Ceramics for High-Performance Applications III

RELIABILITY

ARMY MATERIALS TECHNOLOGY CONFERENCE SERIES

Volume 6 Ceramics for High-Performance Applications – III: Reliability
Edited by Edward M. Lenoe, R. Nathan Katz, and John J. Burke

A Continuation Order Plan is available for this series. A continuation order will bring delivery of each new volume immediately upon publication. Volumes are billed only upon actual shipment. For further information please contact the publisher. Volumes 1 to 5 of this series were published by Brook Hill Publishing Company, Chestnut Hill, Massachusetts, and are available from the Metal and Ceramics Information Center at Batelle Columbus Laboratories.

Published in cooperation with the Metals and Ceramics Information Center at Battelle Columbus Laboratories
(A Department of Defense Information Analysis Center)

Ceramics for High-Performance Applications III

RELIABILITY

Edited by
EDWARD M. LENOE
R. NATHAN KATZ
AND
JOHN J. BURKE
Army Materials and Mechanics Research Center
Watertown, Massachusetts

PLENUM PRESS • NEW YORK AND LONDON

Library of Congress Cataloging in Publication Data

Army Materials Technology Conference (6th: 1979: Orcas Island, Wash.)
Ceramics for high-performance applications, III.

(Army Materials Technology Conference series; v. 6)
"Published in cooperation with the Metals and Ceramics Information Center at Battelle Columbus Laboratories."
Includes bibliographical references and index.
1. Ceramic materials—Congresses. 2. Heat-engines—Materials—Congresses. I. Lenoe, Edward M. II. Katz, R. Nathan. III. Burke, John J. IV. Metals and Ceramics Information Center (U.S.) V. Title. VI.
TA455.C43A75 1979 666 82-13234
ISBN-13: 978-1-4684-3967-0 e-ISBN-13: 978-1-4684-3965-6
DOI: 10.1007/978-1-4684-3965-6

Proceedings of the Sixth Army Materials Technology Conference, held July 10–13, 1979, at Orcas Island, Washington, cosponsored by the United States Department of Energy Office of Transportation Programs

A Division of Plenum Publishing Corporation
233 Spring Street, New York, N.Y. 10013

Army Materials Technology Conference Committee

Co-Chairmen

J. J. BURKE
Army Materials and Mechanics Research Center

J. MUELLER
University of Washington

Executive Committee

A. E. GORUM
University of Washington

E. S. WRIGHT
Army Materials and Mechanics Research Center

International Advisory Board

W. BUNK
Deutsche Forschungs- und Versuchsanstalt für Luft- und Raumfahrt E. V.

D. J. GODFREY
Admiralty Materials Technology Est.

K. INAMORI
Kyoto Ceramics Company, Ltd.

S. KRONOGARD
United Turbine AB

Program Directors

R. N. KATZ
Army Materials and Mechanics Research Center

E. M. LENOE
Army Materials and Mechanics Research Center

Program Committee

R. ASHBROOK
National Aeronautics and Space Administration

C. F. BERSCH
Naval Air Systems Command

C. BLANKENSHIP
National Aeronautics and Space Administration

M. BÖHMER
Deutsche Forschungs- und Versuchsanstalt für Luft- und Raumfahrt E. V.

S. BORTZ
Illinois Institute of Technology Research Institute

W. BRYZIK
U. S. Army Tank-Automotive Research and Development Command

A. DINESS
Office of Naval Research

W. DUCKWORTH
Battelle Columbus Laboratories

J. GANGLER
National Aeronautics and Space Administration

G. E. GAZZA
Army Materials and Mechanics Research Center

H. GRAHAM
Air Force Materials Laboratory

W. McGOVERN
U.S. Army Mobility Equipment Research and Development Command

A. F. McLEAN
Ford Motor Company

D. R. MESSIER
Army Materials and Mechanics Research Center

R. RICE
Naval Research Laboratory

R. SCHULZ
Department of Energy

W. SIMMONS
Air Force Office of Scientific Research

R. SPRIGGS
National Materials Advisory Board

N. TALLAN
Air Force Materials Laboratory

G. THUR
Department of Energy

E. C. VanREUTH
Defense Advanced Research Projects Agency

Program Coordinator

J. A. BERNIER
Army Materials and Mechanics Research Center

Conference Coordinator:

J. M. AYOUB
Army Materials and Mechanics Research Center

DEDICATION

This volume is dedicated to the memory of Dr. Alvin E. Gorum; colleague, mentor, and friend. Al Gorum was a unique phenomenon in the world of materials science and technology. His personal gifts for research, teaching, management and organizational leadership enabled him to play key roles in many of the major materials advances of the past three decades. Characteristically, Al was a strong advocate of the application of ceramics to heat engine technology -- as early as the 1960's. Indeed, one of his first actions upon becoming Director of AMMRC in 1970 was to suggest that we initiate research on silicon nitride for turbine applications. Thereafter, he played a major role in the initiation of the DARPA Brittle Materials Program, and later as the program's technical monitor he had the key role in its success. Throughout his career, Dr. Gorum provided inspirational and astute leadership in his many enterprises. Subsequent to retiring from AMMRC, he was instrumental in developing a unique interdisciplinary educational program in structural ceramic technology at the University of Washington. With Al's premature death, the high performance ceramics community, worldwide, has lost a dedicated advocate; and those of us privileged to have worked closely with him have lost a friend.

The Editors

FOREWORD

The Sixth Army Materials Technology Conference, "Ceramics for High Performance Applications-III-Reliability", was co-sponsored by the Army Materials and Mechanics Research Center and the U.S. Department of Energy, Office of Transportation Programs. The program highlighted all issues relevant to the reliability of ceramics in advanced systems. The conference emphasized programmatic reviews of the major efforts on ceramic gas turbine technology, on an international basis. The conference showed how ceramic design, materials development, materials processing, NDE, and component systems testing are being integrated and iterated in specific engine development programs. Further, the conference promoted interchange among the various technical disciplines working in the advanced turbine and heat engine areas.

This volume will join its earlier companions, Ceramics for High Performance Applications (1974), and Ceramics for High Performance Applications-II (1978), in chronicling the rapid progress being made in the applicaton of ceramics to the very demanding service environment of gas turbine and piston engines. At the last meeting of this series at Newport, RI, in March 1977, successful high temperature tests of ceramic components in test rigs were described. Only two and one half years later we were shown films of a partially ceramic configured gas turbine engine powering a truck on the highway and enduring the rigors of a test track; given a report of a small gas turbine with an all ceramic nozzle producing electric power; told of a successful diesel piston cap demonstration; and even told of ceramic components in racing car engines. Yet, these are all one of a kind demonstrations. Now that we know ceramics will work, the question focuses on the issues related to assuring reliability of ceramic materials and components in operating systems; thus, the focus of this volume -- Reliability. It is also gratifying to see the greatly expanded international involvement in developing this technology, potentially so important to our energy and materials scarce world.

We wish to thank all of the participants at the Conference for contributing to this important event. The editors wish to thank the authors and members of the various committees for the privilege of having worked with them in producing what we believe will be an important volume in the area of high performance ceramic technology.

Special acknowledgement is due to Dr. D. R. Messier and Miss M. M. Murphy of the Army Materials and Mechanics Research Center, for their efforts in updating the Annotated Bibliography on Si_3N_4, which was distributed at the conference. The dedicated assistance of Mrs. J. Ayoub of the Army Materials and Mechanics Research Center throughout the stages of conference planning and final publication of this book is deeply appreciated. We are especially appreciative of the participation of our colleagues at the University of Washington and their students for conducting the Workshop Session.

Army Materials and Mechanics Research Center The Editors
Watertown, Massachusetts 02172

CONTENTS

KEYNOTE ADDRESS

SESSION I
HEAT ENGINE APPLICATIONS
Chairman: Dr. A. E. Gorum, University of Washington
Vice Chairman: Dr. R. Taggart, University of Washington

SESSION I (Continued)
HEAT ENGINE APPLICATIONS
Chairman: Dr. W. Bunk
Deutsche Forschungs-und Versuchsanstalt für Luft-und Raumfahrt E.V.
Vice Chairman: Dr. W. Scott, University of Washington

SESSION Va
TEST, EVALUATION, AND DEVELOPMENT OF COMPONENTS
Chairman: Mr. V. N. Saffire, General Electric Company
Vice Chairman: Dr. R. Kamo, Cummins Engine Company

SESSION Vb
TEST, EVALUATION, AND DEVELOPMENT OF SPECIMENS
Chairman: Prof. W. Love, University of Washington
Vice Chairmen: Prof. A. S. Kobayashi, University of Washington
Dr. R. Chait, Army Materials and Mechanics Research Center

SESSION VI
CRITICAL ISSUES WORKSHOP
Chairman: Prof. J. Mueller, University of Washington

KEYNOTE ADDRESS

Dr. E.M. Lenoe

Army Materials and Mechanics Research Center

RECENT ACCOMPLISHMENTS AND RESEARCH NEEDS IN STRUCTURAL CERAMICS

Dr. Edward Mark Lenoe

Army Materials and Mechanics Research Center
Watertown, Massachusetts 02172

The purposes of this talk are rather straightforward. The first objective is to reiterate and re-emphasize the theme of this Conference: namely, reliability issues in structural ceramics. Hopefully, this will further encourage our speakers to directly focus their attention and particular expertise on the various aspects of ceramic structural integrity. A second objective is to briefly review accomplishments in the field, and finally, to outline research needs in the technology. Hopefully, this will serve to stimulate more productive discussions during this Conference.

Let us begin with an outline of recent progress. Figure 1 lists some important technology accomplishments. Consideration of the tabulation reveals that significant and steady progress is evident in the numerous engine and component durability programs conducted during the past few years. Other encouragement includes the general trend of improving ceramic properties, along with the proliferation of varieties and producers of ceramics both here in the United States and abroad. The required mechanical properties data base continues to grow, although admittedly not as rapidly as desired. Finally, there is an increase in international cooperation, as attested by this meeting.

SOME ACCOMPLISHMENTS OF U. S. CERAMIC ENGINE PROGRAMS

		Achievement	Comment
ARPA/Westinghouse	1973	2200 F Static Rig Tests of Ceramic Stator Vanes, 100 Peaking Cycles	
	1975	2500 F Static Rig Tests of Ceramic Stator Vanes, 100 Peaking Cycles	Temperatures 2600 F
ARPA/Ford Gas Turbine Test Engine	1974	50-hr Demonstration of an Engine Ceramic Stationary Components and Metal Rotors (1930 F TIT)	First Integration of Multiple Ceramic Components into an Operating Engine
	1977	36.5-hr Engine Demonstration 2200-2500 F + TIT 40,000-50,000 rpm with Ceramic Rotor	First Integration and Successful Demonstration of an Uncooled, All-Ceramic Flow Path at Temperatures Beyond Superalloy Capability
DOE/Ford Regenerator Program	1977-1978	200-hr Rig Demonstration of all Stationary Hardware to 2500 F - Rotor to Approximately 2200 F	200-hr Capability Demonstration
	1978	10,000-hr Regenerator Life Demonstrated in Engines (1900 F TIT)	Regenerator Problem Solved
Army/Solar 10-KW Turboalternator	1978	45-hr Demonstration of Engine with Hybrid RBSN/HPSN/RSSiC Nozzle (1750 F TIT)	First Ceramic Configured Engine Providing Electric Power at Required Performance Levels
DOE-NASA/Detroit Diesel Allison GT-404 Test Engine	1978	1200+ hr Demonstration of KT SiC Stator Vanes in Engine (1900 F TIT)	No Obvious Deterioration in the Engine Environment 1st 1000 hr+ Engine Test of Flow Path Component
Army/Cummins ADIABATIC Diesel	1978	250-hr Full Engine Performance Demonstration of HPSN Piston Cap (Max. Temp. Approximately 1750 F)	First Successful Demonstration of a Dynamic Component in a Heat Engine to Meet Performance Goals
NAVAIR/GARRETT	1978	Integration of Over 100 Ceramic Components into an Engine. Short Engine Run at Design Conditions 2200+ F	Multi-Stage Ceramic Rotor, Largest Size Engine Run to Date

TIT = Turbine Inlet Temperature
RBSN = Reaction Bonded Silicon Nitride
RSSiC = Reaction Sintered Silicon Carbide
HPSN = Hot-Pressed Silicon Nitride

Figure 1

Let us consider some specific accomplishments of the U. S. Ceramic Engine Programs in more detail. Under DARPA sponsorship, Westinghouse investigated short-time behavior of fairly large ceramic stator vanes. Ford accomplished the first integration of multiple ceramic components into an operating engine as well as short term durability testing of an all-ceramic rotor at 2500°F. The U. S. Department of Energy and Ford contributed significantly via their 200-hour durability of all stationary hardware to 2500°F. In 1978, 10,000-hour regenerator life was demonstrated. During this Conference we will have an update of current achievements of various programs listed here, as well as description of newly initiated activities.

It is encouraging that awareness of ceramic technology is increasingly evident in the popular press. For instance, the July 5th edition of the New York Times contains an article by Peter Schuyten titled, "Using Ceramics in Auto Engines". Mr. Schuyten reported that Detroit Diesel Allison Division (DDA) has run their truck gas turbine with ceramic heat exchangers and stationary vanes for 1000 miles over the G. M. Test Track in New Milford, Michigan. This afternoon, Gene Helms of DDA is expected to give us the latest facts.

In addition to ceramic component demonstrations, numerous researchers have documented progress and made optimistic estimates of potential improvements in properties. They are confident that all varieties of structural ceramics will be improved. Figure 2 lends support to their confidence, since it illustrates the progressive improvement achieved in both silicon nitride and carbide.

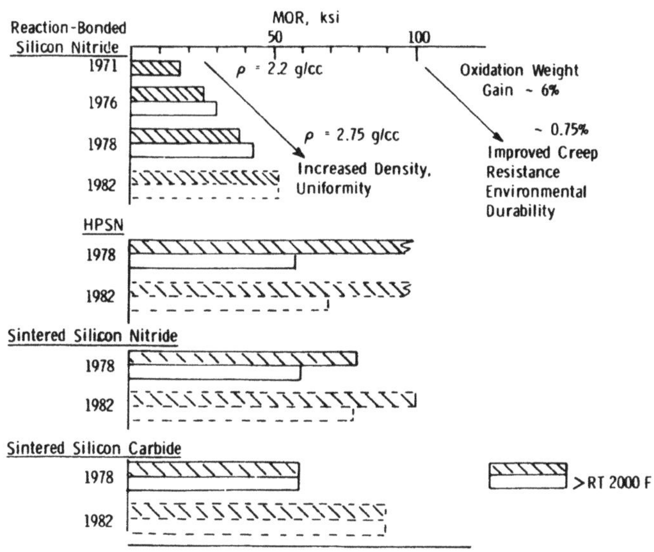

Reference: D. L. Mann, AiResearch and R. N. Katz, AMMRC

Figure 2

Now let us change tack and briefly consider problem areas to be confronted. Figure 3 is a general listing of such areas. These cover a wide range of subjects requiring different levels of effort to solve. Actually, these numerous problems were motivating factors in organizing this Conference. During the week we will have various experts addressing reliability issues from a number of viewpoints. Now, I want to briefly consider several of these items.

GENERAL PROBLEM AREAS IN
STRUCTURAL CERAMICS TECHNOLOGY

- Lack of reproducibility in many available ceramics.
- Neglect of process improvement potential.
- Ignorance of aspects of phenomenological behavior.
- Limited full-scale correlative tests.
- Inadequate design properties data base.
- Limitations of design/analysis capabilities.
- Need for better definition of service environments.
- No universal test standards.
- Nondestructive evaluation - need to define defects and acceptance/rejection criteria.
- General inconsistencies in experimental procedures/data interpretation.

Figure 3

It is self evident that processing of ceramic components and their associated properties variation is a complex process. The manufacture of structural ceramics generally requires many processing steps. Often detailed understanding and precise control of each step is lacking. Some factors influencing reproducibility of hot pressed materials illustrate important aspects. For instance, use of fine powders of the order of microns size, and the necessity for dispersion of pressing aids and additives requires careful procedures for uniform dispersion since trace elements greatly affect high temperature properties. It is useful to recall that many structural ceramics are still in early stages of development. Thus, we must improve existing materials, develop new materials and must, in addition, investigate new fabrication approaches. Some possibilities for improved nitrides and carbide ceramics include:

- Vapor phase addition of densification aids
- Fluidized bed powder treatment and pre-reaction of densification aids
- Sintering plus "cladless" hot isostatic pressing
- Devitrification of N_2 glasses
- Bulk ceramics from polymeric precursors.

As far as nondestructive evaluation (NDE), the relatively limited experience of correlative service tests has not permitted adequate definition of types and importance of flaws in hardware. Presently, this lack of information hinders meaningful development of appropriate accept/reject criteria. Much NDE development is aimed at inspection of microstructural features which may or may not be relevant. Currently, relatively little effort is being directed to development of equipment to improve inspection of complex components, and this is problematical in terms of adequate screening of structural components.

Regarding observed mechanical properties variability, George Quinn [1] of AMMRC, under U. S. Department of Energy sponsorship has evaluated thirteen nitride and carbide ceramics at room temperature for durability effects. His data are of some interest (see Figure 4). (Note the data are for silicon nitrides and silicon carbides of varying degrees of maturity.) This chart of room temperature flexure strength indicates mean values, maximum and minimum, summarizes coefficient of variances (CV); i.e., δ/σ_m, as well as two parameter Weibull moduli estimated via maximum liklihood estimating techniques. Along each bar, the upward arrow denotes mean strength, the line shows minimum value, the solid triangle the mean value after high temperature exposure, the CV is shown at the bar end, the density at the beginning (ρ) and the Weibull m, just beneath each bar. For silicon carbide, one observes CV varying from 9.5 to 20.5%, Weibull m from 6.8 to 11.9 and multimodal Weibull parameters. Therefore, simple statistical models do not fit all materials behavior. It is also noteworthy that in spite of high quality and careful machining of specimens, machining damage was the strength controlling defect in HPSC.

Silicon nitride (Si_3N_4) showed similar behavior in Quinn's studies -- fairly wide range of CV ranging from 8.4 to over 15%, some instances of bimodal statistical distributions of strength, and fairly high instances of machining damage in HPSN. Later in this Conference, Quinn will report on portions of his durability studies based on high temperature static environmental exposure as well as a moderate thermal cycling exposure.

Regarding further documentation for materials variability, there are several fairly large data bases for ceramic materials. For instance, under the DARPA/Ford/Westinghouse contract, 90 billets of NC-132 were characterized. Figure 5 is a histogram of characteristic strength and Weibull parameter, based on two parameter models. NC-132 is a rather expensive and difficult to machine structural ceramic. However, the material is of interest since several major projects are currently evaluating rotor concepts which use HPSN. Potential users include Ford, AiResearch, and Daimler Benz. Note the range in this quality assurance data.

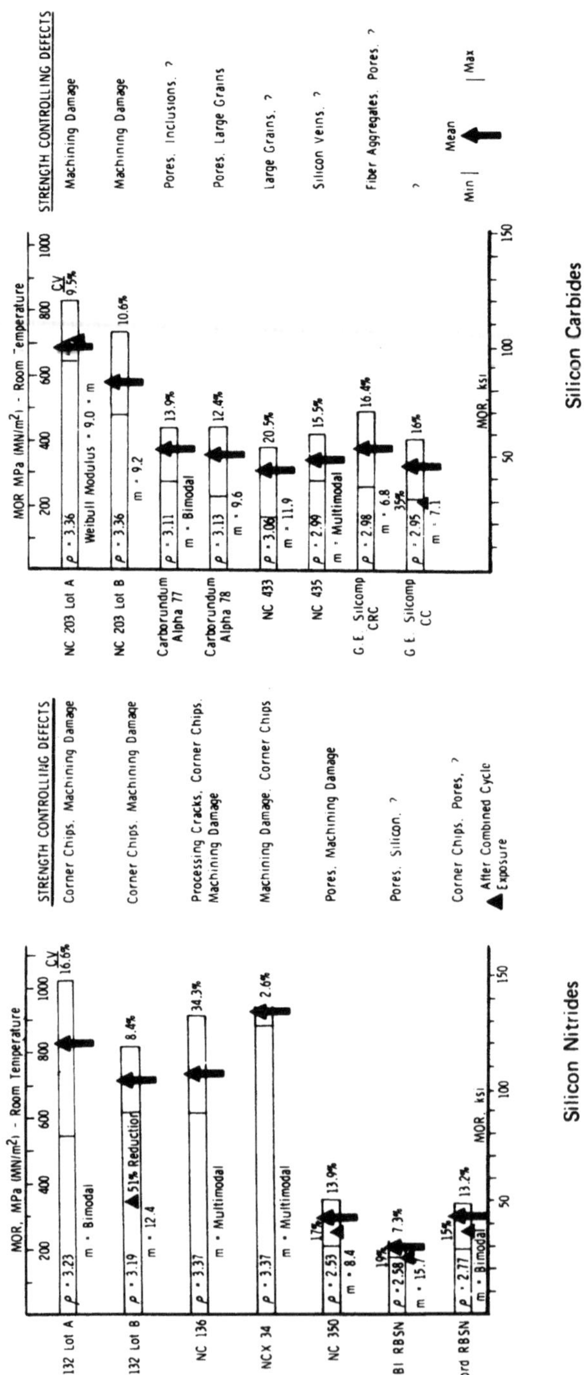
MOR, MPa (MN/m²) - Room Temperature
STRENGTH CONTROLLING DEFECTS
NC 132 Lot A
NC 132 Lot B
NC 136
NCX 34
NC 350
KBI RBSN
Ford RBSN
Corner Chips, Machining Damage
Corner Chips, Machining Damage
Processing Cracks, Corner Chips, Machining Damage
Machining Damage, Corner Chips
Pores, Machining Damage
Pores, Silicon, ?
Corner Chips, Pores, ?
After Combined Cycle Exposure
51% Reduction
MOR, ksi
Silicon Nitrides
MOR MPa (MN/m²) - Room Temperature
Weibull Modulus = 9.0 = m
NC 203 Lot A
NC 203 Lot B
Carborundum Alpha 77
Carborundum Alpha 78
NC 433
NC 435
G.E. Silcomp CRC
G.E. Silcomp CC
Machining Damage
Machining Damage
Pores, Inclusions, ?
Pores, Large Grains
Large Grains, ?
Silicon Veins, ?
Fiber Aggregates, Pores, ?
Min Mean Max
Silicon Carbides

Figure 4

For each billet, this QA data consisted of a minimum of eight flexure tests. It is useful to discuss statistical evaluation of the test results. When the three-point flexure QA data is used together (i.e., 759 tests), a fairly narrow range (in 90% confidence limits) is observed. Figure 6 illustrates probabilities of failure versus flexure strength. In the right corner of the figure, root mean

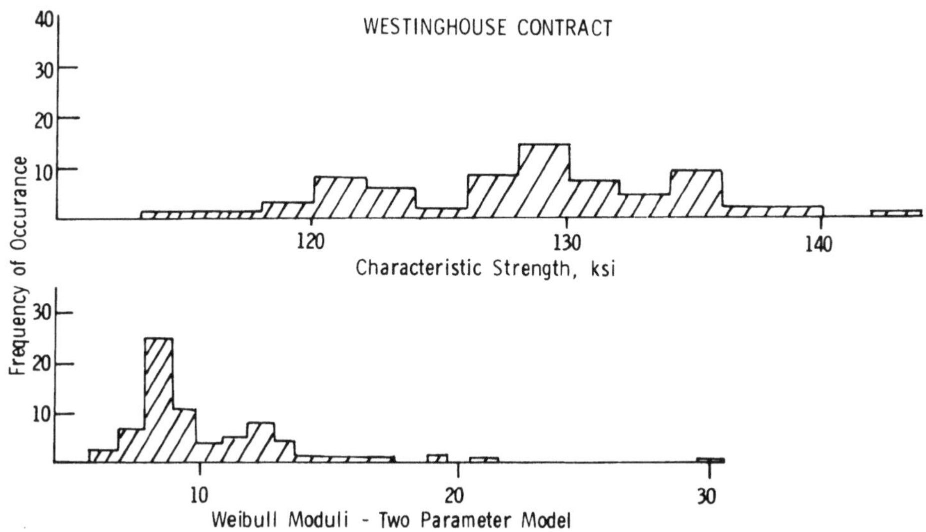

DISTRIBUTION OF CHARACTERISTIC VALUES AND WEIBULL MODULUS
90 Billets, NC-132

Figure 5

square (RMS) results for three distributions are listed. The two parameter Weibull distribution yields the lowest RMS value. Also listed are confidence limits for mean value and standard deviation (90% confidence intervals) for the normal distribution, as well as associated limits for Weibull slope and characteristic value. Note that Ri are various empirical ranking methods.

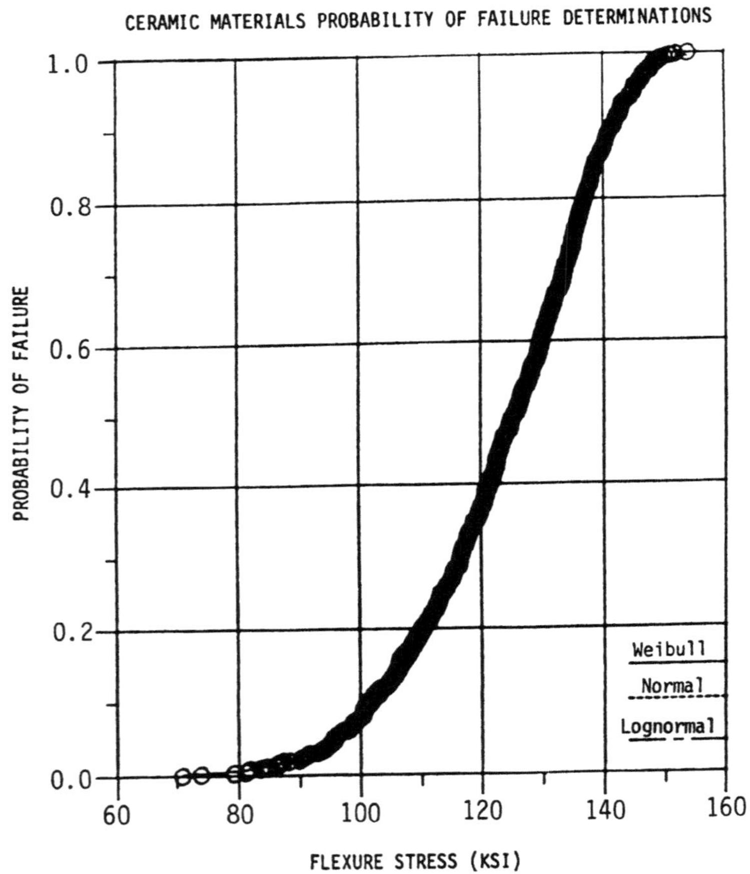

RMS (error)

	Normal	Lognormal	Weibull
R1	.043	.144	.016
R2	.043	.144	.016
R3	.043	.144	.016

		90% CONFIDENCE INTERVAL	
MEAN (KSI)=	123.559	122.625	124.494
S DEU (KSI)=	15.165	14.553	15.836

WEIBULL PARAMETERS

		90% CONFIDENCE INTERVAL	
SLOPE M=	9.927	8.762	11.067
CHAR VALUE=	13C.023	127.957	132.109

MATERIAL : WESTINGHOUSE NC132

NO. OF SPECIMENS : 759

MISC. : TOTAL DATA

Figure 6

It is quite interesting to note the effect of sample size. Table I tallies characteristic values and Weibull moduli with confidence limits in brackets. Consider what happens in comparing the total data set versus 30 and 60 randomly selected data points. Note the 30 data points result in the highest value of Weibull m -- but also has the widest range in 90% confidence limits.

Table I

INFLUENCE OF SAMPLE SIZE - WESTINGHOUSE VANES
NC-132 HPSN

Number of Data	Characteristic Value (Ksi)	Weibull Modulus
759	130.023 [127.96 - 132.11]	9.927 [8.76 - 11.07]
30	133.286 [129.16 - 137.6]	10.616 [7.96 - 12.95]
60	133.41 [130.45 - 136.48]	10.217 [8.46 - 11.84]
Lowest Values		
Vane	Characteristic Value (Ksi)	Weibull Modulus
79	131.743 [116.16 - 150.24]	5.967 [2.961 - 8.29]
63	126.682 [112.741 - 143.08]	6.441 [3.2 - 8.95]
Highest Values		
66	143.204 [139.63 - 147.03]	29.71 [14.74 - 41.26]
65	135.37 [130.70 - 140.42]	21.397 [10.62 - 29.72]
All Data (759)	130.023 [127.96 - 132.11]	9.93 [8.76 - 11.07]

THREE POINT FLEXURE, 1/8 x 1/8 x 3/4" BEAMS

Results for sets of the lowest and highest observed strength values are also presented. Here individual billets (vane qualification tests for instance were based on 8 QA replicates) with Weibull moduli as low as 6 and as high as 30 were observed. Having to account for such variabilities obviously would impose restrictively conservative design limitations. These data suggest the wisdom and necessity for a fairly large number of replicates and lead to concern over the source of such variability and also to question what can be done to improve the situation. During the next few days we will hear, in some detail, of the various factors which affect reproducibility of properties in ceramic component properties for the various fabrication processes.

Regarding advanced design, it is evident that we need a better design data base, particularly for the more recently developed materials. Of primary importance is the necessity for increased design iterations as well as correlative tests to failure which will undoubtedly lead to deeper understanding of the sources and relative importance of defects.

The design data base is inadequate in several respects: first, there remain gaps in many areas of mechanical properties characterization, and secondly, probabilistic data evaluation is still in its relative infancy in structural ceramics. We have generally relied on very simplified models, perhaps overly emphasizing two parameter Weibull statistics and elementary models of time and temperature dependent failure processes. There are still many instances where a ridiculously small number of replicate experiments are conducted and their results fitted to incorrect statistical models. Furthermore, there are no generally accepted test techniques and in many cases the test specimens are not representative of component properties.

The flexure test is a worthwhile case to discuss. There are a range of specimen geometries used by different organizations in the United States. Oftentimes each organization uses different types of test fixtures and experimental techniques which further

AMMRC STUDIES - LIFE PREDICTION METHODOLOGY

ACCOMPLISHMENTS TO DATE

- HIGH TEMPERATURE (1000-1400 C) MOR

  STRESS RUPTURE (150)
  STEPPED STRESS RUPTURE (150)
  DURABILITY TESTS (STATIC) (75)
  DURABILITY TESTS (COMBINED CYCLE) (75)
  BASELINE STRENGTH (176)

- PREFLAWED (INDENTED BEAM) MOR

  FRACTURE TOUGHNESS, K_{Ic} AND
  CRACK VELOCITY (150)

- TENSILE CREEP (12 AT 1205 C)

  CREEP EXPONENT ⟶ 4
  NONRECOVERABLE STRAIN ON UNLOADING

- PRELIMINARY TESTS BY OTHER METHODS

  IN-PLANE BENDING
  DOUBLE TORSION
  SHORT ROD
  NOTCHED BEAM

- PRELIMINARY FRACTOGRAPHY AND THEORETICAL ANALYSIS

Figure 7

influence data scatter. Our German colleagues are better organized, having adopted a more widely used specimen standard. Our German colleagues also have reached more unanimity regarding test fixtures -- only several types prevail. During this Conference, particularly at the Workshop Session, we hope to address the problem of agreeing on test standards, at least for the flexure test.

Environmental effects have not been completely evaluated for many important structural ceramics. For these materials we must determine the basic phenomena as well as develop methods to cope with them. Under U. S. Department of Energy sponsorship, AMMRC has been attempting to contribute to such requirements. To date, thirteen types of ceramics have been evaluated to varying degrees via flexure tests and by tension creep as well as various fracture mechanics tests. HPSN-NC-132 has been studied via flawed beam tests in some detail. This has included development of experimental procedures and statistical data generation. Figures 7 and 8 summarize AMMRC studies of this material.

AMMRC FLAWED BEAM EXPERIMENTS

DEVELOPMENT OF EXPERIMENTAL PROCEDURES

- INDUCING, ANNEALING, CHARACTERIZING KNOOP INDENTATION FLAWS (15 Tests at 25, 1200, 1300, 1400 C)

PRELIMINARY SLOW CRACK GROWTH DATA

- UNFLAWED BASELINE STRESS RUPTURE (8 at 39 ksi and various temperatures)
- MECHANICALLY (STRESS) CYCLED FLAWED BEAMS (4)
- THERMALLY CYCLED FLAWED BEAM
- FLAWED BEAM (2.6 Kg LOAD) (18 at various temperatures and 39 ksi)

STATISTICAL DATA GENERATION - FLAWED BEAMS

Knoop Load	Temperature, deg C 1200	1225	1250	1275	1300
0.6 Kg	Unsuccessful - failed from other flaws				
1.0 Kg	X	2	2	X	X
1.6 Kg	X	11	10	4	8
2.6 Kg	8	8	9	4	5
3.5 Kg	5	9	8	X	8

Figure 8

It is instructive to discuss the flawed beam experiments. Figure 9 shows typical fracture surfaces of Knoop indented flexure specimens. Note the roughly semi-elliptical flaw growth patterns and the times to failure shown beneath each fracture face. For this set of experiments, the hours to failure ranged from .0827 to .045 for the ratio R = .0827 ÷ .045 = 1.84.

A summary of lifetime distributions for various indent loads and temperatures is provided in Figure 10. These flawed beams were loaded in four-point flexure under constant load conditions (38 Ksi) until failure occurred. Note that for 1.6 kg loads at

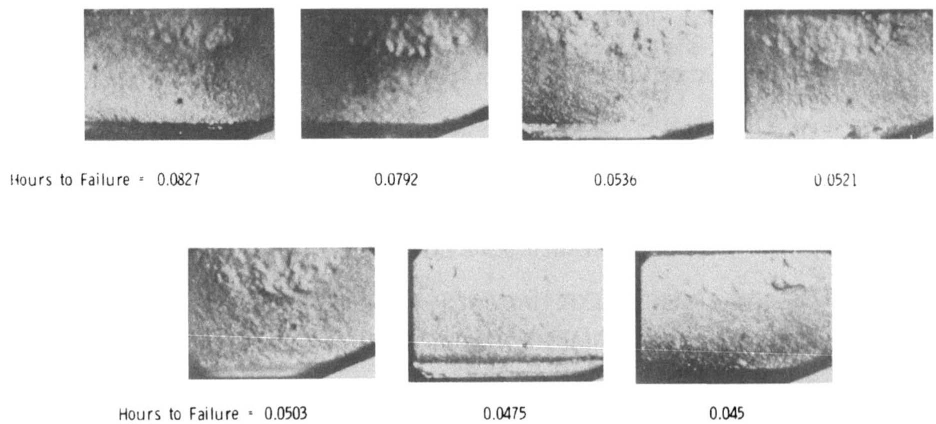

KNOOP INDENTED FLEXURE CREEP, 38.6 KSI, 3.5-KG LOAD, 1300 C
AMMRC TESTS - NC-132 HPSN

Figure 9 - Typical Fracture Surface of Knoop Indented Flexure Creep Specimen

1225°C, the ratio of maximum to minimum time to failure was 14, whereas at 1300°C the ratio dropped to 5.2. For 2.6 kg loads the ratio of maximum to minimum time to failure was about 5 at 1200°C and 1.9 at 1300°C. The flawed beam tests conducted at 3.5 kg Knoop indent load showed a similar trend of reduction in the ratio with increasing temperature, the ratio at 1225°C being 4.65 whereas at 1300°C, R = 1.83. All told, there is convincing fractographic evidence for less scatter in times to failure at the high temperature, perhaps due to the increasing predominance of a single failure mode. One might expect that mixed failure modes would tend to lead to more widely scattered life distributions.

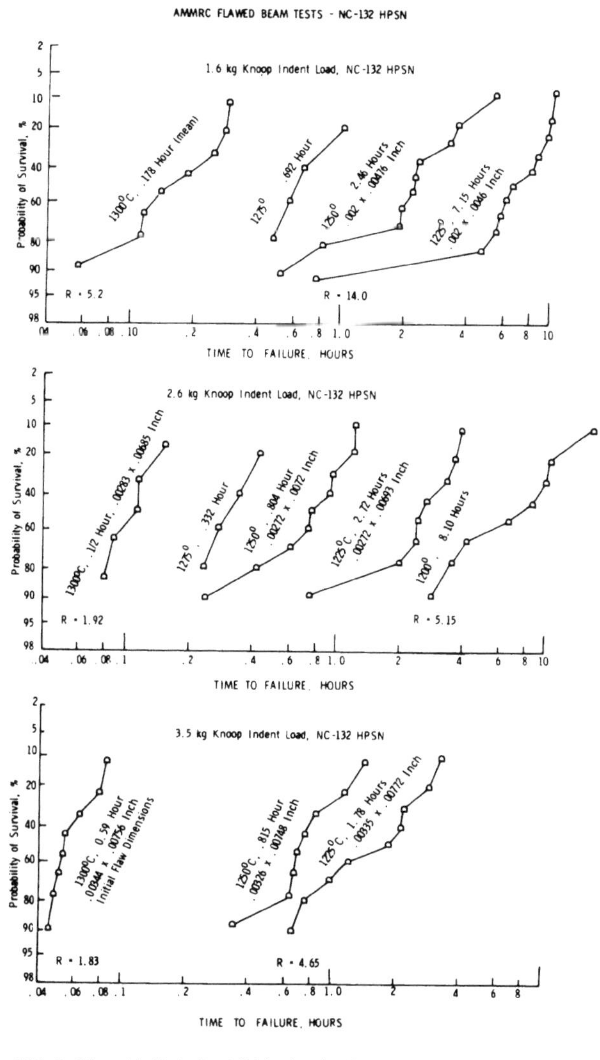

Figure 10 - Lifetime Distributions for Flawed Beam Experiments

This matter is worthy of further discussion. Refer to Figure 11 where for 2.6 kg indent load and 1200°C temperature, three specimens; i.e., about 25% of those tested, did not fail during the experiment. The longest surviving specimen lasted 1062 hours, and was subsequently fractured at room temperature to examine the indentation site. Surface deposits were evident and very little crack extension occurred. In this same specimen gross deformation occurred under the flexural load, obviously

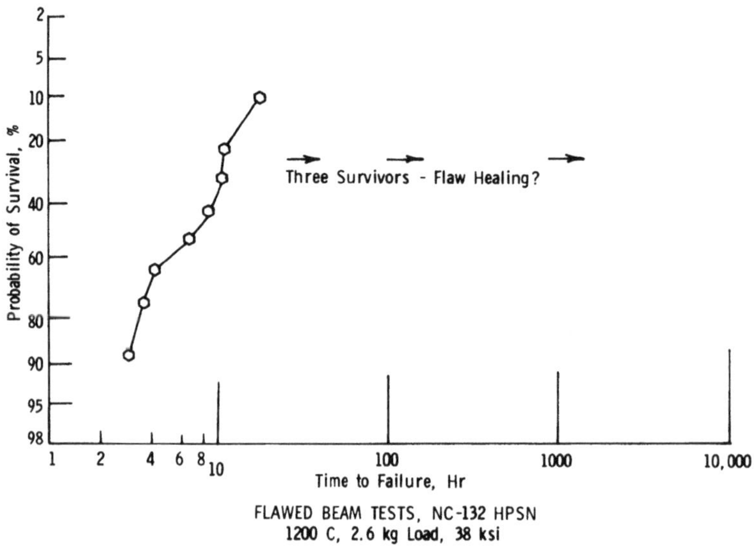

Figure 11 - Lifetime Distributions for 1200°C Flawed Beam Tests

TABLE II

TYPICAL INITIAL AND FINAL FLAW GEOMETRIES
NC-132 HPSN Test Temperature 1225°C

Indent Load Kg	Initial b/a	Approximate Final b/a	Initial b/h	Final b/h
1.6	0.855	0.896	0.0217	0.395
2.6	0.788	0.785	0.0295	0.385
3.5	0.852	0.868	0.0363	0.365

modifying the stress distribution by shifting the neutral axis of the beam. This would also tend to diminish slow crack growth.

Oftentimes these indented beam tests are used to obtain fracture toughness estimates, or for approximate value of constants to be utilized for power law description of crack growth behavior.

Behavior of the initial and final flaw sizes is of interest. Examination of Table II indicates that the initial ratio of flaw width/depth, b/a, was fairly constant for the various Knoop loads at 1225°C. Note that the final depth of critical flaw b/h tended to decrease, on the average, with increasing indentation load.

It is of fundamental importance to realize that the stress intensity factor varies along the crack front. Calculations of stress intensity factors also assume ideal perfectly elastic behavior. These facts, as well as numerous experimental observations indicate that improved analysis and interpretation of fracture mechanics specimens are a prerequisite to further improvements in life estimating procedures. Sufficient evidence has been accumulated to indicate that simplified models of life estimating based on single flaw-based failure, and single mode of response are not valid for high temperature ranges.

In order to make significant progress in life prediction methodology, we must be able to clearly delineate, for important classes of structural ceramics, the regimes of classical material behavior, namely, perfectly linear elastic, and the nonlinear and inelastic ranges of response. This is illustrated by Figure 12. Presently it is not possible to clearly delineate these various regimes of behavior for many important structural ceramics. In general, valid constitutive equations for simple uniaxial tension, compression and shear behavior have not been obtained. At their highest useage temperatures, many ceramics undergo severe microstructural, chemical and physical changes which are not currently taken into account in life estimating procedures. The mechanisms of these effects also remain to be fully understood.

In summary, perhaps problem areas have been over emphasized in this introductory presentation. Therefore, it is appropriate to recall all that has been accomplished. There is no doubt that ceramics are making important contributions to improved engines. In addition, we must all remember that the range of applications of these new generation of ceramics is much broader. Now we are laboring mighty in heat engine technology; in the future there will be broad expansion of other important application areas. And so, I conclude and wish you all a productive and enjoyable conference.

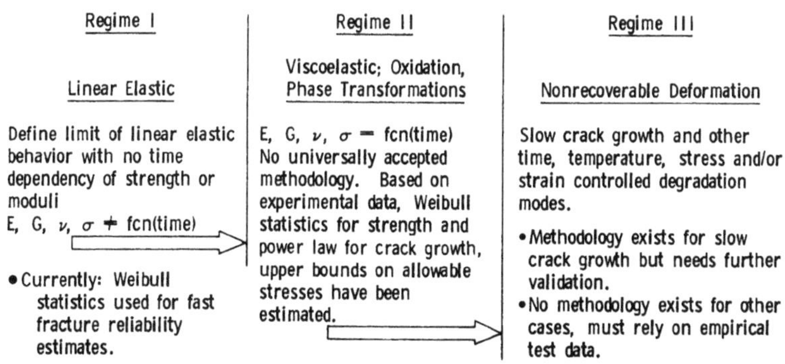

Figure 12 - Idealized Regimes of Behavior

REFERENCES

1. George D. Quinn, "Characterization of Turbine Ceramics After Long-Term Environmental Exposure", AMMRC TR 80-15, April 1980.

SESSION I

HEAT ENGINE APPLICATIONS

Chairman: Dr. A.E. Gorum
University of Washington

Vice Chairman: Dr. R. Taggart
University of Washington

OVERVIEW OF U.S. DEPARTMENT OF ENERGY

CERAMIC GAS TURBINE PROGRAM

Robert B. Schulz

U.S. Department of Energy
Assistant Secretary of Conservation
and Solar Applications
Office of Transportation Programs
Washington, D.C. 20585

INTRODUCTION

The U.S. Department of Energy (DOE) has embarked upon a major automotive gas turbine development program that depends for its success on the application of ceramics to the engine's hot section program and a ceramic gas turbine program.

NATIONAL NEED AND PROJECT OBJECTIVES

The need for an automotive gas turbine program comes from our country's increasing dependence on foreign sources of petroleum for highway vehicles. Presently, major downsizing of vehicles and all practical propulsion system improvements to the conventional internal combustion engine (ICE) are being made to meet federally mandated fuel economy and emission standards. These demands are pushing the limits of spark ignition and diesel engine technology. For the future there is a need for new heat engine technology to decrease petroleum comsuption during the 1990's when increased numbers of vehicles and miles traveled are projected.

In recognition of the national need for Government action, on February 25, 1978, the President signed Public Law 95-238, Development Act of 1978. This law specifically mandates the Secretary of Energy to undertake "... expanded research and development of new automotive propulsion systems."

Specific requiremtnes of Title III include:

- Develop and demonstrate advanced systems within 5 years.
- Supplement, not supplant, industry R&D.
- Establish new projects and accelerate existing ones.
- Intensify research in key basic areas necessary for developmental success.
- Give priority attention to fuel-flexible systems.

Therefore, the Advanced Automotive Heat Engine Development Project was established within the DOE Office of Transportation Programs to develop and demonstrate propulsion systems in automobiles having these overall objectives:

- At least 30-percent improvement in fuel economy compared to the ICE.
- Ability to meet the stringent Clean Air Act emission standards or better.
- Capability to use a wide variety of combustible fluids as fuel.
- Competitive production costs.

Under the Heat Engine Project both gas turbine and Stirling engines are being developed because they are considered the two best options to advance beyond the diesel and stratified charge engines.

DOE/NASA PROJECT DESCRIPTION

The Lewis Research Center (LeRC) of the National Aeronautics and Space Administration (NASA) has been delegated responsibility for accomplishing the project objectives. The Army Materials and Mechanics Research Center (AMMRC) supports the project in the ceramics area.

Project Plan

The Heat Engine Project major milestones and assessments are shown in Fig. 1. The project objectives will be pursued through major systems contractors for the gas turbine and Stirling. There will be one new engine generation (Mod I) and one engine upgrading (Mod II). In FY 1982 the results of Mod I engine performance mapping on a dynamometer, together with results of rig and in-engine component tests, will allow an assessment of the system development progress and a decision made by DOE on whether or not to converge to a single engine type (gas turbine or Stirling).

To satisfy Title III, in FY 1983, a Mod I vehicle will be demonstrated and dynamometer tests of the Mod II engine will be completed. Final demonstration of the project objectives will be accomplished by the end of FY 1984 in Mod II powered vehicles. The fuel economy and emissions objectives will be demonstrated in Mod I and Mod II vehicles tested at the Environmental Protection Agency (EPA) facilities in FY 1983 and FY 1984, respectively.

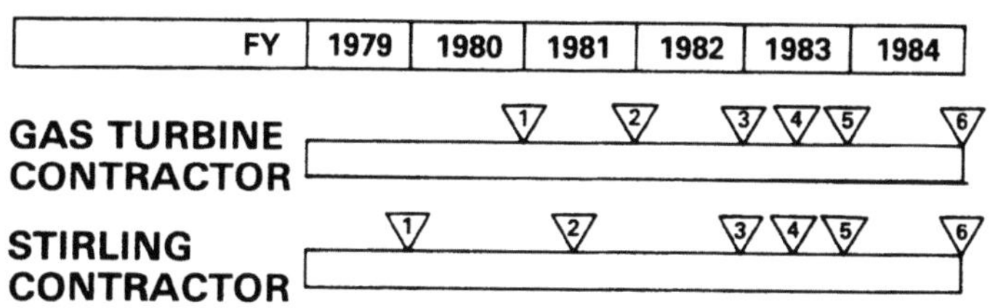

1. Design Freeze And Assessment
2. First Build Mod I Engine On Test And Assessment
3. Major Program Assessment (Iterated Mod I Engine On Dyno Test)
4. Mod I System Vehicle Test (EPA)
5. Mod II System Dyno Test
6. Mod II System Vehicle Test (EPA)

Fig. 1. Heat Engine Project - Major Milestones and Assessments

Two auto industry teams have been chosen for negotiations that will lead to the selection of the gas turbine contractor. One of the teams selected is composed of two divisions of General Motors: Detroit Diesel Allison Division (DDA) and Pontiac Motor Division. The other team is AiResearch Division of Garrett Corporation in conjuction with Ford Motor Company. Negotiations with both teams are taking place simultaneously, and two contracts are expected to be awarded as a result.

Ceramic Applications

Both gas turbine contractor teams have systems that include one or more ceramic turbine wheels, as well as hot section ceramic stationary components and a ceramic heat exchanger core. This need for a ceramic gas turbine was shown in conceptual design studies of an improved automotive gas turbine reported on at the DOE Highway Vehicle Systems Contractor Coordination Meeting, Dearborn, Michigan, April 24-26, 1979. There the results of the three studies by Ford, DDA, and Chrysler Corporation all concluded that the use of a ceramic turbine wheel was necessary to achieve the project fuel economy objectives. Fig. 2 summarizes the main reasons for a ceramic turbine wheel for automobile engines.

The structural ceramics being developed for the automotive applications are primarily silicon nitride, silicon carbide, and silicates. These ceramics are composed of elements including silicon, nitrogen, and carbon that are available in abundance. By using ceramics, the use of scarce metals such as nickel, cobalt, and chromium is avoided, and the net result is a potential for lower cost in production. Also, these expensive imported metals affect the U.S. balance of payments and are subject to cartel price and supply manipulations, just as petroleum is now.

To achieve the project fuel economy objectives, high-temperature materials are needed that can operate at the inlet temperatures and blade tip speeds required. Inlet temperatures up to 1370°C (2500°F) are needed, which is well beyond the capability of metals in small, uncooled integrally bladed wheels.

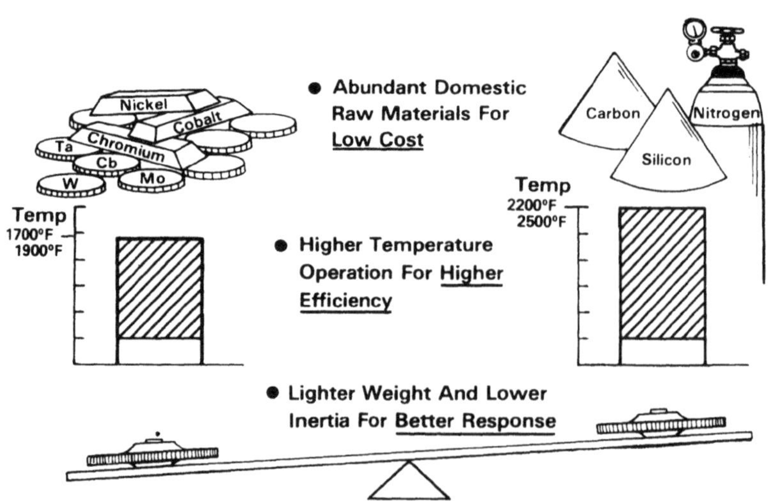

Fig. 2. Ceramic Turbine Wheels for Advanced Automotive Turbine Engines.

A ceramic component can weigh considerably less than a metal part for the same application because of its lower density. For the automotive gas turbine it is important to reduce the turbine wheel rotating inertia to improve engine transient response. A light-weight ceramic turbine improves engine response and vehicle driveability, a well-known problem for automotive turbines with metal wheels.

While the ceramic turbine wheel may be necessary to achieve the fuel economy objectives of the automotive turbine engine, it is not sufficient by itself. Fig. 3 shows the approach in applying ceramics to an existing metal idustrial gas turbine engine, such as what DDA is doing in the Ceramic Applications in Turbine Engines (CATE) program for DOE/NASA. The approach taken for the automobile engine components should be the same, but a new design (Mod I) is expected early in the program. The increased temperature would require the use of ceramics in most hot section components such as the combustor, ducting, turbine nozzles, and turbine wheels. The increased regenerator operating temperature is important for automotive gas turbines because it allows turbine inlet temperature to remain high at part power, which improves part-power efficiency and fuel economy.

Several examples of the interaction between component efficiciencies and the use of ceramics follows: any lowering of overall engine aerodynamic efficiency is not acceptable because it would offset the gains made elsewhere. The low thermal conductivity of

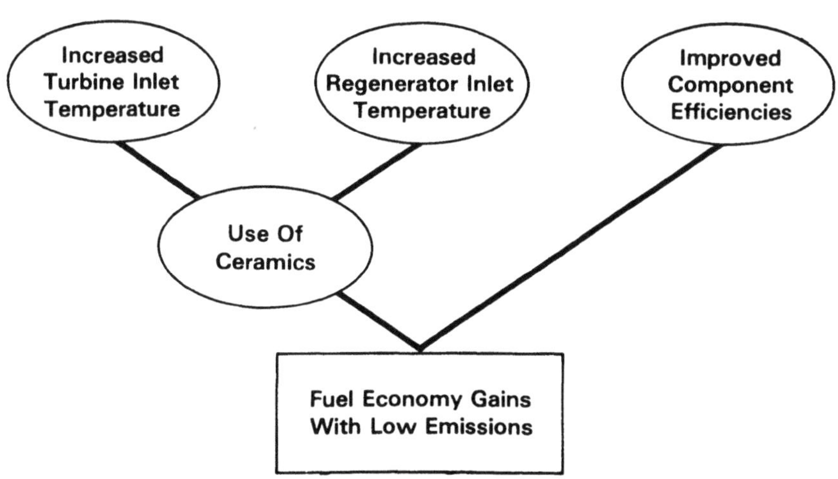

Fig. 3. Component Development Logic for Improved Automotive Turbine Engine

silicate materials (LAS, MAS, AS) can be used to reduce heat loss from the engine and reduce axial conduction loss in the regenerator material. Also, ceramic shroud rings can reduce turbine wheel tip clearance losses.

Problem Areas

Three major ceramics technology problem areas that are being addressed by DOE/NASA follows:

- Material Properties - Strength, Creep and Oxidation Rates, Etc.
- Component Durability - 3500 Hours Turbine Environment.
- Validated Design Method - Slow-Crack-Growth, Oxidation, Etc.

First, a ceramic may not be available today that combines the material properties needed for an integrally bladed axial or radial turbine wheel. The desired material would combine high strength for the hub and low creep and oxidation rates for the blades. Multidensity construction is one approach, but improvements in the properties of sintered silicon carbide and sintered silicon nitride are also promising.

Ceramic component durability data in the automotive turbine operating environment is lacking. The component testing to date by Ford and DDA is a beginning, but certainly more testing is needed. The one exception seems to be the ceramic regenerator that has demonstrated a B_{10} life of over 3500 hours at 800°C.

Life prediction methods that account for slow crack growth and oxidation effects have not been validated. Ford has conducted tensile stress rupture tests of NC-132 ceramic material at 132MPa (19,200 psi) and 1220°C (2200°F). The time to failure for the 14 specimens tested ranged from 4 hours to 21 hours. This was an order of magnitude faster than the analytically predicted time to failure based on strength parameters as determined using flexure-stress rate measurements.

Project Elements

The elements of the Advanced Gas Turbine (AGT) Project are outlined below:

- Advanced Gas Turbine System Development
  - System Studies
  - System Development

- Early Technology Demonstration - CATE
  - DDA 404/505 IGT
  - Selective Use of Ceramics
- Supporting Technology
  - Ceramic Materials
  - Combustors
  - Bearings/Seals
  - Turbomachinery
  - Analysis

Notable is the CATE Project, which was started several years ago in order to gain experience with ceramics in an automotive gas turbine. DDA under NASA contract is applying ceramic components to the 404/505 Industrial Gas Turbine (IGT). This activity, which is described in Session I, will be continued because of its promise for commercialization in the near future.

The Ceramic Materials Supporting Technology Program that is currently addressing some of the previously mentioned problem areas is outlined below:

- Material Characterization
  - Durability
  - Life Prediction Methodology
- Improved Materials
  - High Density RSSN
  - Sinterable Ceramics
  - Heat Exchanger Materials
- Component Technology
  - Regenerator Durability
  - Integral Stator Development
  - Ceramic Rotor Fabrication
  - Nondestructive Evaluation (NDE)
  - Hot Isostatic Pressing (HIP)

Again, much of this work is presented elsewhere in the Conference Proceedings.

Future Directions

In the future the system contractors will be responsible for component development necessary to achieve the project performances objectives and milestones. The supporting technology effort will accordingly be reduced in scope. Plans are to continue in Fiscal Year 80 the AiResearch durability testing, Ford stator evaluation, AMMRC life prediction methodology, and LeRc in-house NDE and HIP activities.

In addition LeRC and AMMRC would provide materials support to the system contractors, as needed. On this basis the sintered silicon nitride work at AMMRC and General Electric would be continued.

To expand our technology base, DOE has signed an International Energy Agency Agreement with the German Aerospace Research Establishment (DFVLR) for a cooperative program in Ceramics for Automotive Gas Turbines.

CONCLUSIONS

The first conclusion is that national need for highway vehicle petroleum conservation have caused the Government to establish challenging project objectives. The Government is also providing the funding needed to do the job.

Secondly, ceramics progress is the likely result of this project because of the resources and efforts that will be applied. The resulting ceramics technology will likely find widespread applications.

Finally, the eventual commercialization of an automotive/ceramic gas turbine is uncertain because this is a high risk technology. However, the possible benefits to society of such an automobile engine and its resulting technology make this project worthwhile.

OVERVIEW OF THE GERMAN CERAMIC GAS TURBINE PROGRAM

Wolfgang Bunk[1], Ernst Gugel[2], Peter Walzer[3]

[1]DFVLR, Institute for Materials Research, D-5000 Köln 90
[2]Annawerk, D-8633 Rödental
[3]Volkswagenwerk, D-3180 Wolfsburg

1. CONCEPT AND FUNDING

In 1974 the German government initiated a vehicular gas turbine program emphasizing high strength - high temperature structural ceramics. From the beginning it was a cost-shared program bringing together government, industrial and university facilities and teaming gas turbine engineers with ceramicists and material scientists. The team was not organized as one main contractor with numerous sub-contractors. Instead, there were more than ten partners of equal status and several laboratories concerned with the more basic scientific problems. Very real communications barriers can exist during introduction of new materials and design techniques. In referring to such difficulties in structural ceramics, we coined the term the materials-design-canyon. This partnership of equals was an attempt to bridge the materials-design-canyon.

In the first part of this paper the program funded by the BMFT, (the German Department of Research and Technology), and monitored by the DFVLR, will be described. The objectives of the gas turbine producing companies in this program are indicated in Fig. 1. MTU is designing components for a 250 kW truck turbine using a hybrid rotor with metal hub and ceramic blades. Daimler-Benz is engaged in developing an all ceramic 125 kW passenger car gas turbine. VW is funded for designing a 50 to 100 kW passenger gas turbine (GT). Research and development activities at these companies started in 1974.

COMPANY	G O A L	ROTOR-CONCEPT
MTU	Truck-GT, 250 kW	Metal-hub with ceramic blades (hybrid-rotor)
DB	Car-GT, 125 kW	Monolithic HPSN-rotor
VW	Car-GT, 50-100 kW	1st stage: duo-density-rotor 2nd stage: hybrid-rotor

Figure 1. Goals of the Gas Turbine Industry.

Also in 1974, some ceramic materials producing companies started to develop materials and processing techniques for GT components, as shown in Fig. 2. H. C. Starck as a powder producer contracted to manufacture silicon nitride powder of high purity. Annawerk addressed hot pressing, post hot pressing, and other techniques. Degussa and Feldmühle preferred injection molding. Rosenthal was asked to manufacture heat exchanger components by extrusion. In 1976, silicon carbide as an alternative material to silicon nitride was introduced and some companies already mentioned added these materials to their activities. In addition, SIGRI agreed to slip cast and to injection mold and ESK to hot press SiC (Fig. 3).

COMPANY	ACTIVITY
H.C. Starck	Si_3N_4-powder
Annawerk	Hotpressing of Si_3N_4 (HPSN), injection moulding and slip casting of Si_3N_4 (RBSN), post-hot-pressing
Degussa	Injection moulding of Si_3N_4 (RBSN) and joining metal-Si_3N_4
Feldmühle	Injection moulding of large size components (RBSN)
Rosenthal	Extrusion and slip casting

Figure 2. Activities of Si_3N_4 Producers.

COMPANY	ACTIVITY
H.C. Starck	SiC-powder
Sigri	Slip casting, injection moulding, siliconized SiC (SiSiC)
ESK	Hotpressing of SiC (HPSiC)
Annawerk	Sintering of SiC, slip casting
Rosenthal	Extrusion and slip casting

Figure 3. Activities of SiC Producers.

The activities of all companies involved in the program and their objectives are summarized in Fig. 4. In order to organize the team work and to guarantee and adequate flow of information, a joint working group of materials engineers and designers was formed for frequent exchange of results and critical review of the status.

Regular attendees at these working group meetings were scientists of universities and research institutions. Fig. 5 describes their activities in the area of basic research taking advantage of their expertise and capabilities.

The goals of the last years of the program are:

a. Stationary Components: (combuster, nose cone, stator, rotor shroud) - Production of original components from various materials and 200 h-tests in a simulated duty-cycle with a max combuster outlet temperature of 2500°F and an inlet pressure of 73 psi.

b. Heat Exchanger: Production of Si_3N_4-recuperator with a wall thickness of 4mm (second effort .2mm), a pressure ratio of 5, an allowable leak rate of 5% and a max temperature of 2015°F (later 2200°F) inlet temperature, 10 h-test, eventually experience transfer to SiC.

c. Rotor:

- Metal-ceramic rotor (cars and trucks)
  200 h-test in duty cycle
  (inlet temperature 2285°F)

- All ceramic rotor (car)
  50 h-test in duty cycle
  (inlet temperature 2500°F)

The funding for the six year period of 1974-1980 is shown in Fig. 6. One observes a balanced situation with funds mainly divided between GT producers and ceramic material manufacturers.

In 1978 the German program was reviewed in a national status seminar at Bad Neuenahr. A book containing all published results is available from the Springer Publishing Company (1). The BMFT intends to continue the program until 1982. During the next years a number of unsolved problems will be tackled following the program goals.

---

(1) Keramische Komponenten für Fahrzeug-Gasturbinen, W. Bunk and M. Böhmer, ed., Springer-Verlag, Burlin, 1978.

In May of 1979, an implementing agreement was signed to allow more international cooperation. This cooperation of US-DOE and German BMFT under the auspices of IEA, Paris, is open to other nations who are invited to join.

	COMBUSTOR	STATOR	ROTOR/BLADES ROTORRINGS	RECUPERATOR
MATERIAL	Annawerk Rosenthal Sigri	Annawerk Rosenthal Sigri	Annawerk Degussa ESK Feldmühle	Rosenthal
DESIGN	MTU VW	MTU VW	DB MTU VW	DB

Figure 4. Activities of Participating Companies.

INSTITUTE	ACTIVITY
Max-Planck-Institut Stuttgart	Basic research
TU Berlin	Powder technology
TU Clausthal-Zellerfeld	Joining technique
IzfP Saarbrücken	N D T
TU Karlsruhe	Creep properties
Universität Erlangen	HT-fracture-mechanics
DFVLR Köln	- Material characterisation - Project management

Figure 5. Activities of Research Institutes.

	Part of BMFT	Total amount
Industry	29.686	59,372
Institutes	4,932	4,932
Σ	34.618	64,304

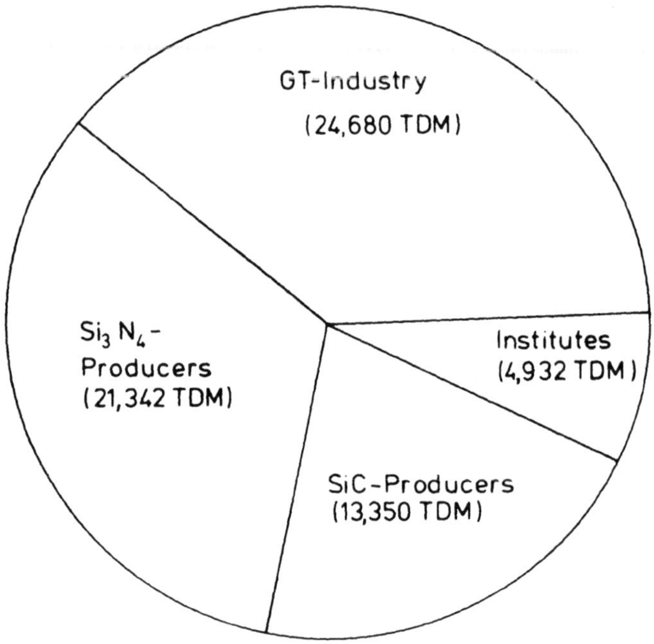

Fig. 6. Funds for the program 'ceramic components for vehicular gas turbines' during 1974-1980 (TDM)

2. MATERIALS AND COMPONENT DEVELOPMENT

2.1 Reaction-Sintered Silicon Nitride

Average and peak values of strength here continued to be substantially increased, but, unfortunately assured reproducibility is still a problem. This is due to the unreliability of silicon sources, but also to the problems arising in controlling the nitri dation of large charges.

The highest average strength values of up to 300 N/mm^2 with peak values of up to 400 N/mm^2 are measured at a bulk density of

2.5 g/cm^3. This proves that very high densities are not required to attain high strength. It is worth noting that materials with a bulk density of as much as 2.8 g/cm^3 have been produced. In general, materials with a high level of homogeneity are more desirable than highly dense materials, where major problems can arise due to lack of reproducible manufacturing. The 2.5 g/cm^3 material has a good homogeneity (Fig. 7) resulting in Weibull moduli of 15-20 and above. Up to 1400°C, this material shows a slight increase in strength rather than a decrease, except for a temporary strength loss below 1000°C, which according to latest investigations, is due to a destruction of the surface oxide layer. At higher temperatures the oxide layer heals again. A more serious problem is the considerable strength loss after long term oxidation treatments. Density and pore size distribution have proven to be the key factors influencing this behavior.

The creep resistance of fine structured homogeneous materials is considered to be sufficient. Fig. 8 shows the improvement in creep resistance during this period of development. Investigations of the microstructure of creep specimens clearly show that improvements of the creep properties can be explained by the refinement of the pore structure - the same factor also influencing oxidation.

The above-mentioned property data mostly apply to reaction-sintered silicon nitride produced by isostatic pressing, injection

Fig. 7. Scanning micrograph of a fracture surface of reaction sintered silicon nitride, ρ=2.5 g/cm^3, σ_B=320 N/mm^2 (Annawerk)

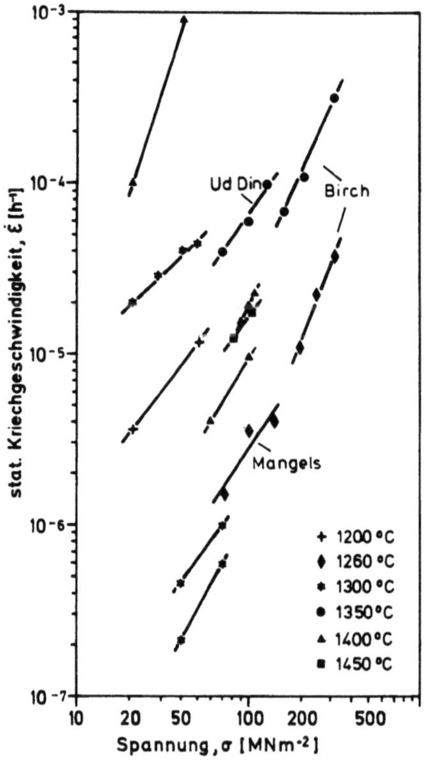

Fig. 8. Steady state creep rate as a function of stress for various reaction sintered materials (Annawerk, TU Karlsruhe)

molding, warm molding or slip casting. For the manufacturing of initial test components, isostatic pressing with subsequent machining is still recommended. For moderate production batches, slip casting offers advantages, since tool costs involved are comparatively low. A method for slip casting has been developed, which allows the manufacturing of complicated components. Rather critical, however, is the liberation of gases in the slip, which if done improperly may cause the formation of larger pores. For larger production volumes, injection molding is the only possible economical way. In this sector, considerable progress has recently been achieved.

Reaction-sintered silicon nitride is used for fabricating heat exchangers, combustion chambers, stator vanes, which are composed of isostatically pressed rings and injection-molded blades (Fig. 9) or cast in one piece, nose cones, also cast in one piece, single blades for rotors, and blade rings for the duodensity rotor. The rotating components have reached 65 000 and 35 000 r.p.m., respectively, in the spin test.

Fig. 9. Reaction-sintered Si_3N_4 stator. The ring is isostatically pressed, the blades are injection molded (Rosenthal)

2.2 Silicon Nitride Powder

One company concentrated especially on the processing of silicon nitride powder on a technical scale. It is now available in a constantly good quality and sufficient quantity. At present,a powder with 93 % α-Si_3N_4, 3-4 % β- Si_3N_4, >38 % N, <1 % O, <0.3 % C, and a sum of <1 % of all impurities is produced. Specific surface is 8-10 m^2/g, with a maximum (medium) grain size of 5-6 (1-2) µm. In order to meet their own needs, the companies doing hot-pressing produced their own powders also.

2.3 Sintered Silicon Nitride

Investigations of sintered silicon nitride are only done by research institutes. The influences of sintering aids, grain size, embedding aids and gaseous atmosphere are studied.

A mixture of boron nitride and silicon nitride with an addition of silicon, also silicon oxide and silicon dioxide has proven to be the best embedding aid. Nitrogen pressure should be as high as possible, but partial pressures of silicon must also be allowed to build up.

A high specific surface area of silicon nitride powder is favourable. Among magnesium compounds, MgO has the most advantageous effect. Best results were obtained at a density of 2.9 g/cm^3, with

3 % MgO and high α containing powder (20 m^2/g surface, attritor milled), sintered at 1750 °C in a closed boron nitride crucible, with a weight loss of 4 % being registered.

Further sintering tests with powders developed in the scope of the program are carried out with $BeSiN_2$, CeO_2, Y_2O_3, and Y_2O_3 mixed with MgO and SrO.

2.4 Hot-Pressed Silicon Nitride

Concerning hot-pressed silicon nitride, the powder as well as the post hot-pressing were further developed. Strength values of more than 1000 N/mm^2 at room temperature and 350 N/mm^2 at1350 °C could be reached. To achieve this, a needle-like microstructure (Fig. 10) has to be developed. But again the component and its reliability in homogeneous properties and their reproducibility seem more important, so that a strength level of 600 N/mm^2 with a Weibull modulus of more than 20 in a component and in a lot is a satisfying result.

The surface condition has a tremendous influence on strength (Fig. 11). If the grinding direction is prependicular to the tensile axis of the sample, strength losses of 40 % are noted. Considerable increase in strength of these samples could be achieved if the surface was treated with an abrasive free grinding medium. Special grinding of the surface of the roots of single rotor blades for the hybrid rotor gave excellent results in spin tests.

Fig. 10. Needle-like microstructure of a fracture surface of hot-pressed Si_3N_4, SEM, 0.5 wt.% MgO (Annawerk)

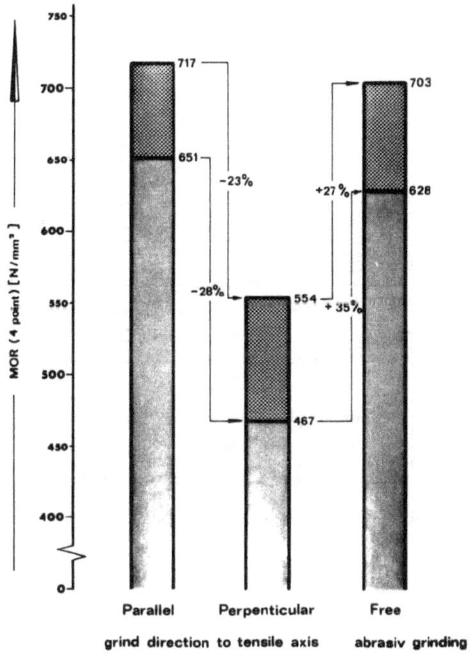

Fig. 11. MOR of hot-pressed Si_3N_4 as a function of grinding condition (Annawerk)

In ground condition a tensile strength of 400 N/mm^2 is guaranteed in the cold spin test for parallel and profiled rotor discs in the hole and neck.

The measured K_{Ic} factor varied between 7 and 8 $MN/m^{3/2}$. Intensive investigations are presently being made in the field of long-term strength. The results of delayed fracture, which are of extraordinary significance for technical application, are shown in Fig. 12 for some qualities. It should be possible to decrease slow crack growth rates by varying type and quantity of additives. Lifetime measurements on non-cracked bend specimens showed 2100 h at 1250 °C and a stress of 250 N/mm^2.

2.5 Duodensity Concept

The post hot-pressing approach is used to produce duodensity turbine wheels. The main problem here is to safely remove the rotor blades from the hot-pressing equipment. To avoid this difficulty, bonding outside the hot-press would be advantageous, but up to now there has been no success. Fig. 13 shows a duodensity rotor, which failed by the break of one blade at 50 000 r.p.m.

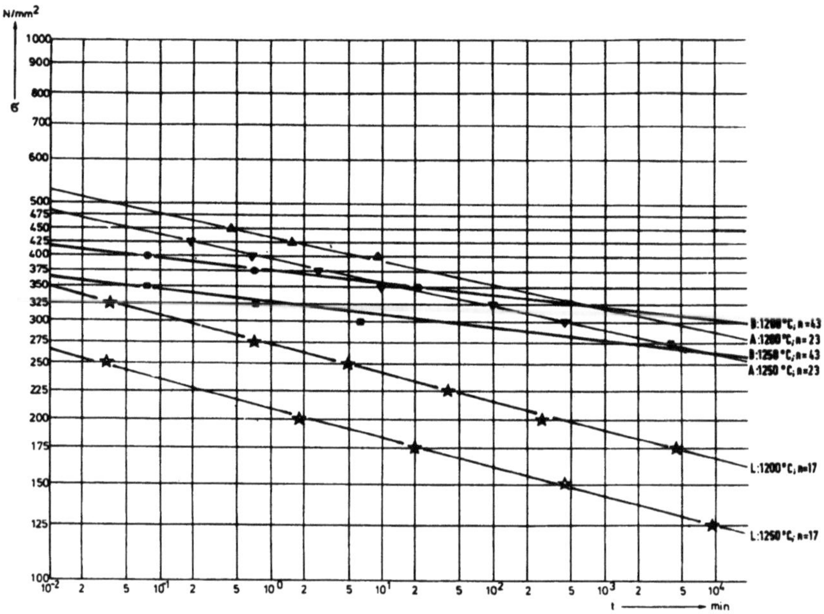

Fig. 12. Stress rupture times for different qualities of hot-pressed Si_3N_4 (Annawerk, TU Karlsruhe)

Fig. 13. Si_3N_4 duodensity rotor (Annawerk)

Fig. 14. Slip cast nose cones (Annawerk) (a) and isostatically pressed combustion chambers (SIGRI, MTU) (b) manufactured with siliconised SiC

2.8 Pressureless Sintered Silicon Carbide

In the future, pressureless sintered silicon carbide might become increasingly significant. At first only β-silicon carbide powder had been considered suitable for sintering. Meanwhile, α-silicon carbide proved to be appropriate for the process also, and even advantageous, so that powder production concentrated on the processing of commercial Acheson SiC.

Furthermore, fundamental investigations were carried out to combine silicon nitride and silicon carbide with metals. Using transition layers, quite acceptable components with metals such as superalloys and tungsten could be obtained.

2.6 Silicon Nitride Mixed Crystals

Investigations of phase equilibria in extended non-oxide ceramic systems were continued. In the Si-Al-Be-C-N system, several new com pounds as well as the possibility of mixing binary and ternary compounds were found. Investigations in the Si_3N_4-AlN-Mg_3N_2-SiO_2-Al_2O_3-MgO system showed that Mg^{2+} is not soluble in β Cialon.

Finally, investigations in the Si_3N_4-ZrN-SiO_2-ZrO_2 (Si-Zr-N-O) system exhibited that a stable ZrO_2 dispersion in nitride materials can only be maintained, if a temporary liquid phase allows rapid densification, and gaseous phase reactions are inhibited.

2.6 Siliconised Silicon Carbide

The material preferably used is reaction-sintered silicon carbide containing pores filled with silicon. This material has the same advantage reaction-sintered silicon nitride offers, i.e., its shape stability is very good during firing. A silicon content of 10% and even less can be reached, but properties do not really require it, since there is no strength increase involved. Bending strength at room temperature ranges between 300 and 400 N/mm^2. There is no perceptible change in strength with changing SiC grain size and up to contents of 20 % silicon. Usually, strength increases with temperature to peak values of 500 N/mm^2, but beyond 1300 ^oC it decreases again due to the softening of silicon.

Interest in this material is caused by its high thermal conductivity (≈70 W/mk at room temperature), which affects resistance to thermal shock in a favourable way. Especially here, a material with moderate silicon content should be appropriate.

Manufacturing of components can be done by dry-pressing, isostatic pressing, extrusion, slip-casting and injection molding. Fig. 14 shows some components presently being tested.

2.7 Silicon Carbide Powder

The manufacturing of SiC powder, which can be sintered and hot-pressed on a technical scale is another task within the scope of the gas turbine program. Doping with additives to support sintering is possible during the process.

In extensive high temperature dilatometer experiments the sintering mechanism was studied and parameters influencing the sintering behavior were investigated. It can be assumed that the process involved is liquid phase sintering.

With small specimens up to 98 % of theoretical density could be achieved. With larger specimens or even components, however, it has not been possible so far to exceed 92 % of theoretical density.

High shrinkage has proven to be a disadvantage impairing the attainment of a high degree of dimensional accuracy of the component.

In mass production of components, which is economically useful only with injection molding, the powder has to meet severe requirements. In order to obtain uniform components highly constant densification behavior is required. To obtain high strength, complete densification, accompanied by an optimum microstructural development, is necessary. To what extent this can be realized in larger components has not yet been clarified.

2.9 Hot-Pressed Silicon Carbide

Hot-pressed silicon carbide has been developed in two qualities, both having a room temperature strength of more than 600 N/mm^2. One quality shows intergranular fracture in a rupture test, the other transgranular fracture. The latter is maintaining its high strength up to 1400 ^{o}C. Hot-pressed SiC is tested as a ceramic rotor material also. The blade ring in this case is machined ultrasonically.

Compared to hot-pressed silicon nitride, dense SiC doubtlessly has the advantage of better high-temperature strength and also of a more favourable oxidation behavior. When silicon carbide is used for the rotor, the high thermal conductivity of 75 W/mK at room temperature certainly presents a problem. Furthermore, it has not yet been proven, if a silicon carbide is a viable alternative to the less brittle silicon nitride.

Results of creep tests on hot-pressed SiC are shown in Fig.15. The improvement of quality is clearly visible. The low and, in all cases, linear dependence of the creep rate on stress at temperatures above 1500°C is worth mentioning.

3. DEVELOPMENT STATE OF THE COMPONENTS

3.1 Combustion Chamber

With respect to low pollutant formation a chamber with a rather large volume is required. Economical reasons demand a single piece component and a simple production method like slip casting.

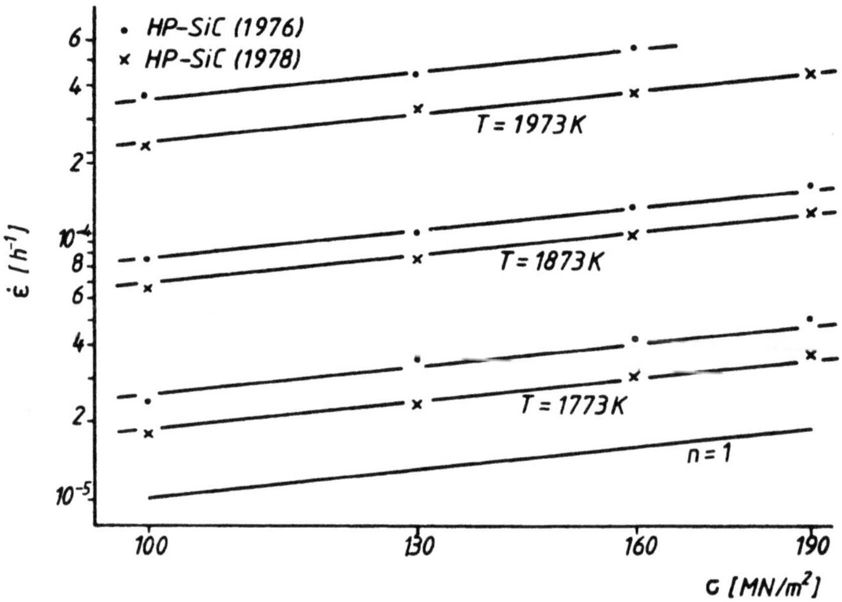

Fig. 15. Creep rate of hot-pressed SiC as a function of stress (ESK, TU Karlsruhe)

Because of the regenerative working process cooling air at the outer surface is not available and the wall might therefore reach a temperature as high as 1800 K near the reaction zone. Between the hotter primary zone and the cooler secondary zone wall temperature gradients as high as 100 K/10 mm wall length might occur. Additional thermal shock is generated during cold start-up and hot shut-down of the engine.

Si-infiltrated SiC flame tubes and turbine inlet nose cones with a wall thickness of 4 mm, developed by the MTU, have been tested. So far this components ran for 20 hours at full load condition, including 300 cycles, simulating rapid load changes between idle and full load.

3.2 Turbine Nozzle Rings

In a high temperature gas turbine the nozzle rings have to withstand gas temperatures up to 1625 K and gas temperature changes of 500 K/s. Although the stresses resulting from gas forces are moderate, rather high thermal stresses can result in a closed nozzle ring from non uniform gas temperatures. In addition, the nozzle blades have to be resistant to high temperature corrosion. The shape of the blade itself is complicated, with a trailing edge of 0.5 mm thickness.

In the development program nozzle blades made from reaction sintered Si_3N_4 and Si-infiltrated SiC in fabrication processes like injection molding or slip-casting, have been investigated. Different designs of Si_3N_4 nozzle rings were tested. The blades are always injection molded and the rings are isostatically pressed. In one design the blades are bonded with the outer ring in the green stage. In other designs the blades are assembled and mechanically attached between inner and outer tip shrouds. The designs differ in the assembling method. Such nozzle rings have been tested in Volkswagen's hot gas test rig. So far nozzle rings with mechanically assembled blades have survived up to 33 hours of cyclic testing. The complex stress situation in integral fabricated nozzle rings, however, still leads to earlier failures.

3.3 Turbine Rotors

The highest stresses wil occur in the turbine rotors. In these components mechanical stresses due to rotation are added to the thermal stresses. Tensile stresses can reach 450 MN/m^2 in the rotor hub and 120 MN/m^2 at the blade root. The question whether a rotor can at all be developed successfully from the available ceramic materials is therefore still the critical question for a high temperature gas turbine.

The goal of the German Ceramic Program is a rotor with 130 mm outer diameter. This rotor should be able to work at gas temperatures of 1550 K with circumferential speeds of 400 m/s at mean blade diameter. In addition, the capability to survive cold start-up cycles should be demonstrated. Since the development of this component is extremely difficult, each of the participating companies is working on this component and investigating special fabrication concepts.

Fig. 16 shows a metal-ceramic hybrid rotor (VW). In this concept the disc is metallic and the ceramic blades are individually inserted in axial rim slots. This concept uses the higher tensile strength of the metallic material in the area of the lower thermal stresses and applies ceramic only in the high temperature area. A further advantage of this concept is that in the case of a fabrication failure only one blade has to be exchanged and not the whole rotor. On the other hand this concept will only allow running temperatures as high as 1450 K, and some reduction in aerodynamic efficiency has to be expected from a reduced number of blades.

The highest load of the ceramic blade occurs in the blade root area. Therefore the development work is concentrating on designing this root area with special care. Some designs try to use soft metallic inserts between blade and disc, in other designs the root gets a special surface treatment. The main development effort on this concept is being done by the MTU. Injection molded blades have

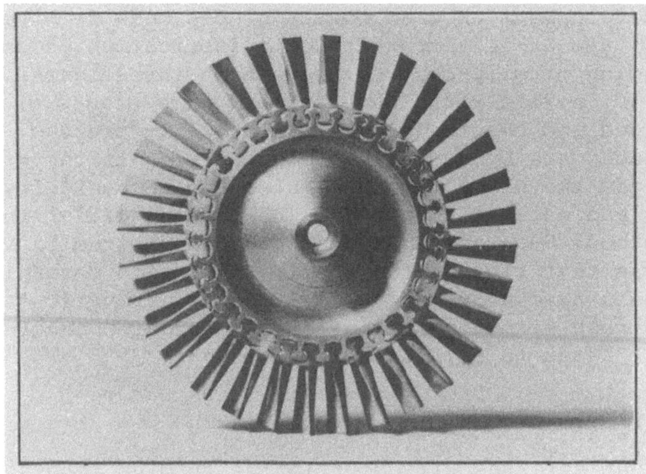

Fig. 16. Metal-ceramic hybrid rotor (VW)

attained circumferential speeds of 400 m/s and hot-pressed blades as high as 500 m/s under conditions of frequent load changes. These results have been obtained in cold spinning tests. It is thought, however, that it is possible to transfer these results to hot operational conditions since this blade attachment should not result in additional thermal stresses.

An all ceramic rotor would offer even more advantages than the metal-ceramic rotor described above. Such an all ceramic rotor would allow even higher gas temperatures. In addition, the lower density of the ceramic disc would give a rotor with a lower inertia, which would result in a shorter response time of the engine. However, in the case of the all ceramic rotor, the high tensile stresses in the hub have to be carried by the ceramic material.

At Mercedes-Benz, prototype rotors have been fabricated from flat hot-pressed Si_3N_4 discs by ultra-sonic erosion and diamond grinding. Fig. 17 shows a gas generator set in which the turbine rotor has been produced by this fabrication method. Continuing progress has been made with these rotors. In the best results, mean blade speeds of 300 m/s at gas temperatures of 1475 K have been attained. The test time has been 22 hours.

Since it is doubtful that machining of a hot-pressed disc will ever lead to an economical product, Volkswagen is investigating a concept in which the all ceramic rotor is built up as a ceramic hybrid. This concept uses reaction sintered Si_3N_4 in the area of the blade ring where the stresses are lower. It is thought that in the

Fig. 17. Gas generator set. The Si_3N_4 rotor has been machined from hot-pressed Si_3N_4 by ultra-sonic erosion (Mercedes-Benz)

final developmental stage such blade rings could be produced by injection molding and that a larger number of the simple shaped hubs could be produced in one single operation in an isostatic hot-press machine. If the development of a bonding technique outside the hot-press is successful, an economical product would result.

Fig. 18 shows different development steps of this rotor. The hubs on the left side have survived speeds of 115 000 r.p.m. The two bonded prototype rotors in the middle have demonstrated that thermal cycles with 500 K/s temperature changes can be tolerated. So far, however, the transfer of these good results to the rotor on the right side has not been successful. Fig. 19 shows typical rim cracks during spinning. In cold spinning 50 000 r.p.m. have been attained, in hot duty cycles only 20 000 r.p.m.

3.4 Heat Exchanger

In a regenerative gas turbine the gas temperature at turbine inlet is also dictated by the maximum temperature capability of the heat exchanger. In the case of a high temperature gas turbine, a heat exchanger has to be developed which can work with gas temperatures as high as 1450 K. In the USA high temperature heat exchangers of the regenerative type are being developed. In the German Ceramic Program a recuperative type is being investigated by Mercedes-Benz. So far this work has been aimed at recuperator elements from reaction sintered Si_3N_4. Several production methods have been investigated. Matrixes made by straight and waved layers did not attain sufficient strength.

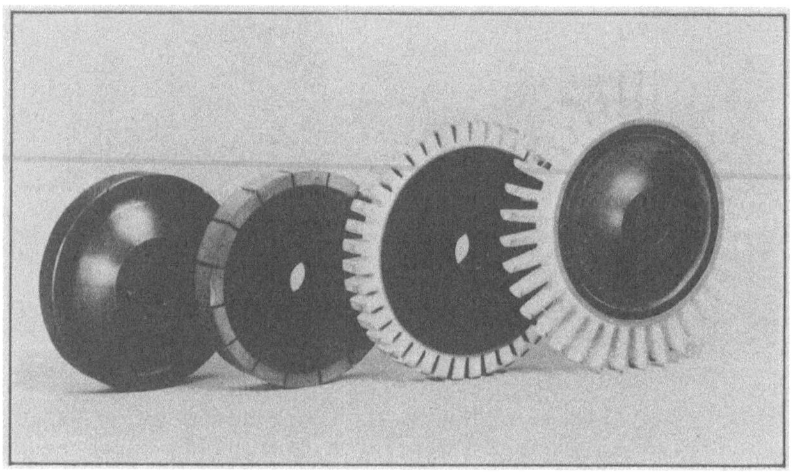

Fig. 18. Development stages of a Si_3N_4 duodensity rotor (VW)

Fig. 19. Si_3N_4 duodensity rotor at fracture

Matrixes made by extrusion had enough strength and low enough porosity. However, no complex shapes could be fabricated. Fig. 20 shows the element of the present recuperator concept in which a package is bonded from several single prepressed layers.

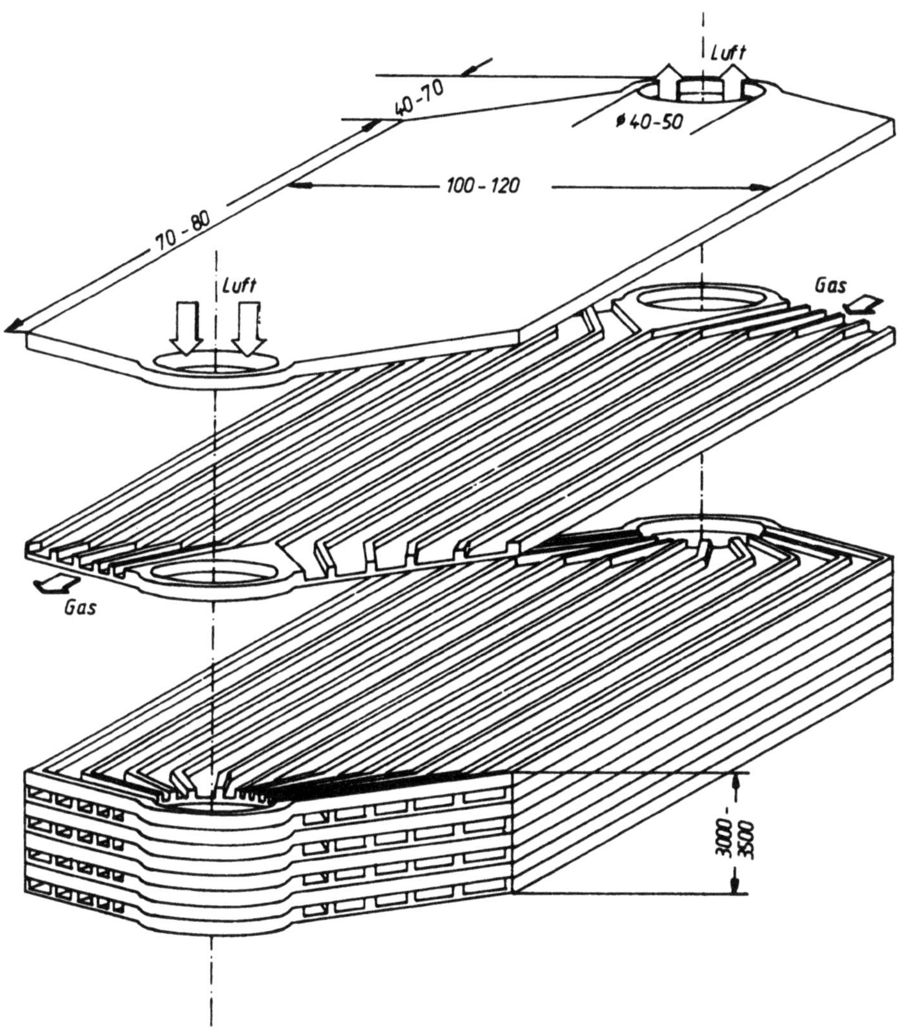

Fig. 20. Concept for a reaction sintered Si_3N_4 recuperator (Mercedes-Benz)

4. FUTURE WORK IN THE CERAMIC PROGRAM

The work reported here represents the results of 5 years of development. Although starting from very limited knowledge, it appears that many of the initial goals of this program can be accomplished, or at least that solutions to the remaining problems seem to be in sight.

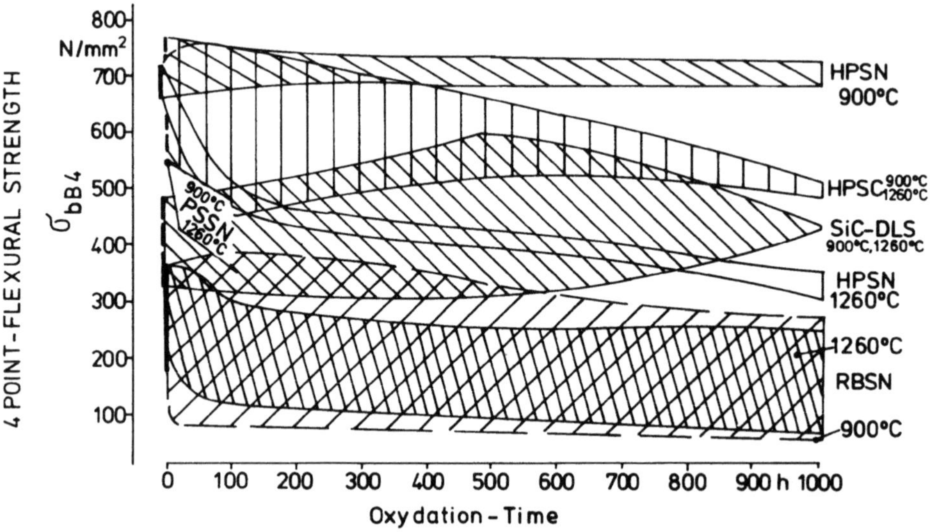

Fig. 21. General data bands of RT strength after oxidizing heat treatment of nonoxide ceramics (state of May 1979), (VW)

In Fig. 21 a problem is touched on where still no solution can be seen. This is the problem of strength degradation with time which was found in all the ceramics under investigation by Volkswagenwerk. If this situation cannot be improved considerably, ceramics will be of no use in highly stressed gas turbine components. This shows how high the risk of these development problems still are and how important worldwide cooperation is to solve these basic problems.

CERAMIC ENGINE RESEARCH AND DEVELOPMENT IN SWEDEN

S. O. Kronogard, and

L. Malmrup

United Turbine

INTRODUCTION

Development of ceramic materials for heat engines is actively going on in all countries which have a viable domestic engine industry.

The reason for this is that ceramic materials have great potential to improve a number of important characteristics of heat engines. Most important is the improvement of efficiency and fuel economy resulting from the higher temperature which can be used in an engine if ceramics were utilized.

Obviously the potential advantages of ceramic materials vary for different applications. The list in Fig.1 shows some of the more general characteristics of ceramics in heat engines.

Both higher working temperature and lower cooling requirements lead to improved efficiency. The first is especially applicable to engines with continuous combstion where the allowable temperature is particularly limited by the material strength at high temperatures.

Examples of such continuous combustion engines are gas turbines and stirling engines.

MAIN ADVANTAGES OF CERAMICS IN HEAT ENGINES

- HIGHER WORKING TEMPERATURE
- LESS COOLING REQUIREMENTS
- LIGHTER
- LOWER INERTIA
- LOWER COST
- NON STRATEGIC MATERIALS

Figure 1. Advantages of Ceramics in Heat Engines

The lower heat loss and reduced cooling requirement on the other hand is of value for all water cooled engines of which the conventional Otto and Diesel engines are examples.

The lower weight and substantially reduced inertia are both due to the lower specific weight of ceramics compared to metal. Typically there is a relationship of 1:2,5 between ceramic and steel. However a lower weight of the complete engine is also a result of the higher temperature used because this dramatically increases the specific power of an engine and thereby reduces the engine weight and volume for a given power. Again this is more pronounced for the continuous combustion engines. A smaller and lighter engine results of course in decreased engine cost. So does the fact that the ceramic materials itself in most cases have a potential to be much less costly than the metal material it replaces, typically a so called super alloy.

Furthermore the ceramic materials are non strategic materials which is important looking ahead at the need to save important and limited mineral resources on the earth. In Sweden we have a rather important and viable vehicle and engine industry. This automotive industry plays an important role in the Swedish employment and balance of trade.

Volvo of course manufactures engines for its own cars, trucks, buses and boats, and so does SAAB-SCANIA.

Volvo Flygmotor manufactures jet engines for Swedish fighter airplanes. The company is also involved in international cooperation on civil jet engines. Stal Laval develops and manufactures gas and steam turbines for power generation, industrial use and ship propulsion.

United Turbine, which is a subsidiary to Volvo, is engaged in the development of a small automotive gas turbine. United Stirling belongs to a government owned section of the industry. The company is developing Stirling engines for different applications such as total energy systems, small delivery vans, etc.

CERAMIC MATERIAL DEVELOPMENT

As regards the Swedish ceramic industry, ceramic companies typically are engaged in manufacturing of electrotechnical components such as large insulators, resistors and sanitary goods but also refractory materials. This mainly gives a background in such manufacturing methods as slip casting and to a certain extent injection moulding of ceramic materials.

SWEDISH CERAMIC INDUSTRY

COMPANY	PRODUCT LINES	TURNOVER (Millions)	EMPLOYEES
IFÖ	ELECTROTECHNICAL SANITARY ENAMEL	$70	1500
GUSTAVSBERG	SANITARY HOUSEHOLD ENAMEL BRICK-WORKS	$70	1300
HÖGANÄS	BUILDING REFRACTORY BRICK-WORKS	$60	900
RÖRSTRAND	HOUSEHOLD	$20	650
FORSHAMMARS BERGVERK	REFRACTORY	$18	200

(The Figures apply to these Specific Product Lines.)

Figure 2. Some Swedish Ceramic Companies

As specified in Fig. 2 the approximate numbers mentioned on turn-over and employees are for the specified product lines only. The companies otherwise belong to much larger units with rather large development budgets.

Swedish Board for Technical Development

With this background in engines, especially gas turbines and stirling engines and with a ceramic industry within the country, the Swedish Board for Technical Development (STU) in 1976 decided to financially support the work in ceramic materials for heat engines. This board for technical development is a government agency. It was decided to support two parallel activities - one aimed at material development, and the other at engine development.

The material development which as a long term goal has a domestic production of ceramic heat engine parts should be carried out at R&D Institutes to start with. The work to introduce ceramic materials in heat engines shall on the other hand be performed by the engine companies. The support to the engine companies shall help the engine developer to obtain ceramic components from abroad for test in engines at an early stage before the Swedish ceramic industry can produce experimental ceramic parts.

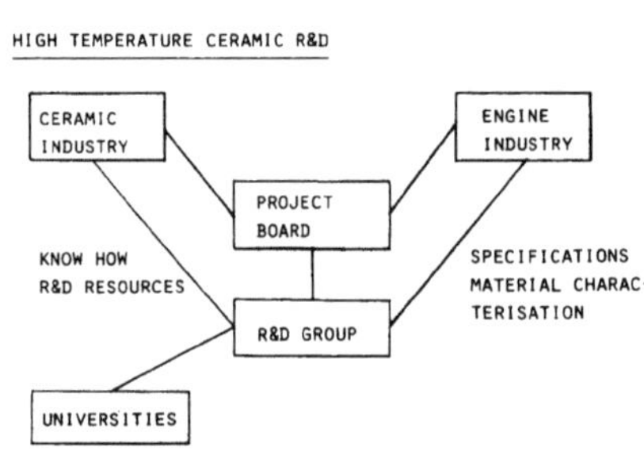

OBJECTIVES

- LONG TERM DOMESTIC PRODUCTION OF CERAMIC PARTS FOR HEAT ENGINES
- MATERIALS FOR HEAT ENGINES
- MANUFACTURING TECHNOLOGY
- MATERIAL CHARACTERISATION
- SINTERED SiC
  SINTERED Si_3N_4

Figure 3. Objectives of Swedish Institute for Silicate Research

Swedish Institute for Silicate Research

It was decided to locate the initial development work on high temperature ceramics at the Swedish Institute for Silicate Research. The objectives of this research and development R&D project are listed in Fig. 3.

As stated earlier the long term goal is to create a knowledge base which can be drawn upon in a future Swedish production of high temperature ceramic components. The work is concentrated on materials and processes that can be used for manufacturing components for small vehicular engines regardless of type.

Materials to be studied as first priority are pressureless sintered SiC and Si_3N_4. The work will include material characterisation and experimental studies on manufacturing methods that can be used in volume production.

The work of the ceramic R&D group is guided by a project board. This board has representatives from the engine industry, the ceramic industry and universities. Through this project board a close cooperation between the R&D group and industry is ensured. For example the projects can use equipment from the ceramic industry and also gain from the know-how already available there.

The engine industry can present specifications for different kind of components and thereby provide important guideline for the work. The R&D group can also function as a service organisation for the engine industry by carrying out for example material characterisation on materials being tested in engines.

Although this project will be separate from the ordinary work at the institute, it can when needed use base resources in the form of equipment and also people from the institute. Today a small team of research people are employed in the project.

Among the equipment available the following ought to be mentioned - scanning electron microscope with X-ray spectrometer, X-ray diffractor and X-ray fluoroscan equipment, thermo balance equipment, high temperature bend test fixture, equipment for powder characterisation and powder handling, carbon resistance furnaces and injection moulding equipment.

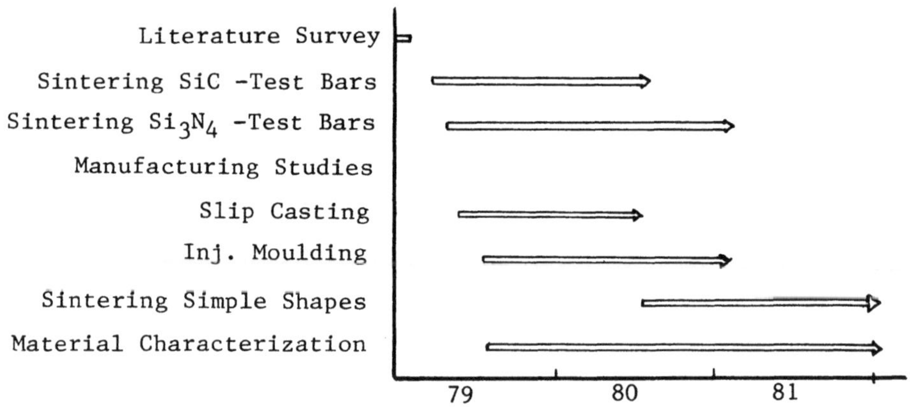

Figure 4. Time Schedule for High Temperature Ceramic R&D

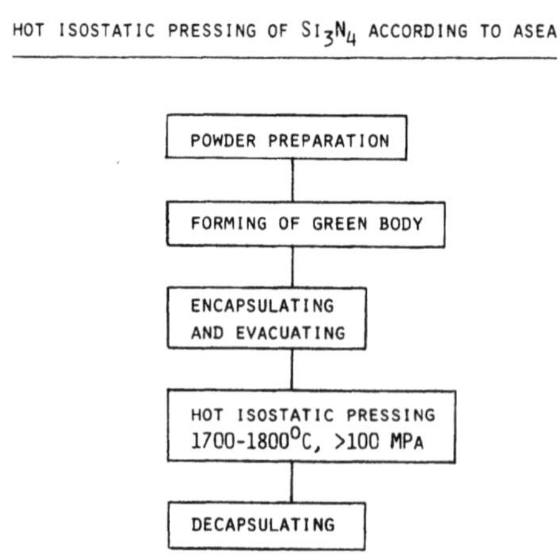

TYPICAL PROPERTIES

$\sigma_{BRT} = 700\ MN/M^2$

$\sigma_{B1370^\circ C} = 550\ MN/M^2$

$M = 20$

CREEP RATE = $1.6\times10^{-6}\ H^{-1}$ AT $70\ MN/M^2$, $1300^\circ C$

Figure 5. Hot Isostatic Pressing Process

The time chart in Figure 4 gives a condensed picture of the work being planned within this high temperature ceramic research program the next couple of years. The activities include sintering of test bars, studies on slip casting and injection moulding but also the manufacturing of simple shapes.

The first goal is to be able to manufacture simple heat engine components within the present frame of this initial project phase.

HIPing of Ceramics at ASEA

At ASEA in Sweden R&D work on hot isostatic pressing of ceramics is carried out. Besides being a manufacturer of electrical equipment and power stations ASEA is also producing equipment for hot isostatic pressing. To develop the high pressure technique further, ASEA in 1965 set up a high pressure laboratory at Robertsfors in Sweden. At this laboratory a new HIP technique to process Si_3N_4 has been developed.

The main advantages in using a HIP process is the possibility to get fully isotropic materials of complete density with sintering aids. The process is summarized in Fig. 5.

Silicon nitride powder is mixed and milled to appropriate particle size. Any sintering aid may be added to the mixture but the amount can be reduced substantially compared to what is needed for the hot press process (HP).

A green powder body is made by any desired process giving uniform density distribution and reasonable strength after burning off any temporary binding agents. As the green body contains a communicating porosity it is necessary to apply a temporary barrier for the high pressure gas before hot isostatic pressing. The body is therefore put into an oversized capsule of glass. The capsule is evacuated at an elevated temperature before sealing off.

The encapsulated body is then compacted during a hip cycle. During the first phase the temperature will be increased to the softening temperature of the glass. Pressure is applied and the capsule will completely fill out the space around the green body. The temperature is then increased to the final operating value, 1700-1800°C.

The capsules are kept at the operating temperature under a constant pressure of at least 100 MPa. (14500 PSI). Actual temperature, pressure and time during the HIP cycle will depend upon type and mixture of silicon nitride to be compacted.

The shrinkage of the preformed part is totally uniform and can be predicted for a given powder and pretreatment. The glass capsule is removed after the HIP cycle by sand blasting.

The properties obtained depend very much on the powder quality. But typical data for a good powder is
σ_{BRT} = 700 MPa (100,000 PSI) and
$\sigma_{B1370^{\circ}C}$ = 550 MPa (80,000 PSI) with a Weibull modulus of 20.

This method with the oversized glass is of course not suitable for mass production of complicated shapes such as a complete turbine wheel for automotive engines.

A new technique suitable for mass production and based on a method where the glass envelope is formed by glass particles has therefore been developed. This technique is described on Fig. 6.

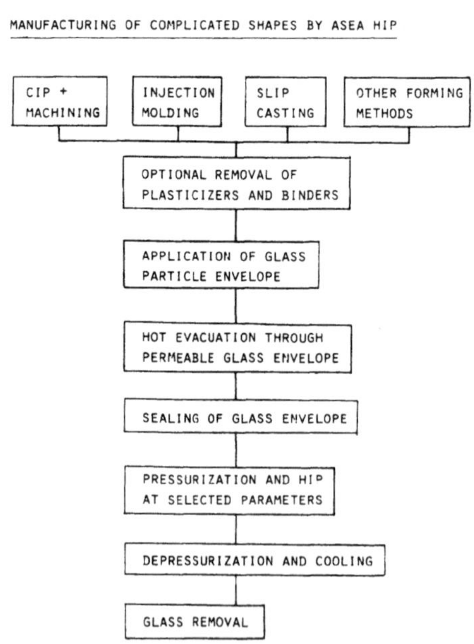

Figure 6. HIP Process Suitable for Mass Production

The preforming can, as before, be made in a number of different ways. Typically injection moulding will be used in mass production.

After preforming the plastic and binders are removed. The component is then covered by glass particles. The temperature is raised so that a glass layer is sintered around the body. Before the envelope is thus completely "sealed-off" the final evacuating of the component has been made. The pressure is then raised further and so is the temperature and the HIP process is completed. After lowering of pressure and temperature the glass can be removed. Fig. 7 shows some components that have been manufactured with this technique. A reproducability of better than 0.1 mm on a 100 mm diameter part has been obtained.

ENGINE DEVELOPMENT

On the advanced heat engine side the major Swedish ceramic activity is on the development of gas turbine parts and to a certain extent parts of Stirling engines.

As regards conventional Otto and Diesel engines both Volvo and SAAB-SCANIA are following the development with great interest. They are also both conducting some experimental R&D work themselves.

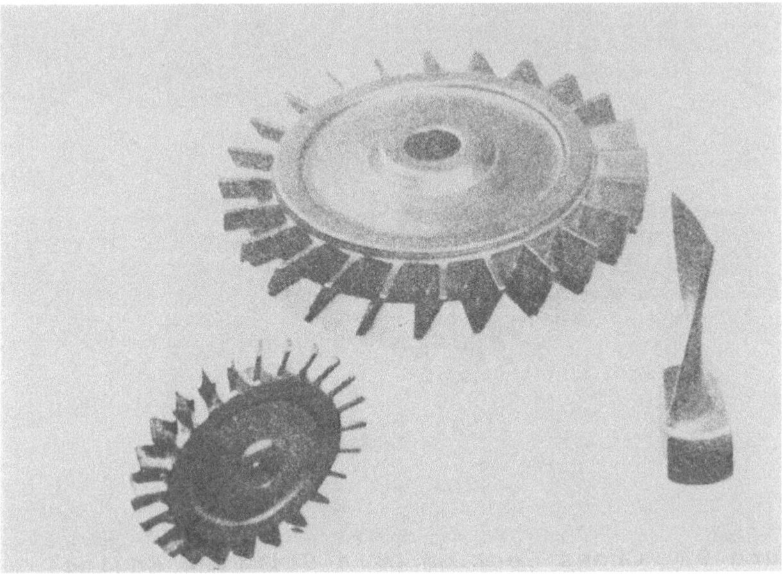

Figure 7. Ceramic Components Made by the HIP Process

CERAMIC ACTIVITIES AT UNITED STIRLING

United Stirling has as a first objective to develop and market the Stirling engine. In Fig. 8, a metallic version is shown. However, since the power and efficiency will be significantly increased with an increased operating temperature, Fig. 9, the use of ceramic materials for the heat system is of very high interest. Furthermore, some components in the "metallic" Stirling engine can be replaced by ceramic components, for instance the pistons, the combustor and the air preheater. Thus it is possible to introduce ceramics in the Stirling engine part by part as they are developed and successively change it to a "ceramic" engine.

The key element and the component that necessarily has to be ceramic to utilize the high temperature potential of the ceramics is the heater, which is the heat exchanger between the combustion gases and the working medium (H_2), hydrogen, in the engine. This is also the most difficult component to design and manufacture in ceramic materials. The projected material temperature levels of a ceramic heat system for the Stirling engine are shown on Fig. 10.

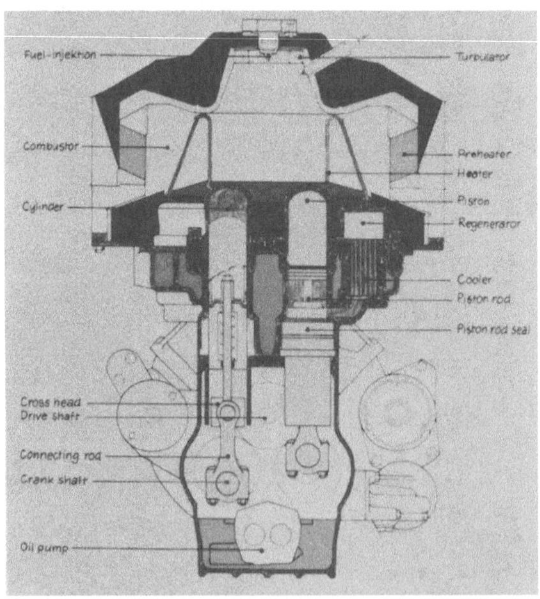

Figure 8. Cross Section of a Stirling Engine

United Stirlings philosophy for a ceramic engine is to simultaneously work with different components in ceramic materials and introduce them into the engine as they are developed.

The air preheater is one of the components that probably is possible to introduce in the relatively near future. United Stirling has been working on adaption of the same type of regenerative air preheater as is used for gas turbines to the Stirling engine. The regenerative heat exchanger tested has been a glass-ceramic one. From the beginning lithium alumina silicate (LAS) and later the alumina silicate (AS) version developed by Corning was used.

A number of LAS cores were tested and failed in less than 100 hrs in a test rig. One AS core has been tested for about 500 hrs and was still intact. United Stirling has also started the development of a recuperative air preheater. This is a cross-flow Si_3N_4 heat exchanger designed in segments surrounding the heater. The first version is designed, heat exchangers are manufactured and testing will soon start.

The combustor is another component that could be made of ceramic material in the near term. A number of combustors have been tested in separate test rigs. Combustors of different types of SiC and Si_3N_4 were tested, all of them with rather discouraging results. They all failed in less than 100 hrs. However, all the combustors tested were of a design similar to that of the metallic combustor. Further analyses show that a redesigned version of the combustor is needed. Such a redesign will be done together with testing of the influence of the combustor shape on the combustion process.

The pistons in the Stirling engine are elementary shapes, much simpler than in an internal combustion engine, and should be a comparatively easy component to make of ceramic materials.

In addition, the temperature distribution is symmetrical and the working conditions can be met by various materials. Ceramic pistons could be used in Stirling engines at present, but since there will be little performance to gain until one has a ceramic heater, pistons are not one of the first ceramic components to be considered.

P40 STIRLING ENGINE

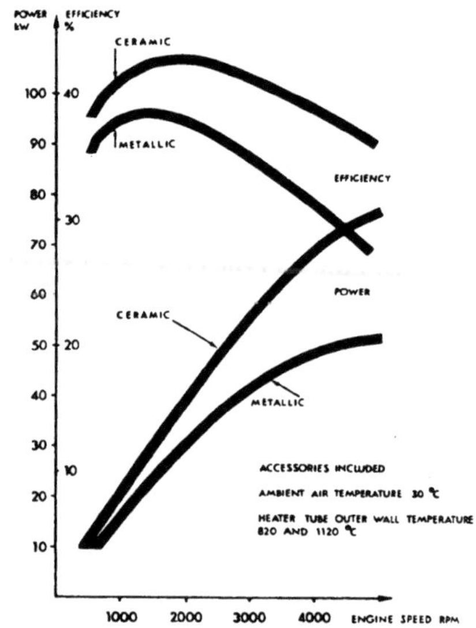

Figure 9. Efficiencies and Output Power of Metallic and Ceramic Stirling Engines

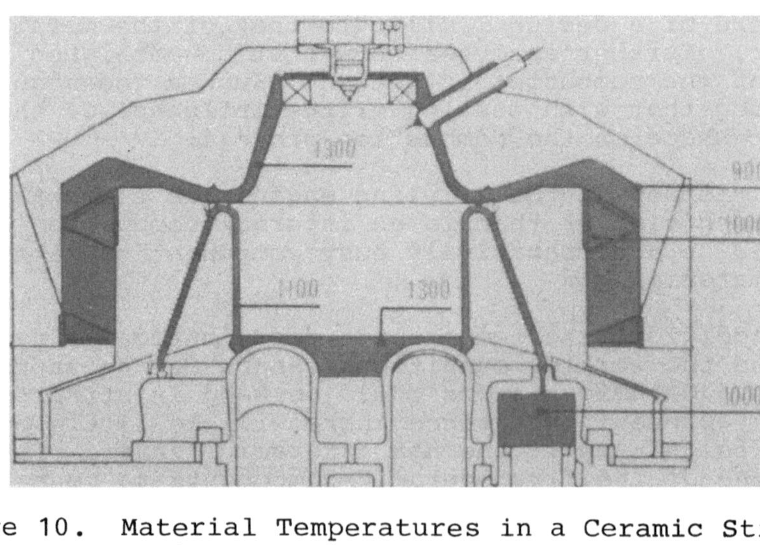

Figure 10. Material Temperatures in a Ceramic Stirling Engine

As mentioned before, the heater is the most important component to make in ceramic materials, but it is also the most difficult and complex one. It is built-up of a heat exchanger joined to cylinders and regenerator housings. The whole unit has to be completely redesigned in a ceramic version but there are some general areas to be investigated independent of (the evolution of) the final version.

The engine contains a certain amount of hydrogen gas that has to be kept in it for a sufficiently long time. Therefore, the hydrogen diffusion rate through candidate materials is an important parameter. United Stirling has tested the diffusion through Refel tubes at 1.100°C. The diffusion rate at that temperature was somewhat lower than through metallic tubes at 800°C. This indicates that loss of working gas can be handled at the raised temperature level by the right material choice.

It is likely that there must be some joints in the heater. Consequently, some joining tests have been performed and the results are promising. Cylinders and regenerator housings will be subjected to cyclic temperatures at the top, heated up at engine operation and cooled down at shut off, and at the same time be attached to a water cooled engine block at the bottom. Fig. 11 shows a design of a ceramic regenerator housing which will be tested.

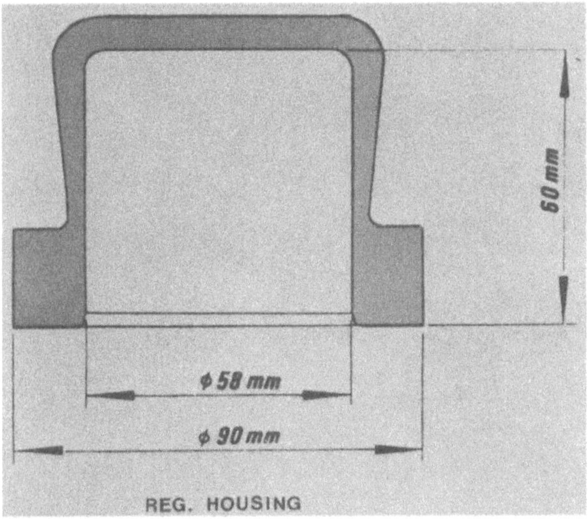

Figure 11. Ceramic Regenerator Housing

Tests of the heater parts mentioned are steps towards the design, manufacture and testing of complete heaters that will follow as soon as the results of the experiments allow.

DEVELOPMENT OF CERAMICS FOR GAS TURBINES AT UNITED TURBINE

Fig. 12 shows the automotive gas turbine now being developed at United Turbine. As you can see from the picture this experimental version uses two heat exchangers located symmetrically on each side of the engine. This is to minimize thermal distortion of the housing. Some design data for this engine are

turbine inlet temperature	1,040°C
pressure ratio	5:1
power	130 HP

The gas turbine we are developing is based on a special system called the KTT system (Kronogard Turbine Transmission System).

Figure 12. Experimental Gas Turbine - KTT MkI

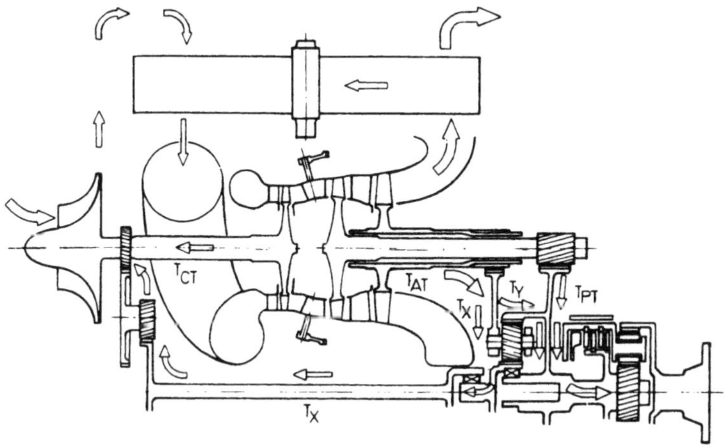

Figure 13. The KTT Gas Turbine Principle

KTT SYSTEM FEATURES

- EFFICIENT AND HIGH PERFORMANCE
- DESIGNED FOR COMPACT VEHICLES
- INTEGRATION - ENGINE & TRANSMISSION
  - HIGH FUEL ECONOMY
  - LOW WEIGHT & VOLUME
  - LOW COST & MAINTENANCE
- THREE TURBINE STAGES
  - HIGH FUEL ECONOMY
  - FAST RESPONSE
  - HIGH SPECIFIC POWER
- SUITABLE FOR CERAMICS - EARLY USE
- HIGH DEVELOPMENT POTENTIAL
- EXTENSIVE PATENTS

Figure 14. Advantages of the KTT System

The KTT Gas Turbine

Fig. 13 shows the principle of this gas turbine system. As can be seen the system includes the same components as a two shaft gas turbine, that is compressor, burner, heat exchanger, compressor turbine and power turbine. But the KTT system also has a third turbine stage on a separate shaft, the auxiliary turbine, which is located after the power turbine. This turbine delivers power both to the compressor, the accessories and to the output shaft. This split of power is done by a planetary gear system.

The KTT Gas Turbine system has a number of advantages compared to a conventional two shaft gas turbine as can be seen in Fig. 14. One of the main advantages is the favourable torque characteristic which means that only a very simple transmission is needed for car propulsion.

Another important feature is that the three turbine stages of the KTT system make the engine very suitable for the use of ceramic materials. Fig. 15 shows a comparison between the turbine system in a two shaft and in a three shaft gas turbine with the same output power and the same shaft speeds. One can see that the diameter of the turbine wheels are smaller in the three shaft case. This means lower blade stress, disk stress, engine volume and rotor inertia as illustrated by the bars.

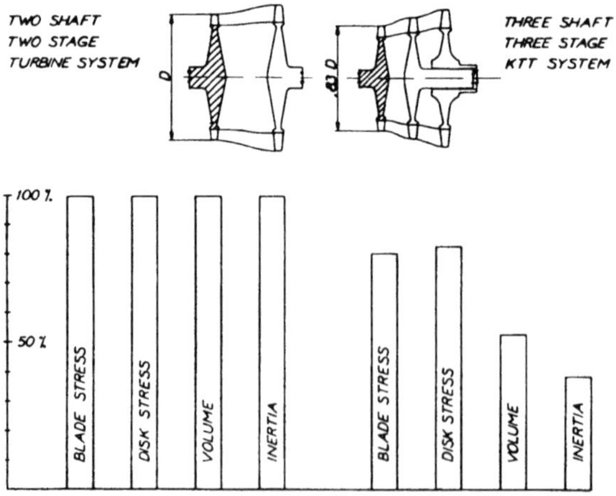

Figure 15. Relation Between Turbine Rotor Stress, Volume and Inertia for Two and Three Shaft Systems.

It is of special interest to note the lower stress levels of the three shaft system both in the turbine blades and in the turbine disks. This is an important advantage when introducing ceramic materials since the main problems with ceramics in turbine wheels are the stress levels.

It is also important to recognize that when going to higher turbine inlet temperatures the specific work increases, which regardless of turbine system leads to the use of more turbine stages, e.g. from two to three stages, not only to reduce stress but to reduce turbine loading in order to keep the turbine efficiency at an acceptable level. In the KTT system we already have the three stages and they are used for other reasons too, such as high torque multiplication, for high efficiency and low stresses.

Ceramic Components

Much has been said about how the higher turbine inlet temperature increases the total efficiency of the gas turbine. In the following, the results of a calculation made for our specific engine are illustrated. Fig. 16 shows the fuel consumption in MPG calculated for a given car over the combined Federal Driving Cycle.

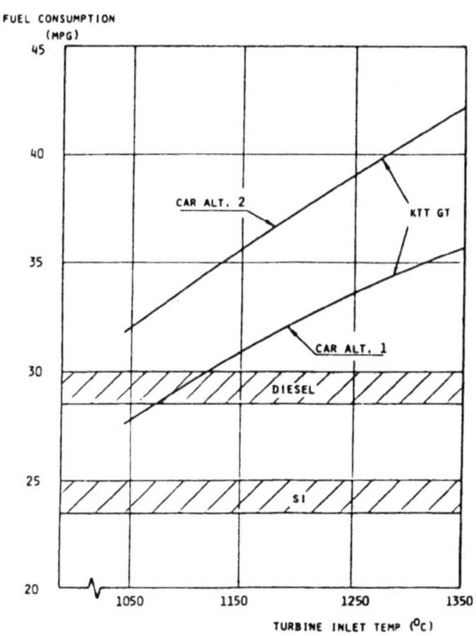

Figure 16. Influence of Turbine Inlet Temperature on Fuel Consumption for the KTT System

Today with 1040°C as turbine inlet temperature the KTT system is somewhat better than the Otto engine. By going to 1150°C it compares very favourable with the Diesel engine. The figure 1150°C is the temperature level we think can be used without ceramic turbine rotors, with the use of powder metal first stage rotor and ceramic stationary parts. By going to 1370°C we have gained almost 30% in combined MPG compared to the 1040°C gas turbine. The absolute fuel economy levels of Fig. 16 can be increased further by improvements in component efficiencies and more use of variable geometry in the flow path.

Fig. 17 shows some hot metallic parts in our present experimental automotive engine, designated KTT MkI. It shows combustor, vortex, compressor turbine nozzle and compressor turbine wheel. The turbine wheel has a diameter of 110 mm. The four parts mentioned are the hottest and therefore the first which will be made in ceramics. A ceramic heat exchanger is also shown. In our development project we plan to increase the turbine inlet temperature stepwise according to the following discussion.

Today at 1040°C we only use ceramic material in insulation and heat exchangers. However, at 1150°C the combustor, scroll and compressor turbine nozzle need to be manufactured in ceramics as well. For the compressor turbine wheel a better metal material is needed than presently used 713LC. Powder metal wheels manufactured according to a method being developed by P&WA, Florida will be tried.

The next temperature level will be 1290°C. Besides the earlier listed ceramic components, the compressor turbine wheel and the power turbine nozzle then also need to be of ceramics. In this case the variable nozzle is moved to a position behind the power turbine, the reason being that it is still not clear how difficult it will be to make a ceramic variable nozzle. In the KTT system it is possible to place the variable nozzle in front of the auxiliary turbine instead of in front of the power turbine and still get about the same control function and overall engine characteristics.

The next temperature level listed is 1370°C. Then all hot components except the auxiliary turbine need to be made in ceramic materials. This limit of 1370°C is set by the ceramic material if this is Si_3N_4. With SiC the potential maximum temperature is higher.

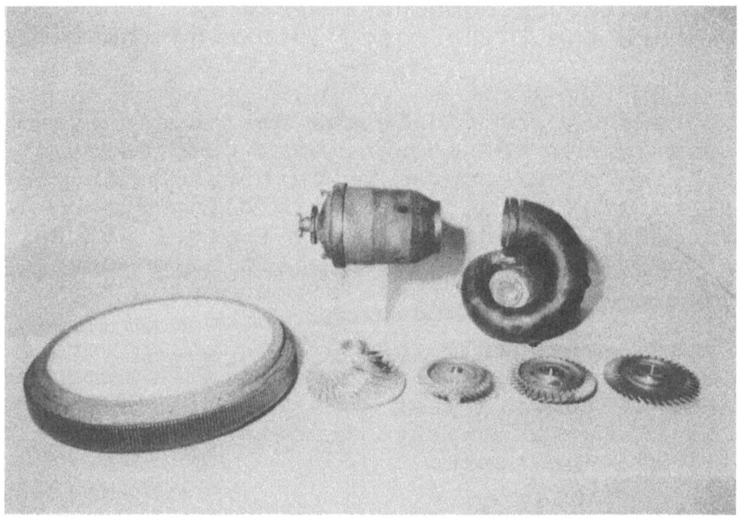

Figure 17. Some Hot Parts in the KTT Gas Turbine Engine

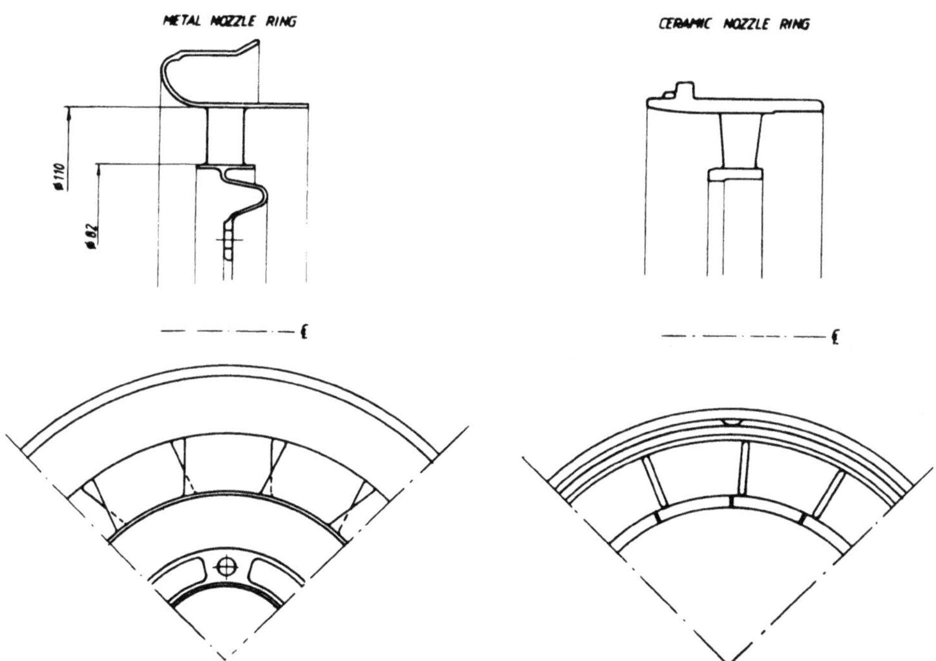

Figure 18. Metallic and Ceramic Compressor Turbine Nozzles for the KTT Engine

High temperature component R&D work is now actively going on within the KTT project involving the following components.

The ceramic scroll is on the design stage and the first vendor is not yet chosen. The compressor turbine nozzle is being manufactured by Carborundum in their α-SiC material. Turbine wheels are being forged in powder metal for United Turbine by Pratt & Whitney (P&WA). A ceramic compressor turbine wheel is being developed together with ASEA using their HIP process.

The heat exchanger cores earlier shown were made from LAS material with a maximum allowable temperature of about 800°C. Engine test cores made from thin wall AS material by Corning with a temperature capability of 1000°C will also be tested.

Ceramic Turbine Nozzles

Fig. 18 shows a comparison between the present metal compressor turbine nozzle and a suggested design for a ceramic nozzle. The shape of shrouds of the ceramic nozzle is as can be seen much simpler. The blading is changed to allow an axial retraction of the injection moulding tool. Thus, there is no overlap of the blades in the ceramic nozzle, as exists in the metal one.

The inner shroud of the ceramic nozzle is slotted. This is to lower stresses due to differences in thermal expansion of inner and outer shroud.

Fig. 19 shows the result of a temperature calculation of a SiC compressor turbine nozzle vane during a start up cycle. This particular plot shows the temperature eight (8) seconds after light up. As can be expected the hottest portion of the blade is at the trailing edge where we have the darkest area. The analysis is made by a three dimensional finite difference computer program on the complete nozzle including the shrouds.

Fig. 20 shows the corresponding stress picture for the nozzle vane. The stress plotted is the maximum principal positive stress. In this case the analysis is made by a two-dimensional finite element model, using elements which allow a variation of the blade thickness.

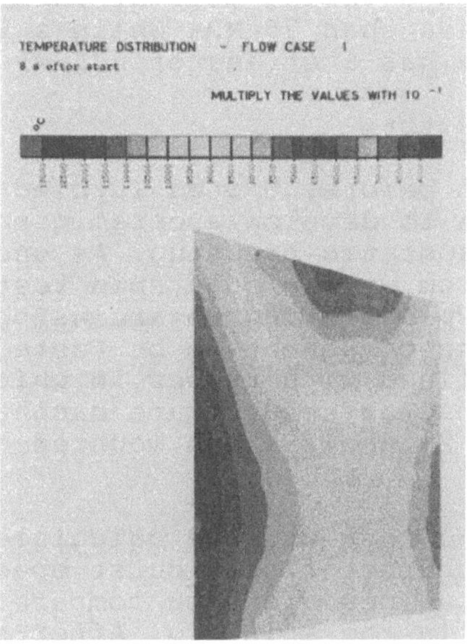

Figure 19. Temperature Distribution in a Nozzle Vane
8 Seconds After Start

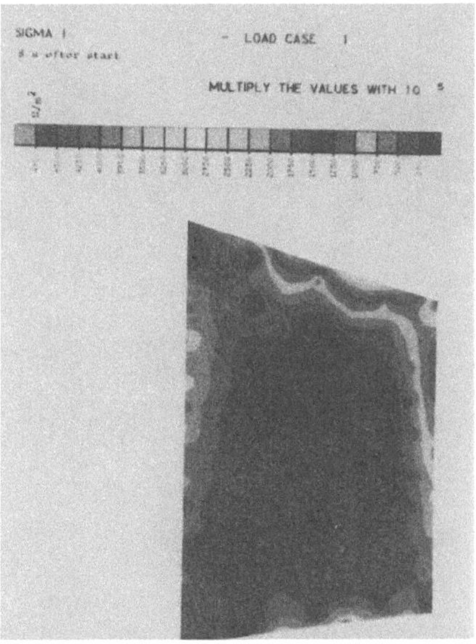

Figure 20. Stress Distribution in a Nozzle Vane
8 Seconds After Start

From the scale to the right of the plot one can see that the dark area covering most of the blade corresponds to a stress of less than 25 MPa while a local point at the trailing edge has a stress of 275 MPa.

Ceramic Turbine Rotor

As mentioned before, United Turbine has a joint venture with ASEA to develop a ceramic turbine wheel by the use of hot isostatic pressing. As one part of that development program we are cold spin testing simple hub shapes. For the spin testing we are using an air turbine, which can spin the component to be tested to 100,000 rpm. The spin pit is lined with rubber in which some of the pieces coming loose at a burst are caught and can be analysed. Figure 21 shows a hub mounted on its shaft in the spin rig.

From the bend test data we calculate the Weibull modulus and the characteristic burst speed of the hubs. These calculated values are then compared to values actually reached in the spin rig. After some initial

Figure 21. Simple Ceramic Turbine Hub Mounted in the Spin Rig

problem with manufacturing errors on the hubs which resulted in lower characteristic burst speed than calculated, we have now got a fairly good agreement between calculated values and test values.

The hubs are manufactured by cold pressing of billets of Si_3N_4 from which the final shape is machined while still in a green form. Thereafter the hubs are HIPed and the only machining made after HIPing is grinding of the shaft section.

Besides spinning simple hubs we are also spin testing hubs with simple shaped blades. This is for testing of among other things the area where the blades join the hub. Fig. 22 shows different hub/blade configurations manufactured for this type of test. The figure also shows some blades which have been manufactured for us by ASEA. Blades with a trailing edge thickness down to .3 mm have been manufactured. All the components shown have been preformed by cold pressing followed by machining to shape of the green body.

Figure 22: Different Hiped Ceramic Parts from the KTT Gas Turbine Ceramics R&D Program

Fig. 23 shows a simple disk before and after spin test. As you can see the pieces are rather small which always seems to be the case when the hub bursts at high speed. In some cases one can trace the starting point of the rupture on one of the bigger pieces which was located near the center of the hub.

The heat transfer and stress analyses of the complete turbine wheel are carried out with the same computer program as used for the nozzle ring. The result of a transient stress analysis is shown in fig. 24. The stress plotted is maximum principal positive stress 10 s after start. The dark areas in the neck area and on the side of the disk somewhat above where the shaft joins the disk are the areas with the highest stresses. Fig. 25 shows a calculation made at the steady state operating condition of 70.000 rpm and 1370°C TIT. The highest stress in the lower portion of the disk is now 285 MPa to be compared to the 335 MPa in the transient case. One can also see that the stress levels in the major parts of the disk are lower in the steady state case.

With a Weibull modulus of 20, a MOR of 700 MPa measured by three point bending of a 3x3x20 mm test bar the probability of failure for this particular experimental turbine wheel in the transient case is .0005 (1:2000).

To be able to use injection moulding as the preforming method the blading of the turbine wheel has to be designed to allow the tool to be retracted in a near radial direction.

Figure 23. Ceramic Disk with Simplified Blades Before and After Spin Test.

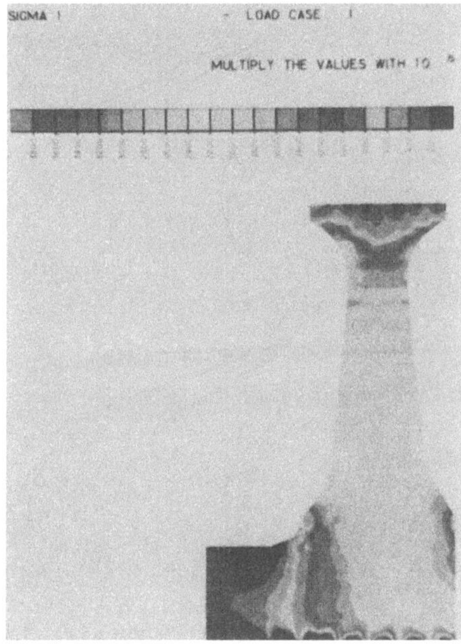

Figure 24. Stress Distribution in a Ceramic Rotor 10 Seconds After Start

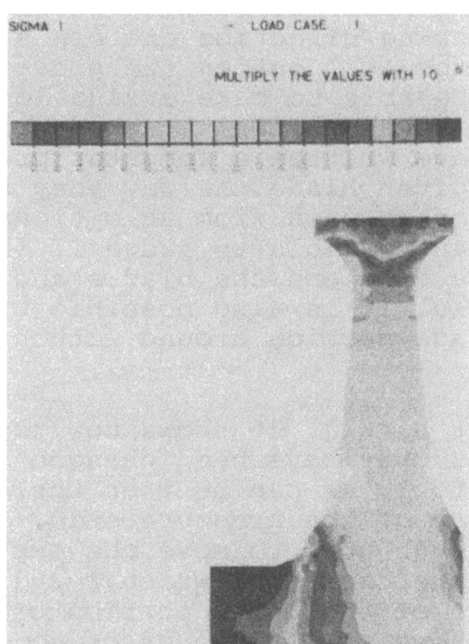

Figure 25. Stress Distribution in a Ceramic Rotor at Steady State

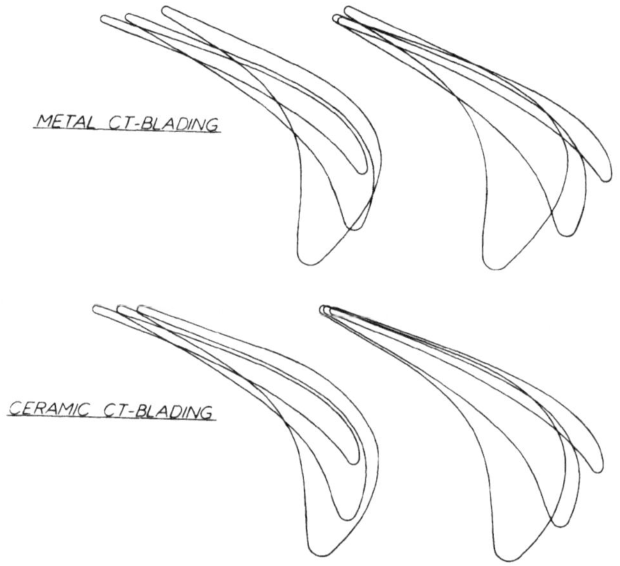

Figure 26. Shape of Metallic and Ceramic Compressor Turbine (CT) Blade

On the upper half of Fig. 26 the profiles of our present cast metal compressor turbine wheel are shown. In this particular viewing direction one can see that it is impossible to get the tool out on the pressure side at the trailing edge. To be able to make evaluations of different blading designs as regards stress, aerodynamics and manufacturability we have written a computer program which plots the sections of a given blade the way they stack on each other when being looked upon from an optional angle. From this plot it is then possible to judge if there is interference between the tool and the blades and what changes are needed to avoid this. It is also possible with this program to move a given blade section around without changing its profile shape.

The lower part of Fig. 26 shows how the profile sections look after they have been changed to allow the retraction of the tool. As can be seen there is no overlap at the trailing edge of the pressure side. In this case it was not enough just to move the sections. We had to redesign the two inner sections which resulted in a somewhat smaller throat area of the inner portion of the blade. We have calculated a loss in efficiency of about 1% due to this modification. The pulling direction of the tool for this particular blade is 8 degrees towards the leading edge and 1.5 degrees towards the suction side of the blade.

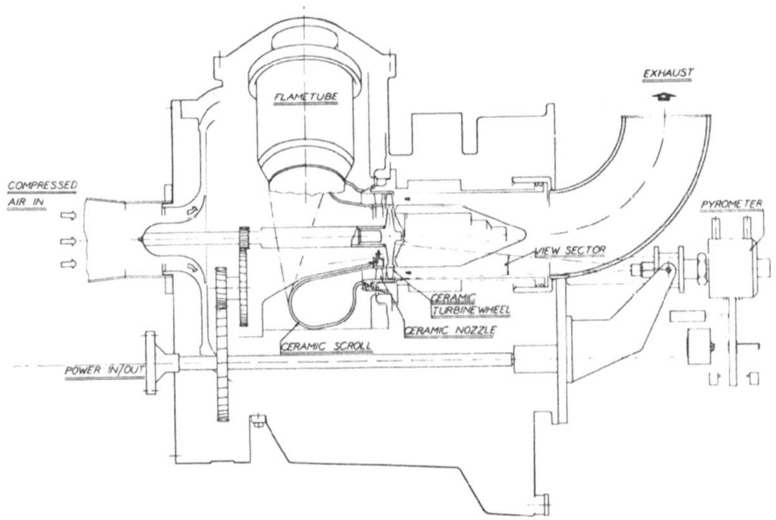

Figure 27. Hot Spin Rig Using the KTT Engine Gas Generator and Main Housing

Hot Rig Testing of Ceramics

Complete ceramic parts will be hot tested in a special test rig. This rig, which is a modified gas generator rig, as shown in Fig. 27, also shows how the ceramic scroll, the ceramic nozzle and the ceramic turbine wheel are mounted in the rig. The special cooling and insulation system used to protect the metal parts against the high temperature is not shown on the sketch. The compressed air needed for the test is taken from the central laboratory air supply and is then led to the combustor which is metallic in this particular picture.

Depending on what type of components are being tested the rotor can either be run as a turbine by hot gas or driven by an engine via a power take off at the front end of the rig. A pyrometer will be used to monitor the temperature on the turbine disk. The pyrometer can be set so it can monitor the temperature all the way from the hub of the disk to the tip of the blades.

The rig will be used for studies of both transients and steady state running conditions. The tests start at rather low temperatures and speeds and then during the course of the test program these parameters will be increased successively.

In this rig different fastening principles will be evaluated for attaching the ceramic wheel to the metal surroundings. The sketch does not show how the wheel is connected to the shaft but we are especially studying principles with a turbine wheel having a solid disk and a short stub shaft as part of the wheel as shown on earlier slides. This ceramic stub shaft is then elasticly mounted inside the metal shaft. We are working with different patented solutions to keep the ceramic stub shaft under compression and to cool the metallic shaft.

Volume Production of Ceramics

Injection moulding of complete turbine wheels seems to be a very attractive manufacturing method. This is especially true when it comes to high volume production. United Turbine and ASEA are therefore working with the adoption of injection moulding as a preforming method for the HIP process.

Fig. 28 shows to the right an injection moulded turbine wheel before the plastizers have been removed. The turbine wheel to the left is a complete turbine wheel after it has been HIPed by use of the technique with a glass particle envelope described earlier. However, this piece is preformed by cold pressing with the subsequent machining of the green part. It is put side by side with the injection moulded green part to illustrate the linear shrinkage which is about 14%.

Figure 28. Turbine Wheel before and After the HIP Process. Also shown is Separate Blade Test Element Made by same Procedure

Reliability Analyses

At United Turbine we are at present also working with new statistical methods in connection with stress and reliability analyses of ceramic components. This work is made in cooperation with Professor W. Weibull, also from Sweden, after whom the Weibull modulus is named.

Fig. 29 shows some areas where R&D investigations can be very rewarding in order to get a better understanding of the reliability of ceramic components. We are studying a new technique to calculate the parameters of the Weibull distribution. This technique is called the "MCZ-method" and we think it is more efficient than the most commonly used "maximum-likelihood-principle". We are also looking at three parameter Weibull distributions where the so called threshold strength, which more generally is called the "location parameter", is assumed not to be equal to zero. We have studied a number of test samples of ceramic materials presented in the literature where the threshold strength does not seem to be zero. One example is presented in the figure. On the X-axis we have two scales of probability of failure one for each of the two distributions and on the Y-axis we have stress.

RELIABILITY

- IMPROVE STATISTICAL METHODS

  MINIMIZE THE SAMPLE SIZE BY ADVANCED PARAMETER ESTIMATION TECHNIQUE

  USE THE THREE PARAMETER WEIBULL DISTRIBUTION

  METHODS FOR RECOGNIZING IF A SAMPLE CONTAINS SUB-SAMPLES AND OUTLIERS

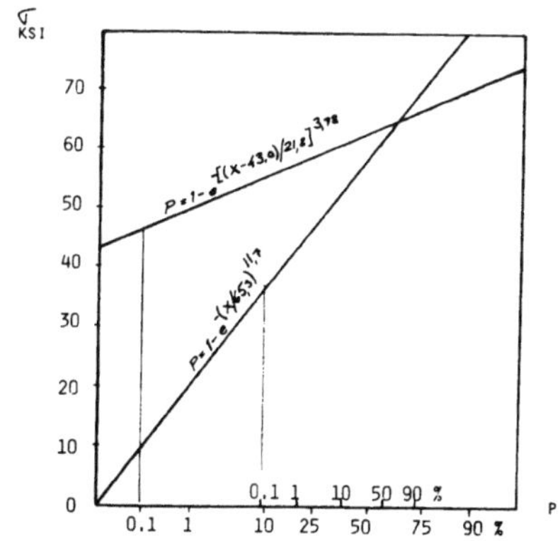

Figure 29. Influence of a Two Parametric Respectively Three Parametric Weibull Distribution on Probability of Failure for the same Test

If we assume the threshold strength to be zero we get the lower curve while if we assume a three parameter distribution we get the higher curve, which indicates a threshold strength of 43 ksi. Below this stress level no piece would fail. If we look at a probability of failure of 0.1% the three parameter distribution gives an allowable stress of 46 ksi while the two parameter distribution only allows 36 ksi, which shows a difference of about 25%.

At United Turbine we are also working with methods to recognize if a sample contains "sub-samples" or "outliers". The "sub-samples" could be the result of one surface dependent distribution and one volume dependent. The "outliers" could for example be a surface defect from machining.

Future Work at United Turbine

At United Turbine tests of ceramic components will now first be done in our present experimental KTT engine, MkI. In parallel we are also working with the design of a new KTT engine called MkII which is shown on Fig. 30. This MkII engine will be designed for ceramics from the beginning and will include at least stationary ceramic parts in the first phase and probably also a ceramic first stage turbine rotor. As can be seen from the figure this engine has only one heat exchanger. It is cross mounted and fits very well into the engine compartment of an ordinary Volvo car.

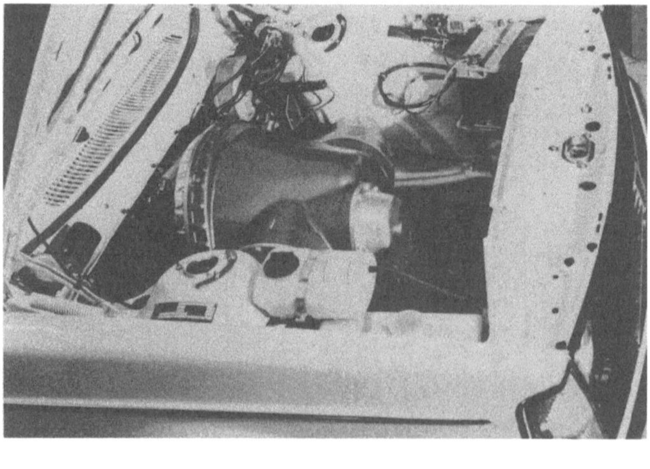

Figure 30. Mock Up of the KTT MkII Gas Turbine Engine in a Car

RESEARCH ON THE USE OF CERAMICS IN DIESELS

D. J. Godfrey*, D. A. Parker†, M. L. P. Rhodes†
and R. F. Smart†

* Admiralty Marine Technology Establishment
Holton Heath, Poole, Dorset, U.K.
† Associated Engineering Developments Limited
Rugby, Warwickshire, U.K.

INTRODUCTION

Exploration of the use of ceramics in diesel engines in the UK has so far shown that in comparison with the gas turbine situation the potential advantages of ceramics are harder to identify, but the practical difficulties are considerably less. The feasibility of using a reaction-bonded silicon nitride (RBSN) 50 mm piston in a Villiers 0.9 kW gasoline engine was demonstrated in 1970[1], as was the use of a 108 mm piston in a Gardner 9 kW diesel engine in 1973[2,3]. In a further and more detailed investigation a highly rated 15 kW Petter diesel piston had a 50% survival rate, as partly described[4] and reviewed[5] in previous presentations. An analysis of this work is given below; subsequent work has been with a 4 cylinder vehicle engine, for which RBSN pistons have been designed and are currently being tested.

The prime objective of these investigations has been demonstration that inductile ceramics could survive the high thermal and mechanical loads and severe dynamic stresses of the diesel engine. Future investigations will undoubtedly concentrate more on attempting to realize the advantages to be gained from using ceramics for overall heat loss reduction. It is probable that their low thermal conductivity will allow the cooling system size to be reduced, and permit improvements in efficiency when waste heat diverted from the engine and coolant to the exhaust is utilized by a turbocharger. Noise reductions could result from cooling fan miniaturisation and the use of low expansion ceramics.

Such materials are also of interest for other engine components e.g. pre-combustion chambers[6] and turbochargers[7] because of their potential for cost savings and/or higher temperature (and more efficient) operation.

SINGLE CYLINDER ENGINE TESTS

The investigation was carried out, using all-ceramic pistons in a highly rated engine; the aim was firstly to determine the ability of RBSN to survive such a stress regime and secondly to assess the benefits that might be expected from the use of ceramics.

Single cylinder supercharged Petter AVB diesel engines, capable of speeds up to 2000 rev/min and power ratings up to 1586 kPa b.m.e.p., were used for the test. The piston design generally followed that of the conventional aluminium-silicon piston (Fig.1) but alterations were made to take account of differences in the material and method of manufacture (Table 1); the chosen design was substantiated by finite element stress analysis.

Pistons were prepared from commercial silicon powder, isostatically pressed to billets 150 mm diameter by 420 mm long, of density 1.6 Mg/m^3. The billets were sintered in argon, machined to piston shape and nitrided. Dimensions with tolerances greater than $\pm$ 0.05mm were machined to final size in the sintered silicon billets; with tighter tolerances, diamond grinding after nitriding was necessary and this was specified for piston crowns, skirt and ring lands and grooves, and for the gudgeon pin hole. The loose surface layer of material was removed from all unground areas.

Test pieces were included with the piston blanks; materials characterisation tests gave the results in Table 2. The weight gains, coupled with microstructural evidence, indicated good conversion to silicon nitride. In view of suggestions that diamond grinding can seriously impair the strength of RBSN, tests were carried out to check this but there was no evidence of strength reduction due to either longitudinal or transverse grinding (see Table 2).

The pistons (see Fig.2) were vacuum impregnated with oil before running; they were engine tested with normal cast iron piston rings and liners, and with a steel gudgeon pin. Endurance tests comprised steady state running and thermal cycling and gave the results summarised in Table 3. Of the eight pistons tested, four survived their respective test schedules (in one case, this required nearly 60 hours running). Four of the pistons failed during testing. In three cases, cracks had propagated through the piston, but had not shattered it; the reasons for these failures were deduced (see Table 4) and it was felt they could be avoided in future designs. Only in one case

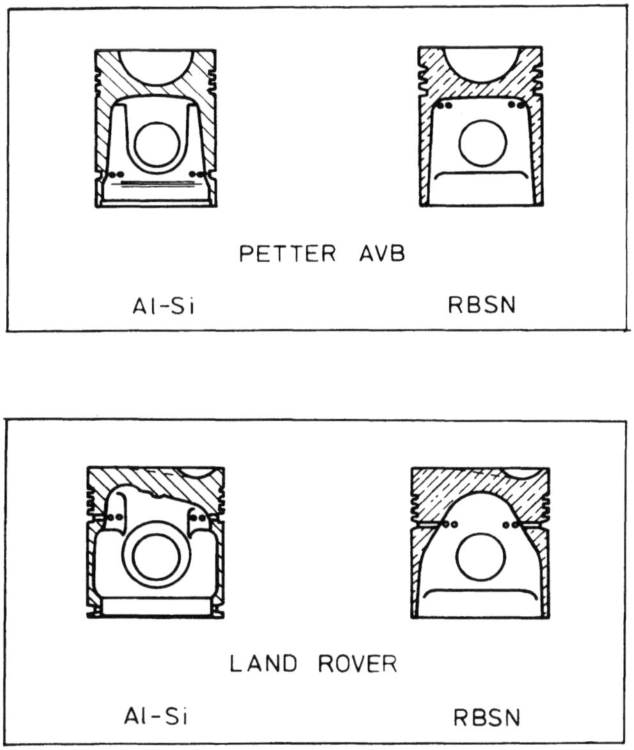

Figure 1. Comparison of Al-Si and RBSN Piston Designs.

Table 1. Design Changes in Petter RBSN Pistons.

Batch No.	Changes in Design Compared with Aluminium Piston
Batch 1	Interior geometry simplified Number of rings reduced to three Ring groove land widths increased All oil drillings eliminated Gudgeon pin located by end pads Internal and external corners radius Skirt clearance specified as 0.013 mm
Batch 2	As above except: Oil holes restored Compression ring grooves lowered (in an attempt to reduce their operating temperature)

Table 2. Properties of Petter RBSN Piston Material.

	Batch 1		Batch 2
Piston density	2.55 Mg/m^3		2.55 Mg/m^3
Testpiece density	2.51 Mg/m^3		2.49 Mg/m^3
Piston weight gain %	63.1		63.1
Mean flexural strength*	186 MPa		176 MPa
Weibull modulus	18		45
No. of tests	20		10
<u>After diamond grinding</u>	Transverse	Longitudinal	
Mean flexural strength*	194 MPa	199 MPa	
Weibull modulus	18	16	
No. of tests	8	10	
Thermal conductivity (20-400°C)	$14.55-9x10^{-3}T$ W/mK		$18.35-13.6x10^{-3}T$ W/mK
Thermal expansion (20-1400°C)	$3.27 \times 10^{-6}/K$		-
Fracture Toughness K_{1c}	2.3 $MN/m^{3/2}$		-

* 3 point bend

Figure 2. RBSN Petter Engine Pistons.

Table 3. Petter Engine Tests with RBSN Pistons.

Piston No.	Ceramic Finish	Total Engine Hours	Ceramic Failure	Remarks
2D*	Ground	$6\frac{1}{2}$	No	No visible deterioriation. Top ring stuck.
2C*	As nitrided	$6\frac{1}{2}$	No	No visible deterioriation. Top ring stiff.
		$59\frac{3}{4}$	No	Stopped - persistent fouling and ring sticking.
3E*	Ground	$56\frac{1}{2}$	No	No visible deterioriation. Top ring stuck.
		$64\frac{1}{4}$	Yes	Extensive fracture above pin.
1B*	Ground	$15\frac{1}{2}$	Yes	Extensive fractures.
12A†	Ground	$1\frac{1}{2}$	Yes	Extensive fracture above pin.
11C†	Ground	$19\frac{1}{2}$	No	Stopped - excessive fouling.
		$22\frac{3}{4}$	Yes	Shattered during injection timing tests.
11A†	Ground	$3\frac{1}{2}$	No	Thermistors fixed under crown. Linkage fitted. No visible deterioriation.
11B†	Ground	$2\frac{3}{4}$	No	Thermistors fixed in drilled crown. Linkage fitted.
		$3\frac{3}{4}$	No	
		$9\frac{1}{4}$	No	No visible deterioriation.

Table 4 Analysis of Failures in Petter RBSN Pistons

Piston No.	Engine Running Time Before Failure (hours)	Failure Type	Comments
3E	$64\frac{1}{4}$	Fracture initiated in outer edge of valve pocket and spread via crown edge and top ring groove, to pin holes.	Before fracture, fuel accidentally switched off and cold compressed air taken in; cause of failure probably thermal shock. Fracture spread via high stress areas in outer crown edge region.
1B	$15\frac{1}{2}$	Cracking started in top ring groove.	Fracture initiation in regions of high hoop stress as predicted by stress analysis. Did not pass through oil drain holes.
12A	$1\frac{1}{2}$	Failure from gudgeon pin hole area.	Failure attributed to sharp corners at inner edge of pin hole, since substantial subsurface cracking around crack initiation point. Sharp corners could have caused high load concentration resulting from pin bending.
11C	$22\frac{3}{4}$	Piston shattered.	Cause not known. Material structure no different from other pistons.

did failure lead to disintegration of the piston and consequential damage to the engine.

Instrumented piston assemblies were tested to obtain data on engine performance. The availability of a mechanical telemetry system, designed and developed by AED[8], gave a unique opportunity to obtain accurate data about the operation of the ceramic piston. The first of the instrumented pistons was equipped with four thermistors, attached under the crown in the positions shown in Fig.3; this provided initial temperature checks without incurring additional hazard to the piston. The second was equipped with four further thermistors fitted within deep drillings and a fifth repeating one of the undercrown positions on the first (see Fig.3).

Wiring from the thermistors was attached to the piston interior by ceramic and epoxy resin cements in the hotter and lower regions respectively. It was conducted from the piston to the connecting rod in a spring steel Bowden sheath, anchored at one end within a bracket attached near the bottom of the skirt with epoxy resin cement, and at the other end by another bracket screwed to the connecting rod immediately below the small end boss. Electrical connections across the big end joint were achieved by gold plated plugs and sockets. The leads were taken to the engine crankcase along a 2-beam mechanical linkage, shown in Fig.4. In this, a bracket fixed over the big end cap, the two beams and a stationary bracket bolted to the crankcase were pivoted together by hollow hinge pins, through which the leads were passed prior to being fixed under cover plates to the webs of the I-section beams. The outer beam passed through a hole cut in the crankcase inspection plate, and a metal box was placed over the hole and the stationary bracket to collect oil splash. The thermistors were ultimately connected to a digital voltmeter, via a switching unit, to measure resistance and hence, from prior calibration, temperatures.

The results of the instrumented tests are shown in Table 5 and the measured piston temperatures are included in Fig.3. The piston stress levels were originally determined by finite element analysis, from predicted temperature levels derived from known heat transfer data for aluminium pistons. The availability of the measured temperatures allowed the stress calculations to be refined and this yielded the stress levels shown in Fig.5. The gas load stresses proved to be very small, the thermal stresses being much greater; the most significant were the hoop stresses induced in the outer edge of the piston crown. A rather high tensile stress was found near the top ring groove, which was near the origin of one of the piston fractures (see Table 4).

It is apparent from Table 5 that the crown temperatures with the RBSN piston are very much higher than those with conventional metallic

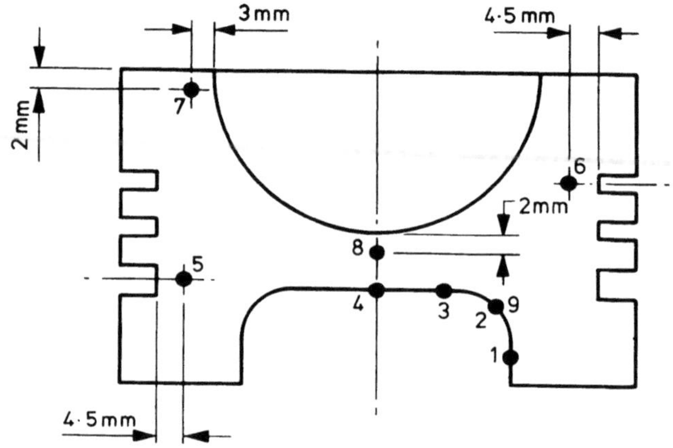

MEASURED TEMPERATURES °C			
POSITION	IDLE	1000 REV/MIN FULL LOAD	2000 REV/MIN FULL LOAD
PISTON 11A			
1	98	152	178
2	120	190	233
3	125	198	250
4	130	202	244
PISTON 11B			
5	110	170	236
6	150	260	395
7	173	362	646
8	150	276	397
9	118	170	218

Figure 3. RBSN Piston Temperatures in Petter Tests.

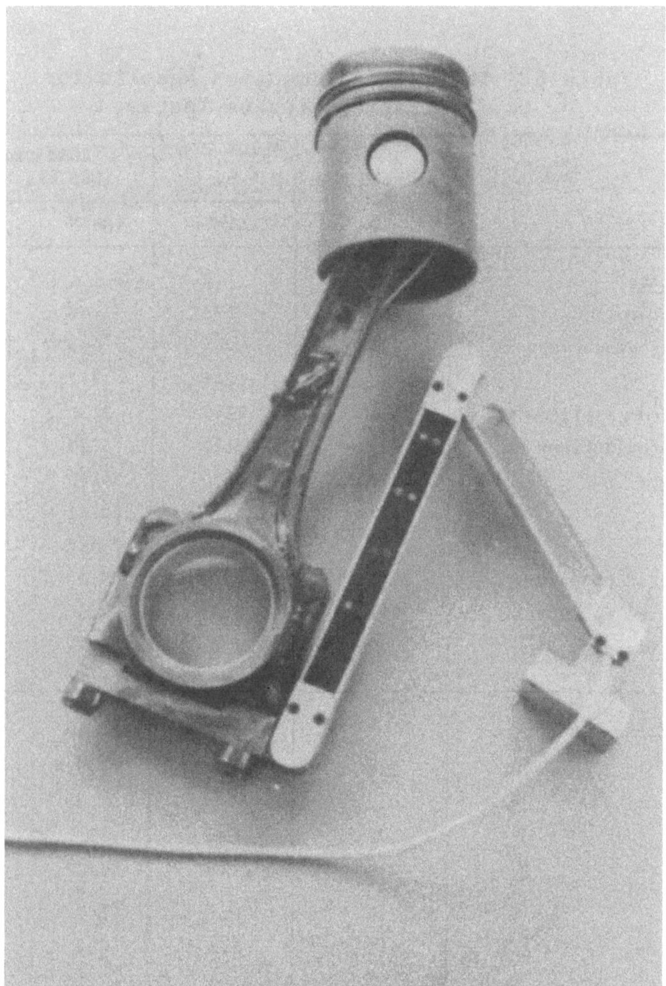

Figure 4. RBSN Petter Piston Showing Telemetry.

Table 5. Thermal and Emission Results for Petters Engine Tests.

	1000 rev/min 620 kPa b.m.e.p.		2000 rev/min 1586 kPa b.m.e.p.	
	RBSN	Al-Si	RBSN	Al-Si
Thermal Results				
Max. piston temp. °C	510	173	748	226
Max. cylinder head temp. °C	267	244	405	366
Exhaust temp. °C	401	346	771	648
Heat flow through piston W.	1460	1140	2130	1900
Heat flow through liner W.	2950	2620	6790	5060
Fuel flow W.	12290	9870	54160	49490
Air flow g/s	5.54	5.71	22.61	23.80
Blowby g/s			0.562	0.239
% heat rejected to water	27.3	24.3	19.3	14.8
% heat rejected to exhaust	23.1	25.0	42.5	42.7
% thermal efficiency	23.95	29.2	26.2	29.4
Emission Results				
CO_2 %	8.7	9.8	9.5	10.8
CO %	0.55	0.54	0.28	0.20
HC p.p.m.	102	136	63	92
NO p.p.m.	2180	2240	1330	1420
O_2 %	5.8	6.7	5.7	5.6
Bosch smoke no.	4.2	3.2	4.2	3.1

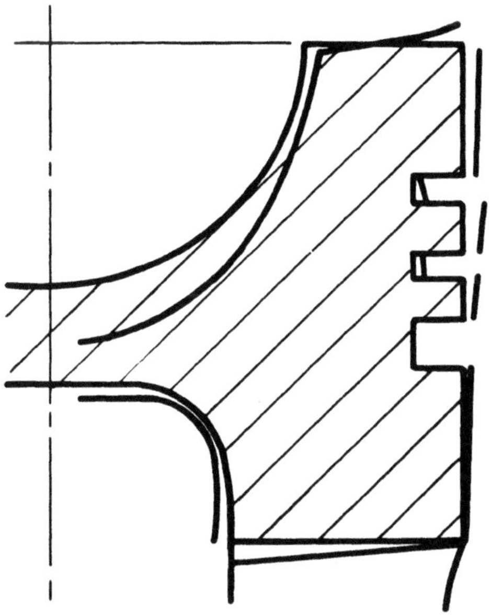

Figure 5. Hoop Stresses in RBSN Petter Piston.

Compressive Stresses	:	inside piston;
Tensile Stresses	:	outside piston;
(2000 rev/min	:	1586 kPa B.M.E.P.)

pistons. Measured temperatures of cylinder head, liner and exhaust were also higher with the ceramic. The greater temperature drop across the ceramic caused a greater heat flow through the piston, despite the lower conductivity of RBSN compared with aluminium. When the heat rejected to exhaust was calculated as a percentage of the fuel supplied, the figure was rather less for the RBSN piston, due to the considerable increase in the quantities of fuel but, even on this basis, the heat rejected to water was still higher with the ceramic piston.

The data show no improvement in thermal efficiency for the ceramic piston. It was initially suggested that this might have arisen from poorer sealing with a significant loss of chamber pressure; however, calculations based on measured blowby discounted this hypothesis. A more likely explanation is that the different thermal conditions when running with RBSN pistons have induced a fundamental change in the nature of the combustion process and the combustion conditions. The measurements indicated a reduction in the flow of air entering the engine and this was also shown by calculation, using a cycle simulation programme. It therefore seems likely that the higher temperatures of the combustion chamber have hindered aspiration by heating the incoming air charge. This has led to ignition delay and burning later in the cycle. The pressure diagrams revealed these characteristics - a lower peak pressure and higher pressures later in the cycle - while the emissions data also indicate later, less efficient burning.

It is interesting that Griffiths[9], in a computer simulation of a turbocharged engine in which the piston crown temperature was assumed to be raised from 283°C to 616°C, predicted a 1.22% increase in thermal efficiency when the trapped air mass flow was increased by 2.4%; in this case the additional energy had been recovered from the exhaust gases by a turbocharger, which supplied the extra air flow to the cylinder head (see Fig.6).

Although more efficient running operation was not achieved with ceramic pistons, the results are particularly promising in that four of the pistons survived to the end of their test schedule; this highlights the important conclusion that, if suitably designed and operated, RBSN pistons - and possibly other ceramic parts - can endure severe operating stress levels.

MULTICYLINDER ENGINE TESTS

In view of the high rating, very low power/weight ratio and generous cooling of the Petter engine, further tests are in progress, in a 2¼ litre Land Rover indirect injection unit with rating, geometry and cooling more typical of the transport field.

The RBSN piston is shown in Fig.7 and the principal differences

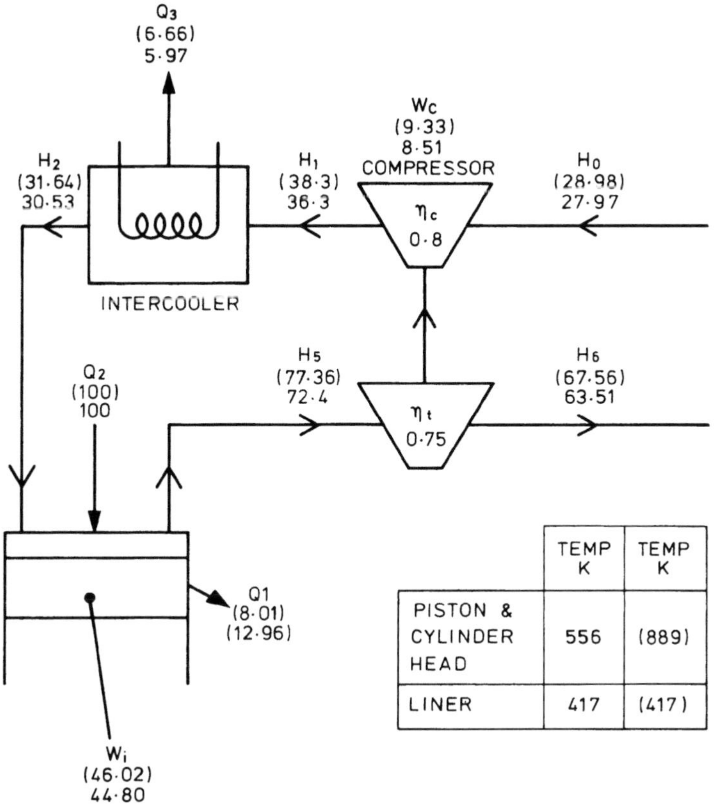

	TEMP K	TEMP K
PISTON & CYLINDER HEAD	556	(889)
LINER	417	(417)

- H AIR ENTHALPY AT AMBIENT TEMPERATURE.
- H_1 AIR ENTHALPY AT TEMPERATURE AFTER COMPRESSOR.
- H_2 AIR ENTHALPY AT TEMPERATURE AFTER INTERCOOLER.
- H_5 GAS ENTHALPY AT TURBINE INLET TEMPERATURE
- H_6 GAS ENTHALPY AFTER TURBINE
- Q_1 HEAT LOSS FROM GAS TO COMBUSTION TO CHAMBER WALLS (HEAT REJECTION)
- Q_2 HEAT RELEASED BY FUEL
- Q_3 HEAT TO INTERCOOLER
- W_i INDICATED WORK
- W_t TURBINE WORK
- W_c COMPRESSOR WORK
- η_c ISENTROPIC EFFICIENCY OF COMPRESSOR
- η_t ISENTROPIC EFFICIENCY OF TURBINE

Figure 6. Effect of Piston/Cylinder Head Temperature on Thermal Efficiency and Heat Rejection. (Griffiths Simulation)

Figure 7. RBSN Land Rover Pistons.

from the standard aluminium piston are included in the sketch in Fig.1. Two pistons have been instrumented to provide temperature measurements at a total of 5 positions within the crown and 6 under the crown; temperatures in the cylinder liner and head will also be recorded. Since the use of a Bowden sheath is likely to be unreliable at the high engine speeds (4000 rev/min), the gudgeon pin has been pegged to the rod and the wiring taken through the pin and peg directly.

During initial testing of the ceramic pistons, a problem has been experienced due to lubrication breakdown between gudgeon pin and piston, especially with the pins pegged to the rod. The poor capability of the porous RBSN material to support an adequate high pressure oil film over a small area, along with the influence of pin boss surface finish on friction coefficient, may have contributed to this problem, and work is in progress to solve it, so that the engine tests can be completed.

CONSEQUENCES OF WORK

Since its invention in 1893 the diesel engine has been subject to continuous development using metal pistons. It is therefore perhaps hardly surprising that the single cylinder tests did not show an immediate advantage when a conventional metallic diesel piston was replaced with one of RBSN. Owing to the higher temperatures of the combustion chamber with the silicon nitride piston, the incoming charge of air also became hotter, and therefore more viscous, resulting in a lower trapped mass with consequentially reduced rates of rise of temperature and pressure during adiabatic compression. Thereafter, more fuel had to be injected to compensate for the more difficult combustion conditions and the combustion pressure peak was both lower and later. To utilise the full potential of a ceramic piston in reducing rejected heat, it is therefore suggested that the re-optimisation of the engine must be effected in a number of areas, such as:

(1) Increasing air flow into the cylinder to compensate for greater viscosity.

(2) Altering the timing and nature of injection, including the position and angle of the fuel spray, to allow for the different combustion conditions.

(3) Altering the shape of the combustion system, including parts contained in the cylinder head, piston or pre-chamber.

(4) Introduction of additional heat barriers in the piston to ensure that the additional heat available passes into the exhaust pipe rather than into the body of the piston or liner (and hence to water) as in the single cylinder tests.

(5) Introduction of means in the exhaust pipe to recover the additional energy diverted thereto, e.g. turbocharging or turbocompounding.

The use of a significant amount of insulating material in the walls of the combustion chamber increases the surface temperature very considerably, e.g. the maximum temperature of the Petter piston from 226°C to 748°C (Table 5). These high and localised operating temperatures can have serious consequences in terms of operating stresses, e.g. in Fig.5 the peak tensile stresses are seen to be high and predominantly thermal. As more ceramic components are added to the combustion chamber in an effort to provide still greater insulation, this problem is likely to get worse; it also renders more difficult the operating conditions of the remaining components such as the valve gear, so that valve life may become a limitation.

Another problem experienced in the single cylinder engine work was that of sticking rings, due to decomposition of the oil at the high operating temperatures. If other engine constraints allow, this problem might be alleviated by placing the rings lower down the piston. If normal and insulating pistons are compared at the same rating, the crown temperature of the insulating piston would be higher than that of the normal and the skirt temperature would be lower. It therefore follows that there is a surface through the body of the piston for which the temperature is the same in both materials. If the rings are placed below this surface they may be expected to run cooler. If this solution is precluded by emissions or other considerations, it may be necessary to resort to the use of synthetic or even gaseous lubricants.

It may be that composite pistons, comprising a ceramic crown and a metallic body, could have advantages over all-ceramic pistons and some work on composites has been reported[4,10]; however the problems of sustained attachment of RBSN to metal has by no means been solved.

The potential of ceramic pistons in diesel or other engines will depend on the extent of the additional efficiency they bring about coupled with their acceptability to engine users. The acceptability in turn will depend upon the additional degree of complexity introduced, with its consequent effect upon price, reliability, and servicing needs.

CONCLUSIONS

The engine tests reported in this paper confirm that all-ceramic RBSN pistons endure high operating stresses and temperatures, provided care is taken in their design, manufacture and fitting to recognise the characteristics of the ceramic. That the potential benefits in improved engine efficiency have not been realised emphasis that, with ceramic pistons (or indeed other engine components) re-optimisation of the engine conditions is necessary to take account of the quite different thermal regime that occurs in the engine when ceramic is substituted for metallic pistons. Ways to achieve this re-optimisation are outlined in the text. The commercial potential of ceramic pistons will clearly depend upon the extent to which these changes are acceptable to diesel engine users.

ACKNOWLEDGEMENT

The work described has been carried out with the support of the Procurement Executive, Ministry of Defence.

REFERENCES

1. D. J. Godfrey and E. R. W. May, "The Resistance of Silicon Nitride Ceramics to Thermal Shock and other Hostile Environments" in Ceramics in Severe Environments, Ed. W.W. Kriegel and Hayne Palmour III, Plenum Press, New York (1971), 149.
2. D. J. Godfrey, "Silicon Nitride Ceramics for Engineering Applications", Trans SAE, 83; (2): 1036 (1974)
3. B. D. Gibson and R. B. Stone, Private Communication.
4. D. A. Parker and R. F. Smart, "An Evaluation of Silicon Nitride Diesel Pistons", Proc. Brit. Ceram. Soc. 26: 167 (1978)
5. D. J. Godfrey, "The Performance of Ceramics in the Diesel Engine", in "Ceramics for High Performance Applications - II", Ed. J. J. Burke, A.E. Gorum and R. N. Katz, Brook Hill Publ. Co., Chestnut Hill (1978), 877.
6. D. McLaughlin, "Ceramic Applications in the Diesel Engine", Proc. Brit. Ceram. Soc., 26: 183 (1978)
7. J. Panton, "A Turbine Rotor in a Low Cost Ceramic - Keramos" Project", Proc. Brit. Ceram. Soc., 26: 53 (1978)
8. N. A. Graham and M. Kendrick, "Instrumentation for Measurement in Operating Engines", Associated Engineering Symposium (1978), Paper No.19.
9. W. J. Griffiths, "Thermodynamic Simulation of the Diesel Engine Cycle to show the Effect of Increasing Combustion Chamber Wall Temperatures on Thermal Efficiency and Heat Rejection", Wellworthy Topics, 63, (1976-7)
10. R. Kamo, "Cycles and Performance Studies for Advanced Diesel Engines", in loc. cit. reference 5 above.

SESSION I (Continued)

HEAT ENGINE APPLICATIONS

Chairman: Dr. W. Bunk
Deutsche Forschungs-und Versuchsanstalt
für Luft-und Raumfahrt E.V.

Vice Chairman: Dr. W. Scott
University of Washington

CERAMICS POTENTIAL IN AUTOMOTIVE POWERPLANTS

A. F. McLean

Engineering and Research Staff
Research
Ford Motor Company
Dearborn, Michigan 48121

ABSTRACT

The paper addresses the potential that ceramic materials can play an important role in future automotive powerplants — both advanced heat engines and advanced battery systems. A number of related experimental programs are reviewed including ceramics for gasoline and diesel piston engines, gas turbine and Stirling Engines and sodium-sulfur batteries. A strong integrated program to develop ceramics technology is recommended.

INTRODUCTION

Today's automotive powerplant relies on the heat engine and this will likely remain the predominant situation for several decades. With the continued and critical depletion of fossil fuel resources during this period, there will be continued incentive and pressures to reduce the size and fuel consumption of auomotive heat engines for trucks and passenger cars. In the long run, at least for small passenger cars, the electric vehicle is perhaps the most likely alternate to the automotive heat engine, and this hinges on development of advanced, high energy density, rechargeable battery systems. Ceramic materials with potential for low cost and ready availability can play an important role in both advanced battery and heat engine development.

Historically, metals have dominated the field of engineering, especially "engine" engineering. The reasons are many. Probably the most important are cost, reliability, and the ready ability to be formed into many varied shapes. While ceramic bodies can be readily generated, they have not gained widespread use for engineering applications primarily because of their lack of strength and ductility. In the past decade, however, activity in the mechanical structural application of ceramics has increased, the primary reason being that newer, higher strength and lower thermal expansion ceramics have been emerging. These ceramics are primarily nitrides, carbides and oxides, including in particular, silicon nitride, silicon carbide and aluminum silicates. In addition to high strength and low thermal expansion, other important characteristics in selecting ceramics over metals are their corrosion resistance, their electrical characteristics, their ability to withstand higher temperature and thermal shock, and their retention of corrosion resistance and high strength at temperature.

AUTOMOTIVE HEAT ENGINES

The high temperature properties of certain ceramics are particularly desirable in materials for heat engines. The heat engine predominates the means by which man produces energy, the most well known being the piston engine (spark ignited and diesel) used in cars, trucks and ships, the steam engine and turbine used in stationary power plants and the gas turbine used in aircraft. In the field of automotive power plants, research is underway on all of these different powerplants as well as another heat engine known as the Stirling Engine.

A limitation of fundamental importance in the science of thermodynamics, pointed out by Carnot in the year 1824, is that the ideal efficiency of a heat engine (more generally known as Carnot efficiency) is uniquely related to the maximum and minimum temperatures of the thermodynamic process. While it is impractical to achieve Carnot efficiencies in an actual engine, the important factor is that, for given minimum or sink temperature conditions, the higher the maximum temperature of the process (sometimes called cycle temperature), the higher the potential efficiency of the machine. The questions then become what materials can be used to contain the high temperature in a practical heat engine and can ceramics be beneficial in achieving this potential improvement?

In the case of conventional gasoline and diesel piston engines, iron can be used for the cylinder block and cylinder head, where combustion takes place, since combustion is intermittent, and since the block is water cooled which both prevent excessive cylinder wall material temperatures. This cooling of the block, however, does represent a heat loss from the cycle and the potential for ceramics in these engines is based on the opportunity of reducing or eliminating the coolant and associated heat loss. The material temperature problem becomes much more severe with continuous combustion heat engines such as the gas turbine. Here the best available high temperature nickel-chrome superalloys have to be utilized. These are expensive metals and still impose a limit on uncooled turbine inlet temperature of about 1000°C. Ceramics such as silicon nitride and silicon carbide offer potential to increase turbine inlet temperature to values as high as 1600°C for the vehicular gas turbine and, at the same time, replace expensive, strategic metals with low-cost ceramics. The material problem in the Stirling Engine is similar to that of the turbine, and ceramics could facilitate an increase in the engine cycle temperature and replace costly metals.

Most of the research and development to date on ceramics for heat engines has been applied to the gas turbine. This will be reviewed first, followed by a review of ceramic applications to piston engines and to the Stirling Engine.

Ceramics in Gas Turbines

The gas turbine is an internal combustion engine where combustion is continuous and compression and expansion are accomplished by high speed, continuously rotating turbomachinery. The automotive gas turbine engine differs from the aircraft engine in that it produces output shaft power as opposed to jet propulsion and it incorporates a heat exchanger to recoup much of the exhaust heat thereby improving engine efficiency.

As mentioned earlier, the motivation to apply ceramics to the gas turbine is to increase turbine inlet temperature and replace costly nickel-chrome superalloys with potentially low cost ceramics. In studies conducted at Ford, a 1600°C 250 horsepower ceramic automotive gas turbine engine using high-efficiency aerodynamic components could have a very attractive specific fuel consumption, approaching 0.3 lb/hp-hr. It also appears that such an engine could have low exhaust emissions, based on projections from laboratory tests of an experimental turbine combustion system. Further claimed attributes of the turbine engine include multi-fuel capability, smooth, vibration-free production of power, low oil consumption, good cold-starting capabilities, and rapid warm-up time. Because of this potential for the ceramic gas turbine, automotive companies, turbine engine manufacturers, and material companies are active in developing ceramics technology. The following is a brief summary of some research programs which have been underway at Ford to develop technology for ceramic designs, materials, and fabrication processes and to evaluate these by actual rig and engine testing at turbine inlet temperatures up to 1370°C.

Figure 1 shows a schematic of the hot flowpath of the Model 820, high temperature regenerative gas turbine engine being used for this research and development. All of the high temperature components in this flowpath were designed for ceramics. In the case of the regenerator, shown in Figure 2, there had been a number of previous investigations into the use of ceramics. For example, twin Corning ceramic regenerators made of very low expansion lithium aluminum-silicate had been used in an earlier industrial turbine program at Ford and had accumulated thousands of hours of operating experience. Under certain conditions, however, the lithium-aluminum-silicate material underwent chemical attack associated with the sulfur content in the fuel or sodium (salt) ingested into the engine. This problem now seems to have been solved by changing to a chemically resistant, still low expansion, aluminum silicate material. Approximately 300 hours per month are being accumulated on each of several Corning aluminum silicate regenerator cores and to date, five have survived testing for over 10,000 hours at 800°C regenerator inlet temperature, and another four have survived for over 3500 hours at 1000°C indicating its suitability for a high temperature gas turbine engine.

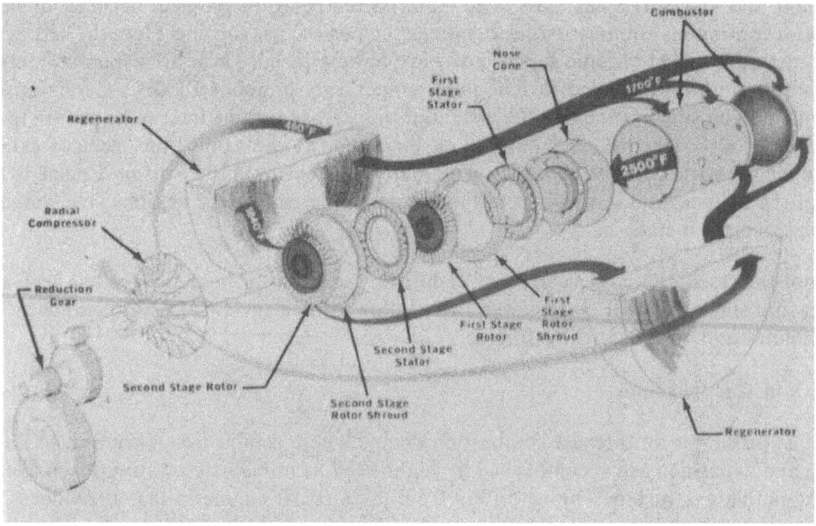

Figure 1. Ceramic Flowpath Schematic — Ford Model 820 Turbine Engine.

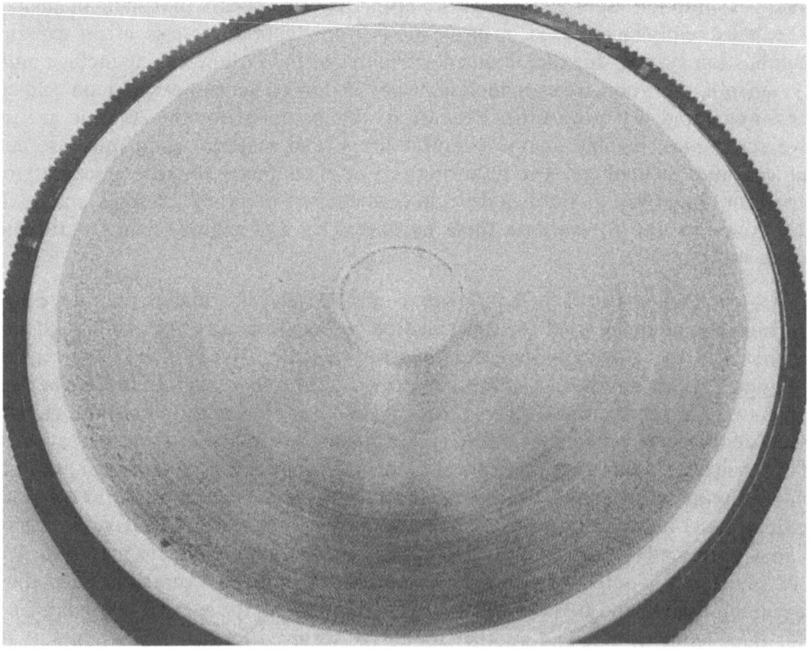

Figure 2. Ceramic Regenerator.

The stationary hot flowpath components which are subject to the peak cycle temperature of 1370°C, as shown in Figure 1 and separately in Figures 3 through 6, comprise the combustor, turbine inlet nose cone, turbine stators, and rotor tip shrouds. As previously reported, each of these ceramic components has been under development with respect to design, material, fabrication process and testing technique.

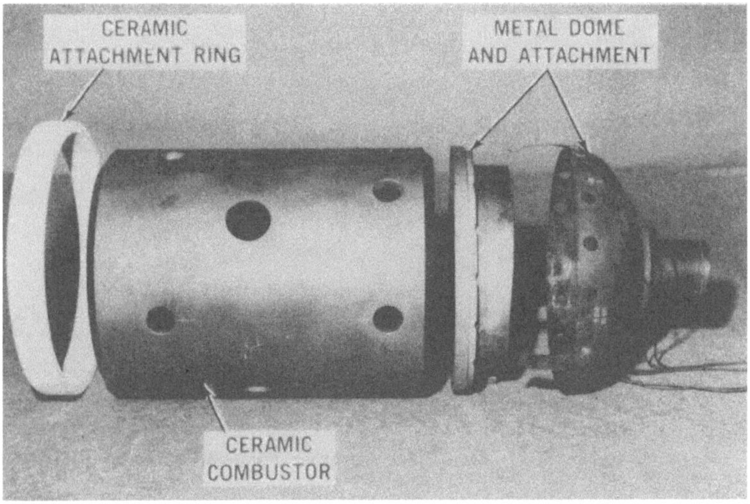

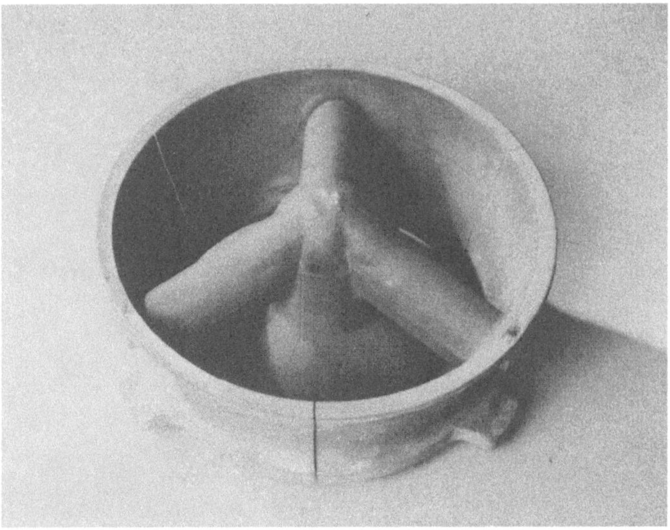

Figure 4. Turbine Inlet Nose Cone.

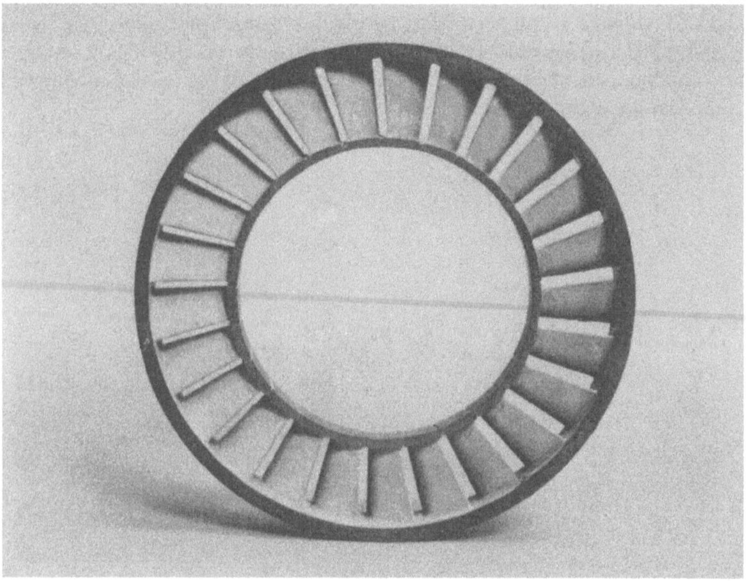

Figure 5. One-Piece Ceramic Turbine Stator.

Figure 6. Ceramic Rotor Tip Shrouds.

An important goal, in terms of component development, was to achieve 200 hours durability under conditions of start-up and shut-down transients and cycle operation based upon an engine operating duty cycle as shown in Figure 7. All of the stationary hot flowpath components, as follows, were shown to have met this goal of 175 hours at 1050°C and 25 hours at 1370°C:

- one-piece RBSN* stators made by injection molding
- one-piece RBSC** stators made by injection molding
- one-piece RBSN nosecones made by injection molding
- RBSC ("Refel") combustor made by isostatic pressing
- RBSN rotor tip shrouds made by slip casting

*Reaction bonded silicon nitride.
**Reaction bonded silicon carbide.

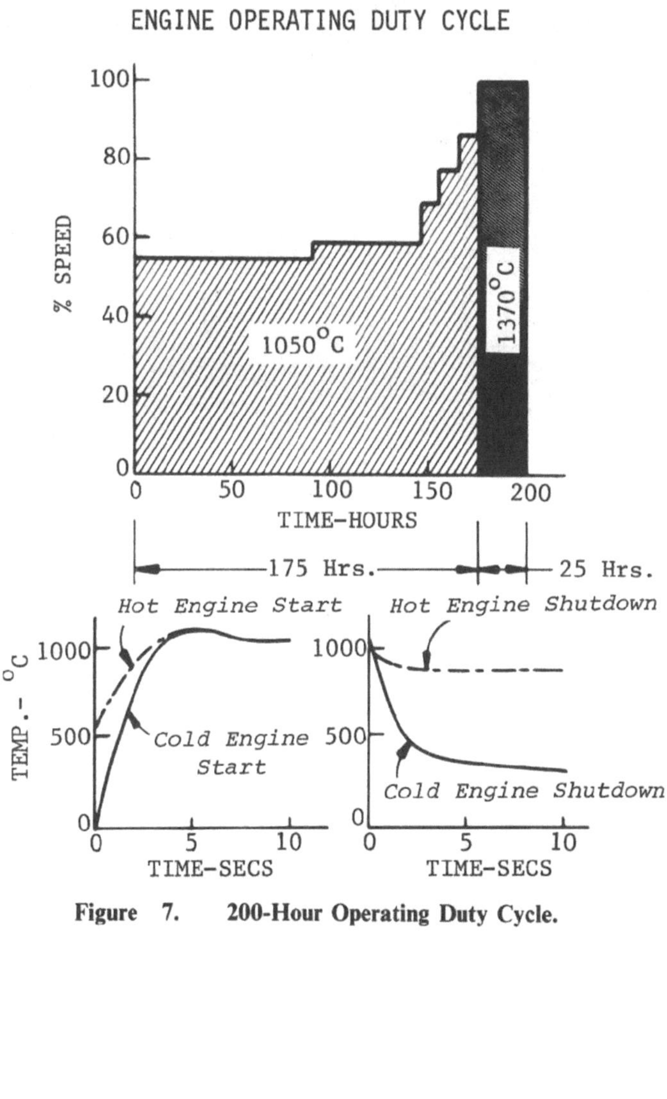

Figure 7. 200-Hour Operating Duty Cycle.

As discussed in more detail later at this conference, recent testing of one-piece ceramic stators has focused on their evaluation under conditions of an extremely transient duty cycle where temperatures are cycled each minute from 700°C to 1120°C, 1180°C, or 1200°C as shown in Figure 8. One injection molded RBSN stator successfully accumulated almost 16,000 cycles of such testing before fracture occurred, believed due to failure of another part in the test rig. Further testing over this duty cycle is planned, not only to evaluate RBSN stators (Ford), but also sintered αSiC stators (Carborundum) and RBSC stators (Norton), all of the integral, one-piece design.

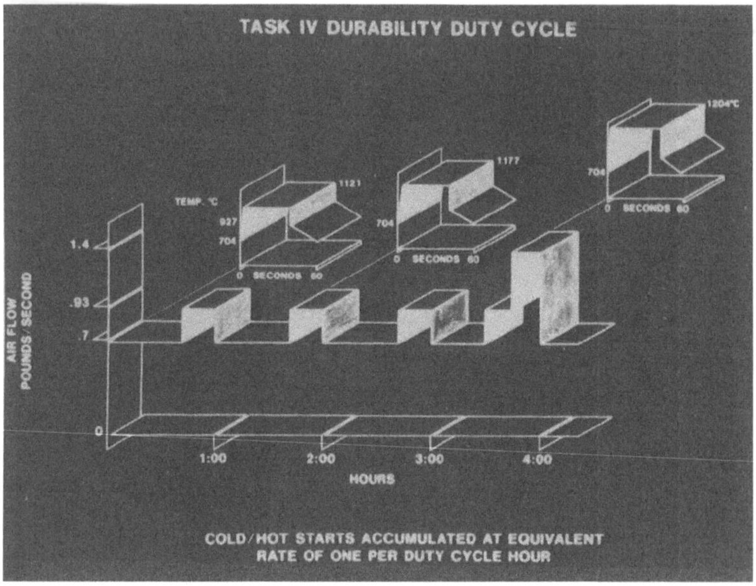

Figure 8. Severely Transient Duty Cycle.

It is generally acknowledged that the most difficult task is development of the ceramic turbine rotor, and that success of a high temperature ceramic turbine engine rests with success of the ceramic turbine rotor. Efforts at Ford have focused on the all-ceramic, "duo-density" silicon nitride rotor as shown in Figure 9, in which an injection molded RBSN blade ring is diffusion bonded to a rotor hub made of hot pressed silicon nitride (HPSN).

As previously reported, the duo-density rotor concept was selected for continued development following experimental investigations on some ten approaches to make a ceramic turbine rotor. Many detailed iterations were explored during process development of the duo-density rotor and at one stage the process was held fixed for the fabrication of several test rotors. Results of the tests on these rotors are shown in Table 1.

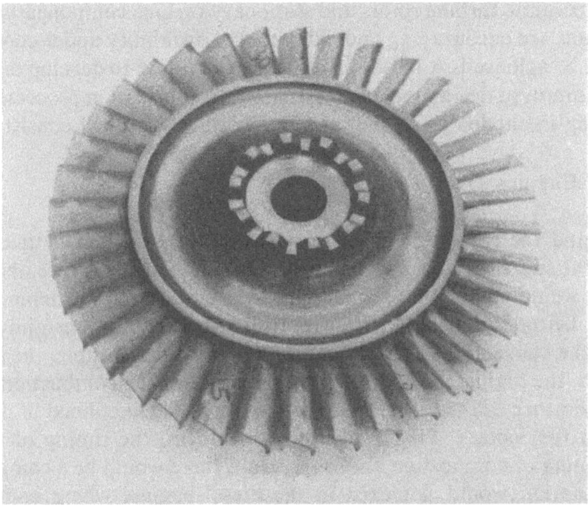

Figure 9. Duo-density Silicon Nitride Turbine Rotor.

Table I. Ceramic Rotor Hot Spin Test Results.

ROTOR	ROTOR RIM TEMP, (°C)	RPM	CUMULATIVE TIME (HR.)	SURVIVED RUN	REMARKS
Six-Rotor Test					
1	1000	32,000	-	No	Simplified bolt and mounting system
2	1000	24,000	-	No	"
3	1000	50,000	25	✓	Failed in curvic on disassemby
4	1000	50,000	25	✓	"
5	1000	50,000	25	✓	
6	1000	50,000	25	✓	
200 - hour Durability Test					
6	1000	Duty Cycle	200	✓	

CERAMIC ROTOR HOT SPIN TEST RESULTS

The first test series was to evaluate each of six rotors for 25 hours at a given speed (50,000 rpm) and rim temperature (1000°C). While two rotors failed prematurely due to an attachment problem, the remaining four each survived the 25-hour test. In addition, to meet one of the major program goals, a duo-density rotor was tested to 200-hours at 1000°C rim temperature over the range of speeds shown in the duty cycle of Figure 7. It is encouraging that this first attempt at a 200-hour run was successful. More complete information on the development of the duo-density rotor will be presented later at this conference.

These results on ceramic turbine rotors and stationary turbine components of potentially low-cost, one-piece designs are encouraging and indicate that durability under conditions beyond the range of metals can be achieved. A major effort is now required to develop technology to assure reliability through improved design techniques, materials, fabrication processes and nondestructive and proof testing methodology so that acceptable durability will consistently be achieved.

Ceramics in Piston Engines

The spark plug and the catalytic converter substrate are well-known areas where ceramics are used in or with the spark-ignited, gasoline piston engine. For the future, the combustion zone of the piston engine would seem a likely area to incorporate ceramics to achieve higher temperature capability and better insulating characteristics. However, this certainly introduces problems in the case of the conventional, spark-ignition engine. For example, the greater the degree of the ceramics used, the higher the operating temperature of the combustion surfaces. In turn, the higher the temperature of these surfaces, the greater is the likelihood of detonation or auto-ignition for a given fuel octane. Under such circumstances, the timing of the spark may no longer define the timing of combustion and, as a result, there would be a complete loss of engine control. Such a situation would not exist in the diesel engine, where combustion timing is controlled by the injection of fuel and there is no spark plug. A further problem arises in the case of an "all-ceramic" combustion zone, i.e. ceramic piston, cylinder liner, and cylinder head. Here, the cylinder walls would run so hot that oil lubrication would become a problem. A possible solution would be elimination of oil and use of a piston/cylinder air gap but this then introduces questions of adequacy of sealing the combustion zone and of durability of the air gap or rubbing surfaces. While development of an all-ceramic system is being pursued and might be considered an ultimate goal, the experimental application of ceramics to certain key areas in the piston engine is also being pursued. These developments are discussed in the following examples:

Figure 10 shows a picture of a sintered α-SiC prechamber cup (Carborundum) for a small diesel automotive engine. While this ceramic replacement offers improved temperature capability and lighter weight, its final selection for this application may well be on the basis of lower cost.

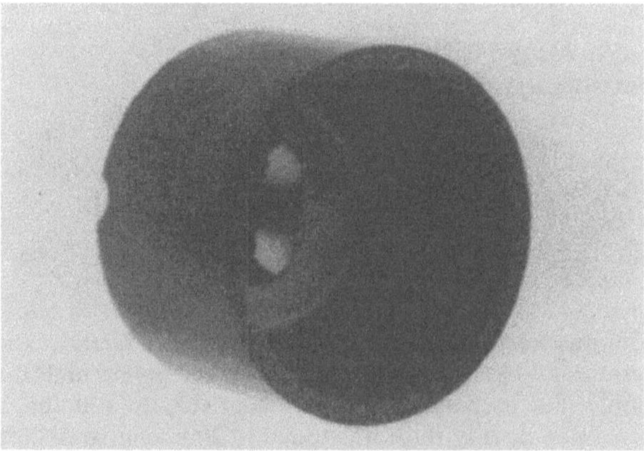

Figure 10. Silicon Carbide (Carborundum) Prechamber Cup.

Pistons used in direct/indirect injection engines are commonly of the bowl-in-piston type. Such engines depend upon high-induced swirl within the bowl for desired fuel mixing and vaporization for combustion. Figure 11 shows a picture of a machined HPSN bowl for experimental work on Ford's PROCO (programmed combustion) engine. Hot pressed silicon nitride was selected purely to maximize reliability during early experimental work. Clearly, for production, a sintered or reaction bonded, directly formed part would be necessary.

Figure 11. Silicon Nitride Piston Bowl.

One difficulty encountered with the conventional gasoline piston engine has been associated with hydrocarbon emissions during cold starts. To circumvent this problem, experiments have been made to form a ceramic exhaust port liner as shown in Figure 12 which would serve to insulate the exhaust gas stream. This would then minimize loss of exhaust gas heat to the cold exhaust structure and promote thermal oxidation of hydrocarbons upstream of the catalyst. It would additionally provide for faster warm-up time of the catalyst itself.

As reported by Kamo at this conference, an all-ceramic, "adiabatic" diesel engine is under development with potential for ceramic pistons, cyliner liners, head liners, valves and possibly unlubricated ceramic ball bearings. Figure 13 shows a RBSN (Ford) piston cap made for this experimental program. The ceramic components in this concept are primarily for the purpose of insulation. By minimizing heat loss from the cylinder, exhaust gas temperature will rise, increasing the potential energy recovery via turbo compounding. Such an engine not only offers improved fuel consumption, but simplification by elimination of the water cooling system.

Figure 12. Ceramic Exhaust Port Line.

Figure 13. Silicon Nitride Piston Cap.

Turbocharging of spark-ignited, automotive piston engines is now becoming as familiar a concept as the well-known turbocharging of diesel engines. The automotive industry now offers several turbocharged car models and this trend will likely continue due to the continued pressure of reducing engine size for better fuel economy. While the primary purpose of the turbocharger is to boost engine power, there is concern at the lack of response of the turbocharger in providing this power boost. Figure 14 shows the influence of turbocharger inertia on boost pressure response for a given engine operating speed.

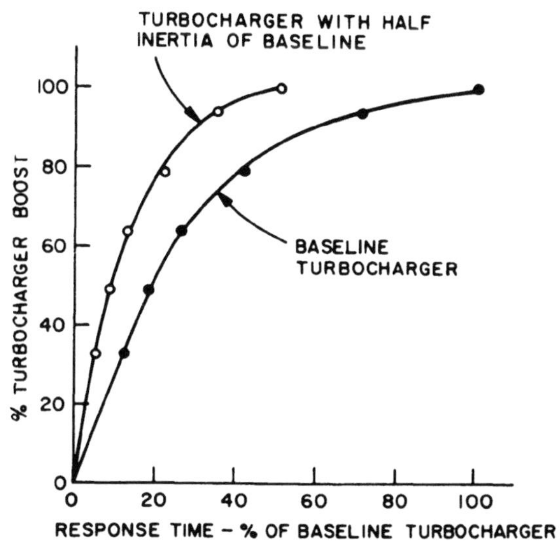

Figure 14. Turbocharger Boost Pressure Response.

Replacing the superalloy turbocharger rotor with a silicon carbide or silicon nitride ceramic rotor will reduce inertia of the rotating assembly by about 50% and, thus, improve turbocharger and vehicle response. Figure 15 shows a RBSN turbocharger rotor currently under investigation. A number of these have been made and cold spin tested during development of the fabrication process.

The highest spin test speed reached to date is 140,000 rpm, which is considered acceptable for continued experimental work. Further work is planned to improve rotor quality and to evaluate rotors through both cold and hot spin testing. Besides the rotor, other parts of the turbocharger are under consideration for ceramics. For example, the turbine housing shown in Figure 16 is a relatively heavy casting made of nodular iron. The application of ceramics for this function can provide an opportunity for significant weight savings and possibly simplified design. In addition, adoption of lower expansion materials for the turbine rotor and its housing facilitates closer control of turbine tip clearances with associated gains in operating efficiency.

Figure 15. Ceramic Turbocharger Rotor.

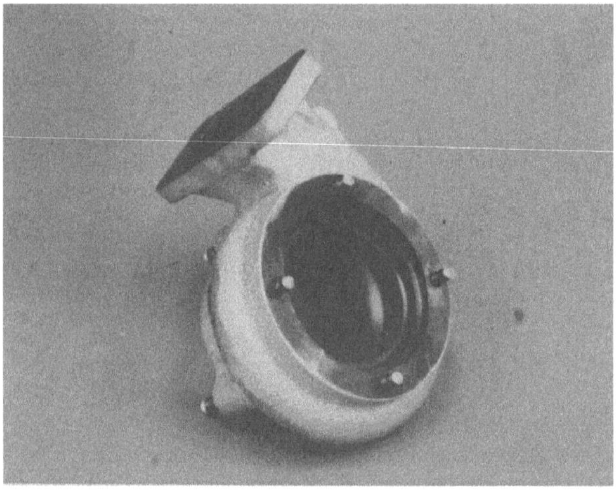

Figure 16. Turbocharger Turbine Housing.

Ceramics in Stirling Engines

The Stirling Engine was conceived in 1816 by its namesake Robert Stirling. It is an external combustion, closed cycle engine and, basically, its principle of operation is to alternately heat and cool an entrapped reciprocating volume of gas. The heated gas is expanded by a piston mechanism which delivers power to the output shaft. When the gas is cooled, it contracts so the piston returns for its next power stroke. Claimed attributes of the Stirling Engine include attractive specific fuel consumption, low exhaust emissions, multi-fuel capability, quiet operation and potentially good reliability and maintenance since it is a "sealed" engine.

Stirling Engine development is being pursued by several companies, worldwide, and is also being supported through various government contracts. Ford Motor Company has in the past worked on development of experimental automotive Stirling Engines, including, in conjunction with the Philips Company of Holland, a 170 hp four cylinder unit, shown schematically in Figure 17.

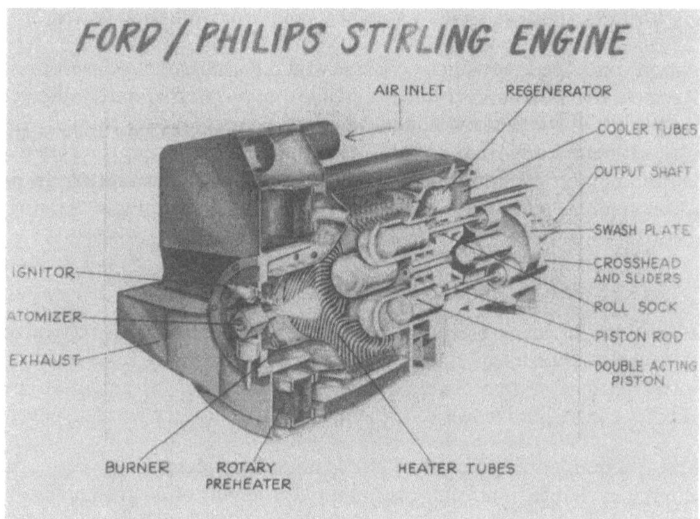

Figure 17. Stirling Engine Schematic.

In this engine, hydrogen was used as the working fluid within the closed cycle, to minimize engine size and fuel consumption. In addition, for packaging reasons, a swash plate mechanism was used to transfer the reciprocating motion of the piston to the rotating output shaft. The hydrogen pressure used in this experimental Stirling Engine was up to 250 atmospheres with maximum hydrogen temperatures of 750°C. The heat source comprised an external continuous combustion system similar to that used with the gas turbine engine. To reduce fuel consumption, intake air was passed through a preheater and then mixed with fuel for combustion. The hot combustion gases were directed over heater tubes attached to a heater head containing the hydrogen working fluid, which was heated to temperatures of about 750°C. To minimize the volume of the external combustor/preheater system as well as the heater tubes, an external combustor exit temperature of over 1650°C was used.

Considering the potential for ceramics in this engine, the external combustor/preheater problem is similar to that of the gas turbine combustor/regenerator system. In fact, aluminum silicate regenerator technology developed under gas turbine programs, has been used in experimental Stirling Engines. Other components of the Stirling Engine that might be made of ceramics are the heater tubes and heater head, and the combustor and hot air ducting. Ceramics for the heater tubes represent the greatest risk, but offer the greatest potential benefit by reducing cost and increasing temperature capability. Such a material for the heater tubes would have to have the following requirements:

- Low Cost
- Adequate conductivity to promote efficient transfer of heat from the combustion gases to the hydrogen working fluid.
- Economical fabrication means for the tube bundle.
- Durability and strength to withstand high internal pressures at high temperatures for the engine life.
- Ability to contain hydrogen at high pressures and temperatures.
- Ability to withstand thermal shock on engine start ups and shut downs.

Past investigations indicated that silicon carbide would perhaps come closest to meeting these requirements. As such, reaction bonded silicon carbide, pressureless sintered silicon carbide, and chemically vapor deposited silicon carbide should be considered first for this very risky ceramics application. However, before any final material selection, design concepts which exploit each of the candidate ceramic materials and fabrication methods must be evaluated. In particular, an assessment must be made as to whether the crucial problem of attaching a ceramic heater head can be solved.

AUTOMOTIVE BATTERY SYSTEMS

While the advantages of using electric power for vehicles have been recognized for many decades, there is now, as a result of the oil crisis, increased urgency to develop a high energy density, automotive battery system. Table 2 shows a comparison of potential energy storage systems and indicates the attractiveness of the sodium-sulfur battery from an energy viewpoint.

Of course there are many other factors to be considered besides energy requirements, including charge/discharge cycle life, materials availability, problems of corrosion and safety, system complexity and reliability and battery cost. A critical aspect of Ford's research program on the sodium-sulfur battery is the development of the solid, ceramic electrolyte.

Table II. Comparison of Potential Energy Storage Systems.

BATTERY SYSTEM	PROJECTED ENERGY DENSITY WATT-HRS/KG.	PROJECTED POWER DENSITY WATTS/KG
Low temperature systems(20-50°C)		
Lead acid	40	100
Lithium-tithanium sulfide	130	-
Nickel zinc	80	150
Zinc cholorine	150	70
High temperature systems(200-400°C)		
Sodium-antiminy tricholoride	100	50
Lithium-iron sulfide	130	160
Sodium sulfur	120	160
Estimated Minimum Requirements for Electric Vehicles	100	150

Sodium-Sulfur Battery

The sodium-sulfur battery uses a solid, ceramic, ionically-conductive membrane as the electrolyte to separate the reactants. The battery operates at 350°C at which temperature the reactants are in a liquid state. In fact, with liquid reactants, it is the solid ceramic membrane that makes the concept of the sodium-sulfur battery a practical one.

The ionically conducting material developed for the electrolyte is a β''-alumina which is basically a sodium-lithium-alumina ceramic. Initial experimental sodium-sulfur cells have utilized a β''-alumina closed-end tube as the electrolyte as shown in Figure 18. The problem of supporting and sealing the β''-alumina was overcome through development of an alumina/β''-alumina seal as indicated in the figure.

Continued development of the sodium-sulfur battery and of other high temperature battery systems is necessary to solve problems of corrosion and to develop refinements to achieve adequate reliability and durability. For this reason, the low temperature batteries (advanced lead acid and nickel zinc) will likely be the first to appear in the market place. However, with the promise of higher energy and power densities, continued development of high temperature batteries such as the sodium-sulfur system and related materials technology is recommended.

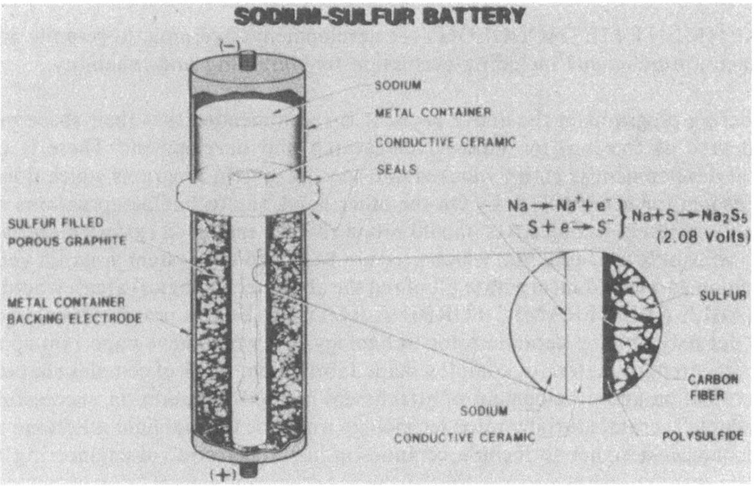

Figure 18. Sodium-Sulphur Cell Schematic.

CERAMICS TECHNOLOGY RECOMMENDATIONS

In the past decade, the possibility that advanced, low cost, available ceramics can be applied to improved heat engines and battery systems has been demonstrated. While this is, of course, of paramount importance for automotive applications, it also has much broader implications, covering the whole spectrum of the power and energy fields. With this degree of importance combined with the encouraging developments to date, an integrated program to develop ceramics technology is strongly recommended.

In assessing which aspects of ceramic technology should be pursued, it is clear that the major current problems center around the question of "reliability." Considerable research and development is needed to take us from the stage of "ceramics can work" to the stage of "ceramics won't fail." The following represent suggestions of areas of R & D which will enhance the reliability of using ceramic materials.

- CERAMIC MATERIALS — development of material compositions and microstructure for improved characteristics (e.g., strength, Weibull modulus, electrical characteristics).

- CERAMIC PROCESSES — development of processes such as molding, slip casting, sintering and machining to practically form complex shapes while retaining quality material characteristics.

- NONDESTRUCTIVE EVALUATION — development of nondestructive evaluation techniques to detect strength-controlling flaws including, in particular, relatively gross fabrication flaws in complex-shaped parts. High frequency ultrasonics and x-ray tomography are promising ceramics NDE techniques requiring further development.

- LIFE PREDICTION METHODOLOGY — development of methodology for ceramic lifetime reliability through experimental work involving flexural testing, tensile testing, spin testing and controlled testing of actual ceramic components.

- ATTACHMENT METHODOLOGY — development of ceramic-to-ceramic and ceramic-to-metal attachments including evaluation for durability and reliability.

In considering programs in the above areas, it is recommended that their scope provide the necessary degree of freedom for innovative research and development. There is concern if technological development is entirely locked into specific system programs which don't provide adequate freedom for needed R & D. On the other hand, the technology programs need to be kept generally on track, so objectives should relate to heat engine or battery material requirements. A significant R & D program which need not be locked to a system program yet would be very challenging and would incorporate all of the the above technological areas, would have as a goal: RELIABLE ALL-CERAMIC TURBINE ROTORS. Such a program would be valuable as it would not only develop ceramic rotor technology, but would have important spin off with respect to high strength materials, complex shape fabrication, NDE of complex shapes, lifetime reliability evaluation and development of attachment methods. In addition, successful development of "reliable" ceramic turbine rotor technology would be the highlight milestone to encourage the mechanical designer to include ceramics in his "repertoire" of engineering materials.

SUMMARY AND CONCLUSIONS

In the past decade, ceramics with promise of low cost and ready availability have been experimentally applied to potential automotive power plants — silicon nitride, silicon carbide and aluminum silicate to heat-engines and β''-alumina to advanced battery systems. Successful operation of such ceramics has been demonstrated, at least for modest durability, thus indicating that "ceramics can work."

An integrated program to research and develop ceramics technology is recommended to take us to the next stage of "ceramics won't fail" — a stage necessary to really establish ceramics as a class of engineering materials. Specifically a multi-discipline, research-type program is recommended to develop "Reliable All-Ceramic Turbine Rotors."

ACKNOWLEDGEMENTS

The author wishes to acknowledge the Advanced Research Project Agency, the Army Materials and Mechanics Research Center, the Department of Energy, the National Aeronautics and Space Administration and the Office of Naval Research for their support of a number of programs referred to in this paper.

ARPA/NAVAIR CERAMIC GAS TURBINE ENGINE DEMONSTRATION PROGRAM

J. E. Harper

Project Engineer
AiResearch Manufacturing Company of Arizona
A Division of The Garrett Corporation
Phoenix, Arizona 85010

ABSTRACT

This paper presents an overview of the ARPA/NAVAIR Ceramic Gas Turbine Engine Demonstration Program being conducted at Garrett, with emphasis on the component rig and full-scale engine development phases of the program. Program accomplishments and problems encountered in applying ceramics to gas turbine engines are discussed.

INTRODUCTION

The ARPA/NAVAIR Ceramic Gas Turbine Engine Demonstration Program was initiated in February 1976 to identify problem areas and demonstrate the use of ceramic hot flow path components in a gas turbine engine. Only through full-scale engine testing could the full range of ceramic technology be identified for application of ceramics to gas turbine engines. Included in this program are the following goals:

- Fifty-hour demonstration test consisting of twenty-five 2-hour cycles
- 2500°F peak temperature with 2200°F average temperature
- Potential for low specific weight and cost
- Decreased use of strategic materials

The program consists of six phases, which contain four major tasks.

- Design - employing 3-D probabilistic analysis for brittle materials
- Materials - material characterization of reaction-bonded and hot-pressed silicon nitride (RBSN and HPSN).
  - process development to establish fabrication methods
  - nondestructive evaluation (NDE) to develop inspection techniques and accept/reject criteria
- Rig Development - develop ceramic hardware in component rigs prior to conducting engine tests
- Engine Tests - full-scale engine development tests consisting of instrumented and endurance testing

Effort on the Design, Materials, and Rig Development tasks has been successfully completed as shown in the program schedule of Figure 1. Current effort is being concentrated on engine development testing with the 50-hour demonstration test scheduled for late 1979. The engine has successfully demonstrated capability to full design conditions, and no barriers that would preclude conducting the 50-hour test have been noted.

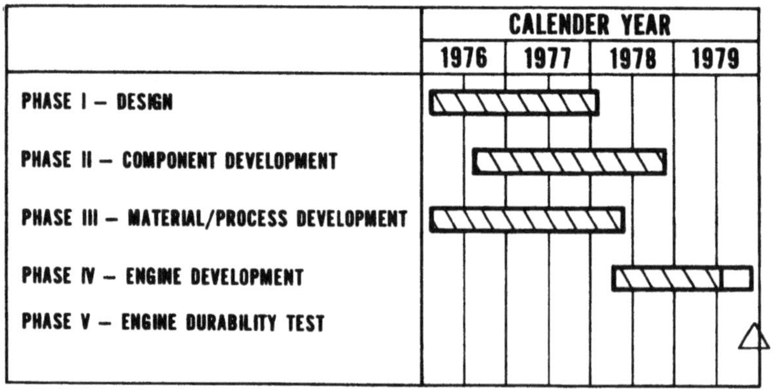

Fig. 1. ARPA/NAVAIR ceramic engine demonstration program.

DISCUSSION

The engine selected for this program is the AiResearch Model T76 Turboprop Engine shown in Figure 2. The various derivatives of this engine are used in 50 different military and civilian aircraft. The gearbox and compressor sections are unchanged; however, the hot-end has been redesigned to employ ceramic gas-path components. This permits turbine-inlet temperature to be increased 350°F to 2200°F, thereby increasing power from 715 SHP to 1000 SHP and decreasing SFC by 10 percent.

The turbine section includes 102 separate ceramic components with RBSN material in the static parts, and HPSN material in the blades. A cross-section of the hot end that identifies the ceramic hardware is shown in Figure 3. The rotating group includes 28 ceramic blades in each of the first- and second-stage turbine rotor assemblies. The rotor assembly with ceramic blades installed can be seen in Figure 4. Figure 5 shows a complete engine set of 102 ceramic parts and indicates the large number of ceramic parts that are included in each engine test.

Tasks leading up to rig and engine tests have been documented in previous progress reports and presentations; however, a brief review of the design and materials tasks will be presented.

The design philosophy has been to minimize program risk by designing within the limitations of ceramic materials. At the start of the program in 1976, RBSN and HPSN materials were chosen because they were the most characterized of the candidate ceramic materials. Analytical techniques employed 3-D finite-element and probabilistic analysis to determine cumulative probability of success (CPS) of each ceramic component. Design iterations between the materials, aerodynamics, and mechanical design disciplines were required to assure satisfactory CPS prior to design release. Originally, the turbine was to include three ceramic stages. However, this iterative process indicated a ceramic third stage to have an unacceptably low CPS. Therefore, metal components were substituted for this lower temperature stage. This approach also indicated a monolithic combustor to have an unacceptable CPS, and resulted in a major redesign to a stacked-ring configuration as shown in Figure 6. This delayed fabrication of the combustor and initiation of the ceramic combustor rig test program. As a result, the combustor has been deleted from engine tests.

The design and materials efforts included effects of fabrication methods to assure that actual hardware properties were reflected in the analyses. Manufacturing techniques were developed for slip-cast and hot-pressed components to optimize hardware quality and material properties. The material and process development approach is schematically shown in Figure 7. An example of the machining sequence used to fabricate HPSN

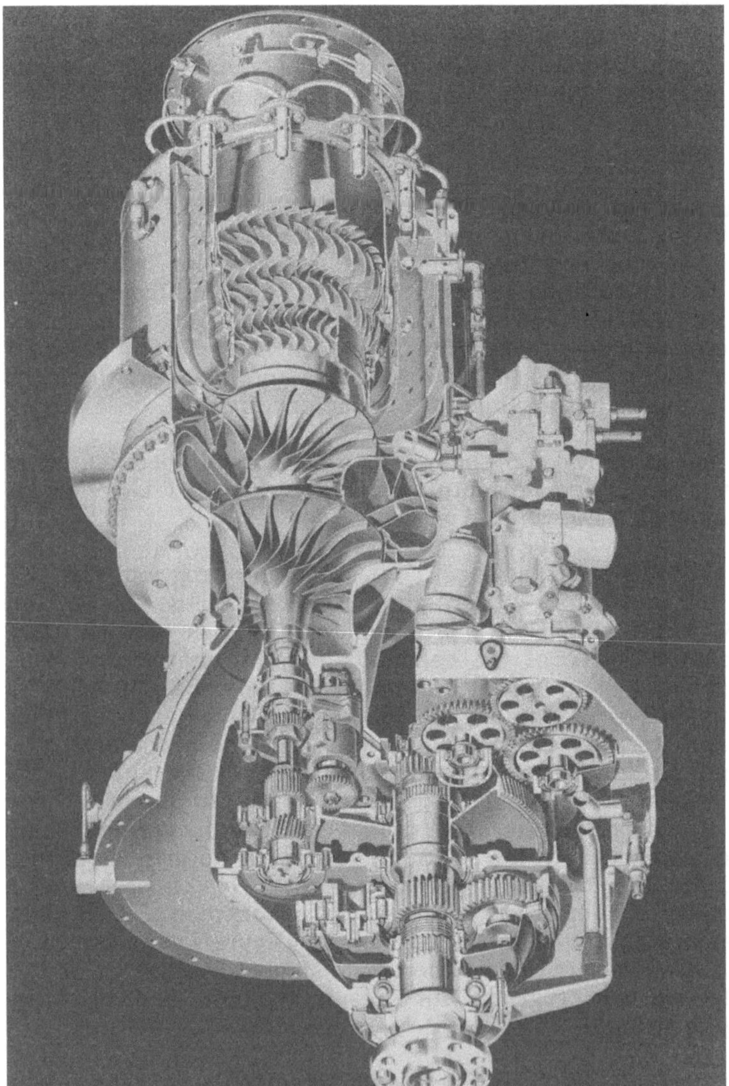

	T76 ENGINE	CERAMIC ENGINE GOALS
POWER OUTPUT	715 SHP	1000 SHP
SPECIFIC FUEL CONSUMPTION	0.60 LB/HP-HR	0.54 LB/HP-HR

Fig. 2. Demonstration engine.

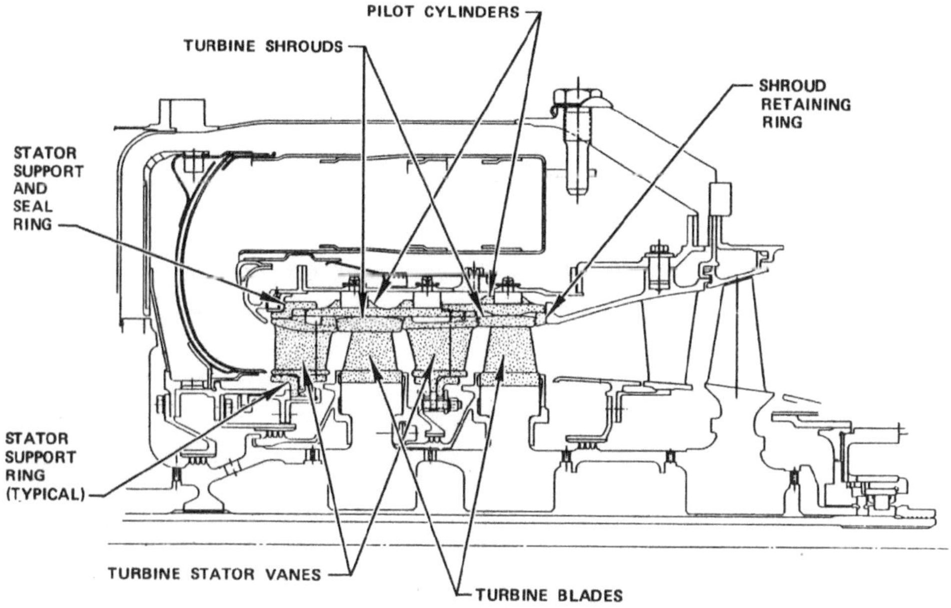

Fig. 3. TSE331C-1 ceramic turbine cross section.

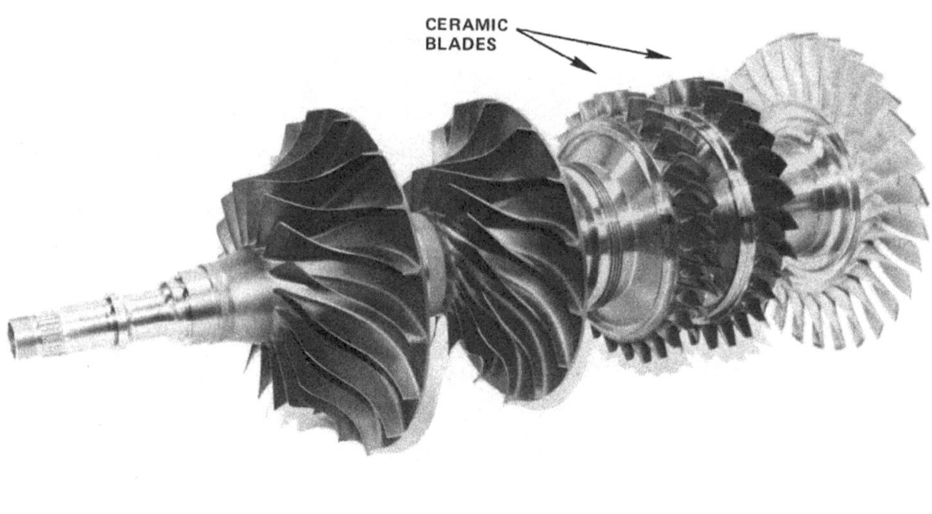

Fig. 4. Ceramic engine rotating assembly.

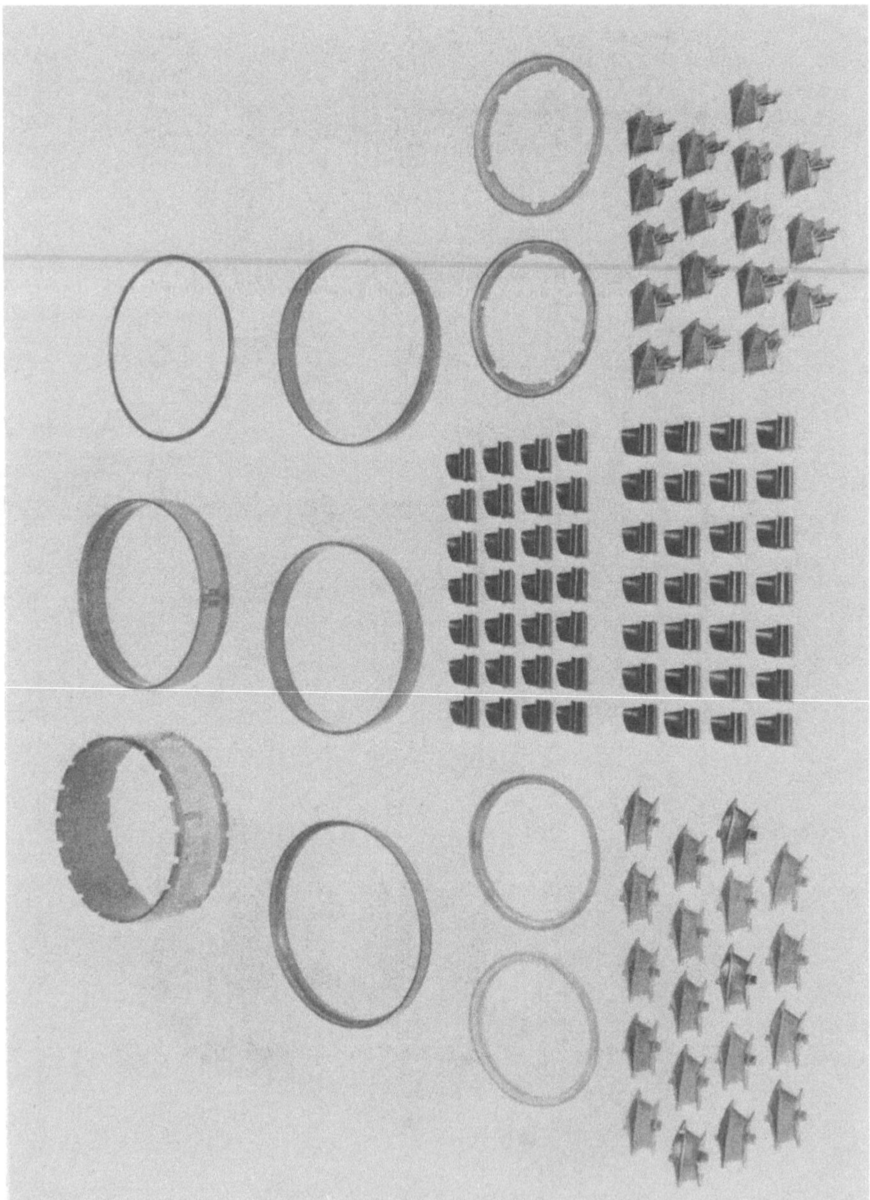

Fig. 5. TSE331C-1 ceramic turbine components.

Fig. 6. Stacked metal ring combustor.

blades is shown in Figure 8. Extensive use is made of ultrasonic trepanning to reduce fabrication time and tool wear during the final diamond grinding operation.

Effects of surface finish from machining of HPSN blade and material properties and post-machining oxidation treatments are shown in Figure 9. This illustrates the results of the iterative process described in Figure 7 wherein the design, materials, and manufacturing disciplines work together toward a final design optimization. In this example, the effects of machining operations upon blade strength were eliminated, which resulted in more predictible properties and reduced fabrication costs.

Proof testing of blades validated the design and analytical techniques. For example, the predicted rejection rate of 3 percent for second-stage blades subjected to proof tests was demonstrated as shown in Figure 10. This verification of analytical predictions demonstrated that the components were ready to begin the rig test phase of the program.

DEVELOPMENT TEST PROGRAM

As of June, 1979, the ARPA/NAVAIR Ceramic Engine Program has completed the following rig and engine tests.

- o Turbine Static Rig
  - 21 separate tests conducted
  - full temperature capability demonstrated
  - 20-percent overpressure capability demonstrated
- o Turbine Rotating Rig
  - 7 separate tests conducted
  - capability to 108-percent speed (hot) demonstrated
  - cyclic life of blade compliant layer demonstrated
- o Ceramic Engine
  - 24 separate engine tests conducted
  - 36 hours engine testing completed
  - Successfully demonstrated capability to full temperature and design speed with no evidence of distress in ceramic components

The rig and engine development programs have provided valuable insight into the behavior of ceramic components. The simulated engine

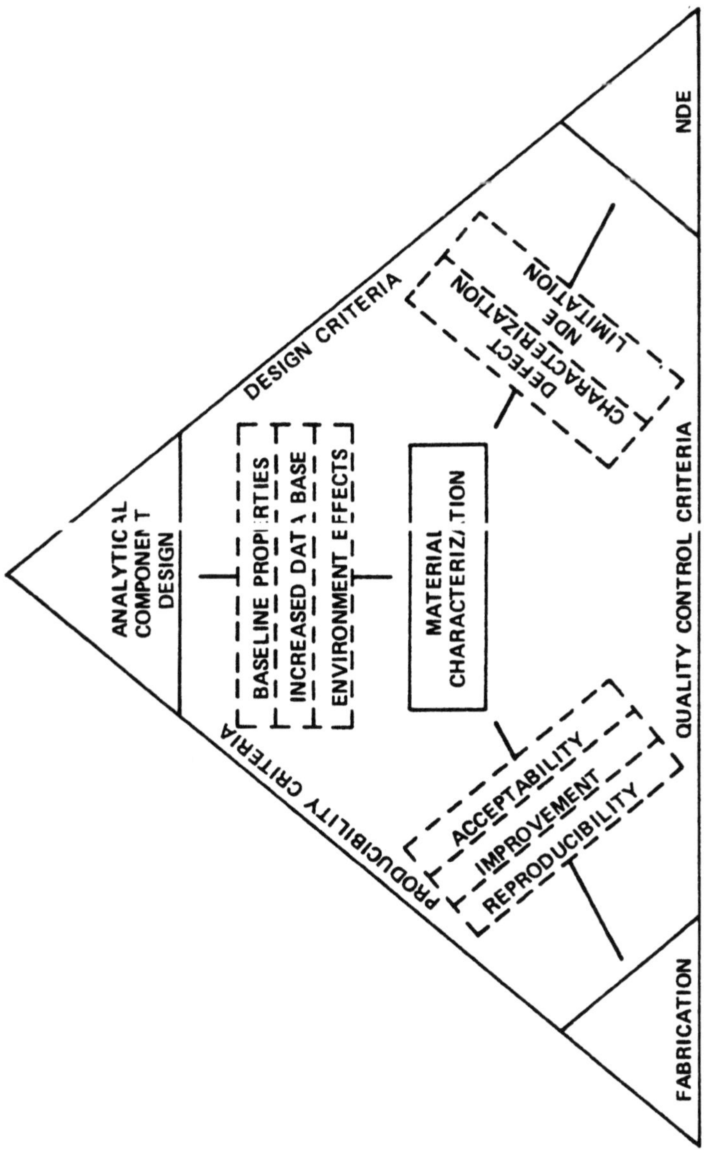

Fig. 7. Material and process development approach.

Fig. 8. Blade machining using ultrasonic trepanning and pantograph profile grinding.

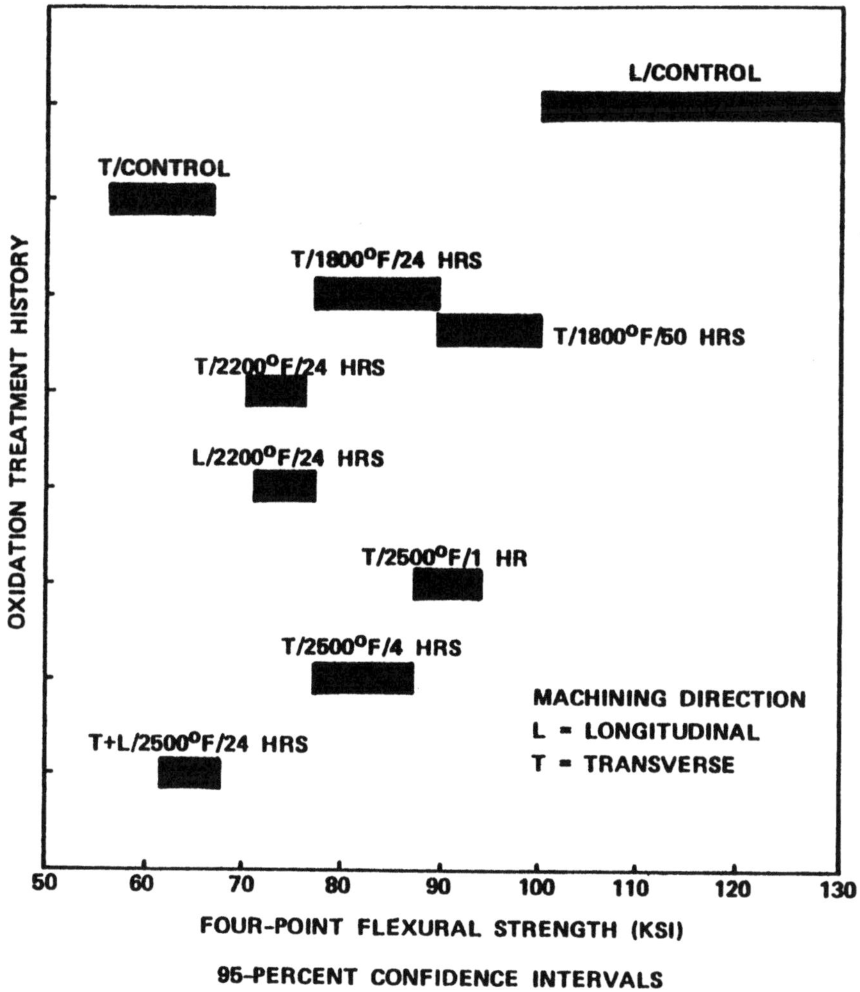

Fig. 9. Strength of NC-132 after static oxidation.

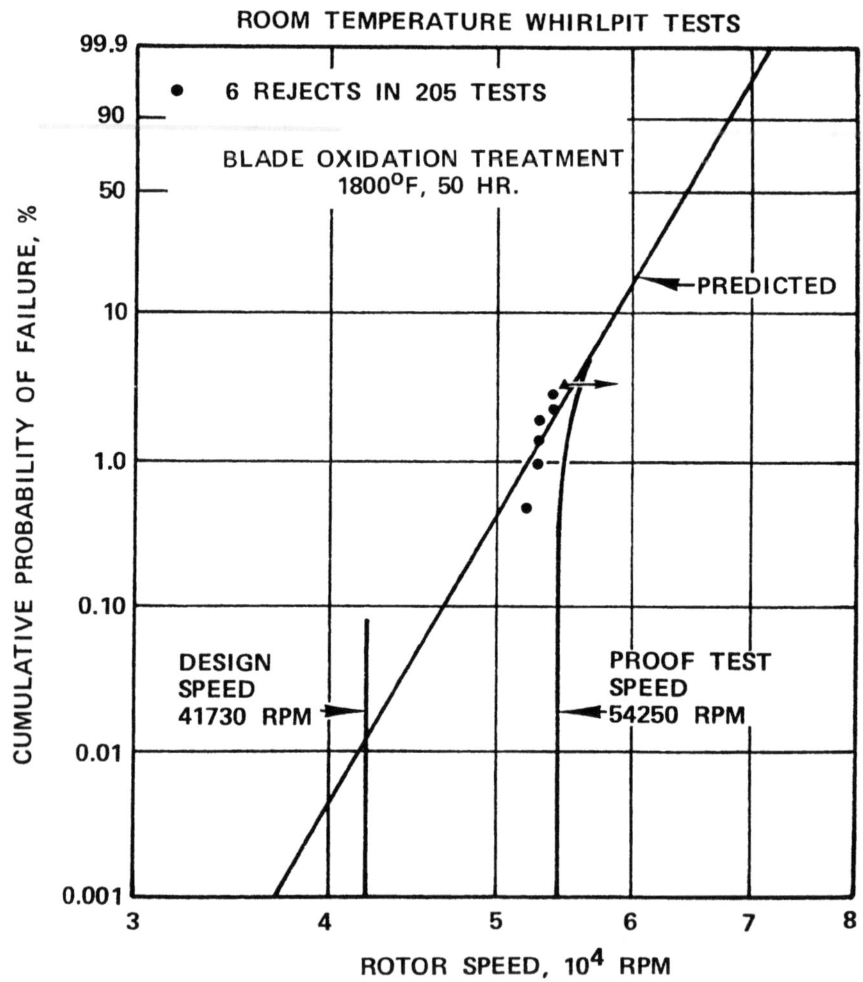

Fig. 10. Rejection rate for second-stage turbine blade.

environment of turbine rigs permitted tuning of the design, and verification of component capability to survive an engine environment prior to subjecting the hardware to an engine test. Rig testing indicated the critical areas that affect ceramic component reliability to be:

- o Contact zone behavior
- o Friction
- o Load Sharing
- o Thermal Shock

Engine testing verified the capability of the ceramic components that were developed under the rig program. The engine has also indicated the above items to be critical to the reliability of ceramic components, especially thermal shock and contact zone behavior.

The remainder of this paper describes methods of improving ceramic component reliability that have been developed in the rig and engine test portions of the ARPA/NAVAIR Ceramic Engine Program.

Component Rig Development

Extensive component rig testing was conducted to verify hardware capability prior to initiating engine tests. This minimized risks to engines and also allowed fracture origins to be identified and corrected, since fractures that occurred in the rig would not cause extensive secondary damage. The ability to identify fracture origins permitted detailed analysis to be performed on the critical area, thereby providing insight into ceramic component behavior. Lessons learned were used to evaluate other critical areas and to increase the reliability of the design as demonstrated in the following examples.

Combustor

During rig testing of the ceramic combustor, fracture of the inner forward ring occurred as shown in Figure 11. Investigation revealed the fracture to have originated at the surface of a dilution air orifice located within the ring. A detailed stress analysis was performed on the area immediately adjacent to the orifice shown in Figure 12 to determine if effects of localized cooling could have initiated the fracture. Analysis indicated a peak stress of 29 KSI at the surface of the orifice. A design study was then conducted, which included effects of localized cooling, to optimize the orifice location and to investigate other design modifications that would increase the component CPS. Results of this study, shown in Figure 12, are modifications that also include effects of scalloping the lip of the forward ring (Find No. 6 in the cross-section) on the calculated CPS. In this example, CPS was increased from zero in the original design, to greater than 94 percent in the redesigned ring.

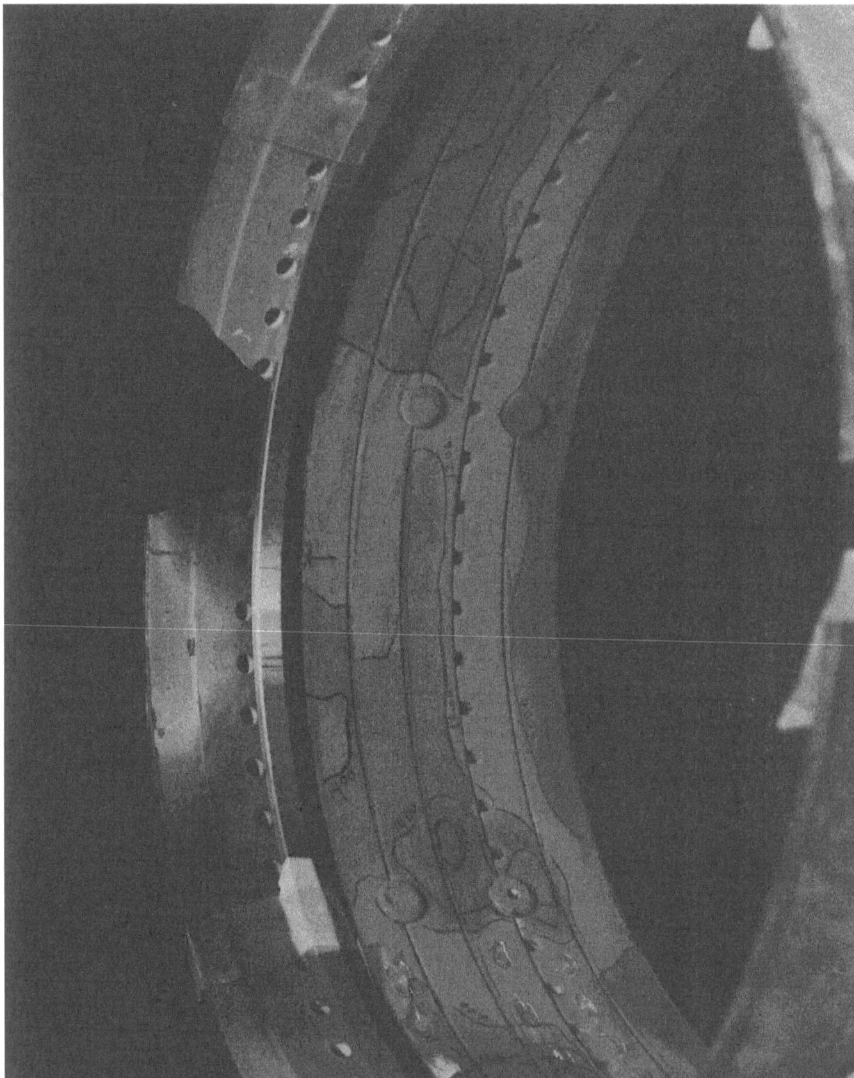

Fig. 11. Ceramic inner liner following combustor test.

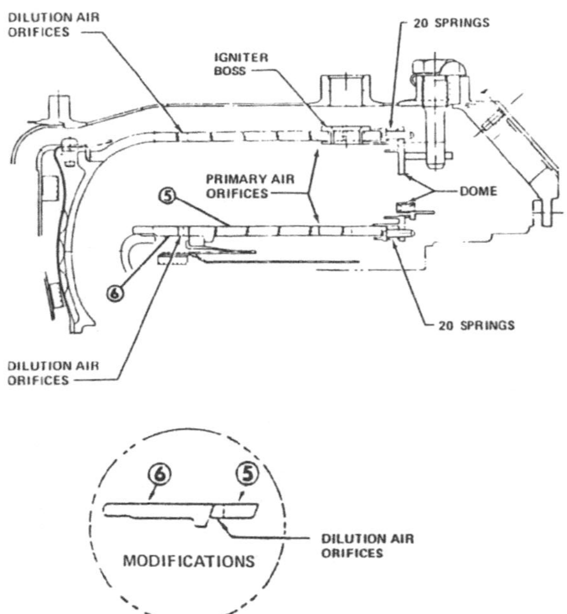

	FIND NO.	DESIGN STRENGTHS, KSI		PEAK STRESS, KSI	C.P.S. %
		TENSILE	FLEXURE		
COMBUSTOR SEGMENT, INNER WALL <u>WITH</u> DILUTION ORIFICES	6	19.0 13.0	34.0 23.0	29.0	
COMBUSTOR SEGMENT, INNER WALL <u>WITHOUT</u> DILUTION ORIFICES - S.S. CONDITION, <u>EXCLUDING</u> LOCAL COOLING AT AFT FACE	6	19.0 13.0	34.0 23.0	9.2	99.89 97.67
COMBUSTOR SEGMENT, INNER WALL W/O DILUTION ORIFICES. S.S. CONDITION, INCLUDING LOCAL COOLING - <u>10 EQUALLY SPACED RADIAL SLOTS.</u>	6	19.0 13.0	34.0 23.0	14.3	99.08 81.80
COMBUSTOR SEGMENT, INNER WALL W/O DILUTION ORIFICES - S.S. CONDITION, INCLUDING LOCAL COOLING - <u>10 EQUALLY SPACED RADIAL SLOTS.</u>	6	19.0 13.0	34.0 23.0	17.2	99.43 88.27
COMBUSTOR SEGMENT, INNER WALL W/O DILUTION ORIFICES - S.S. CONDITION, INCLUDING LOCAL COOLING - <u>60 EQUALLY SPACED RADIAL SLOTS.</u>	6	19.0 13.0	34.0 23.0	12.7 - 15.5	99.93-99.76 98.42-94.84

Fig. 12. Analysis summary, combustor forward ring segment.

This investigation pointed out the importance of considering effects of localized heating or cooling of ceramic components on component reliability. Critical attention must be paid to areas where leakage of hot gases or cooling air can impinge upon ceramic parts with resulting thermal stresses that may initiate fractures.

In the ARPA/NAVAIR ceramic engine, no cooling air is impinged upon ceramic components. Instead, cooling air is introduced as a film to eliminate localized stresses resulting from high thermal gradients. Where hot gases can impinge upon a component, such as leakage between stator segments impinging upon the pilot cylinder, the cylinder has been removed from the stators, thereby minimizing the effects of such leakage. This can be seen in the turbine cross-section of Figure 3.

Turbine Static Rig

Twenty-one separate tests have been conducted on the turbine static component rig to establish static component clearances and mechanical design refinements that assured maximum probability of success in engine testing. Rig testing permitted ceramic contact zone behavior, thermal shock capability, and clearance requirements to be observed and corrected. Test experience indicated chipping and fractures from contact stresses to be the most critical items limiting static component life.

Contact zone behavior can be affected by a number of variables, including:

- friction
- relative movement
- vibration
- unequal load sharing
- interface design

It was evident from early rig test results that these problems would have to be resolved before acceptable static component life could be realized. The rig test effort consisted of a systematic program of test--analysis of results--modification--test. Where problem areas were evident, testing was not continued until the problem was understood and corrective action taken.

In many instances, material data was not available to aid in analysis of test results. In these cases, data had to be generated. The effects of temperature upon friction, for example, were not available and data had to be generated to understand the effects of friction upon component behavior. Figure 13 shows the effects of temperature upon coefficient of friction between machined RBSN/RBSN surfaces. Note that friction increases with temperature on bare RBSN surfaces. One method of

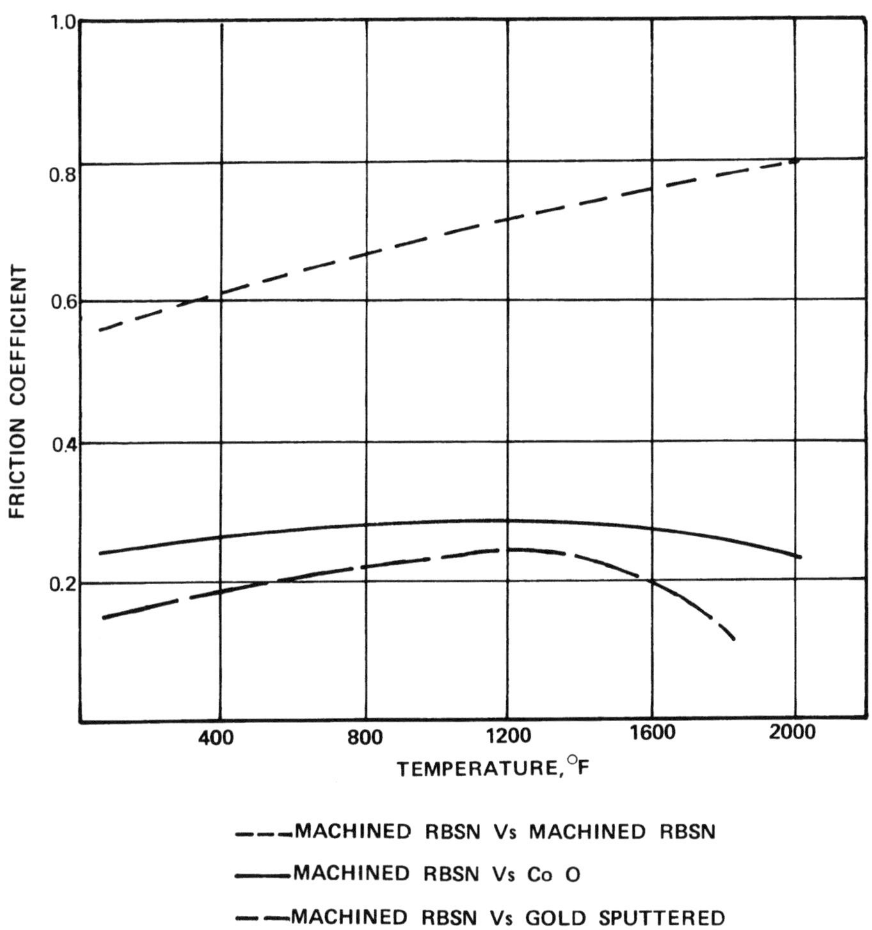

Fig. 13. Effect of temperature upon friction.

reducing friction is to use a lubricant or lubricating strip that reduces friction without significantly affecting material properties. Use of cobalt oxide (CoO) lubricant or gold sputtered ceramic surface resulted in a substantial reduction in friction, especially at elevated temperatures typical of engine operation.

The choice of a lubricant or lubricating strip for a particular application depends upon a number of factors, such as:

- level of friction
- staying power of lubricant
- effects of lubricant on material properties

Relative movement and vibration appear to go hand-in-hand with friction in contributing to contact zone distress. In early rig testing, stator segment anti-rotation lugs suffered from chipping at loads that were much lower than maximum loads. Since relative movement was evident on the contact surface adjacent to the fracture zone, specimen testing was initiated to determine effects of relative movement on material behavior. The effect of relative movement upon load carrying capability of RBSN specimens was significant. While no specimens fractured when subjected to a given static load, fractures occurred at load levels below 20 percent of static levels when relative motion was present.

The use of lubricants and compliant layers at ceramic interfaces serves to reduce the effect of relative movement by lowering the biaxial stress state that contributes to this phenomenon.

Proper choice of compliant layer material and perforation pattern will serve to:

- permit compliant layer to act as a lubricating strip
- tailor the "spring rate" of the compliant layer
- act as reservoir for lubricant

Perforated compliant layers have been successfully employed in the ARPA/NAVAIR Ceramic Engine in such areas as anti-rotation lugs. As an example Figure 14 shows a first-stage stator anti-rotation lug positioned in the pilot cylinder. In one view the lug is positioned without a compliant layer, while in the other the compliant layer is installed. This shows how the compliant layer serves as a cushion between the stator lug and the pilot cylinder.

Unequal load sharing in which one component is subjected to a disproportionate loading is a real problem in ceramic static structures. Since ceramics do not redistribute loads, load sharing must be assured by

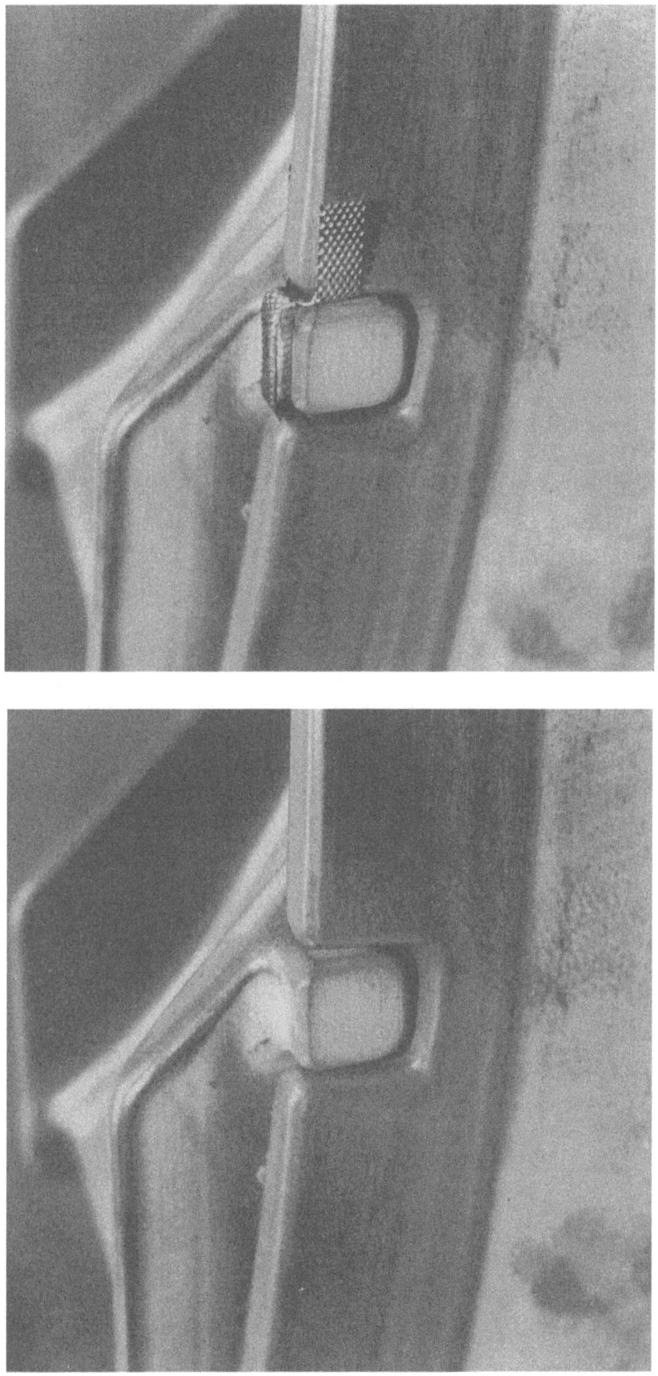

WITH COMPLIANT LAYER

NO COMPLIANT LAYER

Fig. 14. First-stage stator anti-rotating lug.

proper control of locating surfaces and design techniques that work within the material capabilities.

The following techniques have been applied in the ARPA/NAVAIR Ceramic Engine Program to assure acceptable load sharing.

- Locating surface control - Machining of components in an assembly eliminates effects of tolerances on stack dimensions between locating surfaces. This match machining of critical locating surfaces relative to each other, assures against adverse stack dimensions that can result in unequal loading.

- Compliant Layers - Use of compliant layers between contacting ceramic surfaces serves multiple purposes. In the case of load sharing, compliant layers act as cushions to absorb minor surface irregularities, as well as serving to distribute loads over a greater contact area.

- Redundant Load Paths - Adverse stack conditions, improper clearances, and distortions of components can result in redundant load paths that impose severe loads upon components. Significant attention has been paid in the ARPA/NAVAIR Ceramic Engine to assure against such redundant load paths.

It was evident in observing load patterns on rig tested hardware that loading was occurring close to the edges of some ceramic components. Proper ceramic interface design results in controlled, predictible contact, and extends contact zone life. An example of interface design is shown in Figure 15 where crowning of one contact surface results in a controlled line contact removed from the edge of the mating parts. Control of the contact zone through proper interface design has been employed in all ceramic interfaces of the ARPA/NAVAIR Ceramic Engine.

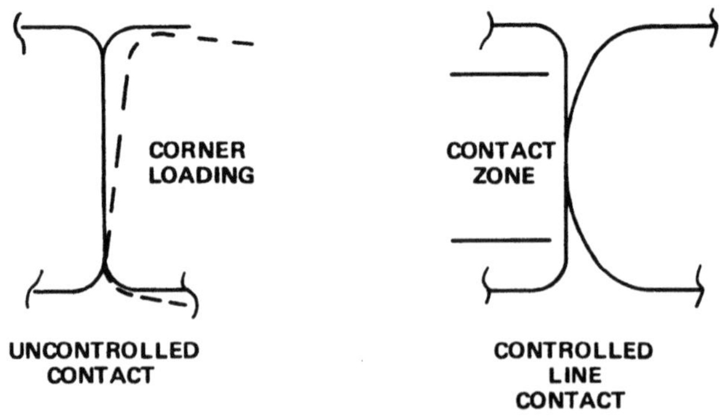

Fig. 15. Ceramic interface design.

TURBINE ROTATING RIG

Use of the turbine rotating rig has provided valuable information about the behavior of ceramic blades in a dynamic environment. This program demonstrated the critical effects that blade attachment zone has upon blade vibration.

The first dynamic rig test ended in a blade fracture at the top of the attachment zone as shown in Figure 16. This occurred at a speed of 36,280 RPM (100-percent speed is 41,730 RPM). Investigation revealed that a combination of a "hard" HS-25 material compliant layer and irregularities of 0.001-inch in the dovetail slot contributed to a lowering of the blade fundamental bending frequency into the engine operating range. This combination of hard compliant layer and irregular surface also served to lower the vibration tolerance of the blade by acting as a stress concentration at the high point of the dovetail slot, shown in Figure 17. This high point correlated with the blade fracture origin seen in Figure 16.

The effects of slot irregularities and compliant layer hardness were evaluated on a static frequency test rig. A blade was tested using both platinum and oxidized HS-25 compliant layers with and without simulated irregularities. The test results, Figure 18, show the increase in frequency that resulted in going from HS-25 compliant layers in irregular surface slots to platinum compliant layers in smooth slots.

It appears that ceramic blades are much more sensitive to attachment zone irregularities and stress concentrations than their metal counterparts. This is because ceramics do not redistribute stresses or offer any degree of compliance. Where inserted ceramic blades are used, great care must be taken to assure that proper "seating" of the blades takes place.

As a result of this effort, all turbine wheel dovetail slots are lapped to remove positive surface irregularities. Platinum compliant layers are also used in the blade attachment area since they offer greater compliance than HS-25 and have demonstrated acceptable cyclic life.

Engine Development

Full-scale engine testing was initiated in April 1978 following a successful rig test of the ceramic static and first-stage rotor components to engine conditions. Figure 19 shows a TSE331C-1 engine on the test stand.

Thermal shock proved to be a problem during early engine testing, since the measured engine lightoff temperature levels were higher than predicted. Where the parent T76 metal engine of the TSE331C-1 could survive such thermal shock during lightoff, the ceramic stators exhibited

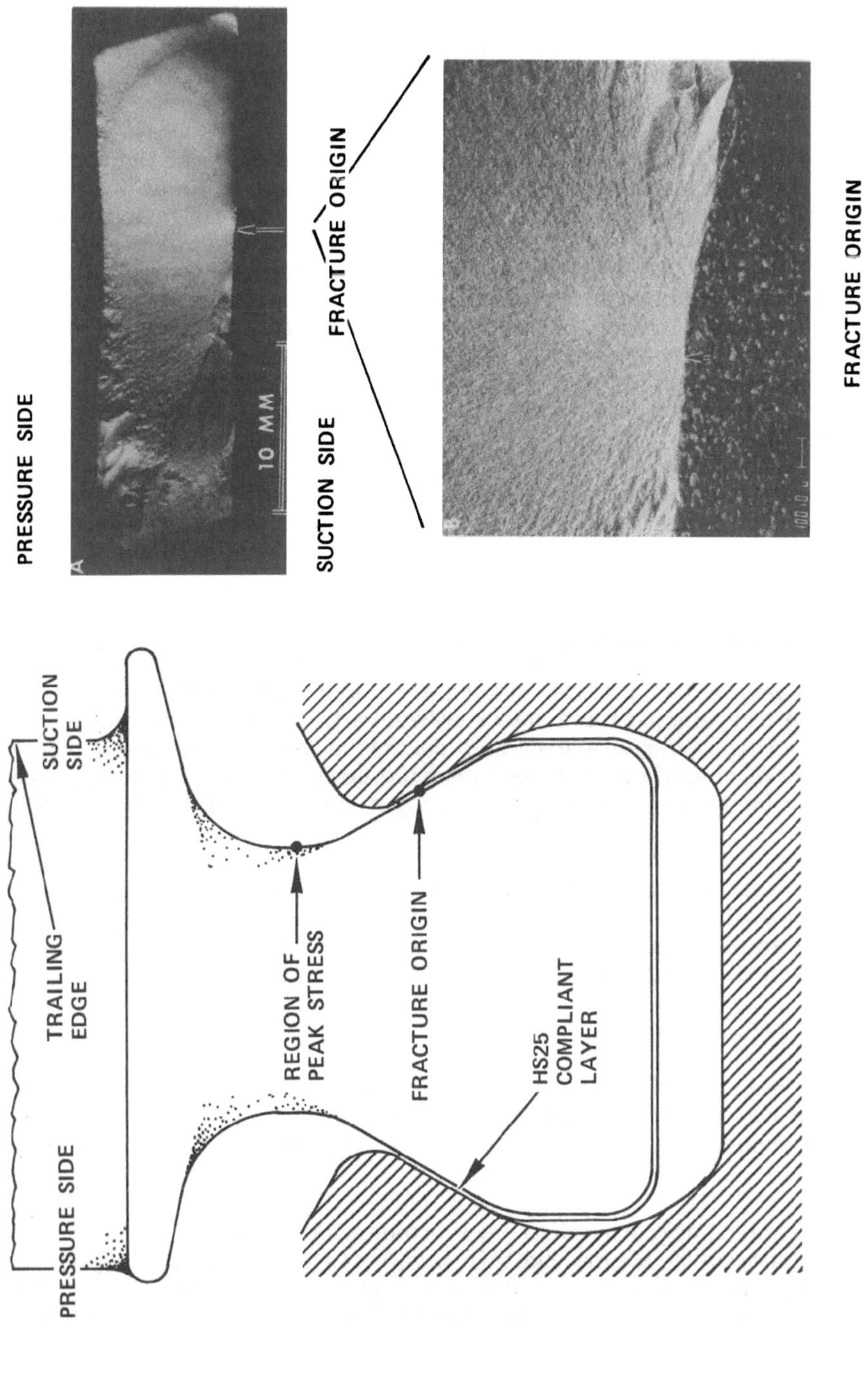

Fig. 16. Blade fracture investigation.

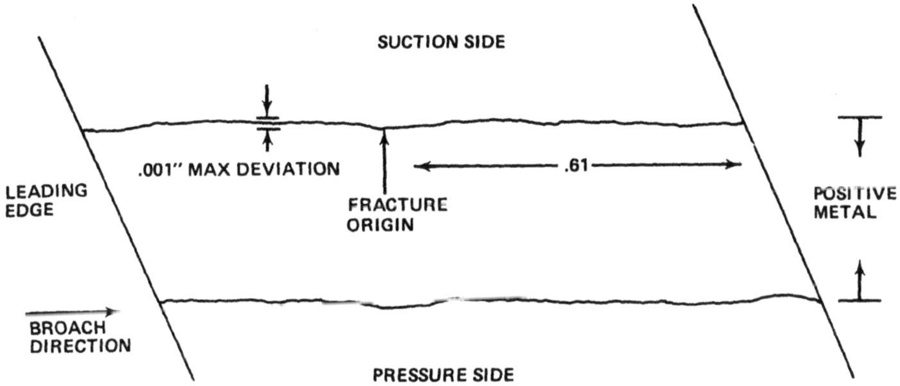

Fig. 17. Disk slot contour.

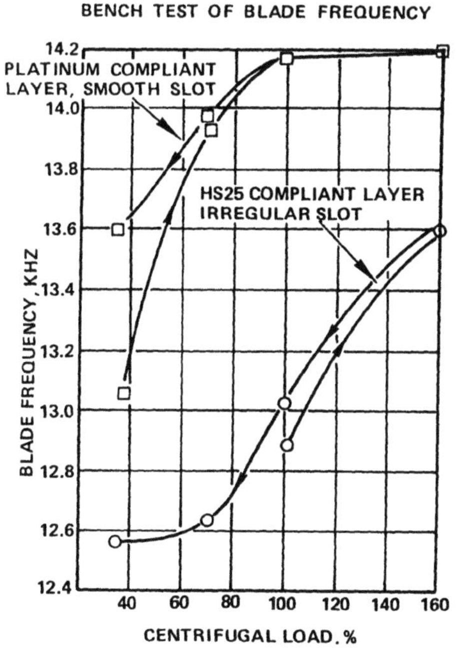

Fig. 18. Effect of dovetail slot and compliant layer on flexural frequency of second-stage blade.

Figure 19. TSE331C-1 engine installed on test stand.

cracks that initiated at the airfoil trailing edges. A program to reduce the severity of lightoff thermal shock resulted in changing the fuel from JP-5 to JP-4, which has more consistent ignition characteristics. Ignition speed was also increased from 10- to 15-percent speed. Figure 20 shows the effects of these changes upon measured turbine inlet temperature (T_{t4}).

No incident of stator airfoil fractures as a result of lightoff thermal shock have been noted since JP-4 fuel and 15-percent ignition speed have been incorporated into the engine test program.

This experience indicates that use of ceramics in an engine requires assessment of the effects of ceramic behavior at a systems level as well as at the component level.

The engine development program has shown ceramic static structure contact zone distress to be the problem area requiring most attention. Engine test experience in the ARPA/NAVAIR engine has shown that

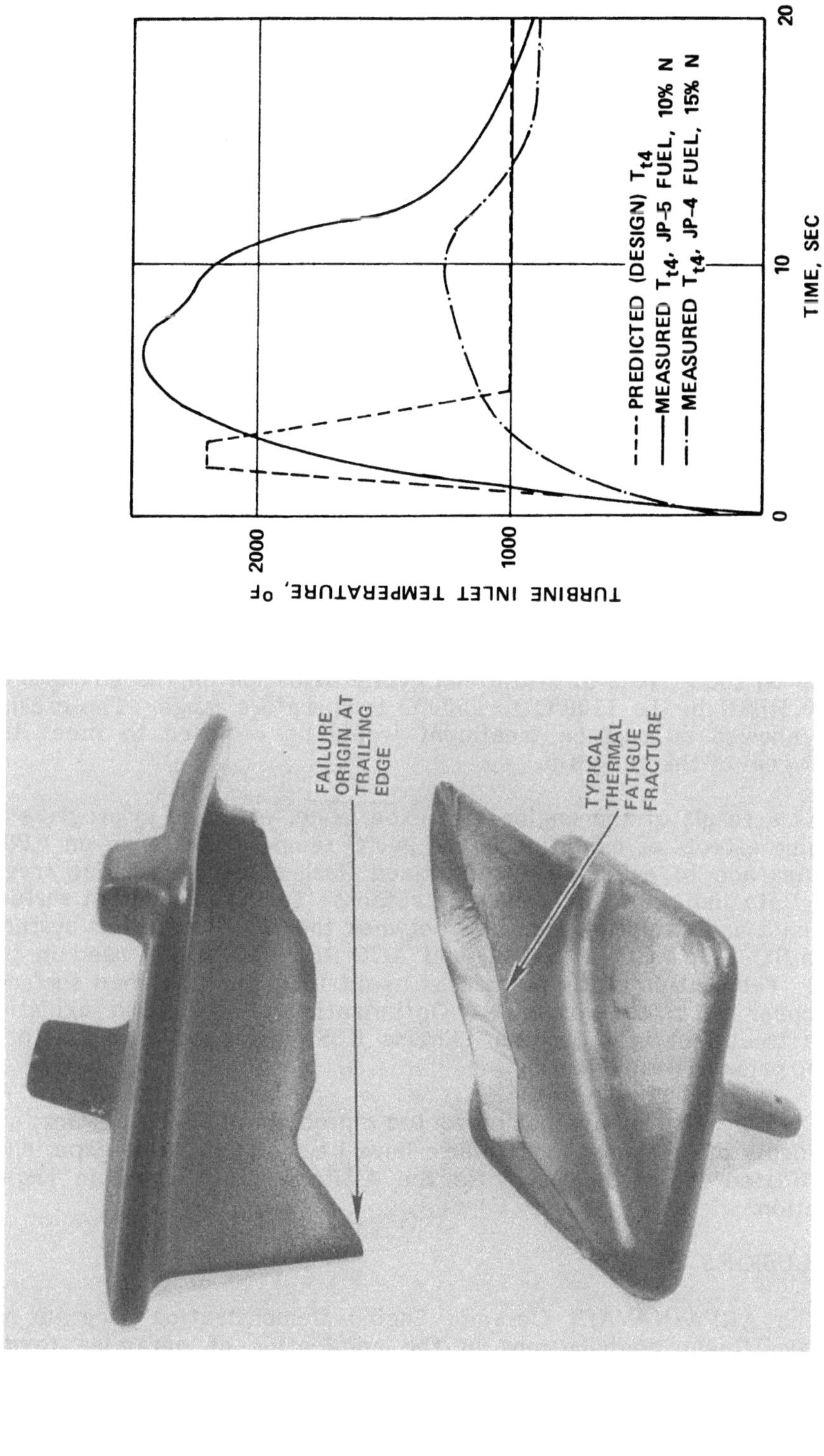

Fig. 20. Lightoff thermal shock.

compliant layers are required at all ceramic interfaces in the static structure to improve reliability and reduce contact zone distress.

The compliant layer is perforated HS-25 material, as previously shown in Figure 14, which is pre-oxidized before being installed in the engine. During assembly, cobalt oxide lubricant is added to reduce friction during engine operation. The compliant layer is designed to perform a number of critical functions. For example:

- HS-25 material work hardens after taking an initial set. This extends the life of the compliant layer after it has conformed to the contours of the mating ceramic components.

- Pre-oxidizing results in a cobalt oxide surface layer which serves as a lubricant to reduce friction at the ceramic interface.

- Perforations serve to tailor the spring rate of the compliant layer as well as acting as a lubricant reservoir.

A contribution to contact zone distress is the effect of oxidation upon RBSN strength. As engine test time and temperature increased, evidence of degradation of RBSN properties were noted at relatively low temperatures in the 900°C range. Work was conducted early in the program on the effects of static and cyclic oxidation on the strength of NC-350 RBSN in the 1100°C to 1500°C temperature range. These early results showed no surface treatment would be required to meet the 50-hour goal of this program.

As a result of the engine test experience, effort is in progress to determine effects of oxidation in the 900°C temperature range on RBSN properties and to establish an optimized flash oxidation surface treatment. Data indicates differences in response to flash oxidation surface treatments and oxidation exposure between the RBSN materials systems (Norton NC-350, ACC RBN-101, and ACC RBN-122) being used in the engine. Fabrication methods such as as-nitrided and machined surfaces also appear to affect response. Optimization of the flash oxidation surface treatment is in process. Engine RBSN components will be pre-oxidized prior to future testing.

Ceramic blades have not presented a problem in engine testing, and no incidents of primary blade failure have been noted. This experience has validated the blade design for the ARPA/NAVAIR Ceramic Engine application.

CONCLUSIONS

The ARPA/NAVAIR Ceramic Engine Demonstration Program has made significant contributions in the application of ceramics to gas turbine engines. Under this program, material characterization of HPSN and RBSN was expanded; fabrication techniques for ceramic components developed, and inspection methods and acceptance criteria for finished parts were established. Analytical tools using 3-D finite-element modeling with probabilistic material properties were developed and verified by rig and engine tests.

Engine testing is now indicating areas where future ceramic effort should be concentrated. Understanding of contact zone behavior and improvement of material resistance to long-term engine exposure are two areas that will offer the most immediate payoffs to increasing the reliability of ceramics. As test exposure of the ceramic engine expands, additional areas will be uncovered where ceramic effort should be applied. Only through full-scale engine testing can the full range of required technology be identified for application of ceramics to gas turbines. The ARPA/NAVAIR Ceramic Engine Program being conducted at Garrett is providing that knowledge.

CERAMIC APPLICATIONS IN TURBINE ENGINES

H. E. Helms

Detroit Diesel Allison
Division of General Motors Corporation
P. O. Box 894
Indianapolis, Indiana 46206

ABSTRACT

The objective of the CATE program is to apply ceramic components in a highway vehicle gas turbine engine and thus reduce engine fuel consumption. Ceramic components permit increased engine operating temperatures which improve engine cycle efficiency.

An initial program milestone has been achieved as ceramic components are operating over the road in a gas turbine powered truck which is available for demonstration. In preparation for this demonstration, the silicon carbide nozzle vanes and aluminum silicate regenerator disks were operated in a turbine powered truck over city streets, interstate highways, road hazards, Belgian block and rough road courses. Both components emerged from this testing in completely serviceable condition.

Overall, a total of 4484 operating hours have been accumulated in this program, representing nearly 250,000 miles of vehicle operation at 1900°F turbine inlet temperature. Included are over 400 hours of successful transit coach durability subjecting silicon carbide vanes and alumina silicate regenerator disks to far more accelerations/decelerations (thermal shocks) than the composite truck cycle used for the other engine durability testing. Alumina silicate regenerators experience now totals 4484 hours and reaction bonded silicon carbide vanes 1829 hours.

Design of the 2070°F engine continues with the ceramic gasifier nozzle, turbine tip shroud and gasifier turbine blades already being fabricated. Supporting this design and fabrication, has been process development using a prototype blade and compliant layer testing using a ceramic coupon duplicating the blade attachment.

Materials characterization activities have resulted in initial design data definition of reaction bonded silicon carbide, reaction bonded silicon nitride, sintered alpha silicon carbide and sintered silicon nitride. Further, the design to be used with ceramic materials has been defined including analysis techniques and statistical treatment of materials strength data. Evaluation of second and third generation ceramic parts and test bars has indicated that process development activities are increasing the realizable strength of ceramics. Testing of several abradable layer samples has been very encouraging with parts now being fabricated for full scale evaluation. Oxidation resistance tests have been initiated on candidate materials for 2070°F operation indicating operation to be feasible without application of resistant coatings.

CATE DEMONSTRATION VEHICLE

Ceramic parts developed under the CATE program are now in a DDA 404-4 gas turbine powered truck (Figure 1).

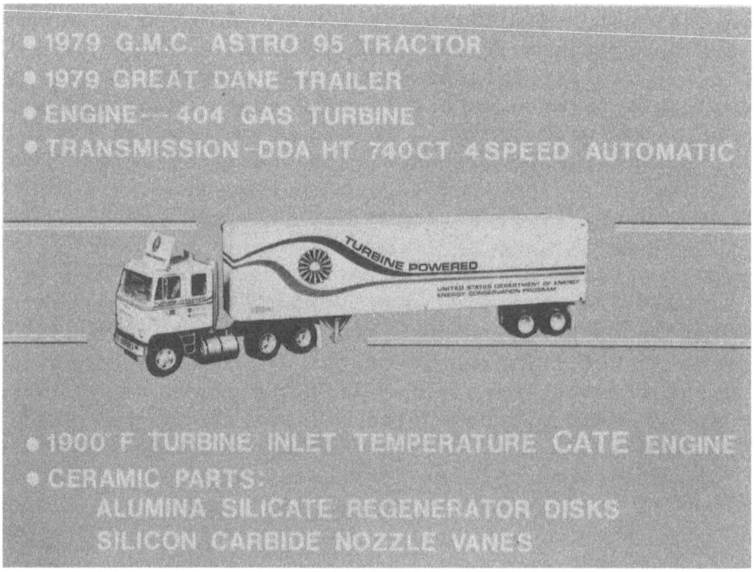

Figure 1. CATE demonstration vehicle.

The truck serves to demonstrate ceramic components operating in a turbine engine in a real operational environment. One notes the smoothness, quietness, and clean exhaust inherent in the turbine powered truck.

Ceramic gasifier turbine nozzle vanes and regenerator disks are utilized in the truck engine which operates at a turbine inlet temperature of 1900°F, 65°F above the baseline all-metal engine. The nozzle vanes are made of reaction bonded silicon carbide. Over 1800 hours of successful laboratory durability have been demonstrated on these kind of vanes in an engine operated to typical truck and transit coach operating duty cycles. The pair of alumina silicate regenerator disks represent a component with over 4400 operating hours of laboratory durability on truck and transit coach duty cycles. This initial 1900°F configuration engine demonstrates ceramic parts and their chemical and structural stability in the 404-4 gas turbine. This is believed to be the first public demonstration of structural ceramic turbine flow path nozzle vanes in an engine which produces useful power in a vehicle that is a candidate for commercial highway usage. Static ceramic components have been shown to be feasible for vehicular gas turbine applications.

The new demonstration truck consists of a 1979 GMC Astro 95 tractor and 1979 Great Dane Trailer. The DDA 404-4 engine is the CATE program's initial engine which has been operating at 1900°F turbine inlet temperature with ceramic components since September 1977. The engine is complimented with a DDA HT740CT four speed automatic transmission which is a commercially available transmission adapted to the turbine engine. The tractor has been customized to accommodate five observers comfortably for demonstration rides. Ceramic materials have truly been put to work in turbine trucks and the CATE Program has begun to prove the credibility of ceramics for turbine engines.

CERAMIC COMPONENT TRUCK DEMONSTRATIONS

A fundamental question which had to be answered before proceeding with a CATE program demonstration truck was "Can ceramic parts withstand the thermal environment and the mechanical shock and vibration encountered while operating over roads and highways?" To help answer this question, the above truck was equipped with CATE engine C-1 incorporating reaction bonded silicon carbide gasifier nozzle vanes and alumina silicate regenerator disks for environmental road testing. The truck was operated over streets, highways, road hazards and severe test courses that should produce extremes in operational loadings and demonstrate engine transient and steady state operation in a highway truck. This test would suggest potential capability of mechanical success during the intended operation of the Demonstration Vehicle if ceramic parts emerged without damage. Both the ceramic vanes and regenerator disks emerged in excellent condition and completely serviceable for continued use.

The initial truck tests were designed to place the engine under full control of a highway truck driver under various road conditions and subject the engine to the mechanical shock and vibrations of both normal operation and the extremes of road surface irregularities. Truck driver control would result in the engine experiencing a complete spectrum of temperature and thermal transients inherent in the application. The shock and vibration exposure would be a test of the inherently brittle ceramic materials over the wide range of mechanical loadings conditions of operation which are nearly impossible to produce in laboratory component vibration and shock tests. The intended objective was 50 hours of test on the General Motors Proving Ground along with operation on city streets, interstate highways, over rough railroad crossings and other normally encountered truck environments.

Overall, the engine was operated in the truck 92 operating hours accumulating some 1840 miles. Of these, 780 miles were over streets and highways and 1055 miles over Belgian block and truck rough road test courses at the General Motors Proving Grounds in Milford, Michigan. In addition, five miles of towing with the engine off were conducted over streets and road hazards.

Finally, trailer coupling was conducted to test capability for this shock environment. The truck operation was split between tractor alone (Bobtail) operation and tractor-trailer operation at a gross weight of 40,000 lbs. The Bobtail configuration produced the more severe shock loadings. During operation in each of these conditions, on board instrumentation recorded the truck frame and engine vibratory conditions for comparison to known loadings in trucks with diesel powerplants.

Tear down inspection following this testing concluded that each ceramic part was intact (free of damage) and completely serviceable. Weather conditions (snow) produced truck operation in a severe road salt environment. There was no evidence of chemical reaction of the salt with ceramic materials. It was noted that some unprotected engine metal parts showed significant corrosion from this salt (especially external brackets and chassis parts). The "milestone demonstration" of ceramic vanes and regenerators is the first of many demonstrations required to evolve ceramic materials technology from the the exploratory development state-of-the-art to fully useful and fully developed production engine materials.

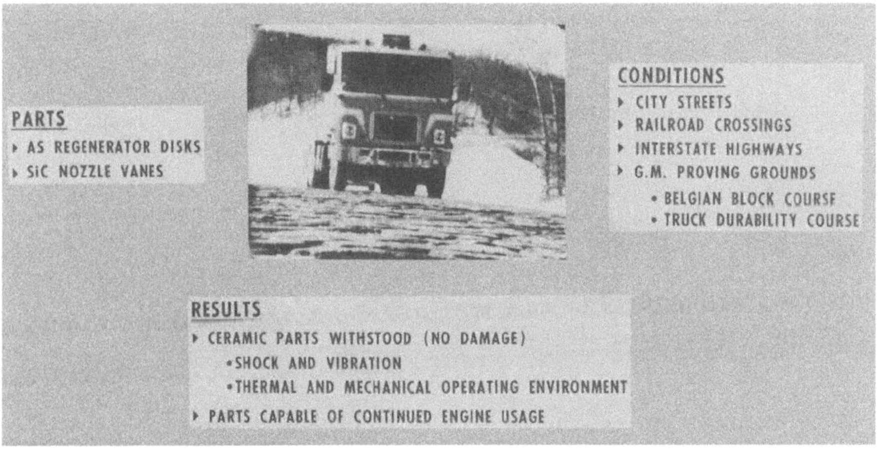

Figure 2. Ceramic component truck demonstration.

TRUCK DEMONSTRATION RESULTS

The shock and vibration environment experienced in the initial C-1 engine truck application is shown in Figure 3. Accelerometers were located on the truck frame close to engine mounts, and on the engine at its mounts. The shafts which support each of the regenerator disks were strain gaged to reflect regenerator loadings. The engine mounting system consists of 3 mounts--two to the rear carrying vertical, lateral and fore and aft loads, and one in the front carrying vertical and lateral only allowing freedom for axial thermal growth of the engine.

Figure 3 is an overall composite of the highest excitation level (0 to peak) in G's recorded throughout the test. It is observed that an engine acceleration of 4.8 G's was recorded at the front mount while operating in the Bobtail condition. This proved to be the most severe loading experienced during the truck testing. This particular reading was experienced on the Belgian block course at the maximum safe speed for the course (25 mph). The average of the 10 most severe loadings over each course was 3.15 G's, again in the Bobtail configuration over the Belgian block course. It is noted that the peak G's encountered at the regenerator disk was about 1.5 G's, considerably below that found at the engine mounts. Observation of the frequency spectrum over the Belgian block course showed frequencies in the range of 2 to 20 hertz with 10 hertz tending to predominate. Close examination reveals the presence of several discrete frequencies none of which are coincident, thus no resonance in the mount system.

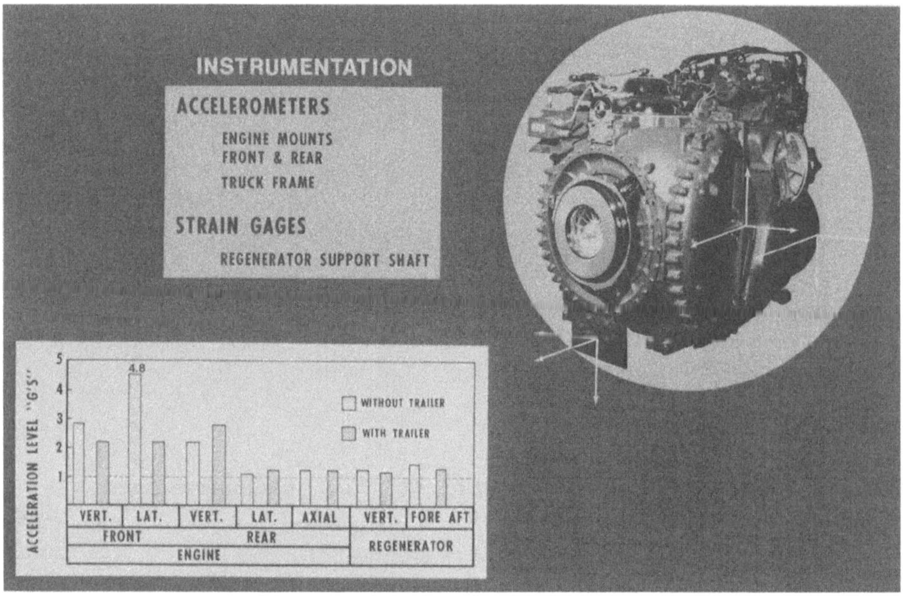

Figure 3. Truck demonstration results.

A few specific conditions are worthy of note. Rough trailer coupling shocks peaked at 1.5 G's. Railroad crossing (particularly poor track condition) between 8 and 25 mph resulted in 1.8 G's. It is noted that the regenerator disk under engine operation is supported by aerodynamic pressurization of its seals as well as the spindle. Thus towing without the engine running is a particularly critical condition for regenerators. The regenerator loadings recorded represent a range of .2 to 1.6 G's, the worst case being towing over railroad tracks.

It must be kept in mind that the data shown is the highest reading recorded and the level prevailing is significantly below these peaks. In general, the data generated compares with accepted design data for trucks operated under the same road conditions.

CERAMIC COMPONENT DEVELOPMENT

Detroit Diesel Allison Division of General Motors has been under contract to NASA Lewis Research Center with funds from the Department of Energy, Automotive Technology Development Division, Office of Transportation Programs, since July 1976, for application of ceramic components to demonstrate improved cycle efficiency by raising the operating temperature of the existing 404 vehicular gas turbine engine (see Figure 4). Early efforts studied the improvement potential and approaches to ceramic application to the 404 engine (.45 SFC) and resulted in a plan to apply ceramics incrementally at successively higher operating temperatures leading to an SFC goal of .35. Initial evaluations continue at 1900°F, which is 65°F above the baseline 404 engine with all metal components, and has led to the first vehicle installation of initial configuration ceramic components. Concurrently design and fabrication activities are directed at the second ceramic configuration to be operated at 2070°F. In the future a 2265°F configuration will introduce additional ceramic components for assessment of the SFC attained by introduction of ceramics. At each technological step, emphasis is placed upon durability of ceramic components with engine operation test stands to road usage cycles. In addition, ceramic component process development, ceramic material characterization, component non-destructive inspection and rig qualification testing support the program's planned intent to promote the understanding of design with ceramic materials and establish a technology base for automotive and other gas turbines.

FY 79 funding permits continued evaluation of initial configuration ceramic components at 1900°F while concentration on completion of design, ceramic part process development and fabrication of ceramic parts for the 2070°F configuration engines.

CURRENT ACTIVITIES

As illustrated in Figure 5, engine durability testing continued in this reporting period along with conduct of three significant engineering data tests to guide the design of the 2070°F configuration. Silicon Carbide nozzle vanes and alumina silicate regenerator disks were tested extensively at 1900°F T.I.T. in the

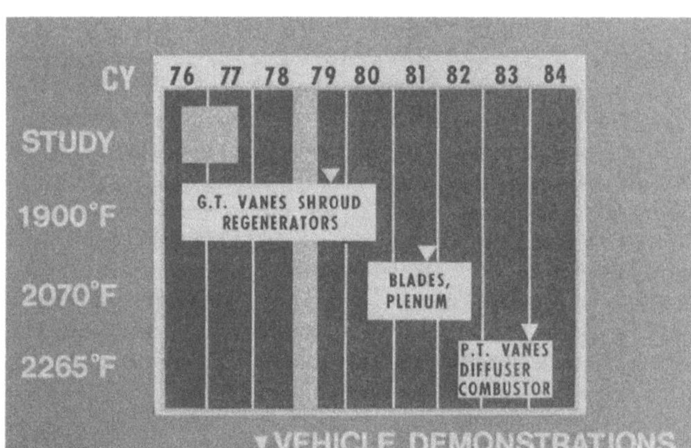

Figure 4. DOE/NASA ceramic component development.

engine with continued success displaying chemical and structural stability as well as operating durability. Durability was conducted in programmed dynamometer test cells to composite truck and transit coach cycles as well as the previously cited over the road truck operation. The introduction of the transit coach durability cycle (404 hours) testing with 55 severe thermal transients per operating hour evaluated the thermal shock resistance of the silicon carbide vanes and alumina silicate regenerators.

Supporting the component test program, ceramic material characterization continued utilizing test bars representative of both silicon nitride and silicon carbide products from four sources. Elastic, thermal and strength characteristics have been established for reaction bonded silicon carbide and silicon nitride and sintered alpha silicon carbide and sintered silicon nitride. This data is currently being used for design data. Strength tests results coupled with failure mode characterization have been statistically treated to portray strength in Weibull statistics used by the designer. Oxidation testing of current materials is in process. Abradability samples based upon both controlled densification and coated ceramics are yielding preliminary but encouraging results. A study was completed by Technology Associates, Purdue University, identifying viable near term NDE approaches applicable to ceramic materials. The long term goal is to establish critical flaw sizes associated with given stress levels in a given ceramic material and to be able to identify critical flaws in ceramic components used for gas turbine applications. Experimental NDE investigations based upon identified technologies have been initiated.

Design and analysis of the 2070°F configuration ceramic components are nearing completion with process development and part fabrication in progress. Alternate materials, processes and sources for each are being pursued. Process development in sintered alpha silicon carbide of a prototype blade yielded a steady improvement of dimensional and material quality. Blade coupon tests are underway to select the optimum compliant layer design by static loading and verify the design of the dovetail attachment by spin pit testing. Special instrumented runs utilizing 1900°F engines have been completed with evaluations leading to engine controls based upon exhaust gas sensing (T_6 vs current burner outlet temperature T_4); engine block crossarm thermal gradients and deflection assessments; and high temperature T_6 regenerator seal temperature measurements resulting in seal temperature assessments. Seal temperatures measured indicate graphite peripheral seals are usable and temperature gradients of the seal metal structures are within platform capabilities.

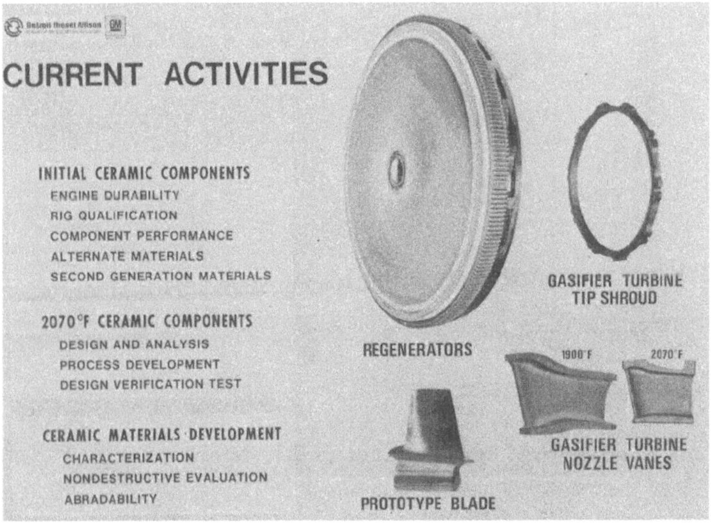

Figure 5. Current activities.

REGENERATOR SEAL DEVELOPMENT

The seal shown in Figure 6 is a metal structure faced on one side with applied wear face materials with leaf springs on the opposite side and is known as the hot or inboard regenerator seal. It is held against the rotating ceramic regenerator disk by leaf spring and gas pressure forces forming a running seal around the periphery and diametrically across the face of the regenerator disk.

The seal across the diameter, known as the crossarm seal, has shown small radial distortion during fabrication and distortion from the thermal environment of the engine. This distortion results in increased leakage during start up and idle conditions when gas pressure clamping the seals is minimum, making starting difficult. On early seals the wear face on the crossarm was either 100 percent nickel oxide or 85% nickel oxide and 15% calcium fluoride which was plasma sprayed on the metal substrate.

The first phase testing, directed to minimizing distortion, consisted of making a series of plasma sprayed specimens varying materials, spray intensity, clamping, and heat treatment. The series of tests on these samples concluded that plasma spraying under controlled conditions, a 70%/30% nickel oxide/calcium fluoride (NiO/CaF_2) material on inconel 625 prebent metal structure, fol-

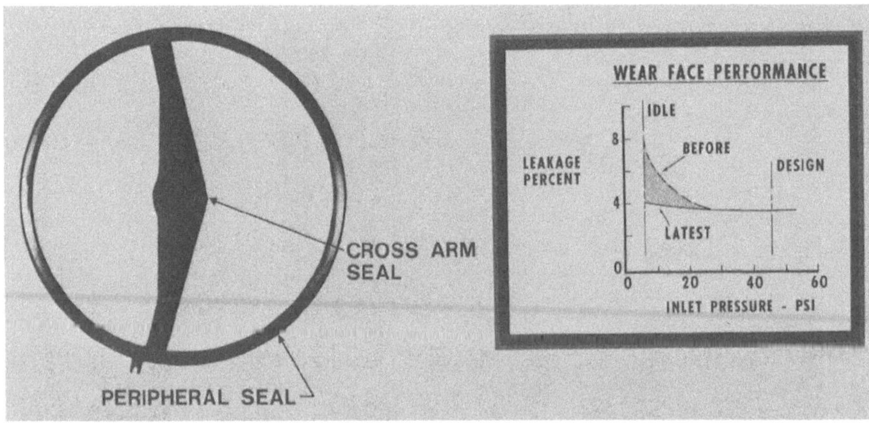

Figure 6. Regenerator seal development.

lowed by a thermal cycle treatment could result in minimum seal crossarm distortion. The second phase of testing scaled up the approach to the full scale crossarm. After several iterations to the process, a 70%/30% NiO/CaF_2 crossarm was put into a seal assembly and subsequently rig tested with the results shown in the chart. With less than 4 percent leakage at the higher pressures, this seal leakage profile yields an excellent engine running characteristic. The 8 percent leakage is excellent for the starting condition with 4 percent at idle very exceptional. This seal has been run 100 hours in an engine in addition to rig testing and indicates no performance degradation with engine environmental exposure.

CERAMIC BLADE SHAPE FABRICATION DEVELOPMENT

It was recognized early that the fabrication of complex shapes such as rotor blades of ceramic materials without extensive airfoil machining would require an iterative approach. It was desired to produce parts to the dimensional requirements of turbine aerodynamics and have them possess the material strength required for the engine application. Efforts at dimensional requirements focused on structuring and controlling each step of the fabrication process to produce repeatable airfoil shapes. Material strength was to be achieved by basic material development, process and tool modification to minimize flaws and their size, process control to minimize material variability and in-process utilization of non-destructive evaluation techniques.

A 2070°F prototype blade configuration was defined resembling the attachment shape, airfoil size and material thicknesses envisioned for the blade to be designed. This blade was fabricated of sintered alpha silicon carbide where shrinkages of up to 20 percent from the as molded green state to the final sintered part were observed.

To evaluate dimensional uniformity, dimensional measurements of airfoil geometry have been monitored. The setting angle indicates the control of the airfoil positioning on the mounting axis of the attachment dovetail. The contour deviation portrayed the trueness of the airfoil shape. (See 0 and X on the facing chart.)

As shown in Figure 7 processing progress has been made with more than 150 parts having been produced. At the present time, blades are being made which comply with the dimensional requirements imposed by aerodynamics.

Continuing effort is reducing variability and improving part yield at each step of the fabrication process. To achieve the current blades, processing variables investigated include mold changes, mix variation, molding parameter variation, and sintering parameters.

CERAMIC MATERIALS CHARACTERIZATION

Materials activities have been concentrated upon characterization of candidate ceramic materials in support of the 2070°F configuration engine design and analysis, evaluation of second and third generation ceramic parts of the initial 1900°F configuration and study of non destructive evaluation technology applicable to ceramic parts. In addition, effort has continued defining ceramic design methodology for ceramics and the statistical treatment of ceramic test bar strength data. The ceramic materials include sintered alpha silicon carbide, reaction bonded silicon carbide, sintered silicon nitride and reaction bonded silicon nitride. An initial set of materials properties design data sheets has now been compiled for these materials.

Materials characterization is providing ceramic material elastic properties, thermal properties, strength, fracture toughness, abradability (tip shrouds) and chemical/structural stability assessment. These properties are augmented by microstructure and fracture surface topography analysis to relate properties to the structural and chemical make-up of the materials.

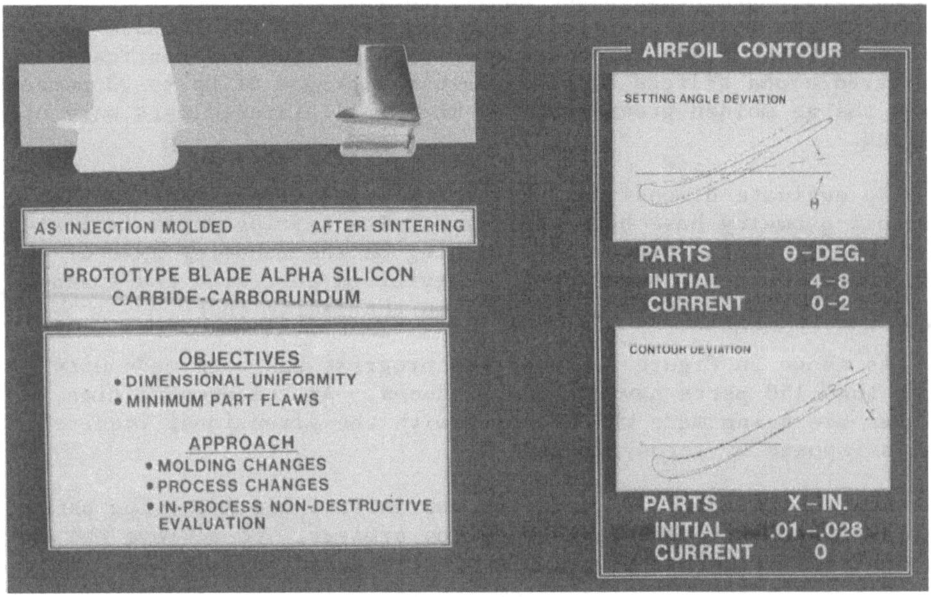

Figure 7. Ceramic blade shape fabrication development.

As an example of characterization activities, the results of thermal properties testing are shown in Figure 8. Linear thermal expansion was measured in an argon atmosphere with a recording quartz dilotometer up to a temperature of 1472°F. At higher temperature, an optical dilatometer was used as marks on the sample were viewed with twin telemicroscopes capable of defining length changes of less than .00010 inch per inch. Specific heat was derived utilizing a drop (ice) calorimeter. A specimen was preheated to a predetermined temperature, wrapped in tantalum, and placed in the closed calorimeter wall where its heat melts the ice. Mercury was introduced as a measure of the volume change of the melting ice. Enthalpy was derived from the weight of mercury introduced as the specimen cooled from its introduced temperature to that of ice. Specific heat was established from the data collected for six temperatures from the relationship Cp=dH/dt. Thermal conductivity was determined from thermal diffusivity measurements using the flash laser technique. The technique utilizes a laser impulse to heat a thin specimen in the isothermal zone of a furnace with the atmosphere preheated to the data point temperature. Conductivity is derived from recording the thermal response of the back face of the specimen of known thickness.

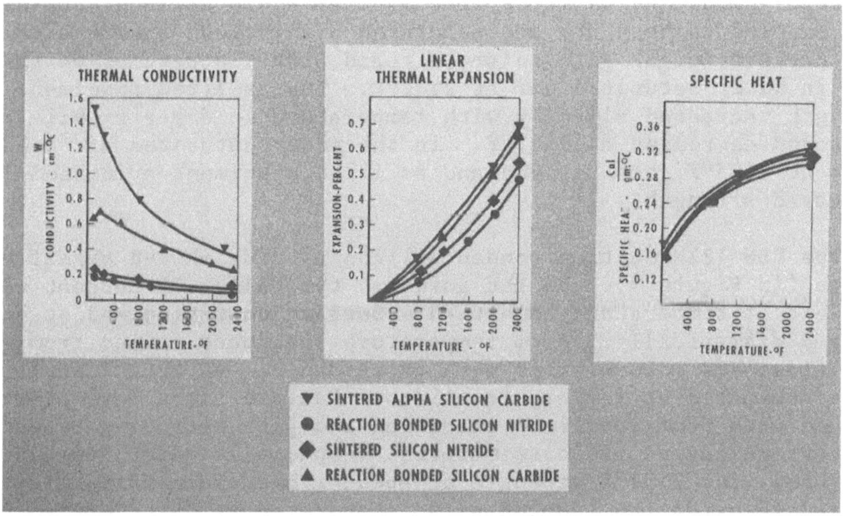

Figure 8. Ceramic material characterization--thermal properties.

CERAMIC MATERIAL OXIDATION TESTS

Oxidation tests have been initiated which expose ceramic specimens to an oxidizing environment for varying lengths of time over a range of temperatures. This particular test series was run at constant temperature. These tests are used to establish the effect on ceramic material strength that exposure to elevated temperatures for extended periods of time may have. Tests are conducted both in dry air and water saturated air because of the known accelerating affect of water vapor on the oxidation kinetics of both silicon nitride and silicon carbide materials. Materials will be evaluated from room temperature through 2500°F for lengths of time to 1000 hours.

During these initial tests, both sintered alpha and NC430 silicon carbide specimens displayed excellent resistance to oxidation and no strength degradation. Strength of RBN 122 reaction bonded silicon nitride remained unchanged through 1922°F but displayed a marked decrease in strength after exposure at 2292°F.

The strength of ground specimens of alpha silicon carbide remained at room temperature strength through 2507°F with failure traced to both surface and volume sites where undensified structure proved to be the strength controlling defect. Weight gain after 100 hours at 2507°F was approximately .10 percent indicative of excellent oxidation resistance. Weight gain remained under .5 percent when exposed to temperatures of 2282°F in water saturated air, without apparent loss of strength.

Both ground and as fired specimens of NC430 silicon carbide were exposed to both dry and saturated air through 2507°F with .125 percent weight gain in dry air and with .10 percent weight gain in water saturated air at 2282°F. The as fired specimen strength increased slightly with temperature in dry air while the as ground decreased at 2507°F. In the water saturated air conditions at 2285°F, both ground and as fired specimens evidenced increased strength.

The RBN 122 reaction bonded silicon nitride showed significant change (in Figure 9). Weight gain was typically .33 percent regardless of temperature condition. Surface condition and strength changed dramatically. Room temperature specimens failed from surface flaws and volume sites were in evidence. At 2282°F surface flaws dominated at failure locations. Failure sites were clearly surface pits from localized oxidation and pits extending beneath the surface layer into base material (especially after 500 hours of exposure). At 2507°F strength dropped with surface failure sites related to oxidation induced pits. Saturated air accelerated the oxidation and resulted in strength loss at 2282°F. All failure sites were surface flaws at glass puddles with the presence of such impurities as iron, iron/chrome, calcium, sodium and aluminum.

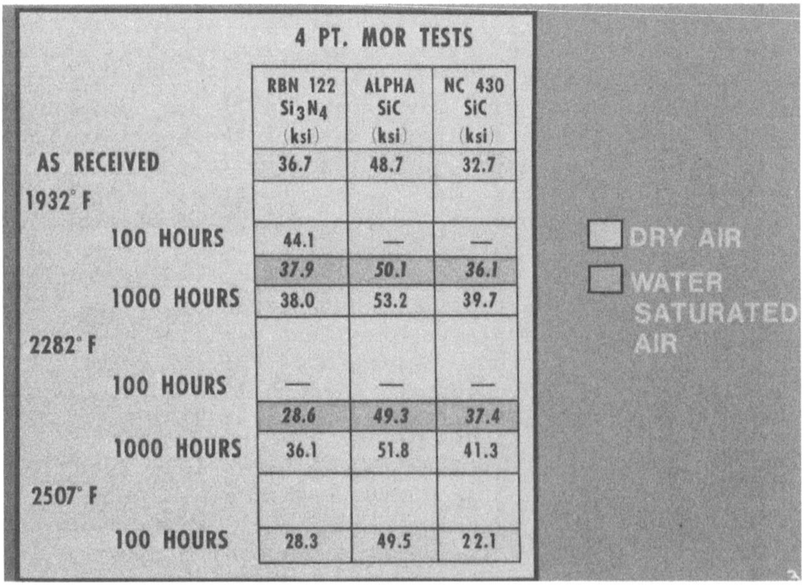

4 PT. MOR TESTS

	RBN 122 Si_3N_4 (ksi)	ALPHA SiC (ksi)	NC 430 SiC (ksi)
AS RECEIVED	36.7	48.7	32.7
1932° F			
100 HOURS	44.1	—	—
	37.9	50.1	36.1
1000 HOURS	38.0	53.2	39.7
2282° F			
100 HOURS	—	—	—
	28.6	49.3	37.4
1000 HOURS	36.1	51.8	41.3
2507° F			
100 HOURS	28.3	49.5	22.1

Figure 9. Ceramic material oxidation tests.

CERAMIC MATERIAL STRENGTH - MACHINING RELATIONSHIP

Finish grinding on critical surfaces such as the turbine blade attachment can produce surface damage which may lower strength. The strength of ground test bars of sintered alpha silicon carbide is used herein to illustrate results obtained (see Figure 10). Modulus of rupture bars were subjected to strength test by 4 point loading with machining surfaces lying along the plane of maximum tensile stress and transverse to it. It was observed that the strength of transverse ground bars is reduced significantly from longitudinal ground bars confirming the relationship between strength and machining direction.

In the case of transverse ground bars, fracture proceeds initially along grinding grooves. Failure origin was at the surface in 75 percent of the longitidinal ground bars. These results clearly establish the necessity of proper orientation of finish grinding and further confirm the extremely urgent requirement to either produce parts without machining or to establish machining which minimizes surface damage.

MODULUS OF RUPTURE RESULTS
ALPHA SILICON CARBIDE-CARBORUNDUM

MACHINING DIRECTION	MEAN STRENGTH KSI	STD. DEV. KSI	CHARACTERISTIC STRENGTH KSI	WEIBULL MODULUS
LONGITUDINAL GRIND	45.6	6.5	48.3	8.2
TRANSVERSE GRIND	39.9	4.7	42.0	9.9

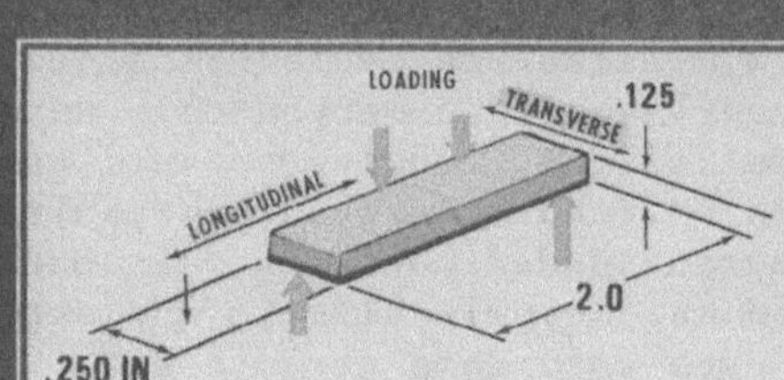

Figure 10. Ceramic material strength--machining relationship.

CERAMIC MATERIAL DEVELOPMENT

One of the objectives of the CATE program is to work with the ceramic industry in producing ceramic parts with strengths advertised for the basic material. An iterative approach to fabricating and testing the parts has been underway to encourage process development, process control and the effective utilization of nondestructive inspection techniques. Currently, the third generation parts and test bars for the initial 1900°F part configurations have been fabricated and are under evaluation.

Figure 11 portrays this progress with reaction bonded silicon carbide. Significant progress has been made but the program demands continued progress. The third generation parts represent a 25 percent improvement in strength over initial parts and the near term 55 ksi strengths are deemed achievable with the current process improvements. The 2070°F ceramic components will profit from these improvements.

CERAMIC PART DESIGN ANALYSIS

Ceramic parts require a design methodology which encompasses both analytical methods and statistical portrayal of material properties. The meaningful output of design analysis is expressed as probability of survival for the operating conditions stipulated. The analytical approach being pursued follows Weibull statistic methods.

Several elements combine to permit Weibull analysis. First, a design goal is established with a quantifiable value. To arrive at this goal, a design reliability is established based upon a competitive warranty cost objective in a mass production environment typical of automotive products. Reliability apportionment of the entire engine established reliability goals for each component, including the ceramic parts such that reasonable failure modes and affects analysis would conclude engine warranties would be within the established goal. Next, the engine duty cycle is defined based upon composite truck utilization cycles recorded for the application. This then translates into mechanical, aerodynamic, and thermal loads. For practicality, conditions were assessed and the worst case condition defined. Analytical results point out the maximum principal stress under most adverse conditions. Material characteristics such as thermal, elastic and strength are established by test for each candidate ceramic material. Since ceramics are flaw sensitive, strength data must be presented statistically and acknowledge the probability of the existence of flaws both at the surface and within the volume. Data is generated over the range of temperatures anticipated with fast fracture assumed. Strength data on sintered alpha silicon carbide is shown as an example (note the variability displayed with part finished surface condition). Construction of 3-D finite element models of ceramic component are generated (Figure 12) and mechanical and thermal stresses are calculated for every element. Using the material data in Figure 12, the probability of survival is calculated for each

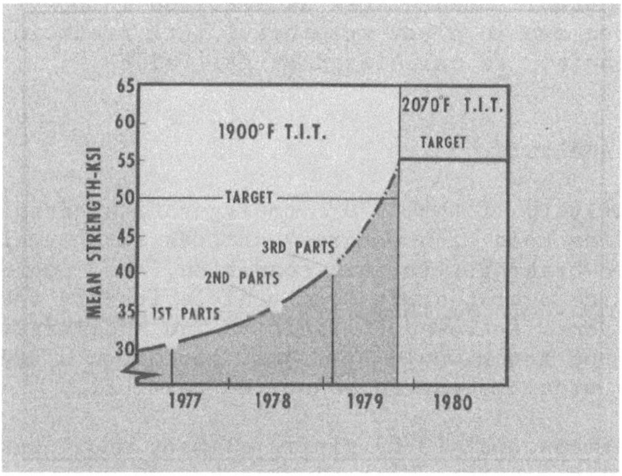

Figure 11. Ceramic material development--reaction bonded silicon carbide-carborundum.

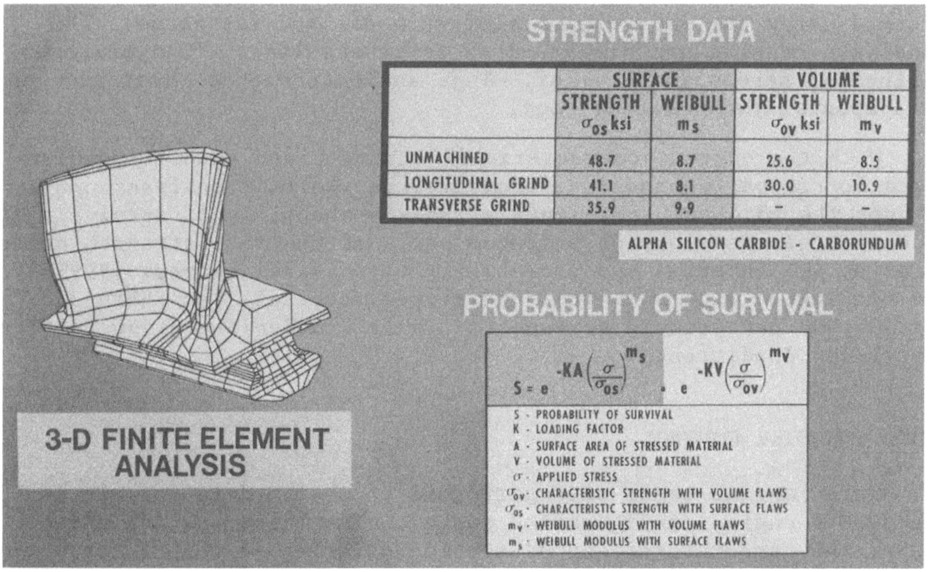

Figure 12. Ceramic part design analysis.

element and a cumulative probability of survival is obtained. This result is then compared to the design value established.

3-D stress analysis provides the tools to understand part failures which are encountered during rig qualification or engine testing. It also provides a design tool which offers initial design geometry selection. Much effort is required to prove the theory of failure assumed and to place meaningful interpretations on materials data comparison to calculated stress values.

CERAMIC VANE ANALYSIS

In the analysis of the 2070°F configuration ceramic vane, the critical loading condition occurs 2 seconds into acceleration following dynamic braking. In this condition, the compressor is driven by the engine output shaft from vehicle inertia forcing cooler air over the vanes followed by ignition and acceleration. In operating terms, the truck would have just gone down a steep hill and the driver is accelerating to go up the next hill.

A three dimensional (3-D) finite element model was constructed containing 452 elements. Each element has 20 node points. Upon applying the thermal transient conditions from the operating conditions, the thermal analysis depicts the isotherms (see Figure 13). From these it is readily apparent that thermal gradients exist in both the radial and axial directions. Stress analysis is then performed using the same finite element model and isotherms. The analysis results are displayed as isostress lines. The maximum principal stress if 53.4 ksi. This is located at a local area at mid span of the trailing edge.

With the current characteristic strength and Weibull modulus data for reaction bonded silicon carbide, Weibull analyses concludes the 53.4 ksi represents a .9859 probability of survival. When process development produces parts at the full strength potential of the material the probability of survival becomes .9999737. In production with this fully developed material, less than 3 out of 100,000 parts could be expected to fail under the given critical operating condition.

2070°F ENGINE CONTROL

Three engine tests were performed to obtain data for use in the 2070°F engine configuration design. The first testing established the temperatures and distortion existing in the crossarm area of the ductile iron block. Temperature levels must not produce creep and temperature gradients can produce distortion in this critical regenerator seal area. Runs were made with a block crossarm cooling concept, showing it is effective in producing proper block crossarm temperatures.

A second engine test series was conducted with regenerator inlet temperatures of 1450°, 1500°, and 1600°F. Instrumented hot regenerator seals recorded the temperature environment and permitted projection of seal temperatures to the 1800°F temperature obtained in the 2070°F engine. Significant conclusions were derived from the data obtained. The maximum temperatures at the hot seal periphery should be within the limits of the current graphite seal material. The radial gradients in the seal metal substrate were within that of existing regenerator experience showing that redesign required is minimal.

The third test is illustrated in Figure 14. This test provided verification that the engine could be controlled by sensing T_6 temperature (power turbine exhaust) rather than the current approach of sensing T_4 (combustor out) temperature. Current temperature measurement limits of 2000°F are associated with commercial thermocouples. The T_4 temperature will exceed this thermocouple limit in the 2070°F configuration. With work having been taken out of the gas stream by the two turbines, T_6 temperatures are considerably lower than T_4. The test established the temperature response in both the T_4 and T_6 locations in various steady state and transient conditions. T_6 followed T_4 within 10 degrees during the tests which is well within the limits of the engine control system.

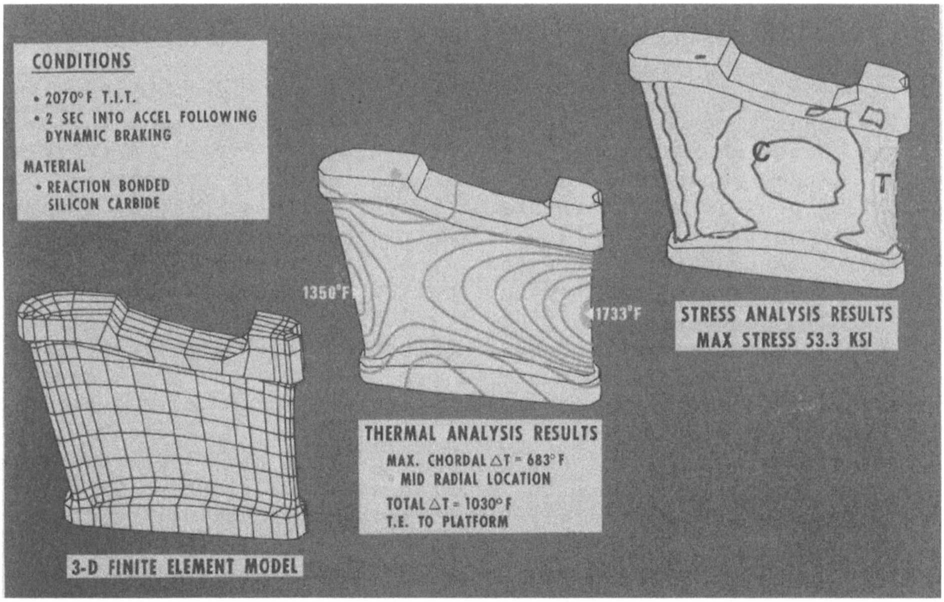

Figure 13. Ceramic vane analysis.

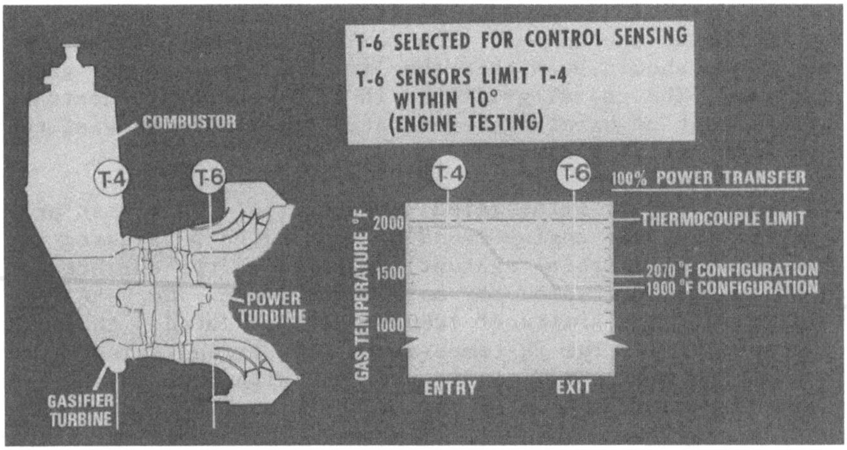

Figure 14. 2070°F engine control.

ENGINE TESTING

An additional 862 hours of engine durability at 1900°F were recently accumulated bringing the program total to 4471 hours. 404 hours of dynamometer durability to a transit coach operating cycle were run and 92 hours were run in truck operation. The balance of the engine testing was to a typical truck operating duty cycle.

The transit coach cycle presents quite different operating conditions to the ceramic components than the truck cycle. A truck cycle includes departure from a loading point, city streets out of town, long periods on interstate highways, followed by city streets to the terminal. This cycle has few transients with a high percentage of the running at constant speed and relatively high power output. The transit coach cycle consists of frequent starts and stops as the coach stops to pick up and discharge passengers operating exclusively on city streets. Fifty-five full acceleration/ deceleration cycles are encountered per hour in the transit coach cycle rapidly accumlates thermal transients on parts of the engine. (See comparison of truck vs. transit coach cycle in Figure 15.) Alumina silicate regenerator disks and reaction bonded silicon carbide nozzle vanes withstood the transit coach cyclic test experience and emerged in completely serviceable condition.

In the overall program engine testing, thin wall regenerators have experienced 2232 operating hours and thick wall regenerators have accumulated 2767 operating hours. The high time thin wall alumina silicate regenerator disk has 2087 hours and is still operating.

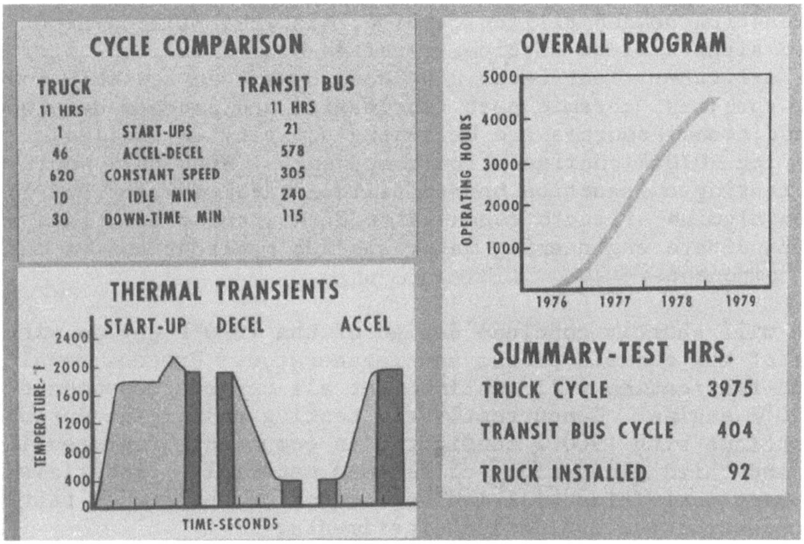

Figure 15. CATE engine testing.

Reaction bonded silicon carbide nozzle vanes have accumulated 1829 engine operating hours with the high time parts at 1247 hours and still in use. Reaction bonded silicon nitride vanes remain at 81 hours of experience as parts with improved strength are awaited.

561 hours have been accumulated on reaction bonded silicon carbide and 102 hours on reaction bonded silicon nitride shrouds. Specimen testing of abradable ceramic materials has been encouraging and shrouds of abradable materials are anticipated early in the next period for resumption of tip shroud evaluations in the engine. During the next report period third generation vanes, new source vanes and abradable tip shrouds of several materials will be available for rig and engine evaluations.

SUMMARY

Ceramic components are in the demonstration truck application with reaction bonded silicon carbide nozzle vanes and alumina silicate regenerator disks emerging fully serviceable from initial road operation at the hands of the driver and after exposure to special road hazard shock and vibration testing. These same components survived the rigorous thermal shock environment of the transit coach operating cycle on the test stand. These parts/materials are now installed in the CATE program truck for further road demonstrations at the 1900°F operating temperature.

Characterizaton, rig qualification and engine operation testing have been conducted on five of the most promising ceramic materials: reaction bonded silicon carbide, sintered silicon nitride, sintered alpha silicon carbide, reaction bonded silicon nitride and alumina silicate. Four ceramic sources have been actively involved in test specimen, ceramic part fabrication and process development. Three additional sources are beginning activity as candidate sources for 2070°F configuration components. With 1829 hours of engine testing on reaction bonded silicon carbide vanes and 4484 hours on alumina silicate regenerator disks, these materials stand out as candidate engineering materials for applications in turbine engine components.

We will shortly conclude design of the 2070°F engine with release of the ceramic plenum and regenerator. Process development and fabrication will continue for all ceramic components of the 2070°F engine. Concurrently rig testing and engine durability will continue with 1900°F configuration components representing second and third generations of ceramic materials. Materials development activities will include oxidation resistance testing and non-destructive evaluation experiments.

It is clear that significant progress is being made in ceramic part development, process control and producibility. The current ceramic component development dictates that process control must receive emphasis and ceramic part users must provide the ceramic suppliers with feedback on factors affecting part quality. The engine test results with initial ceramic components continues to encourage and support the application of ceramics in turbine engines for highway vehicles including trucks, buses and automobiles.

CERAMICS FOR SMALL RADIAL INFLOW POWER GENERATORS

Arthur G. Metcalfe and James C. Napier

Solar Turbines International,
an Operating Group of International Harvester
2200 Pacific Highway
San Diego, CA 92138

INTRODUCTION

Small radial inflow power generators are widely used by the U.S. Army for field operation. Experience has shown that the nozzle vanes are the most vulnerable component of engines in service. Degradation of vanes appears to result from erosion by ingested dust and by carbon from imperfect combustion, particularly with heavier fuels (e.g., Diesel #2). Teardown of affected engines shows that the nozzle vane trailing edge has been worn back to an extent that is in line with the calculated decrease of performance plotted in Figure 1. Replacement of the metal vanes by hot pressed silicon nitride solved the erosion problem as reported at the first meeting of this group in November 1973.

Work since this time has been directed along three lines:

1. Develop a ceramic vane design and demonstrate by engine tests.

2. Extend application of ceramics to both vane and shroud to allow higher inlet temperatures and higher output.

3. Select effective manufacturing methods for implementation of ceramic vanes in engines.

Initiation of a manufacturing methods program by the U.S. Army provides support for the view that the theme of this conference reliability - may be closer to attainment for this very

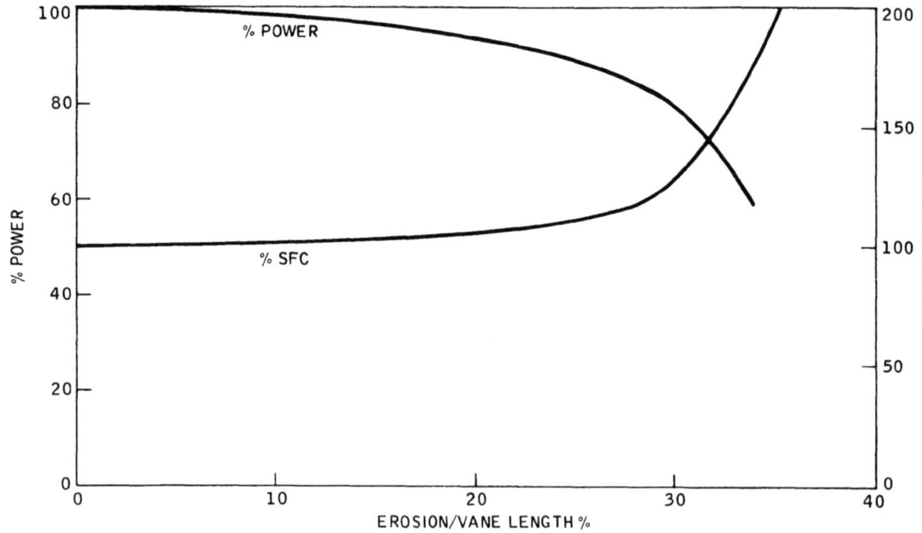

Figure 1. Relative Performance Loss With Nozzle Vane Erosion in a Small Radial Gas Turbine

special application than for other engines. The reasons for this include the favorable statistics derived from the small component size, the low structural stresses in the ceramic component, the use of the relaxing joint concept and other special conditions. These will be reviewed in the body of this paper.

THE MERADCOM 10 kW AUXILIARY POWER UNIT

The initial demonstration that ceramic vanes solved the erosion problem was made on the larger, 60 kW output Titan APU. A comparison of the erosion of ceramic vanes with the conventional metallic vanes is shown in Figure 2. Based on these results, the smaller 10 kW Gemini APU was proposed as a test bed to introduce ceramics economically into radial gas turbines and establish design approaches for introduction to larger engines. The 10 kW APU is shown in Figure 3. The radial compressor and radial turbine are mounted back-to-back and operate at a design speed of 93,500 rpm. The pressure ratio is 3.5/1 and mass flow 0.45 lb/sec. The nozzle and turbine wheel are cast in Mar M421.

DESIGN OF EROSION RESISTANT CERAMIC NOZZLE

The erosion shown in Figure 2 does not result from low angle attack as dust laden gases flow through the nozzle. This is because the dust particles are not accelerated to the gas velocity in the 1-2 inches of travel. As a result they are overtaken by

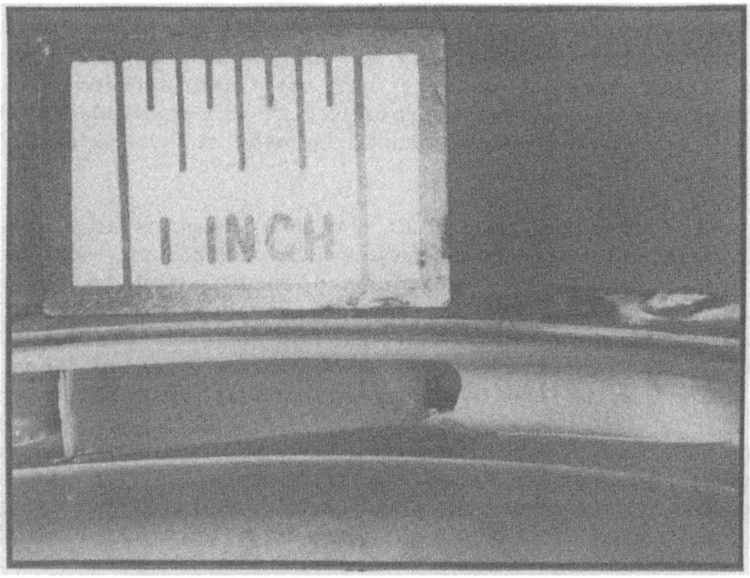

Figure 2. Silicon Nitride (left) and N-155 Vanes (right) in Titan Engine After Testing

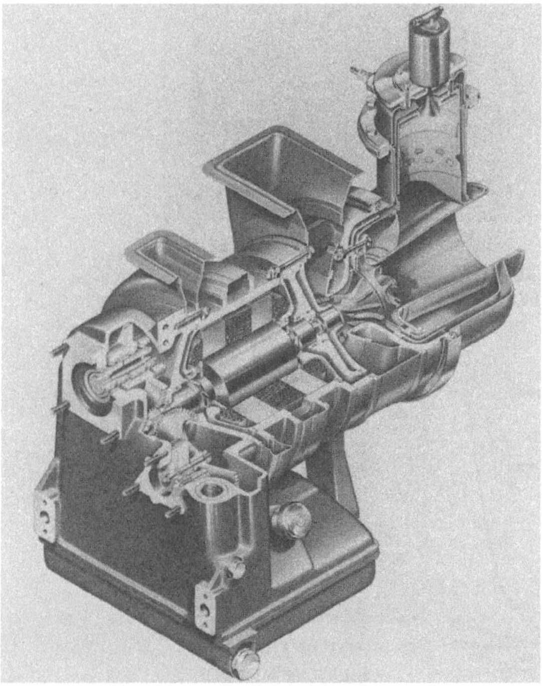

Figure 3.

10 kW Turboalternator Set

the turbine wheel and batted forward to hit the vanes at a high angle. The relative velocities at the turbine wheel impact are low and little erosion occurs; but they are extremely high at vane impact and cause severe erosion. It was concluded that high velocity, high angle impacts should be used to evaluate candidate nozzle vane materials.

Figure 4 compares Inco 713C with two ceramics at a series of particle velocities up to 840 feet per second at room temperature. Higher impact velocities were anticipated in the engine but could not be attained in the room temperature test rig. The 50X improved erosion resistance of hot pressed silicon nitride and the 13-fold improvement of NC-430 over that of Inco 713LC led to selection of the hot pressed material for initial tests.

The metallic vanes were partially machined away and replaced with the hot pressed silicon nitride material. A glass adhesive was used that provided a gas seal yet permitted some flow to relieve thermal stresses, as shown in Figure 5. Five hundred thermal shock cycles showed the design to be sound (Fig. 6). This was followed by a 10-hour dust erosion test of a nozzle

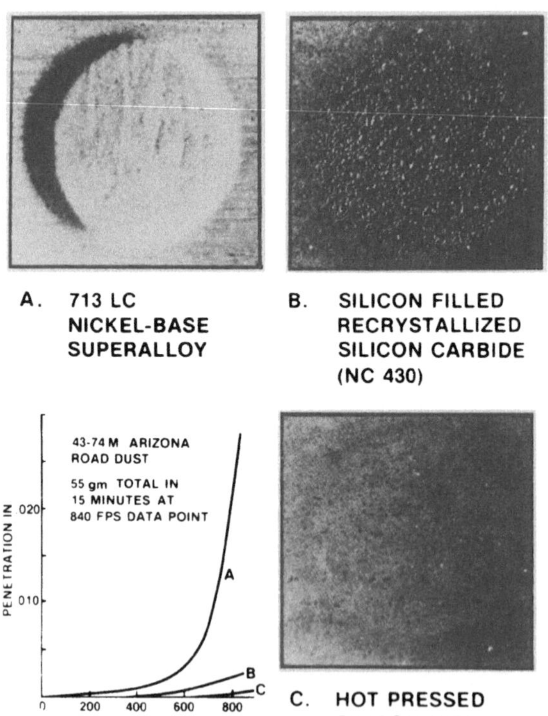

A. 713 LC NICKEL-BASE SUPERALLOY

B. SILICON FILLED RECRYSTALLIZED SILICON CARBIDE (NC 430)

C. HOT PRESSED SILICON NITRIDE

Figure 4. Erosion of Vane Materials

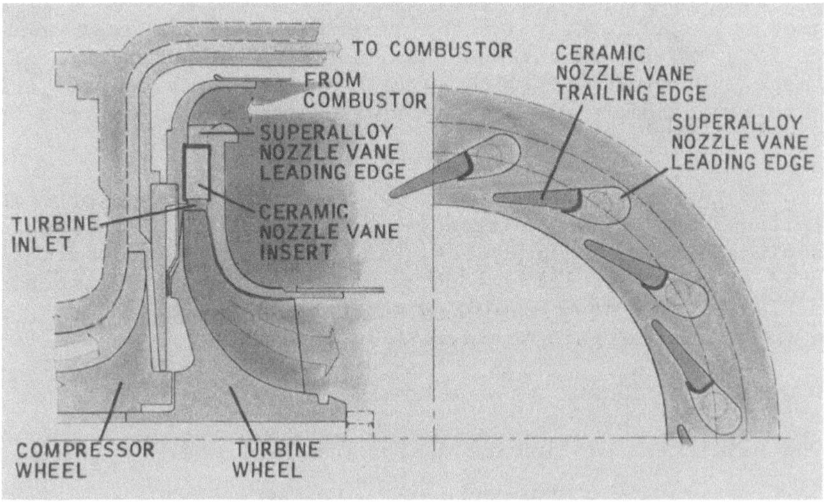

Figure 5. Ceramic Nozzle Vane in 10 kW Turboalternator

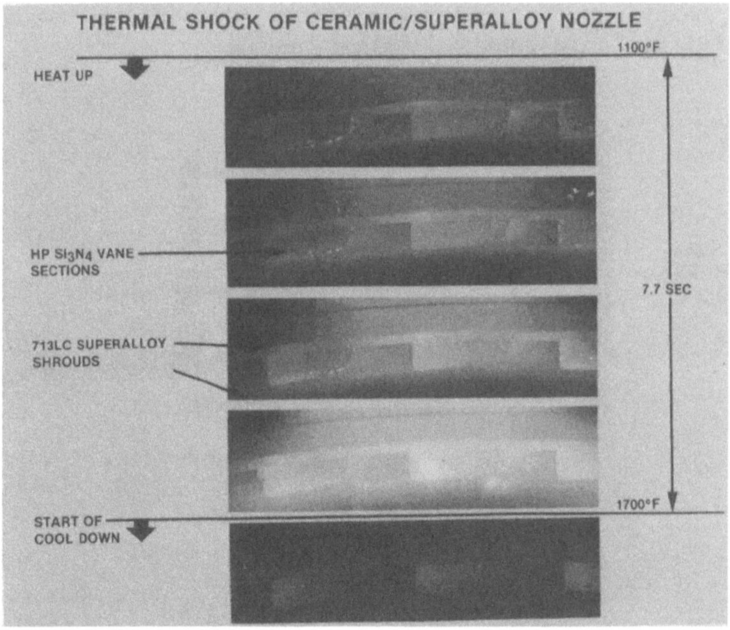

Figure 6. Thermal Shock of Ceramic/Superalloy Nozzle

assembly to compare with a metal vane, as shown in Figure 7, and confirmed the superiority of the ceramic. Another test was to compare loss of section under hot corrosion conditions. This test lasted 70 hours at 1700°F with 3 ppm of simulated sea salt. Attack on the ceramic was small and was estimated to be less than 4% of that on the metallic vane.

Two engine tests were made to check the design approach. Each included 24 hours of steady-state running at various load levels plus 50 start/stop cycles. In the first test 6 of the 15 vanes had suffered trailing edge fractures; these were traced to deformation of the inner lip of the metallic forward shroud. Redesign to eliminate this problem gave a successful engine test.

DESIGN OF AN ALL-CERAMIC VANE ASSEMBLY

The major ceramic turbine design problems are:

1. Accommodation of expansion mismatch between metallic and ceramic members.

2. Reduction of localized contact stresses.

3. Accommodation of misfits from normal manufacturing tolerances.

4. Maintenance of service stresses (mechanical and thermal) within safe design limits.

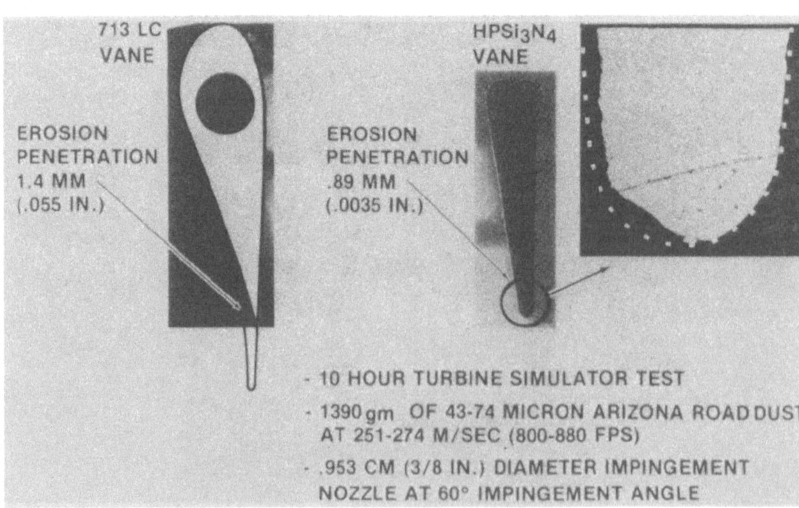

Figure 7. Results from Engine Simulator Erosion Tests

It will be noted that only Problem #4 relates to the ceramic component; the other problems relate to intercomponent interactions.

Expansion mismatch is minimized through combination of component temperature and material selection for suitable expansivity so that the parts move largely together. All-ceramic Design #1 shown in Figure 8 reveals such a problem at the seal plate (metallic) with the forward ceramic shroud. Review of temperatures and materials led to choice of Inco 903, a low expansion alloy, for the seal plate.

Reduction of contact stresses has been achived by several approaches including the relaxing glass adhesive, use of spring washers and a Rene' 41 leaf spring to retain the forward shroud to diffuser housing. These approaches are used also to accommodate typical manufacturing misfits. These concepts were tested first in a simulated all-ceramic nozzle using flat plate shrouds.

The Design Concept #1 (Fig. 8) used $RBSi_3N_4$ shrouds with air-cooled Inco 718 bolts. These bolts are piloted at the compressor housing and accurately locate the nozzle rear shroud to matching turbine wheel contour. The rear shroud has radial slots at the bolt penetration points, and the forward shroud has clearance holes to allow for radial growth. The seal plate provides accurate centering of the nozzle forward shroud.

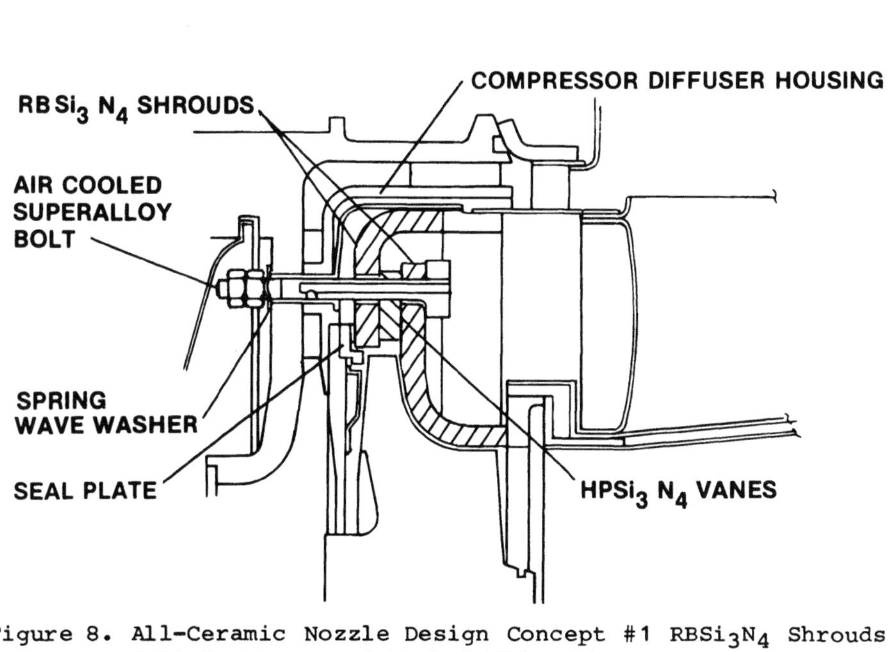

Figure 8. All-Ceramic Nozzle Design Concept #1 $RBSi_3N_4$ Shrouds, $HPSi_3N_4$ Vanes and Cooled 718 Bolts

Design Concept #2 (Fig. 9) used $HPSi_3N_4$ in place of the air-cooled Inco 718 for the bolts. Stress attenuation is achieved by the spring wave washers at the cold end and a 0.004 inch platinum interface layer at external spring loaded collets.

Design Concept #3 (Fig. 10) used silicon filled reaction bonded silicon carbide shrouds with hot pressed silicon carbide vanes. The vane/shroud joints were bonded with a relaxing glass adhesive to relieve contact stresses and to provide an integral unit for ease of handling.

The rear shroud-to-exhaust scroll joint is maintained under pressure in all cases by an axial load at the exit of the exhaust scroll provided by a bellows.

Engine Simulator Tests

Each design concept has been tested in an engine simulator shown in Figure 11 to evaluate factors such as leakage at seal interfaces and turbine tip clearances under full rated engine operating conditions.

Results have shown that face seal interfaces at ceramic section inlet and exit positions function well without special materials applied. The design provisions which allow compati-

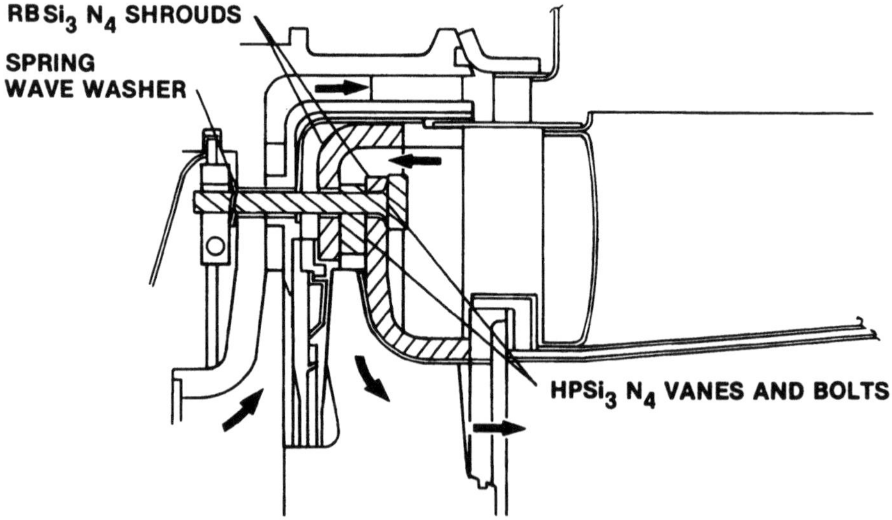

Figure 9. All Ceramic Nozzle Design Concept #2 $RBSi_3N_4$ Shrouds, $HPSi_3N_4$ Vanes and $HPSi_3N_4$ Bolts

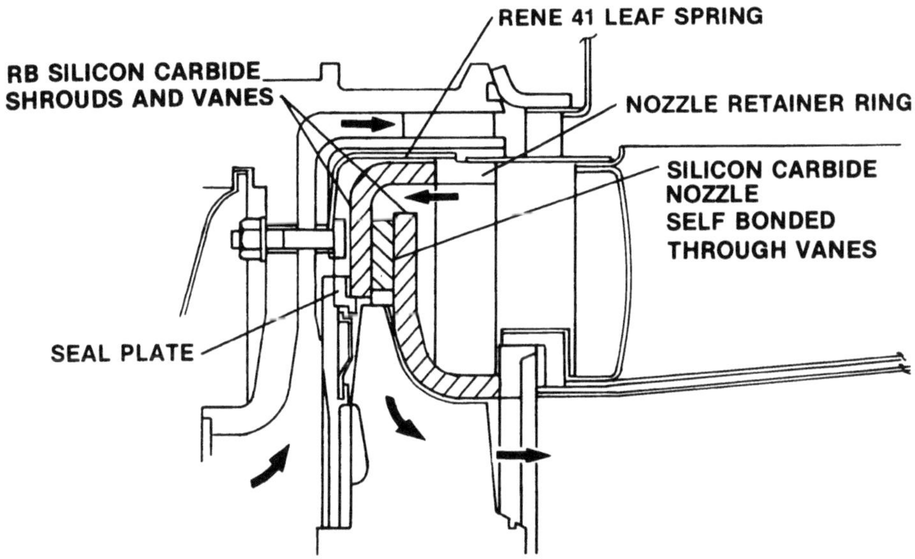

Figure 10. All-Ceramic Nozzle Design Concept #3 RBSiC Shrouds, HPSiC Vanes and Vanes Bonded to Shrouds

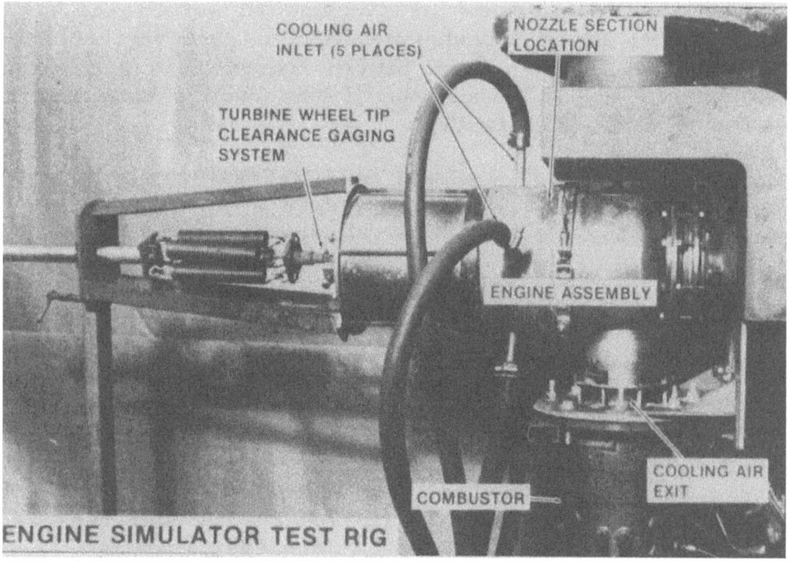

Figure 11. Engine Simulator Used in All-Ceramic Nozzle Testing

bility of large differential expansions between ceramics and metal alloys have been proven to function well. Accurate rear shroud centering critical to turbine tip clearance has been demonstrated with each of the concepts in the engine simulator (shown in Fig. 11) at 1700°F and 2100 fps nozzle throat velocity.

Engine simulator testing, which included 500 thermal shock cycles to 2000°F peak temperature, and some steady state testing in which turbine tip clearances were checked, consistently showed the reaction bonded silicon nitride forward shroud to fail despite design measures taken to avoid failure. The silicon filled reaction bonded silicon carbide forward shroud of design concept #3 had no difficulty in surviving the same test sequence.

From this result it was concluded that reaction bonded silicon carbide should be used for the forward shroud in the engine test of the nozzle. Design Concept #1 was selected over the others because it offered the most conservative design of the three considered. The use of a silicon carbide forward shroud and silicon nitride rear shroud was found to be a thermally coordinated design since the higher expansion coefficient SiC at the cooler forward shroud position would match thermal growth of the Si_3N_4 rear shroud.

No features specific to Design Concepts #2 or #3 were found to be a problem in engine simulator tests. The ceramic bolts of Concept #2 survived steady state testing in the simulator and both silicon carbide shrouds of Concept #3 were found to be in perfect condition after 500 thermal shocks. However, the Concept #2 ceramic bolts had not been proven through the entire engine performance sequence and location of the rear shroud in Concept #3 remained an uncertain issue.

The silicon carbide forward shrouds fabricated for Concept #3 were modified to meet the requirements of Concept #1. The clearance holes were cut by electro-discharge machining in the Si/SiC material.

Engine Tests

Two 50-hour engine tests have been performed under the conditions given in Table 1. The first was made on an engine modified by Design Concept #1, except that the forward shroud was reaction bonded silicon nitride, with hot pressed silicon nitride vanes and a reaction bonded silicon nitride rear shroud. The RB silicon carbide forward shroud survived the 50-hour test with no flaws or cracks visible by NDE after the test (see Fig 12). The rear shroud sustained one fracture at a support bolt location after 33 hours of test as shown in Figure 13. Between the 33rd and 50th hour, the metallic exhaust scroll collapsed (Fig. 14)

Table 1

Engine Test Sequence for All-Ceramic Vane Nozzle

Operating Condition	Time at Load (Hours)
Calibration	5
50% Load	25
0	4
100	10.5
75	6.5
	50 hours total

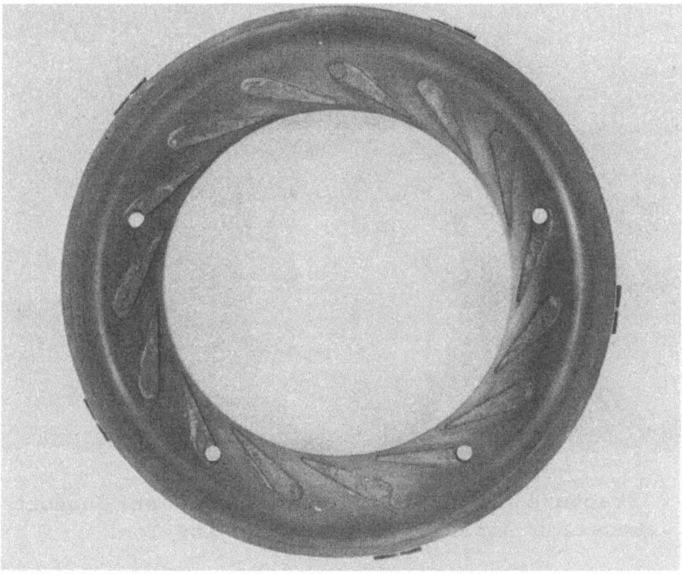

Figure 12. RBSiC Forward Shroud After 50 Hours Engine Test

due to an error in design on modification of the engine, and this led to a more substantial fracture of the rear shroud (Fig. 15). However, the fracture in the first 33 hours of testing could not be attributed to this cause.

The reaction bonded silicon nitride was replaced by reaction bonded silicon carbide for the second engine test in view of this

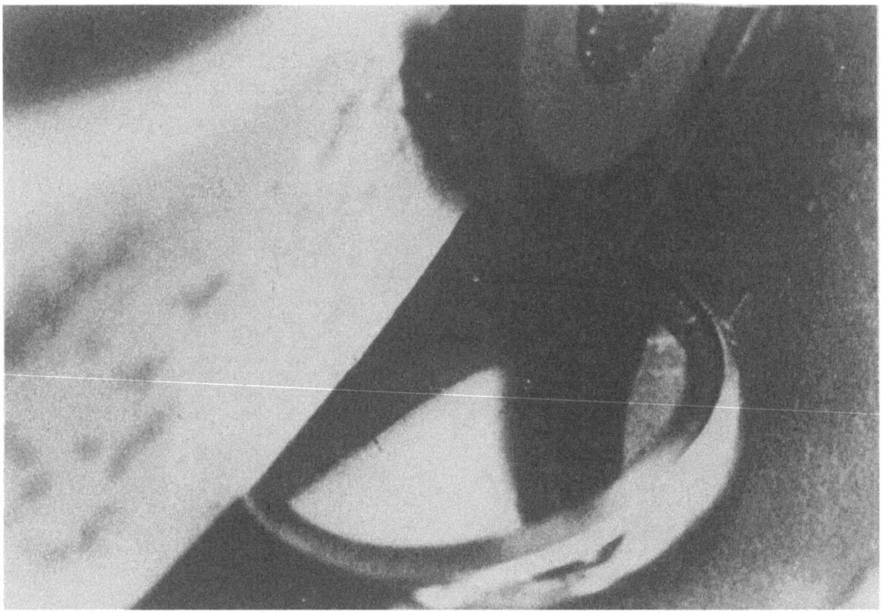

Figure 13. Fracture in $RBSi_3N_4$ Rear Shroud at Support Bolt Location After 33 Hours of Engine Test

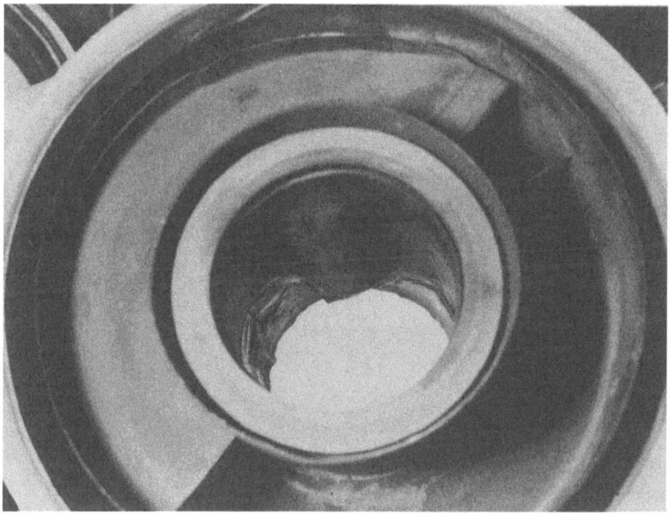

Figure 14. Collapsed Exhaust Scroll.

Figure 15. All-Ceramic Nozzle After 50-Hour Engine Run (Note fracture at bottom section of $RBSi_3N_4$ rear shroud)

unexplained failure. This eliminated the partial expansion compensation between the cooler, higher expansion carbide forward shroud and the hotter, lower expansion nitride rear shroud. The metallic exit scroll was rebuilt to avoid the overheating and buckling. The engine test with this configuration has completed 108 hours without problems at the time of this meeting.

DISCUSSION OF RESULTS

The principal conclusions from this work are:

1. Ceramic components can be successfully substituted in production engines and be qualified using standard qualification engine tests.

2. Interaction between components is more critical than the stress within components in an engine of this type.

3. Reaction bonded silicon carbide (Norton NC-430) performed better than reaction bonded silicon nitride (Norton NC-350). This was contrary to expectations based on strength, expansivity and elastic modulus.

CERAMICS FOR ADIABATIC TURBOCOMPOUND ENGINE

Roy Kamo

Director - Advanced Engines and Systems
Cummins Engine Company, Inc.
Columbus, Indiana 47201

W. Bryzik

U.S. Army Tank Automotive R&D Command
Propulsion Systems Division
Warren, Michigan 48090

ABSTRACT

Material limitation is no longer a barrier to higher temperature engine operation with consequent higher engine performance. High performance ceramics offer a new dimension in diesel engine design for improved fuel economy, engine output, and reliability. This paper presents an adiabatic diesel engine system adaptable to the use of high performance ceramics which will enable higher operating temperatures, reduced heat loss, and turbocompound exhaust energy recovery. The engine operating environments as well as the thermal and mechanical loadings of the critical engine components are covered. Design criteria are presented and techniques leading to their fulfillment are shown. The present shortcomings of the high performance ceramic design in terms of meeting reliability and insulation targets are discussed, and the need for composite designs is shown. A ceramic design methodology for an insulated engine component is described and some of the test results are shown. Other possible future improvements such as the minimum friction-unlubricated engine through the use of ceramics are also described.

INTRODUCTION

Improvements in diesel engine efficiency, maintenance, multifuel capability, emissions and durability can be expected as a result of the adiabatic turbocompound engine program which is being jointly funded by Cummins Engine Company and the Army Tank Automotive R&D Command (TARADCOM) of the U.S. Army.

Mechanical properties of high performance ceramics have been improved to the point where their use in heat engines is possible. The high temperature strength and low thermal expansion properties of high performance ceramics offer an advantage over metals in the development of a non-water-cooled engine. The removal of the watercooling system and the increased operating temperatures will provide additional exhaust energy available for recovery. The lack of a cooling system will also improve reliability and maintenance problems. The adiabatic turbocompound diesel engine concept (1) emerges as a simple and direct approach to take full advantage of these new developments.

The program on the adiabatic turbocompound engine is now in the fourth year. Widespread interest and a high degree of cooperation have been shown by various industrial, educational, and governmental organizations. The progress has been encouraging, and the targets set forth for the first demonstration in 1979 appear achieveable.

This paper describes the objectives of the program, its targets, and its implementation. This program represents the beginning of the use of high performance ceramics in the diesel engine. Future areas where high performance ceramics can be applied are also described.

Adiabatic Turbocompound Diesel Engine

The adiabatic turbocompound diesel engine being developed by Cummins and TARADCOM insulates the combustion chamber. The combustion chamber is permitted to get "hot" to minimize heat transfer from the combustion chamber. The engine has:

- no water cooling
- no air cooling

Thus, the exhaust energy from the insulated adiabatic diesel engine is quite high. In order to improve its cycle efficiency, an exhaust heat recovery device is necessary. This can be accomplished by a turbocompound device. The turbocompound presents a simple, lightweight, low cost, and a potentially low maintenance engine configuration. A schematic of the adiabatic turbocompound engine concept is shown in Figure 1. In the adiabatic turbocompound engine concept, the cooling system is completely eliminated.

These cooling system components are responsible for some of the troublesome maintenance and warranty problems in the commercial diesel engine business and their elimination is highly desirable.

The need to consider the turbocharged diesel engine, rather than the naturally aspirated version, was described earlier (1, 2). Thus, an adiabatic turbocompound diesel engine is simply a non-watercooled turbocharged diesel engine fitted with an insulated combustion chamber and a turbocompound unit. The turbocompound unit is a second stage turbine which is located after the turbocharger and is geared to the crackshaft.

Engine Benefits

The chief advantage of the adiabatic turbocompound engine over the basic water-cooled engine is its high thermal efficiency. Other important power plant considerations that are significantly improved are the specific weight, specific size, and specific output.

Although not thoroughly investigated at this time, it is believed that the following advantages could accrue:

- multifuel capability
- combustion efficiency improvement
- reduced combustion and mechanical noise
- white smoke elimination
- lower compression ratio operation
- reduction in maintenance

Program Objective and Strategy

The objective of the Cummins-TARADCOM joint program is to demonstrate the technical feasibility of adiabatic turbocompound engines of advanced ceramic design to meet all foreseeable performance and sociability goals on a commercial Cummins NTC-400.

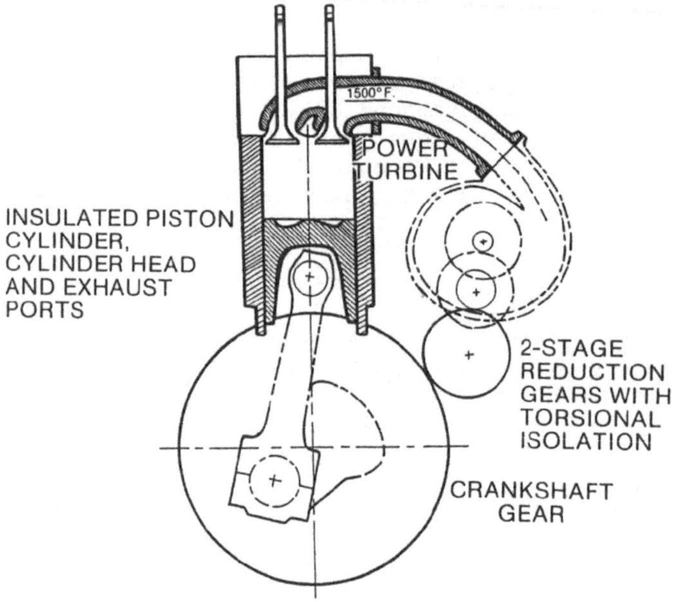

Figure 1. Total energy recovery via Cummins adiabatic turbocompound engine.

The targets are:

BHP @ Rated	500 HP
RPM @ Rated	1900 RPM
Emissions (NOx+HC)	5 grams/bhp-hr
BSFC	0.28 lb/bhp-hr
Life	250 hours

The basic adiabatic engine which was designed for testing in the laboratory is illustrated in Figure 2. The basic engine consists of the following components:

1. Insulated piston cap
2. Insulated "Hot Plate" for cylinder head
3. Insulated top liner "deck-spacer"
4. High temperature lubrication package
5. Insulated exhaust ports
6. Others: exhaust manifold insulation, valves, injectors

Three builds of the basic adiabatic engine were made: namely, (1) metallic, (2) glass ceramic, and (3) hot pressed silicon nitride. These three builds are illustrated in Figure 3.

The metallic adiabatic engine was fabricated and assembled because of expediency. Valuable operating data regarding engine performance, temperatures, etc., were quickly obtained for designing of subsequent ceramic builds.

The glass ceramic adiabatic engine design components were fabricated from lithium alumina silicate. The excellent insulating properties of glass ceramic with its extremely low thermal coefficient of expansion greatly simplified the design. The glass ceramic offered the best chance for a low cost, simplified engine design. The major disadvantage of the glass ceramic was its low material strength.

The hot pressed silicon nitride (HPSN) adiabatic engine design was quite similar to the metallic build. Insulating metal shims were required to reduce heat loss. The high strength, high temperature capability of HPSN offered the best chance of success.

Operating Environment

The operating temperature within the insulated combustion chamber of the adiabatic engine depends greatly on the load, air-fuel ratio, intake air temperature, injection timing, etc. Figure 4 shows a least square curve fit of the thermocouple probe temperature within the combustion chamber for the peak

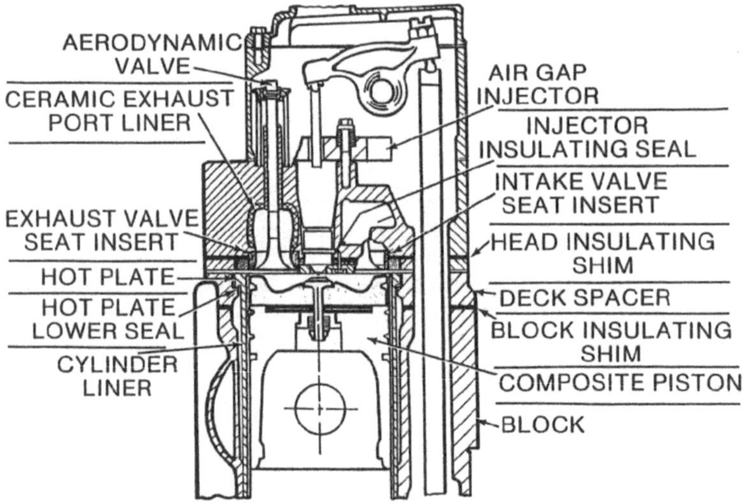

Figure 2. Cross section of Cummins basic adiabatic (insulated) diesel engine.

Figure 3. Three builds of Cummins basic adiabatic-insulated engine design with insulated liner deck.

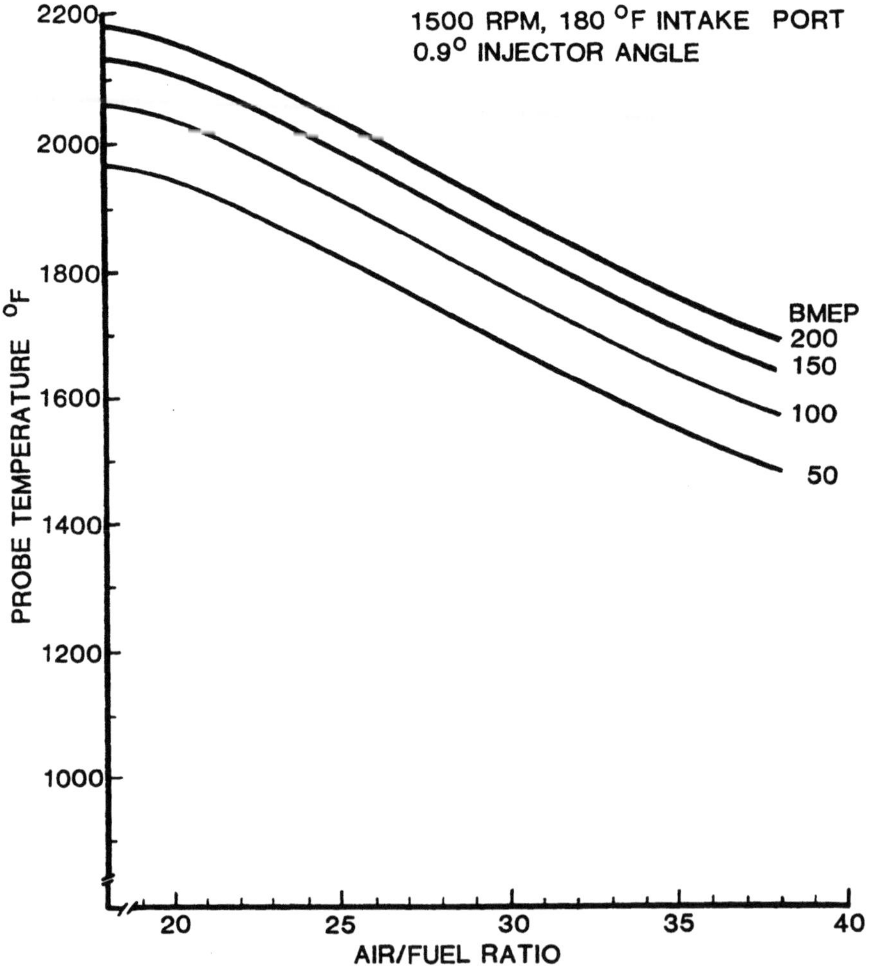

Figure 4. Least-square fit of probe temperature.

torque condition of the engine. The thermocouple probe temperature shown in Figure 5 is thought to provide a reasonable approximation of time-mean surface temperature of an adiabatic engine (6) Thus, the measured temperature was used in all piston, liner, and head analyses. For design analysis purposes, the peak cylinder pressure is 2000 psi. A typical combustion diagram as a function of crank angle is shown in Figure 6.

The engine under investigation had the following operating specifications:

Bore x Stroke	5 1/2 x 6 inches
Engine Speed	1900 rpm
Engine Configuration	In-Line Six Cylinder
Engine Cycle	4-Cycle
Combustion Chamber	Direct Injection Quiescent Chamber
Peak Cylinder Pressure	1800 psi
Lubrication	Oil Lubricated
Brake Mean Effective Pressure	195 psi
Air-Fuel Ratio	27:1
Intake Air Temperature	140°F
Overspeed Capability	25% Burst Margin

Reliability and Durability Requirements

Although the feasibility requirements for durability are only 250 hours at rated load and speed, the requirements for a viable production engine are considerably more demanding. The real life engine must meet the following mature engine life goals:

- Life to overhaul; 10,000 to 15,000 hours
- 90% reliability on major engine components
- Service interval, 500 to 1,000 hours
- Life of major components, 5,000 to 7,000 hours
- Warrantable features:
  - 70% reliability through 100,000 miles or 2,000 hours

Design Methodology

Designing reliable engine components with ceramics is considerably more difficult and unquestionably different than designing with ductile materials. Figure 7 shows the probability of failure of a given component at different stress levels when designed with metal and ceramics. The probability of failure of a metallic design is quite predictable; whereas, ceramic designs are considerably more unpredictable. In order to avoid

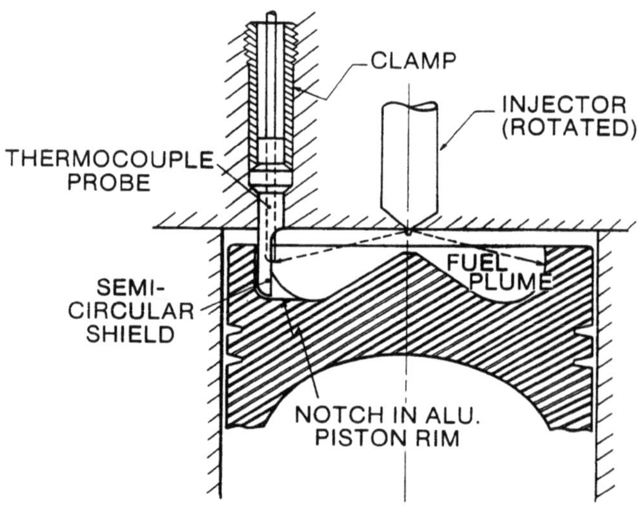

Figure 5. Installation of thermocouple probe in cylinder head.

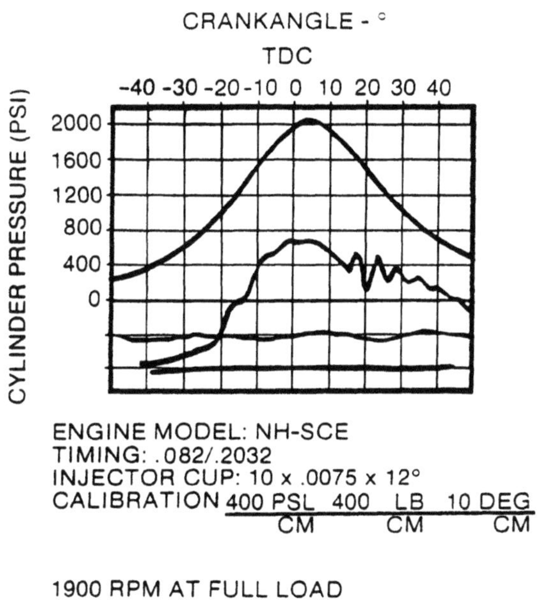

Figure 6. Combustion diagram of an uncooled 5½"x 6" (bore x stroke) diesel engine operating at full load and speed.

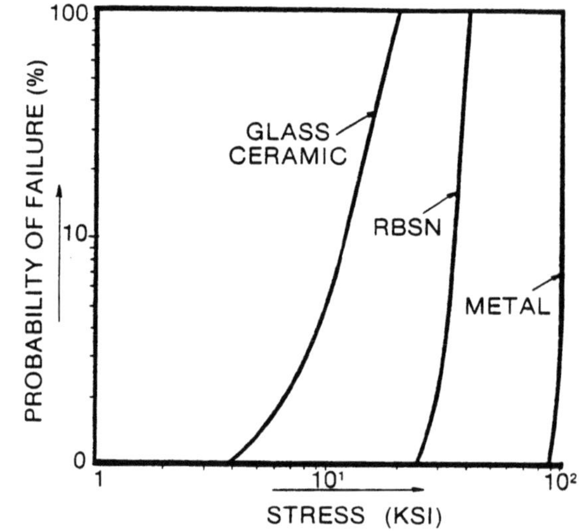

Figure 7. Estimated probability of failure as a function of stress for various materials.

failures, ceramic components must be stressed very low and ceramics of good quality and strength must be used.

The previous sections of this paper have dealt with the design requirements and the operating environments of an adiabatic turbocompound diesel engine. These inputs coupled with an initial design concept will lead to a preliminary design which is an iterative result. Upon a satisfactory preliminary design, it is subjected to detailed analyses by a team of specialists. Interplays between disciplines are necessary for completing the iterative loop. The final detailed design is subjected to computer test. If the detailed design is unacceptable, the design will be subjected to reiteration. If acceptable, the next procedure will be followed by a procurement. The design methodology process is shown in Figure 8, (5).

The procured component will subsequently be subjected to non-destructive testing, temperature, and stress measurements, and finally dimensional checks. In our next step leading to final acceptance, the component design will be screened in a rig or a laboratory bench test. Failure of ceramic components in an operating engine is catastrophic in terms of damage to other useful engine components and much time is required for cleaning up debris and salvaging other good components. Such a bench screening rig is shown in Figure 9 for screening engine components. It consists of a single cylinder engine block which is motored by a variable speed drive. The cylinder head is replaced by a piston retainer assembly; the test piston is stationary. While the primary purpose of the rig is to test pistons, it has many other uses. The rig only simulates mechanical loads and speed (7).

Once the design component has passed the bench screening rig, the component is now proof tested in a single cylinder fired engine under actual operating conditions. Proof testing is continued for 250 hours. If the final component design survives the single cylinder proof testing, it is now ready for the multicylinder performance testing. Upon satisfactory completing of the performance testing, the component is ready for the demonstrator engine. The testing procedure is shown in Figure 10.

All major adiabatic engine components have been, or will be, subjected to the design and test methodology presented above.

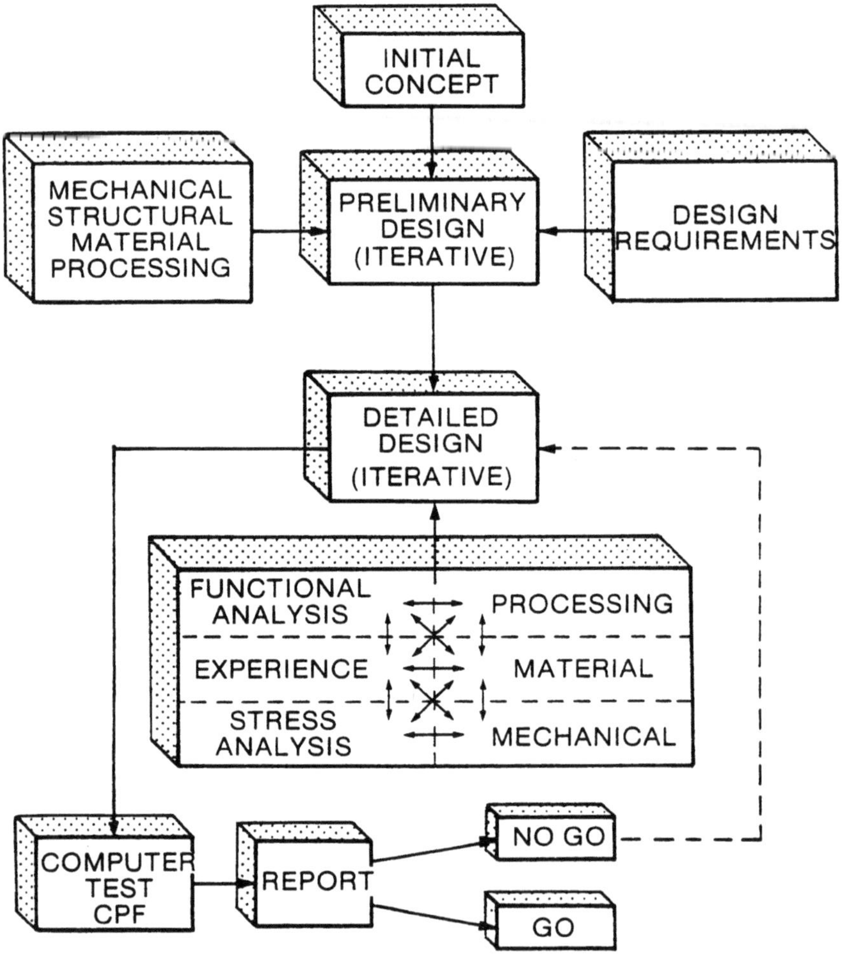

Figure 8. Methodology for designing with ceramics through team effort (University of Washington).

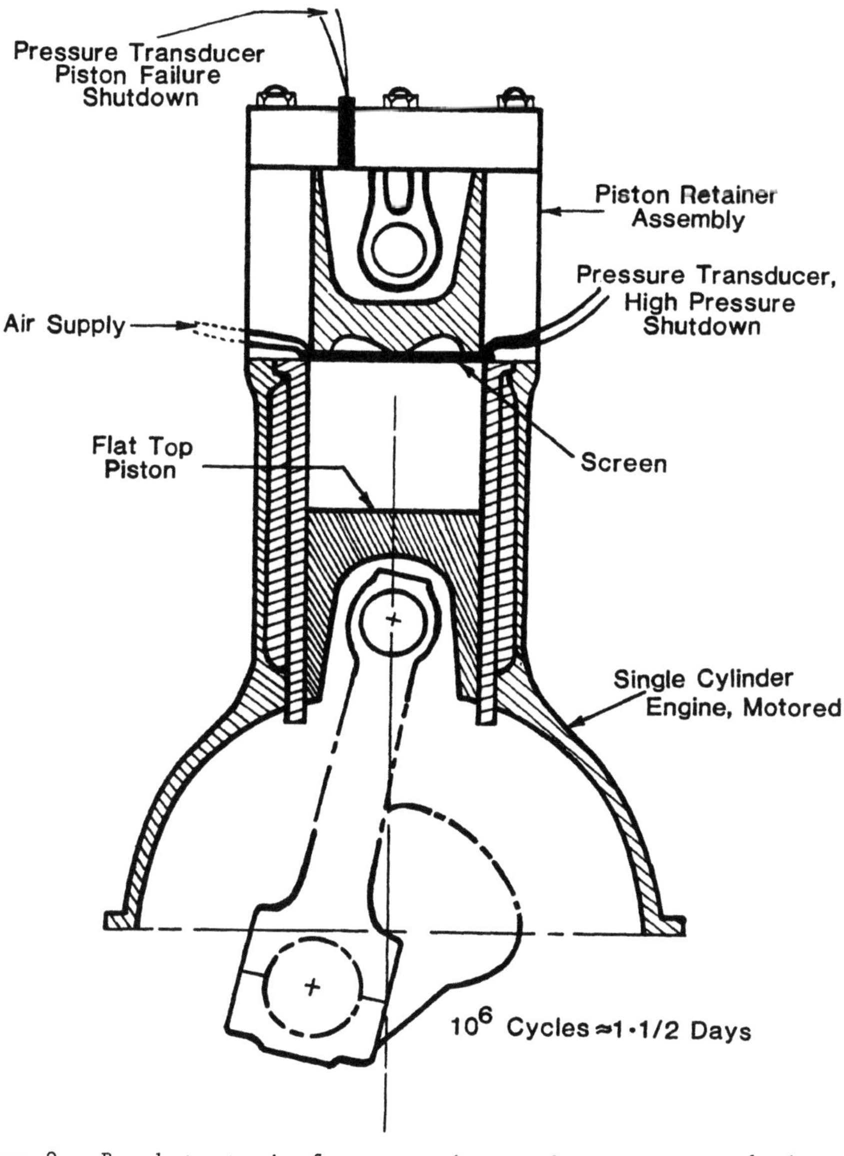

Figure 9. Bench test rig for screening engine component design parts.

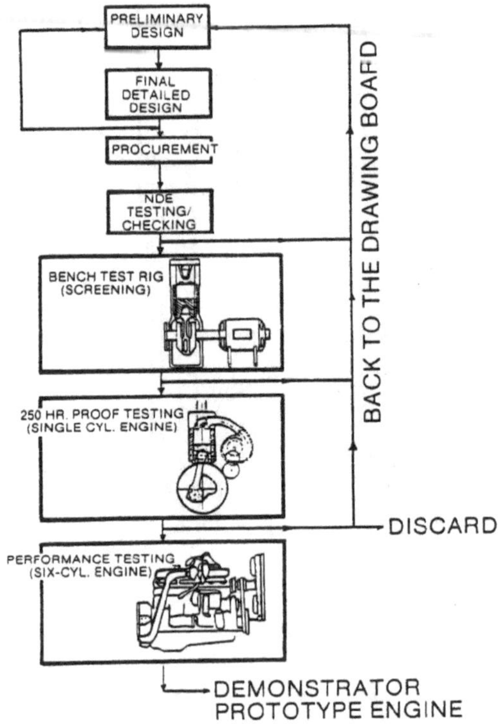

Figure 10. Testing sequence and iterative process.

These major components are as follows:

- Composite ceramic piston cap
- Cylinder liner
- Cylinder head system (including valves, and ports)

Only one of the above components; namely, the composite piston ceramic cap, will be discussed in detail in the following sections to illustrate the design and test methodology used.

The Ceramic Piston Cap

Our earlier attempts to design a monolithic piston design were stopped because of limited success and the inability of the high performance ceramics to provide necessary thermal insulation property. Figure 11 shows the latest types of composite pistons with HPSN caps. The piston base is an aluminum full-skirted design. The HPSN piston caps are fastened to the piston base by means of a Waspalloy bolt and Vellville washers. Piston insulation is achieved by surface roughened metallic discs (3). Other piston caps besides HPSN were designed, fabricated and tested. They were:

- Reaction bonded silicon nitride (RBSN)
- Sintered Silicon Nitride (SSN)
- Lithium Alumina Silicate (LAS)
- Stainless Steel (SS)

A drawing of the circular optimized bowl composite piston cap is illustrated in Figure 12.

Based on the operating environment data of an adiabatic turbo-compound engine, the temperature distribution in the SSN piston cap was calculated and the isotherms are shown in Figure 13. The normal stresses were then calculated from the temperature data. The isostress lines are shown in Figure 14. Similar calculations were made for other high performance ceramic materials, glass ceramics, and metals. Table I shows the results for other materials.

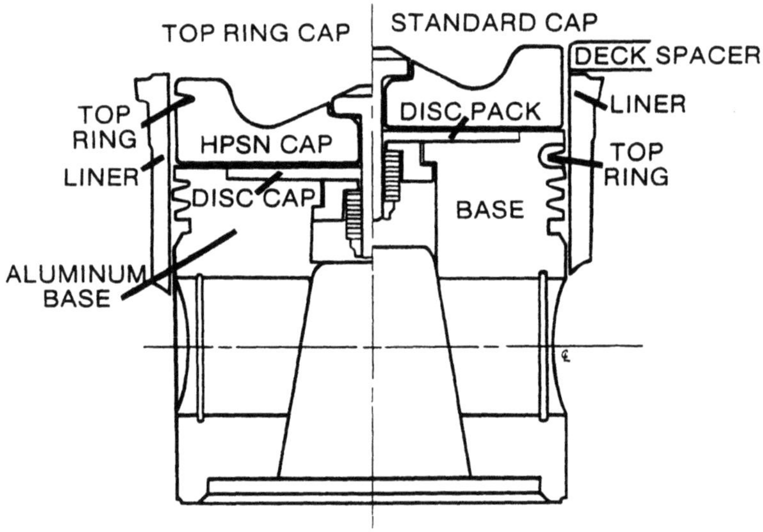

Figure 11. Two types of composite pistons with HPSN caps under investigation.

PART NO.	MATERIAL	A DIA.	B
Y-3013754	HPSN	5.464/5.459	1.248/1.242
Y-3013755	SSN, SIALON	5.462/5.457	1.248/1.242
Y-3013756	LAS	5.472/5.467	1.248/1.242
Y-3015505	STAINLESS STEEL	5.388/5.386	1.003/.997

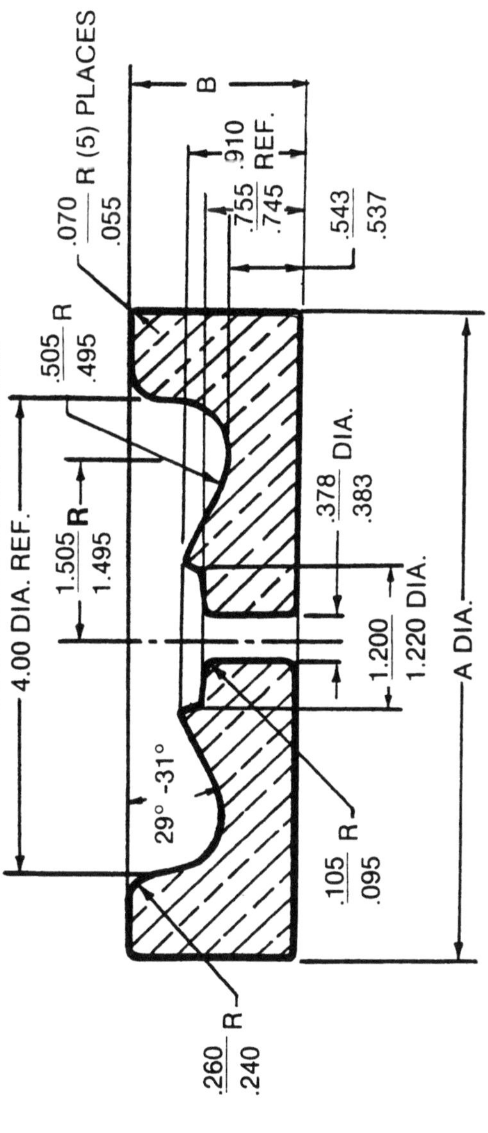

Figure 12. Composite piston cap (Optimum bowl).

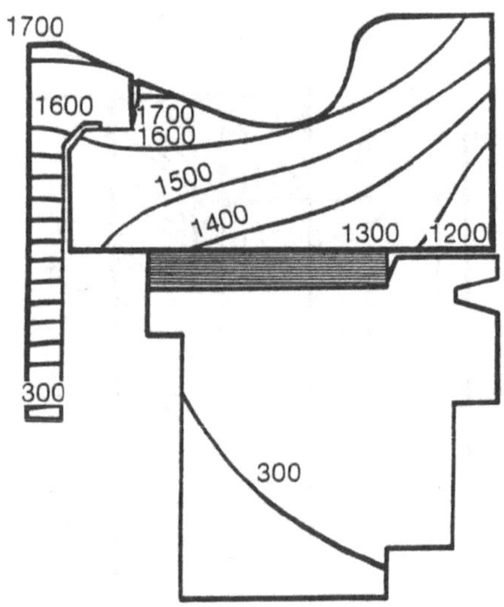

Figure 13. Temperature distribution in a composite piston with an SSN cap.

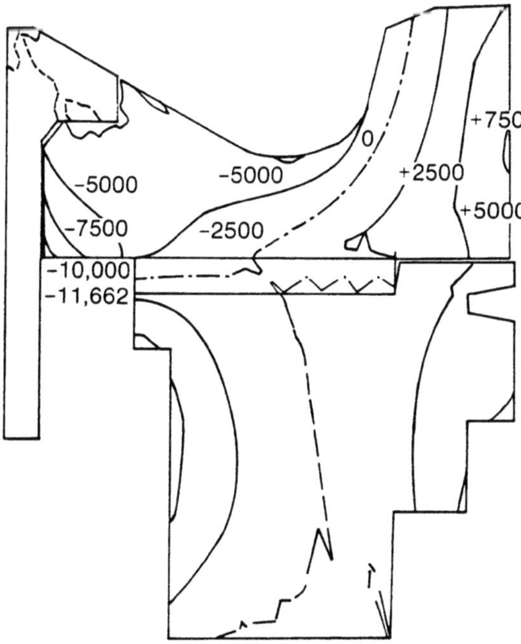

TGAS = 1910°F, HGAS = 122 BTU/HR-FT² - °F
PRESSURE = "O" PSI, INERTIA = 500g
DISC PACK CONDUCT = .41 BTU/HR-FT - °F

Figure 14. Normal stress of composite piston, SSN with optimized bowl.

Table 1 - Composite Piston Cap Stress

Material	Av. Coef. Thermal Expansion ($x10^{-6}$/F)	Thermal Conductivity (Btu/hr-ft-F)	Elastic Modulus (psi x 10^6)	Max. Tensile Stress (psi)
310 S.S.	10.0	18.0	15	15,800
HPSN	1.2	10.5	44	6,744
SSN	1.7	9.7	40	7,674
LAS	0.40	0.97	11.5	2,627
SN-201	1.33	7.25	34	7,400
SiAlON	1.78	9.7	35	5,920
RBSN	1.2	6.4	23.1	5,250

The next consideration for the insulated composite piston is the heat transfer analysis. To achieve the indicated performance target of the adiabatic turbocompound engine, it was assumed that 80% reduction in heat rejection was necessary. Table 2 shows the results of the piston thermal analysis for various materials. Surface roughened metallic shims were used as thermal barriers for metal, HPSN, and SSN. The LAS required no artifical thermal barrier and it shows the best insulation characteristics.

Table 2 - Insulated Composite Piston Thermal Summary

Cap Material	Heat Lost From Gas to Piston BTU/MIN	% Reduction in Heat Rejection
Metal	110	78
HPSN	98	80
SSN	94	81
LAS	47	90

Each of the above pistons were procured and subjected to the testing procedure outlined in Figure 10. Some of the piston caps failed during assembly. Some made it to the screening test rig. Others made it to the single cylinder engine. However, only one ceramic cap made it through the 250 hours proof testing on the single cylinder engine. The successful piston cap was the HPSN cap made by the Army Materials and Mechanics Research Center in Watertown, Massachusetts. At the time of this writing, AMMRC is fabricating six similar caps for the six cylinder demonstrator engine. The summary of our composite piston engine test is shown in Table 3.

Usually the failure of ceramic components in the fired engine test results in broken pieces in the combustion chamber and the crankcase. It is very difficult to trace the origin of failure by shifting through the debris. However, in Test Nos. 9 and 10 (sintered silicon nitride caps) the failures were caught during the 50-hour inspection period. The piston caps were cracking but experienced no catastrophic failures. The cap in Test No. 10 was sent to the U. S. Naval Research Laboratory (Ceramics Branch of Materials Science and Technology Division) for failure analysis. The failure mode was analyzed as being caused by extremely small flaws and machine marks at the outer edge during processing. A photograph of failed sections is shown in Figure 15. It appears that the wide spread in the probability of failure as a function of stress can be attributed to poor quality control, complex designs, finishing process, or even in the handling of parts.

Other Components

From the liner temperature and heat transfer relation, the temperature distribution in the cylinder liner can be calculated and checked experimentally. Figure 16 shows such a plot for an uncooled composite iron cylinder liner. Overall stress analysis can now follow with the additional input of cylinder pressure vs. time.

Cylinder Head and Valves: Although roughly 50% of the heat rejection of a conventional diesel engine is rejected through the cylinder head cooling system, it must be remembered that more than 47% of the cylinder head area is valve head surface areas in a four-valve system. Similar design analyses were conducted for the cylinder head and valve. Figure 17 shows the temperature distribution for the exhaust valve and seats.

Engine Accessories, Manifold, Etc.: In a truly adiabatic engine, it is advisable to insulate the exhaust manifold and even the turbocharger housing to retain the exhaust energy from the engine for optimum operation.

Problem Areas

In an adiabatic engine, the ceramic combustion chamber material must possess:

1. High strength at high temperatures
2. Good insulating properties

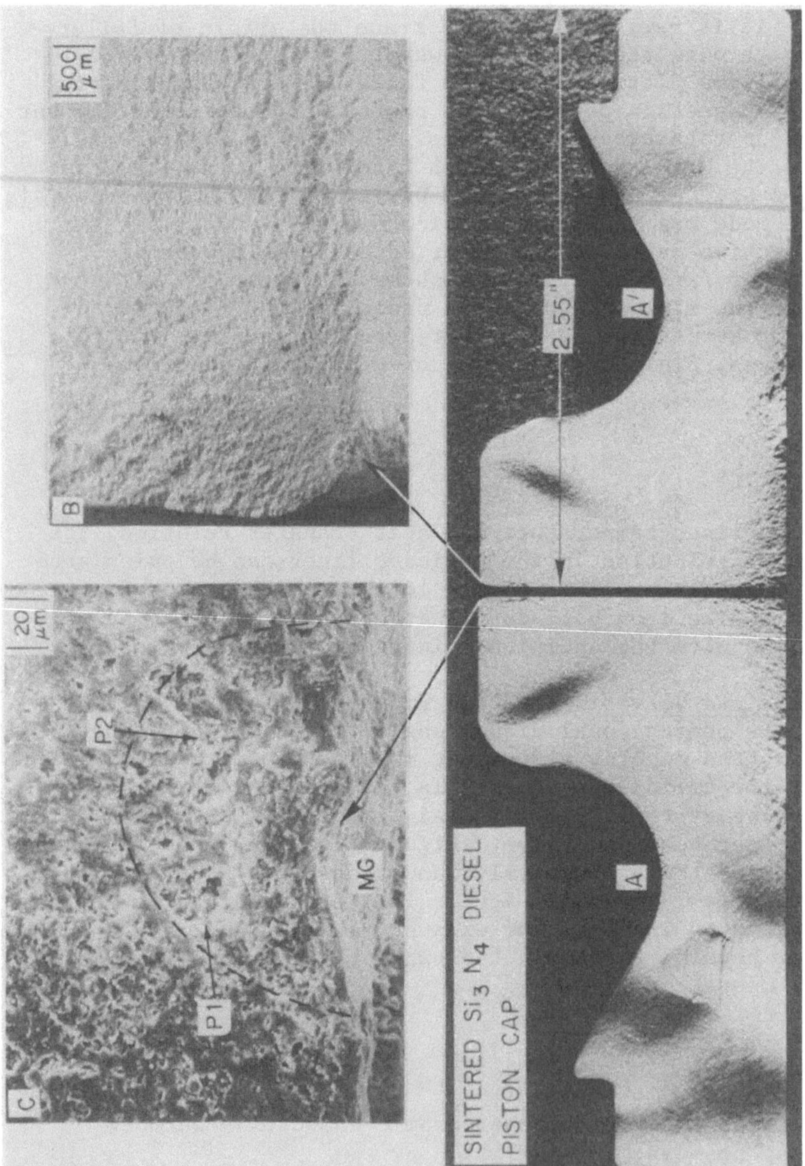

Figure 15. Failure analysis of mating sections from sintered silicon nitride diesel piston cap for Cummins test engine (courtesy of U.S. Naval Research Laboratory).

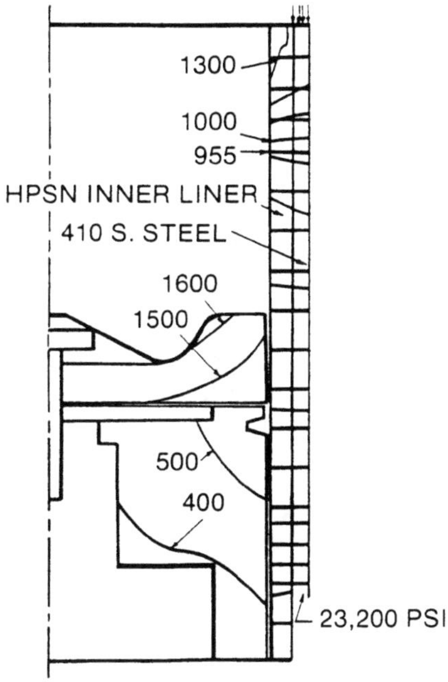

Figure 16. Temperature prediction for HPSN composite liner.

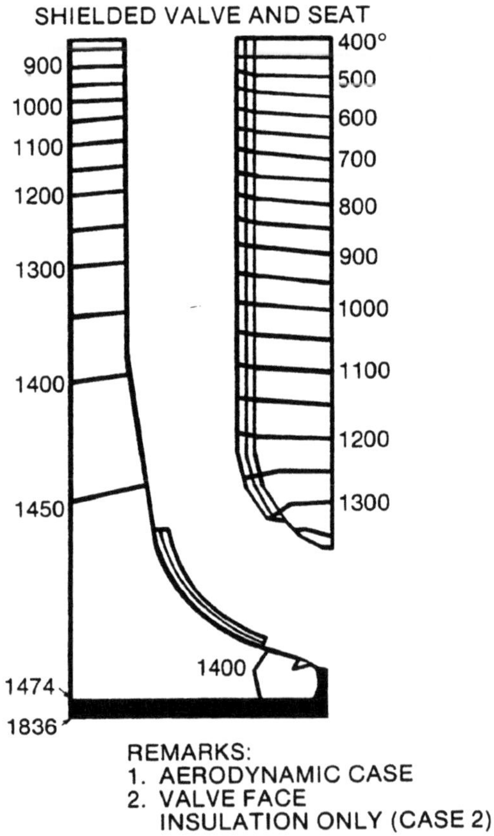

Figure 17. Valve and insert temperatures.

Table 3

Composite Piston Engine Test Summary

TEST NO.	YEAR	CAP	BOWL SHAPE	TOTAL HOURS OPERATION	COMMENTS
1	1976	310 SS	SHALLOW	350	ESTABLISHED NEW BASELINE
2	1976	RBSN	SHALLOW	11	FAILURE - HIT DECK EDGE
3	1977	HPSN 1	SHALLOW	90.8	FAILURE DUE TO SNAP RING
4	1977	HPSN 2	SHALLOW	164.2	FAILURE DUE TO LOOSE BOLT
5	1977	SSN	SHALLOW	172	FAILURE, CAP MATERIAL RELATED
6	1978	316 SS	DEEP	20	FAILURE, BASE FLANGE
7	1978	SSN	DEEP	17	CAP FAILURE, CONTACT LINER WALL
8	1978	SSN	DEEP	5	DISCONTINUED, CAP-LINER CONTACT
9	1978	SSN	DEEP	36	DISCONTINUED, CAP CRACKED
10	1978	SSN	DEEP	200	DISCONTINUED, CAP CRACKED
11	1978	HPSN	DEEP	250	GOAL ACCOMPLISHED
12	1979	HPSN	DEEP W/TOP RING	37	STILL IN OPERATION
13	1979	LAS	DEEP	7	FAILED AT HALF-LOAD

3. Low coefficient of expansion
4. Low cost

The LAS glass ceramic possesses (2), (3), and (4), but not (1). Hot pressed silicon nitride possesses (1), but not (2), (3), and (4), and so on. Hopefully, a more suitable material is in the making.

Although the bending strengths of advanced ceramic materials have made significant advances, higher strength is desirable in designing for high confidence level.

Another shortcoming of the present ceramic material is that the useful applicable or design stress is considerably lower than the data obtained from the laboratory specimen. Nitriding of thick sections over 3/4 inch, flaws, inclusions, cracks, and scratches all tend to lower the useful stress by a considerable margin. Could we not get more consistency in manufacturing quality? Possibly by accelerating the development of fiber ceramics or CVD.

There are many new developments coming on stream in the ceramic field. Complete and consistent physical properties are totally lacking in many cases. This can probably be attributed to testing not keeping up with the pace of development. Tribiological data, chemical properties, etc. are sorely needed.

The need for high temperature lubricants was singled out as one of the major stumbling blocks of an adiabatic turbocompound diesel engine during the initial period of the program, 900°F on the liner in Figure 16 as compared to 350-400°F on the cooled engine counterpart. High temperature lubricants providing adequate lubrication are difficult to find. The lube oil temperature capability must be consistent with viscosity requirements for low friction.

Future Developments

The next significant improvement in the advanced diesel-based power plant, after the adiabatic turbocompound engine, is friction reduction. A 50% reduction in the overall adiabatic engine friction could reduce the brake specific fuel consumption from 0.28 (lb per brake horsepower hour) in the adiabatic turbocompound engine to 0.25 for the minimum friction adiabatic turbocompound engine. In order to meet these reductions in overall engine friction and brake specific fuel consumption, the following engine component design approaches have been given careful study and scrutiny:

A. Gas lubricated ceramic piston-ring-liner combination
B. Unlubricated ceramic roller bearings for the wrist pin, crank pin, and main bearings
C. Solid lubricant coatings for gears, valve guides, rocker arm and push tube assembly

Again, all of the above component designs involve the use of advanced high performance ceramics technology. Not only would the engine be uncooled, but if successful, a good possibility exists for the development of an unlubricated diesel engine.

CONCLUSIONS

A research and development program for demonstrating the technical feasibility of an adiabatic turbocompound diesel engine using ceramics has been presented. Although most of the high performance ceramics are good thermal conductors, a viable approach to achieve the desired insulation as described in Ref. 3,4 has been fabricated, assembled and engine tested.

Cummins and TARADCOM feel confident that the original objectives and targets set forth can be demonstrated. The minimum friction engine approach is still in the very early stages of development. Upon successful demonstration of these studies, we can expect:

1. Uncooled diesel engine design with a 48% thermal efficiency in conjunction with turbocompounding.
2. Unlubricated engine design with a 55% thermal efficiency with significant exhaust energy still available for a bottoming cycle.
3. If widespread use of performance ceramics in advanced diesel based power plants becomes a reality, new dimensions in engine design must be considered.

ACKNOWLEDGEMENTS

The authors wish to acknowledge the following for their contribution in making this paper possible: Cummins Advanced Engine Team, the U. S. Army Materials and Mechanics Research Center, the U. S. Naval Research Laboratory, and the U. S. Army Tank and Automotive Research and Development Command.

1. Kamo, R. and Bryzik, W., "Adiabatic Turbocompound Engine Performance Prediction", SAE Paper 780068, February 1978, Detroit, Michigan.
2. Kamo, R. and Bryzik, W., "Ceramics in Diesel Engine", Societe les Ingeneures les L'Automobile Conference, Paris, France, February 1979.
3. Kamo, R., Wood, M., and Geary, W., "Ceramics for Adiabatic Diesel Engines", CIMTECH 4th, June 1979, St. Vincent, Italy.
4. Layne, J. L., "Thermal Contact Conductance of Piston Insulating Wafers", Cummins Engine Company Technical Report No. 0750-78003, January, 1978.
5. Mueller, J. I., Kobayashi, A. S., Scott, W. D., "Designs with Brittle Materials", University of Washington, Seattle, 1979.
6. Stang, J., "Designing Adiabatic Engine Components", SAE Paper 780069, February 1978, Detroit, Michigan.
7. Stang, J., Johnson, K., "Advanced Ceramics for Diesel Engines", U. S. Army Tank-Automotive Development Center Technical Report No. 12131, January 1976.

CERAMIC TECHNOLOGY REQUIREMENTS FOR

1425°C (2600°F) UNCOOLED POWER GENERATION APPLICATION

D. W. Richerson and J. M. Wimmer

AiResearch Manufacturing Company of Arizona
A Division of The Garrett Corporation
Phoenix, Arizona

S. M. Wander
DOE Division of Power Systems
Department of Energy
Germantown, Maryland

INTRODUCTION

The limited supply and high cost of oil used by the utility industry have resulted in the realization that an alternative fuel must be developed that would be both economically and environmentally acceptable. Since coal is the most readily available fuel- with the highest reserves in the United States, recent emphasis has been on a return to coal-fired steam power plants. However, net-system thermal efficiencies of these plants, including the cost of pollution control systems, are only about 33 percent[1]. In contrast, the combined cycle utility gas/steam turbine systems burning oil have reached thermal efficiencies of 43 percent. Uncooled components in gas turbine systems operating at 1425°C (2600°F) have projected efficiencies of over 50 percent.

Unfortunately, the technology base for direct, coal-fired gas turbines does not exist; hence, the need to convert coal to a liquid or gaseous fuel. Although current conversion efficiencies are poor, the combined cycle system has the potential to outperform the direct coal-fired steam system. In addition to improving coal conversion efficiencies, turbine inlet temperatures must be raised to 1425° to 1645°C (2600° to 3000°F). Clearly, new materials are required to meet this goal.

In October 1976, the Committee on the Use of Ceramics in Industrial Gas Turbines published the results of an analysis entitled, "Concept Feasibility Study of Ceramics for Use in Electric Utility Gas Turbines Fired with Coal Derived Fuels." One of the major conclusions of the study was that ceramics provide a viable technological opportunity for the development of uncooled components in high-power, high-temperature, long-lived utility gas turbines. As a result, the Department of Energy (DOE) contracted for Phase I of a four-phase Ceramic Technology Readiness (CTR) Development Program--a program expected to require 10 years to reach full-scale demonstration of a ceramic, utility gas turbine hot section. A main goal of the CTR program is to provide a ceramic utility gas turbine design compatible with low-Btu coal gasification fuel systems expected to be in use in the near future. One innovative design developed during Phase I was the High-Temperature Ceramic Augmentation Turbine (HiCAT) concept.[2,3]

HiCAT

The HiCAT concept makes use of a small, high-speed, high-temperature ceramic turbo-compressor coupled to a large, low-speed (3600 rpm) utility gas turbine that drives a conventional 60-Hz generator. The power system concept, using a 70 MW_e Brown-Boveri, Type 11, Utility Gas Turbine, is illustrated in Fig. 1. The design is based on the concept of destaging the compressor of the Brown-Boveri unit and performing a portion of the compressor work in a separate 1425°C (2600°F) high-speed ceramic turbo-compressor and then expanding the turbo-compressor exhaust through the utility gas turbine at the original design inlet conditions. The higher efficiency, derived from increasing the cycle pressure ratio and operating with a 1425°C (2600°F) ceramic turbine inlet temperature, results in a system that delivers 48-percent more power than the original turbine, while providing a 31-percent net-system thermal efficiency improvement. A major advantage of this concept is that the high-temperature ceramic turbine can be designed specifically for small ceramic components with minimum component stresses, surface area, and volume--all critical factors in achieving a reliable design. A schematic of the complete system is shown in Fig. 2.

Two basic ceramic turbine designs are being considered. In one, the ceramic blades are attached to a rotating disk, and centrifugal forces cause blade stresses that are principally tensile in nature. In the second design (Fig. 3), the blades are attached to the inside of a rotating cylinder, and centrifugal forces cause the blades to become compressively loaded--a state of stress that takes advantage of the inherent compressive strength of ceramic materials.

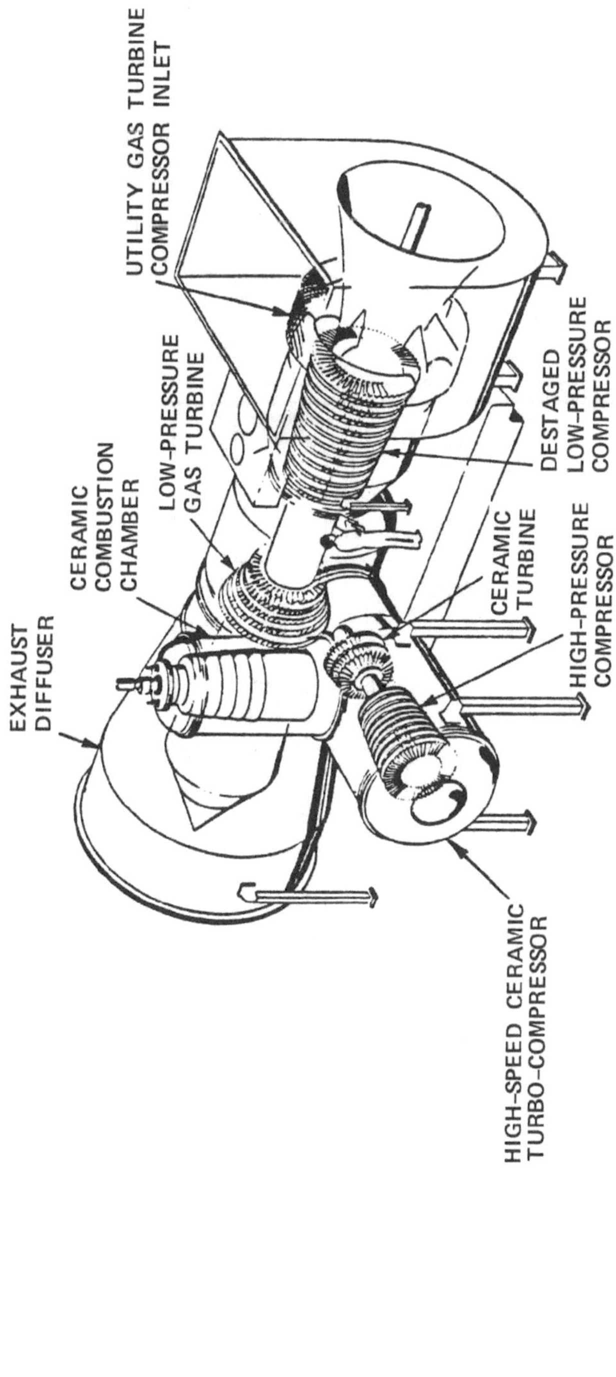

Figure 1. High-Temperature Ceramic Augmentation Turbine (HiCat).

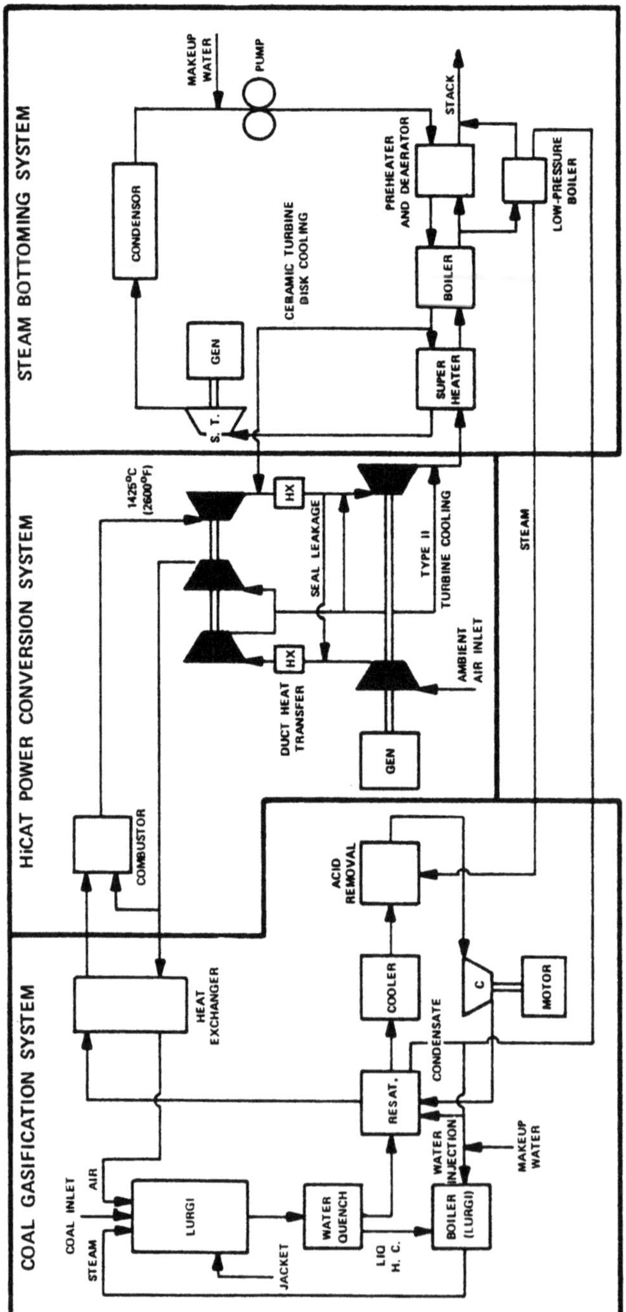

Figure 2. System Schematic.

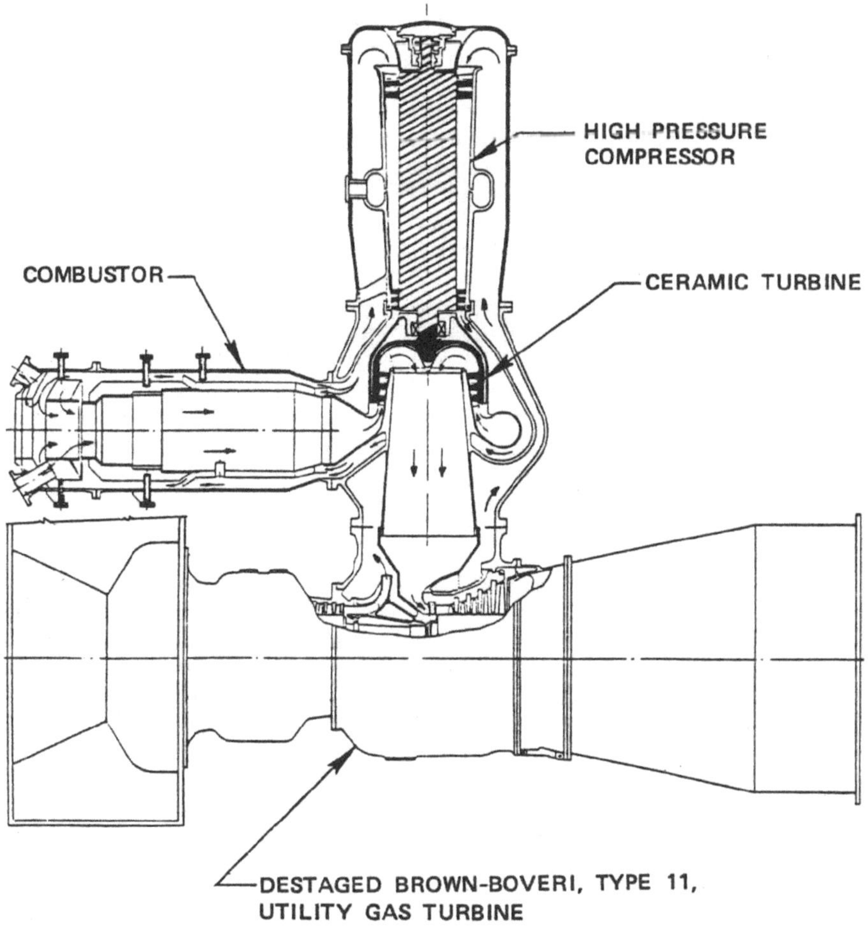

Figure 3. HiCat Concept II, Compressively Stressed Turbine.

Although this design is very attractive, several problems must be solved: (1) the high stresses in the rotating metal cylinder to which the blades are attached are estimated to exceed 200 ksi, and (2) the radial growth of the metal rotor, due to thermal expansion and centrifugal loading, is estimated to exceed 0.5 in., two-thirds of which is due to thermal expansion and one-third to centrifugal loading. The growth increases blade-tip clearance, resulting in a decrease in turbine efficiency unless mating shrouds are mechanically or thermally controlled to follow rotor growth.

In both the tensive and compressive cases, aerodynamic and thermal stresses combine with the centrifugal forces to produce stress fields that have both compressive and tensile components. Although the centrifugal stresses are the dominating factor, the combined stress fields must be considered in the ceramic turbine design and life-prediction efforts.

Table 1 is a comparison of tensive and compressive rotor systems. The relatively low net-system thermal efficiency in both cases is due to the Lurgi gasifier, which operates only at 71 percent efficiency. This gasifier was chosen for the system study due to the published data and field experience available. The choice of an 80-percent efficient gasification system would improve overall efficiency by about 5 percent. If the gasifier were eliminated and a liquid fuel burned directly, the net-system thermal efficiency would be greater than 52 percent.

The major penalty incurred by the Lurgi system is the loss of heat energy in the gas clean-up process. As illustrated in Fig. 2, raw gas leaving the gasifier is water-quenched to remove particulate matter and liquid hydrocarbons. The latter is then used to fuel the boiler that delivers steam to the gasifiers. The low-Btu gas is further cleaned in an acid gas-removal process. The composition of the gas leaving the purification train is given in Table 2. In addition to these major constituents, significant amounts of a number of impurities would be present, e.g., Al, Ca, Cu, F, Fe, Mg, P, K, Na, Si, and Ti. At this time, the quantity of particulate matter to enter the combustor is not known.

Table 1. Comparison of Tensive and Compressive Rotor Turbine Systems (Low-Btu Lurgi Coal-Gas and Steam Bottoming)

Parameters	Tensive	Compressive
Net-System Thermal Efficiency, %	36.4	35.9
Net-Power, MW_e	150.4	148.7
HP Spool Speed, rpm*	5675	4250
Number of HP Compressor Stages	25	31
Last-Stage Blade Height, in.*	4.92	5.00

*Note: Parameters for later designs differ somewhat.

Table 2. Cleaned Low-Btu Gas Composition.

Compound	Percent	Compound	Percent
H_2O	0.2	C_2H_6	0.05
H_2	24.8	C_2H_4	0.04
CO	17.2	N_2	42.8
CO_2	10.7	H_2S	0.09
CH_4	4.1	COS	0.02

The primary ceramic components to be used in the HiCAT turbocompressor are the combustor, transition liner, stator vanes,rotor blades, and associated static structure. The temperatures and stresses experienced by these components vary along the hot-gas flow path and dictate material requirements. The combustor liner will experience the highest temperatures but relatively low stresses. The transition liner and first-stage stator vanes will experience the 1425°C (2600°F) average inlet temperature, and the trailing edge of the vanes will experience a relatively high stress (20 to 30 ksi) occurring at peak temperatures of 1535°C (2800°F). The rotor blades will withstand the maximum stresses; e.g., 35 ksi at 760°C (1400°F) in portions of the third-stage blade of the tensive design. As indicated in Table 1, maximum blade heights are approximately 5 in.

Material Reliability Development Requirements

Major emphasis during Phase I of the CTR program, in addition to evaluating innovative design concepts, was placed on definition of critical requirements for development in the following areas:

- Design Methodology
- Life-Prediction Methodology
- Environmental Effects
- Material Development
- Manufacturability
- Accept/Reject Criteria

These are important segments of technology where advancement is required; Fig. 4 shows how these segments interrelate.

Design Methodology

Current design methodology for ceramics may not be adequate for high-temperature, long-life utility requirements. For example, most ceramic engines currently under development rely on compliant layers to distribute loads at ceramic/ceramic and ceramic/metallic interfaces.[4,5] These compliant layers may be acceptable for a 50- or even 1000-hr life but are not likely to be stable for 30,000 hr. Therefore, a new design methodology must be developed and demonstrated.

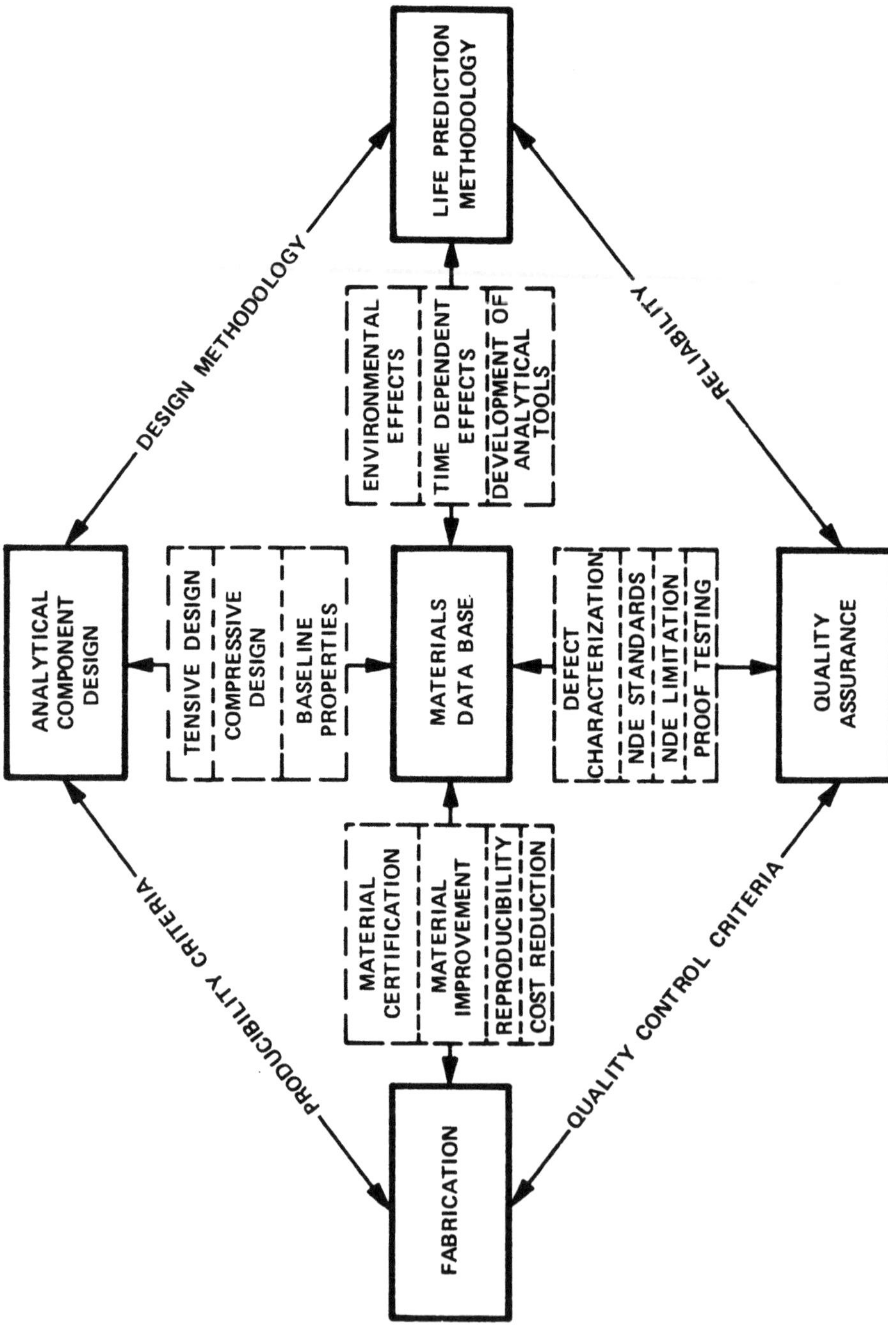

Figure 4. Technology Readiness Plan Integration.

Attachment of rotor blades is a good example of the use of compliant layers. To ensure minimal blade attachment stresses and provide enough blades to obtain reasonable aerodynamic efficiency, a dovetail angle of approximately 60-deg has typically been selected, with a ductile metal compliant layer between the ceramic blade and metal disk to avoid stress concentrations. If the compliant layer cannot be used, the 60-deg dovetail angle may no longer be optimum.* Inability to use a compliant layer may require compromises in allowable blade-tip speed, or the change in dovetail angle may limit the number of blades. Both would have an adverse affect on performance and would necessitate a change in design methodology in order to achieve the desired system efficiency.

The HiCAT concept and the compressive rotor configuration were conceived to provide better reliability for critical ceramic components than could be projected for a design methodology integrating much larger ceramic components directly into current utility turbines. To directly replace the large blades and vanes of current turbines would not only require scale-up of ceramic manufacturing procedures but also a scale-up of ceramic manufacturing facilities. This would require testing components in a full-scale utility turbine or a comparably large test rig, either of which would be expensive and require long turnaround time between tests.

The HiCAT design methodology provides a much simpler problem statement for component manufacture and development testing. The ceramic components are small enough to be manufactured on a prototype basis with existing facilities and shaping technology. Development testing of ceramic components can be conducted with the HiCAT unit, which can operate as a gas generator without requiring hookup to the Brown-Boveri turbine. The HiCAT approach permits concentration of initial effort and funding on ceramic component development in an incremental program that will demonstrate the feasibility of critical ceramic technology before committing large expenditures to the errection of full-scale utility systems.

The HiCAT concept also influences other design methodology areas, such as maintainability. The HiCAT portion of the utility system can be designed for easy disassembly for inspection, repair, or replacement.

Life-Prediction Methodology

Currently, the data base for ceramic materials is quite limited. Figure 5 illustrates the problem this causes in the attempt to design and predict the life of a reliable component. The initial flaw size distribution in a material can be characterized by

*The current design of the HiCAT rotor utilizes a 45-deg dovetail to reduce contact stresses.

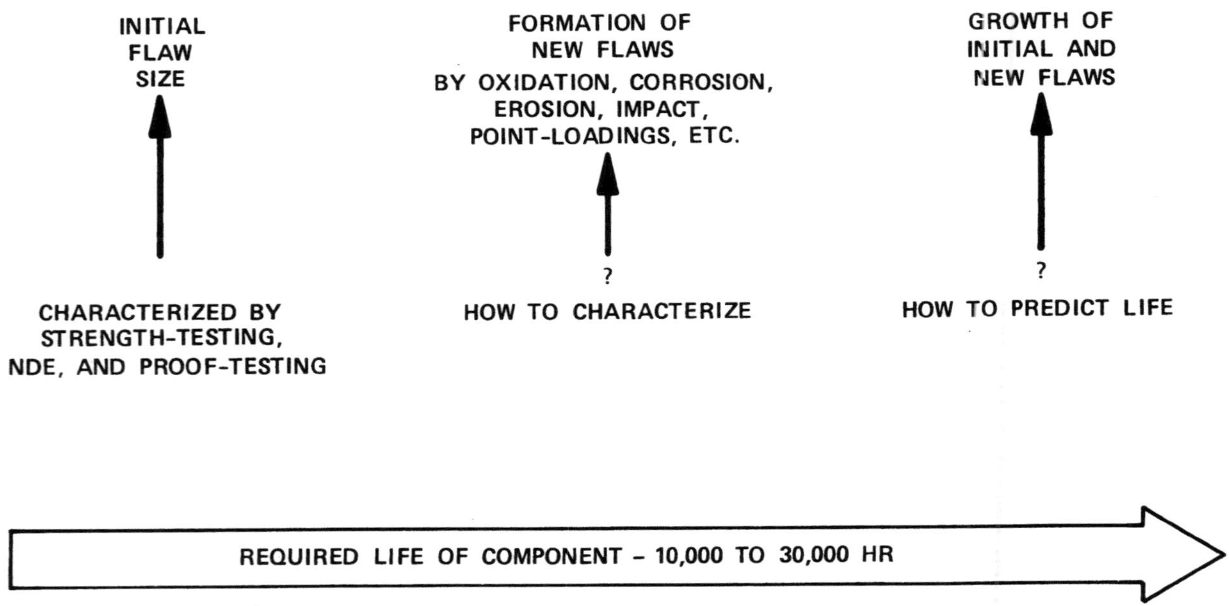

Figure 5. Requirement for Improved Life-Prediction Methodology for Long-Life Application.

strength testing, nondestructive inspection, and proof-testing. To a lesser degree, the growth rate of these initial flaws under a static load in oxidizing environment can be predicted. Growth predictions of flaws in the engine environment and the formation of new flaws, resulting from this environment, cannot be accomplished accurately.

For long-life high-temperature applications, a new life-prediction methodology must be developed that requires a vastly expanded data base to include the probability and kinetics of flaw formation due to oxidation, corrosion, erosion, contact stress (point loading, impact, friction), stress corrosion, and cyclic fatigue. This methodology will also have to cover growth of initial and new flaws due to engine/environment-induced mechanisms and to microstructural related mechanisms, such as creep, grain growth, and impurity diffusion.

Environmental Effects

Engine environmental effects are defined broadly as the following:

- Oxidation
- Corrosion
- Stress Corrosion
- Erosion
- Fouling
- Contact Loading

These have the potential for decreasing the performance or load-bearing capability (and thus the reliability) of ceramic components.

Oxidation. Si_3N_4 and SiC are not thermodynamically stable in a high-temperature oxidizing environment. Their usefulness is dependent upon the formation of a protective surface layer of SiO_2 formed by passive oxidation. Once a coherent layer of SiO_2 is formed, further oxidation is controlled by diffusion through this layer.

The oxidation rate increases as temperature increases. The oxidation kinetics of Si_3N_4 and SiC have been studied[6,7,8] from 900° to $1400^{\circ}C$ (1650° to $2550^{\circ}F$) but not at temperatures above $1400^{\circ}C$ ($2550^{\circ}F$). However, considering the potential impurities in gases from the coal gasification process, corrosion rather than pure oxidation is more likely to be the controlling factor.

Corrosion. Corrosion can be defined as the combined chemical effects on the ceramic components of oxidation and gaseous, condensed, or particulate impurities, either introduced via the gas stream or present in the material. These impurities can increase the rate of passive oxidation, cause active oxidation (direct formation of SiO vapor without forming a protective layer), or produce compositions that can chemically attack the ceramic. Some of the corrosion mechanisms that must be considered are:

- Change in the chemistry of the SiO_2 layer, increasing the oxygen diffusion rate
- Disruption of the protective SiO_2 layer, allowing increased oxygen access
- Decreased viscosity of the protective surface layer, which is then swept from the surface by the high velocity gas flow
- Formation of a molten composition at the ceramic surface, with solubility for the ceramic
- Localized reducing conditions, resulting in low oxygen partial pressures and active oxidation
- Formation of new surface flaws, such as pits or degraded microstructure, that decrease the load-bearing capability of the component

Corrosion testing in flowing-gas combustion environments has been conducted and data compiled during various studies[9-13] at temperatures of 900° to 1370°C (1650° to 2500°F). These studies have indicated that fully dense SiC may be more stable than the current commercial products of Si_3N_4, also, that corrosion for both material systems is highly dependent upon the impurities present in combustion gases and the materials.

The studies referenced above indicate that the effects of corrosion and the possible limitations on component life cannot be predicted for the 1425°C (2600°F) coal-gas-fired utility system based upon current data. Long-term exposure testing in a realistic combustion environment that simulates the application will be required, followed by strength, microstructure, and dimensional and weight change evaluations.

<u>Stress Corrosion</u>. To date, very little testing has been conducted to evaluate the simultaneous effects of applied stress and corrosive environment on the crack growth and life of candidate ceramic materials. Rowcliffe and Huber[14] conducted stress-corrosion studies in a simulated gas turbine environment and compared the results with duplicate tests in static air. Hot-pressed HS-130 Si_3N_4* showed significantly accelerated crack growth in the turbine environment.

*Norton Company material

The rate was about the same at 900°C (1650°F) in the turbine environment as in air at 1100°C (2012°F). Hot-pressed NC-203 SiC* also showed accelerated crack growth in the gas turbine environment, but the effect was less than the Si_3N_4.

High priority must be placed on the evaluation of the effect of stress corrosion on candidate materials for the 1425°C (2600°F) coal-gas-fired utility turbine application.

Erosion. Coal-derived fuels typically contain particulate material that cannot be completely eliminated prior to entering the gas turbine. These particles may result in erosion that could limit component life in the long-term utility application.

Erosion can affect the reliability of turbine engine components at least three ways. First, dimensional change of a component may cause aerodynamic or leakage losses and thus reduce engine performance. Second, surface flaws or reduced cross sections can decrease the load-bearing capability of the component. Third, a roughened surface or abrasive particles can increase friction and limit relative motion, resulting in abnormally high localized stresses and fracture or chipping.

Erosion resistance of hot-pressed Si_3N_4 under simulated turbine conditions, as shown by Metcalfe[15] and Pettit[16], is superior to metal turbine alloys by a factor of 5 to 10. Direct erosion of Si_3N_4 or SiC does not appear to be a major concern; however, a coupled erosion/oxidation mechanism must be considered. The surface layers formed on Si_3N_4 and SiC due to oxidation or corrosion are much softer and are likely to be more susceptible to erosion. Removal or thinning of these layers would then expose fresh Si_3N_4 or SiC to further oxidation or corrosion, resulting in a synergistic degradation significantly accelerated compared to either mechanism operating independently.

Fouling. The opposite of erosion is fouling--the buildup of material on a component surface. By providing a protective layer against erosion or corrosion, fouling can be beneficial, but blockage of the turbine airflow can occur, resulting in loss of performance.

Fouling is the major problem with direct firing of coal in a gas turbine and is likely to be a problem with any coal-derived fuel in which the slag or ash is not substantially removed prior to entering the gas turbine.

Contact Loading. Although contact loading is not an environmental effect, interaction of the turbine environment with surfaces that

*Norton Company material

are in contact can result in increased stresses and possible formation of critical flaws. For instance, oxidation or corrosion could produce a glassy surface layer or a roughened surface that would increase friction between interfacing components. If relative motion were restricted, a significant increase in stress could result within the ceramic component.

Impact damage can also result from the gas turbine environment. Examples of concern include spalling of combustor and transition liners and the dislodging of combustor carbon deposits.

Material Development

Currently available ceramic materials were not developed for 10,000- to 30,000-hr application in a gas turbine engine operating at an average inlet temperature of 1425°C (2600°F). Although the data base is not adequate to quantify material deficiencies, improvements in creep resistance and oxidation/corrosion/stress-corrosion resistance will be required.

Figure 6 illustrates steady-state creep data from Larsen and Walther[17] for the following currently available Si_3N_4 and SiC ceramic materials:

o	NC-132	Hot-pressed Si_3N_4	Norton Company
o	NCX-34	Hot-pressed Si_3N_4-Y_2O_3	Norton Company
o	KBI RSSN	Reaction-sintered Si_3N_4	Kawecki-Berylco Co.
o	NC-435	Reaction-sintered SiC	Norton Company
o	NC-350	Reaction-bonded Si_3N_4	Norton Company
o	β-SiC	Sintered beta SiC	General Electric Co.
o	SASC	Sintered alpha SiC	Carborundum Co.

These results indicate that Carborundum sintered α-SiC has the lowest creep rate.

Manufacturability

Although the HiCAT design significantly reduces the size of components required for utility scale application, manufacturing development will still be necessary to fabricate the configurations with uniform, reproducible properties at a reasonable cost.

Accept/Reject Criteria

Current accept/reject criteria for ceramic components are based upon the following:

(1) Certification according to a specification that starting materials meet defined standards

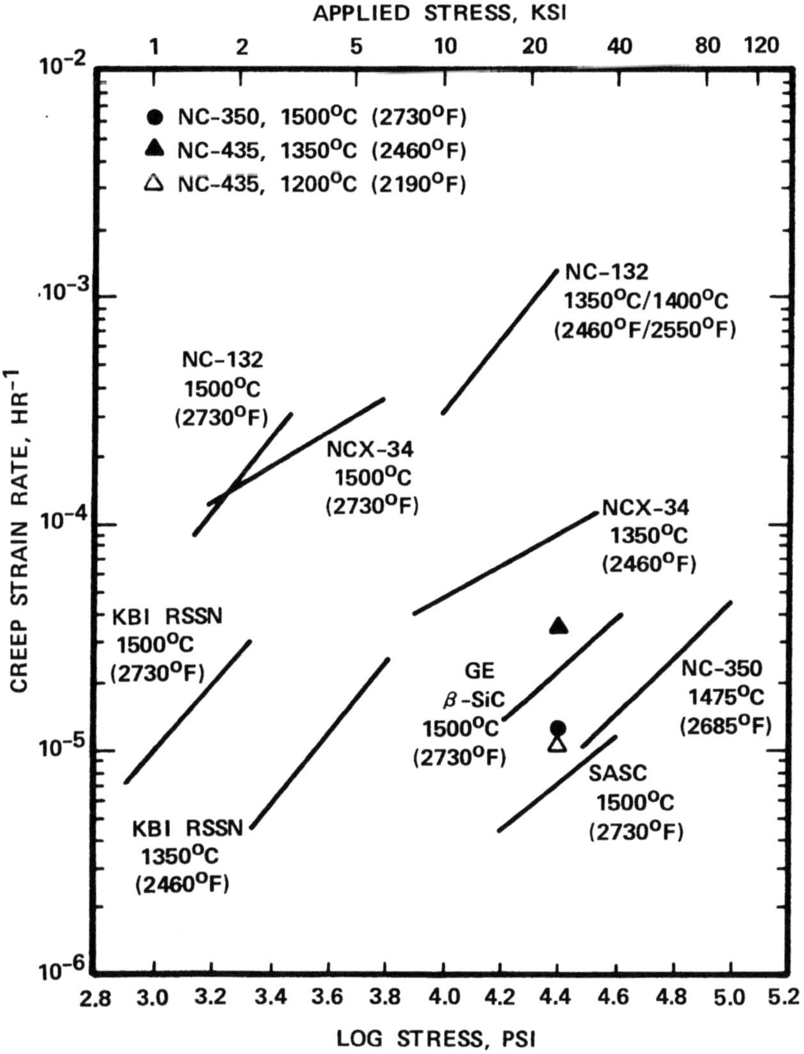

Figure 6. Steady-State Creep Rate vs Applied-Stress for Various SiC and Si_3N_4 Materials.[17]

(2) Property testing of specimens fabricated at the same time as the component

(3) Nondestructive evaluation (NDE) of the component

(4) Component proof-testing or qualification testing

These are being used with some success for prototype testing in short-life applications but require extension and refinement before reliability for either a short-life or long-life application can be assured.

CONCLUSION

The reduction in ceramic component size in the HiCAT concept and the possibility of achieving compressively loaded blades are major steps toward achieving a reliable utility gas turbine operating at 1425°C (2600°F)--a temperature at which the gas turbine system can utilize coal more efficiently than conventional steam power plants. However, consideration of present day materials and the anticipated degradation in combustion products from coal-derived fuels indicate that an extended development effort is necessary to achieve the long life required in the utility application. In the initial phase of this effort, emphasis should be placed on material development, data base extension, development of life-prediction methodology, and a continuation of innovative design approaches.

REFERENCES

1. "Energy Conversion Alternatives Study (ECAS) - Summary Report," NASA TM-73871 (September 1977).
2. CTR Program Interim Report: HiCAT Conceptual Design Study, AiResearch Manufacturing Company of Arizona, Report No. 31-3003 (November 1978).
3. R. T. Elkins, B. B. Heath, and T. Yonushonis, "Innovative Design of Ceramic Utility Gas Turbines," ASME Paper 78-WA/GT-9, presented in San Francisco, CA (December 1978).
4. H. L. Kington, C. R. Dins, L. J. Meyer and D. J. Tree, "Turbine Rotor System Design," in: Proceedings of the 1977 DARPA/NAVSEA Ceramic Gas Turbine Demonstration Engine Program Review, MCIC-78-36 (March 1978).
5. H. E. Helms and F. A. Rockwood, "Heavy Duty Gas Turbine Engine Program Progress Report for Period 1 July 1976 to January 1978," Report No. DDA EDR 9346, prepared under NASA Contract NAS3-20064 (February 1978).
6. W. C. Tripp and H. C. Graham, "Oxidation of Si_3N_4 in the Range 1300°-1500°C," J. Amer. Ceram. Soc., Vol. 59 [9-10] (1976).

7. E. A. Gulbransen, K. F. Andrew, and F. A. Brassart, "The Oxidation of Silicon Carbide at 1150° to 1400°C and at 9×10^3 to 5×10^{-1} Torr Oxygen Pressure," J. Electrochem. Soc., Vol. 113, No.12, 1311-1314 (1966).
8. S. C. Singhal, "Oxidation of Silicon-Based Structural Ceramics," in: Properties of High Temperature Alloys, ed. by Z. A. Foroulis and F. S. Pettit, The Electrochemical Society, Inc., New York (1976).
9. D. W. McKee and D. Chatterji, "Corrosion of Silicon Carbide in Gases and Alkaline Melts," J. Amer. Ceram. Soc., Vol. 59, [9-10], 441-444 (1976).
10. D. W. Richerson and T. M. Yonushonis, "Environmental Effects on the Strength of Silicon Nitride Materials," in: Proceedings of the 1977 DARPA/NAVSEA Ceramic Gas Turbine Demonstration Engine Program Review, MCIC-78-36, 247-271 (1978).
11. S. C. Singhal, "Corrosion Behavior of Silicon Nitride and Silicon Carbide in Turbine Atmospheres," in: Proceedings of the 1972 Tri-Service Conference on Corrosion, MCIC-73-19, 245-250 (1973).
12. K. G. Miller, C. A. Anderson, S. C. Singhal, F. F. Lange, E. S. Diaz, E. S. Kossowsky, and R. J. Bratton, "Brittle Materials Design High-Temperature Gas Turbine - Materials Technology," AMMRC CTR-76-32, Vol. 4 (1976).
13. W. D. Carruthers, D. W. Richerson, and K. W. Benn, "Combustor Rig Durability Testing of Ceramics," Ceramics for High Performance Applications - III - Reliability, in: Proceedings of the 6th AMMRC Materials Technology Conference, Orcas Island, Washington (July 10-13, 1979).
14. D. J. Rowcliffe and P. Huber, "Hot Gas Stress Corrosion of Si_3N_4 and SiC" Proc. Brit. Ceram. Soc., No. 25, 239-252 (1975).
15. A. G. Metcalfe, "Ceramics in Radial Flow Gas Turbines at Solar," in: Ceramics for High Performance Applications, ed. by J. J. Burke, A. E. Gorum, and R. N. Katz, Brook Hill Publishing, Chestnut Hill, MA, 739-747 (1974).
16. F. S. Pettit, DOE, EPRI, GRI, NBS Third Annual Conference on Materials for Coal Conversion and Utilization, Conference 781018, available from NTIS (October 10-12, 1978).
17. D. C. Larsen and G. C. Walther, "Property Screening and Evaluation of Ceramic Turbine Engine Materials," Semiannual Interim Technical Report No. 6, prepared for AFML under Contract F33615-75-C-5196 (July 20, 1978).

STATIONARY CERAMIC COMPONENT CONSIDERATIONS FOR ADVANCED INDUSTRIAL TURBINES

R. J. Bratton

Westinghouse Electric Corporation
Research and Development Center
Pittsburgh, PA 15235

K. L. Rieke

Westinghouse Electric Corporation
Combustion Turbine Systems Division
Concordville, PA 19331

INTRODUCTION

Industry and government in the United States are working together to achieve several specific goals to improve industrial combustion turbine technology. The major goals are improved system reliability, increased fuel flexibility, reduced emission, higher operating efficiencies, and decreased equipment cost. Achievement of these goals will provide the opportunity for wide-scale use of combustion turbine engines in combined cycle power generating systems operating on coal-derived fuels (gaseous and liquid). These highly efficient combined cycle plants can be advantageously applied to base load or intermediate load duty. In addition, the combustion turbine systems alone can continue to serve peak load duty in the simple cycle mode.

The concept of utilizing ceramic components in advanced turbine systems has been continuously studied in recent years[1-7] as one of several concepts to improve operating performance. These studies have shown that if reliable ceramics can be developed for engines, reliable performance can be predicted along with significant performance improvements. However, it has also been shown[3-7] that: a) much work remains in order to achieve ceramic turbine technology readiness, and b) stationary components are likely to find first application in industrial combustion turbines. Stationary components in this case include ceramic parts for the combustion system (combustor/transition piece) and those for the combustion turbine (vanes and shrouding).

Improved combustors are needed, in the near term, to meet exhaust emissions regulations for combustion turbines. Of particular concern is the low nitrogen oxide emission limitations imposed by U.S. and local Environmental Protection Agencies. The combustion problems are further compounded with the current trend toward the use of heavy petroleum fuels and the use of coal-derived liquid and gaseous fuels. Thus, the industrial combustion turbine manufacturer must be forward looking regarding combustor systems development which addresses fuel flexibility.

The present paper reviews the cycle performance benefits that can be obtained by the use of ceramics in industrial combustion turbines while pointing to specific benefits of the stationary ceramic components, and then focuses on advanced combustor concepts that utilize stationary ceramic components.

COMBINED CYCLE PERFORMANCE OF TURBINE SYSTEMS

A combined cycle plant performance improvement realized over advanced air-cooled turbines provides incentive to develop a coal-derived fuel fired engine with turbine ceramic components operating at high turbine inlet temperature (TIT) (2300 to 2600°F). A potential thermal efficiency of above 51% is attainable on coal-derived liquid fuel delivered-on-site and with a 2600°F TIT (technology development goal) as shown in Fig. 1. The coalpile-to-bus bar efficiency potential is about 13% less on low Btu coal gas fuel generated on-site.

The application of ceramics to utility size engines can provide the following benefits:

- Combined cycle plant efficiency approximately 2.8 percent points greater than that of air-cooled engines at 2600°F TIT (see Fig. 1, Ref. 1, 2, 3, 6, and 7).
- Same combined cycle efficiency at 2300°F as at 2600°F TIT air-cooled high temperature turbine (HTT) (see Fig. 1).
- Improved corrosion/erosion resistance for longer life.

Figure 2 shows the contribution to efficiency when adding ceramics to the turbine stationary vanes and rotating blades (combined forming a stage) or to the stationary vanes only.

- 1st Stage, 51% of total improvement or approximately 2.5% in cycle performance.
- 1st and 2nd Stages, 82% of total improvement or approximately 4% in cycle performance.
- 1st and 2nd Vanes, about 57% of total improvement or approximately 2.9% in combined cycle thermal efficiency.

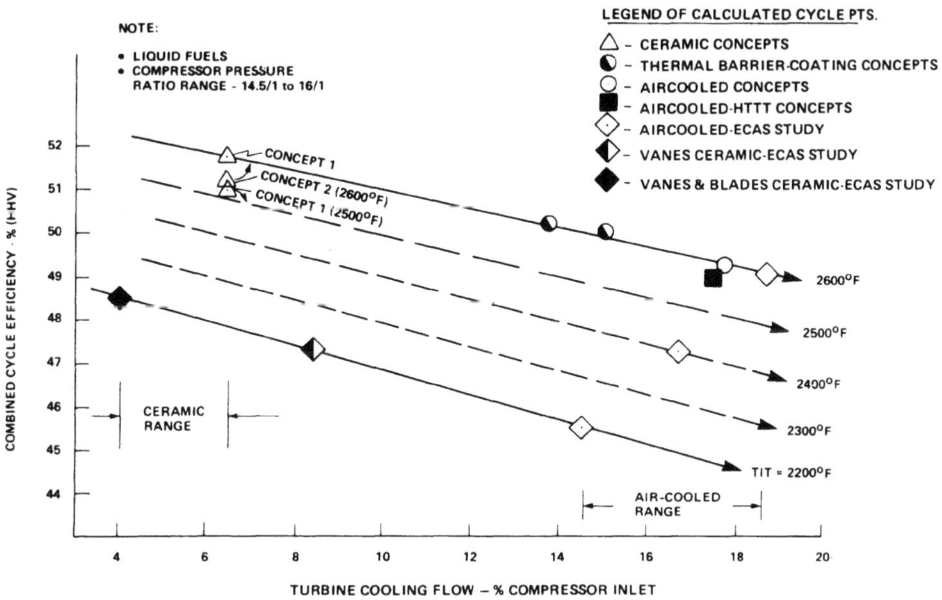

Fig. 1. Combined cycle efficiency vs. cooling flow.

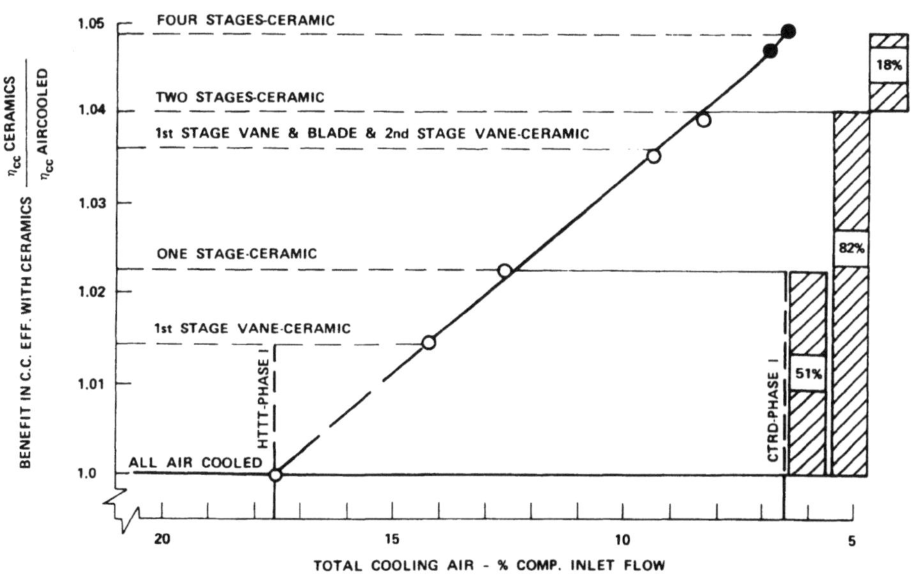

Fig. 2. Benefits in combined cycle efficiency.

This allocation of improvement in combined cycle thermal efficiency becomes a major factor in planning the cost effective development and implementation of structural ceramics to commercial engines for utility power generation systems. The benefits of stationary ceramic vane components in the turbine stages are noteworthy.

ADVANCED COMBUSTION SYSTEM

The combustion system (i.e., combustor and transition components functionally and mechanically integrated as a system) is a major and important element of any heat power engine, be it an internal or external combustion engine. The function of the combustor is to convert chemical energy bound in the fuels to thermal energy which is readily extracted as mechanical energy by the turbine element. The transition component of the combustion system functionally contains the hot gas product and ducts it to the turbine inlet under specific flow, temperature, combustor outlet environment, and imposed design constraints.

The primary objective of an advanced combustion system is to meet the following engine design requirements:

- A high temperature exit gas temperature (technology development goal up to 2600°F).
- A pattern factor (PF) goal = 0.05, where
  $$PF = \frac{(T\ max - T\ avg)}{(T\ avg - T\ combustor\ inlet)}.$$
- Ability to operate on coal-derived fuels (fuel flexibility).
- Ability to meet federal and local EPA emission regulations.

Presently, no known combustion systems exist that provide these high temperature performance goals.

In the case of the combustion turbine system, maximum temperature (i.e., hot spot as defined by T_{max} in the pattern factor (PF) equation above) at the combustor exit is important. At the turbine inlet temperature goal of 2600°F, even a modest local temperature above average may exceed the temperature limit of present state-of-the-art structural ceramic materials applied to the combustion system and the turbine inlet vane components. Thus, the need for establishing a PF goal is evident.

Additional key development issues associated with the combustion system designs are summarized as:

- Structural integrity against vibrational forces.
- Structural configuration to minimize thermal stress and interstitial leakage.
- Attachment interfaces designed to accommodate differential expansions and allow for temperature differences between "hot wall" ceramic and outer metal support system.

- Capability to reliably maintain component integrity under engine transition operation conditions, including severe emergency shutdown conditions.

COMBUSTION SYSTEM CONCEPT DEVELOPMENT

As previously noted, the combustion system function is to burn either gaseous or liquid fuel efficiently with the combustion air. The combustor is also required to provide a near uniform turbine inlet temperature distribution to limit the value of the local hot spot temperature. Two areas of technology development needed to advance the state-of-the-art application of ceramics to utility sized engine combustion systems are:

1. Combustion process.
2. Hot wall combustion system structures.

Combustion Process

The following design requirements and system constraints must be factored into the complete combustion process, i.e., chemical energy release process within the containment structure:

- Steady state performance parameters such as pressure drop, turn down, reference velocity, efficiency, maximum wall temperature, heat release rate, etc.
- Positive ignition and flame stability that meet start-up.
- Steady state and shutdown operation requirements of the engine system.
- A 0.05 pattern factor (PF) goal at the combustion system exit (i.e., 94°F hot spot temperature above 2600°F TIT goal).
- Exhaust emissions that meet EPA regulations.

Combustion process designs in general have the following three regions (zones) of commonality to functionally meet design requirements for the combustion "process" (see Fig. 3).

- Primary Combustion Zone - Involves fuel preparation, ignition and chemical reactions. This zone also provides means of flame stabilization.
- Secondary Combustion Zone - Completes chemical reactions and provides diluent to the hot gases.
- Mixing - Completes mixing of diluent and products of combustion for turbine entry.

The engine operational requirements and integrated engine and combined cycle performance design constraints influence the high temperature combustion process. As an example, the total pressure drop of the combustion system in today's turbine engine is limited

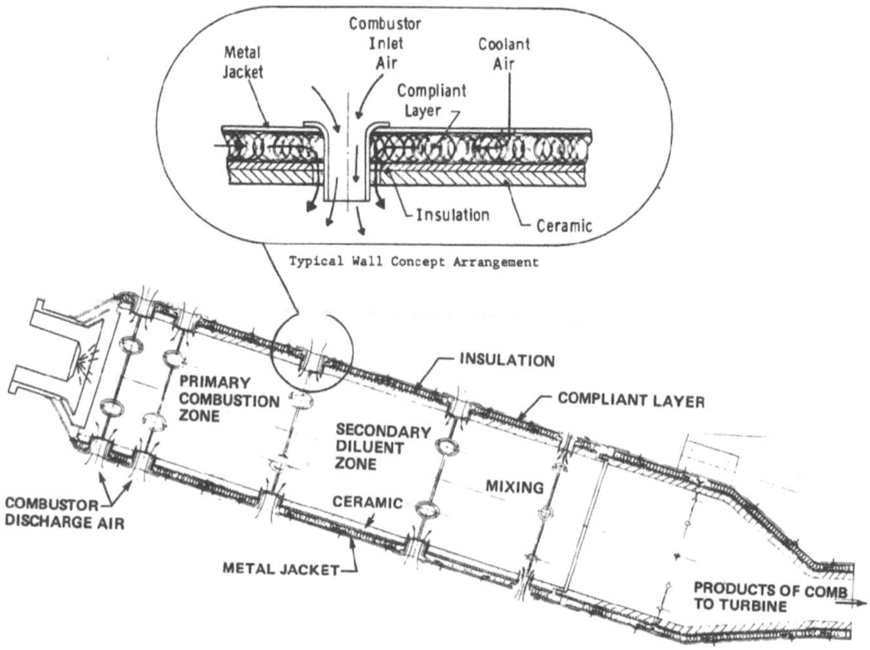

Fig. 3. Combustion process zone allocations.

to about 4% in order to maintain high performance; furthermore, combustor efficiency greater than 99+% is required by its direct relationship with engine fuel economy and exhaust pollutants.

The combustion process design must be integrated with the mechanical configuration in order to assure that positive and reliable ignition occurs. The combustion process must meet transient fuel/air ratio scheduling requirements of the engine for both normal and predictable malfunction operational modes. The combustion process must remain stable, provide thermal energy to the engine system and meet performance goals and emission limits imposed upon it during the steady state and transient modes of operation.

The pattern factor goal of 0.05 is the most stringent single requirement imposed to date on any combustion system. A critical assessment which weights a development goal of up to 2600°F average turbine inlet temperature against the current available ceramic material strength capability provides the incentive to impose this pattern factor goal on the combustion system.

The EPA emission requirements must be factored into the combustion process development plan from the beginning. The pollutants included in the EPA regulatory requirements are sulfur oxide, unburned hydrocarbons, smoke, particulate matter, carbon monoxide, and nitrogen oxide. With the exception of nitrogen oxides (NO_x) the above pollutants can be controlled by careful attention to combustion system design and fuel clean up. However, NO_x control at high temperatures and high pressures requires fundamental changes in the combustion design approach and requires development.

The thermal NO_x reaction occurs at high temperatures for all combustion processes involving air (i.e., $N_2 + O_2$) as an oxidant, and is essentially independent of the fuel. In contrast, fuel bound nitrogen found in many coal-derived liquid fuels forms NO-type compounds during the combustion process. The relative appearance of the two types of NO_x (thermal vs. fuel-bound) is dependent on time and temperature. The NO_x formation can be limited to two processes:

- Low temperature combustion (i.e., lean burn catalytic combustion and others).
- Rich combustion (i.e., a deficiency of oxygen; the relatively inactive nitrogen doesn't compete with carbon and hydrogen for oxygen.)

The use of either approach means moving through the air/fuel vs. temperature plane of Fig. 4. The low NO_x window lies within

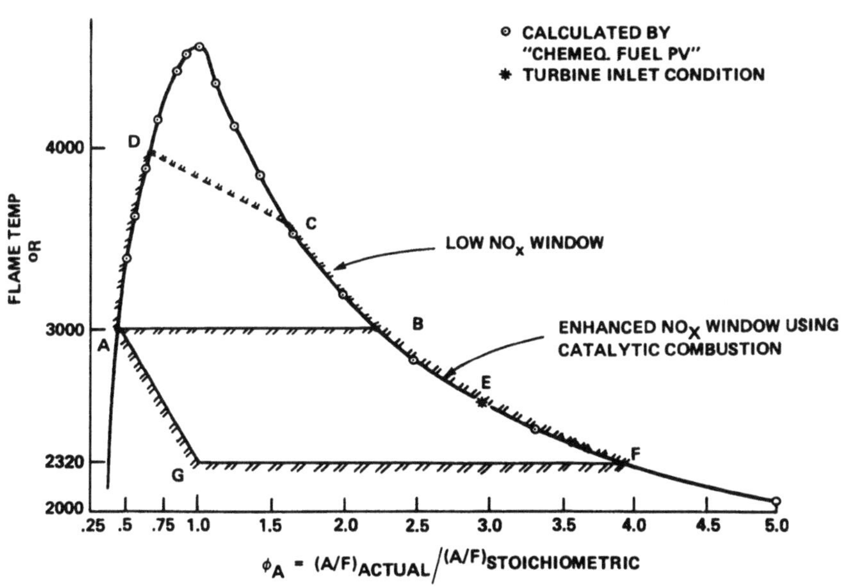

Fig. 4. Low NO_x window in temperature-equivalence ratio plane.

ABCD, where AB represents the lean limit of combustion without a catalyst. Segments AD and CB represent thermodynamic equilibrium conditions. Segment CD connects maximum acceptable NO_x values on the rich and lean side. If a catalyst is used, the operable NO_x window can be extended to ADCBEF. The rich limit for catalytic combustion has not yet been defined; it is assumed to be at least equal to normal combustion at point A. Point G is assumed to define a reasonable shape of the line connecting A and F.

Hot Wall Combustion System Structures

The potential application of hot wall structures in combustion system component designs include the following concepts.

- Catalytic combustor
- Conventional combustor
- Transition section

An assessment of the application of ceramics in hot wall combustion systems results in the following benefits:

- Hot walls provide high probability of attaining 0.05 pattern factor goal (i.e., low turbine inlet local maximum temperature).
- Hot walls increase flexibility in fuel usage (i.e., coal, gas, coal derived liquids, residuals, etc.)
- Hot wall designs (including catalytic) permit burning over wider air/fuel ratios.
- Hot wall designs reduce coolant requirements when operating at high exit temperatures (especially important when operating with low Btu gas).
- Hot wall designs provide flexibility in rich/lean staged combustor concepts to meet NO_x emissions with high bound N_2 fuels.
- Hot wall designs assist in meeting emissions with other combustion process concepts (such as lean burn).
- Hot walls potentially increase life and reliability over cooled metals by providing improved oxidation/ corrosion/ erosion resistance with coal-derived fuels and reducing structural distortions (thermally induced stress).
- Ceramic hot walls reduce the need for high cost, strategic materials.

The application of ceramic hot wall structures to combustor conceptual designs is shown in Figs. 3 and 5 for the conventional combustor and catalytic combustor, respectively. Figure 5 also shows a reference engine used for design concept studies on ceramics in turbine hot section components.[9]

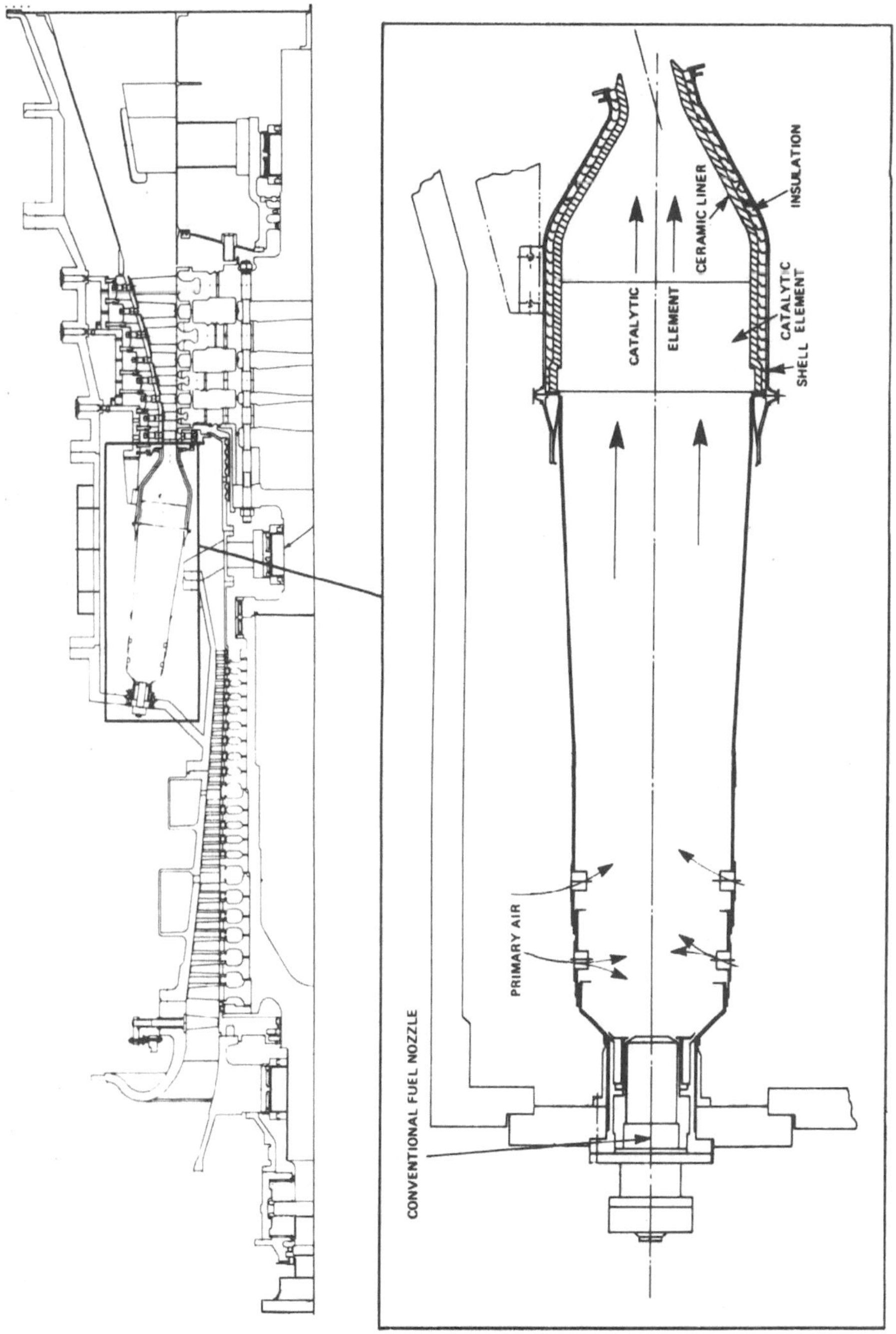

Fig. 5. Arrangement drawing showing catalytic element in hot wall combustion system.

The transition section is common to all the combustion systems concepts. This component directs the combustion products to the turbine and maintains or improves the pattern factor developed in the combustor. The shape of the transition is more complicated than the cylindrical combustor regions (see Fig. 6). Its complex shape causes additional structural design considerations when applying ceramics to the hot wall concepts. The structures have about the same thermal design objectives with high temperature products of combustion as those expressed for hot wall combustors; i.e., operation at high inner surface wall temperatures, with low heat flux. The concept shown in Fig. 7 makes use of a segmented ceramic tile liner concept. The common design elements of such hot wall structures are:

- Inner structural ceramic material.
- Insulation capability to control the hot wall temperature and heat loss to ambient surroundings.
- Cooled attachment layer between the ceramic and outer support layer.
- Outer metal support system.

SEGMENTED APPROACHES TO HOT WALL STRUCTURES

The various combustor and transition concepts are expected to require structural segment designs. The key development issues identified for such structures are:

- Configurations that minimize thermal stress and interstitial leakage.
- Attachment schemes that provide the necessary flexibility to accommodate differential expansions.
- Interface schemes that accommodate thermal gradients within ceramic and metal components, and high temperature differences between them.
- Structural designs that endure high temperature product of combustion without compromising the objective of the 0.05 pattern factor design goal or attainment of the emissions regulatory requirements.
- Structural designs that withstand start-up and shutdown transients and the steady state engine operation.

Examples of a segmented wall structure are shown in Fig. 8a and 8b. In Fig. 8a, a layer hot wall concept is used. It illustrates ceramic-to-ceramic edge treatments, and use of compliant layers between metal and ceramic. Fig. 8b shows a potential hot wall design approach that has been employed in experimental MHD work at Westinghouse. The concept shows a design which features an offset at ceramic edges, a stainless steel plate about the spring clip, and a sealer between clips. The experimental design has successfully contained combustion products approaching 4000°F

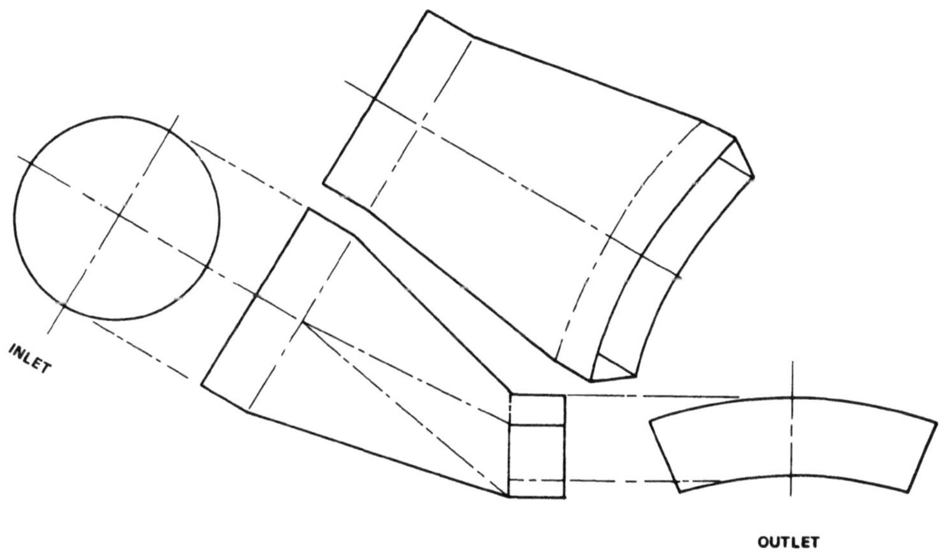

Fig. 6. Typical transition configuration.

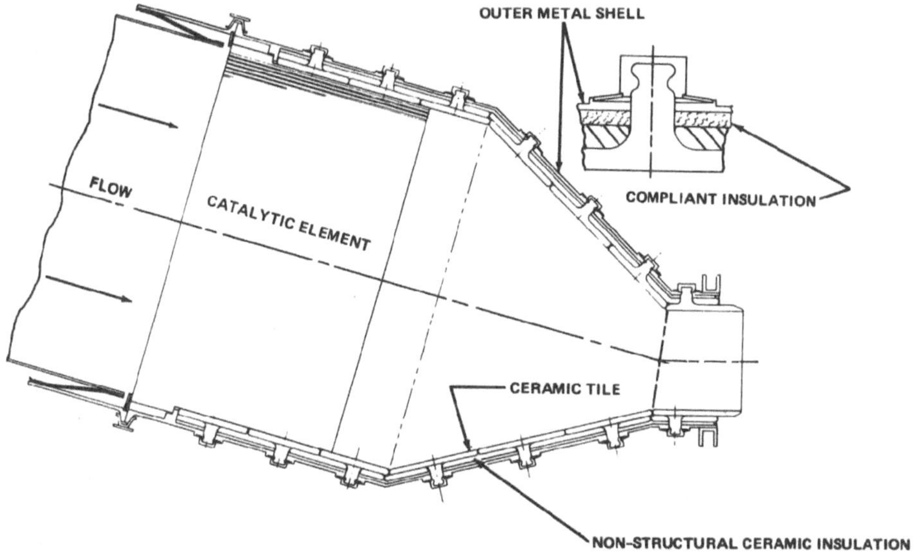

Fig. 7. Multi-piece ceramic tile liner transition concept.

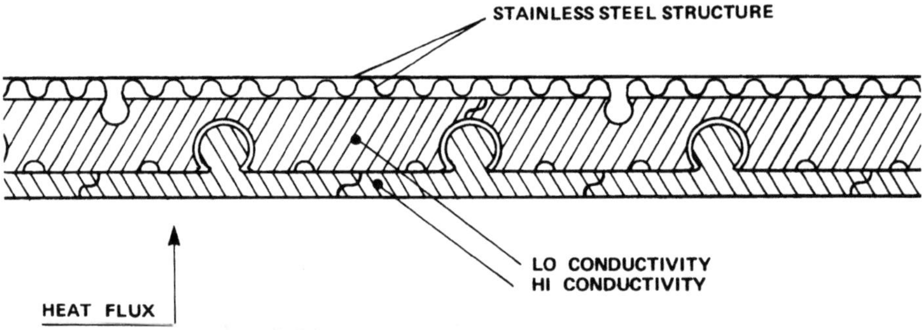

(A) CERAMIC LINER WITH KEYED PLATES AND COMPLIANT LAYERS

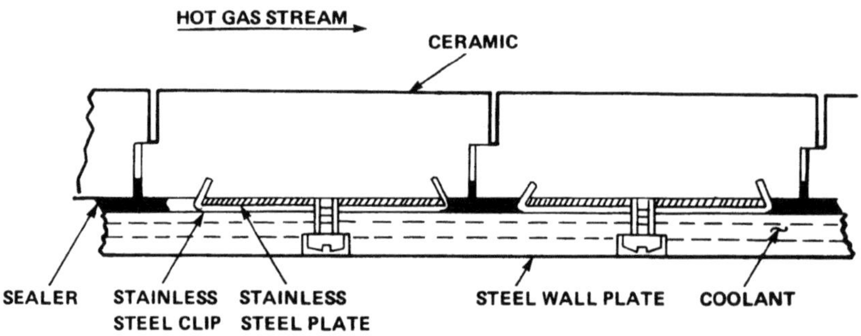

(B) A HOT WALL DESIGN APPROACH

Fig. 8. Segmented wall structures concept.

in the MHD application. The preceeding examples display a multi-concept basis for which ceramic hot wall combustion systems may be evolved through development and test programs.

CONCLUSIONS

Recent studies have reaffirmed that significant cycle performance improvements, engine durability and reliability may be achieved through the use of high performance structural ceramics in combustion turbine systems. Reliable stationary ceramic components are needed in the near term, in particular for advanced combustor concepts, that must be developed for the purpose of meeting emission regulations, while providing the user with the capability of using fuels ranging to heavy petroleum fuels and coal-derived fuels. Ceramic hot wall structures are feasible in the near term for the conventional combustor, catalytic combustor and the combustion transition section.

REFERENCES

1. D. T. Beecher et al, "Energy Conversion Alternative Study - ECAS-Westinghouse Phase I Final Report," NASA CR 134941, February 1976.
2. D. T. Beecher et al, "Energy Conversion Alternative Study - ECAS-Westinghouse Phase II Final Report," NASA CR 134941, March 1977.
3. "Final Report, Concept Feasibility Study of Ceramics for Use in Electric Utility Gas Turbines Fired with Coal Derived Fuels," by Committee on the Use of Ceramics in Industrial Gas Turbines, University of Pittsburgh, Pittsburgh, PA 15260, Report SETEC DO 76-024F, October 1976.
4. R. J. Bratton and D. G. Miller, "Brittle Material Design, High Temperature Gas Turbine, Volume I," Final Report on Contract DAAG-46-71-C-0162, AMMRC CTR 76-32, December 1976.
5. R. J. Bratton, "Ceramic Turbine Components Research and Development," Final Report Summary, EPRI Contract RP 421-1, March 1979.
6. K. L. Rieke, "Ceramic Technology Readiness Development Program, Phase I - Conceptual Designs and Materials Screening, Interim Report," DOE Contract EF-77-C-01-2786, July 1978.
7. K. L. Rieke, "Ceramic Technology Readiness Development Program, Phase I - Conceptual Designs and Materials Screening, Final Report," DOE Contract EF-77-C-0102786, March 1979 (draft not yet released).

ACKNOWLEDGEMENTS

The authors acknowledge the Electric Power Research Institute and the U.S. Department of Energy for their recent sponsored work in the areas reviewed in the present paper.

SESSION II

QUALITY OF MATERIALS AND COMPONENT PROCESSING

Chairman: Dr. R.N. Katz
Army Materials and Mechanics Research Center

Vice Chairman: Dr. O.J. Whittemore
University of Washington

SINTERED SILICON CARBIDE

Y. Hamano, M. Yamaguchi and S. Nagano

Kyoto Ceramic Co.

1 - 1, Yamashita-cho, Kokubu-city, Kagoshima 899-43
Japan

HIGH EFFICIENCY GAS TURBINE DEVELOPMENT PROGRAM IN JAPAN

In Japan we have two government projects for the resolution of the energy problem. The Sunshine Project was started in 1974. The purpose of this project is to develop production technology for new energy sources such as solar, geothermal, coal gasification and liquefaction, and utilization of hydrogen for energy storage. Meanwhile, to promote energy conservation technology, the Moonlight Project was started in 1978. The main tasks of this project are the trial manufacturing of prototype plants with high temperature and high efficiency gas turbine and MHD power generation systems together with the development of advanced materials and production technologies.

The target of the Gas Turbine Development Program is to realize 55% thermal efficiency (LHV) and 1500°C turbine inlet temperature by a prototype plant of 100 MW with the combination of a steam plant. The prototype plant will be built by fiscal year 1984 to realize the target of the program, with advanced materials and technology which will have been developed in the next six years.

An engineering research association was established last year with members of six gas turbine makers, four superalloy makers, four ceramic makers and the Central Research Institute of Electric Power Industry of Japan to coordinate and promote the development program. A reheat cycle industrial turbine of 100 MW class (AGTJ-100A) was designed by the association which had two combustors (high pressure combustor and reheat combustor), two compressors and three turbines. The combined cycle thermal efficiency of the reheat cycle gas turbine was calculated for three different stages of materials technology and production technology, such as precision casting and precision machining. The lowest line in Figure 1 shows the relation between TIT and thermal efficiency using conventional materials and technology. The next line shows the relation with the most advanced materials and technology which are currently available. The upper line shows the calculated efficiency of the plant with rotor blades and vanes made by ceramics or a new alloy with a complicated cooling system. At the TIT of 1500°C, the combined efficiency is well over 55%. New alloy and complicated cooling technology for blades and vanes, ceramic combustors and newly designed metallic combustors, which can withstand TIT of 1500°C, are the main targets of component development of the program. Improved designs of the compressor and turbine with 2% better aerodynamic efficiency are also expected.

The development of a ceramic combustor will provide great advantages for the protype plant because 1500°C TIT is necessary. Kyoto Ceramic Co. is engaged in the development of a ceramic combustor. One of the candidate materials for the ceramic combustor is sintered silicon carbide.

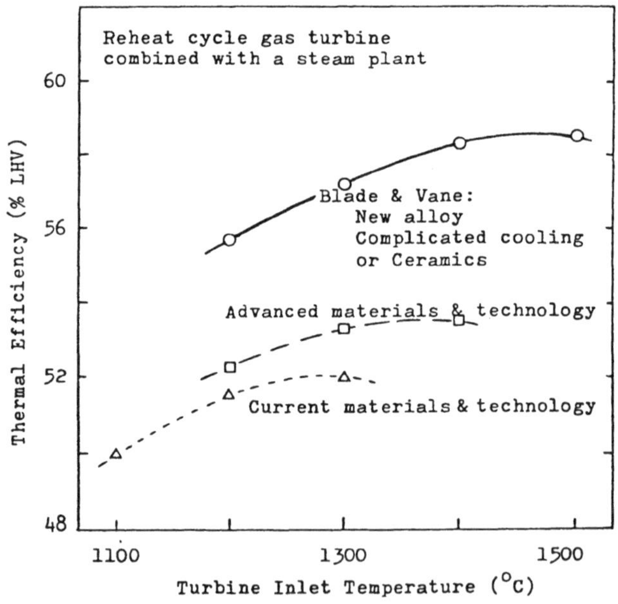

Figure 1. Combined Cycle Thermal Efficiency of 100 MW Reheat Cycle Gas Turbine for Three Different Stages of Materials and Production Technology.

SINTERED SILICON CARBIDE

A high density, high purity and fine grained silicon carbide was developed at Kyoto Ceramic Co. and is now produced and sold in quantity (Kyoto Ceramic Co. designation SC-201). Fabrication technology for complex shapes has been developed at the same time. SC-201 may be formed by any of the conventional ceramic technologies such as die pressing, isostatic pressing and machining, slip casting and extrusion. Selection of forming technology is based on shape and quantity of parts to be manufactured. Each part is ground, if necessary, to shape with given dimensional tolerance. Figure 2 shows several pieces made of SC-201. Some physical properties will be discussed hereafter.

Mechanical Strength

Figure 3 shows flexural strength of SC-201 measured in air by three point bending with a span of 20mm. Specimens had a 3mm square cross section and were ground to the shape after firing. Because of instrumental limitation, the strength was measured at temperatures below 1430°C. No strength degradation was observed at these temperatures.

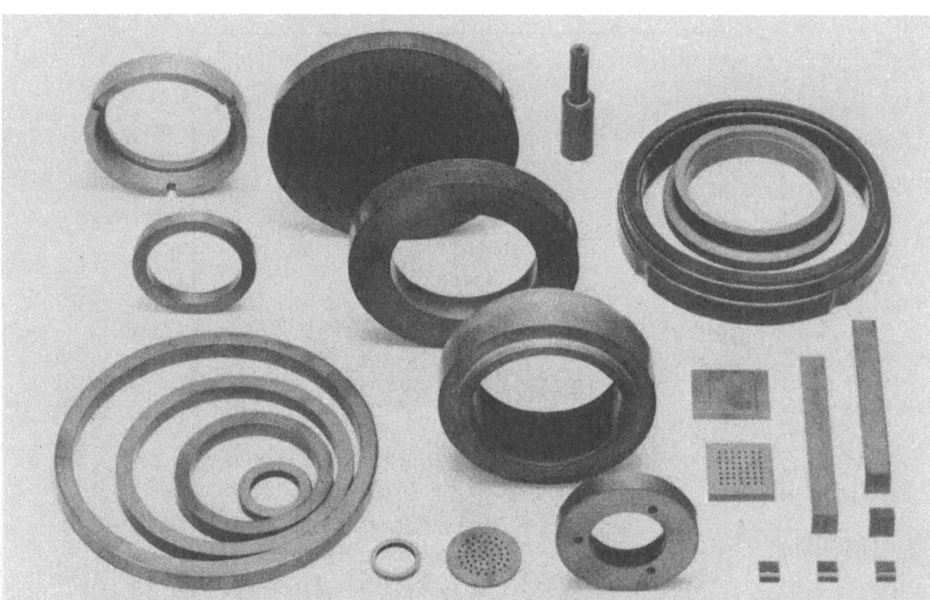

Figure 2. Typical Shape Manufactured from SC-201.

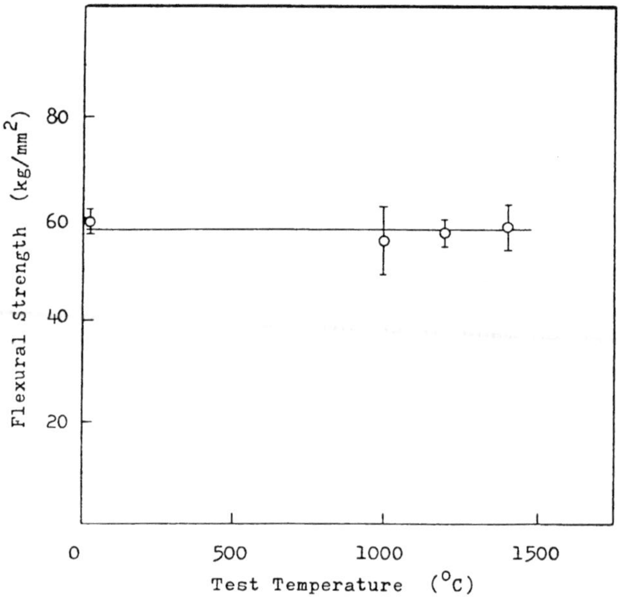

Figure 3. Flexural Strength - Temperature Relationship for SC-201.

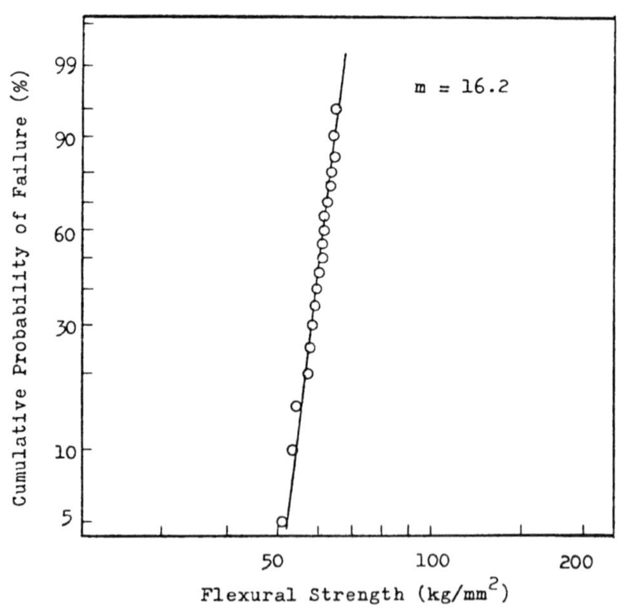

Figure 4. Weibull Distribution of Flexural Strength of SC-201 at Room Temperature.

The distribution of flexural strength of 19 test bars at room temperature is presented in Figure 4, from which the Weibull number is calculated to be 16.2 for this lot of SC-201.

K_{IC} values of SC-201 at room temperature, 1200°C and 1400°C are presented in Figure 5. These were measured with test bars with different notch widths. The K_{IC} value was calculated for 200 notch width.

Figure 6 shows the relation between K_I and crack velocities measured with double torsion specimens at the same temperature intervals as above. n values in the following equation were calculated to be 41, 33 and 15:

$$V = A\,K_I^{\,n}$$

Izod impact test was carried out with specimens and the instruments identical to ASTM test method (ASTM D256-73). The average impact strength of 10 specimens was 2.6kg cm/cm or 25.5 J/m.

Creep Deformation

Creep deformation rate of SC-201 was measured at 1400°C in air. To a test bar measuring 1 x 5 x 100 mm, a steady load was applied with a 1055g alumina block by four point bending alignment with the inner span of 40 and outer span of 80 mm. At the end of each test time and after cooling, the deformation at the center of the test bar was measured with a microscope The deformation was between 100 to 200 for the test time of 80 to 200 hours. The following equations were applied in calculating stress and strain:

$$\sigma = \frac{3\,W\,(l - a)}{2\,b\,h^2}$$

$$\varepsilon = \frac{4\,h\,e}{a^2}$$

Where W is the load, l is the outer span, a is the inner span, b is samples width, h is sample thickness and e is the deformation at the center of the sample. Figure 7 shows the creep strain of SC-201 at 1400°C in air.

Young's Modulus

Young's modulus of SC-201 was measured up to 1300°C by elastic deformation and sonic resonance methods. As is shown in Figure 8,

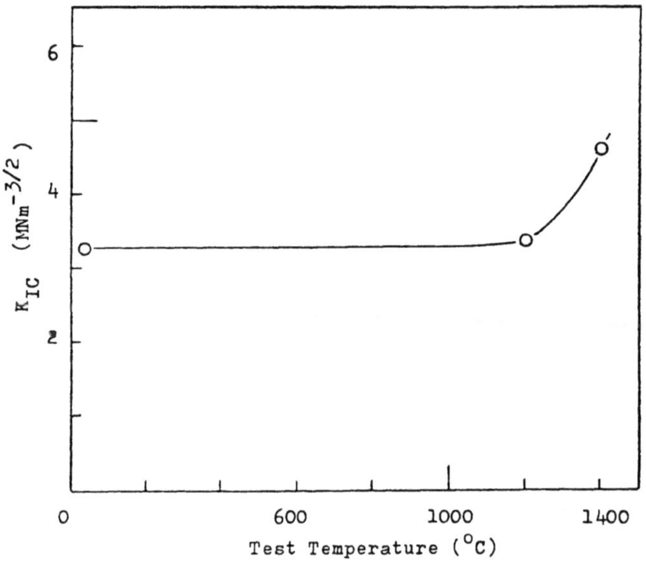

Figure 5. K_{IC} of SC-201 at 25°C, 1200°C and 1400°C.

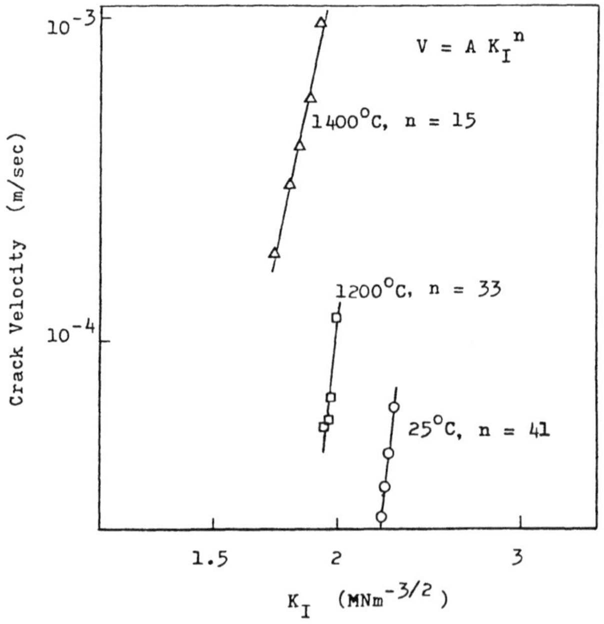

Figure 6. K_I-V relationship for SC-201.

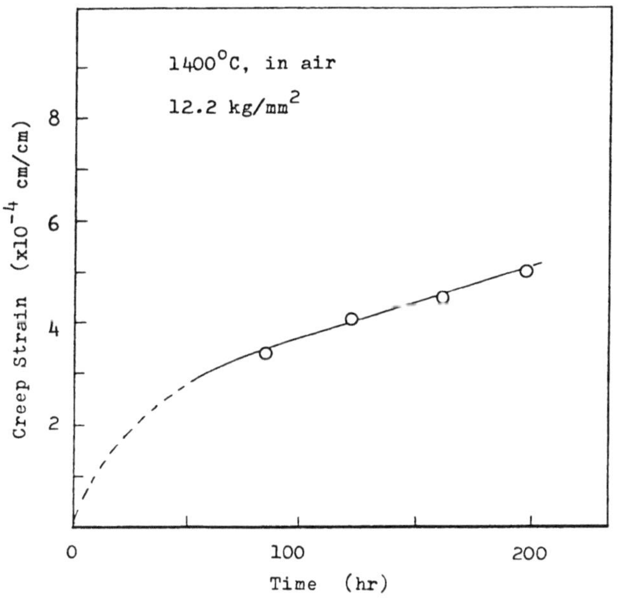

Figure 7. Flexural Creep of SC-201.

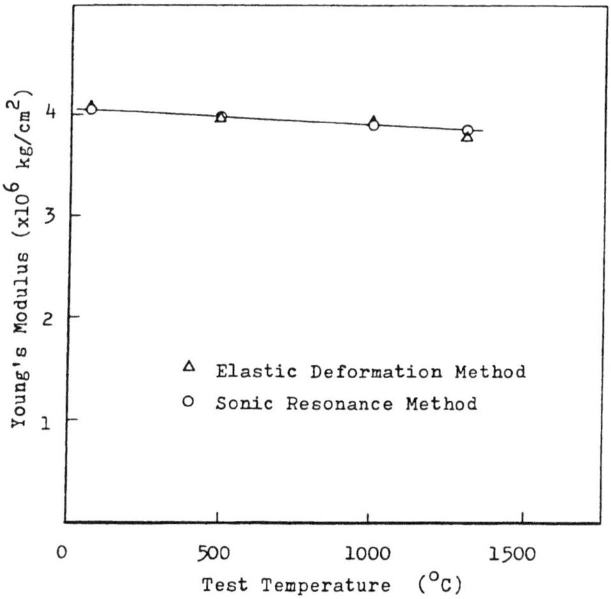

Figure 8. Young's Modulus - temperature relationship for SC-201.

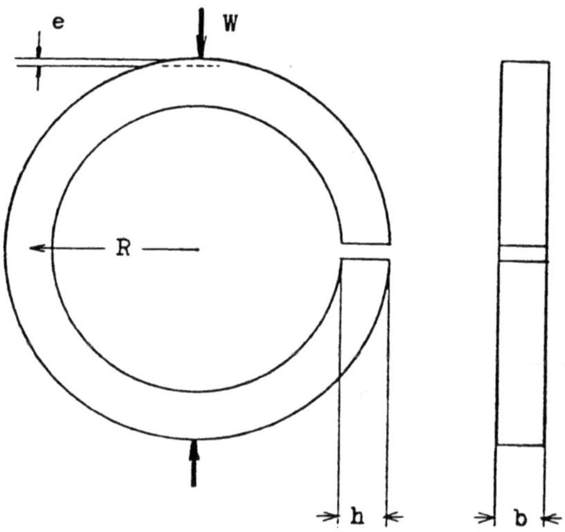

$$E = \frac{21.43}{b}\left(\frac{R}{h}\right)^3 \frac{W}{e}$$

Figure 9. Specimen geometry for Young's Modulus measurement by elastic deformation.

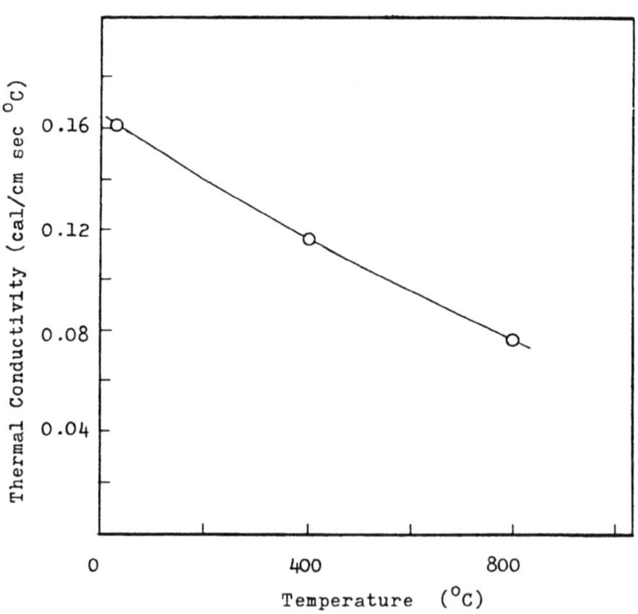

Figure 10. Thermal Conductivity - temperature relationship for SC-201.

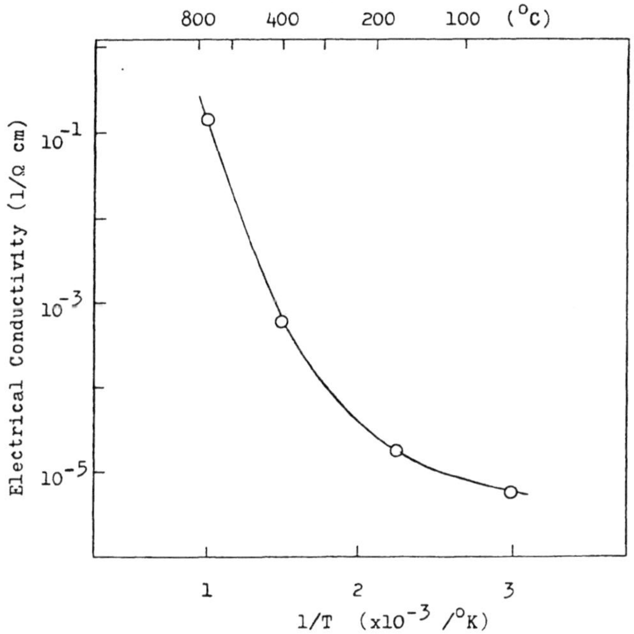

Figure 11. Electrical Conductivity versus Reciprocal Temperature.

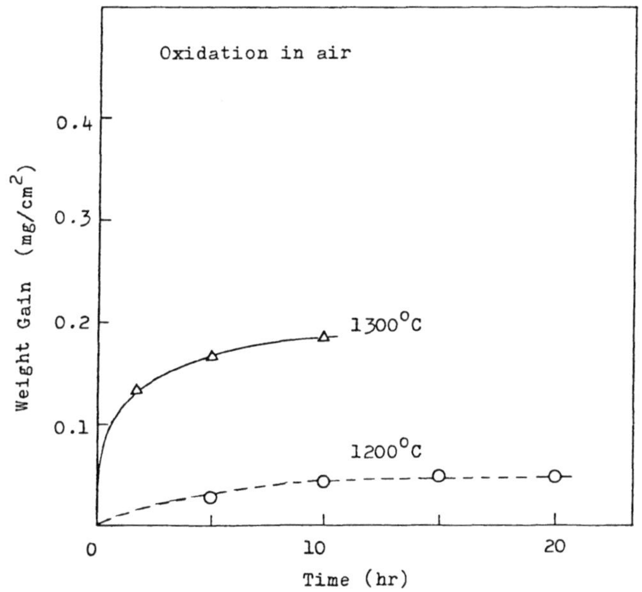

Figure 12. Weight gain of SC-201 in heating in air.

both methods gave identical values which decreased slightly with increasing temperatures. Figure 9 shows specimen geometry for elastic deformation measurement. The amount of deformation was measured with a differential transformer, and the load was measured by a load cell.

Thermal Conductivity

Thermal conductivity of SC-201 was measured by the Xenon flash method. A Xenon flash of t^* duration was applied to one side of the test specimen which was 20mm in diameter and 2mm thick. From the temperature increase on the other side, thermal conductivity was calculated from the following equation where T_m is the maximum temperature increase, $t_{1/2}$ is time required for the temperature increase of $\frac{1}{2}T_m$, h is sample thickness, and Q is an instrument constant representing flash energy.

$$k = \frac{1.37\ h\ Q}{2\ T_m\ (t_{1/2} - 0.36\ t^*)}$$

Electrical Conductivity

Electrical conductivity was measured at 60 Hz from voltage and current relationship in the temperature range from 25°C to 800°C. As shown in Figure 11, log conductivity was plotted against reciprocal temperature for a typical value of conductivity measurement. The conductivity depends very much upon the amount of impurities.

Oxidation

Figure 12 shows weight gains caused by oxidation in air at 1300°C and 1200°C. These were measured by heating in an electric furnace and parabolic curves were obtained.

CONCULSION

SC-201 is the candidate material for combustors for the industrial gas turbine for the Japanese Government project. A detailed design of the cermaic combustor is not obtained yet, but a tubular design incorporating tubes of about 7 inches diameter and stacked ring structure will probably be adapted. Because of great high temperature strength and stability and versatile forming characteristics of SC-201, utilization of SC-201 for the ceramic combustor seems to be promising. The key to success for the ceramic combustor will be the completion of NDE technology of ceramic thin walled parts together with design technology to overcome the brittleness of ceramic materials.

PROCESSING HOT PRESSED SILICON NITRIDE FOR IMPROVED RELIABILITY: HS-110 to NC-132

M. L. Torti

Industrial Ceramics Division
Norton Company
Worcester, MA 01606

INTRODUCTION

In this paper I would like to review our experiences in the evolution of hot-pressed silicon nitride over the past ten or so years, to look at what progress we have made and what progress we haven't made in the hope that this review will help to guide us in our efforts in the future.

EARLY WORK ON HS-110 LEADING TO HS-130

Although Norton has been working on silicon nitride for a number of years previously, HS-110 really began July 30, 1970, with the signing of an agreement between the Lucas Company of England and the Norton Company in which Norton obtained licenses in the United States for the Lucas silicon nitride patents. We then moved rapidly to develop our own powder preparation facilities and to scale-up the hot-pressing technology. These developments coincided with the beginning of ARPA sponsorship of the ceramics in structural applications program which was administered so ably by AMMRC and carried out by ford and Westinghouse.

Our first realization of the need for improvement occurred when we began to check the high temperature strengths of the hot-pressed silicon nitride that both Norton and Lucas had been preparing. The initial testing at Lucas had been carried out in an inert atmosphere. When strength tests were carried out in air the strengths at temperatures over 1000°C were considerably lower than those obtained in the inert atmosphere. These results are shown in Figure 1. These results were a great disappointment to us and we rapidly instigated a program to improve the high temperature strengths of HS-110. That program was carried out by David W. Richerson.(1) The first thing he did was to look at the

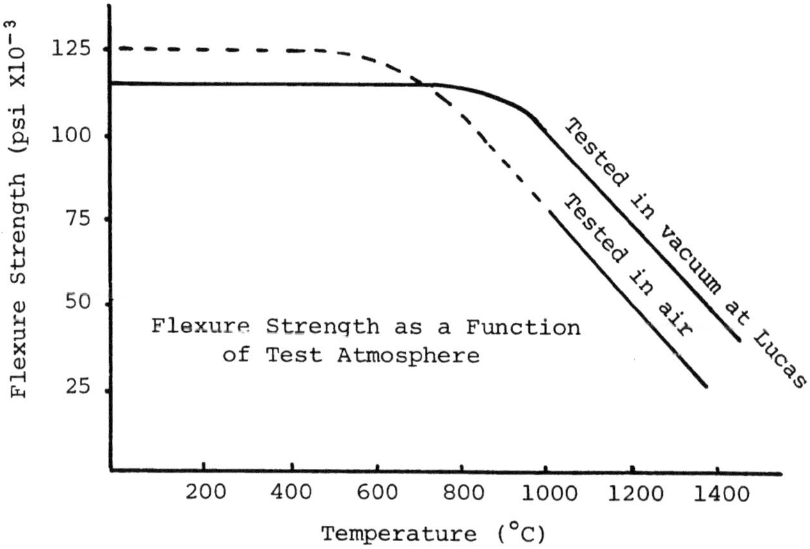

Figure 1. Flexural strength of HS-110 hot-pressed silicon nitride, 3-point bend.

composition of the glass phase in the boundaries of the silicon nitride and to see the effect of the intentional or unintentional impurities upon the melting point of those phases. Figure 2 shows his analysis of the compositions we were working with at that time. Note the low high temperature strength in areas where the bond compositions had a liquidus below 1500°C. The experimental compositions we were developing which had higher high temperature strength were in the area moving towards silica and away from the metal oxides. Following this approach, HS-130 material was developed. Figure 3 shows the chemical analysis for HS-130 compared to HS-110. As can be seen there are reductions in a number of impurities but the principal one is calcium. This resulted in considerable improvement in strength in HS-130 as is shown in Figure 4. Also the strain rate dependence of HS-130 was much less than HS-110, as shown in Figure 5. A somewhat unexpected bonus was the improvement in the oxidation rate as shown in Figure 6. As may be expected from all this the stress rupture life for HS-130 was considerably better than HS-110 as shown in Figure 7.

It is my intention to review this, not so much to show how much better HS-130 is than HS-110 but rather to show in retrospect how poor HS-110 was and to illustrate the order of magnitude improvements that were accomplished by a 1500 to 2000 ppm reduction in the calcium content.

Composition	Flexure Strength at 1375°C
1	16,500
2	20,600
3	45,000
4	62,000
5	25,500

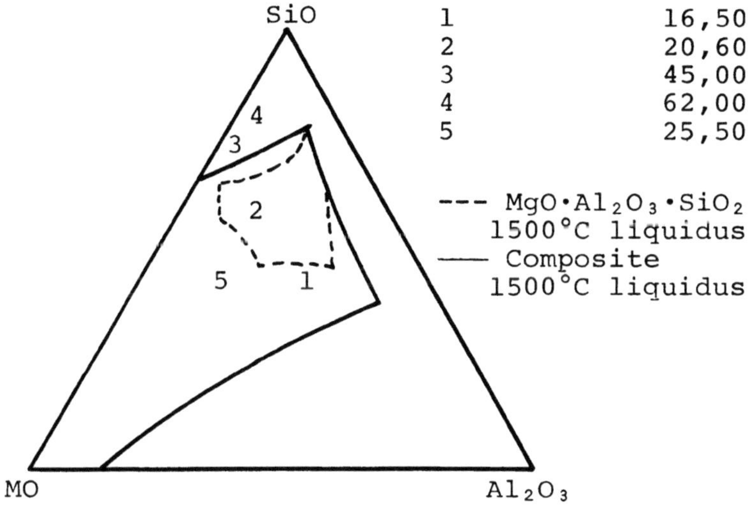

Figure 2. Correlation of High Temperature Flexural Strength with Refractoriness of calculated bond phases.

	Weight Percent	
	HS-110	HS-130
Al	0.62	0.18
Fe	0.50	0.30
Ca	0.26	0.08
Mg	0.56	0.60
Ti	0.06	0.04
Mn	0.07	0.10

Figure 3. Typical billet analysis.

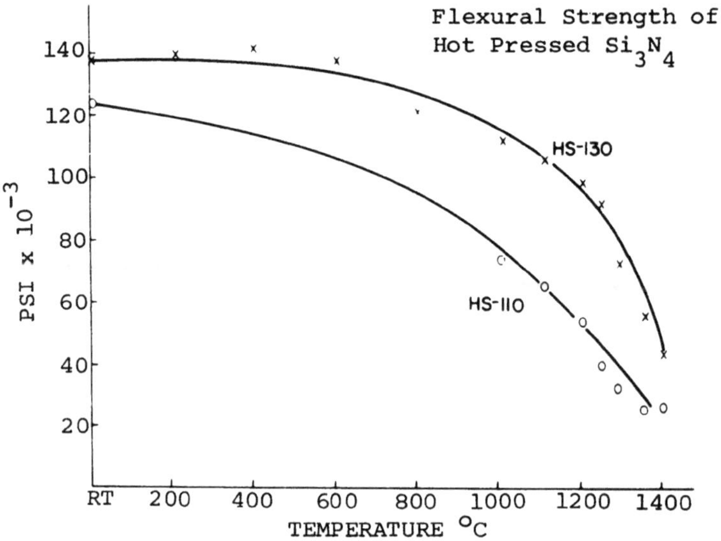

Figure 4. Flexural Strength, 3-point bend.

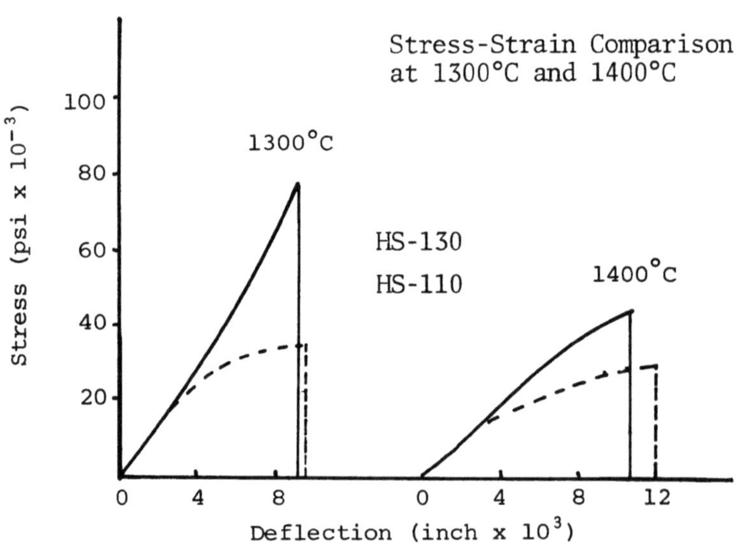

Figure 5. Flexural Strength, 3-point bend.

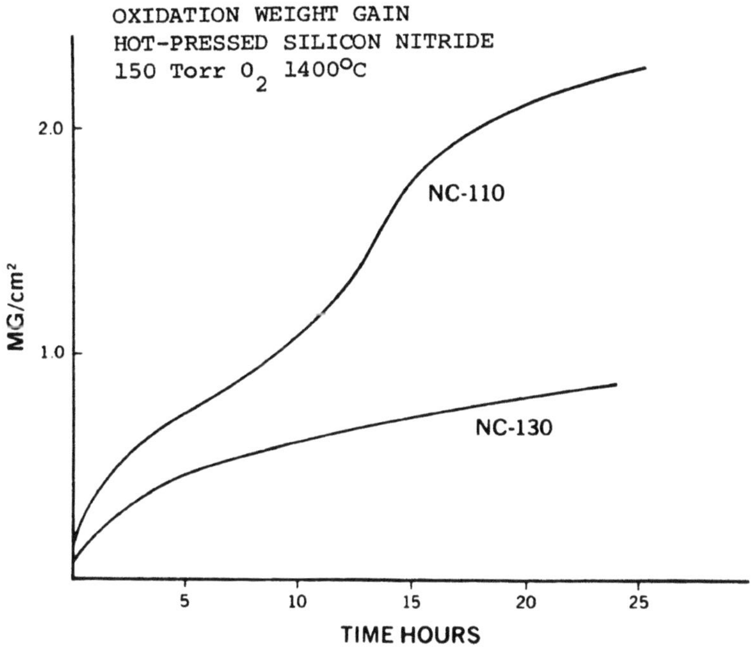

Figure 6.

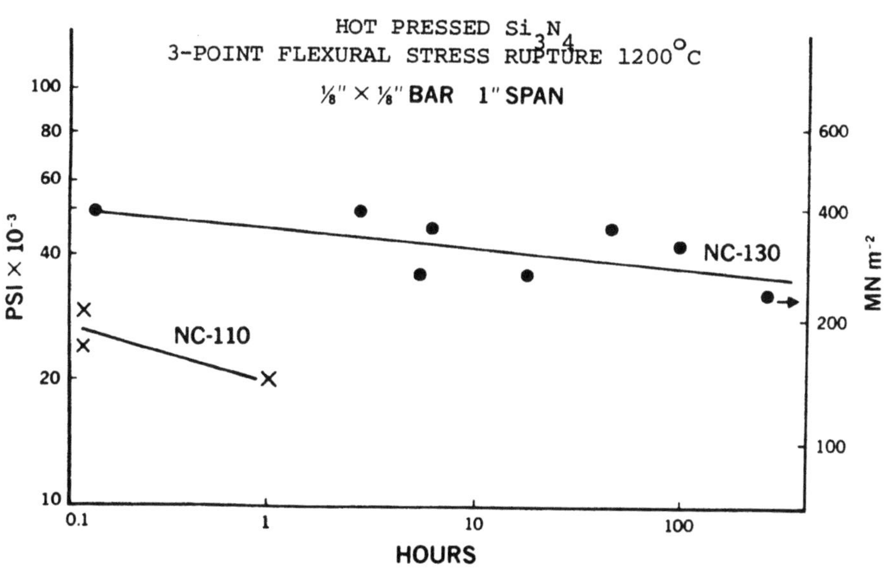

Figure 7.

Throughout the period of compositional improvements we were developing pilot plant facilities for the handling of the fine silicon nitride powder. Our objective was to develop an area with production sized equipment with hot presses capable of producing discs up to 11 inches in diameter and powder preparation facilities capable of several hundred pounds of powder. We structured the organization along research lines so that the emphasis would be upon improvements, reliability and documentation first and cost reduction and production throughput second. I think we have accomplished the former but I am not sure we accomplished the latter because it always seemed that we were never far from the need to ship because of commitments to other programs which depended upon our outputs.

One of the key elements to this pilot facility was the establishment of clean rooms for the powder processing. These were maintained at slight overpressure, there is an ante-room where the technicians change shoes and put on lab coats. The silicon nitride powder is handled only in these facilities just prior to loading into the hot pressing dies.

Development of NC-132

Westinghouse, while evaluating HS-130 material for large gas turbine vane applications under ARPA sponsorship, discovered that some anomalously low stress rupture values could be traced to metallic inclusions. As a result of this we added additional powder purification steps, carried out in the clean room. This move, although not changing the typical chemical analysis of the material, did substantially reduce inclusions as a cause of fracture and resulted not only in improving the high temperature stress rupture properties but also in giving more uniform and higher room temperature strengths. This final evolution was given a designation of NC-132. The change in the room temperature strength can be seen in Figure 8.

These processes were then used to prepare billets from which the large vanes were machined for the Westinghouse gas turbine rig testing under the ARPA program. The as-pressed vane shape is shown in Figure 9 and the vanes machined from those in Figure 10.

NC-132 Quality Assurance Proceedures and Tracibility

At this point, NC-132 was essentially established as a composition. Our emphasis was placed upon defining a fixed process including in-process checks against specifications and maintaining traceability and documentation throughout the processing of prototype orders so that we could be confident that the same route was followed in each case and that individual billets would have their complete written history in case there was a need to review any unusual results.

FLEXURAL STRENGTH
HOT PRESSED SILICON NITRIDE
4-POINT BENDING

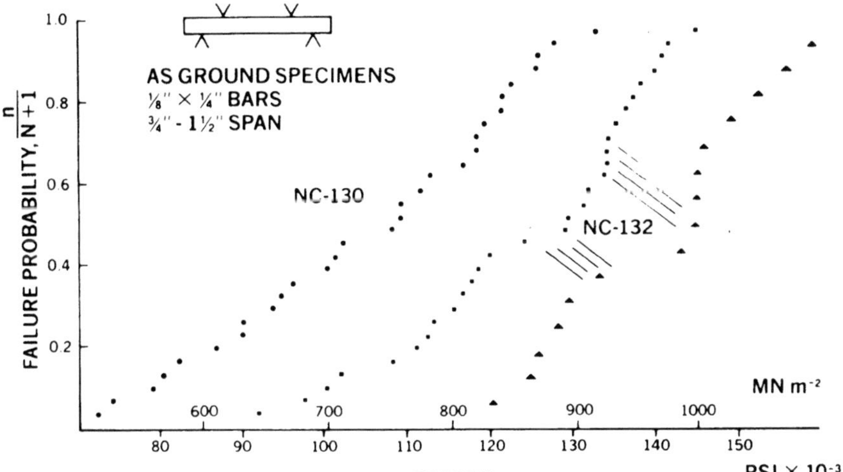

STRESS

Figure 8.

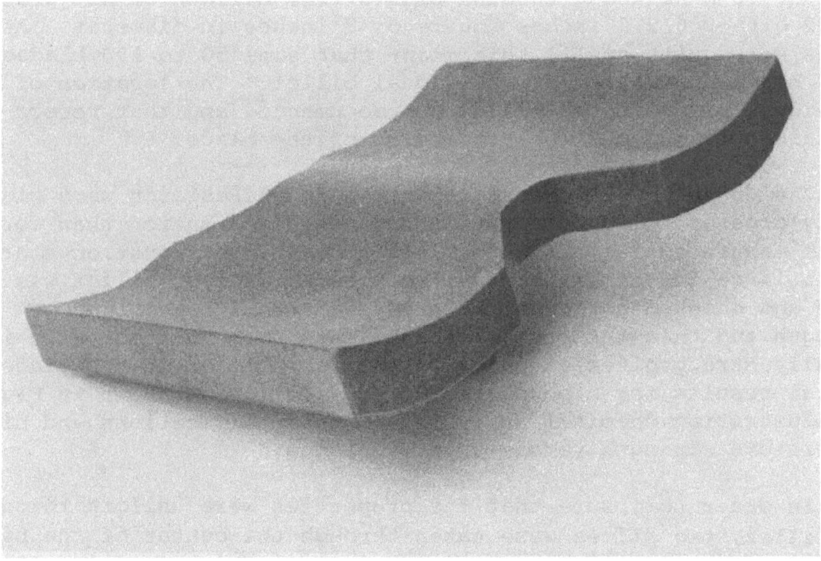

Figure 9. As pressed two vane blank, NC-132.

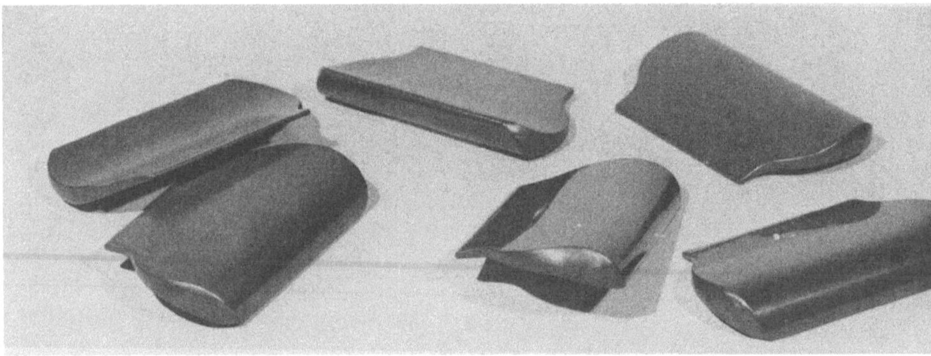

Figure 10. Ground vanes of NC-132.

These procedures and documentations were formalized under our programs with AiResearch and the Marine Turbine Programs sponsored again by ARPA but this time through Navy monitoring. The decision was made to make these turbine blades from billets of hot-pressed NC-132 either 6 x 6 inches square or 9 inches in diameter. As the blades were quite small, this meant that some 50 to 100 blades would be machined from an individual billet. The location of the individual blade in the billet was documented and that record maintained throughout the processing of the blade.

The costs involved both at Norton and AiResearch were significantly greater for documentation and characterization than for the actual manufacturing of the billet. The characterization started with a large lot of silicon nitride powder; a test billet was prepared and checked for chemical analysis and high temperature strength and then the rest of the billets from that lot were individually strength tested prior to the fabrication of the blades. Typical results for a powder lot certification is shown in Figure 11 illustrating chemical analysis versus specifications and high temperature strength versus specifications.

In order to insure that the properties were uniform throughout the billet, two slices were taken through the center of the billet. One slice sent to AiResearch for their evaluation, and one slice, after x-ray at the Norton Company, was cut into test bars along a predetermined pattern. That pattern is shown in Figure 12. Note that bars are cut in both the horizontal and vertical direction.

Hot Pressed Silicon Nitride
NC-132
Lot HN11
ground surfaces

		Specification	Actual
Flexural Strength (K309286898)		psi	psi
Room Temperature	Mean	115,000	143,505
(4-point	Mean -2 std. dev.	90,000	112,825
2500°F	Mean	40,000	63,089
(3-point)	Mean -2 std. dev.	36,000	59,807
Chemical Analysis		%	%
Al		0.5	0.18
Ca		0.05	0.006
Fe		0.75	0.006
Mg		1.00	0.43
W		3.0	2.16
O		-	2.12

Figure 11. Properties of NC-132 billet used for blade fabrication studies.

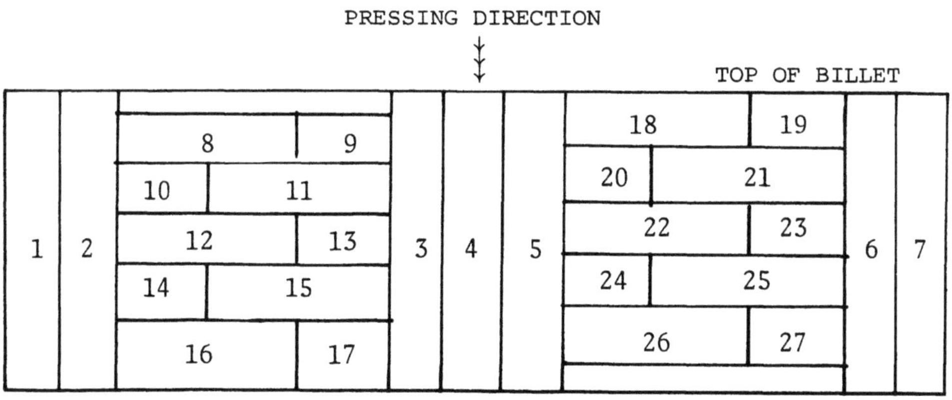

Figure 12. Test bar location in center slice of 9" x 9" x 1¼ billet of NC-132.

NC-132 tends to demonstrate a slightly lower strength when bars are cut parallel to the pressing direction, (in the vertical direction in this figure). We wanted to maintain documentation of that strength as well as the strength in the perpendicular to the hot pressing direction. Figure 13 shows the results of the two groups of test bars which were broken at the Norton Company for this particular billet. No specification was established for the strengths in the vertical direction, that is breaks one through seven, although typically they ran around 90,000 psi. The requirement for the strengths of the bars cut perpendicular to the pressing direction was 115,000 psi minimum mean strength and a mean less two standard deviations of over 90,000 psi was required. One break could be dropped from the average, if a fracture clearly originated at the edge of the bar. Note, this was done for one of the strength determinations in this group. This procedure was carried for all of the blades used in this program. It is in one sense a brute force approach to reliability, but one that appeared to have gotten the job done and which was appropriate for the needs of the program and the state of the art. These billets then went on through the various steps in blade machining shown in Figure 14 and an example of the finished blade in Figure 15.

This brings us to a basic paradox in the current state of the art of hot pressed materials. The utmost reliability and ease of characterization and proof testing occurs in large simple billets which is precisely the approach which tends to be the most costly. Pressing more directly to shape in order to reduce cost makes it more difficult to obtain a meaningful destructive evaluation. This emphasizes the critical need for viable production oriented non-destructive testing techniques.

Norton Company Billet Number HR304785943

For Record Only		For Billet Certification	
Test Bar Number	Strength (psi)	Test Bar Number	Strength (psi)
1	84,096	9	93,380*
2	82,231	10	109,064
3	70,048	13	146,134
4	126,114	14	115,723
5	111,387	17	147,499
6	109,091	19	142,400
7	104,031	20	148,227
		23	104,322
		24	142,036
		27	128,925

*not used, bevel break - machining flaw

Figure 13. Flexural strength 4-point bend NC-132 billet.

Figure 14. Steps in machining turbine blades from NC-132 billet.

Figure 15. Finished turbine blade of NC-132.

Bearings

Another area which can only briefly be cited here but which is a very demanding one for high reliability in NC-132 is that of high strength ceramic bearings. Norton, Federal-Mogul, and SKF have been working for a number of years under sponsorship from Naval Air Systems Command and more recently the Air Force Materials Lab to evaluate hot-pressed silicon nitride roller and ball bearings such as bearing components shown in Figure 16 and the assembled bearing in Figure 17. This is a requirement in which the maximum tolerable inhomogeneity or impurity may be even smaller in size than required for a turbine component and it is also an application in which cost requirements would dictate more and more emphasis on hot pressing nearer to the final configuration.

Conclusions

I think the point that I would like to make at this junction, is that the further development efforts in hot-pressed silicon nitride will need to be directed more specifically toward a specific application. Further improvements will come harder than the initial steps and need to be directed by close cooperation between the suppliers and the people who are evaluating materials for an application. In short, we need feedback as quick as possible on what areas are not adequately reliable, whether it be the minimization of impurities in existing systems or the development of new alloying systems. We need to improve the strength at elevated temperatures and to improve the retention of strength at those elevated temperatures.

Figure 16. Disassembled roller bearing, rollers and races of NC-132.

Figure 17. NC-132 assembled bearing roller.

This paper has pointed out the criticality of impurities in affecting the properties of the earlier generations of hot-pressed silicon nitride. I believe, this indicates that the current work looking at other additives such as yttria and zirconia holds the potential for even further improvements.

REFERENCES

1. D. W. Richerson, "Effect of Impurities on the High Temperature Properties of Hot-Pressed Silicon Nitride", American Ceramic Society Bulletin, Vol. 52, No. 7, (1973).

IMPORTANCE OF PHASE EQUILIBRIA ON PROCESS CONTROL OF Si_3N_4 FABRICATION

F. F. Lange

Structural Ceramics Group
Rockwell International Science Center
Thousand Oaks, California 91360

INTRODUCTION

In order to understand and develop materials, some material scientists are striving to establish interrelations between fabrication, composition (or phase equilibria), microstructure and properties. Knowledge of these interrelations leads to the anticipation of engineering problems in using the material and to the routes to material improvement. Important interrelations have been uncovered for the family of silicon nitride alloys fabricated with a densification aid. These interrelations will be the subject of this review; interrelation between composition and properties will be emphasized.

POLYPHASE NATURE OF Si_3N_4 ALLOYS

Silicon nitride powder obtained by either nitriding silicon, or other reactions involving silicon components (e.g., $3SiCl_4 + 4NH_3 \rightarrow Si_3N_4 + 12HCl$ and $3SiO_2 + 6C + 2N_2 \rightarrow Si_3N_4 + 6CO_2$) are usually mixtures of α and β crystal structures. As reviewed elsewhere, high α-Si_3N_4 powders (>70%) are required for development of the fibrous microstructure which optimizes fracture toughness and strength (Lange, 1979a).

Impurities in the raw materials, and those purposely added to catalyze the nitriding process, are usually carried through to the Si_3N_4 powder. Major cation impurities in Si_3N_4 manufactured by nitriding commercial grade silicon are Fe, Al, and Ca. Iron catalyzes the nitriding reaction, and it can be added to speed the reaction. Oxygen is a major impurity in most silicon nitride powders; its weight fraction varies between ~0.003 to ~0.03 depending on the powder source and batch-to-batch variability. If it is assumed that the oxygen associated with silicon is SiO_2, the molar content of SiO_2 can vary from 2 mole % to 12 mole %. Other impurities can be picked up during powder processing; for example, in milling the powders prior to densification, small amounts of milling media such as WC or Al_2O_3 are introduced as contaminates.

Fabrication of Si_3N_4 powder currently requires a densification aid; a variety of metal oxides and nitrides serve this purpose. These densification aids promote a liquid phase by the general reaction above the solidus temperature:

$$Si_3N_4 + SiO_2 + \text{metal oxide} \rightarrow Si_3N_4 + \text{Liquid.} \qquad (1)$$

It should be noted that this general reaction involves <u>all</u> constituents in the composite powder including the impurities not shown in Eq. (1). The equilibrium fraction of solid Si_3N_4 particles left by this reaction is densified through solution-reprecipitation mechanisms within the liquid phase. Equilibrium at the densification temperature is achieved when all of the initial α-Si_3N_4 is transformed (through solution-reprecipitation) to the more stable β-Si_3N_4 phase (Iskoe and Lange, 1976; Bowen et al., 1978).

If equilibrium was also achieved during cooling, the crystalline phase would be expected to partition from the liquid upon cooling to the eutectic temperature where the remaining liquid with the eutectic composition would crystalize into the two or more end-member components. Thus, with the exception of specific compositions in two systems discussed below, Si_3N_4 fabricated with a densification aid is a polycrystalline, <u>polyphase</u> material. As will be shown, all properties are strongly affected by the type, chemistry and amount of the secondary phases. Optimization of properties thus relies on optimizing secondary phases through compositional control.

If it were again assumed that equilibrium was achieved during cooling, the chemistry and amount of the secondary phases could be predicted with knowledge of the components in the starting powder and knowledge of the phase relations within the system of interest. During the last several years phase relations have been established for many important Si_3N_4 alloy systems. The subsolidus phase relations of four of these systems, viz., Si-Mg-O-N (Lange, 1978), Si-Al-O-N (Gauckler et al., 1975) and Si-Y-O-N (Lange et al., 1977) and Si-Zr-O-N (Lange, 1979b) systems are shown in Fig. 1. Thus, knowledge of the three constituents Si_3N_4, SiO_2 and metal oxide (or metal nitride), and the phase relations

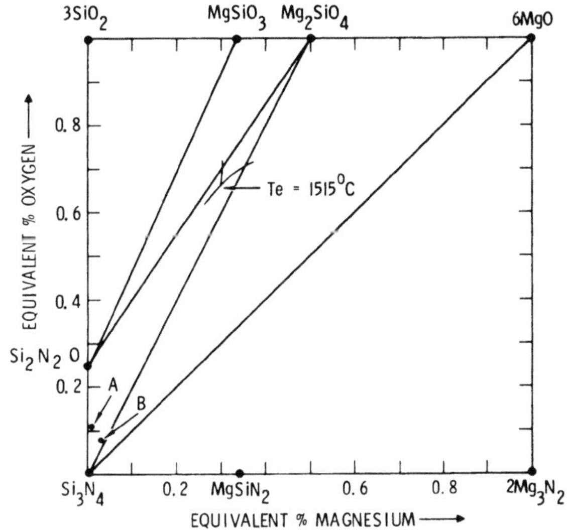
3SiO2
MgSiO3
Mg2SiO4
6MgO
0.8
0.6
0.4
0.2
Si2N2O
A
B
Te = 1515°C
EQUIVALENT % OXYGEN
Si3N4
0.2
MgSiN2
0.6
0.8
2Mg3N2
EQUIVALENT % MAGNESIUM

Fig. 1a

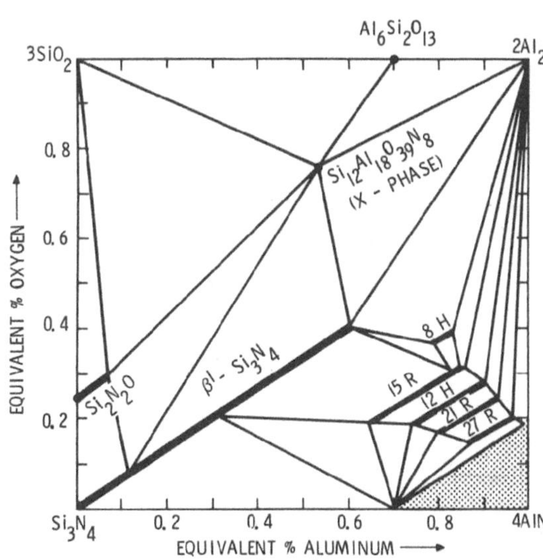
Al6Si2O13
3SiO2
2Al2O3
0.8
0.6
0.4
0.2
Si12Al18O39N8
(X - PHASE)
8 H
15 R
12 H
21 R
27 R
β' - Si3N4
Si2N2O
EQUIVALENT % OXYGEN
Si3N4
0.2
0.4
0.6
0.8
4AlN
EQUIVALENT % ALUMINUM

Fig. 1b

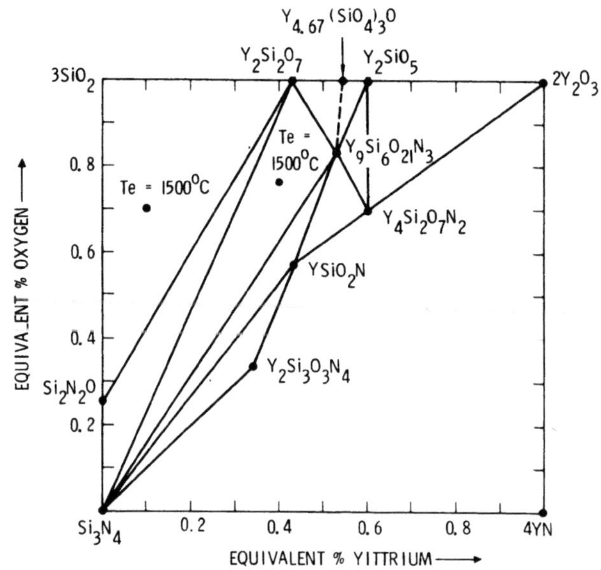

Fig. 1c

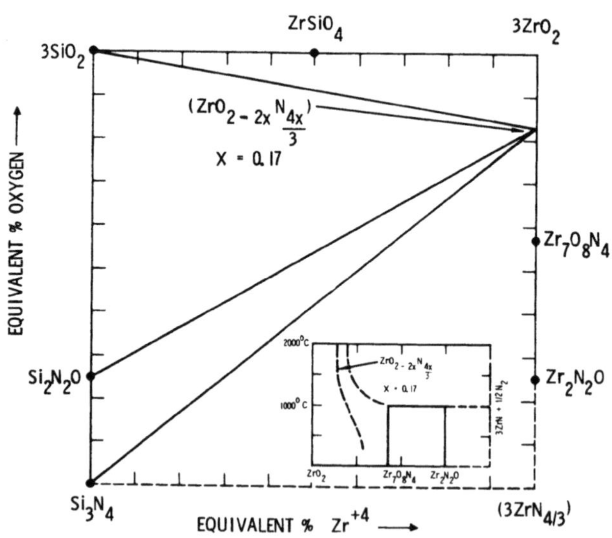

Fig. 1d

Figure 1. Sub-solidus phase relations in the (a) Si-Mg-O-N, (b) Si-Al-O-N, (c) Si-Y-O-N and (d) Si-Zr-O-N systems.

shown in Fig. 1, is required if the equilibrium phases (chemistry and content) in polyphase Si_3N_4 alloys are to be predicted.*

It should be noted in Fig. 1b that an extensive solid-solution range exists in the Si-Al-O-N system which has a general composition $Si_{6-x}Al_xO_xN_{8-x}$ ($0 < x < 4.2$). Theoretically, a composition formulated in this system with starting constituents that satisfy this formulation would be single phase materials. A similar solid-solution range also exists in the Si-Be-O-N system ($Si_{6-x}Be_xO_{2x}N_{8-2x}$) (Husby and Petzow, 1975). Thus, specific compositions that satisfy the above formulations are the two theoretical exceptions to the polyphase nature of Si_3N_4 alloys.

In practice, equilibrium has yet to be achieved during cooling. Namely, all materials examined to date contain a glassy phase between the crystalline grains and/or triple points. Figure 2a illustrates the glassy "grain boundary" phase in Si_3N_4 alloy densified with MgO (Clarke, 1978). In this system the Mg-silicate phase compatible with Si_3N_4, viz., Mg_2SiO_4, is never observed to be crystalline for compositions close to the Si_3N_4 end-member. But as the material's composition is shifted toward the Si_3N_4-Si_2N_2O tie line (MgO/SiO_2 molar ratios <1.5), Si_2N_2O is observed as a crystalline phase by both x-ray diffraction and high resolution electron microscopy.

Crystalline and glassy secondary phases are also known to co-exist in other systems. Figure 2b illustrates the $Y_2Si_3O_3N_4$ phase located at the triple point between three Si_3N_4 grains (Clarke) for a composition close to the Si_3N_4-$Y_2Si_3O_3N_4$ tie line (see Fig. 1c). The location of this phase is consistent with the view that it had crystallized from the liquid during cooling. A glassy "grain boundary" phase also exists in this material between the Si_3N_4-$Y_2Si_3O_3N_4$ and Si_3N_4-Si_3N_4 grains. These examples illustrate a common feature of Si_3N_4 alloys, viz., the co-existence of both equilibrium and non-equilibrium secondary phases.

The exact composition of the glassy phases in Si_3N_4 alloys has yet to be determined. The hypothesis most consistent with facts is that the glassy phase is similar to the eutectic composition within the compatibility triangle in which the material is fabricated (Lange, 1979c). This view is consistent with the fact that the last liquid to solidify has the eutectic composition. With this hypothesis, the volume fraction of the glassy phase can be estimated by using the lever rule of compositional phase diagrams.

*Precluding the strong possibility of mass loss through volitalization.

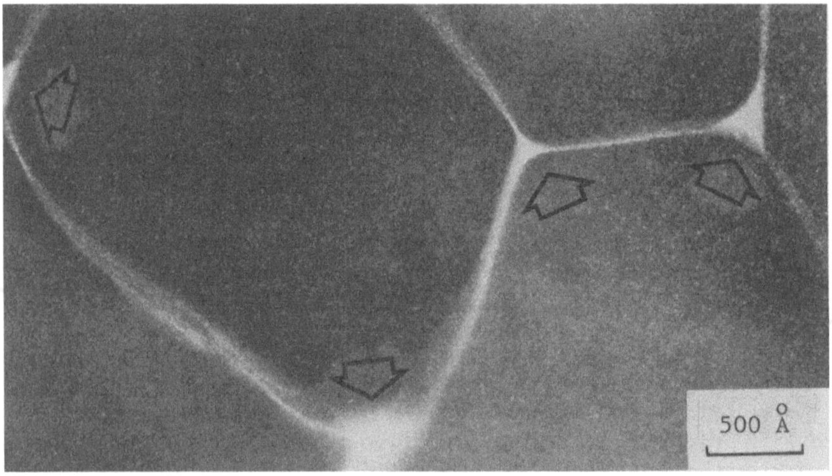

Fig. 2a

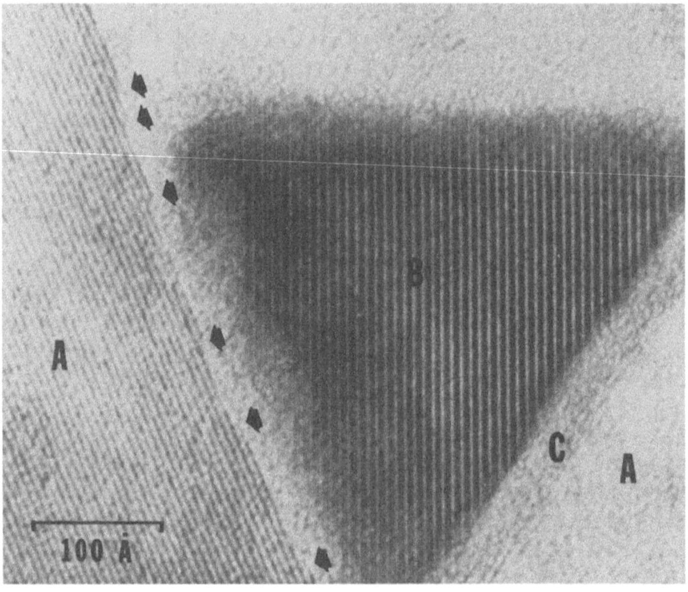

Fig. 2b

Figure 2. Glassy "grain boundary" phase in Si_3N_4MgO alloys, (b) $Y_2Si_3O_3N_4$ crystalline phase, surrounded by glassy phase at Si_3N_4 triple point.

The effect of impurities or additional metal oxide (or nitride) densification aids on the type, chemistry and amount of secondary phases must also be viewed in terms of phase equilibria. Although these additional constituents become unwieldy to represent in compositional space, each impurity or additional aid will influence the phase relations and eutectic compositions/temperature. For example, if CaO is an impurity in Si_3N_4/MgO alloys, the material's composition must be represented in the Si-Mg-Ca-O-N system. Experiments have shown that CaO addition lowers the melting point of eutectics in the Si_3N_4-Si_2N_2O-Mg_2SiO_4 system. Thus, CaO in this system will not only lower the eutectic temperature but will also increase the volume fraction of the glassy phase by moving the composition closer to the eutectic composition. Similar results are expected for all impurities that do not form a solid-solution with one of the crystalline phases.

AVOIDING UNSTABLE SECONDARY PHASES (Lange et al, 1977; 1979b)

In some systems the secondary phase is very prone to oxidize at low temperatures. Oxidation can lead to large molar volume changes, which in turn cause stresses that lead to cracking and, in some cases, the total disintergration of the material. An example of this is the $Y_2Si_3O_3N_4$ phase. The second phase, shown in Fig. 2b, increases its volume by 30% upon oxidation to $Y_2Si_2O_7$ + SiO_2. Material containing this phase will disintergrate upon oxidation at temperatures <1000°C. Many other secondary phases are known to cause the same problem; these are listed in Table I.

The phenomena causing material degradation due to a molar volume change are currently under investigation. Work on Si_3N_4/CeO_2 and Si_3N_4/Y_2O_3 materials show that at higher temperatures the oxidation products extrude to the surface at triple points and between Si_3N_4 grains, which minimizes the stresses developed. At lower temperatures, extrusion does not take place. Thus, at lower temperatures, the increased molar volume of the oxidation products cause compressive stresses that first increase the flexural strength and then cause surface spalling and large, through cracks as oxidation proceeds. These observations have been documented for Si_3N_4/ZrO_2 alloys that contained different volume fractions of the $ZrO_{2-2x}N_{\frac{4x}{3}}$ ($x \sim 0.17$) phase which oxidizes to monoclinic ZrO_2 (molar volume increase of 4-5%) at temperatures as low as 500°C (Lange, 1979b). The degradation phenomena are related to the volume fraction of the unstable phase, its oxidation kinetics and its distribution and to the deformation characteristics of the oxide products. Although some materials do not appear to degrade over short periods with very small fractions of an unstable phase, long term degradation is possible.

Table I. Molar Volume Changes of Unstable Secondary Phases Compatible with Si_3N_4

Secondary Phase	Oxidation Product	% Volume Change
$Y_2Si_3O_3N_4$	$Y_2Si_2O_7 + SiO_2$	+30
$YSiO_2N$	$0.5Y_2Si_2O_7$	+12
$Y_9(SiO_{3.5}N_{.5})_6$	$1.5Y_{4.67}(SiO_4)_3O$ + $0.5Y_4Si_3O_{12}$	+ 5
$Y_5(SiO_4)_3N$	$0.75Y_{4.67}(SiO_4)_3O$ + $0.75Y_2SiO_5$	+ 4
$Ce_5(SiO_4)_3N$	$5CeO_2 + 3SiO_2$	+ 8
$CeSiO_2N$	$CeO_2 + SiO_2$	+14
$Ce_2Si_2O_7$	$2CeO_2 + 2SiO_2$	+ 7
$ZrO_{2-2x}N_{4x}$ $(x = .2)$	ZrO_2 (monoclinic)	+ 5

Table 1 should be avoided by shifting composition of other, stable phase fields.

RELATIONS BETWEEN COMPOSITION AND PROPERTIES - A SYSTEMATIC STUDY

Materials fabricated in the Si-Mg-O-N system were chosen for systematic study since current commercial materials are hot-pressed in this system. Three series of materials each containing 0.91, 0.83 and 0.75 mole fraction of Si_3N_4 and different molar ratios of MgO/SiO_2 were fabricated by hot-pressing. Strength, creep and oxidation resistance were examined as functions of composition for materials in each series. The results of this study showed these properties could be expressly related to phase equilibria concepts that could be used in directing material improvements.

Strength (Lange, 1978; 1979c)

The results of flexural strength testing are shown in Fig. 3a for the series containing 0.83 mole fraction of Si_3N_4. Results for the other two series were nearly identical. At room temperature, strength did not appear to significantly change with

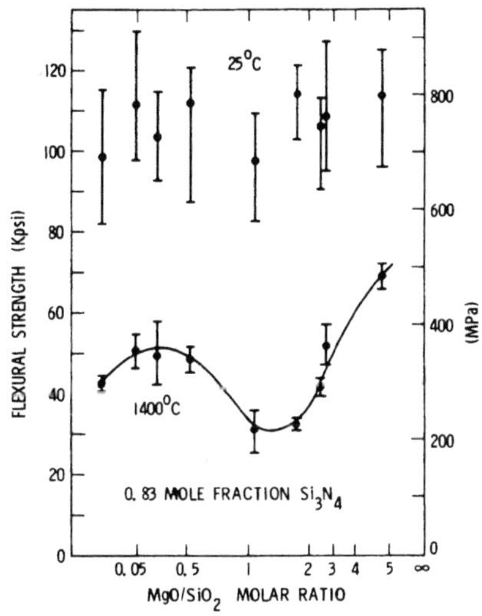

Fig. 3a

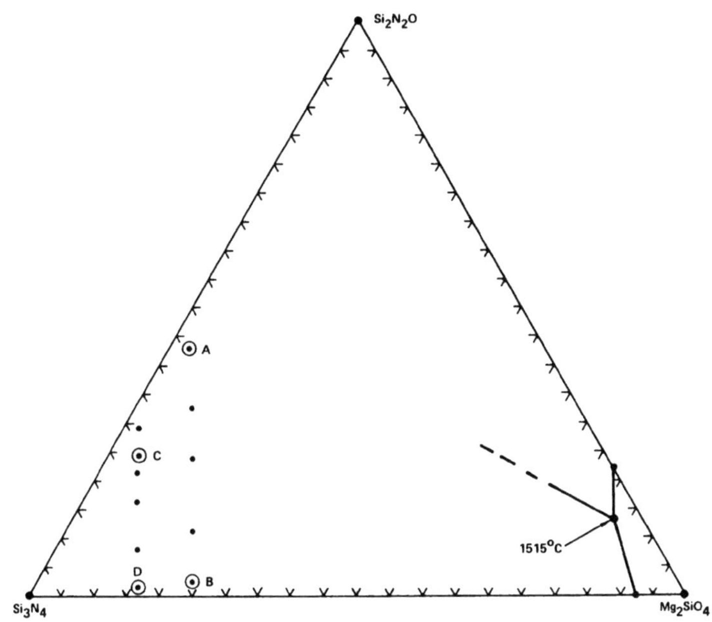

Fig. 3b

Figure 3. (a) Strength results at 25°C and 1400°C for 0.83 Si_3N_4 mole fraction series of materials. (b) Compositions of materials A, B, C and D plotted on molar phase diagram showing eutectics.

the MgO/SiO_2 ratio. But at 1400°C, a strength minimum occurred at a MgO/SiO_2 ratio between 1 and 2. Subsequent eutectic studies showed that the ternary eutectic in the Si_3N_4-Si_2N_2-Mg_2SiO_4 compatibility triangle occurs at a MgO/SiO_2 molar ratio of 1.6. The close correlation between the strength minima and the eutectic composition suggests the following analysis.

All the materials examined contain a glassy "grain boundary" phase. If it is assumed that glass has a composition close to that of the eutectic, its volume content will maximize as the MgO/SiO_2 molar ratio approaches 1.6. As the eutectic temperature is approached or exceeded, the glassy phase becomes viscous. Since the high temperature strength will be proportional to the volume fraction of the viscous phase, all compositions should exhibit strength degration. Strength minimums should exist for compositions containing the maximum amount of viscous phase., viz., within each series, compositions where the MgO/SiO_2 ratio in ≈ 1.6. Thus, maximum strengths should be obtained for compositions furthest from the eutectic composition.

The effect of impurities on high temperature strength can be viewed in a similar manner. Impurities such as CaO that are known to reside within the glassy phase will decrease the eutectic temperature and increase the volume content of the viscous phase. Since the viscosity of the glass will be related to the eutectic temperature, the viscosity of the glass containing such impurities will be lower than that of the glass without the impurities. The effect of increasing the CaO content is not to further decrease the viscosity, but to move the composition closer to the eutectic within the compositional space that includes the impurity as a consitituent. Thus, increasing the impurity content will increase the volume fraction of the viscous phase which will result in decreased high-temperature strength. It is obvious that impurities of this nature should be completely avoided; if this is impossible within cost and fabrication constraints, their level should be kept at a minimum.

Creep (Lange, Davis and Clarke, 1979)

The critical compositions A, B, C, and D in Fig. 3b, all fabricated in the same compatibility triangle, were subjects of an extensive compressive creep study at 1400°C in air. The estimated volume fraction of the glassy phase for each composition is shown in the table on Fig. 3b, the estimation used the level rule of phase equilibria and assumed the glass phase had the eutectic composition. As detailed elsewhere, steady-state creep rates were obtained as a function of stress. Evidence for cavitational creep was obained through precise density measurements ($\Delta\rho = \pm 0.01\%$) and transmission electron microscopy.

As shown in Fig. 4a, compositions with the larger amounts of glass exhibited higher strain rates at comparable stresses. Of major importance is the exponential stress dependence of creep rate. The stress exponent for compositions containing the smallest amount of glass (A and C) was ≈ 1, indicative of diffusional creep, whereas compositions closer to the eutectic (B and D) had stress exponents >1 (~2).* As shown in Fig. 4b, the composition with the smaller glass content (A) did not cavitate, whereas the composition with the large glass content (B) exhibited extensive cavitation.

These results illustrate that diffusional creep and cavitational creep are concurrent deformation mechanisms in Si_3N_4 alloys. Cavitational creep will dominate when the volume fraction of the viscous phase is large, and diffusional creep will dominate for smaller volume fractions. Thus, composition not only affects the creep rate, but also the dominant creep mechanism.

Other findings in this study showed that primary creep was caused by a viscoelastic phenomenon which also resulted in recoverable strain upon unloading. Also, oxidation leads to improved creep resistance as discussed in the latter part of the next section.

Oxidation Resistance (Clarke and Lange, 1979)

The effect of composition on the oxidation resistance of the same series discussed above is shown pictorially in Fig. 5 (only two series shown) for specimens exposed to air at 1400°C for 288 hrs. Specimens which appear to be white are covered with a thick oxide scale, whereas black specimens are relatively resistant to oxidation. The parabolic rate constant decreases by ~4 orders of magnitude as the composition is shifted from the Si_3N_4-MgO tie line to the Si_3N_4-Si_2N_2O tie line. Composition thus has a strong affect on oxidation kinetics.

Experiments have shown that the Mg-containing glassy phase has a strong influence on the oxidation kinetics. As observed by many investigators, Mg (and impurities such as Ca and Fe) diffuses to the surface during oxidation and become heavily concentrated within the oxide scale. This outward diffusion of Mg is the result of the Mg-containing secondary phase not being compatible with the SiO_2 formed on the surface by the oxidation of Si_3N_4. A

*True steady-state creep rates were never observed for materials B and D. Theoretical analysis of cavitational creep suggests that no single exponen- tial behavior exists between creep rate and stress.

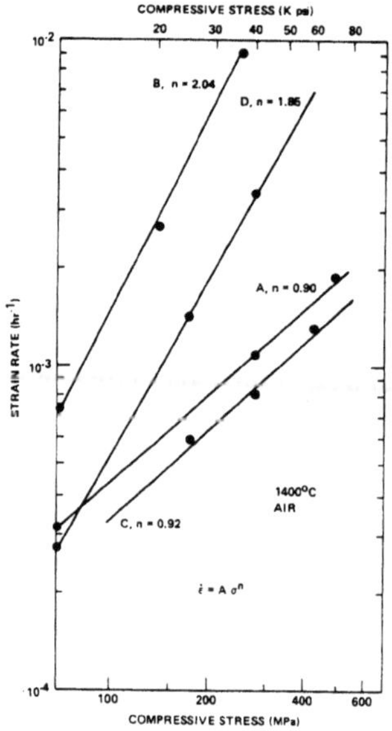

Fig. 4a

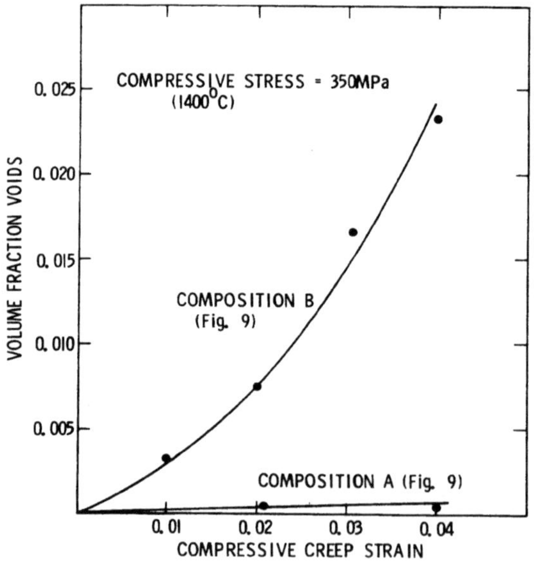

Fig. 4b

Figure 4. Compressive creep rate vs. stress (log-log) plots for materials A, B, C and D (see Fig. 3b). (b) Density changes of materials A and B vs. creep strain.

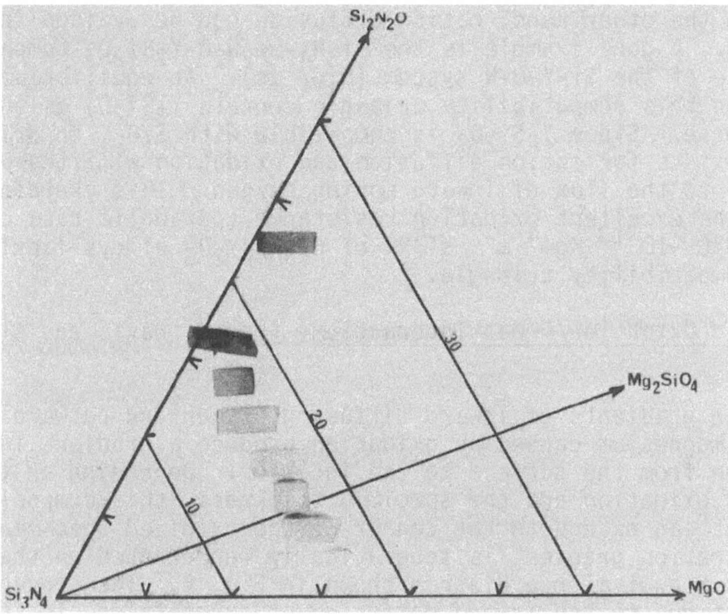

Figure 5. Appearance of two series of Si_3N_4/MgO specimen oxidized at 1400°C for 288 hrs.

diffusion couple is set up between the bulk Mg-phase and the surface SiO_2, and cation diffusion (in this case, the diffusion of Mg) then equilibrates the diffusion couple.

It can also be shown that to maintain charge balance, the outward diffusion of $\frac{2}{3}$ N^{-3}. The oxygen left behind by the outward diffusion Mg^{+2} reacts with Si_3N_4 to form Si_2N_2O (or SiO_2). Thus, the outward diffusing of Mg^{+2} is equivalent to the inward diffusion of oxygen. The oxidation kinetics of these alloys will be proportional to the flux of both outward moving magnesium and inward moving oxygen. Since the flux of outward moving magnesium is proportional to its initial concentration, the oxidation kinetics will be directly proportional to the MgO content, viz., increasing with increasing MgO/SiO_2 molar ratio within a given series.

As indicated above, this outward diffusion of the cations, which strongly affects oxidation kinetics, is caused by the incompatibility of the secondary phase with the SiO_2 formed on the surface. If the secondary phase were compatible, cation diffusion would not occur. In the Si-Mg-O-N system, $MgSiO_3$ is compatible with SiO_2, but the phase relations shown in Fig. 1a preclude $MgSiO_3$ as a secondary phase in Si_3N_4/MgO alloys. Thus, cation diffusion will always occur during the oxidation of Si_3N_4/MgO alloys.

On the other hand, cation diffusion can be avoided in other systems. A good example is the Si_3N_4-Si_2N_2O-$Y_2Si_2O_7$ compatibility triangle of the Si-Y-O-N system (Fig. 1c). At equilibrium, materials in this compatibility triangle contain $Y_2Si_2O_7$ as the secondary phase. Since $Y_2Si_2O_7$ is compatible with SiO_2, no driving force exists for cation diffusion and oxidation will therefore be limited to the flux of inward moving oxygen. This explains, in part, the excellent oxidation resistance (parabolic rate constant at 1400°C ~10^{-13} Kgm^4 m^{-2} s^{-1}*) of Si_3N_4/Y_2O_3 alloys fabricated in this compatibility triangle.

Combined Oxidation/Creep Interactions (Lange, Davis and Clarke, 1979)

The gradients of inward diffusing oxygen and outward diffusing magnesium caused by oxidation produce a gradient in composition from the surface to the interior. Depending on the extent of oxidation and the specimen thickness, this compositional gradient can extend to the center of the oxidized specimen. This concentration gradient is schematically represented on the Si-Mg-O-N equivalence diagram shown in Fig. 6. The concentration gradient due to the inward diffusion of oxygen alone is shown by the broken line A (surface) to X (interior limit of oxygen diffusion). The combined effect of magnesium and oxygen diffusion is shown by the solid line B (surface) to X. Magnesium is concentrated within the surface scale and is thus depleted from the

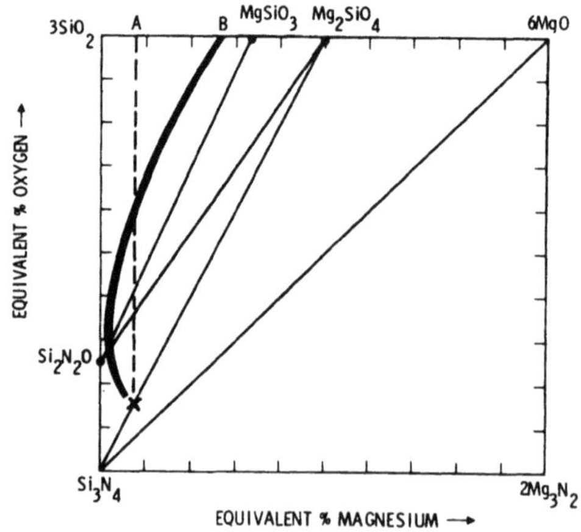

Figure 6. Compositional gradients (surface A, B) to interior (x) shown on equivalence diagram.

*Over one order of magnitude better than the best rate observed for the Si_3N_4/MgO alloys.

interior. The net result of this concentration gradient is that the average composition of the bulk shifts towards the Si_3N_4-Si_2N_2O tie line as oxidation proceeds.

This shift in composition moves the integrated composition away from the eutectic and thus alters properties which strongly depend on the volume fraction of the glass phase; the glass phase is a fugitive of oxidation.

As illustrated in previous sections, both strength and creep resistance strongly depend on the volume content of the glass phase. As shown in Fig. 4a, a significant decrease in the glass content will not only improve the creep resistance, but may also change the dominant mechanism from cavitational to diffusional controlled creep.

Results of specimens oxidized for 100 hr at 1400°C in air and then tested under compressional creep are shown in Fig. 7 along with results from unoxidized specimens. As can be seen, creep resistance was improved for all compositions tested. Of significant interest is the dramatic change in the mechanism for compositions B and D from that of cavitational creep to diffusional creep. Thus, with respect to creep resistance, compositions that exhibit relatively poor resistance can be significantly improved merely by an oxidation treatment.

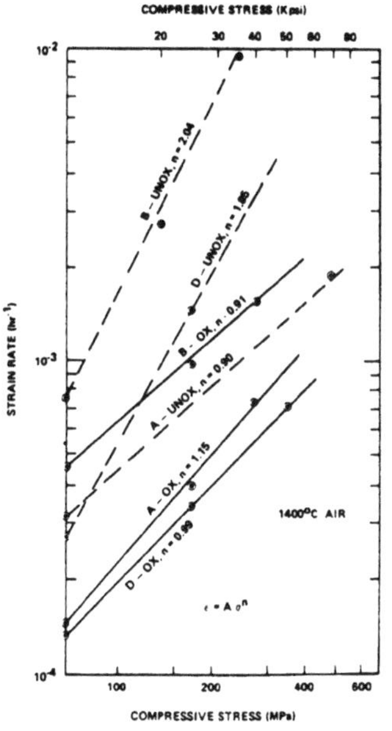

Figure 7. Compressive creep rate vs. stress (log-log) plots for unoxidized and oxidized (100 hr, 1400°C) Si_3N_4/MgO alloys.

CONCLUDING REMARKS

The above results and discussions have shown strong relations between composition and properties. The two major problems that currently exist in maintaining compositional uniformity are control of the powder constituents (SiO_2 content and impurity content) and variable mass losses that occur during during fabrication. When a small amount of the densification aid is used, slight changes in the oxygen content (viz., SiO_2 content) can be critical in determining the location of the composition relative to the eutectic composition and the degradation that will occur at high temperatures. Compositional control is, therefore, essential if product uniformity is to be maintained.

Knowledge of phase relations and eutectic compositions and temperatures is critical for directing material improvements. From a stability standpoint, phase relations must be established so that the chemistry of the equilibrium, secondary phase can be determined. The oxidation resistance of these secondary phases must be established to determine if stability problems might arise during low temperature oxidation.

From the standpoint of high temperature mechanical properties, eutectic temperatures and compositions must be established. High eutectic temperatures are desirable for increasing the viscosity of the residual eutectic glass phase, the presence of which may be unavoidable. Eutectic compositions are required for selecting compositions that minimize the glass phase. Both of these directions appear to be in direct conflict with densification kinetics which indicate that large amounts of the liquid phase are desirable for fabrication. Solutions to this conflict must be sought through fabrication science and innovation.

From the standpoint of oxidation resistance, compositions that will exhibit optimum oxidation resistance should contain a secondary phase that is compatible with SiO_2. Phase relation studies will have required to establish this compatibility question.

In closing, it should be noted that these directions for material improvement are already being put to practice. First, many investigators have already recognized that the single phase Si_3N_4/Al_2O_3 and Si_3N_4/BeO_2 alloys would be desirable in eliminating the polyphase nature of Si_3N_4. Second, compositions in the Si_3N_4-Si_2N_2O-$Y_2Si_2O_7$ system, which contain secondary phases compatible with SiO_2 and have a high eutectic temperature (~1640°C), are also being sought. These efforts have already led to improved Si_3N_4 alloys which should be put to commercial practice.

REFERENCES

Bowen, L. J., Weston, R. J., Carruthers, T. G. and Brook, R. J., 1972, Hot-Pressing and the α-β phase transformation in Si_3N_4, J. Mat. Sci. 13:341.

Clark, D. R., 1979, On the detection of thin intergranular films by electron microscopy, Ultramicroscopy, 4:33.

Clarke, D. R. unpublished TEM micrograph.

Clarke, D. R. and Lange, F. F., 1978, Oxidation of Si_3N_4-SiO_2-MgO System, to be published.

Gauckler, L. J., Lukas, H. L. and Petzow, G., 1975, Determination of the phase diagram Si_3N_4-AlN-Al_2O_3-SiO_2, J. Am. Ceram. Soc. 58:346.

Huseby, I. G., Lukas, H. L., Petzow, G., 1975, Phase equilibria in the system Si_3N_4-SiO_2-BeO-Be_3N_2, J. Am. Ceram. Soc. 58:377.

Iskoe, J. L. and Lange, F. F., 1977, Development of microstructure and mechanical properties during hot-poressing of Si_3N_4, in "Ceramic Microstructure '76," ed. by R. M. Fulrath and J. A. Pask, Westview Press.

Lange, F. F., Davis, B. I., and Clarke, D. R., 1979, Compressive creep of Si_3N_4/MgO alloys: I. effect of composition, II, source of viscoelastic effect; III. effects of oxidation induced compositional change, to be published (Rockwell Science Center Rept. SC5099.2AR).

Lange, F. F., 1978, Phase relations in the system Si_3N_4-SiO_2-MgO and their interrelations with strength and oxidation, J. Am. Ceram. Soc. 61:53.

Lange, F. F., 1979a, Silicon nitride alloy systems: fabrication, microstructure and properties, Met. Rev. Inter., to be published (Rockwell Science Center Rept. #SC-PP-79-20).

Lange, F. F., 1979b, Compressive surface stresses developed in ceramics by an oxidation-induced phase change, J. Am. Ceram. Soc. (in press).

Lange, F. F., 1979c, Eutectic studies in the Si_3N_4-Si_2N_2O-Mg_2SiO_4-MgO system, J. Am. Ceram. Soc. (in press).

Lange, F. F. Singhal, S. C., and Kuznicki, R. C., 1977, Phase relations and stability studies in the Si_3N_4-SiO_2-Y_2O_3 pseudoternary system, J. Am. Ceram. Soc. 60:249.

INJECTION MOLDING OF CERAMICS

T.J. Whalen and C.F. Johnson

Research Staff
Ford Motor Company
Dearborn, Michigan

ABSTRACT

An injection molding process used to fabricate ceramic flow path components for a turbine engine is described. Both Si_3N_4 and SiC are amenable to mold forming and subsequent reaction sintering. Fabrication equipment and technique are presented along with a discussion of major accomplishments and problems encountered during molding technology development at Ford.

INTRODUCTION

The use of injection molding or transfer molding in the fabrication of ceramic articles goes back many years. Karl Schwartzwalder[1] described an injection molding method for producing ceramic articles. Compositions wholly lacking in clay, and thus, lacking in plastic forming characteristics, required the use of plastics to provide the proper molding properties. The development of resinous binders for hot-molding ceramics began about 1932 and large-scale ceramic production was achieved by 1937.

More recently, injection molding has again been called upon to produce complex structural shapes for use in high temperature gas turbine engines. Not only are complex shapes required, but also high strengths and precise, reproducible dimensions are needed. Two types of binder systems are employed for molding turbine parts--a thermoplastic binder[2] used for forming silicon pieces as a first step in making Si_3N_4 parts, and a thermosetting system[3] which has been used primarily for forming silicon carbide-silicon materials. It is the purpose of this paper to discuss these injection-molding processes and to identify potential improvements in reliability by refinements of the injection-molding process.

INJECTION MOLDING OF SILICON

The production of Si_3N_4 is a unique process, by which silicon is converted to Si_3N_4 without dimensional change. The green density of the molded silicon preform completely determines the final density of the reaction bonded Si_3N_4. Although this process is advantageous for making complex shapes to high dimensional exactness, high density material cannot be obtained, since no shrinkage or densification occurs during nitriding.

The molding composition includes metallic silicon powder of a predetermined particle size, Fe_2O_3 as a nitriding aid and a thermoplastic organic carrier to provide flowability during the injection molding process. Mangels[2] has summarized the requirements of volume percent solids, viscosity and silicon particle size distribution needed for an adequate molding composition. To produce a 2.7 g/cc Si_3N_4 material, a composition of 73.5 volume percent of solids ($Si + Fe_2O_3$) and a batch viscosity low enough to be injection molded, are required. It was shown by Mangels[2], that a minimum viscosity is obtained by a very broad range of particle sizes. A silicon powder milling time of 140 hours with a particle size distribution as seen in Figure 1, was sufficient to yield a flow of 10 inches in a spiral flow mold, which is indicative of adequate flow for turbine part fabrication by injection molding. Further discussion of injection molded silicon will imply the use of a standard composition consisting of a thermoplastic organic binder with 140 hour milled silicon and 73.5 volume percent solids.

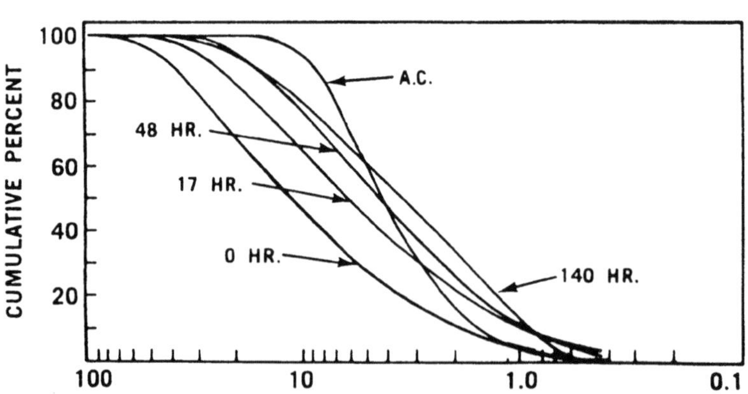

Figure 1. Particle size distribution curves for silicon dry-ball-milled for several times.

For the purpose of this discussion, we have chosen a turbine rotor blade ring shown in Figure 2, to demonstrate how improvements in the injection molding process have improved the quality of this part. Three types of flaws have typically been observed in this part: cracks, voids and inclusions. Cracks in the base of the blades are seen in Figure 3, voids in Figure 4 and inclusions in Figure 5.

Figure 2. Photograph of turbine rotor blade ring.

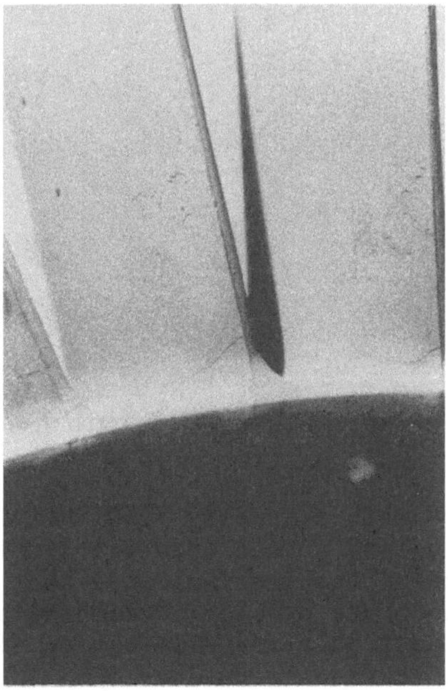

Figure 3. Photograph of cracks at base of turbine rotor blades.

Fig. 4. Photograph of voids in turbine rotor blade ring.

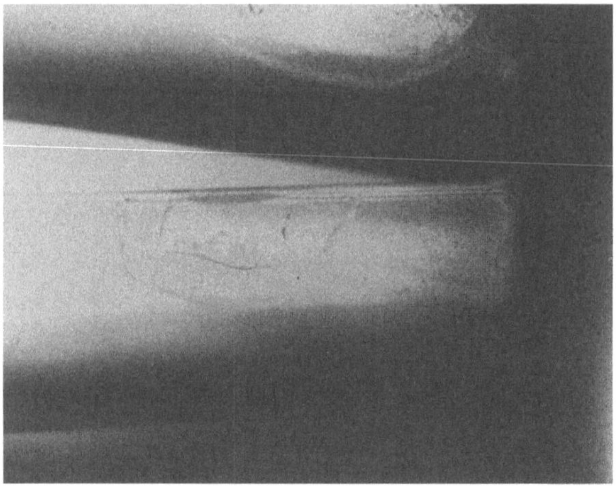

Fig. 5. Photograph of inclusion of turbine rotor blade ring.

Early experience with a Newbury V130T vertical clamp, parting line injection machine and a Reed Prentice 450 TC horizontal clamp, through-platen injection machine indicated that a ram type injection system with highly accurate hydraulic clamping and through-platen injection, was most desirable for elimination of molding induced blade cracks. Specifications were developed for an improved molder and Ford, in conjunction with a machine vendor, designed and built such a molding machine which is seen in Figure 6. This significantly reduced blade cracking.

Fig. 6. Photograph of injection molder.

Temperature control is an important factor in molding crack-free parts with the highly-loaded thermoplastic material. Shrinkage in the die cavity from changes in temperature can impart molding stresses of sufficient magnitude to produce cracks such as those seen in Figure 3. Thermocouples were included in the tool and injection barrel to monitor and control die, material and component temperature. Temperature controls were independent in each half of the die permitting independent temperature control in both stationary and moveable die halves. Figure 7 shows the temperature reproducibility for die, component and material during the molding process. Die actuation was made dependent upon the part temperature level, thus, eliminating reliance upon process time and acceptance of a random temperature. These changes also reduced cracking during molding.

Another factor which caused cracking of the fragile molded pieces was the magnitude of tool deflection during clamp application and release, leading to shifting of the tool inserts. To determine and record die deflections, Bently proximity probes were mounted on the die. Data from these probes, as shown in Figure 8, showed that die and machine modifications were required to reduce deflections below 0.0005", a level where no component damage was observed. Once this was accomplished, blade base cracking was nearly eliminated.

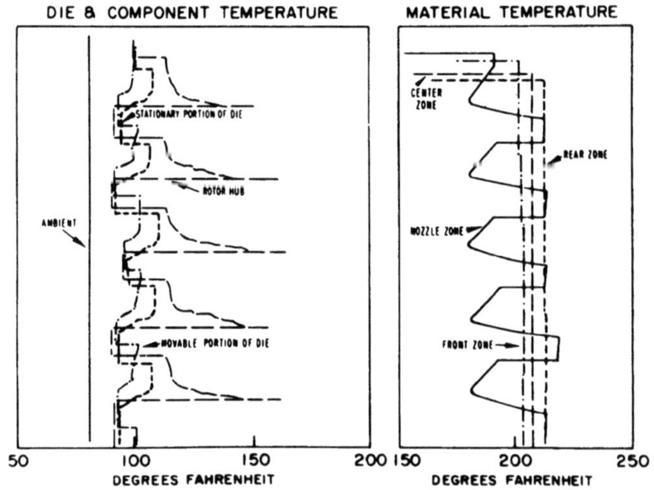

Figure 7. Temperature recording of die, component and material for four consecutive molding shots.

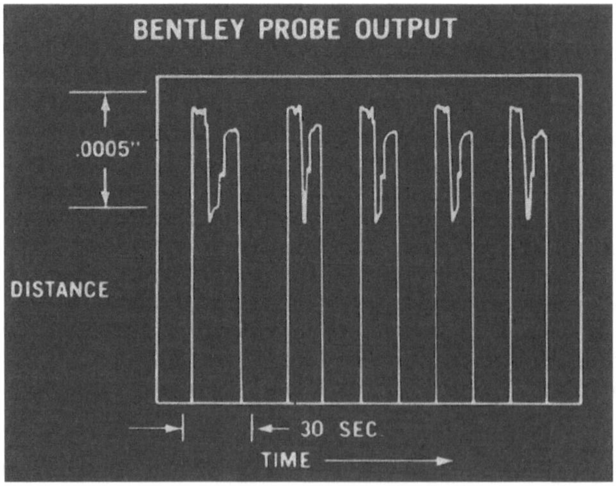

Figure 8. Bentley probe readings for deflection of die during molding.

Voids have been a constant problem throughout the molding program. An example of a void is shown in Figure 4. Their presence can be due to turbulent flow in the die, inadequate plastication with a ram type injection system, entrapped gas in the die due to flow away from the vent system or gas entering with the molding mix.

To attack the void problem, the flow and pressure in the mold cavity during the injection process were studied. A graph of ram position as a function of the pressure in the mold cavity is shown in Figure 9. Control of the shrinkage during hold in the die is critical and is controlled by precise metering of molding pressure. Pressure control over the system is achieved with a Hunker adaptive process control as seen in Figure 10. At ten sensor positions, the unit allows electronic control over hydraulic pressure. Control of

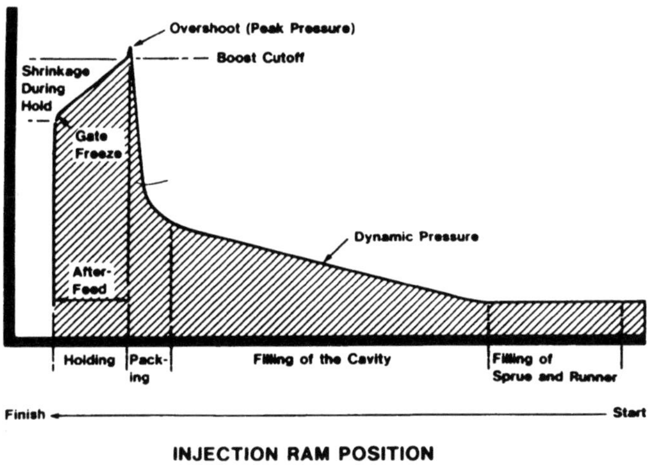

Figure 9. Pressure as a function of ram position during injection molding.

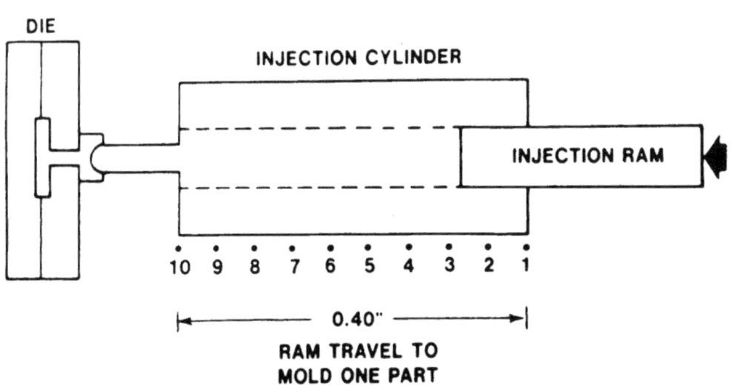

Figure 10. Hunker adaptive pressure control at ten sensor positions.

material cushion and feed is also provided. Based on time, ram position, hydraulic pressure and cavity pressure, the control modulates a flow-divider hydro valve regulating injection and hold pressure.

Before adaptive control of ram position and cavity pressure was initiated, typical results of random cavity pressure for four successive moldings are presented in Figure 11. Maximum cavity pressure and flow, as indicated by ram position, are seen to vary widely from one shot to the next. To coordinate the entire molding system, control, temperature, flow, machine sequence and data acquisition, Ford is developing a microcomputer (Intel 80-80) based control system. This system improves control of molding parameters, thus, improving component quality by reducing unwanted parametric variation. Improvement in reproducibility of cavity pressure as a function of ram position is shown in Figure 12, which is a significant improvement over results seen in Figure 11. Figure 13 shows an even closer control of maximum pressure and flow profile, corresponding to the use of cushion control.

An expanded ram position vs. cavity pressure curve is given in Figure 14 and this highlights the after-feed portion of the molding process; cracks can occur if this part of the curve is not adequately considered. Extreme or inadequate packing pressure can cause component flaws (i.e., voids or cracks).

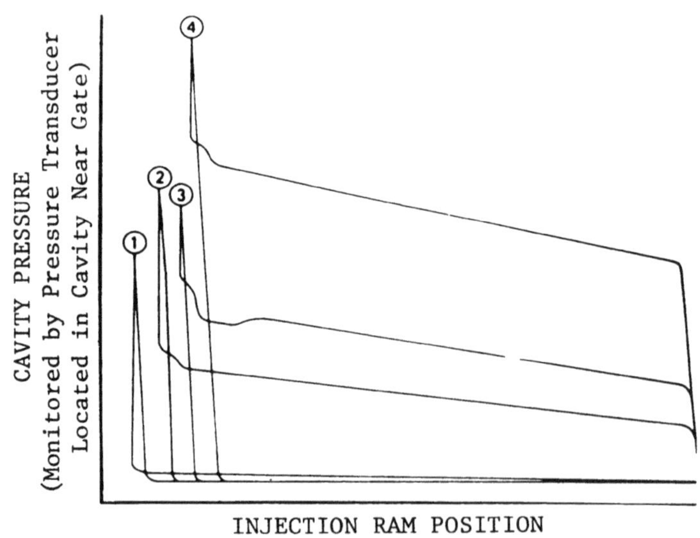

Figure 11. Typical cavity pressure vs. ram position curves for four moldings prior to adaptive controls.

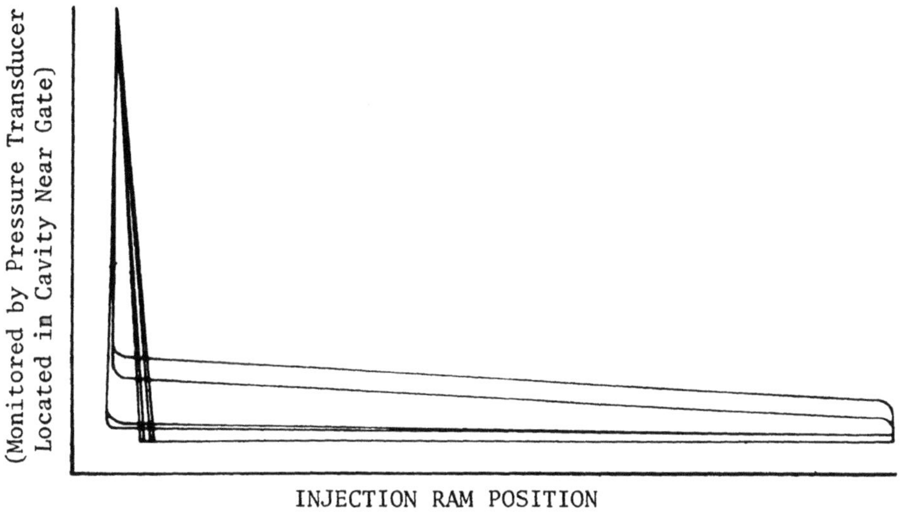

Figure 12. Improved cavity pressure vs. ram position curves for four moldings after adaptive controls.

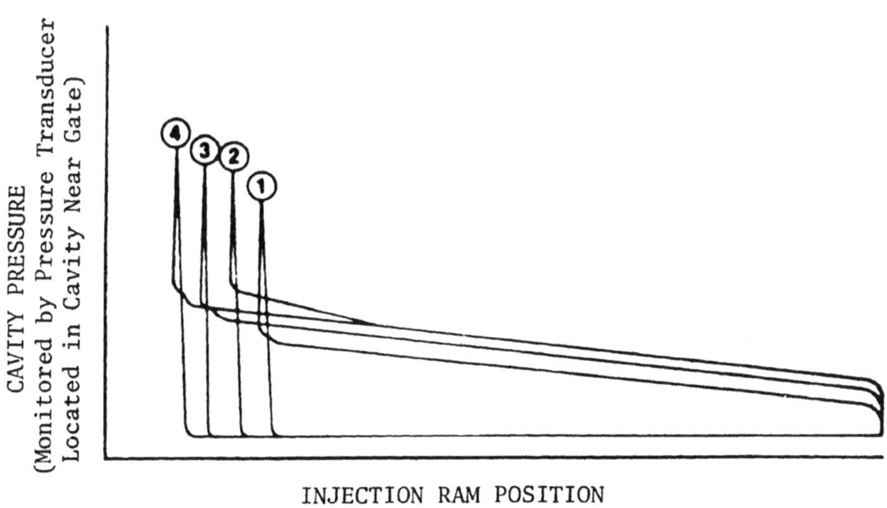

Figure 13. Recent cavity pressure vs. ram position curves with cushion control.

Molded-in stress from improper packing can cause cracks or planar flaws which show no exposed surface. Such cracks are only visible after destructive testing. Such flaws are shown in Figure 15.

Work going on at Cornell University on the prediction of flow in plastic systems using finite element analysis, has shown encouraging results in predicting possible causes for this type of flow. A prediction of flow pattern for polypropylene and the actual flow is shown in Figure 16. Ford is working with Cornell in anticipation of applying this work to more complex ceramic-plastic systems.

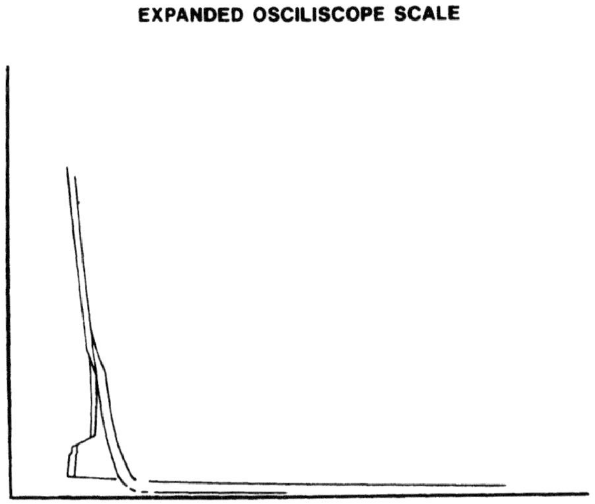

Figure 14. Expanded cavity pressure vs. ram position curves showing details of after-feed portion.

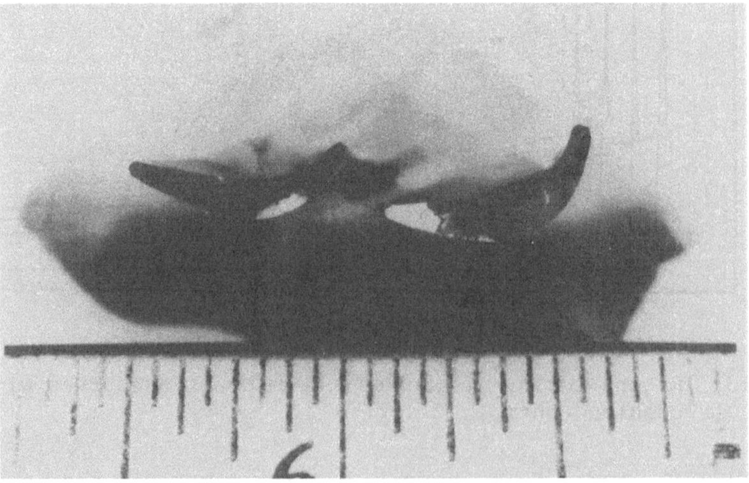

Figure 15. Photograph of planar flaws in rotor blades.

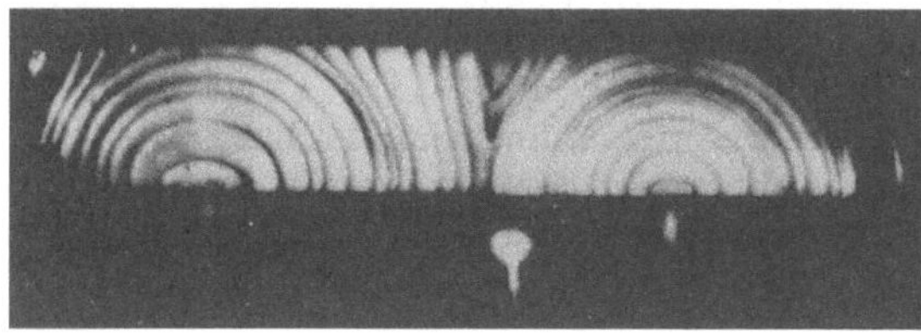

Figure 16a. Short-shot sequence for two-gated rectangular plates for polypropylene molding.

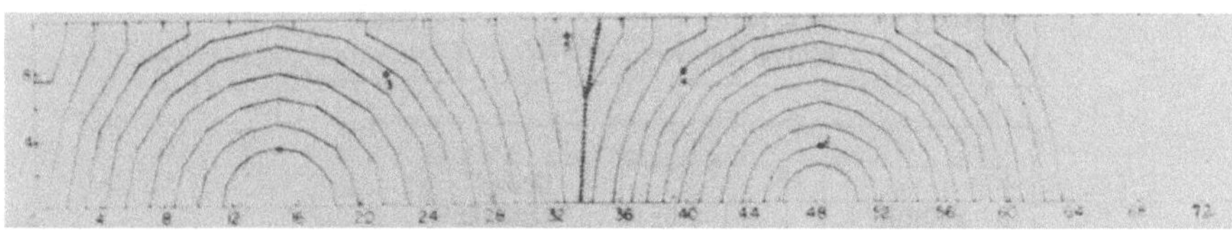

Figure 16b. Predicted melt fronts from Cornell University finite element analysis.

TRANSFER MOLDING OF SiC

Another example of molding complex shapes, which are later reaction sintered, and retain their original dimensions established during molding, is found in the SiC-Si system. A thermosetting plastic is used in the molding process and it possesses the distinct advantage of producing strong bodies in the "green" condition. The plastic discussed in this section is a phenolfurfural phenolformaldehyde copolymer. It also possesses, however, the disadvantage of long cycle times in the molder, generally required of plastics which harden upon application of heat. Thermosetting plastics are generally favored for the SiC process because they provide the large carbon contents needed for reaction bonding. A schematic of the processing steps to produce the SiC-Si materials is shown in Figure 17. Carbon supplied by the plastic is subsequently reacted with Si to form SiC.

As in most molding investigations, a spiral mold was used to evaluate the flowability of SiC-filled thermoset compositions. A typical composition was 47% SiC, 47% polymer, 5% graphite and 1% zinc stearate (vol %). These materials were in powder form and were blended by ball milling prior to molding. Blended powders were then pre-extruded at 116°C to reduce air and moisture contents. The extrudate was ground to -1/4 inch mesh molding powder.

Trials with molding in a spiral mold showed that the flow was influenced by the amount of SiC powder in the mix. Figure 18 shows this effect when powder of a unimodal particle size distribution with an average of 16μm was used. Flow is also affected by the

SiC particle size as shown in Figure 19. Short shots of this mix in simple molds have shown that the material enters the mold as a rope-like strand which forms a coil-like structure until the mold is full. A high pressure is then required at the end of the fill cycle and during the thermosetting plastic cure cycle to knit the strands into an integral part and avoid low density areas at knit lines.

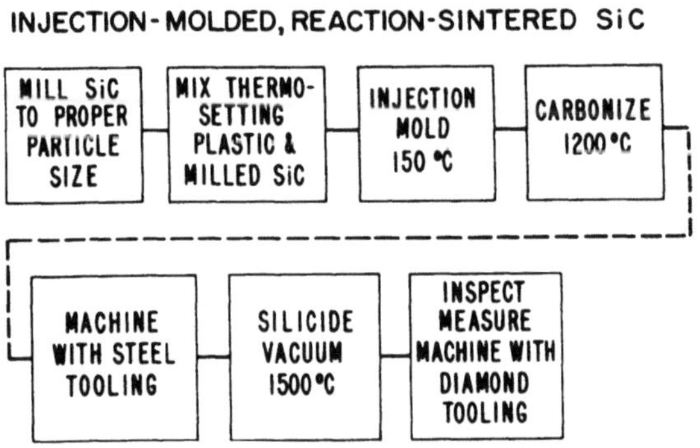

Figure 17. Schematic of processing steps to form SiC-Si materials.

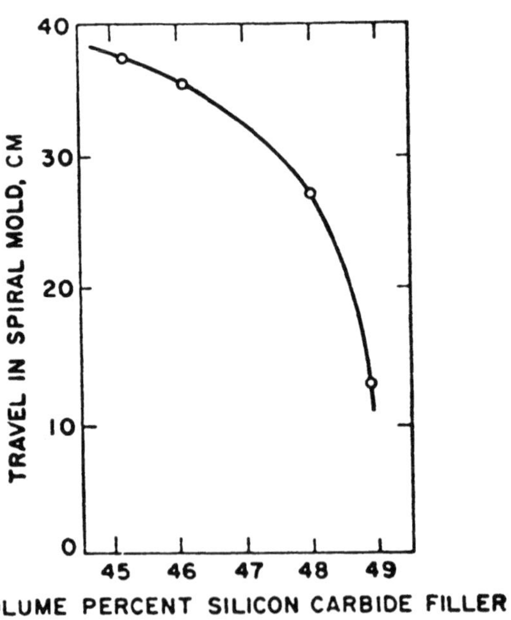

Figure 18. Dependence of spiral mold flow on volume fraction of 16mm average SiC.

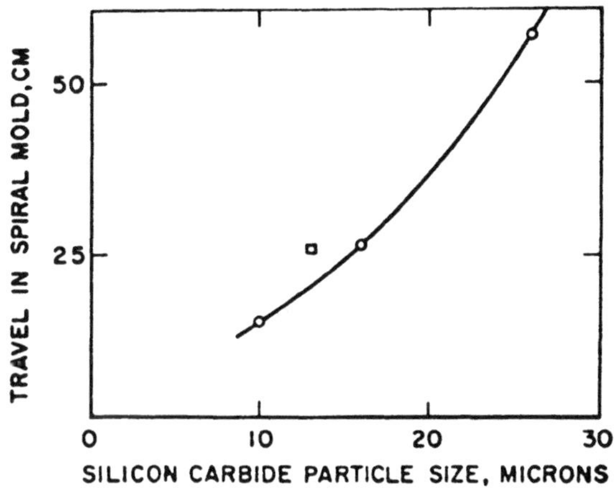

Fig. 19. Dependence of spiral flow on SiC particle size.

Flaws observed in this material, believed to be due to molding problems, are seen as silicon rich planes (shown in Figure 20) and large silicon volumes (seen in Figure 21) in the microstructures of the fully processed material. Mold design, shrinkage caused by compositional and particle size variables, and molding and die clamping pressures, all contribute to these material flaws.

Experience with simple molds has shown that injection should occur at the center line of the mold with equal cavity volume in each half of the mold. Other arrangements, for example, with the entire cavity on one half of the mold, have led to highly stressed pieces manifested by warpage in the carbonized state.

Significant shrinkage during the carbonizing process is a sign of potential problems in the fully processed (silicided) state. Earlier work on SiC-Si material[4] has shown that the shrinkage is a function of average particle size. Linear shrinkages between 1% and 2% can be easily obtained, as shown in Figure 22. If a closely classified SiC powder is used, shrinkages as low as 0.4% can be obtained.

Potential improvements in material reliability by refining the molding process for SiC-Si should be reflected in increased Weibull parameter "m", when parts or test bars are evaluated. For example, groups of test bars containing the silicon planes seen in Figure 20 have m values of 2 to 4. These planes are believed to be cracks formed during the carbonizing step and are attributable to shrinkage or low density planes. Groups containing the less detrimental large silicon pools exhibit larger Weibull m values in the range of 6 to 10. Elimination of these two defects, and careful surface preparation by polishing and grinding of a fine-grained SiC-Si composite may lead to m values of 12 to 15, and strengths in the region of 75 to 100 Kpsi. Individual samples have been measured in excess of 100 Kpsi in several groups of samples. Consistently higher values have been observed in round rods than in rectangular cross section bars probably because of the greater difficulty in machining samples with flaw free edges in the test bars.

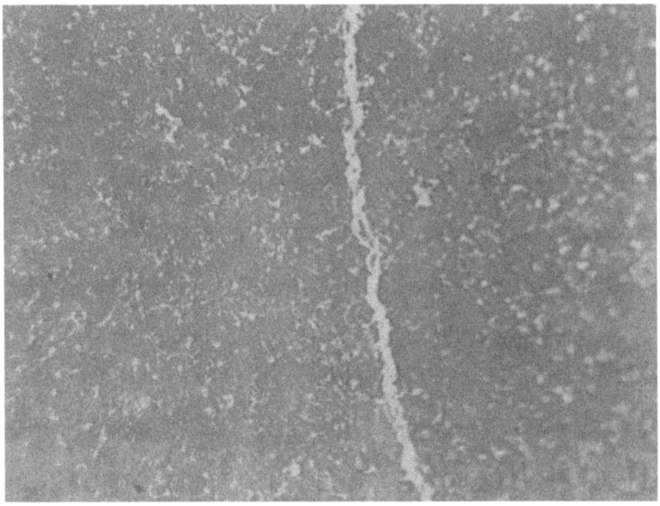

Figure 20. Micrograph of silicon-rich planes in SiC-Si composite.

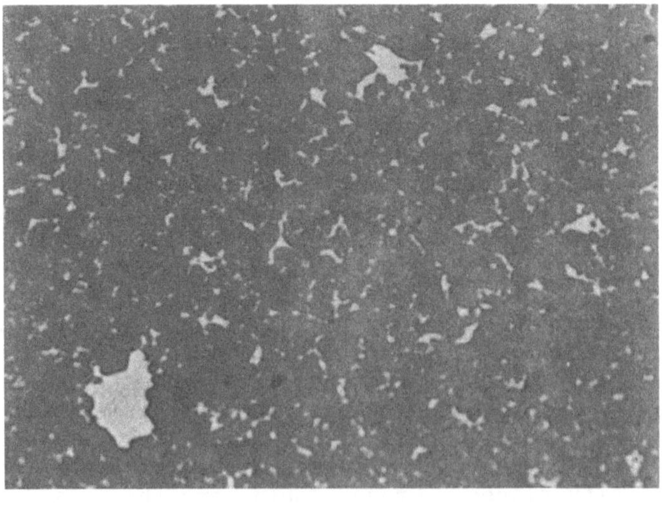

Figure 21. Micrograph of silicon-rich volumes in SiC-Si composites.

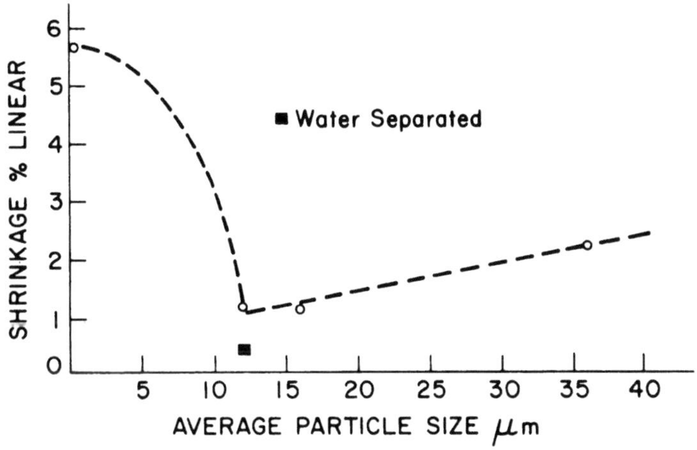

Figure 22. Linear shrinkage of SiC as a function of particle size of SiC powder.

CONCLUSIONS

1. Cracks, planar flaws, inclusions and voids are defects commonly observed in injection-molded parts.
2. Flaws in molded structures can be reduced by use of adaptive controls during molding.
3. Strengths as high as 75 to 100 Kpsi can be expected in flaw-free SiC-Si materials when tested at room temperature.

REFERENCES

1. K. Schwartzwalder, "Injection Molding of Ceramic Materials", American Ceramic Society Bulletin 28 (11), p. 459-462, 1949.
2. J. A. Mangels, "Development of Injection Molded Reaction Bonded Si_3N_4", Ceramics for High Performance Applications -II, Editors J. J. Burke, E. N. Lenoe and R. N. Katz, Metals and Ceramics Information Center, Columbus, Ohio, 1977, p. 113-130.
3. P. A. Willermet, R. A. Pett and T. J. Whalen, "Development and Processing of Injection-Moldable Reaction-Sintered SiC Compositions", American Ceramic Society Bulletin, 57, (8). p. 774-747, 1978.
4. T. J. Whalen, J. E. Noakes and L. L.Terner, "Progress on Injection-Molded Reaction-Bonded SiC", Ceramics for High Performance Applications -II, Editors J. J. Burke, E. N. Lenoe and R. N. Katz, Metals and Ceramics Information Center, Columbus, Ohio, 1977, p. 179-189.

RELIABILITY OF SLIP CAST SILICON NITRIDE COMPONENTS

M.E. Rorabaugh and K.H. Styhr

AiResearch Casting Company

Torrance, California

ABSTRACT

The slip casting process has been applied to several classes of silicon nitride materials systems; however, the majority of effort has been focused on slip casting silicon with subsequent nitriding of net shape components to the final silicon nitride form. This process has many steps with potential difficulties in each, such as raw material control, preparation of the casting slip, preparation of the mold system, the process of casting and solidification, removal of the part from the mold, and its subsequent drying, sintering, finishing and nitriding operations. Process variables in each of these steps are identified along with the current quality control inspection and evaluation methods. Deficiencies are identified and the need for improved techniques for finding critical defects early in the process cycle and tracing them to their source and to the elimination of the cause is stressed.

INTRODUCTION

High performance reaction bonded silicon nitride (Si_3N_4), RBSN, ceramic components have been produced by slip casting silicon powder and nitriding the shaped silicon compact. Figure 1 illustrates the variety of size and complexity of parts that have been formed by slip casting to near net-shape and machining to final dimensions. Under some controlled conditions, slip cast RBSN has provided strength values equal to the highest values measured for RBSN produced by other methods such as injection molding and isostatic pressing. Greater product reliability can be obtained by increasing the average strength or by identifying and eliminating those flaws which result in lower strength components. Finished component inspection must be improved to assure that strength limiting flaws can be detected.

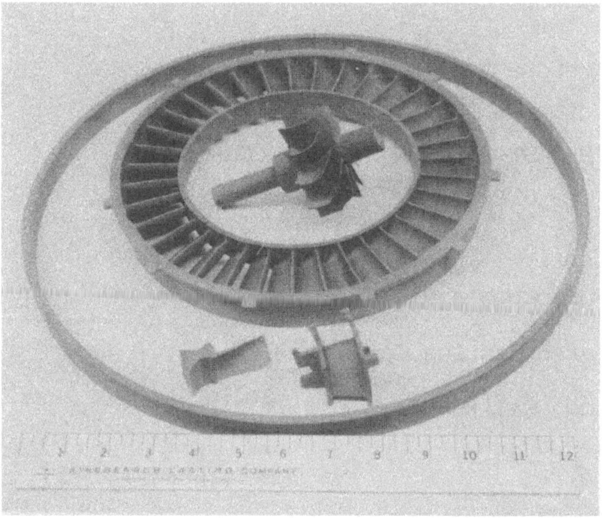

Fig. 1. Typical Slip Casting Components

Slip casting offers many possiblities and even some advantages over some other methods. First, it has been used to produce fully nitrided samples ranging from 2.3[1] to 2.9 g/cm^3. The higher density materials have potential for improved strength and thermal shock and oxidation resistance. Second, a wide variety of particle size distributions (PSD) can be slip cast into test samples. Third, tooling is relatively inexpensive, often costing much less than equivalent injection molding tooling. Finally, the process is more readily adaptable to larger shapes. To take advantage of the possibilities that exist, all the variables in the process steps must be identified, optimized, and carefully controlled.

This paper discusses many process variables. It shows that many process steps can be further optimized. High strengths have been measured in occasional samples, a fact which assures that goals much higher than present capability are reasonable.

PROCESSING VARIABLES

The processing steps listed in table 1 are currently used in producing slip cast components. The variables listed for each of the process steps can influence the final properties. Sintering is included as a process step primarily to provide opportunity for machining large amounts of material from some castings in a soft-prenitrided condition and is not necessary on parts cast to shape.

Table 1. Processing Variables

Process Step	Variables
Starting Powder	Impurities: type and distribution, Particle size distribution (PSD), surface area
Milling	Mill type, media: type and quantity, load, time, speed, aids, packing, temperature, starting PSD
Slip	pH, stability (with respect to age), viscosity, solids content, vehicle, agitation, temperature, deflocculents, wetting agents, antifoam agents
Pattern/Mold	Wax pattern injection parameters: temperature of mold and wax, wax composition, mold release, injection pressure and time, cooling time. Mold: material and preparation conditions, plaster porosity
Casting	Plaster moisture content, slip conditioning, pouring, porous mold to nonporous mold surface area ratio
Sintering	Time, temperature, gas pressure, gas composition
*Machining	Grit size, wheel surface speed, stock removal rate, total stock removed, grinding direction, holding method
Nitriding	Nitriding rate, temperature, time, heating rates, gas pressure, gas composition, nitriding aids, sample geometry, sample condition, furnace load

*Some machining is done on most parts before and/or after nitriding.

Scientific studies have been made on experimental quantities of silicon. Many of these studies have been reported, and some will be referred to here. Process development for prototype hardware using low cost, commercial quality silicon powders, however, provides a basis for the following discussion.

Raw Materials

Significant variations have resulted from changes in starting powder. Figure 2 shows an example of strength variations resulting from using different starting powders all processed in like manner. A significant difference between these powders is the particle size distribution. Powders A, B, and D were nominally -325 mesh and powder C was nominally -200 mesh. Powder A appears to be more difficult to nitride due to particular chemical impurities. Powder B was much finer than the others and had a tendency to agglomerate. Powder C and D both contain many particles >150μm diameter which may establish the ultimate strength limit.

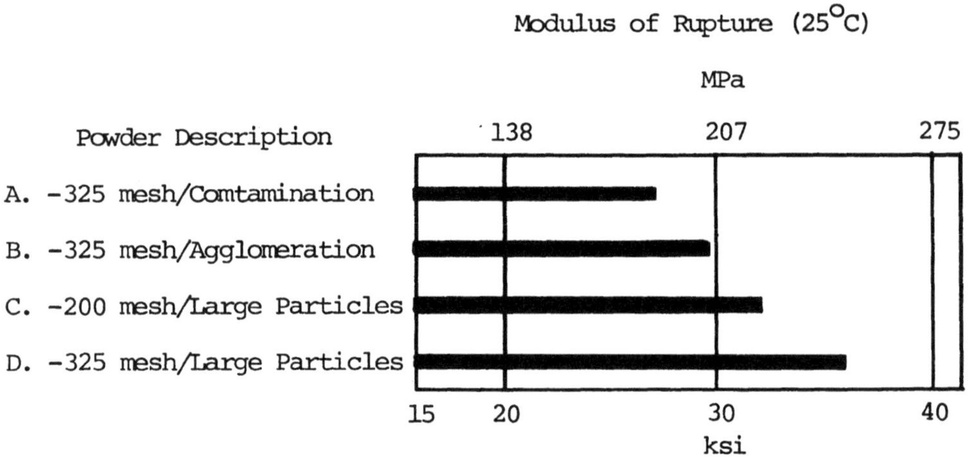

Fig. 2. Modulus of rupture of test bars representative of different powder lots

Particle size distribution data are obtained for each incoming powder lot and representative milled powder batches. Measurements are made by one supplier using a Coulter Counter. Supplemental data are obtained using an x-ray sedigraph. Data for a particular powder lot (as measured by both methods) are presented in table 2. Neither technique can distinguish between individual particles and agglomerates, a limitation which makes sample preparation extremely important in both measurement techniques. Figure 3 shows PSD data for a sample of an agglomerated powder with various degrees of mechancial dispersion. It has been observed that dispersion techniques developed for some

Table 2. Particle Size Distribution Measurements

% Greater Than		Size
Coulter Counter	X-Ray Sedimentation	µm
7.11	1	20
48.63	24	10
85.14	73	5
100.00	93	2.5

oxide materials are inadequate for silicon powder. Another problem is that neither technique gives an accurate indication of the maximum particle size. Large particles must be separated out and individually measured. This can be done by elutriation, separating the coarse and fine material by making use of their different settling rates in water. Measurements have shown that some particles as large as 200µm are present in many nominally -325 mesh (44µm) commercial powder lots.

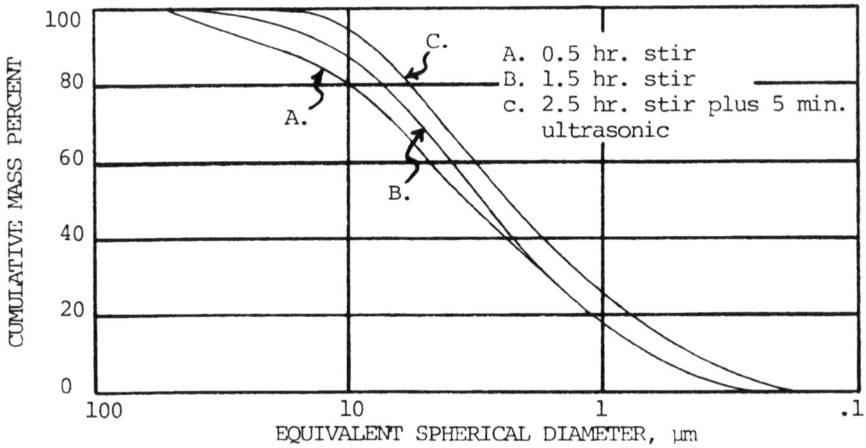

Fig. 3. Particle size distribution measured for a single sample with different mixing conditions

Surface area measurements are made on representative powders. This measurement appears quite reproducible. It is a fair indicator of the PSD below 1µm as the smaller particles make the larger contribution to the surface area. The PSD measurement techniques are practically at their limit at around 0.5µm. The surface area of a powder is very significant in controlling the rheological properties.

Chemical data have been obtained by spectrographic analysis, atomic absorption (AA), and energy dispersion analysis by x-ray (EDAX). Although standards are not readily available for quantitative spectrographic analysis, semi-quantitative analysis appears to be in fair agreement with AA. EDAX has been useful for identifying components of individual features observed. These tests are generally done with very small samples. If a contaminant is in the form of relatively large particles, the result of a given measurement could be misleading. Chemical analysis has, however, helped identify a probable source of magnetic steel contamination as wear in equipment used in reducing silicon to a fine powder. This contaminant is not easily removed from the dry powder because such particles may be trapped in agglomerates which are mostly silicon. It is being removed with some degree of success after the powder is introduced into a slip using a magnetic separator. Present efforts require further evaluation and probably further development.

Powder Processing

Processing conditions must be developed which are suitable for starting powders with varying properties. Maximum particle size must be consistently limited as large particles can result in unreacted silicon, or a large pore [2], which could ultimately be the strength limiting flaws. The PSD must be controlled to assure proper casting properties. Certain chemical impurities must be removed.

Air classification appears to be an effective method of eliminating large particles from a powder lot. An air classifier, however does not make a critically sharp separation and may remove 50% of the 20μm particle to assure that <u>all</u> the 30μm particles are removed. Alternate techniques such as wet or dry sieving are difficult due to the dilatant nature of slips or the tendency toward powder packing, respectively.

Ball milling is effective in reducing the average particle size. It does not assure that all of the large particles are broken down and, therefore, is not an effective means of limiting the maximum particle size. The milling conditions used affect the PSD and, ultimately, the casting and strength properties. Milling can additionally be used to blend two silicon powders of different particle sizes or to blend in a nitriding aid. Figure 4 shows typical changes in a powder by milling and by air classifying and milling.

Table 1 lists many variables which significantly influence the final result of ball milling. Commercial silicon powder has a tendency to pack during dry ball milling. Good casting properties can be obtained even where packing occurs but the resulting unmilled large particles and tightly packed agglomerates can lead to strength limiting flaws in the final product. Milling aids can eliminate packing but must subsequently be removed unless they have no effect on the final properties. Attritor milling[3], wet ball milling and vibration milling are alternate milling methods which should be considered. Attritor milling and vibration milling are higher energy systems, that is they are more effective with shorter milling times. Attrition milling can easily produce a pyrophoric powder or a powder which oxidizes extensively in a water based slip. Wet ball and vibration milling, as well as potentially providing an improved PSD, may be useful in eliminating slip making as a separate operation. Ball milling variables have been discussed by a mill supplier[4]. The effects of mill load and speed and various grinding media are described. The reactive nature of silicon, however, makes such things as milling atmosphere and temperature important variables to consider. A variety of milling aids can be used but their effect on final properties has not been fully explored.

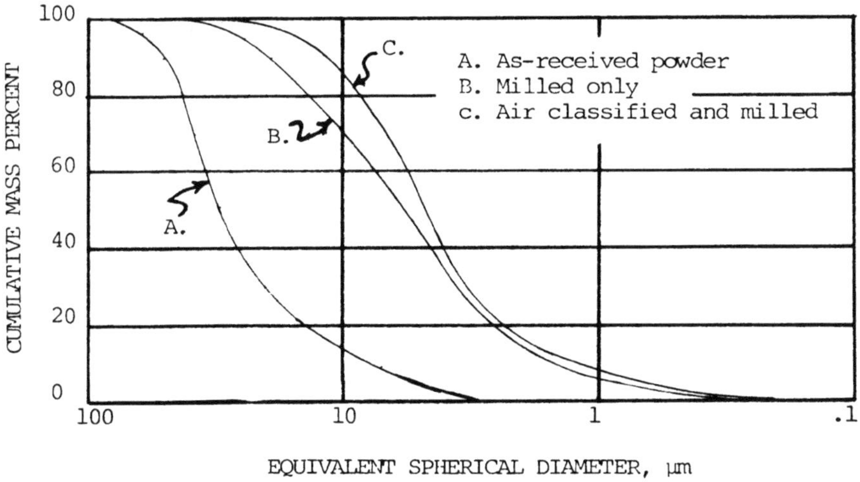

Fig. 4. Air classification reduced the number of particles >40µm

When considering chemical contamination, the type, amount, and form are all significant. Large (>40µ) magnetic steel particles have been identified in several commercial powder lots. Although the total Fe content may be small (0.5%) the nitriding is nonuniform and strengths are reduced with this type of contamination. (Larger quantities of Fe in the form of much finer powder have been added as a nitriding aid and result in improved properties). Improved reliability is possible by the removal of the large magnetic particles. Calcium has been identified as a contributor to creep; however, levels of less than 0.05% appear acceptable[5]. This level has been maintained in recent commercial lots from various suppliers. Vanadium has been identified in several lots of powder in excess of 0.1%; its role in nitriding and final properties is not well established.

Slip Preparation

Many variables in slip rheology have been discussed previously[1]. Slip rheology was controlled by PSD, pH, deflocculants, and solids content. Effect of several of these variables on casting density was discussed, and a linear relationship between increasing density and increasing strength in the nitrided test sample was shown.

Additional considerations must be made for volume production of reliable slip cast parts. The casting rate must be optimized by the control of PSD, electrolytes, solids content, mold condition, etc. Rate optimization is to be based on final properties and manufacturing requirements. Density and part uniformity are affected by some casting rate adjustments. Maximum rate will maximize the production capability of a given space. Slip making and aging processes are being developed which eliminate some of the sources of flaws such as agglomerates, dissolved air, and gassing from the oxidation of silicon.

Prior to casting, slip is "conditioned" to help provide flaw free parts. This process includes filtering (agglomerates and contaminants) and vacuum degassing. Temperature of the slip does influence the slip rheology but temperature controls have not been added to optimize this variable.

Slips with vehicles other than water have been found useful for other powders. Other vehicles may eliminate a source of oxidation of silicon. Other advantages may be realized if alternate systems are further developed.

Molds

After the slip is prepared, it is poured into a mold to produce the desired shape as the vehicle is removed. All mold systems use a porous material, most often plaster, in some portion of the mold to extract the vehicle from the slip by capillary action. Plaster molds in established slip casting production lines (e.g. the whitewares industry) typically are used to produce up to 50 castings.

The mold system used depends greatly on the geometry. All-plaster molds are not desirable for some complicated shapes such as individual stator vanes. For this type of configuration, a process has been used which is very similar to the lost wax process in metals casting[6]. In both cases, a pattern is injection molded, coated with a suitable mold material, and then removed leaving a hollow shell mold. Since a new mold is required for each part, high quality molds must be consistently produced in a cost affordable manner. Variables defined in table 1 must be controlled and regular quality assurance procedures must be established. Simple shapes, however, can be made with entirely reusable molds. Because only small quantities of parts have been required in demonstration programs, to date, the development of reusable molds has not been justified.

The condition of the mold at the time of casting can influence the casting behavior. The long life behavior and the optimum moisture content are not yet established. Scaled-up production will make this knowledge a necessity.

Casting

Pouring slip into a mold, especially one which has fine detail, must be done with extreme care. Air can be mixed into the slip or trapped in corners of a mold quite easily. Vacuum casting can be a benefit; however, too low a pressure can cause porosity by boiling the vehicle. For production of large quantities of parts, automatic equipment should be developed in order to eliminate human variation.

Pouring the slip into the mold is only the start for the slip casting process. Depending on the configuration, slip properties, and plaster condition, the time required for sufficient vehicle to be removed (casting) can be from minutes to hours. During this time some settling does occur. Component orientation can be very significant, especially on complex shapes, to assure that no passages within the mold are closed off prior to complete casting. Where such closure cannot be prevented, reservoirs as shown in figure 5 must be used. As the vehicle is removed, the total volume of casting and slip is reduced. The reservoirs provide material to complete a casting free of cavities on the surfaces. On complex shapes with radial symmetry, slow spinning can assist in assuring complete mold fill.

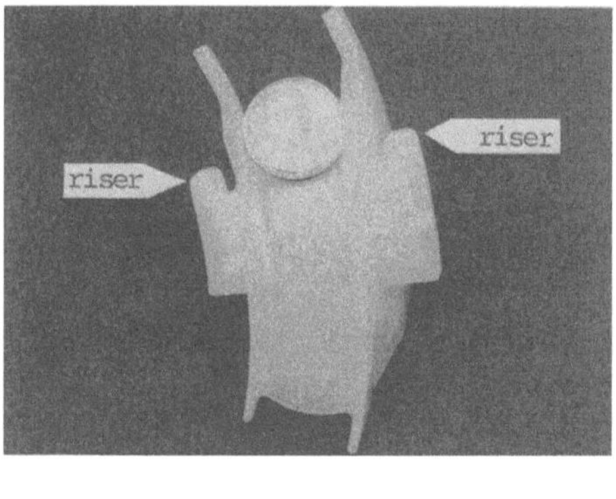

Fig. 5. Risers - slip reserviors at component extremities.

Sintering

When machining is required, it is often desirable to sinter, or partially nitride, the casting in order to give the body handling strength. Sintering of silicon powder compacts prior to nitriding is discussed elsewhere[6]. The referenced report discusses some of the effects of sintering at temperatures from 1050 - 1250°C (1922 - 2282°F) but is not conclusive concerning a strength enhancing mechanism suggested at the higher temperatures. The material evaluated was 99% pure silicon with test samples machined from larger billets.

Sintering of near-net-shape parts containing Fe_2O_3 as a nitriding aid requires additional considerations. Sintering silicon can result in a skin on the component. This skin has a very different appearance as seen in figure 6 as bright regions (the skin appears in different thicknesses at different places due to the angle of cross section). The skin is harder than the core and can lead to fractures when machining with single point tooling. It is not yet clear if the microstructure of the skin enhances or is a detriment to the strength of as-cast parts. Sintering can also result in the growth of large pores (up to 40μm) which can be strength limiting. This pore growth may be partly the result of eutectic melting between iron, silicon, and their oxides. Iron additives help nitriding but possible side effects of Fe should be explored.

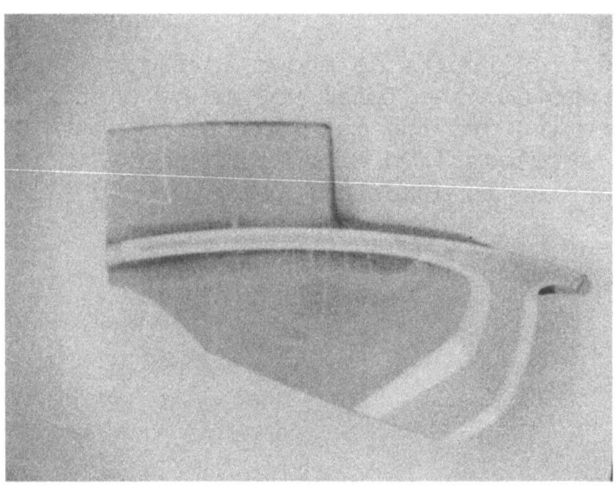

Fig. 6. Skin resulting from sintering (accented by lighting).

Sintering has also been shown effective in lowering the oxygen content which may be helpful. Various temperatures and atmospheres result in a variety of sintering weight losses. Chemical additives also influence the loss of oxygen and even silicon in vacuum sintering.

Nitriding

Nitriding is the chemical reaction of combining silicon and nitrogen and is the final necessary step in producing RBSN components. Considerable discussion has been given on nitriding silicon[2,3,5,7,8].

The literature and experience indicate that much can be learned to improve this process.

Until recently, it has been felt that 2.7 g/cm^3 is the highest density which could be fully nitrided[7]. More recently, 2.8 g/cm^3 material has shown promise of being fully nitrided[8]. AiResearch Casting Company has produced some samples up to 2.9 g/cm^3 with only well dispersed, small metallic inclusions which appear to be the result of a few large particles rather than inherent limitations due to density.

Some durability testing is completed on 2.8 g/cm^3 slip cast RBSN. This material does have traces of unreacted silicon. The material has maintained 93% of its room temperature strength after cycling to 1205°C (2200°F) for 1050 hours. More details of this work are reported in this volume[9]. Further optimization of the nitriding is necessary on the high density materials and will result in improved properties.

Nitriding cycles typically are programmed to last between five and fifteen days, and occasionally even a wider range. Many cycles have been used which attain a reactive temperature early in the cycle and hold at that temperature until a significant portion of the reaction is complete. This method is very sensitive to temperature control. Since the nitriding reaction is exothermic, this method is also sensitive to load in the furnace as a large mass can result in an overtemperature condition. Dilution of N_2 with inert gasses such as argon or helium is another technique sometimes used to control the exothermic reaction.

Rate controlled nitriding has been evaluated[7] and appears to be an improved nitriding technique. By this method, the reaction rate is kept constant for most of the reaction and the temperature increases at a variable rate. This allows nitriding to occur at a maximum, safe (from an uncontrolled exothermic reaction) rate and is somewhat self adjusting to the effects of nitriding aids. Experience has shown that, with Fe_2O_3 as a nitriding aid, the temperature increase is slow at the Fe-Si eutectic where the liquid phase can enhance nitriding; very little temperature increase is necessary to maintain a constant reaction rate. A single, optimum, overall nitriding rate can not be defined. The rate must be established according to the constraints of density, cross section thickness, chemistry, furnace response rate to self heating, and particle size distribution.

Further study of the nitriding cycle may reveal new desirable techniques. Early cycle conditions might be controlled to enhance optimum nucleation conditions. Later cycle conditions may require special consideration as all of the ongoing reaction is concentrated in a small amount of remaining unreacted silicon.

Chemical additions have been made to the silicon and nitrogen to enhance nitriding. Hydrogen and hydrogen sulfide[10] have been shown to enhance nitriding or improve properties. One potential problem of gaseous additives is separation due to density differences. Iron[11] and calcium[5] have also been shown to increase nitrideability of a silicon compact. The full effects of these and other additives are still being explored. The interaction between the additives is even less well known.

Machining

Machining is often necessary when extremely tight tolerances are required or the forming process is not developed to produce net-shape parts. No industry wide standards are established for general application which cover machining of sintered silicon or the nitrided product. These two processes may be very different and must be addressed as such. Machining has been discussed elsewhere[12] and some of the areas requiring further work were pointed out.

Novel approaches are being evaluated. One example is laser cutoff of gating and riser of slip cast stator vanes. Early efforts show promise for easy removal of the bulk of gating considerably reducing the requirement for conventional machining.

COMPONENT EVALUATION

Nondestructive evaluation (NDE) has been developed to the point that many of the flaws in finished parts which could contribute to component failures can be detected. These flaws often occur in areas which do not see peak stresses but are still considered cause for rejection. It can be assumed, though, that undetectable flaws in maximum stress areas could ultimately contribute to failure and that improved NDE techniques would be useful.

Proof testing may provide additional assurance that components will not prematurely fail in use[13]. Proof testing at maximum acceptable stress levels (levels which are maximum conceivable in a particular application) would assure reliability at normal, much lower stress, conditions. Any component which is degraded significantly in such test conditions is probably not suitable for the application. Proof testing may be considered on NDE evaluation in that it is nondestructive on acceptable components. An example of proof testing is in the glass container industry where some bottles are pressurized to assure dependable field performance. Components containing harmful defects are quickly weeded out by this process.

Many ceramics component users are involved in the development of the NDE techniques. When the best methods are developed and there is a demand for large quantities of reliable components, suppliers will be in a position to use the techniques developed. Until that time, a maximum amount of information must be relayed from those making evaluations to those making the parts so that sources of failure due to materials and processing can be eliminated.

Material characterization by test bars is useful, however exact representation of complex shapes using simple test shapes may never be realized. Material strengths resulting from test bars prepared in a variety of ways and tested under different conditions are reported throughout the literature. Often the test conditions are described but the differences in testing are not always fully appreciated. The correlation between test results depends on the process and material and cannot be modified by simple conversion factors. For example, for a particular group of slip cast bars, several values could be reported: 232 MPa (33.6 ksi) for rough quality control bars, 261 MPa (37.9 ksi) for machined bars and 269 MPa (39 ksi) characteristic Weibull strength for machined bars. Another slip cast RBSN[1] has had bend strength values of 286 (41.5) and 365 MPa (53 ksi) reported for different test conditions.

SUMMARY

Although the current state-of-the-art can produce components usable in some applications, it is strongly suggested that there is much room for improvement for RBSN. Even though process controls have been related to slip casting, it should be recognized that many of the strength limiting factors, process steps, and evaluation techniques are similar to what is found in other processing methods. Many improvements in slip casting will have further applications.

Reliability of parts will come from two factors. Consistent, high quality processing and dependable evaluation will result in reliable components for engine building. Everyone involved from the powder supplier to the component user must participate in the development of reliable components.

To date, only a small portion of the development effort has gone into producing better materials. A larger amount has been invested in characterizing today's materials. Larger amounts have also been invested in component fabrication by fixed processes. Improvements in materials will continue slowly if this approach continues. On the other hand, many potential areas of improvement are currently recognized and are waiting to be explored.

REFERENCES

1. A. Ezis, "The Fabrication and Properties of Slip-Cast Silicon Nitride," in J.J. Burke, A.E. Gorum, and R.N. Katz, Editors, Ceramics for High Performance Applications, pp. 207-222, Brook Hill Publishing Company, 1974.

2. D.R. Messier and P. Wong, "Kinetics of Formation and Mechanical Properties of Reaction-Sintered Si_3N_4," pp. 181-193, ibid.

3. T.P. Herbell, T.K. Glassgow, and H.C. Yeh, "Effect of Attrition Milling on the Reaction Sintering of Silicon Nitride," DOE/NASA/ 1040-78/2, NASA TM-78965, May 1978.

4. "Jar, Ball, and Pebble Milling, Theory and Practice," Norton Bulletin P-291.

5. J.A. Mangles, "Development of a Creep-Resistant Reaction Bonded Si_3N_4," pp. 195-206 in ref. 1.

6. AiResearch Report No. 76-212188(7), Ceramic Gas Turbine Engine Demonstrator Program, Interim Report Number 7 (Quarterly), Prepared under contract N00024-76-C-5352 pp. 4-49 to 4-52 November, 1977.

7. J.A. Mangels, "Development of Injection Molded Reaction Bonded Si_3N_4," in J.J. Burke, E.N. Lenoe, and R.N. Katz, Editors, Ceramics for High Performance Applications - II pp. 113-130, Brook Hill Publishing Company, 1978.

8. J.A. Mangels and R.M. Williams, Development of Moldable, High Density Reaction Bonded Silicon Nitride, Interim Report Number 12 (Quarterly), Prepared under contract DEN 3-20, April, 1979.

9. D. Carruthers, D. Richerson and K. Benn, "Combustion Rig Durability Testing of Turbine Ceramics," elsewhere in this volume.

10. P.E.D. Morgan, Activated Nitridation of Silicon and the α/β Ratio," Energy Materials Coordinating Committee (EMACC) Conference on Structural Ceramics, Knoxville, June, 1979.

11. W.A. Fate and M.E. Milberg, "Effects of Fe and Fe Si_2 on the Nitriding of Si Powder," Journal of the American Ceramic Society, V. 62, no. 11-12 pp. 531-2, Nov.-Dec., 1978.

12. R.W. Rice, "Machining of Ceramics" pp. 287-343 in ref. 1.

13. S. Wiederhorn, "Reliability, Life Prediction, and Proof Testing of Ceramics," pp. 635-664 in ref. 1.

LARGE SCALE PRODUCTION TEST OF GAS TURBINE COMPONENTS BY INJECTION MOLDING

E. Lange and N. Müller

Degussa Wolfgang
D6450 Hanau 11

INTRODUCTION

As a participant of the German program for the development of ceramic gas turbines, Degussa is concentrating on the fabrication of injection molded silicon nitride turbine parts. The most important parts are turbine blades and blade rings. The design for these parts originates from the companies developing the turbines.

The essential part of the Degussa program is the development of an economical manufacturing technology and an acceptable quality of the fabricated product. To achieve this, Degussa had to install considerable fabrication equipment. To demonstrate the attained technological level, large test runs have been necessary.

After discussing the development strategy, the manufacturing technology will be described briefly. Thereafter, the results of large test series and the attained quality will be discussed and finally a short outlook on our future work will be given.

DEVELOPMENT STRATEGY

Figure 1 shows the three main directions of the development strategy. The silicon nitride turbine parts have to be cheap, reproducible, and of high quality. To be able to estimate the fabrication costs, manufacturing technology for a mass production has to be developed. This technology should enable us, for example, to form and to nitride 10,000 blades per day.

A satisfying reproducibility can be reached only with a high standard of automation. Rather large investments are necessary to get experience in this field. The standard deviation of the important quality parameters has to be low, or, in other words, the Weibull modulus m has to be quite high ($m \geq 20$).

To guarantee a high average quality, appropriate fabrication parameters have to be found. This can only be achieved by varying these parameters in larger test series. The σ_0 value of the bending strength, for example, has to be ≥ 300 N/mm^2 to get the desired quality.

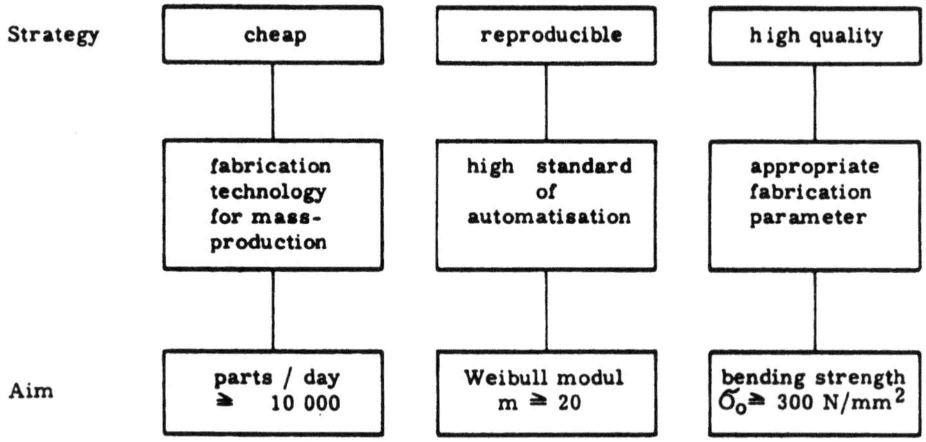

Figure 1. Development Strategy

A future design for ceramic gas turbines will probably specify a minimum strength and the probability with which all measured values are equal or higher than this defined minimum. Figure 2 gives some explanation of this situation. In this case the accepted defect probability is 0.1, that means one out of 1,000 values can be lower than the specified minimum. Assuming a minimum strength of 200 N/mm^2 as shown in Figure 2, a given σ_0 value is followed by a definite Weibull modulus. Since σ_0 values much higher than 300 N/mm^2 are not realistic today, a Weibull modulus near 20 has to be reached to meet the expected specifications. Weibull values much lower than 20 are, therefore, not acceptable.

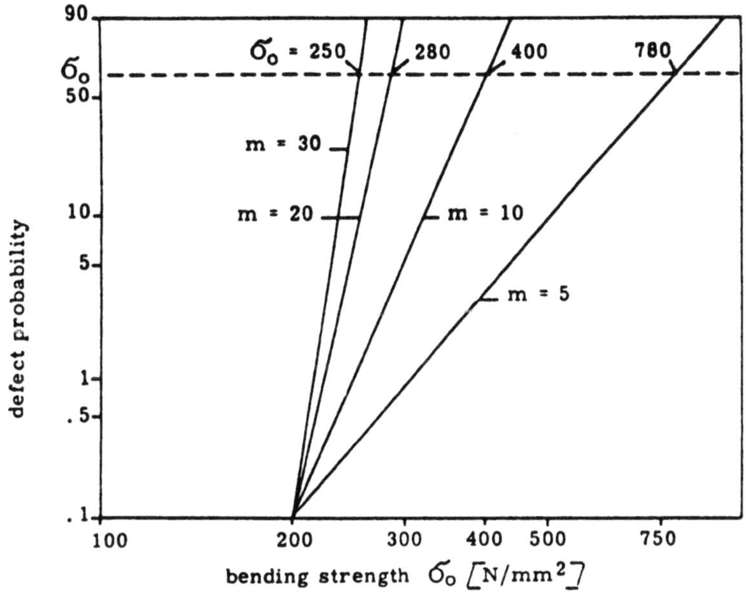

Figure 2. Correlation of Bending Strength, Weibull Modulus and Defect Probability

For that reason high average strength values and small standard deviations (high Weibull values) are repectively the most important aim in this development program.

MANUFACTURING TECHNOLOGY

Figure 3 shows the applied manufacturing technology. The silicon powder with a particle size of $\leq$ 60µm or $\leq$40µm is mixed with thermoplastic material which can be removed from the injection molded part at temperatures $\leq$ 400°C. The nitriding operation has to be carried out at the lowest possible temperature. Temperatures above 1400°C are only allowed in the last stage of the nitriding cycle after 90-95% of the silicon is converted into silicon nitride.

The quality of the silicon nitride product is not only influenced by the quality of the silicon powder, but it also depends strongly on the reproducibility of each single production step. In an automatic injection molding procedure 50-100 pressing operations are usually required to reach equilibrium conditions with sufficiently low standard deviations or high Weibull values.

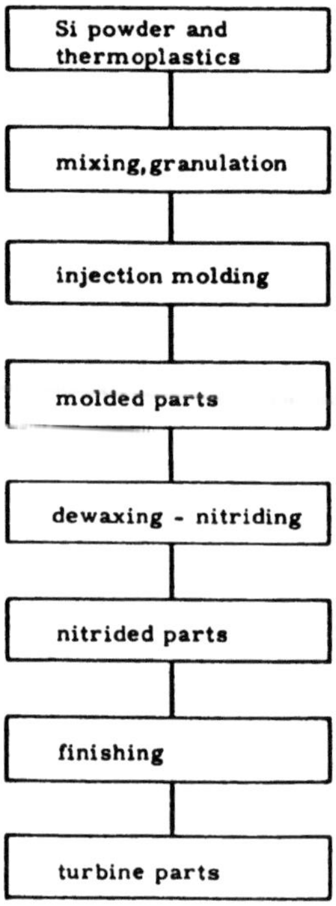

Figure 3. Fabrication Process for Injection Molded Si_3N_4 Turbind Parts

In order to assure high Weibull values in a component, it will be necessary to assure constant nitriding conditions even in larger furnaces. This can only be reached if the nitriding rate can be accurately controlled. The exothermic reaction has to be kept under control, which means the nitriding rate should not be too high. On the other hand, too low nitriding rates would also be unfavorable and lead to products of poor quality.

Important installations in a production line for silicon nitride turbine parts are; a blending machine for getting a homogeneous press feed, an injection molding machine, and furnaces for dewaxing and nitriding.

The pressing machine for the injection molding operation has to be strong and large enough to work continuously over several hours with pressing cycles of 30 seconds and less. Pressing pressures up to 2,000 bar are quite often necessary to form a so called "dry" press feed with plastic contents of 30 vol. % or less.

The nitriding furnace should be suitable for work under vacuum and different gas pressures and atmospheres. To get the desired nitriding rate, one can control the temperature of the product or the consumption of the nitrogen gas. According to our experience both ways are acceptable. Sometimes we had problems with the thermocouples. Therefore, we tend to use the method of controlling the consumption of nitrogen when producing larger quantities in bigger furnaces. Nitriding with flowing nitrogen led to similar results as the nitriding process with only one open valve for filling in the nitrogen gas. There are, however, other experimental conditions which have more influence on the quality of the end product than the question of flowing or not flowing nitrogen gas.

FABRICATION OF TURBINE PARTS

Several types of turbine blades for different turbine manufacturers have been produced and tested. Also, some production experience exists on blade rings. In the following, the results of a large scale production test of turbine blades are described. Later some details concerning the fabrication of blade rings are given.

To control the efficiency of the production equipment and the quality of the produced silicon nitride blades, 5,000 blades were fabricated under realistic fabrication conditions. Within 8 hours, that means 1 shift, 1,000 blades were injection molded, mostly without stopping of the press.

The silicon powder used was a commercially available type. Only a sieving step to eliminate all particles $\geq$ 60μm was added.

To determine both the density and the strength of the pressed material a test bar was produced in one shot together with a blade. Thus, enough test bars were available for a statistical determination of the strength of each pressing and nitriding series.

Two hundred blades and test bars of each of the five 1,000 piece pressing series were statistically selected and put together as a new 1,000 piece series in a dewaxing and sintering furnace. Altogether, 4 sintering cycles were made under almost identical conditions.

Figure 4 shows the result of the determination of 4 point bending strength σ_0 and the Weibull modulus m. As can be seen there is no significant difference between the individual groups. Each group contains 30-40 single values. Concerning the total 5,000 series a σ_0 value of 269 N/mm^2 and a Weibull value of m=13 was obtained.

Visual examination of the fracture surface of the test bars showed in 5-10% of all cases, one relatively large single pore with a diameter of 100μm and more. These pores are large enough to probably be detected by a NDT test. Therefore, we inspected all bars of the nitriding series 4 and selected all samples with these large defects.

Figure 5 shows the result of a new strength calculation as well as a Weibull calculation of this nitriding series 4 in comparison with the corresponding values of Figure 4.

nitriding series		injection molding series 1	2	3	4	5	1 - 5
1	σ_o	265	262	268	271	260	265
	m	17	14	17	10	14	14
2	σ_o	262	274	277	274	264	270
	m	10	17	13	12	12	13
3	σ_o	269	270	270	272	269	270
	m	12	10	16	10	16	13
4	σ_o	267	269	287	267	261	270
	m	9	16	12	10	17	12
1 - 4	σ_o	267	269	275	271	263	269
	m	12	13	15	11	15	13

Figure 4. Injection Molding Series 5,000.
Bending Strength σ_o and Weibull Modulus m

Almost no change is visible at the σ_o values as a consequence of this selection. The average σ_o value is raising only slightly from 270 to 272 but the Weibull value m raises remarkably from 12 to 17.

In our opinion this result is quite important. If it is possible to install an efficient NDT method, one can get Weibull values between 15 and 20 under the described fabrication conditions.

nitriding series	4					
molding series	1 - 3		4 + 5		1 - 5	
sample size	all	no large defects	all	no large defects	all	no large defects
sample number	80	74	56	52	136	126
σ_o [N/mm²]	275	277	264	264	270	272
m	11	15	13	22	12	17

Figure 5. Injection Molding Series 5,000.
σ_0 and m with and without large defects

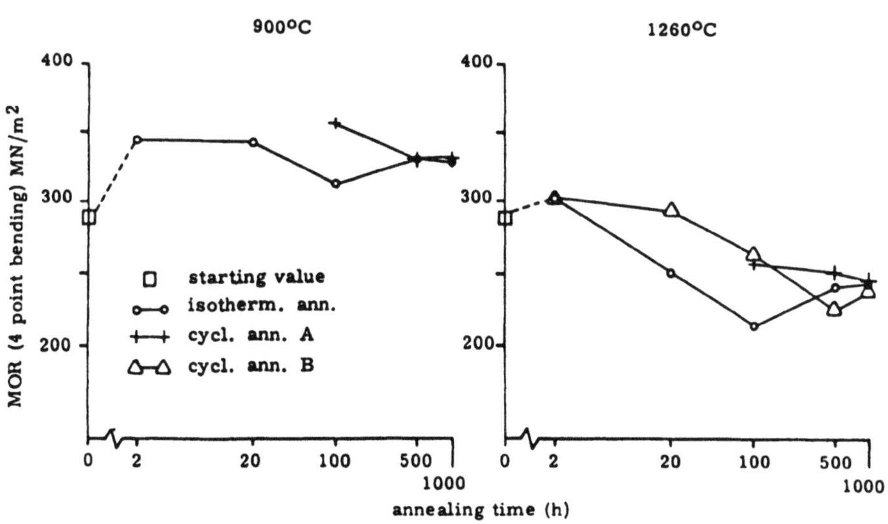

Figure 6. MOR of Injection Molded Si_3N_4 after Isothermal and Cyclic Annealing

Figure 6 shows the effect of isothermal and cyclic annealing on the bending strength at annealing temperatures of 900°C and 1260°C. The sample group for this test was fabricated under conditions similar to the series 5,000 and had a bending strength at the beginning of 287 N/mm^2. At 900°C even after 1,000 hours annealing time the bending strength was always higher than in the beginning. At 1260°C, however, the strength value decreased after a longer annealing time and especially under isothermal annealing conditions. All data of Figure 6 were obtained at the Volkswagenwerk AG.

Figures 7 and 8 show some of the produced blades themselves. Figure 7 shows some blades for gas turbine design of the Volkswagenwerk AG (VW) and the Motoren-und Turbinen-Union (MTU) after nitriding and cleaning and Figure 8 some blades standing in a vertical position as it is necessary for nitriding.

The result of this large scale production test is typical for a quality obtained at present under realistic fabrication conditions with a commercial silicon powder.

Smaller test series with several 100 blades and test bars have been made in the last few months to get higher σ_0 and/or higher Weibull values. For these tests a special silicon powder and a slightly modified manufacturing technique have been used. Two typical results are:

$$\sigma_0 = 320 \text{ N/mm}^2 \text{ with } m=11$$

$$\sigma_0 = 262 \text{ N/mm}^2 \text{ with } m=21$$

These results lead to the conclusion that some further improvements will probably be possible.

The results of the test bar experiments are supported by spin tests of blades at VW and MTU. In both cases, some spin test values above the calculated design values could be obtained. But there are some other test values still far behind this goal.

With the injection molding machine used for manufacturing the blades, blade rings with plane blades were also produced but only in smaller quantities. Molded parts like these blade rings cannot be characterized by a test bar produced in the same shot. In this case spin tests or burst tests are necessary to obtain realistic values of the strength. Both kinds of tests have been performed by VW and shall be discussed briefly. Figure 9 shows the results of these measurements. Spin test values up to 35,500 rpm at a specified value of 36,000 rpm could be obtained. Blade rings of a similar quality reached a burst pressure of up to 195 bar at a specified value of 175 bar. However, in both tests some other blade rings have shown lower values.

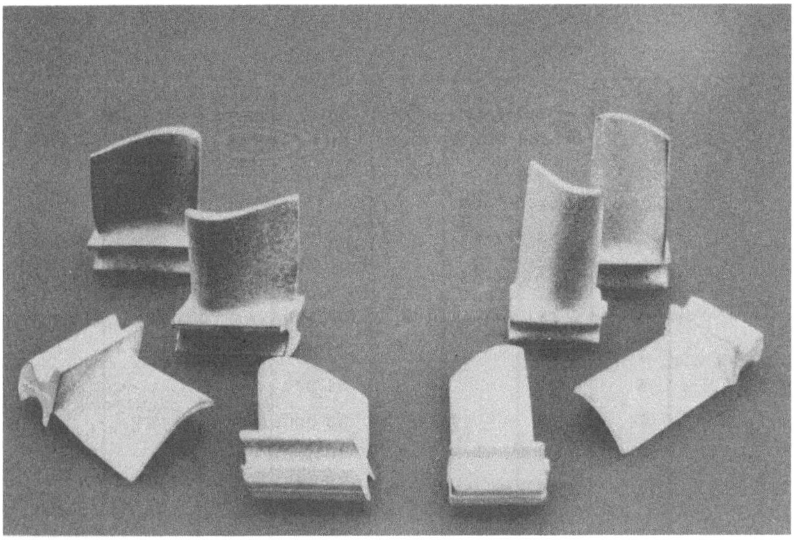

Figure 7. Si_3N_4 Blades of Two Different Designs

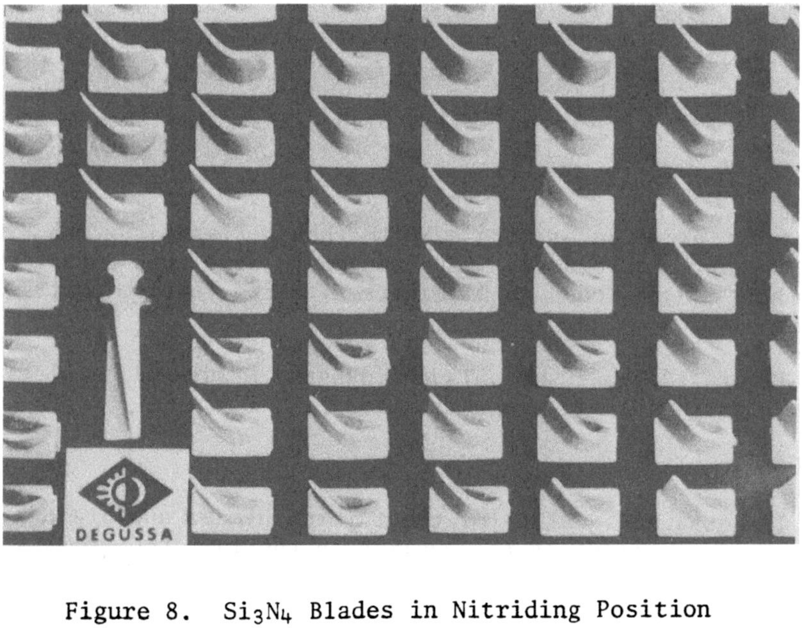

Figure 8. Si_3N_4 Blades in Nitriding Position

blade ring no.	density (g/cm^3) blade ring	blade	ring	spin test (rpm)	burst test (bar)	NDT result
1	2,61	2,62	2,61	-	-	-
2	2,63	2,65	2,62	-	-	-
3	2,62	2,64	2,61	-	-	-
4	2,64	-	-	32 500	-	-
5	2,62	-	-	32 400	-	-
6	2,62	-	-	35 000	-	-
7	2,62	-	-	33 500	-	-
8	2,62	-	-	30 500	-	-
9	2,60	-	-	35 500	-	-
10	2,64	-	-	-	195	-
11	2,60	-	-	-	140	-
12	2,60	-	-	-	150	defect
13	2,60	-	-	-	165	o.k.
14	2,60	-	-	-	175	o.k.
15	2,60	-	-	-	185	o.k.

Remarks:

1. blade rings with plane blades
2. spin test - specified value: 36 000 rpm
3. burst test - specified value: 175 bar

Figure 9. Injection Molded Blade Rings, Test Results

First experiences with a NDT method performed by the Fraunhofer Gesellschaft, Saarbrücken, give reason to hope that a quality test procedure can be found to eliminate nitrided turbine parts of low quality.

Figure 10 shows some blade rings and Figure 11, a section of a blade ring in a larger magnification.

OUTLOOK

The obtained quality of the turbine blades and blade rings and also the improvement of the quality in the last two years are quite encouraging. Values near or above the design values have been reached.

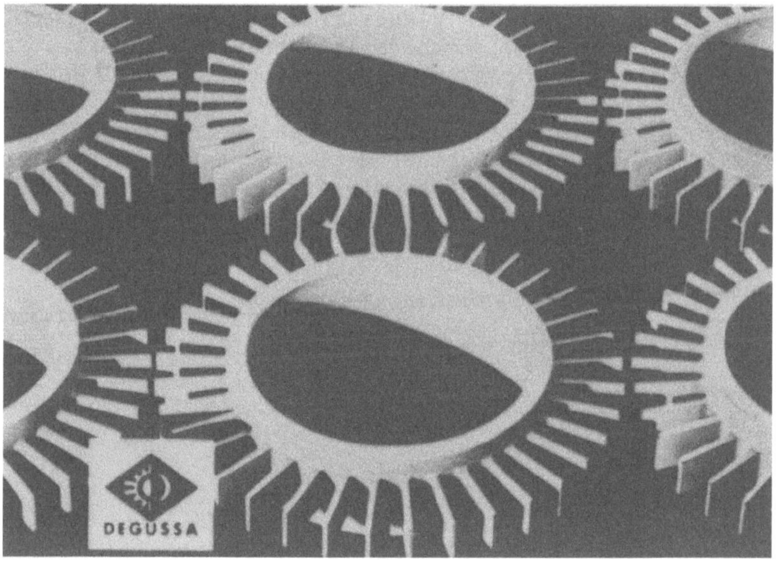

Figure 10. Blade Rings with Plane Blades

Figure 11. Blade Ring Section

However, the Weibull modulus or more generally speaking, the scattering of the single values is a serious problem not really solved so far. NDT methods will be very important to solve this problem in the future. Unfortunately, all Weibull examinations have to be made on large sample groups produced under realistic fabrication conditions and this makes it quite clear that still a lot of further development work has to be done to achieve a cheap manufacturing technology for gas turbine parts with reproducible and high quality.

FACTORS INFLUENCING THE QUALITY OF FULLY DENSE SILICON NITRIDE

G.E. Gazza and R.N. Katz

Army Materials and Mechanics Research Center
Watertown, MA 02172

H. Knoch

Deutsche Forschungs und Versuchsanstalt fur Luft
und Raumfahrt e.V.

INTRODUCTION

Dense silicon nitride, which can be produced by hot pressing or sintering, is adaptable to compositional and/or microstructural alterations which can improve the reliability and performance of the material. The compositional approach involves additive selection, impurity effects, and phase equilibria studies. Microstructural improvement relies on starting material characteristics and optimum selection of process parameters.

Reliability of ceramic materials is strongly influenced by material defects which are introduced by processing, machining and/or applied stress/environmental exposure. The work discussed here focuses on studies aimed at improving material performance and reliability by process control over selected compositions and by development of microstructure.

Considerable experimental effort has been expended in developing dense silicon nitride with high strength for structural applications. It was found in the early development of hot pressed material that additives were required to promote densification and that the phase composition of the starting material was important. Magnesia (1) was first used as a sintering aid while other additives (2-5) have since been found to promote densification. The additive reacts with the Si_3N_4 to form a boundary liquid. Sintering occurs by a solution - reprecipitation mechanism. The liquid acts as a vehicle for the $\alpha \rightarrow \beta$ phase transformation and subsequent grain growth of the beta phase. The grains develop a prismatic or elongated shape after the $\alpha \rightarrow \beta$ transformation with aspect ratios in the range of 10:1. The sintered product may contain glassy and crystalline boundary phases depending on the type and amount of additive used. Some additive cations have high solubility in Si_3N_4, e.g. Be, while others remain principally at grain boundary locations, e.g. Y. Thus, Si_3N_4 may be considered as a composite material of elongated grains in a grain boundary matrix produced by Si_3N_4/additive/SiO_2 reactions. The fracture mode in Si_3N_4 is predominantly intergranular. Therefore, there appears to be two primary approaches to be taken for improving the mechanical behavior of the material from a processing standpoint. The first would be to improve the quality of the grain boundary phase and the bond strength between grains. This concept has been described as "grain boundary engineering" (GBE) (6). Its goal is the deliberate selection and control of composition, structure, and processes occurring in the grain boundary region aimed at specific property modification. The second approach is a structural one. It involves controlling the grain size and morphology in the material. Improvements in both strength and the fracture toughness parameters have been reported (7-9) where microstructures with fine grain sizes and elongated grain shape were developed.

Although fully dense Si_3N_4 may now be produced using a variety of additives, each compositional system exhibits property characteristics which would limit reliability and performance of the material in advanced heat engine applications. The limiting factors on properties generally associated with the Si_3N_4-MgO and Si_3N_4-Y_2O_3 systems are shown in Table I. It is widely recognized that the use of MgO as a sintering aid produces a nitrogen containing glassy phase at the grain boundaries which limits high temperature (>1200 C) strength and creep resistance. In addition, oxidation resistance becomes poorer above approximately 1200 C and significant degradation of strength occurs. For the Si_3N_4-Y_2O_3 system, the development of phases which promote high temperature strength also appear to produce thermal instability of the material at intermediate temperatures (700-1100 C) resulting in structural degradation.

Table I

Factors Limiting Reliability of Fully Dense Si_3N_4 - MgO and Si_3N_4 - Y_2O_3 Systems

Si_3N_4 - MgO	Si_3N_4 - Y_2O_3
• High Temperature Strength	
• High Temperature Creep	• Oxidation 700 - 1100°C
• Oxidation >1200 C	• Processing Difficulty

Some of these limiting factors may be addressed by a combined approach incorporating microstructural, chemical, and process control. Our work on the Si_3N_4-MgO system has emphasized the effects of process variables on microstructural development whereas the work on the Si_3N_4-Y_2O_3 system has focused principally on chemical and environmental effects.

Table II shows the compositional and microstructural control approaches taken in order to improve performance of hot pressed Si_3N_4 (HPSN). Such measures would also be applicable to sintered material.

Table II

Goal	Approach	
	Compositional Control (Grain Boundary Eng.)	Microstructural Control
Improved Performance and Reliability of Si_3N_4 by Process Control	• Additives (MgO, Y_2O_3, etc.)	• Grain Size
	• Impurities (Metal Cations Carbon)	• Grain Morphology (Equiaxed/Prismatic)
	• Phase Behavior (Multi-Component Systems)	• Phase Control (α/β ratio)

Compositional Control

Compositional control techniques, which include GBE concepts, emphasize deliberate selection of additives to promote densification and development of phases which resist high temperature creep and oxidation. The objectives may be to form chemically refractory grain boundary phases, e.g. yttrium silicon oxynitrides (10), use additives with extensive solubility in Si_3N_4 to minimize the amount of grain boundary phase, e.g. BeO, $BeSiN_2$ (11), employ heat treatments to crystallize glassy boundary phases (12), or simply increase the refractoriness of the system by reducing the impurity content (13) which tends to form lower melting compositions. This has been particularly noted for Ca impurity in the Si_3N_4-MgO system (14,15).

The use of Y_2O_3 as an additive has enhanced the high temperature strength of Si_3N_4 by promoting the formation of refractory oxynitrides, producing glassy phases which may be crystallized, and accommodating detrimental impurities, e.g., Ca, into solid solution in these phases which minimizes its effect on high temperature properties. However, the major factor limiting the performance of the Si_3N_4-Y_2O_3 system is material degradation at intermediate temperatures under oxidizing environments. This problem has been initially defined and addressed by Lange (16) using phase equilibria studies. He found that the problem could be alleviated by restricting compositions to within the Si_3N_4-$Y_2Si_2O_7$-SiO_2 compatibility area of the Si_3N_4-Y_2O_3-SiO_2 system. This avoids formation of phases found to be susceptible to the thermal instability problem, i.e., $Y_2Si_3O_3N_4$ (N-melilite), $YSiO_2N$ (K-phase), $Y_4Si_2O_7N_2$ (J-phase), and $Y_{10}Si_7O_{23}N_4$ (H-phase).

Recent work at AMMRC (17) has shown that improvement in intermediate temperature oxidation behavior can be produced by heat treating fully dense, hot pressed Si_3N_4 containing 13% Y_2O_3 in a nitriding environment. Subsequent strength testing at room temperature after oxidation for 100 hrs. and stress-rupture tests at 1000 C in air indicated no material degradation up to approximately 400 hrs. Figure 1 shows the strength vs. time results for heat treated and untreated Si_3N_4 with 13% Y_2O_3 at 1000 C under 275 MN/m^2 applied bend stress. Brennan (18) has noted similar behavior using NCX-34 Si_3N_4 (8% Y_2O_3 additive) which had been subjected to nitriding cycles using both pure N_2 and 96% N_2/4% H_2 mixtures. Although these samples did not gain any measureable weight when oxidized for 190 hrs at 930 C and showed good strength retention, material degradation was noted at a lower oxidizing temperature, 730 C. However, the samples subjected to the nitridation cycle still showed some improvement over the as-received NCX-34 when exposed to the 730 C oxidation treatment. It is apparent that continued studies on the Si_3N_4-Y_2O_3-SiO_2 system are required to further define relationships between processing and phase behavior and their influence on resultant properties.

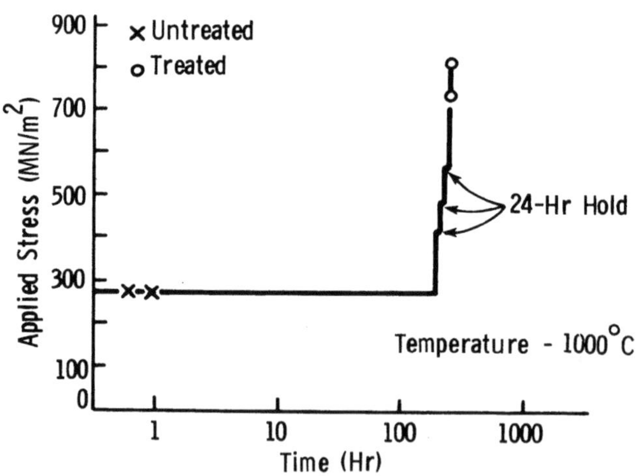

Figure 1. Stress and time dependence of failure in 13 wt % Y_2O_3 doped HPSN.

The effect of metal cation impurities on reducing reliability and performance of Si_3N_4 at high temperature is well known. Another impurity effect, that of carbon in the Si_3N_4-Y_2O_3 system has been observed by Knoch and Gazza (19). It was found that when low amounts of carbon, <0.06%, were present in a hot pressed $Si_3Y_2O_3N_4$ using GTE-SN502 Si_3N_4 powder as starting material, machined specimens would not exhibit cracking after oxidation at 1000 C for >250 hrs. Oxidation kinetics were parabolic. If the hot pressed $Si_3Y_2O_3N_4$ compound was fabricated using a starting Si_3N_4 powder containing 0.6-0.7% carbon, or 2% addition of carbon or silicon carbide were intentionally made to the starting material, the hot pressed specimens would crack severely upon oxidation at 1000 C and oxidation kinetics became linear. Integrated or cumulative oxidation data from 700 C to 1400 C using the low carbon and high carbon containing materials showed cracking and high weight gain only for the high carbon containing N-melilite compounds (Figure 2). Since it is common practice to use graphite dies and/or heating elements to hot press or sinter Si_3N_4, the above findings have important implications on materials reliability.

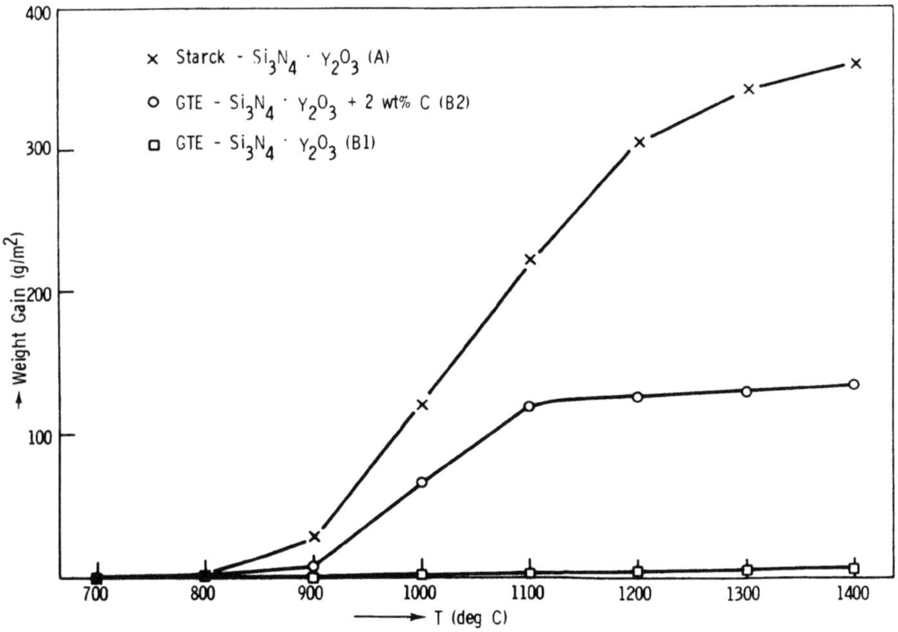

Figure 2. Integrated weight gain of three $Si_3N_4 \cdot Y_2O_3$ compounds as a function of temperature. Holding time at each data point 24 hours.

Microstructural Control

The foregoing discussion has emphasized the identification and attempted resolution of chemical and phase problems affecting performance and reliability of Si_3N_4 materials. Another approach is a structural one. Maximizing reliability and material performance using this approach relies on the development of uniform microstructures with grain size and grain morphology which enhance strength and fracture mechanics parameters. For silicon nitride, it has been shown that the overall grain size and growth of elongated beta grains accompanying densification are important factors influencing the strength (20, 21). The interlocking of growing beta prisms during the α to β phase transformation produces an improvement of bend strength, fracture toughness, work of fracture, and slow crack growth data. However, a degradation of mechanical properties results from grain growth after the phase transformation has been completed. The development of the desired fine grain, fibrous, uniform microstructure in HPSN is primarily a function of the character of the starting powder, i.e., particle size, purity, and α/β phase ratio, and the hot pressing parameters; time, temperature and applied pressure. Studies at AMMRC were conducted using starting powders with 4% and 29% beta phase, adding 5% MgO, and hot pressing at 1500-1700 C for times up to 3 hrs. (22). Applied pressure used was generally 35 MN/m^2 although some runs were made with 70 MN/m^2 pressure. The high beta starting material (29% β) transformed faster to complete β-phase hot pressed product but the average aspect ratio of the resultant grain size was lower than when the 4% β starting material was used. Transformation structures of the low β starting material appear more uniform and more fibrous than those of the high β starting material. Grain growth is observed with increasing hot pressing time and/or increasing volume fraction of transformed β-phase. The influence of the amount of β phase on the strength of the hot pressed product as a function of hot pressing time is shown in Figures 3 and 4. The hot pressing temperature used was 1600 C and the applied pressure was 35 MN/m^2. In Figure 3, the modulus of rupture increases with increasing development of β-phase although complete transformation is not achieved at the time-temperature parameters used. In contrast, when the 29% β material is used, Figure 4, transformation is completed for similar time-temperature parameters. However, modulus of rupture values are lower when starting with the higher β-phase powder and some decrease in strength appears to occur when hot pressing time is extended to 3 hrs. This decrease is attributed to coarsening of the microstructure. The effect may be amplified by using higher applied pressure during hot pressing. Figure 5 shows the effect of hot pressing time on strength and $\alpha \rightarrow \beta$ transformation when 70 MN/m^2 applied pressure is used during hot pressing at 1600 C. It is observed that a maximum in strength is obtained at approximately 1 hr. hot pressing time, then a decrease in strength occurs with further time. The optimum strength is close to the time where the $\alpha \rightarrow \beta$ conversion is nearly complete. Comparison

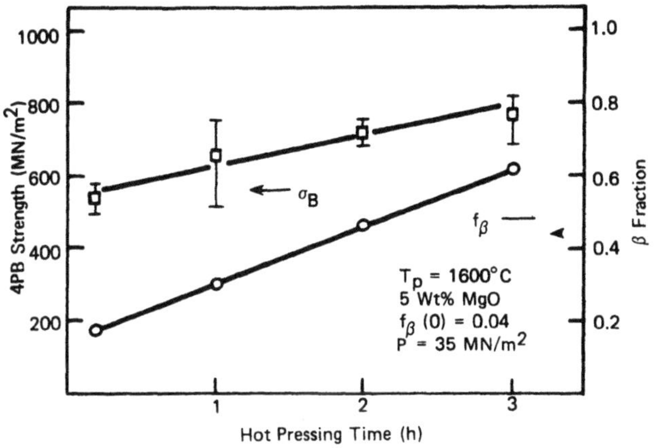

Figure 3. β fraction and 4PB strength of hot-pressed silicon nitride as a function of hot-pressing time for 0.04 f_β powder.

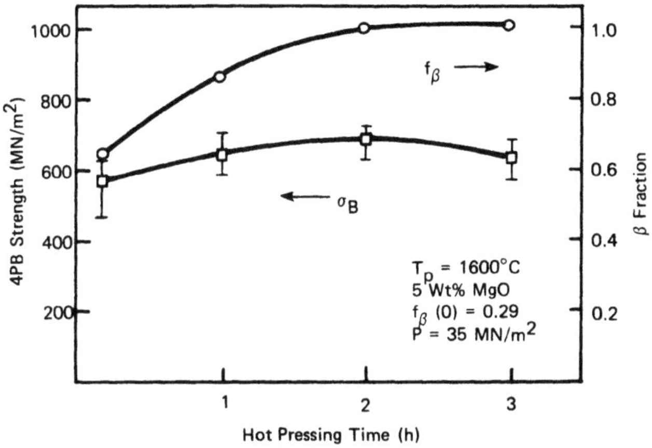

Figure 4. β fraction and 4PB strength of hot-pressed silicon nitride as a function of hot-pressing time for 0.29 f_β powder.

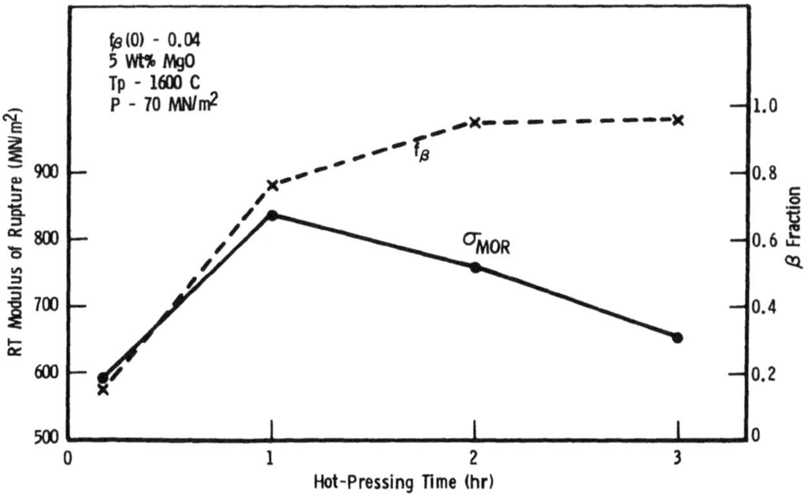

Figure 5. β fraction and 4PB strength of HPSN as a function of hot-pressing time under 70 MN/m^2 applied pressure.

of the microstructures for each processing condition is shown in Figure 6. After 10 min. (Figure 6a), the microstructure is fine and equiaxed. Only limited conversion to β-phase has occurred. With increasing hot pressing time, the grain shape becomes more prismatic as α converts to β. Strength increases considerably with the morphological change in microstructure. After 2-3 hrs. (Figures 6c, 6d), the microstructure is coarsening and the strength decreases until it is only 70 MN/m^2 stronger after 3 hrs. than it was after only 10 min. of hot pressing time. These results may be interpreted by assuming an increase in the nucleation rate with increasing pressure. The formation of new β nucleii increases the number of growing β grains which decreases the average grain size for a given

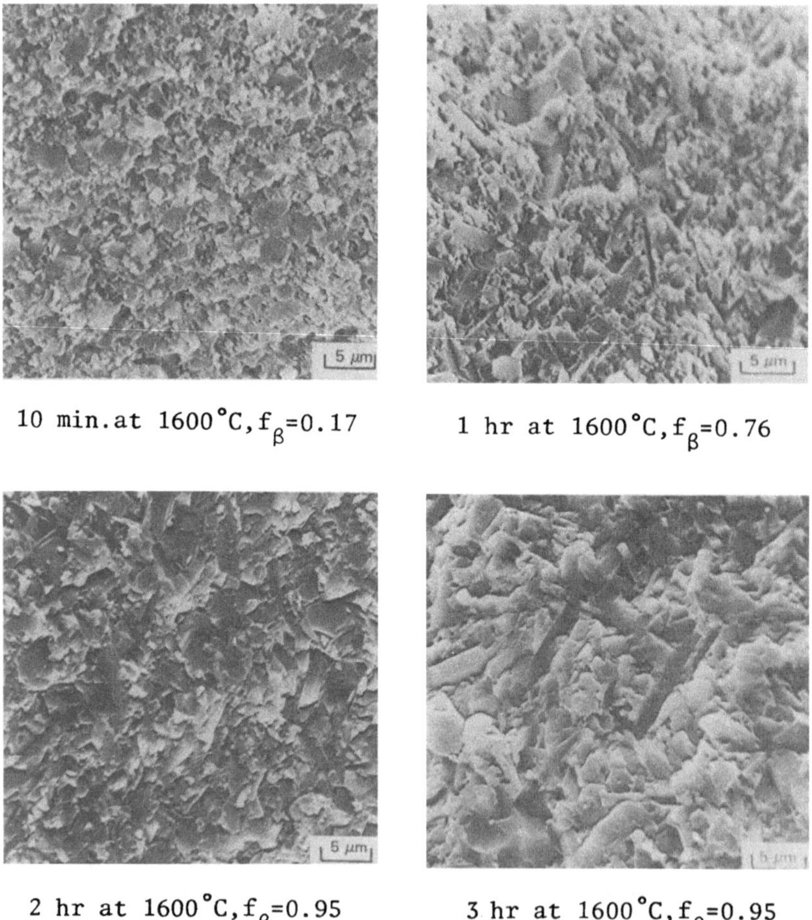

10 min.at 1600°C, f_β=0.17 1 hr at 1600°C, f_β=0.76

2 hr at 1600°C, f_β=0.95 3 hr at 1600°C, f_β=0.95

Figure 6. Microstructure of HPSN as a function of hot-pressing time. Tp - 1600°C, P - 70 MN/m^2, starting powder: f_β = 0.04; 5 wt % MgO.

transformed volume fraction of β-phase. The higher pressure, 70 MN/m^2, increases the transformation rate and produces higher strength material than when 35 MN/m^2 pressure is used at 1600 C hot pressing temperature. However, grain coarsening also occurs in shorter hot pressing time using higher pressure and optimization of process parameters becomes more critical.

In general, the best mechanical properties are achieved at small grain sizes and a high volume fraction of elongated β grains, i.e., when the α→β conversion is complete and no further grain coarsening after completed transformation has occurred. Processing parameters like β fraction in the starting powder or applied pressure during hot pressing are important factors influencing phase transformation and correlated development of microstructure. Optimum microstructures with improved mechanical properties can be achieved by process control of these parameters.

Conclusion

The quality and subsequent performance of fully dense silicon nitride in current areas of interest, i.e., advanced heat engines, bearings, metal processing, etc., is dependent on optimizing process control of material and fabrication parameters. Both compositional and microstructural approaches may be taken to improve material performance for any given application. For silicon nitride, additive selection, impurity control, and knowledge of phase behavior are important compositional factors while starting material characteristics and process parameter selection strongly influence resultant microstructures. The presence of carbon in the Si_3N_4-Y_2O_3 system was found to be particularly detrimental to intermediate temperature properties.

The use of starting material and fabrication parameters which produce a uniform α→β conversion while avoiding grain coarsening will result in superior mechanical properties with improved reliability.

References

1. G. G. Deeley, J. M. Herbert, and N. C. Moore, "Dense Silicon Nitride," Powder Met., 8, 145-51 (1961).
2. G. E. Gazza, "Hot Pressed Si_3N_4," Jour. Amer. Ceram. Soc., 56, [12], 662 (1973).
3. I. C. Huseby and G. Petzow, "Influence of Various Densifying Additives on Hot Pressed Si_3N_4," Powder Met. Int., 6, [1], 17-19 (1974).
4. K. S. Mazdiyasni and C. M. Cooke, "Consolidation, Microstructure, and Mechanical Properties of Si_3N_4 Doped With Rare Earth Oxides," Jour. Amer. Ceram. Soc., 57, [12], 536 (1974).

5. A. W. J. M. Rae, D. P. Thompson and K. H. Jack, "The Role of Additives in the Densification of Nitrogen Ceramics," Proc. 5th Army Materials Technology Conference, Ceramics for High Performance Application - II, Burke, Lenoe and Katz, eds., Brook-Hill Pub. Co. (1978).
6. R. N. Katz and G. E. Gazza, "Grain Boundary Engineering and Control in Nitrogen Ceramics," Proc. NATO Advanced Study Inst. on Nitrogen Ceramics, Nitrogen Ceramics, F. L. Riley, ed., Noordhoff Int. Pub. Co. (1977).
7. F. F. Lange, "Relation Between Strength, Fracture Energy and Microstructure of Hot Pressed Si_3N_4," Jour. Amer. Ceram. Soc., 56, [10], 518-522 (1973).
8. H. Knoch and G. Ziegler, "Influence of Powder Composition and Sintering Temperature on Transformation Kinetics, Microstructure, and Mechanical Properties of Hot Pressed Silicon Nitride," Ber. Dtsch. Keram. Ges., 55, [4], 242-45 (1978).
9. L. J. Bowen and T. G. Carruthers, "Development of Mechanical Strength in Hot Pressed Silicon Nitride," Jour. Mat'l Sci., 13, 684-87 (1978).
10. A. W. J. M. Rae, D. P. Thompson, N. J. Pipkin and K. H. Jack, "The Structure of Yttrium Silicon Oxynitride and Its Role in the Hot Pressing of Silicon Nitride With Yttria Additions," Special Ceramics 6, P. Popper, ed., 347-60, Brit. Ceram. Res. Assoc., Stoke-on-Trent (1975).
11. S. Prochazka and C. D. Greskovich, "Development of a Sintering Process for High-Performance Silicon Nitride," Gen. Elec. Co., Corp. R&D, Schenectady, NY, Final Report AMMRC TR 78-32 under AMMRC/DOE Interagency Agreement EC-76-A-1017.
12. A. Tsuge, K. Nishida and M. Komatsu, "High Temperature Strength of Hot Pressed Si_3N_4 Containing Y_2O_3," Jour. Amer. Ceram. Soc., 58, [7-8], 323-325 (1975).
13. D. W. Richerson, "Effect of Impurities on the High Temperature Properties of Hot Pressed Silicon Nitride," Amer. Ceram. Soc. Bull., 52, (7), 560 (1973).
14. R. Kossowsky, "The Microstructure of Hot Pressed Silicon Nitride," Jour. Mat. Sci., 8, 1603 (1973).
15. F. F. Lange and J. L. Iskoe, "High Temperature Strength Behavior of Hot Pressed Si_3N_4 and SiC: Effect of Impurities," Ceramics for High Performance Application, Chap. 11, J. J. Burke, A. E. Gorum and R. N. Katz, eds., Brook Hill Pub. Co., Chestnut Hill, MA (1974).
16. F. F. Lange, S. C. Singhal and R. C. Kuznicki, "Phase Relations and Stability Studies in the Si_3N_4-SiO_2-Y_2O_3 Pseudoternary System," Jour. Amer. Ceram. Soc., 60, [5-6], 249-52, (May-June 1977).
17. G. E. Gazza, H. Knoch and G. D. Quinn, "Hot Pressed Si_3N_4 with Improved Thermal Stability," Amer. Ceram. Soc. Bull., 57, [11], 1059-60, (Nov. 1978).

18. J. J. Brennan, "Development of Silicon Nitride of Improved Toughness," Final Report NASA CR-159676, Oct. 2, 1979 (United Tech. Research Center, East Hartford, CT, NASA Lewis Contract NAS3-21375).
19. H. Knoch and G. E. Gazza, "Carbon Impurity Effect on the Thermal Degradation of a Si_3N_4-Y_2O_3 Ceramic," AMMRC TR 79-27, (May 1979).
20. F. F. Lange, "Dense Si_3N_4: Interrelation Between Phase Equilibria, Microstructure, and Mechanical Properties," Nitrogen Ceramics, F. L. Riley, ed., 491-509, (Noordhoff-Leyden 1977).
21. G. Himsolt, H. Knoch, H. Huebner and F. W. Kleinlein, "Mechanical Properties of Hot Pressed Silicon Nitride With Different Grain Structures," Jour. Amer. Ceram. Soc., 62, [1-2] 29-32 (1979).
22. H. Knoch, G. E. Gazza and R. N. Katz, "The Influence of Processing Parameters on Development of Microstructure in Hot-Pressed Silicon Nitride," Proc. 4th CIMTEC Int. Meeting on Modern Ceramic Technologies, Saint Vincent, Italy (1979).

SINTERING OF SILICON NITRIDE AT ATMOSPHERIC PRESSURE

A. Giachello(*) and P. Popper (**)

(*) FIAT Research Center S.p.A. - Orbassano (Italy)

(**) British Ceramic Res. Ass.-Stoke on Trent (U.K.)

ABSTRACT

It is well known that highly covalent materials like silicon nitride are difficult to densify by sintering. Even if considerable pressure is applied, as in hot-pressing, sintering aids, e.g.,MgO, Y_2O_3, have to be employed.

Pressureless sintering generally requires a higher temperature than hot-pressing. However, during sintering at such high temperatures a weight loss occurs. This is due to dissociation of the silicon nitride and - what appears so far not to have been fully appreciated - loss of the volatile sintering aid, e.g. MgO. By suitable choice of a powder bed into which the samples are placed, dissociation and volatilization can be suppressed at atmospheric pressure of nitrogen, and high sintered densities can be obtained.

Because of the less drastic decay in strength at high temperatures of samples using Y_2O_3 rather than MgO as sintering aids, a composition 91% Si_3N_4 + 8% Y_2O_3 + 1% MgO has been chosen for the sintering study, High densities were achieved by sintering at 1800°C.

The same sintering aids can also be admixed to silicon before it is subjected to nitridation and final high temperature sintering at 1800°C. This process, which may also be described as "post sintering of reaction sintered silicon nitride", has resulted in very high room-temperature (∿ 1000 MN/m^2) with linear shrinkage of only about 6%.

It is also possible to carry out the post-sintering process on reaction-sintered silicon nitride to which no sintering aids have been added in powder form by transferring the sintering aid in the vapour form into the reaction sintered body. In this way silicon nitride of density 3.05 Mg/m^3 has been formed, which had a higher strength than the precursor reaction-sintered material between room temperature and 1400°C.

INTRODUCTION

Two types of silicon nitride are today commercially available i.e. reaction sintered (RSSN) and hot-pressed silicon nitride (HPSN). Their particularly advantages and disadvantages are well known. In the case of the rotor of a gas turbine both high strength and fabricability of complex shapes are required. This has led to the concept of the "duo-density" rotor which is made up of an injection-moulded RSSN blade ring and a HPSN central disc. The desirability of processing this component in one piece has directed present research effort towards sintering of silicon nitride (SSN). The problems of sintering silicon nitride have been discussed recently[1]; they are i) low green density of highly sinterable silicon nitride powder resulting in high shrinkage, ii) dissociation of silicon nitride at sintering temperature, iii) volatilization of sintering aid. This paper shows how an appreciation of all the problems leads to a pressureless sintered silicon nitride of exceptional high strength at room temperature which has undergone only 6-7% linear shrinkage.

THE DISSOCIATION OF Si_3N_4 AND THE VOLATILIZATION OF MgO

The weight losses occuring during sintering of Si_3N_4 powder to which 5% MgO has been added and the corresponding sintered densities are shown in Figs. 1 and 2 for different sintering temperatures. Curves "a" which are due to Terwilliger and Lange[2] who first reported the resulte of open sintering in nitrogen at atmospheric pressure. The high weight loss and the low density (< 80% theoretical) were ascribed entirely to the dissociation of Si_3N_4. Mitomo[3] was able to obtain higher densities by suppressing the dissociation through sintering at a nitrogen pressure of 10 atm. Similar though slightly lower densities were obtained by sintering in a powder bed[4] consisting of Si_3N_4 and BN (curves "c"). However, such samples were inhomogeneous and exhibited, on sectioning a 1.5 mm thick skin of lower density which by Laser Microspectral Analysis was shown to have a lower Mg content than the bulk of the sample (Fig. 3). This was evidence for loss of MgO from the surface and, in order to decrease it, MgO was incorporated[4] in the powder bed which resulted in homogeneity and further improvement in the densification. The optimum sintering conditions and properties are given in Table 1. The additive Fe was the residue from milling in a steel mill. It is interesting that the iron is contained as an impurity phase which is a solid solution of fayalite and forsterite.

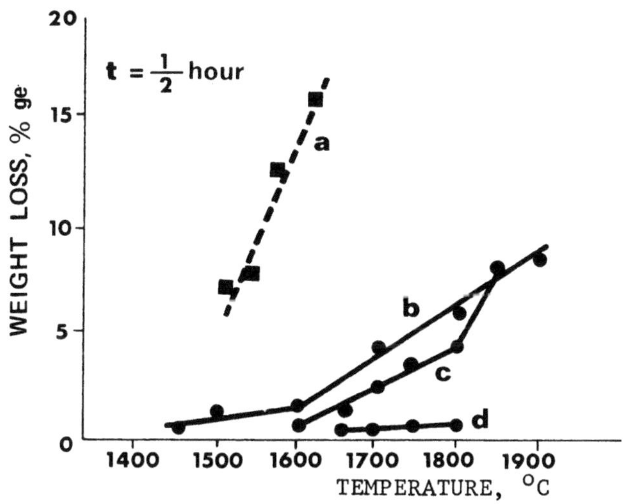

Fig. 1. Weight loss during sintering vs. sintering temperature:
a) obtained in nitrogen at atmospheric pressure by Terwilliger and Lange[2]
b) obtained at 10 atm. nitrogen pressure by Mitomo[3]
c) obtained by Giachello et al. at atmospheric pressure in powder bed[4]
d) as c) but powder bed containing MgO[4].

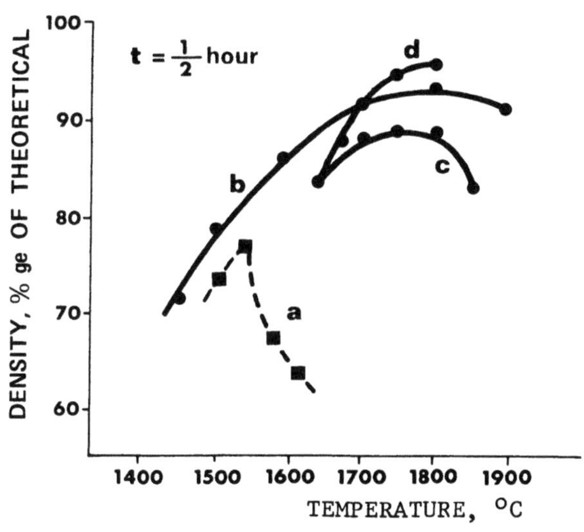

Fig. 2 Density of sintered Si_3N_4 + 5% ge MgO vs. sintering temperature: a), b), c) and d) as in Fig. 1

TABLE I

Pressureless sintered Si_3N_4		
Starting material	:	A.M.E. Si_3N_4 CP85 powder
Additives	:	5 w/o MgO - 2 w/o F_e
Green density	:	2.0 Mg/m^3
Sintering conditions	:	in powder bed at 1650°C for 5 h
Linear shrinkage	:	~ 14 %
Characteristics of the sintered material		
Density	:	3.10 Mg/m^3
Phases by XRD	:	β Si_3N_4
		$(Mg_{0.4}Fe_{0.6})_2SiO_4$
Flexural strength	:	at 25°C = 500 MN/m^2
(3 point)		at 1250°C = 280 MN/m^2
Weibull modulus	:	9.8
Oxidation resistance	:	5 mg/cm^2 (weight gain after 100 hours at 1300°C in air)

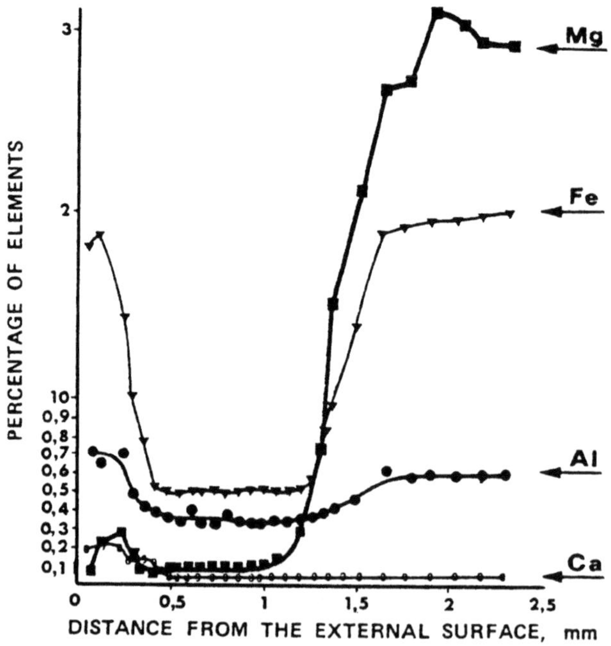

Fig. 3. Concentration profile of different elements vs. distance from external surface

SINTERING OF Si_3N_4 WITH Y_2O_3 AND MgO AS SINTERING AIDS

Because of the well-know deterioration of the strength at high temperatures of materials containing large (∿ 5%) amounts of MgO the following composition was chosen[5]: 91% Si_3N_4 + 8% Y_2O_3 + 1% MgO. The densification vs. temperature when sintering in a powder bed containing MgO is shown in Fig. 4. There was no weight loss but instead a slight gain (∿ 1%) which is thought to be due to nitridation of unnitrided Si in the starting powder. The optimum sintering conditions and the resulting properties are given in Table II.

TABLE II

Pressureless sintered Si_3N_4		
Starting material	:	A.M.E. Silicon Nitride CP 85 powder
Additives	:	8 w/o Y_2O_3 + 1 w/o MgO
Green density	:	2,0 Mg/m^3
Sintering conditions	:	in powder bed at 1800°C for 5 hours
Linear shrinkage	:	∿ 16%
Characteristics of the sintered material		
Density	:	3,25 Mg/m^3
Phases by XRD	:	β Si_3N_4, amorphous
Flexural strength (3 point)	:	at 25°C = 600 MN/m^2 at 1250°C = 380 MN/m^2
Weibull modulus	:	9,3
Oxidation resistance	:	2,5 mg/cm^2 (weight gain after 100 h at 1300°C in air)

The strength versus temperature behaviour is shown in Fig. 5 (curve "a"). There is clear evidence of devitrification Fig. 5 (curve "b")[6]. Oxidation resistance was also improved by the devitrification; there was no evidence of a catastrophic oxidation at 1000°C[7].

POST SINTERING OF REACTION-SINTERED SILICON NITRIDE (SRSSN)

The shrinkage during sintering is to a considerable extent determined by the green density. From a technological point of view it is very desirable to keep it as low as possible. Thus it would be very advantageous if the sintering operation could be carried out on a reaction-sintered body (density ∿ 2.6 Mg/m^3) instead of powder compact (density ∿ 2,0 Mg/m^3). This processing route has been achieved in two ways[8]:

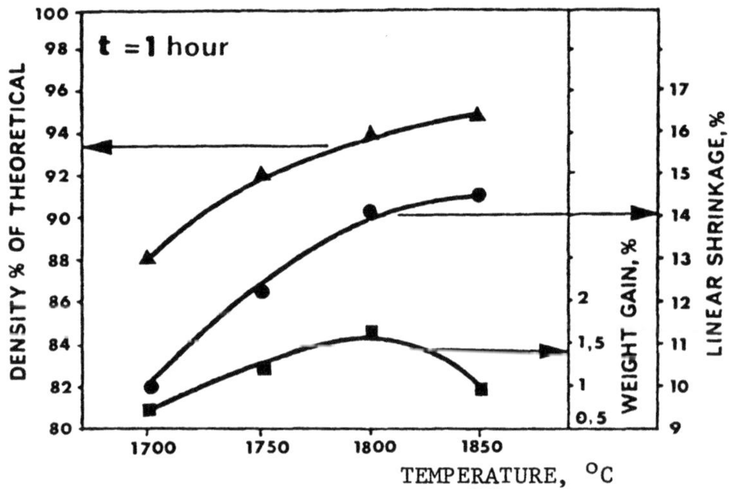

Fig. 4. Density, weight gain and shrinkage vs. sintering temperature for the mixture Si_3N_4 + 8% Y_2O_3 + 1% MgO

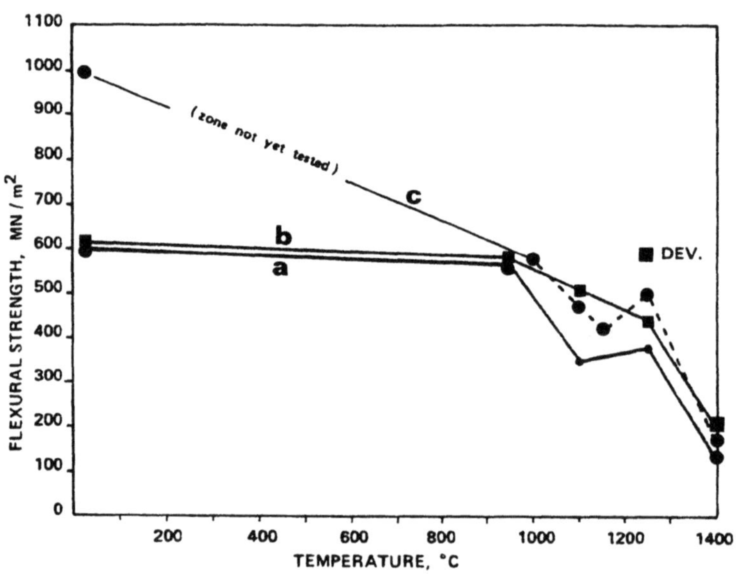

Fig. 5. Tree-point flewural strength for material produced from Si_3N_4 powder (curve "a"), and the same material partially devitrified (curve "b"). Curve "c" in related to the material produced by post-sintering a RSSN containing the same sintering aids and one value (DEV.) is related to partially devitrified material.

i) by adding the sintering aids to the silicon powder before nitridation
ii) by adding the sintering aid (e.g. MgO or Mg_3N_2) after reaction sintering.

POST-SINTERING OF REACTION-SINTERED SILICON NITRIDE CONTAINING SINTERING AID

Two materials (A and B) were investigated. Their starting composition (i.e. Si + MgO and Si + Y_2O_3 + MgO) was calculated to give compositions as described previously under the sintering experiments. Thus after reaction sintering the compositions were as follows: A - 95% Si_3N_4 + 5% MgO, B - 91% Si_3N_4, 8% Y_2O_3 + 1% MgO. Some relevant data are shown in Table III.

TABLE III

Material	Green density Mg/m^3	Weight gain, %	% conv. Si to Si_3N_4	Nitrided density Mg/m^3	Crystalline phases by XRD
A	1,60	54,9	92,4	2,45	α Si_3N_4 (α/β ∿ 75/25) β Si_3N_4 $(Mg_{0.6}Fe_{0.4})_2SiO_4$ FeSi
B	1,64	51,8	94,7	2,52	α Si_3N_4 (α/β ∿ 90/10) β Si_3N_4 $10Y_2O_3.9SiO_2.Si_3N_4$ FeSi

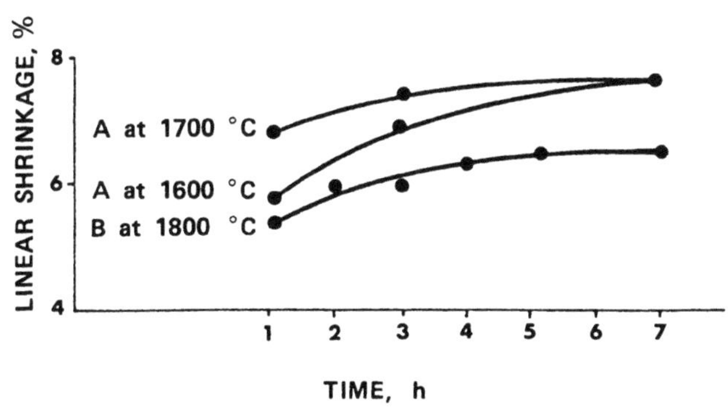

Fig. 6. Linear shinkages for materials A and B vs. sintering time

It can be seen that after nitridation, using the usual type of temperature in time schedule employed for reaction sintering, the sintering aid has reacted with either Si_3N_4 or SiO_2. The latter is an impurity in the Si powder as is Fe which is responsible for the observed trace of FeSi.

Fig. 6 indicates the sintering behaviour and Tables IV some characteristics of the material B sintered to high density.

TABLE IV

Pressureless sintered Si_3N_4 from reaction sintered Si_3N_4		
Starting material	:	RSSN containing 8w/o Y_2O_3 and 1w/o MgO as additives
Density	:	2.52 Mg/m^3
Sintering conditions	:	in powder bed at 1800°C for 5 hours
Linear shrinkage	:	∿ 6.5 %
Characteristics of the sintered material		
Density	:	3.22 Mg/m^3
Phases by XRD	:	β Si_3N_4, amorphous
Flexural strength	:	at 25°C = 1000 MN/m^2
(3 point)	:	at 1250°C = 500 MN/m^2
Oxidation resistance	:	0,10 mg/cm^2 (weight gain after 100 hours at 1000°C in air)
		0.90 mg/cm^2 (weight gain after 100 hours at 1300°C in air)

The strength of 1000 MN/m^2 at room temperature Fig. 5 (curve "c") and the linear shrinkage of only 6.5% are particularly noteworthy. Fig. 5 ("DEV" mark) shows that also Material B can be improved by devitrification of the amorphous phase.

POST SINTERING OF REACTION-SINTERED SILICON NITRIDE

This process which makes use of the volatility of MgO will be illustrated on three materials (C, D and E). Materials C and D were reaction-sintered materials derived from a commercial Si-powder (nominal 97.5% Si, 4μm) which after isostatic pressing was sintered in N_2 at 1 atmosphere by the usual type of schedule (140 h up to 1350°C). Material E was a reaction-sintered material obtained commercially. Samples of both materials were sintered at 1800°C in a powder bed containing MgO. The influence of the powder bed is shown in Fig. 7.

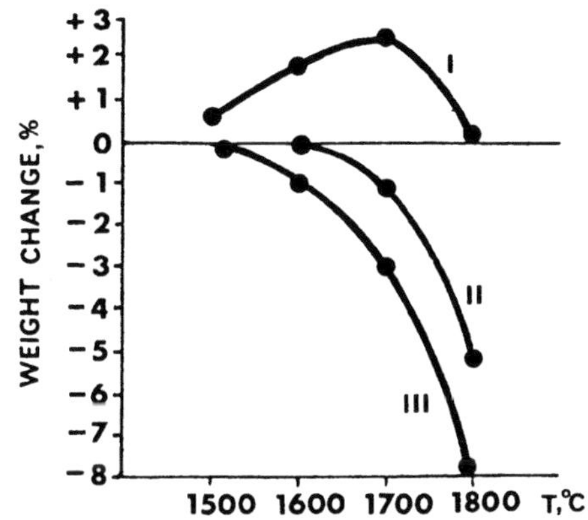

Fig. 7. Ingluence of the powder bed on weight change for RSSN d. 2.35 Mg/m^3. Composition for powder bed for: I) 50% Si_3N_4, 45% BN, 5% MgO -II) 50% Si_3N_4, 50% BN - III) without powder bed.

The sintering results, Table V, again show evidence of the low shrinkages obtained.

TABLE V

Material	Density before sinter. Mg/m^3	Sinter. aid in powder bed	Sintering treatment h	Sintering treatment °C	Linear shink. %	Weight gain %	Mg final content %	Final density Mg/m^3
C	2,35	5%MgO	1	1800	6,9	0,1	0,75	3,01
D	2,64	5%MgO	5	1800	4,7	0,9	0,67	3,05
E	2,47	5%MgO	3	1800	6	0,9	0,90	3,05

Fig. 8 gives the strength vs. temperature behaviour for material E before and after post-sintering. As a result of the take up of ∿ 1% MgO the strenght drops with increasing temperature; however, post-sintering has improved the strength in the material between RT and 1400°C.

SUMMARY

1) Uniform sintering of silicon nitride which contains MgO as sintering aid is prevented by the volatility of MgO at the sintering temperature (∿ 1700°C).

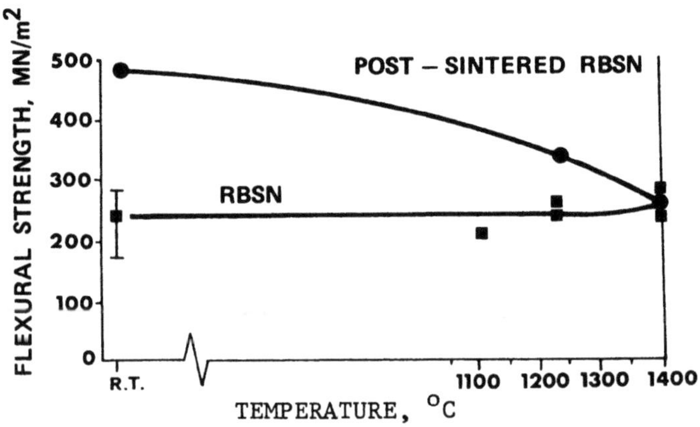

Fig. 8. Three-point flexural strength for a commercial (density 2,47 Mg/m^3) and post-sintered RSSN (density 3,05 Mg/m^3)

2) Loss of MgO can be prevented by sintering in nitrogen at atmospheric pressure using a powder bed containing MgO.
3) Sintering aids (e.g. Y_2O_3 and MgO) can be incorporated with silicon powder before nitridation. Post sintering of reaction-sintered silicon nitride results in material of very high strength (∿ 1000 MN/m^2) having undergone only small linear shrinkage (∿ 6%).
4) Reaction sintered silicon nitride not containg a sintering aid can be post-sintered in a powder bed containing MgO or Mg_3N_2, etc. which results in considerable improvement of strength.
5) The process of post-sintering of reaction-sintered silicon nitride should lead to the production of complex-shaped components to close dimensional tolerances and good mechanical properties by a variety of shaping methods including injection moulding.

REFERENCES

1 P. Popper: "Problems in sintering silicon nitride" Proc. Int. Symp. on Factors in Densification and Sintering of oxide and non-oxide ceramics held in Hakone, Japan, Oct 1978 (to be published)
2 C. R. Terwilliger, F. F. Lange - J. Am. Ceram. Soc. 57 (1974) 25
3 M. Mitomo - J. Mater. Sci. 11 (1976), 1103
4 A. Giachello, P.C. Martinengo, G. Tommasini, P. Popper - J. Mater. Sci. (to be published)
5 A. Giachello, P.C. Martinengo, G. Tommasini, P. Popper - submitted for publication to Am.Cer.Soc.

6 P. C. Martinengo, A. Giachello, P. Popper, A. Buri, F. Branda: "Devetrification phenomena in a pressureless sintered silicon nitride" - as Ref. 1
7 F. F. Lange, S.C. Singhal and R.C. Kuznicki - J. Am.Cer.Soc. 60 (1977) 249
8 A. Giachello and P. Popper: "Post sintering of reaction bonded silicon nitride" - presented at CIMTEC 4 at St. Vincent, Italy (May 1979) to be published in Proceedings and in Ceramurgia International

SESSION III

DESIGN CONCEPTS FOR CERAMICS

Chairman: Mr. A. F. McLean
Ford Motor Company

Vice Chairman: Dr. R. J. H. Bollard
University of Washington

A CERAMIC TURBINE ROTOR - ITS DESIGN EVOLUTION, TEST EXPERIENCE AND FUTURE

A. Paluszny

Ford Motor Company
Scientific and Engineering Lab.
Dearborn, MI 38121

ABSTRACT

Design evolution of a ceramic turbine rotor, developed under the Ford/ARPA Program provides a background for the review and rationalization of conceptual changes that were brought about by the evolutionary developments in ceramic technology as a whole and ceramic design codes in particular. Predictive capabilities of these codes are assessed using results of hot spin - and engine tests of the evolved rotor configuration. The paper then discusses turbine rotor configurations, both axial and radial, that were developed to meet performance and reliability requirements of an advanced vehicular gas turbine of a 100-150 HP class. The analytical results for these future configurations are presented along with material requirements.

INTRODUCTION

Mechanical design of an engine component or structure is a tuning process whereby the designer, using analytical tools, refines the concept, as initially conceived on the drawing board, to assure his design will retain structural integrity under service loading with minimum compromise in performance. For a designer to achieve a mechanically reliable design concept, he must have a detailed knowledge and a thorough understanding of the service environment and loadings and of the component responses (temperatures, strains and deformations) to these loadings. He must, furthermore, be able to accurately define the modes of failure and assess the strength of the component, that is, its capacity to withstand the service loads. Appropriate analytical tools and a reliable material data base are needed for this purpose. Their availability and quality will generally dictate how close and how fast the designer can "tune-in" on the optimum solution.

In the case of the ceramic turbine rotor, which was developed, or more accurately speaking, which evolved during the Ford/ARPA (Advanced Research Projects Agency) Program, the analytical tools available to the designer at the onset of the program were inadequate, lacking the precision required with brittle materials such as ceramics. Furthermore, our understanding of structural ceramics, particularly in respect to their modes of failure in an engine environment, was very limited. Thus the "tuning" of the turbine rotor configuration suitable for ceramics, instead of being an isolated event in the program, continued for its entire duration paralleling the evolution of the ceramic technology. As the design codes for ceramics continue to evolve and our understanding of these materials grew, the rotor configuration was undergoing gradual changes starting with the original "metallic" design and evolving eventually into what one may term a "ceramic" configuration. These changes along with their rationalization are reviewed in a chronological order in the following subchapters. Supplementing the review is a qualitative assessment of the results of engine and rig tests where these changes were corroborated and where the predictive capabilities of the analytical tools used in the rationalization of these changes were evaluated. Finally, the design experiences gained during Ford/ARPA Program are translated into future turbine rotor configurations envisaged for an advanced vehicular turbine engine and the potential of ceramic materials in this demanding application is assessed.

Design Evolution of the Ceramic Turbine Rotor "Metallic" Design Approach

The blade and disk configurations of the ceramic turbine rotors as initially conceived were designed with the help of relatively simple one- and two-dimensional computer codes and followed the general guidelines commonly employed with metallic rotors such as, for instance, uniform stress distribution to affect an efficient utilization of material. Functional features as dictated by either aerodynamic (blade geometry, minimum trailing edge thickness, twist, etc.) or mechanical considerations such as rotor inertia, placement and shape of balancing rings, etc. were retained to see if ceramic fabrication processes could duplicate them with a desired degree of mechanical integrity. The results of this design study have been reported in References 1 and 2 and are included here in Figure 1 showing the contour maps of temperatures and maximum principal tensile stresses in the first stage turbine rotor at 100% thermal and 110% (overspeed) mechanical steady state loading. Material properties used in the study were those of hot pressed

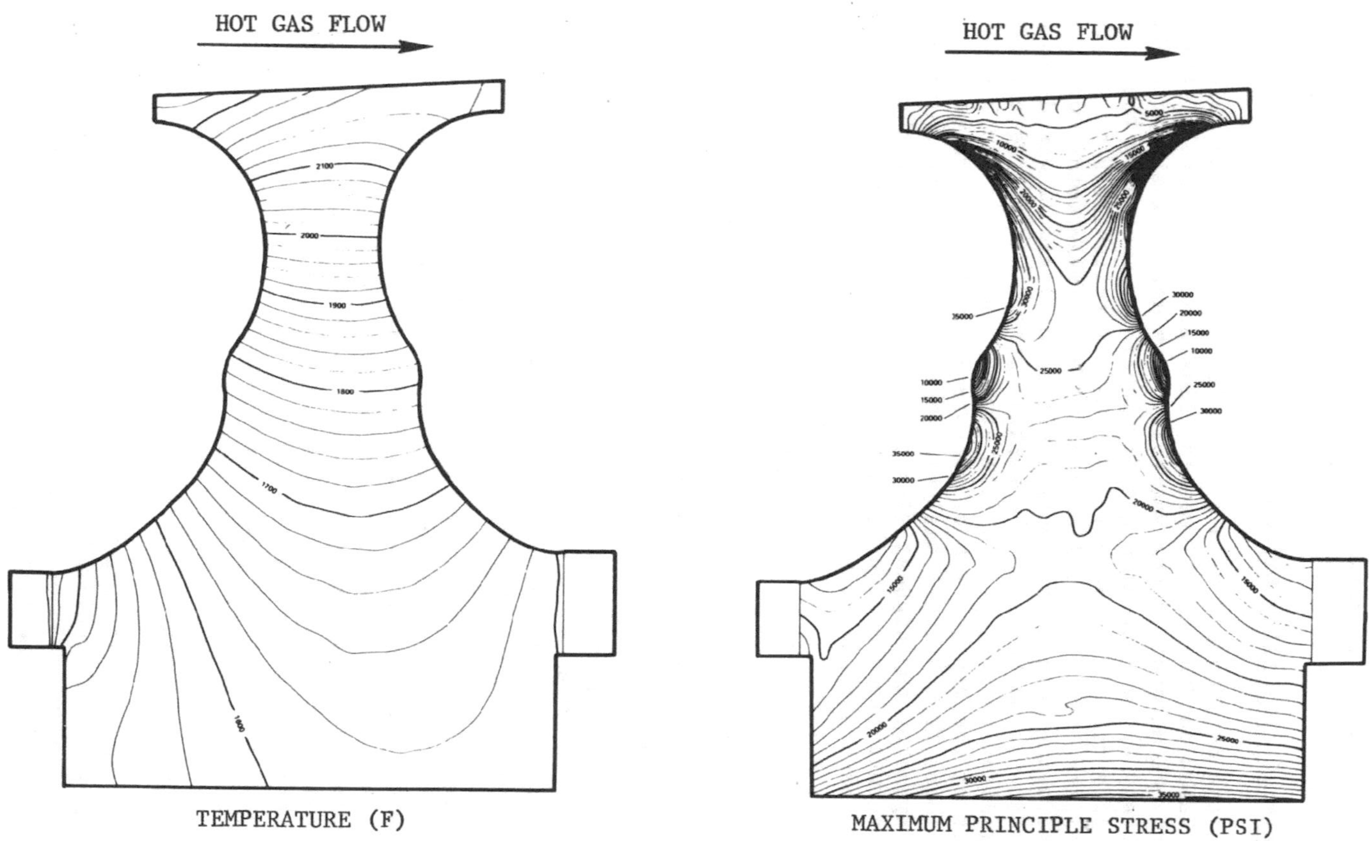

Figure 1. Distribution of Temperatures and Maximum Principal Tensile Stresses in the First Stage Turbine Rotor Disk at Maximum Steady State Loading.

Si_3N_4 (HS-130). The maximum stresses are fairly well balanced and are concentrated in the areas of stress risers such as the central bore, rim to neck transition and around the balancing rings.

Fast Fracture Reliability Effects

With the development of a probabilistic strength analysis, which is based on the Weibull weakest link model of brittle strength and which took account of the variability in material strength (Reference 3 and 4), a rationale was developed for assessing the strength of a ceramic structure and thus its reliability or survival probability under the applied load. The turbine rotors were reanalyzed using Weibull analysis (Reference 3) and it was discovered that the material in the central region was not as effectively utilized as indicated by the stress distribution in Figure 1. A look at the map of risk of rupture intensity generated for the same rotor from Weibull analysis and shown in Figure 2 indicates that the areas of the highest risk of failure coincide with the stress concentrations at the balancing rings where the combined effects of high stress and volume under stress are most pronounced. The risk of failure in the central hub region which is equally severely stressed is however considerably lower.

A subsequent design investigation was conducted which involved 8 disk contour modifications for the purpose of identifying those modifications that will enhance the reliability of the rotor. The results reported in References 3 and 4 showed that in order to improve the reliability of the rotor the stresses in the neck region had to be reduced even at the expense of increased volume indicating that the stress level and not the volume were still the controlling factor. This was to be expected since the stresses appear in the Weibull function in a power-functional relation.

Consequently the disk contours were modified to a reduced stress level in the neck region by increasing neck thickness. At the same time, in order to ease the fabrication the rim thickness was increased to accommodate the duo-density fabrication concept, the balancing rings, which were found to be difficult to fabricate by hot pressing, were removed and the disc geometry was made identical for both turbine stages (Reference 6). The modified profile is shown in Figure 3. The stress analysis for this rotor configuration has been reported in Reference 6.

Effects of 3-D Analysis

As the emphasis in rotor fabrication shifted to the duo-density concept, in which a monolithic reaction bonded Si_3N_4 (RBSN) blade ring is chemically attached to a hot pressed hub, problems associated with the blade geometry and with the blade-ring-hub junction

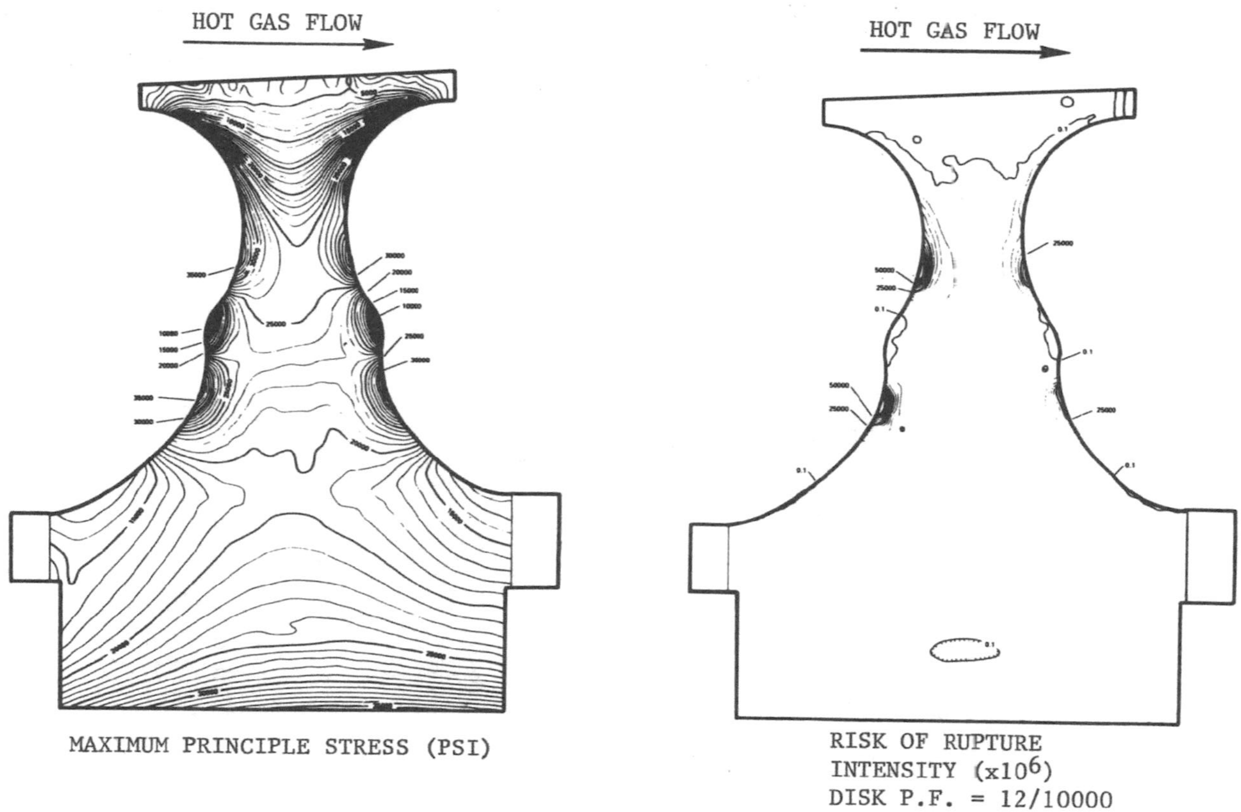

Figure 2. Distribution of Maximum Principal Tensile Stresses and the Corresponding Risk of Rupture Intensity in the First Stage Turbine Rotor Disk at Maximum Steady State Loading.

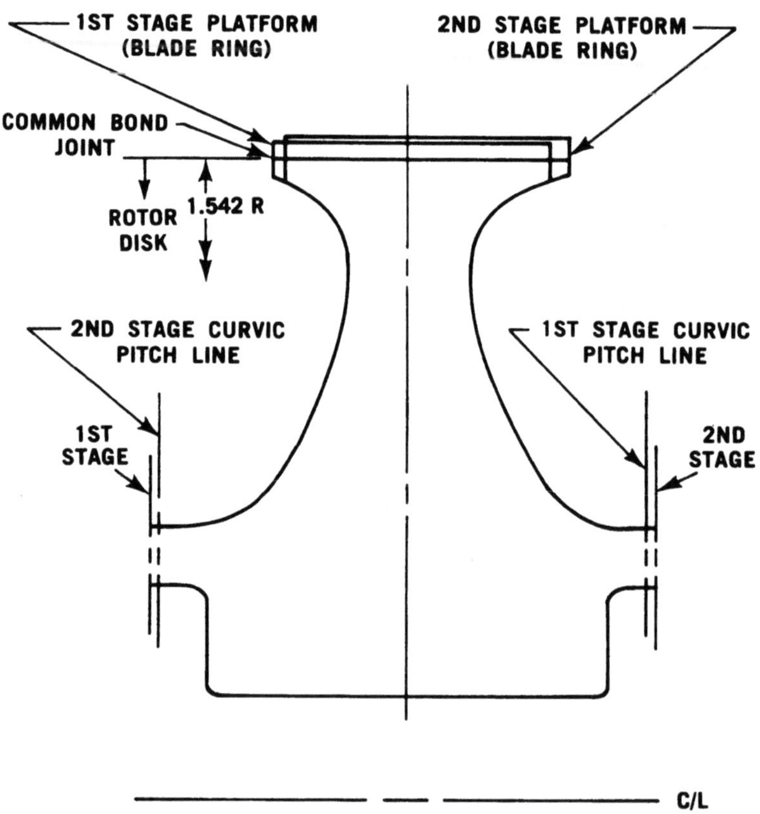

Figure 3. Modified Profile for the First and Second Stage Turbine Rotor Disks.

emerged. In order to gain a better understanding of these problems accurate knowledge of stress distributions in those areas was required. Needed in particular was a better insight into the turbine blade stresses which heretofore were approximated by uniaxial analyses. These complex problems could only be resolved with the aid of three-dimensional analyses and consequently three-dimensional finite element computer codes were acquired and developed.

With the help of these new codes, an indepth analysis was conducted of a turbine disk segment with the original blade geometry using 8-noded isoparametric solids (Reference 7). The analysis revealed that under normal engine loading blade stresses were far from uniform and were reaching (at the leading and trailing edges of the base sections) more than twice the average stress level estimated with the uniaxial analysis. These highly stressed blade regions which in themselves were deleterious to the blade reliability, happened also to coincide with the areas where most of the processing related problems were encountered. The excessive stresses were traced to centrifugal bending and untwisting of the airfoil which is characteristic of the "free vortex" blade design.

In order to eliminate these parasitic stresses the blades were redesigned. The blade sections in the new configuration were radially and centroidally stacked so that the individual camberlines formed a helical surface, similar in concept to the compressor inducer vane and subjected to pure radial loading. The trailing edge radii were increased to improve blade fabrication. A three-dimensional finite element model aided the design of the new blade evolving a configuration with nearly uniformly stressed blade sections free from parasitic stresses found in the "free vortex" design (References 7 and 8). Figure 4 taken from Reference 8 illustrates the distribution of maximum principal tensile stresses, in this case, predominantly radial, on the surface of the first stage blade. Since the distribution of mass in the new blade was similar to that for the original "free vortex" configuration the stresses in the disk were virtually unaffected. Figure 5 shows the distributions of temperatures and maximum principal tensile stresses in the first stage turbine rotor disk with the new blade geometry (Reference 7) at the overspeed condition mentioned earlier - 100% thermal loading and 110% speed.

A Weibull analysis was made for both the original and the new blade configurations using Weibull strength parameters determined for the 2.7 gg/cc injection molded RBSN (S_θ = 36000 PSI and m = 8) from which the blade ring was to be fabricated. As expected the reliability of the new blade ring was considerably improved by the elimination of the parasitic stresses in spite of the fact that the average stresses at the base were 10% higher in the new design.

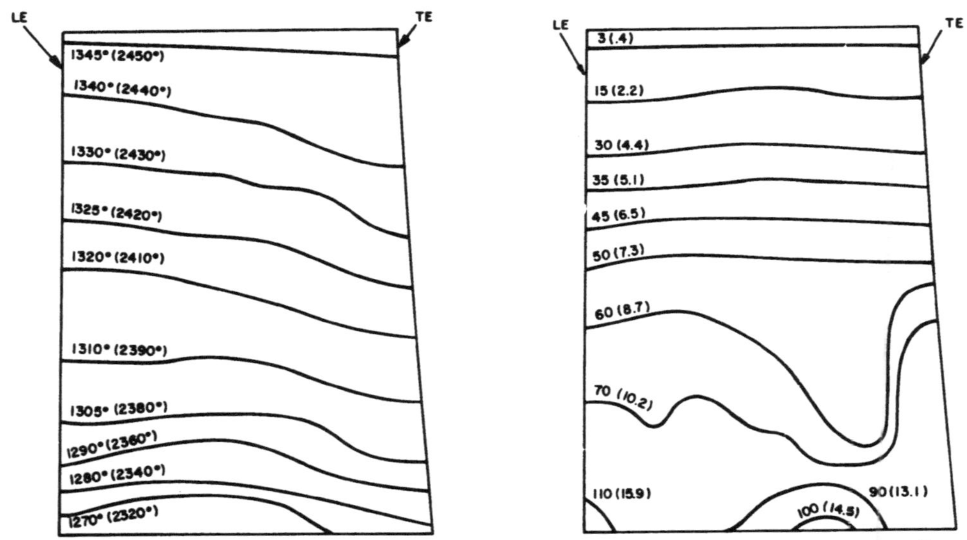

TEMPERATURE CONTOUR MAP IN °C (°F) AT THE CAMBERLINE PLANE OF THE FIRST STAGE TURBINE ROTOR BLADE

100% STEADY - STATE LOADING CONDITIONS (1370°C (2500°F) T.I.T. & 100% CENTRIFUGAL

STRESS CONTOUR MAP IN MN/m^2 (KSI) AT THE CAMBERLINE PLANE OF THE FIRST STAGE TURBINE ROTOR BLADE

100% STEADY - STATE LOADING CONDITIONS (1370°C (2500°F) T I.T. & 100% CENTRIFUGAL)

Figure 4. Distribution of Temperatures and Maximum Principal Tensile Stresses in the Modified First Stage Turbine Rotor Blade at Full Power Steady State Loading.

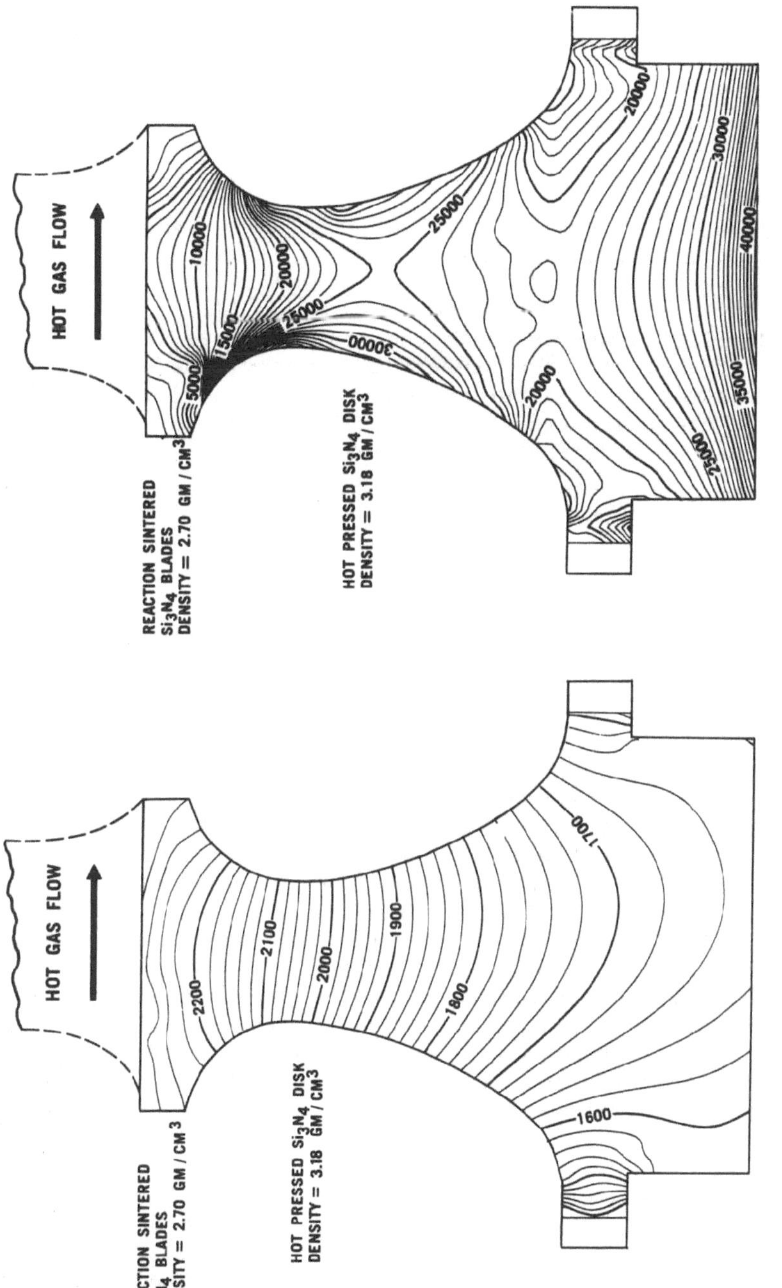

Temperature (°F)

Maximum Principal Tensile Stress (PSI)

Figure 5. Distribution of Temperatures and Maximum Principal Tensile Stresses in the Modified First Stage Turbine Rotor (with Modified Blade) at Full Power Steady State Loading.

Table I

Blade Configuration	Average Stress at the Base Section (PSI)	Reliability of Blade/Ring
Old	11000	.991/.72
New	12200	.998/.93

Time-Dependent Reliability Effects

The reliability analyses conducted thus far were limited to consideration of fast fracture failure modes only, where a ceramic component is assumed to fail catastrophically at the site of a flaw upon reaching a critical stress corresponding to the size of that flaw as given by the modified Griffith relation:

$$\sigma_{critical} = \frac{K_{IC}}{Y \sqrt{a}}$$

K_{IC} - critical stress intensity factor, a material property

Y - geometric factor dependent on shape of the flaw and it's location in the stress field of the component

a - the size of the flaw

The reliabilities thus derived were independent of time and depended only on the severity of the service loading. Although the fast fracture reliability analysis provided good approximation for the blade material (RBSN) within the temperature range involved (2300-2450°F) it failed to account for the time-dependent failure modes to which the disk material such as hot pressed Si_3N_4 was reported (Reference 9) to be subject to in the presence of subcritical crack growth mechanism observed to occur in these materials at high temperatures. As a consequence of subcritical crack growth (SCG) the material was reported to fail with time at stress levels considerably lower than observed in fast fracture tests, giving rise to so called static fatigue failures.

New reliability codes were developed to account for the degradation of strength with time that results from subcritical crack growth. Using dynamic fatigue strength data, obtained from MOR tests conducted at temperature at various constant strain rates, in conjunction with inert strength data, either measured directly or derived (Reference 10), the crack propagation exponent (n) and the Weibull strength parameters (S_θ and m) were determined for the disk material (hot pressed, press-bonded Si_3N_4). A time-dependent reliability analysis was conducted on the first stage turbine rotor to evaluate the effects of subcritical crack growth using as a goal 25 hours of operation at full power condition (2500°F TIT and 64500 RPM), i.e., the objective of the Ford/ARPA program.

The disk configuration which, heretofore, was found to be acceptable on the basis of fast fracture reliability was found completely inadequate for prolonged service when the deleterious effects of SCG were considered (see Table II, 0.3 in. throat). The reliability of the neck region of the rotor, where the service induced temperature and stress conditions were found most conducive to SCG, was shown to deteriorate rapidly with time while the cooler, yet more severely stressed, central region of the disk suffered virtually no loss in strength in the same time.

A study was subsequently conducted for the purpose of optimizing the disk geometry with respect to the time-dependent failure modes while retaining the design constraints set forth by the existing flow path geometry and by the mechanical arrangement, i.e., without affecting either the axial spacing of the rotors (no change in hub width) or their mounting (location and shape of face splines).

A scrutiny of the time-reliability relation (Reference 11)

$$R = \prod_{i=1}^{N} \exp\left[-\left(\frac{\sigma_{1i}^{\,n}\,(n+1)\,\dot{\sigma} t_f}{\left(\bar{S}_{\theta i}\right)_{t=o}^{n+1}}\right)^{\frac{m}{n-2}}\right]$$

Where:

R - Reliability of the rotor

N - Number of finite uniformly stressed elements in the mathematical model of the rotor

$\bar{S}_{\theta i}$ - Characteristic dynamic fatigue strength of the i-th rotor element

σ_{1i} - Maximum principal tensile stress in the i-th element

$\dot{\sigma}$ - Stress rate at which the strength measurements were taken and with which $\bar{S}_{\theta i}$ were calculated

n - Crack propagation exponent

m - Weibull slope of material strength, and

t_f - Time to failure

shows that the most effective way of increasing the reliability of the rotor is either to increase the dynamic fatigue strength ($\bar{S}_\theta$) of the material, a process dependent entity, or to decrease the max. principal tensile stress induced in service. Since only the second entity is design dependent, a simplest way of reducing the stresses in the critical regions is by increasing the load bearing area, i.e., increasing the thickness of the throat of the disk. Three disk configurations were studied: the original with a 0.3 in. thick throat, a 0.4 in. throat and a 0.48 in. throat. The results are summarized in Figures 6 and 7. Figure 6 shows the temperature distribution for all three (3) disk configurations. It was found to be unaffected by the throat modifications. The other Figure shows the maximum principal tensile stress distributions for the three configurations respectively at full power loading condition. The stresses in the neck regions are seen to be progressively reduced at the expense, however, of the bore stresses. Table II summarizes the results of the reliability analysis (Reference 9).

Table II

Throat Thickness (in.)	Max. Throat Stress (PSI)	Max. Bore Stress (PSI)	Reliability	
			Fast Fracture	@ 25 Hrs.
0.3	34000	40000	.99787	.29156
0.4	28000	44000	.99647	.82042
0.48	20000	47000	.98986	.60986

Disk configuration with 0.4 in. throat was found to represent the best compromise between the time-dependent and the fast fracture reliabilities in that the failure probabilities for the neck and the central bore regions were found to be almost equal after 25 hours of operation. The configuration was selected as the optimum for future development.

Test Experience

In parallel with the analytical investigations numerous cold and limited hot spin tests were conducted at various stages of the program. The purpose of these tests was to ascertain the applicability of Weibull probabilistic model of brittle strength to ceramic turbine rotors and other ceramic turbine structures subjected to either or both the fast fracture and the time-dependent modes of failure. Furthermore, it was important insofar future developments were concerned, to corroborate through tests the conclusions drawn

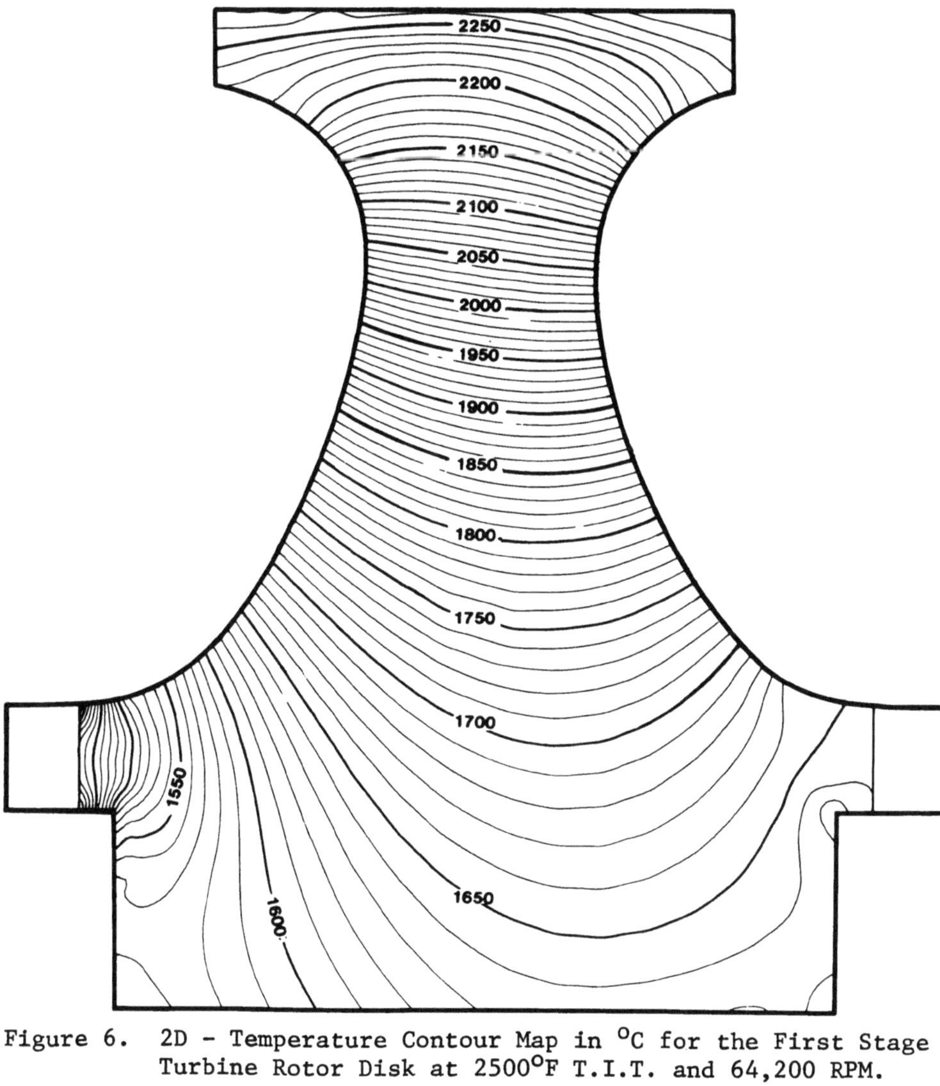

Figure 6. 2D - Temperature Contour Map in ^{o}C for the First Stage Turbine Rotor Disk at 2500°F T.I.T. and 64,200 RPM.

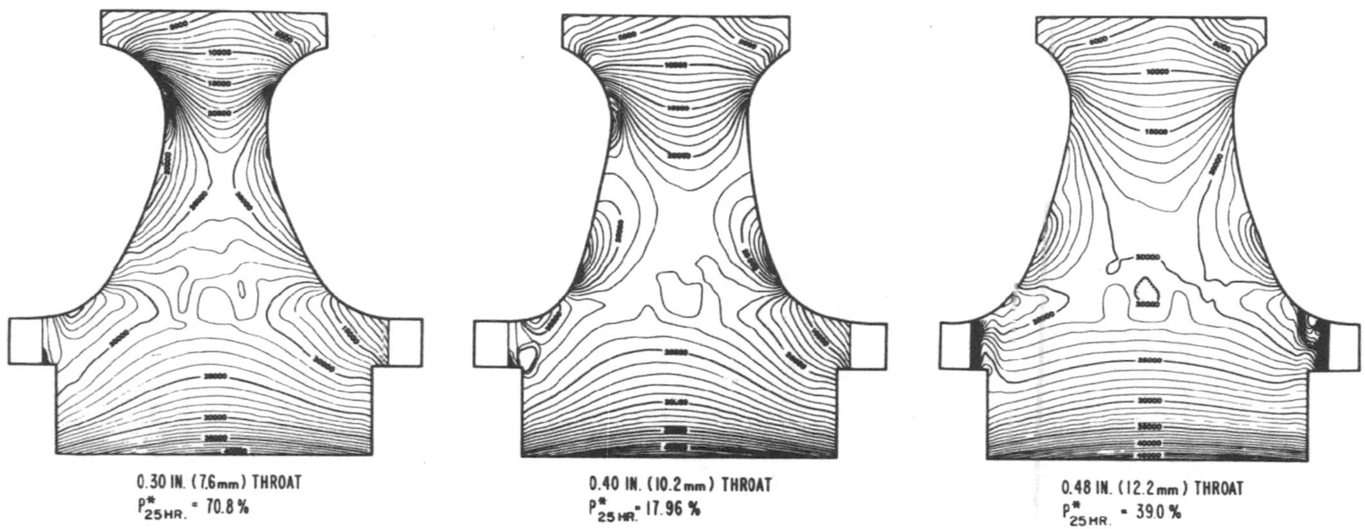

*PROBABILITY OF FAILURE AT 25 HOURS OF OPERATION

Fig. 7. Effect of Disk Geometry on Maximum Principal Tensile Stresses and Their Distribution in the First Stage Turbine Disk for 2500°F T.I.T. and 64,240 RPM (Ref. 10).

from the various studies and validate the resulting design change decisions.

In the category of fast fracture modes of failure destructive spin tests were conducted, initially cold only and towards the end of the program-hot. The cold tests involved simple, flat sided disks 6.0 in. in dia. and 0.5 in. thick (152.4 mm ϕ x 12.7 mm thick) and contoured turbine hubs of actual geometry and size, all made of hot pressed Si_3N_4 (HPSN). The results of these tests were correlated with failure (burst) speed distributions predicted from strength data generated with simple MOR (modulus of rupture) tests of small (.125 x .25 x .75 in. or 3.2 x 6.4 x 19 mm) rectangular bars in four-point bending. As reported in Reference 7 excellent agreement between tests and theory was achieved. For each group of spin tests involving different grades (composition) of HPSN, the theory predicted failure distributions, i.e., both the characteristic burst speed and the Weibull slope (dispersion), well within the 99% confidence interval of actual spin failures.

Hot spin testing in a fast fracture mode of failure involved actual turbine wheels of a duo-density concept with .48 in. (11.8 mm) throat thickness described earlier. The wheels were either fully or partially bladed, depending on whether defective blades had to be removed prior to tests. The tests were conducted in a specially designed hot spin rig[9,11] under carefully controlled thermal and mechanical (centrifugal) loading conditions. The results of these limited tests, as reported in a companion paper by Baker and Swank,[12] are again quite encouraging. The predicted burst speeds fall, also in this case, within the 99% confidence band of actual failures.

Testing for the time-dependent modes of failure involved the same hot spin rig and an identical test set up, except the rotational speed was limited to 50000 RPM in order to maximize the probability of success in reaching the required 25 hours of steady state operation. The 5000 RPM rotational speed chosen was the design speed for a three-stage axial rotor concept being investigated as an alternate candidate for the engine and thus represented, insofar stress levels were concerned, a full power operating point for a practical turbine rotor concept. Detailed results of these tests are reported by Baker and Swank.[12] Figure 8 shows the temperature and the maximum principal tensile stress distributions in the test wheel for loading conditions prevailing in the hot spin rig. This loading was determined from measurements of the rotational speed and of surface temperatures taken at preselected radial locations on the upstream side of the wheel. The rim temperature of 1800°F (982°C), at which these tests were conducted, corresponds roughly to 2150° - 2200°F (1150° - 1204°C) turbine inlet temperature in an actual engine. Material parameters such as the charac-

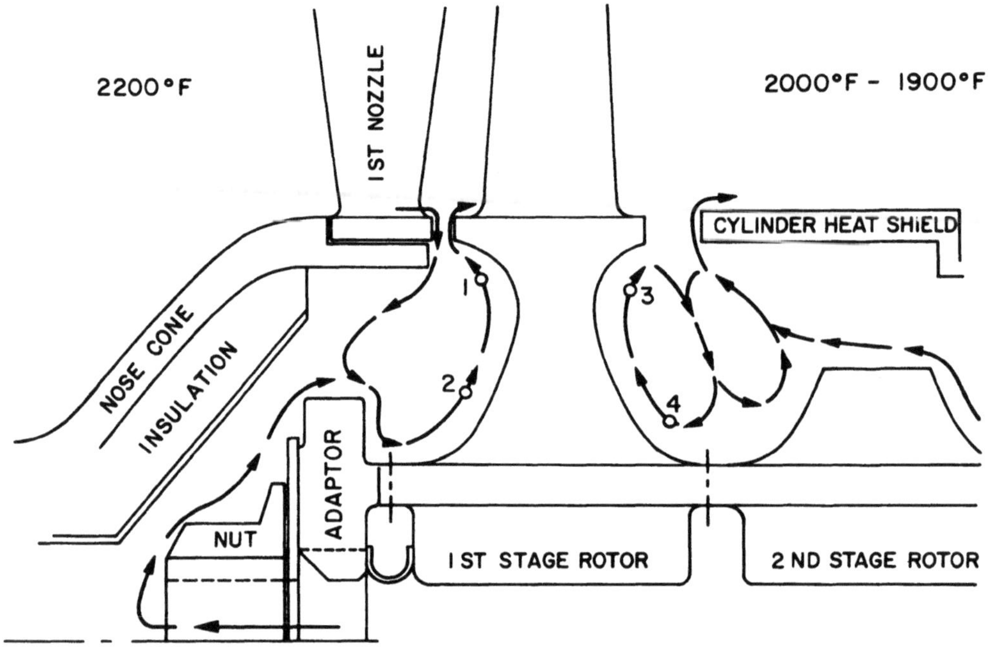

HOT FLOW PATH CONFIGURATION AND ASSUMED RECIRCULATION PATTERNS

TURBINE INLET TEMPERATURE, °F		STATION No. 1	2	3	4
2200°	TEST	1738°	1716°	998°	902°
	CALCULATED	1708°	1718°	972°	901°
2500°	TEST	1896°	1879°	1097°	976°
	CALCULATED	1863°	1870°	1070°	992°

TEST AND CALCULATED AIR TEMPERATURES (°F) @ ROTOR BOUNDARY

Fig. 8. Modified Engine Test of 6-14-77.

teristic MOR strength $S_{\theta}(T)$, measured as function of temperature, the Weibull slope (m) and the crack propagation exponent (n), which were used in both the fast fracture and the time-dependent reliability analyses, were determined in separate tests of coupons cut from actual wheels and are summarized in Figure 9. The crack propagation exponent (n) for the hub material (HPSN) was estimated at 15.9. Using these data as input, the probability of wheels surviving the 25 hour test run at the state of stress and temperature levels depicted in Figure 8 was estimated at 95.5% i.e., a fairly low risk of failure. Altogether four wheels were tested for the time-dependent mode of failure and in line with the analytical predictions (4.5% chance of failure) all wheels survived the 25 hour test runs. Still, no definite conclusions can be drawn, at this point, regards the validity of the mathematical models used in the time-dependent reliability analysis, because the statistical sample size involved in these tests was far too limited. In an attempt to quantify statistically the test results obtained so far, Baker and Swank[12] used the success run theorem and estimated that a reliability level of at least 87.1% at 50% confidence level was demonstrated by the four (4) consecutive successful tests, which compares quite favorably with the predicted reliability of 95.5% computed at the same confidence level.

In addition to the hot spin rig tests an engine test was conducted of the same duo-density turbine wheel concept, fabricated same way and with the same materials, but with the original throat thickness of .30 in. (7.6 mm). As reported in Reference 9 the engine required minor modifications for this test to facilitate testing of a single turbine stage at inlet temperatures up to 2500°F (1371°C). The test wheel accumulated a total of 36.5 hours at turbine inlet temperatures (TIT) ranging from 2200° to 2500°F (1204° to 1371°C) and rotational speeds from 45000 to 50000 RPM. The wheel failed during engine shutdown in the area of the curvic coupling (a self-centering torque transfer mechanism) when the metallic adaptor (See Figure 10) failed to slide freely relative to the ceramic wheel during the cool-down.

A time dependent probability of failure analysis was conducted to determine what chances for success would the theory predict for this particular test sequence assuming it, that is the theory, modeled correctly the modes of failure of the actual test wheel. The analysis and its results have been published in Reference 13. The main highlights of the analysis are presented here for the purpose of reviewing the reliability aspects. Figure 10 shows the mathematical model of thermal boundary conditions reconstructed for this particular engine run from temperature measurements taken in the main gas stream and in the fore (stations 1 and 2) and aft (stations 3 and 4) wheel cavities, with arrows indicating the assumed

3.5 % MgO HPSN CP85	TEMP. °F	WEIBULL SLOPE	CHARACTERISTIC INERT MOR STRENGTH, psi
FAST FRACTURE	78	7	97,465
	500	↓	97,465
	1000	↓	97,465
	1300	↓	97,465
	1500	↓	90,940
	2000	↓	76,000
	2500	↓	60,630

3.5 % MgO HPSN CP85	TEMP. °F	WEIBULL SLOPE m	WEIBULL SLOPE $\bar{m}$	CHARACTERISTICS DYNAMIC FATIGUE MOR STRENGTH, psi (*)
TIME DEPENDENT	78	7		97,465
	500	↓		97,465
	1000	↓		97,465
	1300	↓		97,465
	1420		8.5	93,110
	1500		↓	88,470
	2000		↓	57,430
	2500		↓	25,530

INJECTION MOLDED Si_3N_4	TEMP. °F	WEIBULL SLOPE	CHARACTERISTIC MOR STRENGTH, psi**
	78	10.2	25,410
	1700	↓	23,100
	2100	↓	22,750
	2300	↓	23,240
	2500	↓	22,190

(*) THESE ARE EXTRAPOLATED VALUES FOR A STRESS RATE OF 145 psi/min.

(**) THESE STRENGTH VALUES HAVE BEEN REDUCED BY 30 % TO REFLECT DEGRADATION DUE TO PRESS-BONDING.

Fig. 9. Weibull Strength Parameters used in the Analyses.

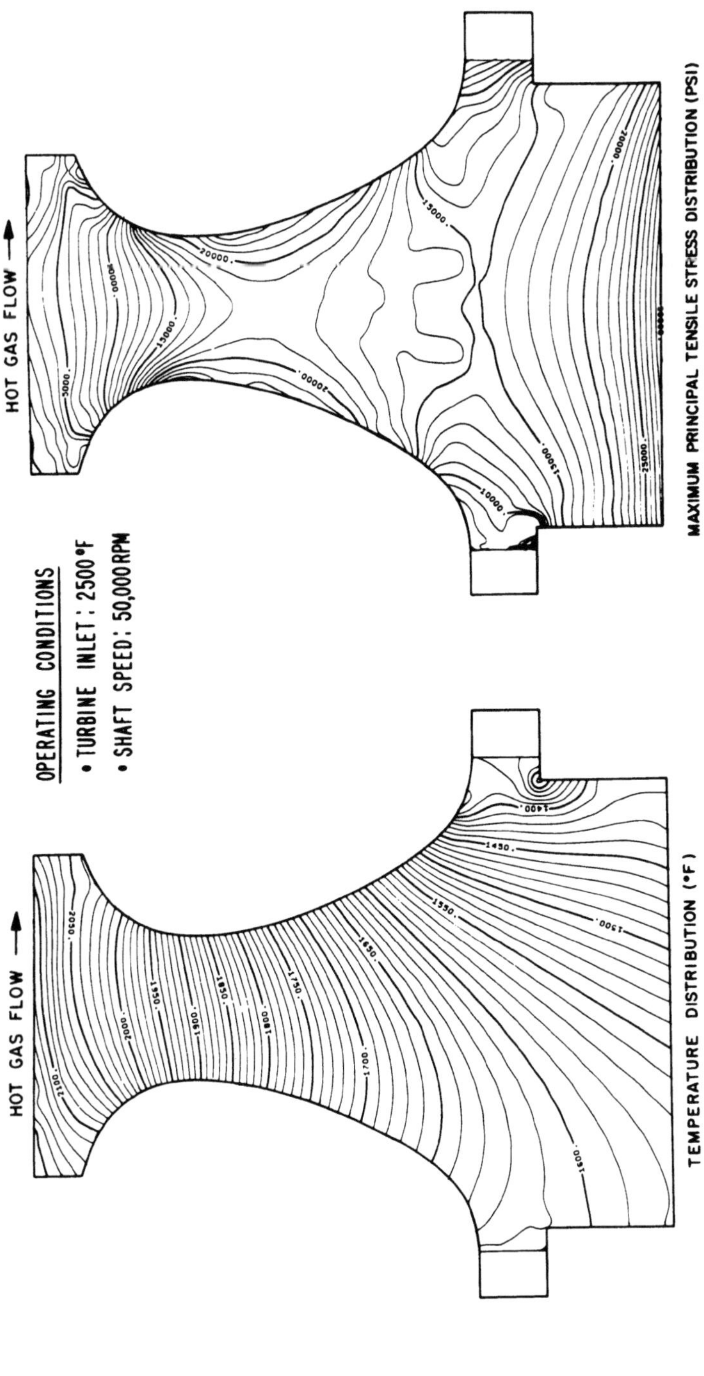

Fig. 10. First Stage Turbine Rotor Analytical Simulation of the Modified Engine Test of 6-14-77 at 2500°C T.I.T.

in the analysis gas recirculation (for more details the reader is referred to Reference 13). The resulting temperature and maximum principal tensile stress distributions are shown in Figures 11 and 12 for the 2200°F (1204°C) and 2500°F (1371°C) TIT engine runs respectively. Both were computed at 50000 RPM rotational speed. The results of a time-dependent reliability analysis, based on material strength parameters discussed earlier and shown in Figure 10, are summarized in Figure 13. It is a tabulation of survival probabilities computed for the individual engine runs and includes the cumulative probability calculated for the combined 36.5 hour engine test. Again, being but a single test, it is difficult to draw definite conclusions regarding analytical predictions which are based on a statistical analysis. In the context of statistics, the only conclusion one can draw at this point is that the predicted probability of success, being a respectable 84%, is consistent with the success of the engine test, since at the relatively low risk of failure (16%) predicted by analysis one intuitively would expect a single test to be a success.

Future Ceramic Turbine Rotor Concepts

The wealth of experience gained in the course of evolution of the ceramic turbine rotor discussed earlier, and during the Ford/ARPA program as a whole, in the area of designing and fabrication of high performance ceramic structures, has spurred considerable interest in exploring the future potential of ceramic materials in heat engine applications. An exploratory study was initiated to this effect at Ford to determine, for instance, whether ceramic materials would meet the strength requirements of the advanced turbine rotor concepts which have been projected for the vehicular gas turbine technology. The results of this study were presented at the 1978 Highway Vehicle Systems Contractors' Coordination Meeting[14] and are reviewed next.

Engine performance analyses conducted over the years and supported by engine design concept studies have identified two different turbine rotor configurations as having greatest potential for the advanced vehicular gas turbine, a high speed, single stage, radial inflow rotor capable of operating at the tip speeds of up to 2500 ft/sec (760 m/sec) and a pressure ratio across the rotor of 4.25, and a multistage axial rotor operating at a moderately high mean blade speed of 1500 ft/sec (457 m/sec.)

The radial inflow rotor operating at a tip speed of 2500 ft/sec (760 m/sec) represents, insofar the strength requirements are concerned, a very advanced state of art even for the best superalloys under development and thus constitutes a true challenge to the ceramic technology. The 1500 ft/sec (457 m/sec) multistage axial concept, on the other hand, although it is very advanced aerodynam-

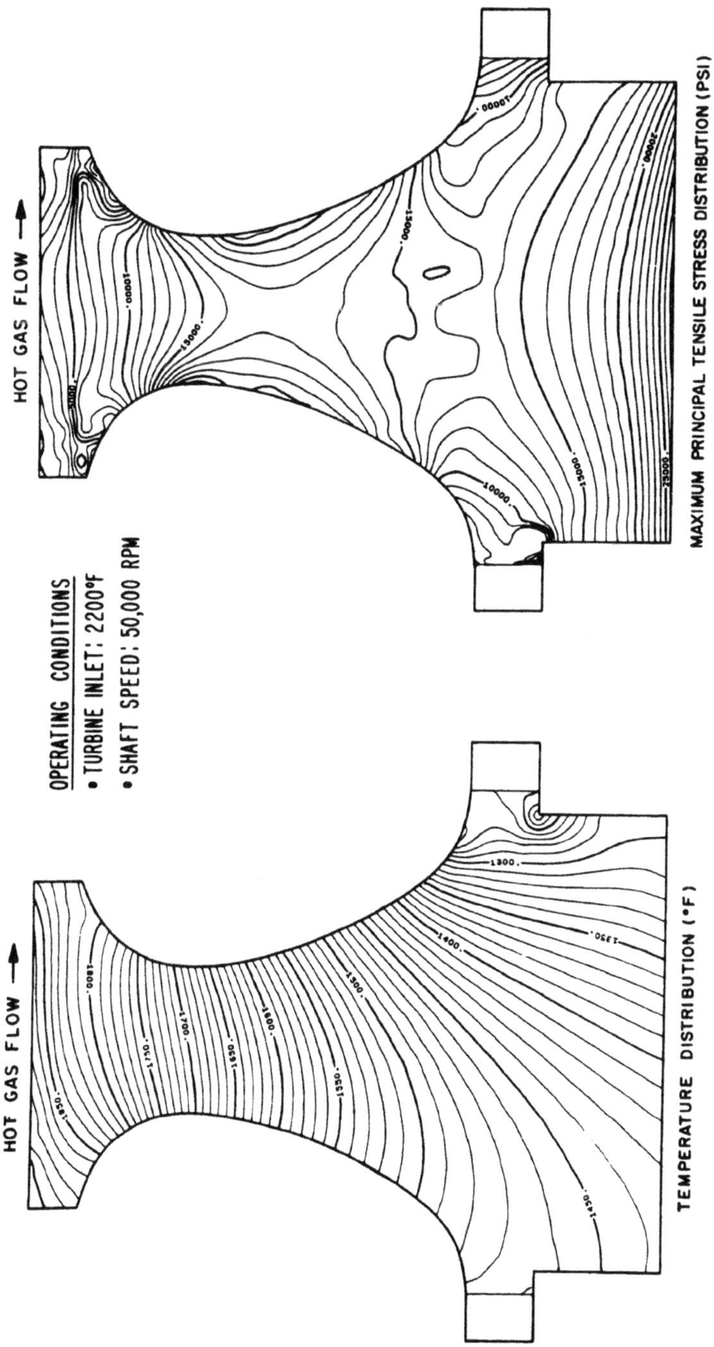

Fig. 11. First Stage Turbine Rotor Analytical Simulation of the Modified Engine Test of 6-14-77 at 2200°F T.I.T.

Loading States	Time (Hours)	Fast Fracture Reliability		Time Dependent Reliability *	
		m = 7	m = 10	m = 7	m = 10
2200°F TIT & 45,000 rpm	10	.98113	.99981	.97443	.99972
2200°F TIT & 50,000 rpm	25	.96145	.99947	.91667	.99845
2500°F TIT & 50,000 rpm	1.5	.89500	.99766	.85830	.99623

* Rotor Reliabilities for the Individual Loading States

Cumulative Time Dependent Reliability for the Total 36.5 hrs of Running Time:

.83843 ($m = 7$, $\bar{m} = 8.5$, $n = 15.9$)

.99547 ($m = 10$, $\bar{m} = 12.2$, $n = 15.9$)

Fig. 12. Predicted Rotor Reliabilities for the Modified Engine Test of 6-14-77.

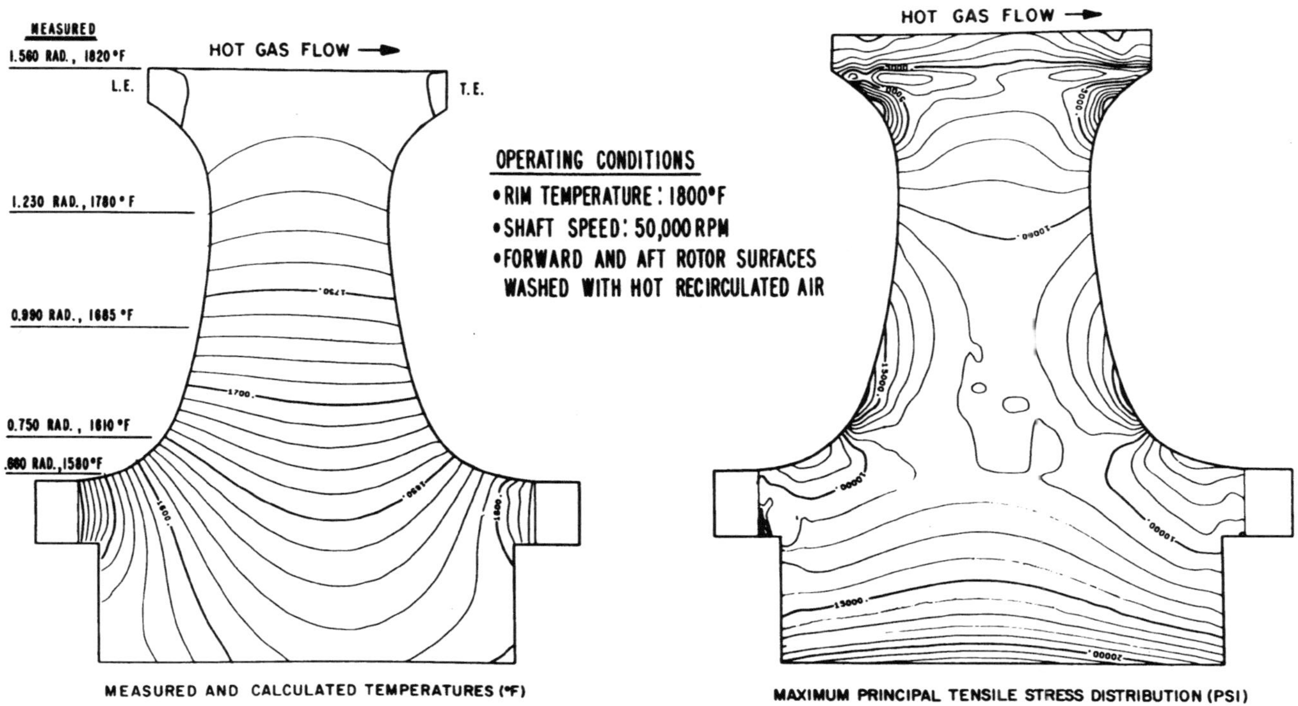

Fig. 13. Analytical Simulation of the Hot Spin Rig Test of the Recontoured 0.40 inch Throat Turbine Rotor.

ically, represents a mechanically more conservative approach because of its considerably lower stresses. It has an advantage over the radial type rotor as it allows full exploitation of the ceramic rotor technology already developed under the current programs utilizing materials of proven ability to be fabricated in complex, close toleranced, aerodynamic configurations. Cost and structural complexity are the main trade-offs in this case, favoring the more demanding mechanically and as yet untried, single stage ceramic radial inflow rotor concept.

The Ford study of these two turbine rotor concepts, being of exploratory nature, was limited to steady state analyses at full power loading conditions, a criterion commonly employed in the sizing and design of automotive turbine rotors. Limited transient analyses involving cold start and acceleration from idle to full power were also conducted in support of the less explored radial rotor concept for the purpose of evaluating its responses to a typical automotive mode of operation. Since considerable experience exists in this area with axial rotors, experience accumulated over years of testing and analysis, it was deemed unnecessary to repeat this exercise for the multistage axial concept.

Figure 14 shows contour maps of temperatures and maximum principal tensile stresses for an eleven-vaned, 2500 ft/sec (760 m/sec) tip speed, ceramic radial inflow turbine rotor at full power loading of 2500°F (1370°C) TIT and 118000 RPM rotational speed. The temperature and stress contours shown have been generated using a 3-dimensional finite-element model of a rotor segment. Physical properties used in the analysis were those of NC-132 hot pressed Si_3N_4. As seen in the Figure, material temperatures in the more severely stressed regions of the rotor (the expeller portion of the hub and the expeller vane fillet) range from 1600° to 1850°F (870°-1010°C). These temperature levels are low enough to neglect the deleterious effects of subcritical (slow) crack growth and the related time-dependent modes of failure such as, for instance, the static fatigue. The strength limiting region is found near the disk center with a stress of 32.4 ksi (223 MPa). The peak stress of 35 ksi (241 MPa) which occurs at the vane fillet reflects local surface stress concentration induced by a rapid change in section. This stress decays rapidly inside the vane to an average value of 25 ksi (172 MPa).

The flow path components in the Ford advanced turbine engine concept (AGT)[15] for which this rotor was designed, have been uniquely arranged with the objective of minimizing radial temperature gradients within critical structural components of the engine. As a result, thermal stresses in this particular turbine rotor have been minimized to the extent that the configuration of the rotor and thus its strength are governed almost entirely by centrifugal

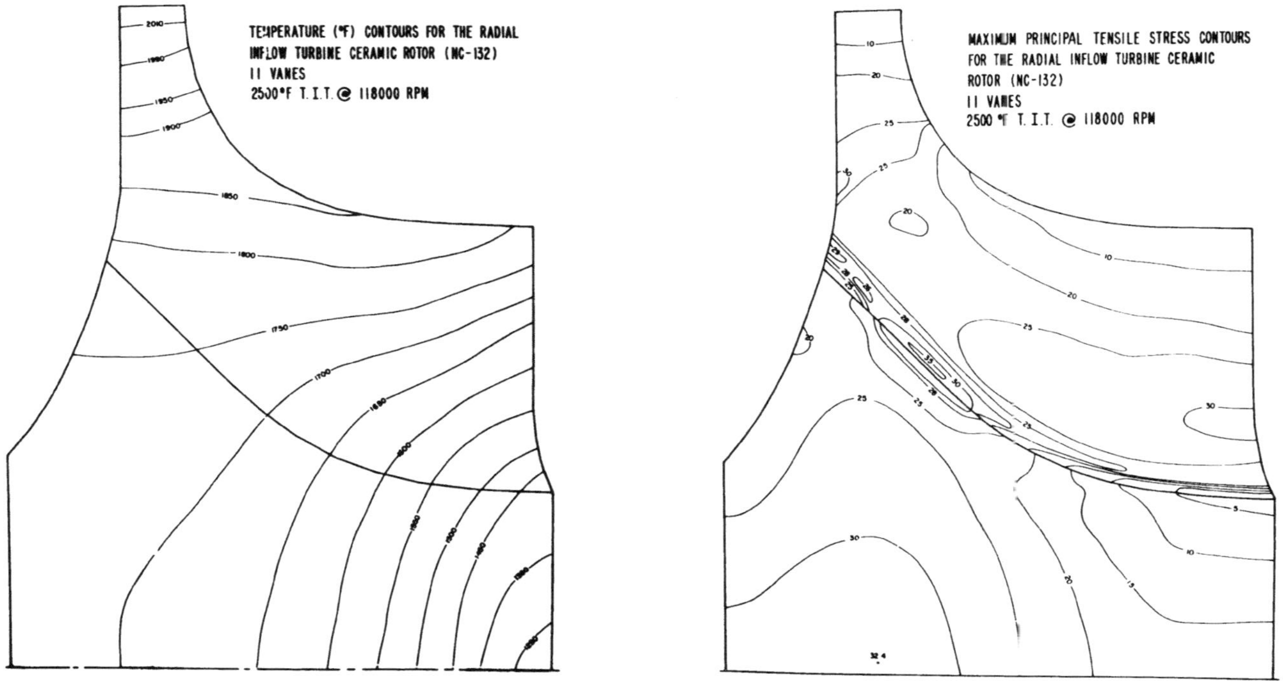

Fig. 14. Distribution of Temperatures and Maximum Principal Tensile Stresses in a Ceramic Radial Inflow Rotor for an Advanced Gas Turbine Engine.

Tip Speed	2500 Ft./Sec.
T.I.T.	2500°F

loading. Hence, one would expect that in the SiC version of the same rotor, keeping in mind that the theoretical density of these two ceramic materials (SiC and Si_3N_4) is equal, the steady state stress distribution at full power loading would be nearly identical to that shown in Figure 14 for the Si_3N_4 rotor. Consequently, the conclusions drawn regarding strength requirements, within the context of this preliminary investigation, will apply to both materials.

The results of the preliminary transient analysis, conducted for the Si_3N_4 version of the rotor, indicate considerably fewer problems than originally anticipated. A 15-25% increase in stress levels over the corresponding steady state values is predicted for the two critically stressed regions described earlier, i.e., the center of expeller hub and the expeller vane fillets. Using Ford test data on NC-132 the probability of failure at peak transient loading was estimated at $2 \cdot 10^{-4}$. Although it has increased by an order of magnitude from the predicted steady state rate of $3 \cdot 10^{-5}$, it was still well within the assumed allowable limit of $1 \cdot 10^{-2}$. It is anticipated that further improvements in rotor design, with respect to its responses to engine loadings and thus its strength, are attainable by way of an optimization process. Of course, the final configuration will need to be verified aerodynamically to insure that no penalties were incurred in rotor performance.

The multistage axial rotor concept studied consisted of three stages. Here again, as in the case of the radial rotor, no attempt was made in the present study to optimize either the geometric configuration or thermal loadings of the individual rotor stages. Therefore, the results presented are considered conservative insofar steady state strength requirements are concerned. Although full steady state analyses were conducted for all three stages, only the results for the first stage disk will be discussed. Past experience has shown this to be the critical stage insofar the temperature and strength requirements. A limited trade-off analysis was conducted to evaluate relative benefits of two different rotor attachment concepts with respect to the effects they will have on rotor responses and the strength requirements. The attachment concepts considered were: a more conventional concept employing a cooled metallic bolt similar to the one developed for a two-stage turbine rotor under FORD/ARPA contract[16] and an alternate approach employing an uncooled ceramic attachment. Steady state distributions of temperatures and maximum principal tensile stresses at the full power point loading of 2500°F (1370°C) TIT and 118000 RPM rotational speed are shown in Figures 15 and 16 for the two attachment concepts respectively. As seen in the figures, the elimination of bolt cooling dramatically reduces the radial temperature gradients and the induced stresses but raises the disk temperatures in the rotor with an uncooled ceramic attachment concept to 2150° - 2320°F (1177 - 1271°C). This range of temperatures and the associated stress levels are for cur-

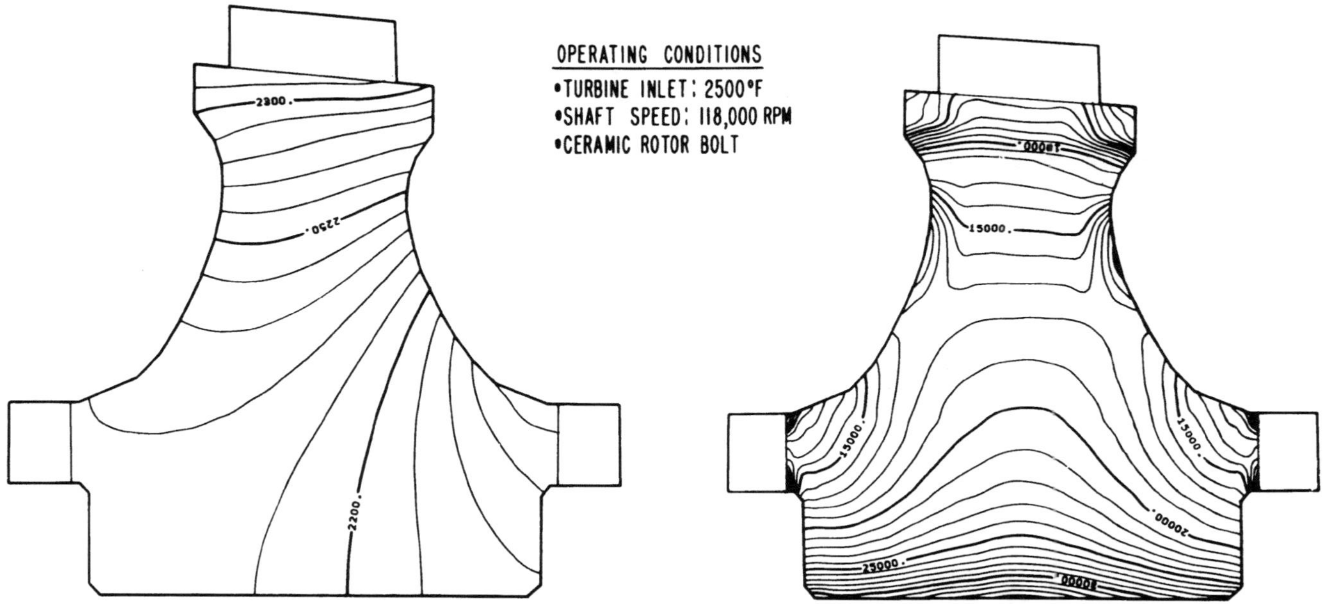

TEMPERATURE DISTRIBUTION (°F)

MAXIMUM PRINCIPAL TENSILE STRESS DISTRIBUTION (PSI)

Fig. 15. First Stage Turbine Rotor of a Cermic Three - Stage Axial Configuration for the Advanced Gas Turbine Engine - Uncooled Bolt.

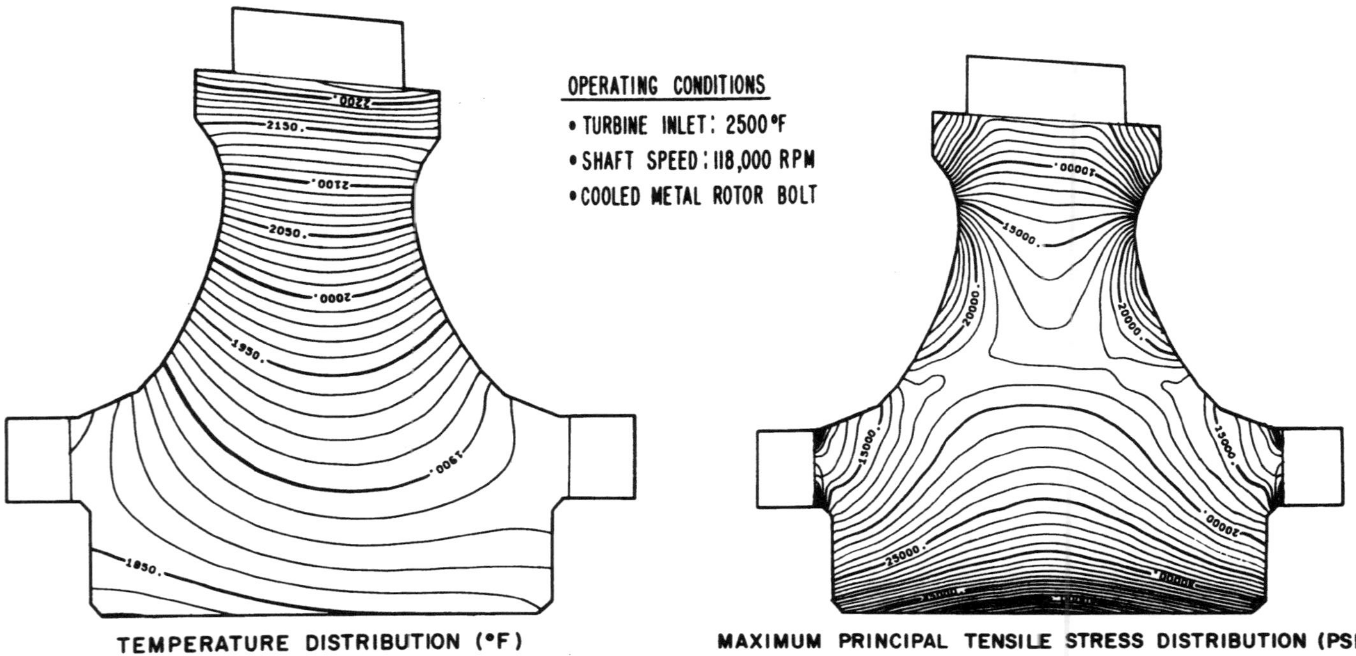

Fig. 16. First Stage Turbine Rotor of a Ceramic Three-Stage Axial Configuration for the Advanced Gas Turbine Engine - Cooled Bolt.

rent grades of HPSN within the regime which requires consideration of time-dependent modes of failure, brought about by subcritical crack growth (SCG), when defining strength requirements or the reliability of a ceramic structure. As a consequence, the strength requirements for the HPSN rotor with an uncooled ceramic attachment concept will increase in order to compensate for the loss of strength resulting from SCG and, as will be shown, will exceed by far the requirements for the same rotor with a cooled metallic bolt. On the plus side of the uncooled ceramic attachment concept are lower operating stresses in the rotor which permit use of ceramic materials which, although not as strong as HPSN, are considerably less susceptible to SCG. This is then where the trade-off lies between these two variants of a multistage axial rotor.

When defining the strength requirements for the advanced ceramic turbine rotor concepts, discussed above, with the objective of determining the potential for the current and future ceramics, a reliability goal of 99% at 100 hours of full power operation was selected. The 99% reliability goal is considered to be in line with rotor reliability levels (98.5%) achieved in commercial aviation.[17] It should be noted that in a typical passenger car application, for which the advanced engine is intended, a relatively small portion, approximately 3% of engine life, is spent at full power and even then it is for very brief time periods only. A probabilistic strength analysis was conducted using stress distributions shown in Figures 14 through 16 under the assumption that all ceramic materials considered in the study can be represented by the Weibull, statistical model of brittle strength. In the Weibull analysis the strength of material is expressed in terms of two (2) parameters: the characteristic strength (S_θ) corresponding to a failure rate of 63.2% and referenced to a specific test coupon and the Weibull slope or shape parameter (m). The test coupon, referenced to in this study, was a .125 x .25 x .75 in. (3.2 x 6.4 x 19 mm) MOR-bar in four-point bending, quarter-point loading (inner span being .375 in. or 9.5 mm).

The results of the strength analysis are summarized in Figure 17. The characteristic, inert (fast fracture) strength requirements (S_θ) for all rotors studied are shown plotted as function of the Weibull slope (m). Material strength requirements have been assumed to be constant throughout the wheel volume[8] and invariant with temperature. Any combination of characteristic strength and slope to the right of a given curve will satisfy or exceed the assumed reliability goal of 99% for that particular rotor (99.7% for the individual stages of a three-stage axial concept). Superimposed upon the strength requirement curves are points representing current and projected early 1980's technology strength levels.

Strength requirements for the first stage of the axial rotor with a cooled bolt attachment version are shown to coincide almost

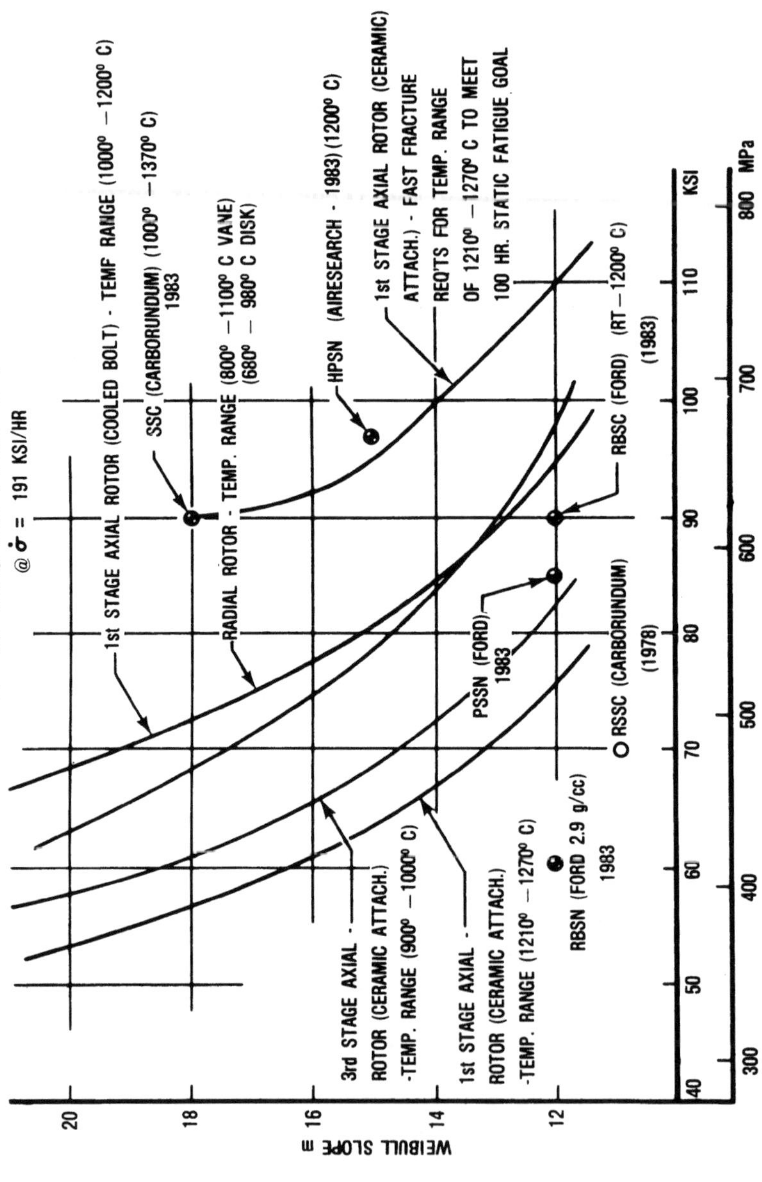

Fig. 17. Projected Requirements of Weibull Strength Parameters for the Turbine Rotors in the Advanced Gas Turbine Engine.

with the requirements for the radial inflow rotor, i.e., there are no relative strength advantages between these two concepts. The curve furthest to the left represents strength requirements for the same first stage disk with bolt cooling eliminated, i.e., a ceramic attachment, without consideration of the time-dependent modes of failure. One must keep in mind that the disk temperatures, in this attachment concept, run considerably hotter and fall into the regime where time-dependent modes of failure occur. The curve furthest to the right represents strength requirements for the same, uncooled version of the disk but with the consideration of the deleterious effects of subcritical crack growth. It was generated for 99.7% stage survival probability (99% for the complete rotor) at 100 hours of full power operation using a crack propagation exponent of n = 17.0 which is representative of hot pressed Si_3N_4. The strength requirements generated in this case correspond to a stress rate of 1317 MPa/hr. The dramatic increase in strength requirements necessary to compensate for the degradation resulting from SCG suggests, as discussed earlier, that the axial rotor with an uncooled ceramic attachment version is more suitable for the reaction sintered SiC (RSSC) or the pressureless sintered Si_3N_4 (PSSN), both of which are expected to be considerably more immune to subcritical crack growth at the temperature levels these rotors will see in service (max. 2320°F or 1271°C). As shown in Figure 17, the strength projections for these materials exceed the requirements, leaving room for possible increases in strength requirements to accommodate extreme transient loadings that will need to be considered in the final design stage. The axial rotor, employing a cooled bolt attachment concept which results in higher disk stresses but considerably lower temperatures, appears, on the other hand, to favour hot pressed Si_3N_4 in either monolithic or duo-density configuration. Again the strength requirements, in this case, are within the projected levels for that material.

The above study, although limited in scope, has shown that ceramics have a real potential in these very demanding applications and that a ceramic turbine rotor of very advanced performance is mechanically feasible. Because ceramics have a definite advantage over superalloys, their development should be pursued in order to realize the potential benefits of the advanced vehicular gas turbine.

Concluding Remarks

The evolution of the ceramic turbine rotor design, the changes made as our understanding of ceramics grew, illustrate how important it is for the designer to thoroughly understand the subject he designs; what is its function, how is it going to be fabricated, what are its failure modes, etc. Only when the designer considers all these factors will his design be material efficient and reliable.

The design tools and fracture models used in this study were novel and perhaps, some may consider, simplistic. They did, however, serve their purpose well for they provided the designer with a rationale upon which he was able to base his decisions and choices. Results of limited destructive tests conducted on actual engine hardware seem to support this rationale. It is true the evidence obtained so far is still far from conclusive. Nonetheless, we must conclude, that a tremendous progress has been made in the science and the art of designing with ceramics. Newer, more sophisticated models for ceramic materials are also on the horizon. These will supplement the models and tools so far developed and, no doubt, improve our understanding of these exciting new engineering materials and make them more palatable to the designer.

Acknowledgement

Partial support for this work was provided by Contract No. DAA6 46-71-C-0162 with the Advanced Research Projects Agency of the Department of Defense and the Department of Energy.

REFERENCES

1. A. F. McLean, E. A. Fisher and R. J. Bratton, "Brittle Materials Design, High Temperature Gas Turbine", Interim Report #3, July 1, 1972 - Dec. 31, 1972. Ford Motor Co., Dearborn, Mich., Army Material and Mechanics Research Center Report, AMMRC CTR 73-9.

2. A. F. McLean, E. A. Fisher and R. J. Bratton, "Brittle Materials Design, High Temperature Gas Turbine", Interim Report #4, Jan. 1, 1973 - June 30, 1973. Ford Motor Co., Dearborn, Mich., Army Materials and Mechanics Research Center Report, AMMRC CTR 73-32.

3. A. F. McLean, E. A. Fisher and R. J. Bratton, "Brittle Materials Design, High Temperature Gas Turbine", Interim Report #5, July 1, 1973 - Dec. 31, 1974. Ford Motor Co., Dearborn, Mich., Army Materials and Mechanics Research Center Report, AMMRC CTR 74-32.

4. P. F. Nicholls and A. Paluszny, "Design of a Small Ceramic Turbine Rotor", Ceramics for High Performance Applications - I, Army Materials Technology Conference Series Edited by J. J. Burke, E. N. Lense, and N. R. Katz, Brookhills Publishing Co., 1974, pp. 63-78.

5. A. F. McLean, E. A. Fisher and R. J. Bratton, "Brittle Materials Design, High Temperature Gas Turbine", Interim Report #6, Jan. 1 - June 30, 1974. Ford Motor Co., Dearborn, Mich., Army Materials and Mechanics Research Center Report, AMMRC CTR 74-59.

6. F. A. McLean, E. A. Fisher, R. J. Bratton and D. G. Miller, "Brittle Materials Design, High Temperature Gas Turbine", Interim Report #8, Jan. 1, 1975 - June 30, 1975, Ford Motor Co., Dearborn, Mich., Army Materials and Mechanics Research Center Report, AMMRC CTR 75-78.

7. A. Paluszny and W. Wu, "Probabilistic Aspects of Designing With Ceramics", Transactions of ASME - Journal of Engineering for Power, Vol. 99, No. 4, Oct., 1977, p. 623.

8. A. G. Evans and S. M. Wiederhorn, "Crack Propagation and Failure Prediction in Silicon Nitride at Elevated Temperature", Journal of Material Sciences (9), 1974, pp. 270-278.

9. A. F. McLean and R. R. Baker, "Brittle Materials Design, High Temperature Gas Turbine", Interim Report #12, Jan. 1, 1977 - Sept. 30, 1977. Ford Motor Co., Dearborn, Mich., Army Materials and Mechanics Research Center Report, AMMRC CTR 78-14.

10. A. Paluszny and P. F. Nicholls, "Predicting Time-Dependent Reliability of Ceramic Rotors", Ceramics for High Performance Applications-II, Army Materials Technology Conference Series Edited by J. J. Burke, E. N. Lense and N. R. Katz, Brook Hill Publication, 1978, pp. 94-112.

11. A. F. McLean and R. R. Baker, "Brittle Materials Design, High Temperature Gas Turbine", Interim Report #10, Jan. 1, 1976 - June 30, 1976. Ford Motor Co., Dearborn, Mich., Army Materials and Mechanics Research Center Report, AMMRC CTR 76-31.

12. R. R. Baker and L. R. Swank, "Duo-Density Ceramic Turbine Rotor Development", Ceramics for High Performance Applications - III, Reliability. Presented on Orcas Island, Washington, July 10 - 13, 1979. To be published.

13. A. F. McLean and R. R. Baker, "Brittle Materials Design, High Temperature Gas Turbine", Interim Report #13, Oct., 1977 - March, 1978. Ford Motor Co., Dearborn, Mich., Army Materials and Mechanics Research Center Report, AMMRC CTR 79-11.

14. Proceedings of Highway Vehicle Systems Contractors' Coordination Meeting, Dearborn, Mich., Oct. 17-20, 1978, U.S. Department of Energy - Fifteenth Summary Report #CONF - 78 1050.

15. Ford Engineering Staff, "Conceptual Design Study of Improved Automotive Gas Turbine Powertrain - Final Report", May, 1979. NASA Publication No. NASA-CR-159580 or DOE/NASA/0037-79/1.

16. A. F. McLean, E. A. Fisher and R. J. Bratton, "Brittle Materials Design, High Temperature Gas Turbine", Interim Report #2, Jan. 1, 1972 - June 30, 1972. Ford Motor Co., Dearborn, Mich., Army Material and Mechanics Research Center Report, AMMRC CTR 72-19.

17. R. A. DeLucia and G. J. Mangano, "Rotor Burst Protection Program, Statistics on Aircraft Gas Turbine Engine Rotor Failures that Occurred in U.S. Commercial Aviation during 1975." Naval Air Population Test Center Report #NAPTC-PE-106 (NASA - CR - 135304), May, 1977.

SILICON CARBIDE COMPONENTS FOR DIESEL AND GASOLINE ENGINE APPLICATIONS

R. C. Phoenix and W. D. Long

High Performance Silicon Carbide Div.
The Carborundum Company
Niagara Falls, New York 14302

ABSTRACT

A recently developed sintered alpha silicon carbide is a candidate material for advanced turbine engine components. Several applications in diesel engines and gasoline engines have been tested with encouraging results. Principal successes are valve lifter, pre-combustion chambers, and turbocharger rotors.

INTRODUCTION

In June of 1976, the Carborundum Company announced the successful pressureless sintering of alpha silicon carbide to 98% of the theoretical density. Subsequent development work has shown that this material has flexural strengths of 80 ksi at both room temperature and up to 1650°C. Some time dependent strength reduction has been identified but it is far less than other advanced ceramic materials. Because of these properties, the material has been selected for use in many U.S. Government advanced turbine engine programs. Since commmerical applications of these engine are many years in the future, the authors determined to evaluate potential uses of alpha silicon carbide components in today's diesel and gasoline engines.

Three different applications were selected for testing with increasing degrees of environmental severity. The first was a solid valve lifter or tappet. The second was the Ricardo Comet design precombustion swirl chamber and the third, the turbocharger rotor. The selection of the components was based not only on the purely technical basis of whether the silicon carbide would survive but also on the economic basis that the fabricated ceramic shape would be more cost effective than the metallic component.

Tappets shown in Figure 1 were first tested in a Chevrolet 350 engine with a new solid lifter profiled camshaft. Various gap settings were used to increase the impact on the ceramic. No wear or cracking of the silicon carbide or the cam was observed after 100 hours at 4000 rpm. Since that first test, ceramic lifters have been used in various stock car races up to and including 500 mile competition. The principal advantage of the ceramic, in addition to its resistance to wear, has been its lighter weight. Improvement in acceleration has been reported. Tests are also under way with several medium and heavy duty diesel engine manufacturers. Tests on a Bedford engine at Ricardo Engineering Ltd. are demonstrating excellent durability. It has been shown that surface finish of the ceramic is critical. A 5 rms of finer finish is required.

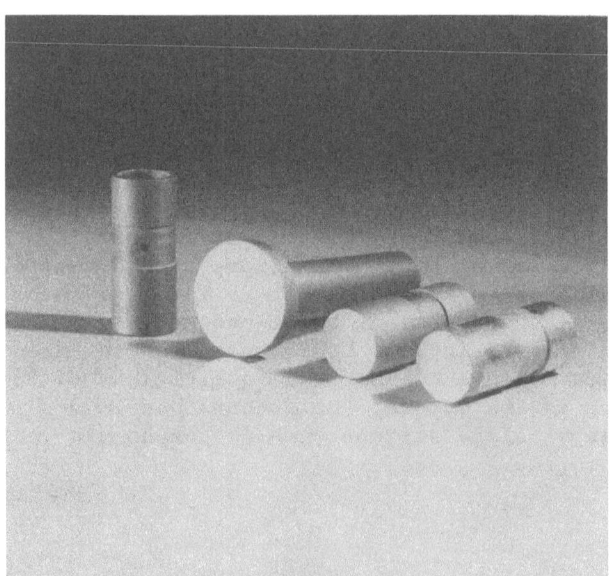

Figure 1

Use as Comet swirl chambers exposes the ceramic to the difficult environment of combustion of the diesel engine. Both sintered alpha silicon carbide and reaction sintered silicon carbide have been tested. In the first tests, a metal retainer rings was used to secure the ceramic to the head. Tests are now under way using a straight sided unit on the right of Figure 2. Units were tested on over the road programs for 2,000 miles prior to a more severe acceleration/deceleration stand test. Both types of silicon carbide units have survived over 500 hours of the test stand cycle. The advantage of the sintered alpha silicon carbide over the reaction sintered silicon carbide is primarily in reduced manufacturing costs. Several design interations are in progress which should enable a ceramic unit to be marketed at a price below that of present nimonic units.

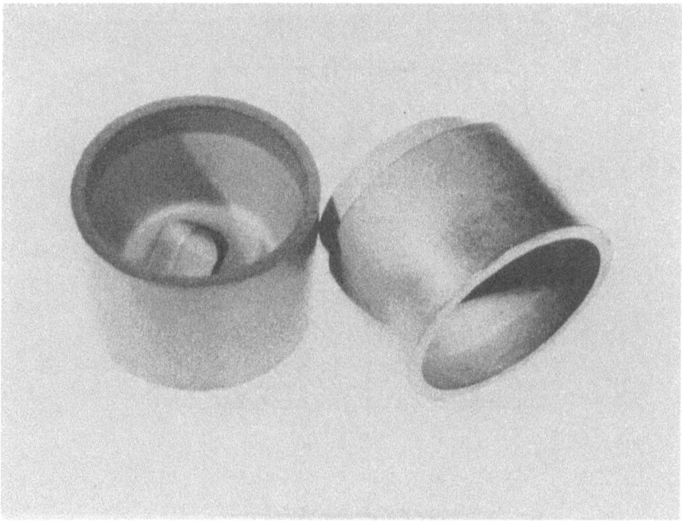

Figure 2

The third, and most difficult, test regimen for silicon carbide is as a turbocharger rotor. Rotational stress added to thermal stress seriously challenges the silicon carbide materials available today. It is our belief that it will be important to demonstrate both technical and economic success in this application as a precursor to the highly stressed rotating components in future ceramic gas turbine engines.

Ricardo Engineering, Ltd. performed a stress analysis on their 3" diameter Keramos rotor using sintered alpha silicon carbide in three different hub configurations shown in Figure 3. The solid is, of course, preferred by the designer; the hollow is best for the silicon carbide injection molder. The webbed compromise is the best current approach. Today it allows a minimum bakeout path for the organics used in injection molding while decreasing the maximum principal stress compared with the hollow design from 30,000 PSI to 23,000 PSI at 100,000 ROM. Other analyses, shown in Figure 4, confirmed that if we attain our goals of 80,000 PSI with Weibul modulus of 18 on injection molded sintered alpha silicon carbide, we can expect to survive design speeds anticipated in automotive gasoline and diesel turbochargers with the webbed hub configuration.

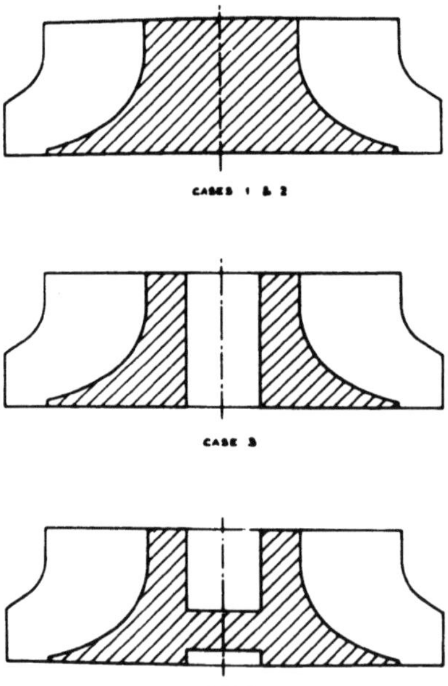

Figure 3

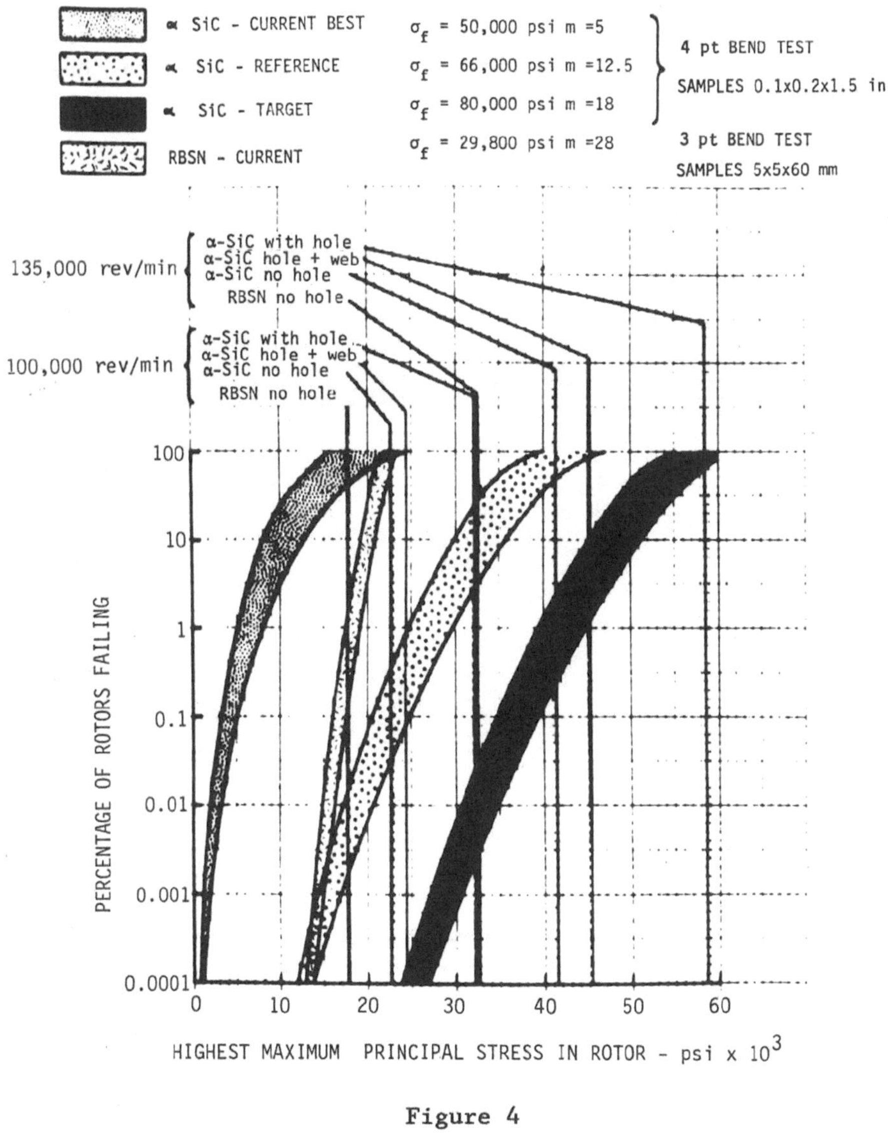

α SiC - CURRENT BEST
α SiC - REFERENCE
α SiC - TARGET
RBSN - CURRENT
σf = 50,000 psi m =5
σf = 66,000 psi m =12.5
σf = 80,000 psi m =18
σf = 29,800 psi m =28
4 pt BEND TEST
SAMPLES 0.1x0.2x1.5 in
3 pt BEND TEST
SAMPLES 5x5x60 mm
135,000 rev/min
α-SiC with hole
α-SiC hole + web
α-SiC no hole
RBSN no hole
100,000 rev/min
α-SiC with hole
α-SiC hole + web
α-SiC no hole
RBSN no hole
PERCENTAGE OF ROTORS FAILING
100
10
1
0.1
0.01
0.001
0.0001
0
10
20
30
40
50
60
HIGHEST MAXIMUM PRINCIPAL STRESS IN ROTOR - psi x 10^3

Figure 4

Figure 5 shows two injection molded silicon carbide turbocharger rotors. Spin test data is being gathered on a variety of sintered silicon carbide material compositions and injection molding tooling iterations. A test vehicle demonstration using a ceramic rotor in an AiResearch TO-3 turbocharger has been made. Cost estimates indicate that production costs for ceramic components in the turbocharger can be lower than presently identified for high temperature resistant metal alloys.

Figure 5

While these demonstrations show that the sintered alpha silicon carbide can survive the challenging environments of a modern diesel or gasoline engine, there is still work to be done. Design engineers must become familiar with brittle design in practice. Material optimization must continue to reach the laboratory proven results in production. It is anticipated that commercial use of an alpha silicon carbide engine component will take place in the 1982 model year. It is hoped that this will lead to a broadening acceptance of new ceramic materials in all types of engine applications.

STATUS OF COMBUSTOR DEVELOPMENT

FROM CERAMIC MATERIALS

G. Kappler

MTU-München GmbH
Dachauer Strasse 665
8000 München 50

1. Introduction

The development of gas turbines with high turbine inlet temperatures and minimized cooling flows will lead to major fuel savings. Engine components fabricated from ceramic materials, have the potential to operate at the required high gas temperatures. In order to overcome their inherent brittleness, research programs were initiated to link improvements in ceramic material properties and fabrication methods with specific ceramic design technologies.

The design of ceramic combustor[1] has to satisfy two main requirements: - high temperature gradients along the flame tube walls, caused by the temperature reduction from the primary combustion zone to the turbine inlet, and thermoshock loads during ignition events or fast changes in operating conditions. Since the load spectra of vehicular gas turbine engines are different for car and truck applications, the development of apropriate ceramic combustors follows distinct technical requirements. For car application, the combustors have to withstand higher cyclic charges and comply with more stringent emission restrictions than truck engines.

2. Selection of ceramic combustors

Some of the ceramic combustors designed and tested at the Volkswagenwerk AG[2] and MTU München[3] are shown in Fig. 1.

The configuration of the ceramic combustors at the Volkswagenwerk AG was dictated by the special requirements of car application. The more complex shape, which

VOLKSWAGENWERK AG

Siliconized SiC Slip-Cast — SiC-Mullit Saturated Isostatic Pressed — Siliconized SiC (Refel) Isostatic Pressed

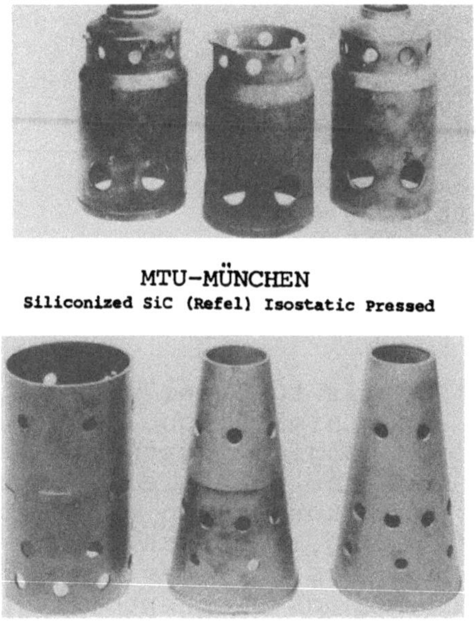

MTU-MÜNCHEN

Siliconized SiC (Refel) Isostatic Pressed

Fig. 1. Combustion Chambers fabricated from Ceramic Materials

was partly derived from existing metal combustors, was intentionally chosen to probe the low-cost design possibilities of low-emission combustors. For low-emission considerations the combustors had a large volume. They were manufactured from siliconized SiC by slip-casting and by isostatic pressing or from mullite-impregnated SiC and isostatically pressed, with walls as thin as technically possible.

At MTU-München, previous ceramic material tests with cylindrical probes had clearly demonstrated the technical advantages of siliconized SiC in comparison with Si_3N_4, especially for thermoshock loadings. Therefore, the combustion chambers were manufactured from siliconized SiC by isostatic pressing. Along with the usual cylindrical combustors, conical flame tubes were designed because of specific ceramic material considerations. Through the conical shape the combustor walls were brought closer to the hot flame zone, thus better suiting the ceramic material requirements which call for high temperature levels but low temperature gradients along the liner.

3. Test equipment

The ceramic combustor tests at the Volkswagenwerk AG were carried out on a combustion test rig which permitted the simulation of typical operating conditions of gas turbines for car utilization. The combustors were fitted in casings which allowed for the same inlct flow conditions as in the engine.

The rig set up used at MTU-München for the ceramic combustor tests is shown schematically in Fig. 2.

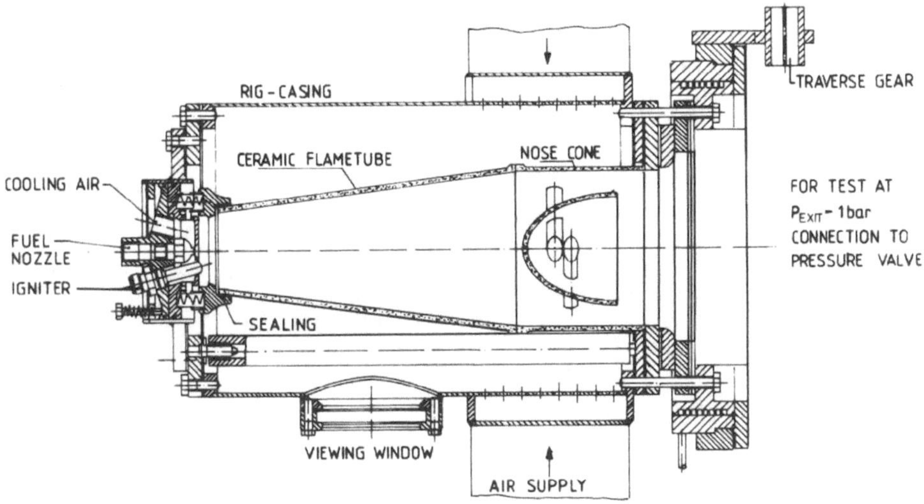

Fig. 2. Schematic of Test Rig

Together with the nose cone, the ceramic combustor was installed in a casing equipped with a viewing window. Air was fed through two adjacent ducts into the casing and uniformly distributed along the combustor. According to the required operating condition, the air mass flow was adjustable between 0.17 and 0.83 kg/s and the air inlet temperature (T_{IN}) between 300 K and 1050 K. The fuel was injected through a commercial spray nozzle at pressures up to 20 bar and ignited by an open spark ignition torch with an electrical energy of 6 Joule at 24 V and a frequency of 2 pulses per second. The maximum fuel mass flow, which corresponds to full power engine operation, amounted to 11.5 grams per second. The additional air fed into the combustor head cooled the nozzle and improved the fuel droplet preparation. A traverse gear was mounted on the nose cone exit permitting full annular exit temperature distribution measurements.

For increased pressure tests above 1 bar a water-cooled nozzle was installed just behind the nose cone. Fig. 3 shows a photograph of the rig set-up during a test

Fig. 3. Rig set up for Pressure Tests up to 5 bars

at a combustion chamber pressure of 4.9 bar. One can see the water-cooled nozzle and the shiny viewing window. A pyrometer was placed in front of the window for combustor surface temperature measurements.

4. Investigations under steady test conditions including cold start trials

The series of ceramic combustor tests commenced under steady operating conditions, with flame ignition initiated only after a preheating sequence. The main goal was to determine which of the selected ceramic materials and combustor shapes withstands the steep gas temperature drop from the primary zone towards the combustor outlet.

For the sequence of steady operating conditions listed in Fig. 4, which was applied as cycle I by the

PHASE	DURATION Min	T_{IN} K	T_{EXIT} K	MASS FLOW kg/s	PRESSURE bar
WARM UP (NO THERMOSHOCK)	10	80 DEGREES/Min		0.5	2.8
IGNITION (WEAK LIMIT)	3	800	≈1000	0.5	2.8
EXIT TEMPERATURE INCREASE	3	800	1230	0.5	2.8
BURNING (STATIONARY)	45	800	1230	0.5	2.8
END	SLOW REDUCTION IN TEST CONDITIONS				

Fig. 4 Stationary Test Conditions VW AG: Cycle I

Volkswagenwerk AG, at ignition the combustor air entry temperature amounted to 800 K and the pressure to 2.8 bar. After ignition the temperature was slowly increased to 1230 K and than kept constant for 45 minutes. The test sequence was ended by slowly reducing the state parameters. The test sequences used at MTU-München were the same in principle, only the maximum exit temperature was higher (T_{EX} = 1600 K). The test results at both companies showed that combustors made from SiC withstand hot steady conditions. However, the cylindrical combustors at MTU-München revealed thin fracture lines around the primary zone air ports and along the transition line from the cylindrical to the concave shape. Previous results gained from ceramic material tests were also confirmed i.e. combustors made from Si_3N_4 fractured at high temperatures.

The ceramic combustors were exposed to higher loads by including cold starts to the steady tests conditions, because of the extreme temperature shock. The temperature increases in fractions of a second from the ambient temperature to the combustion temperature level. Whereas some of the slip-cast SiC combustors fractured after start tests, the combustors from isostatically pressed

SiC withstood the thermoshock. Also, the cylindrical combustors at MTU-München which had already revealed light cracks fractured completely. They were therefore not considered for further tests.

After the cold ignition tests the ceramic combustors were investigated under conditions corresponding to engine operating loads, such as idle and full power. The combustor surface temperature (T_{SR}) was measured during this test sequence, which was designated Cycle II at the Volkswagenwerk AG. The measured temperature distribution along the ceramic combustor surface is shown in Fig. 5.

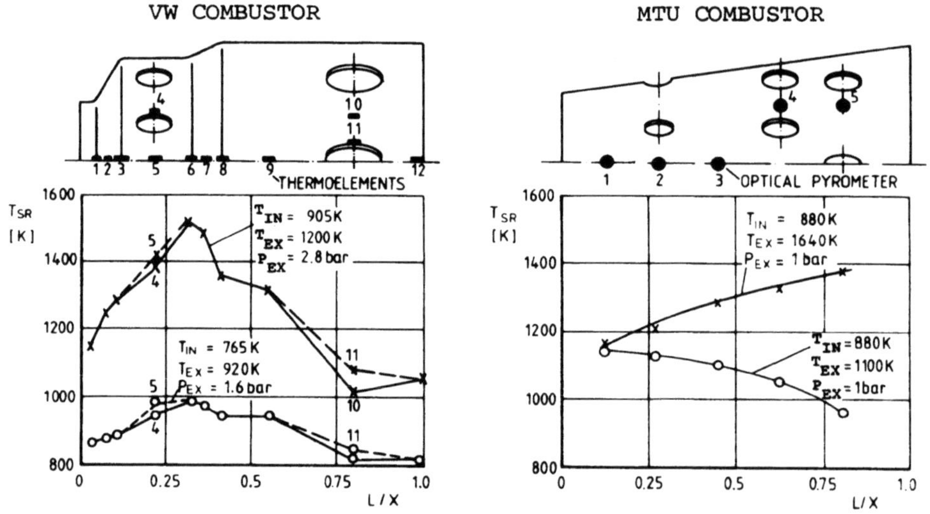

Fig. 5. Measured Surface Temperature Distribution

At the Volkswagenwerk AG the surface temperature was measured by thermocouples glued to the combustor liner. The temperature distribution shows a temperature peak at the combustor enlargment region for both steady test conditions. Since the temperature peak becomes more pronounced with increased loading, i.e. fuel injection, vortex-stabilized combustion can be assumed to occur at the marked enlargment.

The surface temperature along the conical ceramic combustor, which was measured at MTU-München with the aid of an optical pyrometer, reveals a continuous distribution with no temperature peaks. At low exit temperatures the surface temperature distribution follows the familiar drop from the primary zone towards the combustor outlet. At higher exit temperatures, the additional fuel flow leads to a continuous flame stretch which spreads as far as the second row of air injection ports, producing the slight surface temperature increase along the combustion liner. The nose cone may also contribute to the higher outlet surface temperature by improving heat transfer conditions. The conical combustor concept, therefore, permits continuous adaption of the combustion volume to the operating conditions, thus avoiding high local gas temperature peaks.

5. Cyclic Test

After having successfully undergone the investigations under steady conditions including cold starts, the combustors made from siliconized SiC were cyclic tested.

At MTU-München, the tests began with temperature cycles applied at ambient pressure. The cycle depicted in Fig. 6 is made up by an acceleration and deceleration

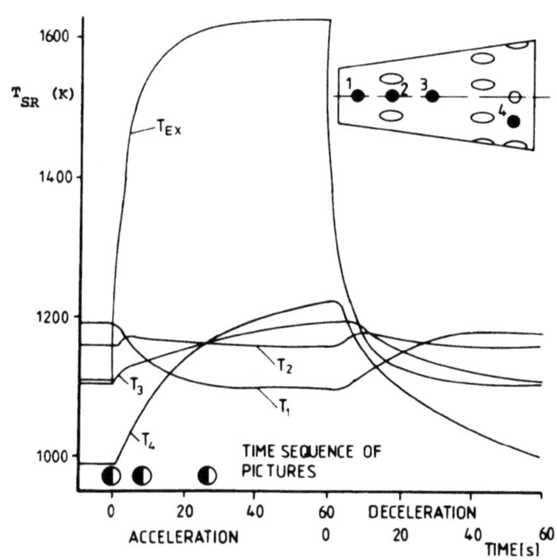

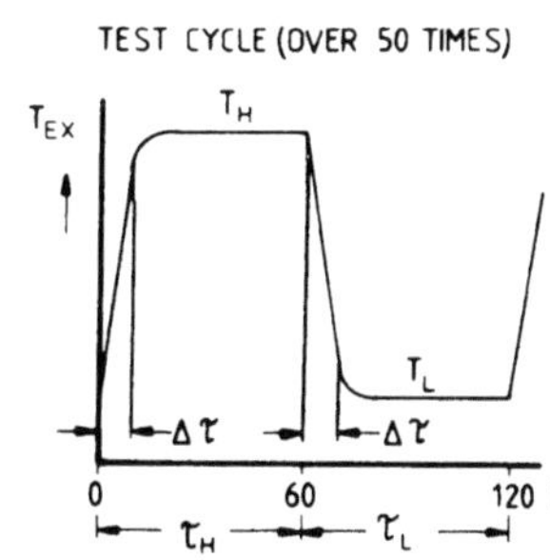

T_H	1600	K
T_L	1100	K
AFR_H	49	
AFR_L	200	
τ_H	60	S
τ_L	60	S
$\Delta\tau$	3	S

TEST CONDITIONS

T_{IN} =880 K

P_{EX} =1.02 Bar

M_{RED} $=5\frac{kg.\sqrt{k}}{s.bar}$

Fig. 6. Cyclic Testing of Ceramic Combustor

phase which was applied more than 50 times. The highest and the lowest temperatures of the cycle (T_H and T_L), measured at the combustor outlet, the corresponding air fuel ratios (AFR) and the time intervals are listed in the table. The air temperature at the combustion chamber entry was 800 K.

The surface temperatures of the liner were monitored at certain locations during the cyclic tests. Together with the variation in the gas exit temperature, they are plotted in Fig. 6. Since with increasing fuel flow the flame spreads downstream of the first row of air ports, the surface temperature at location 1 decreases and increases at location 3, whereas it remains nearly constant between the primary air ports at location 2. The marked increase in the exit surface temperature at location 4 proves the previously explained mechanism of continuous combustion expansion along the conical combustor with increasing fuel flow. During the deceleration phase, the surface temperatures drop steadily towards their starting values without sudden gradients.

The photographs of the glowing surface taken through the viewing window at the beginning of the acceleration phase and at 9 and 27 seconds later, are represented in Fig. 7. The downstream expansion of the combustion zone

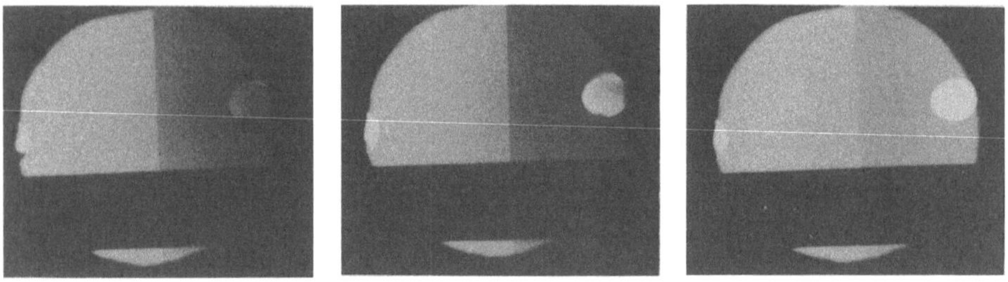

0 Sec 9 Sec 27 Sec

ACCELERATION TIME

Fig. 7. Pictures of glowing Combustor Surface

is quite evident. With increasing fuel flow after 9 seconds the combustion zone has reached the region of the second row of air injection holes making the previous dark ports shine brightly. One can also see the continuous increase in the downstream surface temperature as indicated by the brighter red color, which became nearly uniform after 27 seconds.

At the Volkswagenwerk AG the thermal cycles were part of a sequence of engine-like operating conditions (Cycle III), including cold ignition and steady operating periods. They consisted of 10 temperature leaps from 900 K to 1500 K with a 2 minutes holding period at a combustion chamber pressure of 1.6 bar. Combustors made of isostatically pressed siliconized SiC (Refel) revealed thermal cracks after the tests. The cause of the cracks was traced back to high local temperature gradients and material flaws, such as silicon and other inclusions. Combustors made of mullite-impregnated SiC fractured in the clamping region as a result of mechanical stresses.

The most exacting test conditions the ceramic combustors were subjected to ocurred in Cycle IV, which is shown schematically in Fig. 8. The test cycle included the full power combustor pressure of 4.5 bar and an

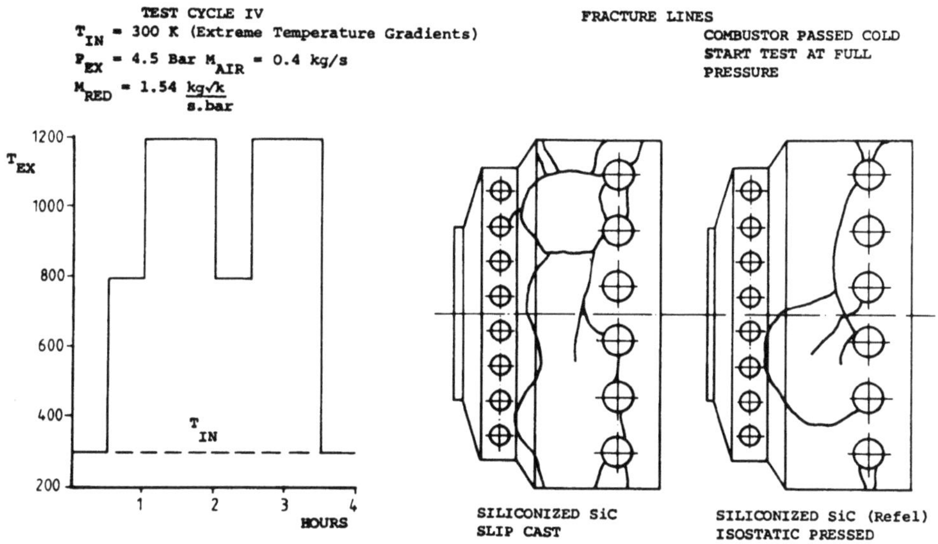

Fig. 8. VW Ceramic Combustor Tests at High Operation Conditions

extremely high temperature difference between 300 K at the combustor entry to 1200 K at the exit. Such a high temperature difference does not occur in the engine over such a long period of time, but only sporadically during cold starts with full power acceleration.

Combustors made of isostatically pressed SiC (Refel) did not fracture after the first period of the test cycle, i.e. the cold start at a combustor pressure of 4.5 bar. At the end of the cycle, however, both the slip-cast and the isostatically pressed SiC combustors fractured. From the fracture lines projected on Fig. 8 it can be seen that the crack in the slip-cast combustor runs along the region of the highest temperature peaks and that in both combustors the cracks were initiated by stress concentrations around the air ports. These findings will be used for a ceramic-oriented redesign of the combustors.

Combustor tests at higher plenum chamber pressures were commenced at MTU-München by increasing the pressure in stages as depicted in Fig. 9, and by keeping the air

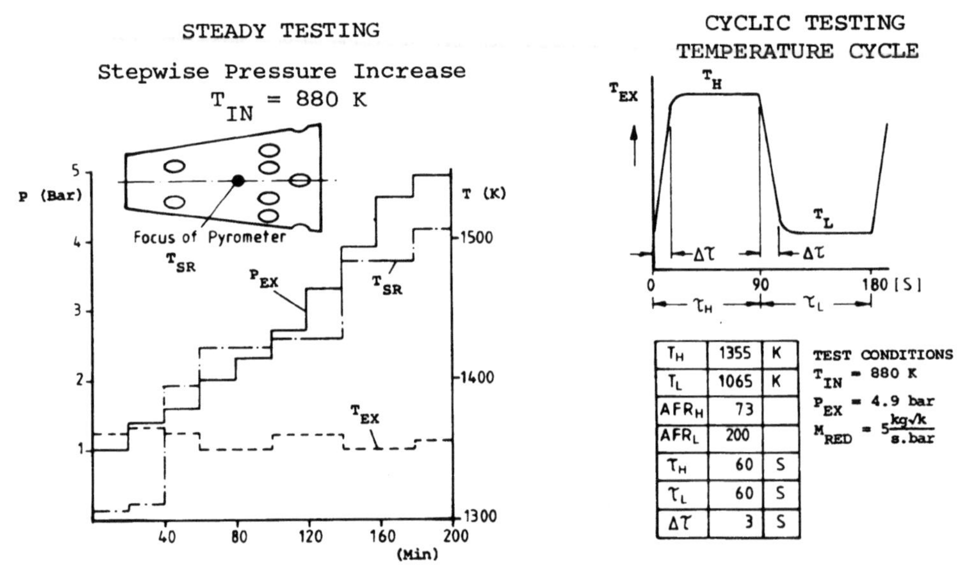

T_H	1355	K
T_L	1065	K
AFR_H	73	
AFR_L	200	
τ_H	60	S
τ_L	60	S
$\Delta\tau$	3	S

Fig. 9. MTU Ceramic Combustor Tests at Elevated Pressures

entry and the gas exit temperature nearly constant. The test data apply to steady conditions. The surface temperature measured at a location just in front of the second row of air ports revealed a pressure-related increase. At the point of measurement in question the surface temperature was higher than the gas exit temperature at pressures over 2 bar.

After the tests, only small silicon secretions were found in the siliconized SiC (Refel) combustors manufactured by BNFL, and the combustors remained undamaged. Similar conical combustors, also made of siliconized SiC but manufactured by NORTON, fractured during the tests. As shown in Fig. 10, the cracks run along the first and second row of air ports, where the material is inherently weakened and where high temperatures prevail. The inner surface became wavy and exhibited a change in color with silicon secretions. Since details of the manufacturing

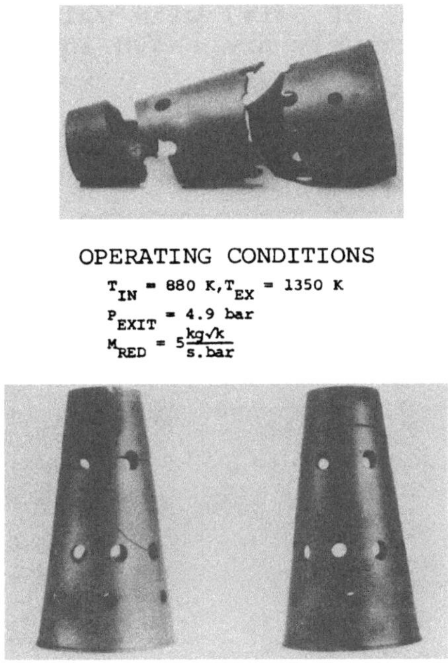

Fig. 10. Fractured Combustor from SiC by Norton

methods of the two companies are not available, no direct conclusions can be drawn.

Various nose cones were also investigated during the tests at elevated pressures. The results were similar to those of combustor tests; nose cones of Si_3N_4 as well as integrally molded SiC nose cones fractured. For cyclic tests at elevated pressures, therefore, only combustors and nose cones of Refel where considered further.

The temperature cycle applied to the ceramic combustors at full power pressure of 4.9 bar is also shown in Fig. 9. The characteristic values of the cycle are listed in the table. The combustors were subjected to 200 cycles, i.e. to temperature leaps of 290 degrees, with an inspection being carried out at the half way stage after 100 cycles. The ceramic combustors made of Refel investigated so far showed no signs of major damage after the cyclic tests. A two-part combustor having successfully gone through the full cycle is pictured in Fig. 11.

The combustor revealed a thin fracture line on the outer surface between two air ports (marked in white) and strong silicon secretion with some fracture lines in the region of high surface roughness on the inner surface. Two unsplit combustors made of Refel with wall thicknesses of 4 mm and 2 mm respectively are shown in Fig. 12 after

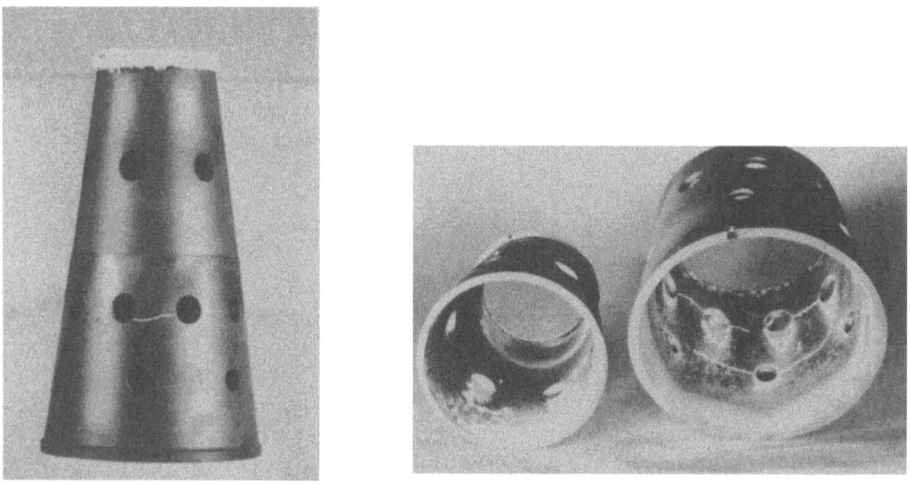

Fig. 11. MTU Ceramic Combustor after Cyclic Tests at Elevated Pressures

NON SPLIT COMBUSTORS

THICK WALLED 4 mm

THIN WALLED 2 mm

Fig. 12. MTU Conical Unsplit Ceramic Combustor

the cyclic tests. Both combustors have no fracture lines and exhibit fewer secretion bubbles. It was concluded, therefore, that some air enters the two-part combustor between the division plane, disturbing uniform combustion and thus producing temperature gradients. This again stresses the requirement for avoiding temperature gradients along ceramic combustion liners.

6. Operation of a hybrid ceramic-metal combustor in a vehicular gas turbine

In order to investigate the behaviour of ceramic materials under full engine operating conditions at MTU-München, a hybrid ceramic-metal combustor was installed in a vehicular gas turbine. A design scheme of the combustor and the installation in the engine is shown in Fig. 13. The combustor consists of an air-cooled metal head end, a ceramic cylinder made of siliconized SiC (Refel) and a metallic transition section. The ceramic cylinder was connected to the metallic parts by a seal and the mechanical loads were transmitted across it through adjustable staves. Such hybrid combustors have

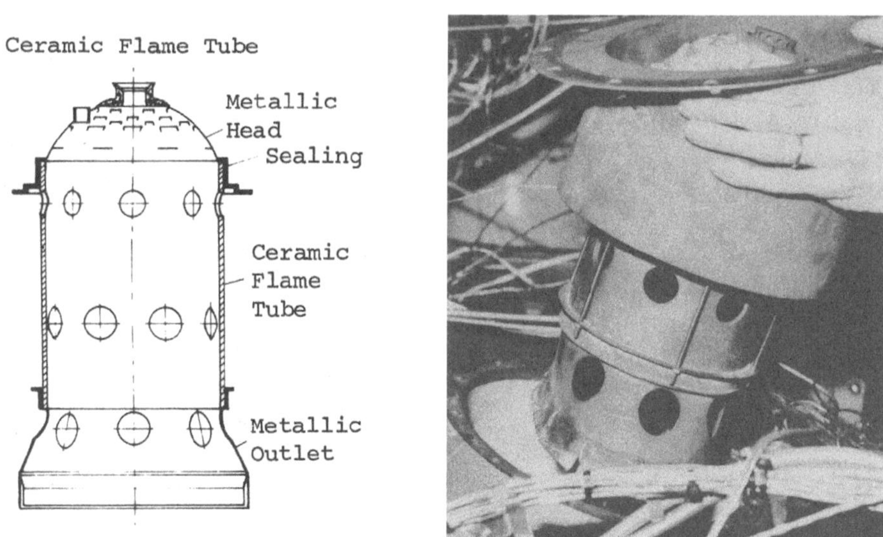

Fig. 13. Ceramic Flame Tube for Vehicular Engine Test

accumulated over 20 hours of testing in a vehicular gas turbine. After test inspections some of the ceramic cylinders revealed cracks originating from the adjoining regions. A typical fracture line, which starts at the head end sealing and crosses the ceramic liner up to the second row of air ports, can be seen in Fig. 14. The transition area from metal to ceramic parts, in general, presents a very demanding design problem because of the diversity in thermal expansion. A new design approach, to be investigated next, omits the insulation, allowing each part to move without direct restrictions. The ceramic cylinder has not yet failed as a result of thermal fatique and thin mechanical cracks have not hampered engine operation.

The hybrid combustor was subjected to a 13-point California test cycle and the exhaust emissions were measured. Under full power the combustor air entry temperature was 891 K, the plenum pressure was 3.67 bar and the exhaust temperature 1255 K. The emissions, plotted in Fig. 14 as brake specific emission factors, in comparison with the emission limits set or to be set for diesel engines in 1979 and 1980, highlight the environmental

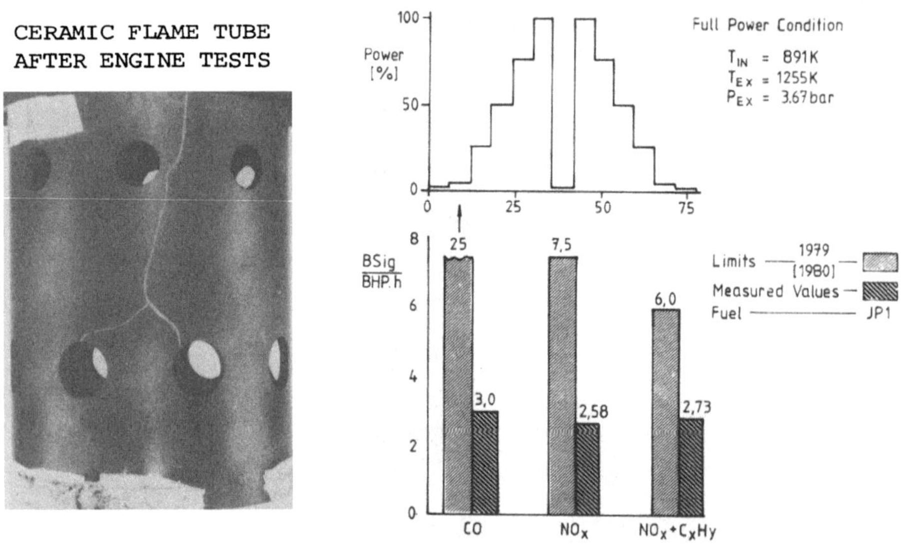

Fig. 14. Results of Ceramic Flame Tube Tests in Vehicular Engine

advantages of the gas turbine. Although the vehicular gas turbine was run on JP1 fuel, because the sulfur in diesel fuel affects the heat-exchanger material, the consequently high NO_x values were still well below the emission limits. Since the emission limits for car engine are much more stringent, greater attention had to be focused on complex low-emission combustor configurations in the ceramic combustor development work carried out at Volkswagenwerk AG.

7. Conclusions

The results of development as it stands show that ceramic combustors can withstand thermoshocks during cold starts and steady operating conditions with high gas temperatures. Conical ceramic combustors designed to avoid high surface temperature gradients have proved their ability to withstand cyclic temperature tests at full power pressure conditions, and investigations with a hybrid ceramic-metal combustor confirmed the potential of ceramic materials to satisfy full engine operating conditions. The technical knowledge gained so far, however, is not sufficient to enable a ceramic combustor to be designed, built and used for complete vehicular gas turbine applications.

Cooperation with the producers of ceramic materials in order to develop better ceramic-oriented design concepts is necessary before the demands of vehicular gas turbine operation can be met. Testing will have to continue with regard to sharper high temperature test cycles and long durability. The iterative experience gained will promote the development of a ceramic technology. To reach the objective, an all-ceramic demonstrator is being set up to demonstrate the low cost and high reliability of ceramic components and the low fuel consumption predictions.

8. Acknowledgements

The author acknowledges the German State Department for Research and Technology (Bundesministerium für Forschung und Technologie) for support of these programs. The combustor development results of the Volkswagenwerk AG were selected and represented by Dr. R. Buchheim.

9. References

1. G. Kappler, K. Giesen, "Stand der Entwicklung von Brennkammern aus keramischen Werkstoffen," Keramische Komponenten für Fahrzeug-Gasturbinen; Springer-Verlag Berlin-Heidelberg-New York 1978.

2. P. Walzer, "Keramische Bauteile für Fahrzeug-Gasturbinen," MTZ Motortechnische Zeitschrift Nr. 10/1978.

3. K. Trappmann, "Development of Ceramic Components at MTU-München," Overview of German Ceramic Gas Turbine Program, ASME Gas Turbine Conference, San Diego, Calif., March 12-15, 1979.

CERAMIC TURBINE NOZZLE RING DEVELOPMENT BY FINITE ELEMENT ANALYSIS

H. Burfeindt and M. Langer

Research and Development Division
Volkswagenwerk AG, Wolfsburg
West Germany

P.M. Stuart

Advanced Analysis AG
Zentralstraße 2A, CH-8160 Uster Zurich
Switzerland

ABSTRACT

This paper describes a finite element analysis procedure adopted at Volkswagen during the development of a turbine nozzle design with a turbine inlet temperature of 1623 K. The analysis considered only the transient stress problem and not long term ceramic material problems such as oxidation. A reference analysis detected two areas of critically high stress, the shrouds and the trailing edge. This agreed with experimental test findings. Independent solutions were found to overcome these problems by analysis assisted redesign cycles.

INTRODUCTION

In the last few years several companies have been engaged in investigating the application of ceramic components to gas turbines. The feasibility has been confirmed in principle [1,2].

Turbine nozzle rings are stationary parts and, unlike blades and discs, stresses due to rotation are not involved. In the case considered here the nozzle is required to withstand gas temperatures up to 1623 K and thermal shocks of 500 K/s. Since the nozzle mounting provides a poor heat conduction path, the heat-up and cool-down transients are the important conditions to be considered during stress analysis.

The ceramic material chosen must be able to withstand attack by combustion gases without loss of strength over many hours of running and must also be suitable for the production of complex shapes with the limitation of trailing edge thickness specifications down to 0.5 mm.

For the analysis of the turbine nozzle ring, steady-state and transient temperature distributions, material properties and boundary conditions were needed for the prototype design so that the high stressed areas could be calculated for all operating conditions. During the iterative development process the relationship and correlation between design, material type, component testing and failure analysis is of particular importance.

The development of ceramic nozzle rings at Volkswagen started with preliminary thermal stress calculations on a 2-dimensional FE model. Fig. 1 shows the design configurations of that early development state. A comparison of analysis and experimental data was not satisfactory. Designing of such a complex structure in ceramics requires a 3-dimensional finite element analysis to get realistic results. For further investigations a 3-D FE model was created to calculate the local thermal gradients that could generate high stressed areas which indicate crack occurance.

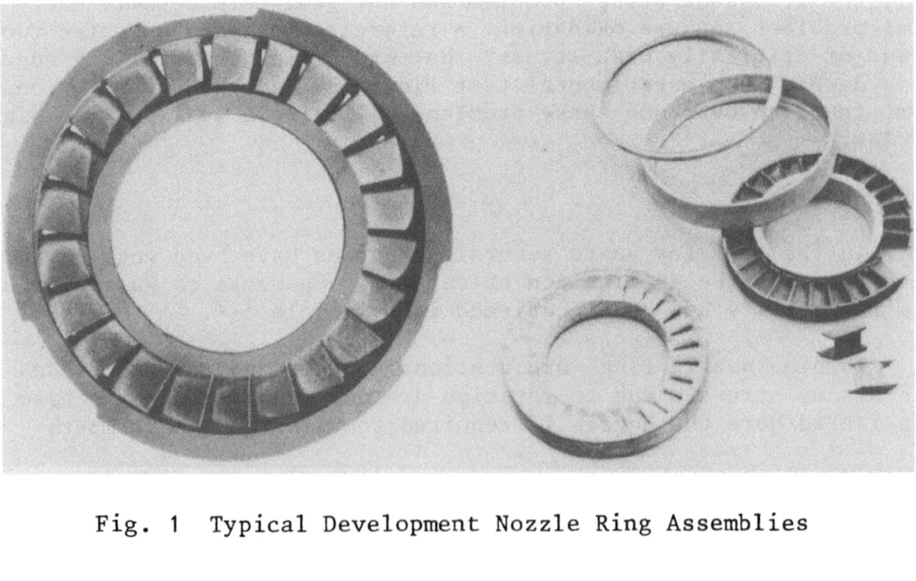

Fig. 1 Typical Development Nozzle Ring Assemblies

The current state of material development indicates that silicon nitride and silicon carbide in various types could possibly be used as suitable component materials.

The thermal stress analysis was performed on a nozzle ring consisting of the materials reaction bonded silicon nitride (RBSN) and sintered silicon carbide (SSC).

The work reported here takes no account of the long term problems of ceramic materials such as oxidation and creep which are the subject of separate investigations.

MODEL DEFINITION

A finite element model was generated using 20 node isoparametric elements. The program MARC was chosen which permits representation of temperature dependence of the thermal and mechanical properties of RBSN and SSC. Further, the time and space dependence of the thermal load, i.e. gas temperature and heat transfer coefficient, were considered. The model shown in Fig. 2 shows the reference configuration of a one-vane sector with proper airfoil to shroud fillet radius representation and a solid shroud section. It consists of 120 elements and 750 nodes. The simulation of the connection of the model into a continuous ring was achieved in the usual way by local transformation and tying of the corresponding perimeter nodes. By this method it was possible to simulate the various inter-vane assembly techniques proposed during the design studies.

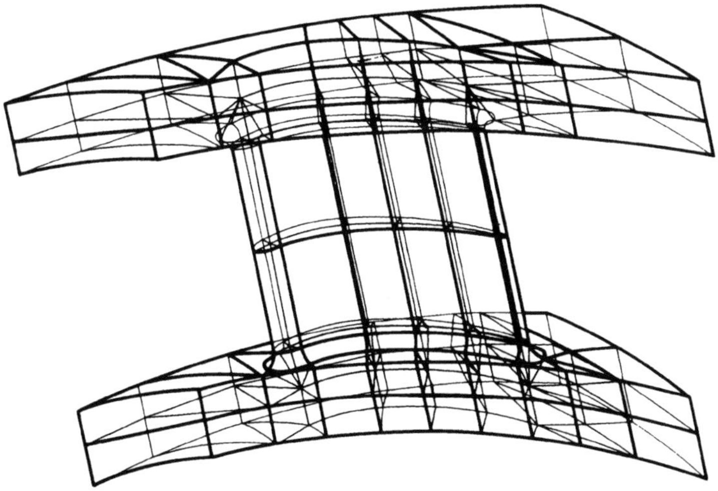

Fig. 2 3-D Finite Element Model

Fig. 3 shows the time dependence of the gas-temperature and heat transfer coefficient during the heat-up and cool-down phases. The design curve shows that the rate of change of gas temperature with time is considerably smaller in the cool-down phase.

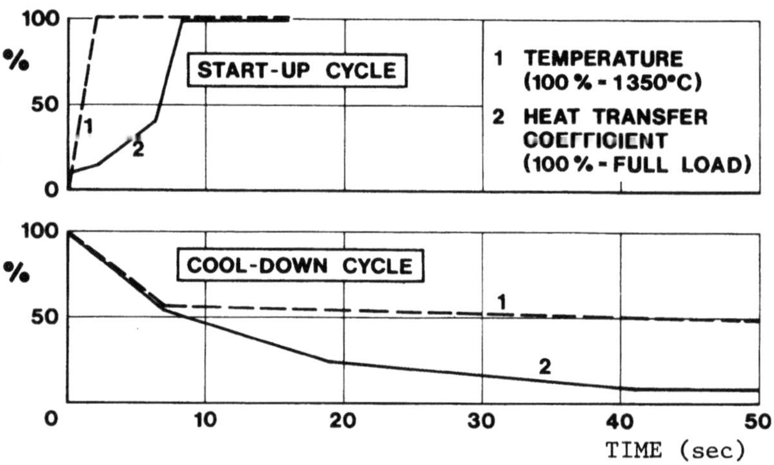

Fig. 3 Transient Load Cycles

Heat exchange between the gas and the structure was restricted to be only through the vane surface and the inner surfaces of the shrouds. The remaining parts of the structure surface were assumed to be insulated. Heat exchange due to radiation was not considered. For parametric comparisons all models representing different inter-vane assemblies and shroud configurations were taken at a selected point in the heat-up transient to observe stresses. The full transient up to steady state was analysed for the model with optimum shroud and trailing edge stresses.

ANALYSIS

Reference Analysis

The reference model was considered to be one in which a one piece integral nozzle ring was represented. The two principal problem regions were immediately revealed as tensile principal stresses in the shrouds in the throat region and in the trailing edge at midspan during heat-up and in the fillet radius during cool-down. These corresponded well with test findings as shown in Fig. 4 and 5 respectively. Interestingly the airfoil/shroud fillets, although regions of higher than average stress, did not show up as problem areas during the analysis as expected except at the one position mentioned. The two problem areas were considered separately as follows.

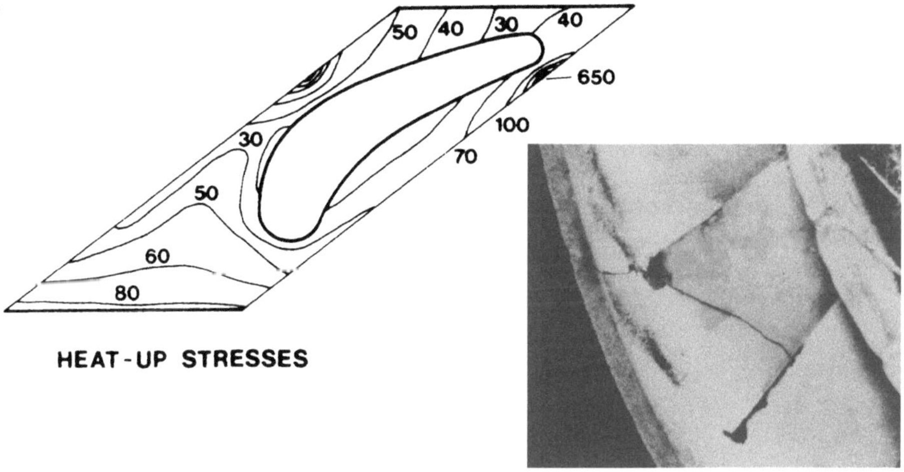

Fig. 4 Typical Shroud Failures

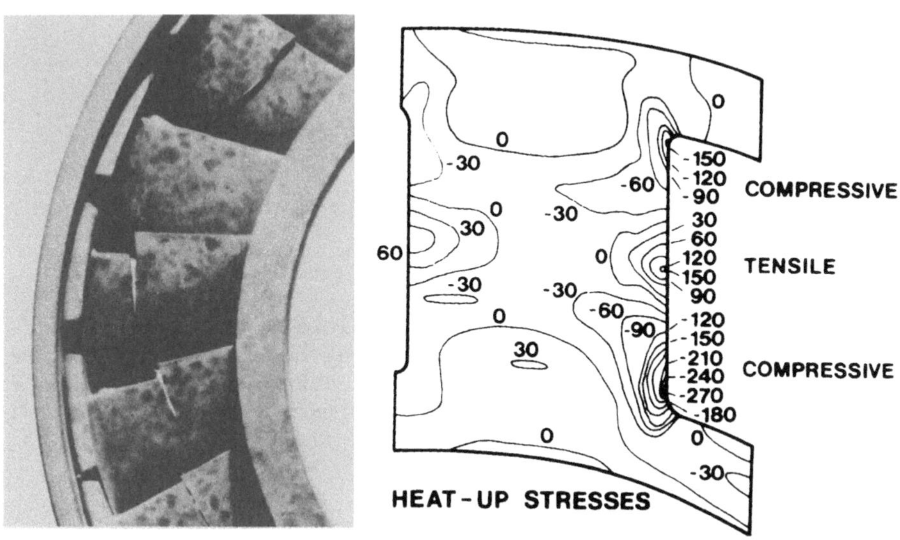

Fig. 5 Typical Trailing Edge Failures

Shroud Problem

The maximum tensile stress in the shroud was tangential and had a value of 650 MN/m^2. The traditional engineering practice of providing a free surface where high stress is found was adopted here. In the case of the nozzle assembly it took the form of disconnecting adjacent vanes in regions of high stress. Connection variations considered included cases where vanes were totally separated from each other and where slots (regions of no bond) were represented, see Fig. 6. Case 5 of Fig. 9 shows how strongly stresses could be affected in the configuration chosen as a satisfactory optimum. Here a shroud stress of 55 MN/m^2 well below the RBSN maximum tensile stress criterion of 130 MN/m^2 was achieved. This shroud stress was calculated at one point in the transient heat-up. At this time the max. tensile stress had not yet been reached. The full transient thermal stress analysis shows that the maximum stress is 20 - 30 % higher.

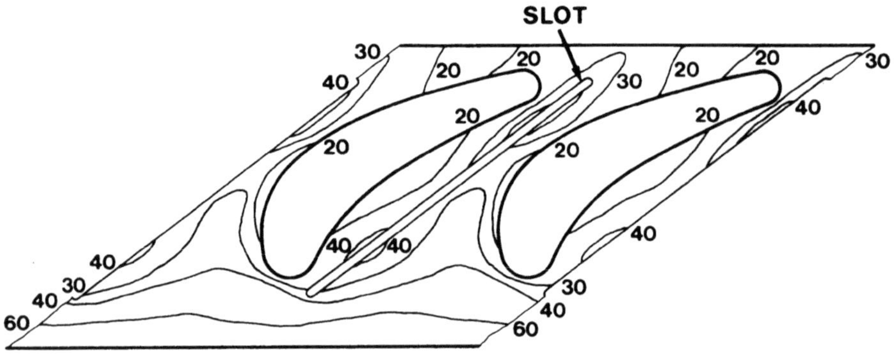

Fig. 6 Solution of Shroud Problem

Trailing Edge Problem

The dominant component of the tensile stress in the trailing edge was noted to be radial. The maximum trailing edge value during heat-up was 190 MN/m^2 and in the fillet during cool-down was 200 MN/m^2.

In searching for the cause it was noted that during heat-up the temperatures at the airfoil midspan were significantly higher than those near the shrouds. The consequence of this was that the thermal expansion of the chord at midspan was greater than that near the shrouds causing the leading and trailing edges to bow outwards at midspan as shown in Fig. 7. This resulted in these edges being stretched to an extent greater than that caused by the thermal expansion of the edges along their own lengths.

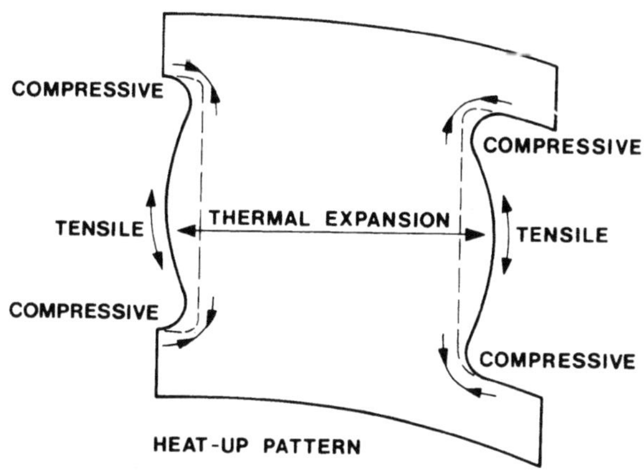

Fig. 7 Trailing Edge Stress Development

During the cool-down cycle the temperature gradient was reversed and was not so strong, resulting in compressive stresses in the trailing edges and tensile stresses in the fillet radii.

If this argument proved correct, a reduction in the stress would be achieved by a reduction in the thermal gradient between midspan and shrouds. A solution to the problem might be found in materials with lower conductivity, such as SSC, but a search resulted in finding a conflict between conductivity, elastic modulus, and expansion coefficient values which prevented a satisfactory result as a full thermal transient analysis on the optimized shroud model showed.

An alternative was the thermal restrictor approach. By restricting the cross-section of the heat flow path leaving the airfoil, the airfoil temperature would rise more rapidly during the critical heat-up period but the gradient between midspan and shrouds would be smaller. Since the condition at the trailing edge was more severe than at the leading edge it was decided to allow the heat to pass from the airfoil to the shrouds in the region of the leading edge, see Fig. 8. This together with a reduction of the cross section of the heat flow path resulted in a reduction of the thermal gradient to an extent which yielded a maximum tensile stress of only 90 MN/m² in the trailing edge giving principal stresses within the specified material maximum of RBSN, see Fig. 9, case 6. For the same reasons, the thermal restrictor approach resulted in a reduction of the high tensile stresses in the fillets during the cool-down phase.

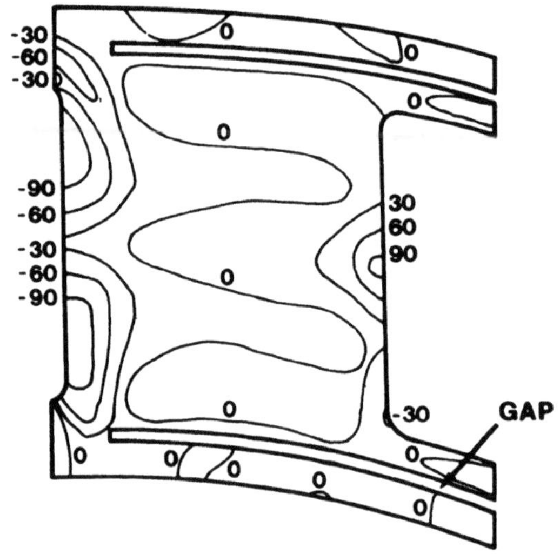

Fig. 8 Solution of Trailing Edge Problem

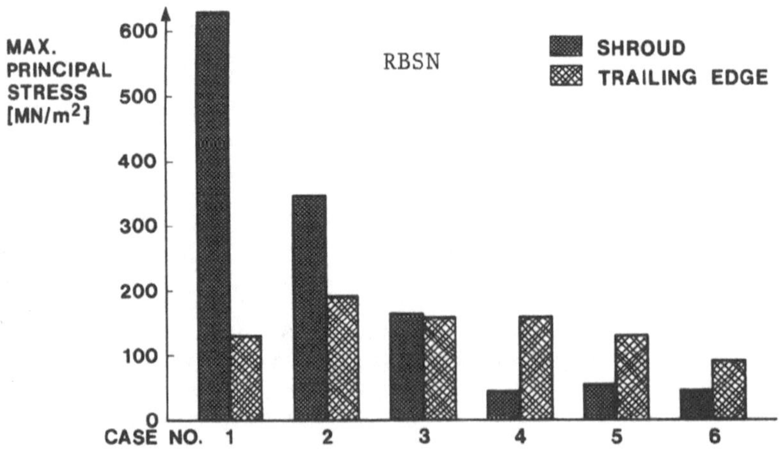

Fig. 9 Maximum Principal Stresses in the Critical Areas

CONCLUSION

Development of ceramic nozzle rings is only possible with the support of FE analysis. The complex geometry demands a 3-D model with a sufficent number of elements to permit detection of critical stress regions. The optimum design is a compromise between aerodynamic and material limitations. Important prerequisites are the proper simulation of different load cycles, a knowledge of material properties and boundary conditions and a suitable reference model configuration.

By the use of finite element analysis it has been possible to gain a good understanding of the stress patterns developed in nozzle vanes and their causes. As a result design changes have been evaluated which show a great reduction of peak stresses compared with the original design concept. With improved bonding techniques developed in parallel with this work, the current plan is to proceed with a nozzle assembled from individual vane mouldings rather than to switch to the integral ring approach [3].

The time required for creating and analysing the reference model and 6 parametric variations was 2 man months. Analysis of the final redesign from a completely new model took an additional month. The total computer time was 150 000 c.p. seconds on an in house CYBER 173. This analytical iterative procedure is therefore considered much quicker and cheaper than the equivalent set of experimental steps. Final confirmation and possible design refinements must still of course be done experimentally.

ACKNOWLEDGMENTS

This work was partially sponsored by the German Ministry of Science and Technology. The authors would like to acknowledge this support and to express their appreciation to Volkswagenwerk AG for the permission to present this paper.

REFERENCES

1. A.F. Mc Lean "Development Progress on Ceramic Turbine Stators and Rotors". ASME Publication 75-GT-11 March 2, 1975

2. E.H. Kraft "Silicon Carbides for DOE/NASA Sponsored Vehicular Gas Turbine Programs". Presented at U.S. DOE Highway Vehicle Systems Contractors' Coordination Meeting Oct. 19, 1978

3. P.H. Havstad and E.A. Fisher "The Development Iteration Process". Ceramics for High Performance Applications - II. Proceedings of the Fifth Army Materials Technology Conference. Newport, Rhode Island, March 21 - 25, 1977

COMPRESSION STRUCTURED CERAMIC TURBINE ROTOR CONCEPT

P. J. Coty

Air Force Aero Propulsion Laboratory
Air Force Systems Command
Wright-Patterson AFB, OH 45433

ABSTRACT

This paper summarizes highlights of a design study conducted by the General Electric Company, Evendale, Ohio, to evaluate a "Novel Ceramic Turbine Rotor Concept" (AF Aero Propulsion Laboratory Contract Number F33615-78-C-2041). The "Novel" feature of this ceramic turbine rotor design involves maintaining the ceramic rotating components in a state of compression at all operating conditions. Many ceramic materials being considered for gas turbine components today display compressive strengths ranging from three to eight times their tensile strengths. Utilizing the high compressive strengths of ceramics in gas turbines for improving ceramic turbine structural integrity has interested engineers in recent years as evidenced by a number of patents and reports issued on Compression Structured Ceramic Turbines with one as early as 1968.[1,2,3,4] Turbine blades designed to be in compression could greatly enhance the reliability of the ceramic hot section components. A design of this nature was accomplished in this contractual effort by using an air-cooled, high strength, lightweight rotating composite containment hoop at the outer diameter of the ceramic turbine tip cooling fins which in turn support the ceramic turbine blades in compression against the turbine wheel. A brief description of the detailed structural and thermal analysis and projected comparable performance between the Compression Structured Ceramic Turbine and an equivalent air-cooled metallic baseline turbine is given.

INTRODUCTION

Ceramic turbine rotor blades are presently being designed so that the ceramic blade is in a tensile state when rotating. Ceramics have a relatively low tensile strength as compared to metals, and designing a ceramic turbine blade with a dovetail to be inserted into a disk, as in the classical metal blade fashion, must incorporate this lesser quality of ceramic materials. To date, there has been little success demonstrating long term reliability and uniform reproducibility of ceramic turbine blades that are designed to withstand the usual dovetail/tensile structure. On the other hand, with the compressive strength of ceramics on the order of three to eight times its tensile strength, a rotor design that would cause the ceramic components to be in a compressive state while rotating should increase survivability significantly over a ceramic turbine designed in the conventional dovetail/tensile manner, thereby increasing the reliability of the turbine rotor system.

This paper covers a brief structural and thermal analysis of a Compression Structured Ceramic Turbine (CSCT) for a small non-man-rated, limited life turbine engine and compares its projected performance with a similar size, air-cooled metal baseline turbine. The study performed on this novel turbine involved two tasks listed below:

Task I - Novel Component Design/Materials Selection/Stress Analysis

Task II - Operational Limits and Performance Analysis

There are many designs capable of effectively producing a Compression Structured Ceramic Turbine Rotor. The design, which is the subject of this paper, fulfills the objectives of maintaining the ceramic blades in a compressive state throughout its entire operating range.

DESIGN AND MATERIALS

A baseline engine assumed for this study is illustrated in Figure 1 with a Compression Structured Ceramic Turbine installed. This conceptual engine was utilized strictly for flow path characterization and to establish structural limits for the novel rotor. In order to achieve the necessary turbine work, a turbine blade tip speed of 1807 ft/sec was established for 100 percent operation. This is approximately 15 to 25 percent higher than most other dovetail/tensile structured ceramic turbines currently being developed.

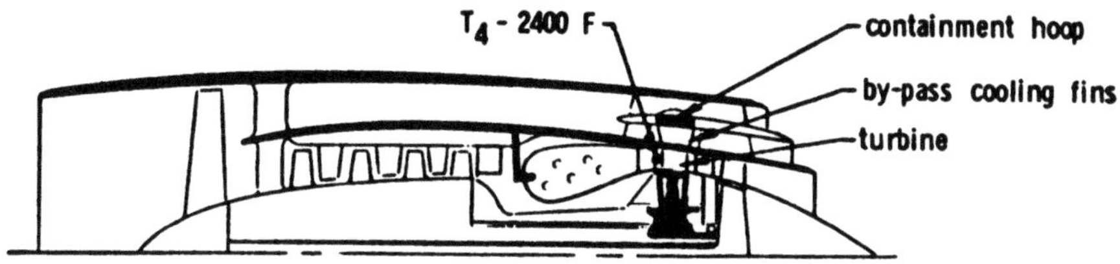

Fig. 1. Proposed Initial Application in a Non-Man-Rated Cruise Missile Engine

The uncooled ceramic turbine temperature (T_4) was set at 2400°F for this study with a turbine life of 50 hours, ten of which are at maximum temperature and speed. The comparable baseline metal turbine was air-cooled to sustain the same 2400°F turbine inlet temperature (TIT). This study was limited to a single design point of 100 percent rotor speed at Mach .7 sea level, with a 111 percent overspeed capability. This overspeed capability allows a 15 percent design safety stress margin. Some design concerns for the Compression Structured Ceramic Turbine are summarized below and will be addressed in some detail throughout the text that follows.

1. Inertia and thermally induced radial mismatch between the disk and tip containment hoop.

2. Hoop integrity versus hoop environment.

3. Hoop and fin drag losses.

4. Rotating seal leakage between the core gas and bypass cooling air duct.

5. Cooling fin work versus aerodynamic cycle match.

6. Compliant interfaces between rotor components.

7. Thrust bearing relationship to seals and engine mount.

8. Lack of material data on compressive strength of ceramics.

Components of a final version of a Compression Structured Ceramic Turbine rotor are illustrated in Figure 2 and are described below:

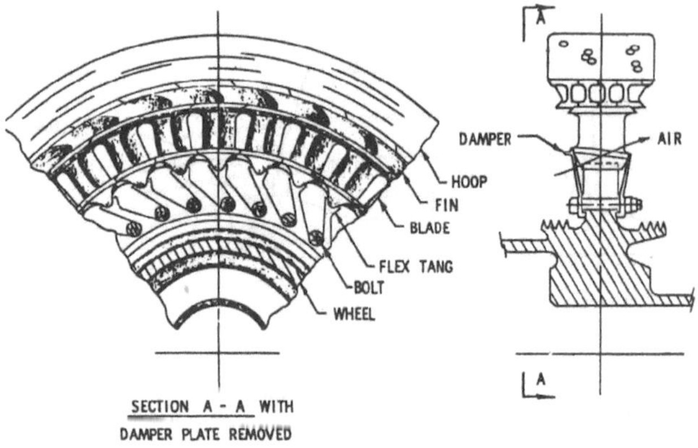

Fig. 2. Compression Structured Ceramic Turbine Rotor

- Hoop - the outer containment hoop is made of high tensile strength filament-wound graphite composite, capable of withstanding tensile stresses of more than 185 ksi.

- Fins - clustered ceramic fins residing against the hoop act to duct a portion of the front fan bypass air as a cooling medium for the hoop.

- Blades - ceramic turbine blades with a uniform extruded airfoil profile matching a section of the baseline turbine blade at 80 percent span, are clustered within a shallow channel generated at the inside surface of the clustered fins. Prospects for refinement of this profile for improved efficiency and optimum work extraction are discussed in a concluding statement.

- Wheel/Flex Tangs - the metal flexible tangs extending from the rim of the wheel are designed to depress radially inward for the axial assembly of the clustered ceramic blades, fins and composite hoop. The tangs are then released against the matching cylindrical profile of the blade roots and act to provide a precise blade indexing and residual radial spring load of about 100 pounds per blade. As the hoop grows from rotating inertia forces, the flexible tangs grow at an equal rate passing through their nominal position with a fairly constant load against the blade root throughout the full operating range as described later. The cylindrical interface between the flexible tangs and the blade root would accommodate the slight relative motion at that joint while the flexible tangs will extract torque and adjust to the difference in growth induced by inertia and thermal gradients between the wheel and the blades.

- Damper Plates - two damper plates assembled to the wheel fore and aft faces act to locate and contain the blades axially and to inhibit windage losses. The tips of the damper plates reside with some opposing axial force against the platforms of the blades and tend to dampen any vibration that may be induced by the flexible tang system.

An enlarged section of the installed Compression Structured Ceramic Turbine is illustrated in Figure 3. It shows a thrust bearing just behind the turbine with its projected path to the engine mounts.

Table I lists a few material assumptions for the main components that were incorporated in this design study. The hoop would incur the maximum temperature of 450°F only at its very inner aft surface. This could be reduced by 100°F or more by inducing more airflow through the fins as described later. It is anticipated that an epoxy system would suffice for the hoop, but the option of a polyamide system may extend the margin of safety if required. A conservative design assumption was used for the compressive strength of the ceramic cooling fins and rotor blades. Since very little data is available on compressive strengths, a

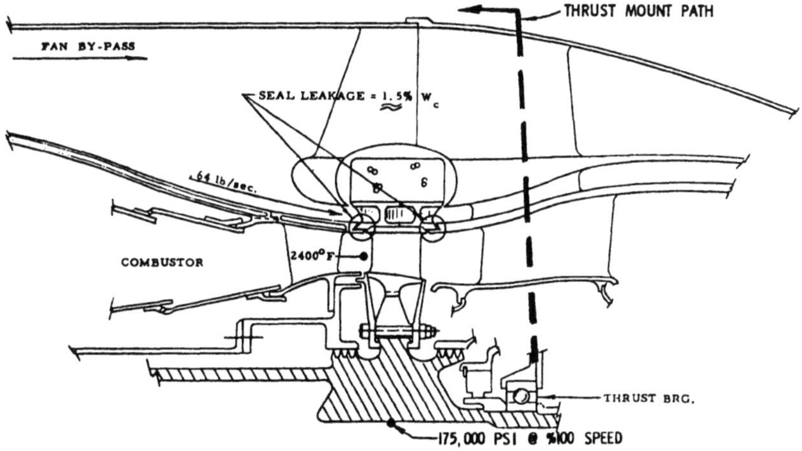

Fig. 3. Installed CSCT, Cross-Section

Table I CSCT Design Assumptions

	Density g/cc	Maximum Temp °F	KSI Ultimate Stress	KSI Allowable Stress
Hoop	1.55	450	210	182
Fin	2.3	1700	80*	74*
Blade	2.5	2300	110*	95*

* Assumed Design Compressive Strength

value of three times the published tensile strength was used for the various ceramic materials.

The ceramic material proposed for the fins and blades was Reaction Bonded Silicon Nitride (RBSN) with variations in reduced density as stresses and environment would allow so that inertia loads into the hoop could be kept to a minimum. RBSN was chosen because of its lower thermal conductivity, lower modulus, sufficient strength at the required temperature, and its density tailorability.

In order to draw as close a comparison as possible between the Compression Structured Ceramic Turbine and the baseline turbine, it was established that the cooling fins would do no work on the induced bypass cooling air. The fins would simply knife the air so the net effect of fin/hoop drag could be isolated without the complexity of modifying the baseline engine aerodynamic system. This does not negate the prospect that some cycle advantage could be gained by having the cooling fins do some work to offset a portion of the drag losses imposed by the fin/hoop system. Figure 4 illustrates the relative relationship between a zero work fin angle and six turbine blades. Working fins would extend over fewer blades depending on the amount of induced work.

The advantage of a shrouded turbine blade tip incorporated in the Compression Structured Ceramic Turbine is somewhat offset by leakage at the fore and aft seals, as shown in Figure 3, that separate the turbine gas from the fin cooling air. Figure 5 shows three seal profiles with estimated leakage rates as a function of effective running clearance. For this study, it was estimated that total leakage at such seals could be restricted to within 1.5 percent of total core air which is less loss than the baseline turbine blade may display with a nominal tip clearance of about

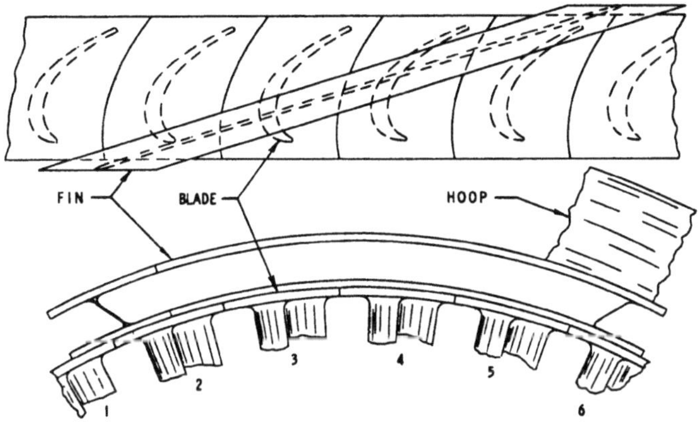

Fig. 4. CSCT Fin/Blade Interface

0.008 inches. By locating the drive shaft thrust bearing in close proximity to the Compression Structured Ceramic Turbine, thermal migrations that could adversely affect seal clearance could be minimized.

HEAT TRANSFER, STRESS AND TIP DRAG ANALYSIS

For the 2400°F T_4 turbine rotor inlet gas temperature used in this study, an initial three-dimensional finite element stress and thermal analysis was made on the blade as illustrated in Figure 6. The right side of the figure shows a maximum compressive

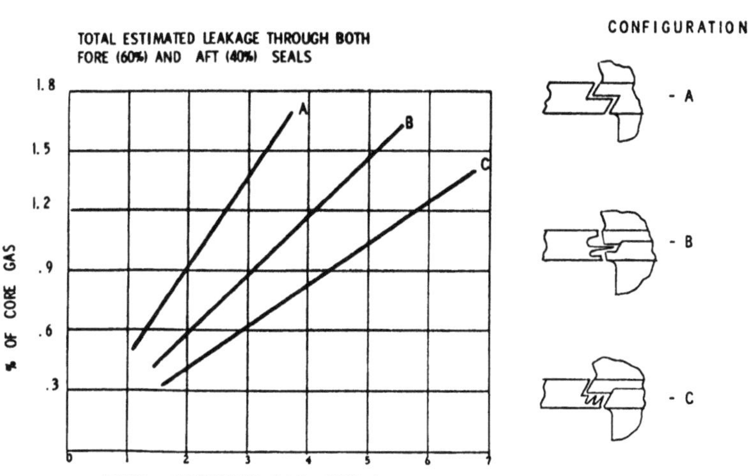

Fig. 5. CSCT Face Seals

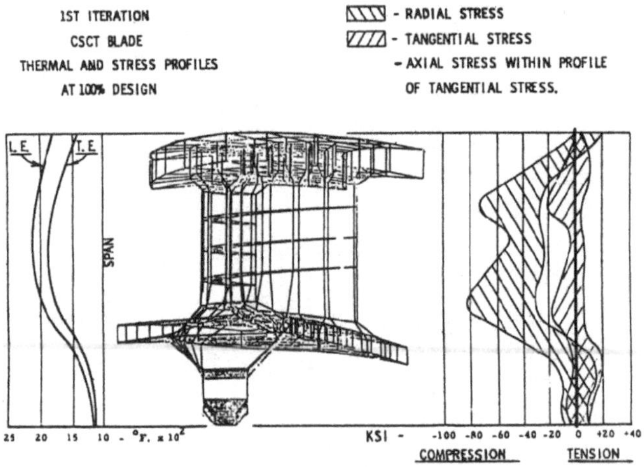

Fig. 6. Temperature and Stress Profiles on the Finite Element Model

stress of 82,000 psi in the airfoil and a maximum tensile stress of 20,000 psi in the platform induced by bending. The leading and trailing edge temperature profiles, 2260°F and 1810°F, respectively, are shown on the left side of the figure. Refinements in the design of the platform could reduce the tensile stress in that area by a considerable amount. However, additional design iterations to reduce these stresses could not be further considered at the time. Temperature distribution in the composite containment hoop illustrated in Figure 7 shows the result of a 0.64 lb/second by-pass cooling airflow rate versus containment hoop temperature for two variations in the number of fins and

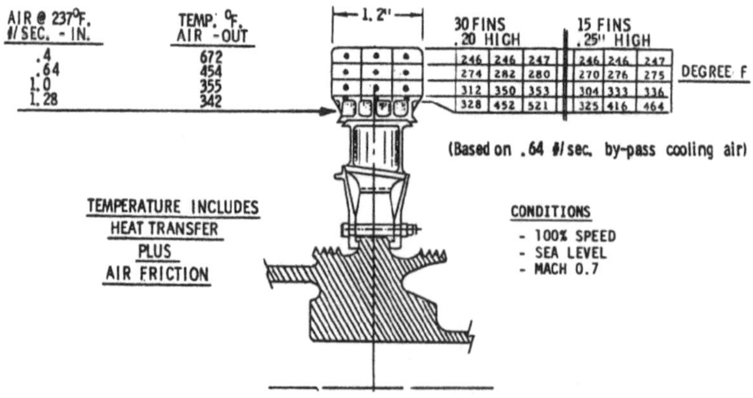

Fig. 7. Heat Distribution in CSCT Containment Hoop and Fins

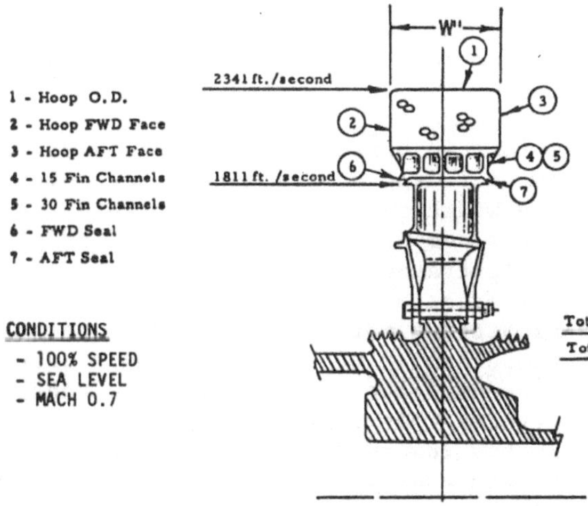

HORSEPOWER	W=1.20"	W=1.40"
1	29	34
2	8	8
3	8	8
4*	45	53
5**	58	68
6	2	2
7	2	2
Total (4)*	94	107
Total (5) **	107	122

Fig. 8. Windage Loss of CSCT

fin radial height. Heat that conducts from the hot turbine blades out through the fins is removed at a rapid rate. In addition, hot leakage gas is directed aft and exits with the fin cooling air before it is fully mixed with the bypass air.

A range of four airflow rates with related air exit temperatures is also listed in this figure to project prospects for more effective cooling of the containment hoop if required. It shows

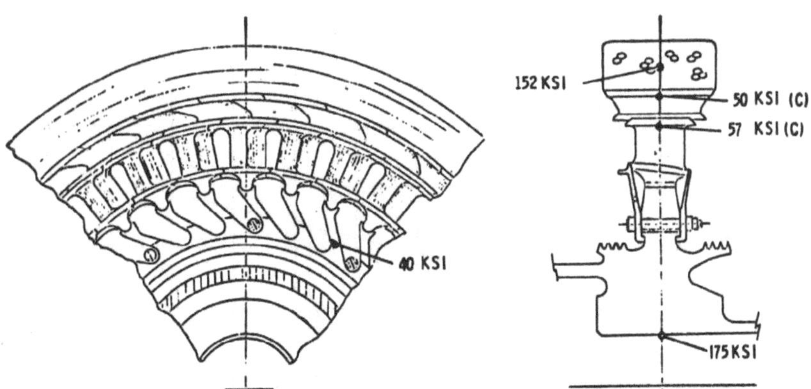

Fig. 9. Average Stress at 100% Design Speed

airflows of 0.4 lb/second, 0.64 lb/second, 1.0 lb/second and 1.28 lb/second, with respective exit air temperatures and assumes continuous fan airflow at the outer radius of the hoop and through the fins.

Figure 8 shows windage loss in horsepower induced by seven drag surfaces totalling a nominal drag of 107 horsepower using a 30-blade configuration. Hoop width was used as a variable. A final design iteration with refinements in the thickness of platforms, shrouds and fins is illustrated in Figure 9 with five areas of maximum stress at critical areas of the CSCT which are well within safe limits of the respective materials utilized. Also listed in this figure are the dynamics of relative growth between the wheel and hoop which is accommodated by the flexible tang. Figure 10 graphically illustrates this relationship. By depressing the flex tangs 0.015 inches inward at assembly, the difference between 0.035 inches hoop growth and 0.024 inches wheel/tang growth is accomplished with 0.011 inches outward deflection of the tip of the tang from centrifugal force. With proper sizing, the tang can be structured to induce a fairly constant load of approximately 100 pounds throughout the entire operating range. The ceramic blades and fins could withstand a much higher load, but the containment hoop would be more severely stressed if a higher tang load were imposed.

Figure 11 illustrates the range of responsive frequency anticipated for the blades of the Compression Structured Ceramic Turbine. This is approximately 600 percent higher than cantilevered metal blade and well beyond the highest excitation range imposed by 24 upstream configuration drivers such as starter holes

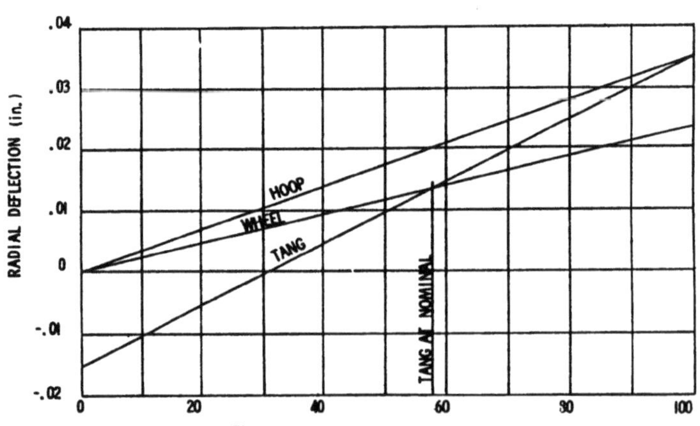

Fig. 10. Rotor System Deflection Relationship

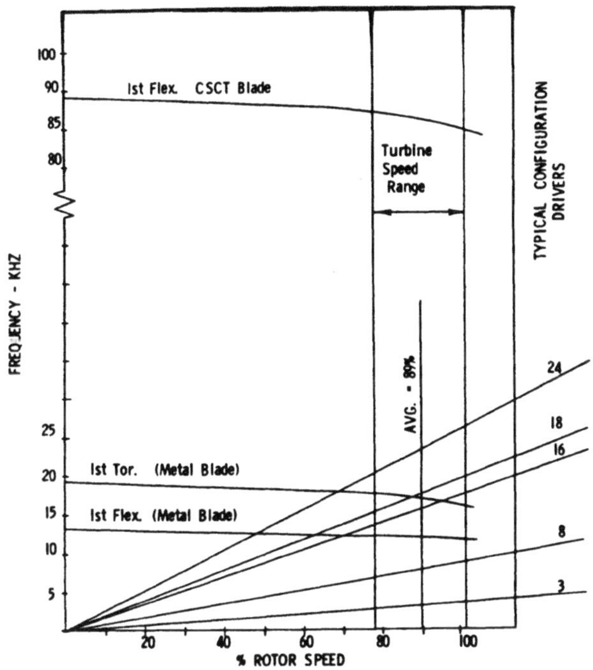

Fig. 11. Campbell Diagram-CSCT Versus Baseline Turbine

or combustion nozzles or stator vanes. The 600 percent higher frequency range relates to natural frequencies observed in cantilevered fixed/free versus fixed/fixed beams. Due to fixed ends of the blades, this first flexural frequency is so high that the first torsional or second flexural is well above the limits of the analysis.

OPERATIONAL LIMITS AND PERFORMANCE ANALYSIS

A summary of the compressor discharge air that is chargeable to the cycle to cool the disk area so as to offset leakage losses and to compensate for equivalent windage/drag losses in the respective baseline and Compression Structured Ceramic Turbine, is illustrated for comparison in Figure 12. The nominal 107 shaft horsepower loss induced by drag of the fin/hoop system on the Compression Structured Ceramic Turbine has been converted from work loss to an equivalent BTU cycle air penalty of 6.4 percent. The summary shows a total of 5.1 percent penalty against the baseline turbine as compared to 9.4 percent against the Compression Structured Ceramic Turbine. However, the 6.4 percent hoop/fin drag penalty could be offset to some extent by improved blade aerodynamics of an inverted taper airfoil as listed in Table II

TABLE II

Potential Aerodynamic Benefits of a CSCT Inverted-Taper Turbine Blade

1. The compression loading of the blades (radially outward) requires the hub airfoil sections be thinner than the tip sections. This is perfectly suited to the usual radial distributions of relative entry Mach number. Hub Mach numbers are high, therefore desired airfoil sections are thin; tip Mach numbers are lower, therefore airfoil sections can be thicker.

2. Blade attachment geometry no longer determines the number of blades that can be accommodated; therefore ideal combinations of airfoil section, annulus flare, and solidity can be selected.

3. The "fixed-end" supported blades permit the application of blade lean to control secondary flow effects to an extent not possible in conventional turbine rotors.

4. At the hub, where the configuration does not require a "radial pull" attachment, leakage and protuberance problems should be less acute than in conventional dovetail designs.

5. Thinner trailing edges should be an advantage, since no cooling-air passages and holes need to be accommodated.

and a redesign of the fin section aerodynamics to produce a pressure rise across the fins. Also, the high energy gas leakage through the face seals of the Compression Structured Ceramic Turbine should produce increased thrust as it mixes with the fin cooling airstream. Therefore, it should not be considered as a total cycle loss as expressed in Figure 12.

Operational limits of a Compression Structured Ceramic Turbine would probably depend more on the reliability of the containment hoop than on the ceramic components. Stress in the graphite composite containment hoop at anticipated temperatures produce less margin of safety than other turbine components, however,

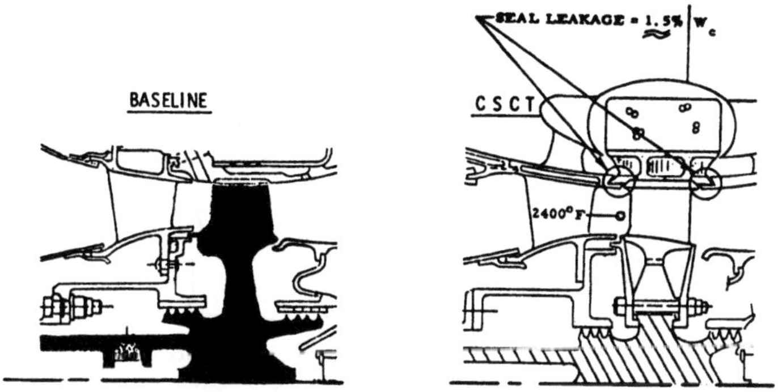

BASELINE	% CORE AIR	CSCT
2.5	BLADE COOLING	0
1.1	TIP CLEARANCE	0
1.5	DISK/DOVETAILS	1.5
0	TIP FACE SEALS	1.5
0	HOOP/FIN DRAG	6.4 (Equivalent to H.P.)
5.1	TOTAL	9.4

Fig. 12. Comparison of Cooling, Leakage and Drag Penalty

emerging technology in Graphite Reinforced Glass Matrix may provide containment hoops capable of full strength at operating temperatures close to 1000 °F[5].

The structural analysis of the Compression Structured Ceramic Turbine selected for this study indicates that design goals can be achieved with a 15 percent margin of safety at 111 percent speed. Aerodynamic optimization of the turbine blades and fins was not of prime concern under this study. Although air loads induce some stresses in the wheel, centrifugal loads impose the greatest stresses and are therefore the major criteria for structural reliability of the rotor system.

ALTERNATE APPROACH

A brief structural analysis of an alternate concept proposed and patented in 1974 by R. Cerrato of Turin, Italy[2], is shown in Figure 13 as it applies to this study. The disk, flex tangs, and blades are common to the basic design. A high temperature composite hoop immediately surrounds the turbine blade end-walls, eliminating the need for the cooling fins. As shown in this figure, the reduction in overall diameter of the wheel considerably reduces the stresses in the containment hoop. The advantages of this concept over the basic design are the lower windage losses of the hoop/fin section and elimination of the leakage out of the hot flow path through the seals. Further study of this design may reveal additional advantages.

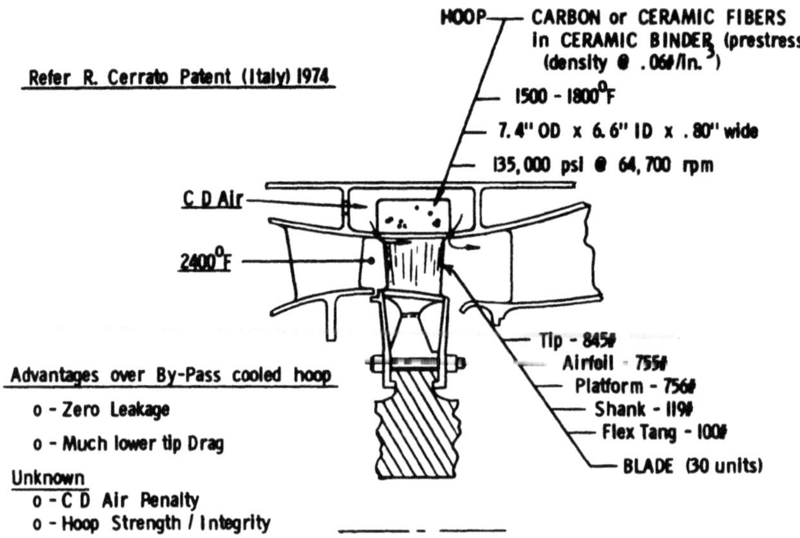

Fig. 13. CSCT in Sealed Hoop Compartment without Cooling Fins

CONCLUSION

A Compression Structured Ceramic Turbine looks feasible. A new engine aerodynamic cycle with effective working fins to offset windage loss, a reduced tip speed to enhance aeromechanics and the possible utilization of leakage gas to augment thrust should be considered. Also, the prospect for more efficient energy extraction offered by inverted taper in the span of the turbine blade should be of prime interest to turbine designers in any future engine utilizing a Compression Structured Ceramic Turbine. Material property data and design refinements based on this data will also have to be seriously considered.

ACKNOWLEDGMENTS

The author wishes to gratefully acknowledge the technical assistance of Mr. L. Stoffer of General Electric Company, Evendale, Ohio, whose contributions greatly enhanced the quality of this publication.

REFERENCES

1. G. T. Davis, "Blades for Mounting in Fluid Flow Ducts", Patent Number 3,378,228, Rolls Royce, Derby, England, 1968.
2. R. Cerrato, "Vaned Rotor for Gas Turbines", Patent Number 3,857,650, Fiat, Turin, Italy, 1973.

3. R. R. Bodman, "Turbine Rotor Construction", Patent Number 4,017,209, United Technology Corporation, Hartford, Connecticut, 1977.
4. L. Stoffer, Patent Disclosure, General Electric Company, Evendale, Ohio, 1977.
5. National Aeronautics and Space Administration Technical Report Number NASA-CR-158946, N79-11126.

SESSION IV

EMERGING TECHNIQUES FOR RELIABILITY PREDICTION

Chairman: Mr. R. Schulz
Department of Energy

Vice Chairman: Dr. B. J. Hartz
University of Washington

RELIABILITY OF CERAMICS FOR HEAT ENGINE APPLICATIONS

S. Bortz

IITRI

Chicago, Illinois

INTRODUCTION

The significant improvements in operating efficiencies that can be obtained through uncooled operation of heat engines at temperatures above those attainable with high-temperature metals and superalloys are well known. Recent advances in the field of ceramics have resulted in materials that have high-temperature capabilities, as well as improved mechanical properties and resistance to chemical attack. However, the structural use of these ceramics introduces problems arising from lack of experience which requires resolution by development of design techniques, materials, and materials processing approaches.

Such problems have been recognized in past programs sponsored by the Department of Defense (DOD), Department of Energy (DOE), and the National Aeronautics and Space Administration (NASA). In this work, ceramic components have been studied for advanced turbine, Stirling, and diesel engines. More recently, the Department of Energy (DOE) has begun research on ceramic turbines directed toward use with coal-derived fuels. The DOD, NASA, and DOE programs have been extremely valuable in demonstrating the technical feasibility of ceramics for these applications and have revealed some of the problems underlying their reliability in service use.

In order to provide the technical community and potential sponsors of this work with a better understanding of the present state-of-the-art for ceramics for heat engine applications, the National Materials Advisory Board of the National Academy of Sciences, established a Committee on Reliability of Ceramics for

Heat Engine Applications. The purpose of this committee was to: identify the principal gaps in knowledge of ceramic materials properties, failure origins, nondestructive testing, life prediction, and proof testing; and to outline how through a more rigorous understanding of this class of materials they can be used in the production of reliable component parts for gas turbine applications. The committee was then to indicate concepts and approaches to a more rigorous definition of the integrity of ceramic engine components and how better integrity of ceramic engines can be attained. The names of the committee members are provided in Table 1 and the government liaison representatives in Table 2.

DESIGN METHODOLOGY

Design methodology has experienced considerable progress over the past ten years as a result of DOD, DOE, NASA and industrial funding. These efforts were primarily focused toward short term, high temperature operation of ceramic hardware in actual engines. They have led to the development of advanced iterative analytical design tools, techniques, and practice for structural ceramics design.

The main features of this iterative process are shown schematically in Figure 1. As shown in the diagram, the design activity is very closely integrated with material sciences, fabrication, nondestructive evaluation (NDE) and testing. This methodology has encompassed design and development of joints, mountings and structural interfaces in general, selection of fabrication processes, tolerances, surface finishes, etc. The state-of-the-art in these areas has progressed in spite of only limited design iterations, a crucial aspect of all design problems. An illustration of this philosophy is shown in Figure 2, which provides the development history of a turbine stator.

The basic concepts of designing with brittle materials are fairly well defined and analytical techniques are available, which allow the designer to make rational assessments of the reliability of complex ceramic engine components when these are subjected to the operating loads. Whether or not the methodology, as it currently stands, will meet the future reliability demands, is difficult to assess at the moment due to lack of adequate test experience with actual engine hardware. The methodology itself is sound. Limited tests conducted so far with simulated or actual turbine engine hardware show very good agreement with theory. Improvements, both in the analytical tools and the underlying theories of brittle failure and fracture statistics need development and these will further strengthen designers' ability to design efficient and reliable ceramic structures.

TABLE 1. - Committee on Reliability of Ceramics for Heat Engine Applications

Seymour A. Bortz Chairman	Donald G. Groves NMAB Staff

MEMBERS

Prof. Richard Bradt	Mr. Winston H. Duckworth
Dr. Frederick F. Lange	Mr. Anthony Paluszny
Mr. David W. Richerson	Prof. John E. Ritter
Dr. Sheldon M. Wiederhorn	Prof. R.J.H. Bollard

TABLE 2. - Committee on Reliability of Ceramics for Heat Engine Applications

LIAISON REPRESENTATIVES

Dr. Richard Ashbrook	Mr. Roy W. Rice
Dr. Edward Lenoe	Dr. Norman Tallan
Dr. Robert N. Katz	Dr. Edward C. Van Reuth
Dr. Henry Graham	

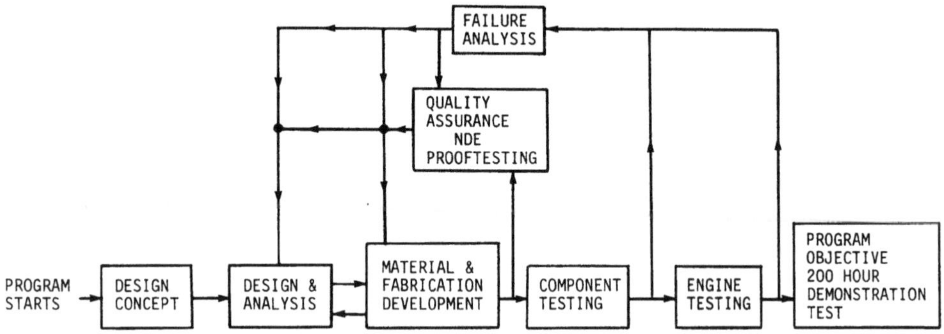

Figure 1. Flow Chart of Major Elements of Ceramic Technology and Iterative Development.

Current analytical tools, sophisticated as they may be, are still far too time-consuming and costly to be used on a scale needed to meet the rising demand on reliability. Future programs should, therefore, stress the adaptation of the more advanced design aids, for instance, advanced computer graphic techniques currently being developed under government funding. These would greatly enhance the designers ability to solve complex loading and stress problems with the detail required for high reliability application.

The reliability based design methodology is, so far, the only viable approach with brittle materials in which strength varies considerably. It will gain acceptance among designers, once successfully demonstrated on actual engine hardware. The basic methodology as outlined by Dukes (1) and subsequently evolved over the past ten years,(2,3) is not likely to change much. Both the concept and the rationale are sound, however, a better design data base is needed. The design data currently available to the designer, are still far from adequate for a

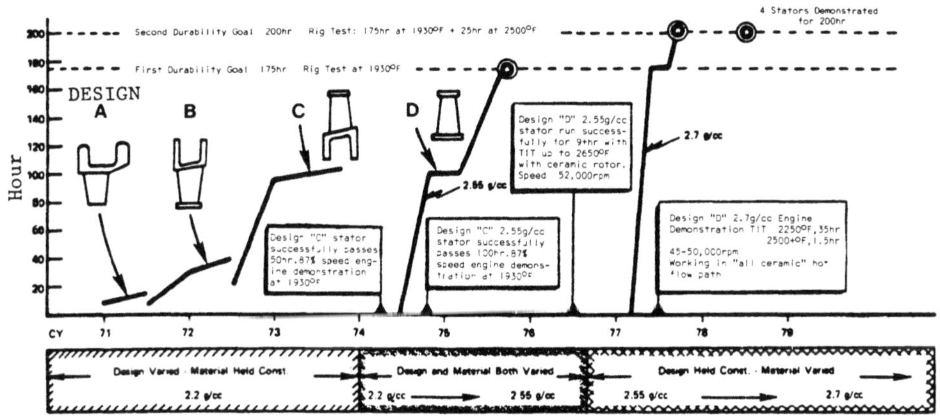

Figure 2. Engineering Development History -- First Stage RBSN Stator DARPA/Ford Engine -- A Case Study of Iterative, Brittle Materials Design.

truly statistical analysis, and this is one of the most serious handicaps a designer, working with ceramics is currently faced with. Improvements and innovations in failure theories of multi-axially stressed brittle structures, in facture statistics, and in statistical tools are also needed.

Presently, the most significant shortcoming of the design methodology for brittle materials, is that too few design iterations have been conducted on any one component; iterations based on test experience from actual engine or rig tests involving the total design cycle from the concept through fabrication, NDE to test evaluation, to truly assess its long-range potential.

More ceramic technology programs that are producted-oriented and yet closely integrated with basic research are needed. Such programs should provide for extensive, statistically-meaningful, correlative testing of ceramic engine components under actual or simulated engine operating conditions, and that also provide for the natural evolution of these design concepts through design iterations.

MATERIALS DEVELOPMENT AND MANUFACTURING

Reliability of ceramic components begins at material selection and is dependent upon all the subsequent process steps. Ceramic materials for heat engine applications are at an early stage of maturity, and reliability can be significantly increased by process improvement and process control. This has been demonstrated for hot-pressed Si_3N_4 at Norton Co. and for reaction bonded Si_3N_4 at Ford Motor Co. In each case, iterative improvement of process parameters and process controls resulted in increased reproducibility, strength, and Weibull Modulus. The effect on component survivability was dramatic: before process improvement, the components typically failed during the first cycle; after process improvement, the components operated for many cycles and hours.

Further process improvement, can and should be conducted for these and other candidate ceramic materials. Depending upon the material and its level of maturity, Weibull "m" values range from 5 to 12. With proper processing, Weibull "m" values of 20 and above with no decrease current strength levels are attainable. This would represent further dramatic increase in reliability.

Ceramic materials in the silicon nitride (Si_3N_4) and silicon carbide (SiC) families are presently the most prominent candidates for heat engine applications. This is due to their excellent combination of high strength, thermal shock resistance, and oxidation resistance. However, the margin between material strength and component design stress is typically not large for currently available materials. Therefore, it is necessary to achieve maximum reproducibility of these materials with a minimum of property variability, and it is desirable to develop improved strength. Attainment of both goals and the resulting component reliability is dependent upon manufacturing process control.

Materials suitable for heat engines have been evolving rapidly. We are now at a point where these materials and the mechanism controlling properties are well enough understood that we should be able to make significant improvements in reliability.

Improved reliability can be achieved by improved process control. Table 3 summarizes specific recommendations. Incorporation of these recommendations will require capital equipment purchase, detailed parametric studies, and extensive physical, chemical and mechanical analysis. Since a large market for heat engine components does not yet exist, it is not likely that ceramic manufacturers will be able to invest the funds and manpower required. Support will have to come from government sources if significant progress is to be achieved.

Table 3. Recommended Processing Improvements to Achieve Increased Reliability.

Raw Material Selection	• Develop powder synthesis techniques that produce high purity submicron powder directly, eliminating the requirement for comminution and thus eliminating a major source of contamination. • Above, with sintering aid included.
Powder Processing	• Incorporate controls into powder processing to minimize pickup of contamination or to positively remove the contamination by techniques such as leaching, screening, or magnetic separation. • Incorporate clean-room procedures or set up closed systems.
Preconsolidation	• Increase effort in preconsolidation to assure uniform distribution of additives and elimination of the types of agglomerates that will end up as flaws during consolidation. • Set up closed system on controlled-environment processing.
Consolidation	• Incorporate instrumentation and/or automation to achieve greater reproducibility. • Incorporate in-process nondestructive evaluation both for process improvement and for process monitoring.
Densification	• Continue development of net-shape approaches such as pressureless sintering, hot isostatic pressing, and chemical vapor deposition. • Incorporate instrumentation and/or automation to achieve greater reproducibility. • Develop solid state or reactive-sintering compositions that yield improved high temperature properties.

Improved reliability can also be achieved by developing improved materials. Table 4 defines three levels of development. The first is short range, consisting of improvement of existing materials. The second is intermediate range, consisting of development of approaches for which some feasibility has already been demonstrated. The third is long range, consisting of development of new approaches or new materials. All three must be pursued if we are to meet the challenges of today's technology where, in so many cases, we cannot progress further until material limitations have been resolved.

MATERIALS CHARACTERIZATION

It has been pointed out that the brittle nature of ceramics imposes the need to use a probabilistic rather than a deterministic approach in structural designing. It is also suggested, that the ability to design reliable structures is seriously limited at present by inadequate characterizations of ceramic strengths.

A valid statistical description of the strength of a ceramic is required for designing with the probabilistic approach. The description is a mathematical formulation of the failure probability of a material in terms of stress and size, (i.e., volume or surface area subjected to tension). It is obtained from an analysis of strength measurements made under appropriate experimental conditions, and is based on modern theories of brittle failure. Practical problems are encountered in describing ceramic strengths for design purposes for a number of reasons, including insufficient basic knowledge of brittle behavior, as will be discussed.

In formulating statistical descriptions of ceramic strengths, it is normally assumed that failure will occur when Griffith's criterion is met. This criterion assumes that failure results from a local condition developing at a flaw site in the material where a crack becomes unstable and grows spontaneously. The criterion states that crack instability occurs when K_I, the stress-intensity factor for crack-opening tension, reaches a critical value, K_{IC}, at any site. K_{IC} is assumed to be a material constant, related to two material properties as follows:

$$K_{IC} = 2 \gamma_f E$$

where γ_f is fracture surface energy and E is Young's modulus. Fundamental understanding of fracture energy, γ_f, is lacking. It is known, however, that the thermodynamic free surface energy of the material can account for only a fraction of the surface energy consumed in fracturing a ceramic. The validity of the assumption that K_{IC} is a bulk material constant requires attention

Table 4. Recommended Materials Development to Achieve Increased Reliability.

PARALLEL DEVELOPMENT PATHS HAVING POTENTIAL

IMPROVE EXISTING MATERIALS	DEVELOP EMERGING MATERIALS	DEVELOP NEW MATERIAL
• OPTIMIZE HIGH TEMPERATURE PROPERTIES	• OPTIMIZE TRANSFORMATION-TOUGHENED CERAMICS SUCH AS Al_2O_3-ZrO_2 AND Si_3N_4-ZrO_2	• EVALUATE AlN AND SIMILAR SYSTEMS
• IMPROVE PURITY	• DEVELOP SIALON TYPE COMPOSITIONS OPTIMIZED FOR HIGH TEMPERATURE ENVIRONMENTS	• DEVELOP TERNARY AND HIGHER COMPOSITIONS
• IMPROVE PROCESSING	• OPTIMIZE UNIFORMITY AND CHEMISTRY VIA HIP	• DEVELOP MIXED COMPOSITIONS WITH SPECIFICALLY ENGINEERED PROPERTIES
• DEVELOP MONOLITHIC CAPABILITY FOR HIGH PURITY CVD MATERIALS	• DEVELOP FIBER AND SECOND PHASE DISPERSIONS TO OPTIMIZE PROPERTIES	• DEVELOP NEW COMPOSITES
• OPTIMIZE CNTD AND/OR CVD COATINGS		• DEVELOP "DUCTILE" CERAMICS FOR IMPROVED TOUGHNESS
		• DERIVE SiC AND Si_3N_4 MONOLITHIC COMPONENTS BY POLYMER PYROLYSIS

in analyzing the strength of a ceramic. Instances are reported where K_{1C} is dependent on local conditions.

When Griffith's criterion is met, the resultant crack extension in the component will usually result in complete separation or serious damage. The extent of propagation of an unstable crack, however, depends on the available stored strain energy, and in the case of highly localized and transient loading, i.e., particle impact or thermal shock, serious damage may not result even though K_{1C} is reached at a local site.

Without change in the failure mode, (i.e., with distinct tensile fracture), the interpretation of stress-state effects requires consideration of three problems:

(1) The fracture mechanics problem of defining the critical condition for crack instability for an arbitrary state of stress and a crack geometry of a form resulting from natural flaws.

(2) The constitutive relation problem of defining the effects of multiaxial loadings on strength; i.e., the problem of formulating strength in terms of principal stresses.

(3) The statistical problem of defining the effect of stress-state on variance of strength data; i.e., the problem of adjusting the Weibull description of the strength of a ceramic to account for the stress-state dependency of failure probability as a function of stress.

Concerning the fracture mechanics problem, it will be recalled that fracture occurs in uniaxial tension when $K_1 = K_{1C}$. For a two-dimensional crack, the effective stress on a crack is dependent on its orientation with respect to the crack-opening tension, owing to a shear-stress contribution. For this reason, K_1 under a given combined stress will differ from that for uniaxial tension to an extent dependent on the shear components, K_{11} and K_{111}. (4-6) Crack shape constitutes a related factor, Cracks that become unstable in ceramics cannot be treated as two dimensional with assurance since, as will be discussed later, their shapes depend on the variable geometry of precursor microstructural features or extrinsic surface defects, and the contribution of K_{11} and K_{111} to crack instability will depend on orientation effects associated with the geometry of the crack.

These sources of variability in effects of K_{11} and K_{111} on crack severity, suggest that a functional relation for σ_{eq} in terms of principal stresses with general applicability is unlikely, and that one can expect a characteristic dispersion of strength values associated with each stress state. However, it is possible that the nature and distribution of critical flaws in a ceramic are such that, for practical purposes the variability in K_{11} and K_{111} effects is insignificant, allowing analytical expressions for failure probability that account for stress-state effects.

Finally, it should be noted that the strength of a ceramic will not necessarily be isotropic even though the material is homogeneous and elastically isotropic. Flaws associated with machining or pressing, for example, tend to have directional severities and will result in different strengths depending on orientation of the tensile stress causing crack instability. Any anisotropy in strength, of course, must be accounted for in the treatment of both fracture mechanics and statistical effects of combined stresses on component failure probability.

The statistical theories of fracture, assume that the scatter in strength values result solely from variability in strength-controlling flaws. Flaws that cause fracture can be of two distinct types: (a) intrinsic features of the microstructure of a ceramic, such as large grains, pores, inclusions, or weak grain boundaries; or (b) extrinsic surface defects introduced in surface finishing or from servide-incurred damage. Sometimes two or more neighboring flaws of either or both types link together and initiate fracture. The intrinsic flaws, being microstructural features, are sensitive to ceramic processing, which again emphasizes the importance of obtaining strength data from speciments prepared identically to the component. In the case of fracture from surface-finishing flaws, it is, of course, necessary that both the strength-test specimens and the component be finsihed alike. Finally, the possibility of strength alteration from surface changes in service cannot be neglected. Service effects could obliterate surface-finishing flaws as fracture-initiating sites with the introduction of a different population of more or less severe flaws in the surface. In any event, investigation of possible service effects on the component surface is imperative in obtaining a valid statistical description of the strength of the ceramic.

Flaws in a ceramic, if not themselves cracks, give rise to the formation of cracks. The severity, s, of a crack is given by the following relations:

$$\begin{aligned} S &= Y\, a/Z \\ &\simeq 0.59\, Y(A)^{1/4} \end{aligned} \qquad (2a)$$

where a is the depth normal to the tensile stress of a surface crack, or in the case of a circular or elliptical subsurface crack, a is the radium or half the minor axis. $Y \simeq 1.99$ for surface cracks and $\simeq 1.77$ for subsurface cracks. Z varies with crack shape, having a value of 1.0 for a long, shallow crack and increasing with depth-to-width ratio. For a circular crack, $Z = \pi/2$. A is the crack area. The approximate form, Equation (2b), introduces a maximum error of less than 5 percent in computing s. (7) This relation shows that the area and location of a crack are sole determinants of its severity.

The photomicrographs in Figure 3 of fracture origins in glass and high-strength ceramics serve to illustrate the point that a range of flaw types can be responsible for a fracture. They also illustrate the fact that the size of a microstructural feature or intrinsic surface defect at which fracture initiates provides only a rough measure of the Griffith crack size. For example, the 70 μm-diameter pore in Figure 3 (b) apparently caused a 150 μm-diameter surrounding crack that ultimately became unstable. Our lack of knowledge of the association between the size of potentially detectable microstructural features or surface defects and the size of Griffith cracks impedes nondestructive assessments of the strength of a ceramic. There is little question that preceding the fracture event environmentally independent stable crack development and growth often occurs at sites of stress intensification. Both fractographs and acoustic emissions give evidence of the occurrence of the phenomenon. The formation of Griffith cracks constitutes a poorly understood facet of the crack instability event in ceramics, and we need to learn more about it in order to predict reliably failure of stressed ceramic components.

No basis exists for extrapolating plits of failure probability as a function of stress. As a consequence, the utility of statistical descriptions of strength is limited to the stress range covered by experimental strength data. Unreliable performance of a stressed ceramic component can be expected, if its failure probability is determined for an applied stress outside this range.

Because of this limitation on statistical descriptions of strength, component size tends to dictate the size of specimens to be strength-tested. If, for example, specimens much smaller than the component of interest are tested, the bulk of the strength data can be expected to fall in the high probability region of the component's failure-stress range, and a very large number of specimens would have to be tested in order to define strength of the low probability region of primary concern in structural designing.

Weibull plots of strength data for an Al_2O_3 ceramic (8) in Figure 4 illustrate the danger of extrapolating failure probability plots. The plots shown are for two sizes of specimens and the effective size of the smaller specimen has been varied by using both 3- and 4-point bending, thus, giving three plots, each covering a different failure-stress range. The Weibull descriptions of

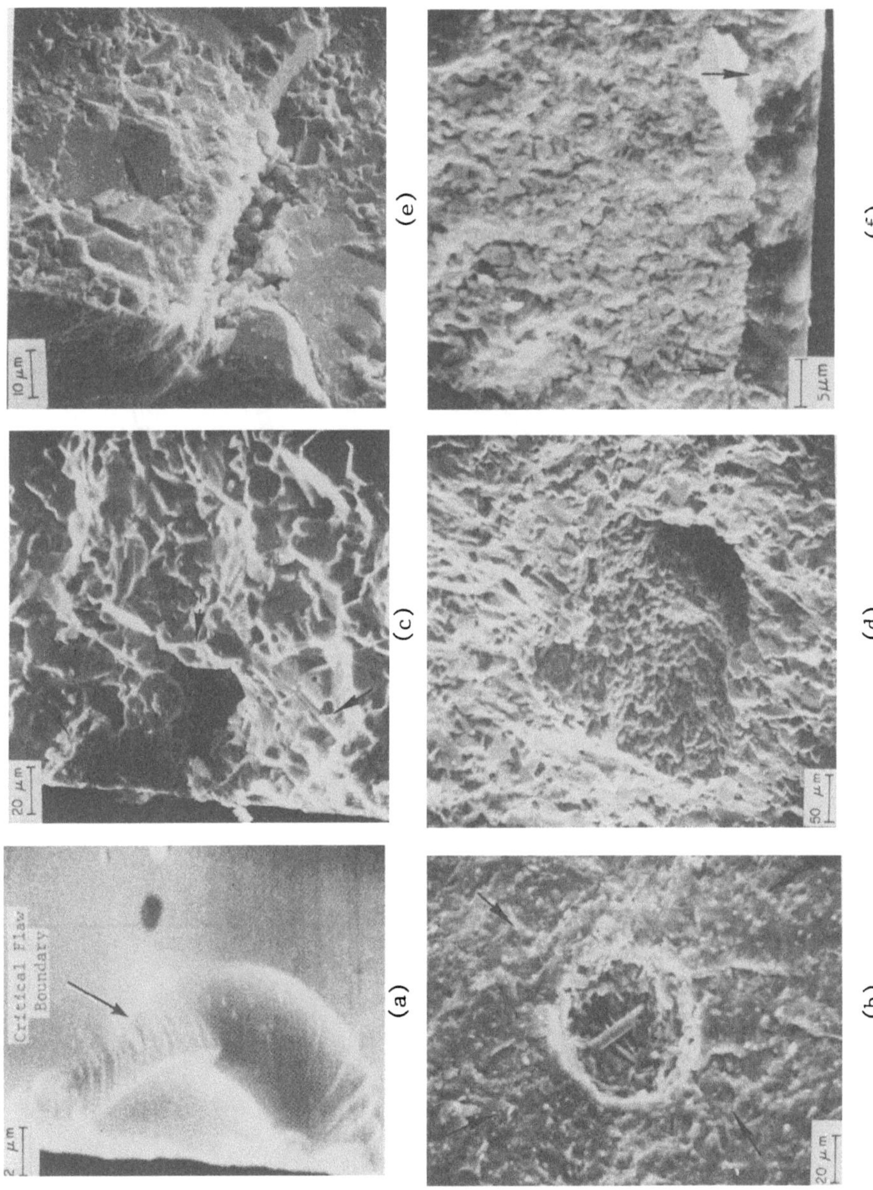

Figure 3. Fracture-Initiating Flaws in Ceramic Materials.

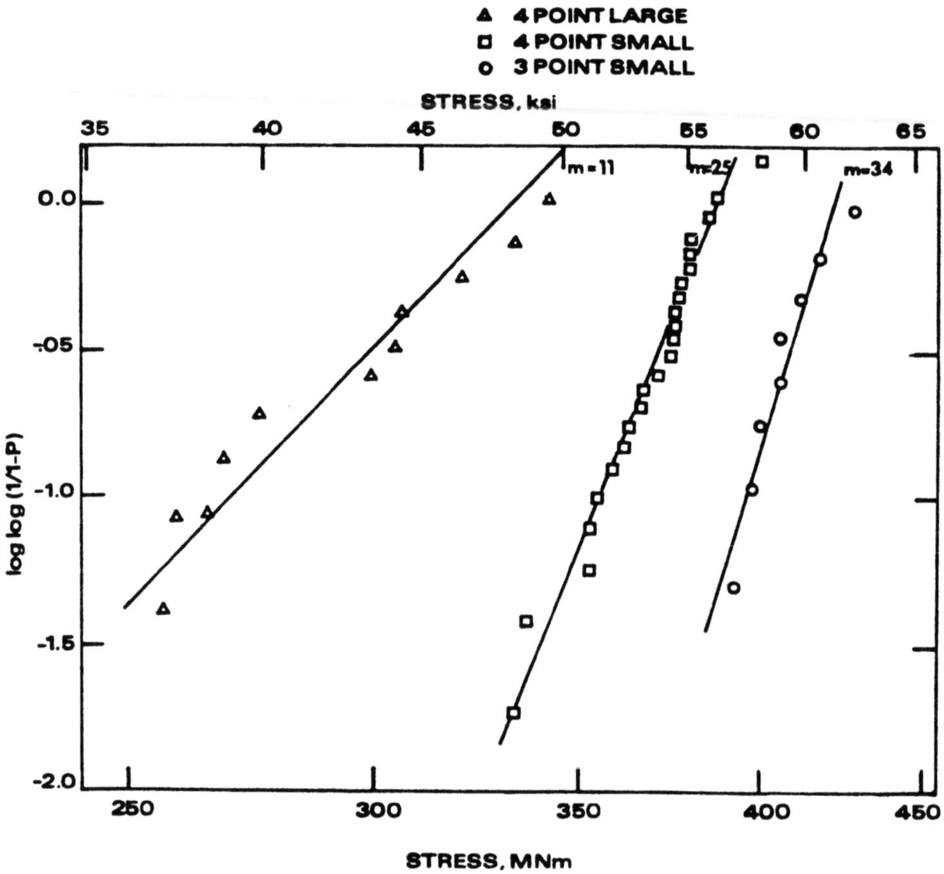

Figure 4. Weibull Plots of Al_2O_3 (Alsimag 614) Strength Data from Specimens of Three Sizes.

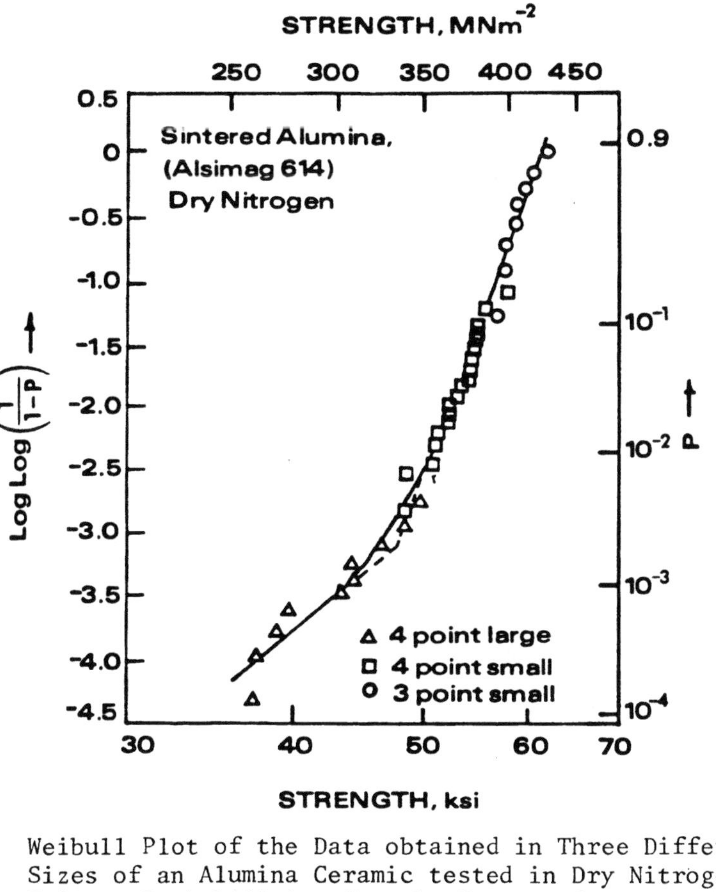

Figure 5. Weibull Plot of the Data obtained in Three Different Sizes of an Alumina Ceramic tested in Dry Nitrogen. Failure Probabilities for the Small and Large 4-Point Bend Data have been Normalized to those of the Small 3-Point Bend Specimen.

strength differ in each range, as indicated by the different slopes, and consequent Weibull moduli, m, with the difference being quite pronounced between the large and small specimens (m = 11 versus m - 26 or 34). Fractography showed that fractures in the small specimens were associated with machining flaws (Figure 3c), while those in the large specimens resulted from intrinsic microstructural defects (Figure 3d). Obviously, in this case, extrapolation of the small-specimen data to obtain a failure probability at stresses below approximately 50 ksi, would greatly underestimate the probability, as reflected by the large-specimen data. Presumably, if sufficient small specimens had been tested in 3-point bending, a Weibull plot as shown in Figure 5, would have been obtained. In this plot, Weibull statistical relations have been used to normalize the data from 4-point small and large specimens to that of the 3-point small specimen. (8,9)

This example of a complex, bimodal strength dispersion, owing to the presence of two statistical populations of strength-controlling flaws, serves to illustrate the importance of using fractography in conjunction with statistical treatments of strength data. One could not expect a single Weibull formulation to describe the failure probability of a ceramic if fractography revealed that some specimens failed from machining flaws and others from intrinsic microstructural defects.

Flaw populations in ceramics apparently can be such that specimen strengths are neither dispersed significantly, nor size-dependent. This fact has been demonstrated to a degree in tests of glass-ceramic. (10) Specimens of widely varying size exhibited no size dependence and small standard deviations (~3 percent) when failure was from extrinsic surface flaws. No abnormal procedures were used in surface finishing the specimens, and the absence of a size dependence was observed in strength tests in which slow crack growth was present as well as absent. The ceramic, however, contained a sparse population of pores which were responsible for some low-stress failures, and there was a size dependence when pore failures were represented in the strength data. The absence of a size dependence indicates a very large Weibull modulus, evaluated to be greater than 50 for the surface flaw failures.

Stable crack growth causes strength to decrease with time, limiting the service life of a stressed ceramic component. Although considerable progress has been made on the problem of life prediction, much remains to be learned about the time dependence of failure probability. From the standpoint of reliability, if a choice exists, it is certainly preferable today to use ceramics for heat-engine components that do not exhibit this phenomenon.

Such growth can result from chemical (water) action even at

low temperatures, or from mass transport (creep) at high temperatures. Also, at suffuciently high K_I levels, stable crack growth is observed in the absence of chemical action or creep. The existence of this environmentally independent stable crack growth has important implications. It suggests that a time dependency of failure is unavoidable at sufficiently high K_I levels, but below K_{IC}, and that a failure criterion less than K_{IC} should be used in designing. Very little research attention has been devoted to this matter. We have no basis today for specifying the K_I range for environmentally independent crack growth or the rate (V) at which it proceeds.

These questions need resolution through research.

The above review indicates several research needs in the area of characterization of ceramics for structural use in heat engines. Some of these concern the further development of a structural design methodology for brittle materials. Others relate to specific effects of engine service on Si_3N_4 and SiC ceramics.

Concerning the general problem, research on the following subjects is highly recommended:

- Establish limitations on the basic assumption of statistical fracture theories that K_{IC} is a bulk property.
- Determine effects of multiaxial stress on failure
- Determine effects of microstructure and temperature on K_{IC}
- Establish relations between microstructural features and crack genesis
- Determine kinetics of environmentally independent stable crack growth
- Improved techniques for statistical descriptions of strength and for relating these to flaw populations, particularly in the case of failures from multimodal populations
- Improve descriptions and basic knowledge of environmentally dependent V-K_I relations, particularly in low V ranges and also when creep is involved.
- Improve techniques for strength measurements to insure both precise fracture-stress values and that variability reflects only effects of "natural" flaws

With respect to Si_3N_4 and SiC ceramics, research on the following is recommended for support:

- Determination of the basic reasons for service-incurred changes in the flaw populations in structural ceramics.

- Development of a design methodology to evaluate long-term life when flaw populations vary with time

- Investigation of the role of oxide formation on creep, crack propagation, erosion, and friction and wear

- Collection of real-time, long-term data on the strength of turbine materials in realistic environments; data to be used as a standard for understanding life-time prediction schemes that are developed for structural ceramics

QUALITY ASSURANCE

Quality assurance of a component can be achieved prior to service by nondestructive evaluation (NDE) and/or over-load proof testing. Fracture models, which pertain to crack size/strength relations and crack growth/stress relations, are required to use either method. Thus, quality assurance not only relies on inspecting and testing components, but also on confidence in the fracture models. The following discussion will review our current understanding of nondestructive evaluation, proof testing, fracture models, and lifetime prediction methodology.

In view of the fact that cracks, solid inclusions and voids are sources of failure in ceramics, nondestructive evaluation (NDE) techniques are being developed to detect these flaws, and to measure their size and severity, so that the reliability of ceramics containing these sources of failure can be predicted. The ultimate goals of nondestructive evaluation (NDE) are to: (a) scan a complex shaped component by automated methods to determine the size, shape, and type of flaws greater than a given size, (b) combine the stress distribution of the component with the flaw distribution in order to determine the correct stress intensity factor to be used for reliable lifetime predictions, and (c) to accept or reject the component based on these predictions. Since powders are normally used to fabricate ceramics, it is important that NDE begin prior to the actual densification and shaping of the component.

Visual inspection (~20X), dye penetrants, x-radiography and ultrasonic mapping are the NDE techniques used in current ceramic heat-engine programs. Table 5 indicates their current limits of resolution as determined by examining 1/4 in. thick billets of hot-pressed IHP) and reaction bonded (RB) Si_3N_4 seeded with known inclusions which were examined under relatively ideal conditions. (11)

Table 5. Ceramic NDE Status -- Conventional Methods.

NDE Method	Material* System	Defect Type	Defect Size (Micron)	Current Limitations
Microfocus X-Ray with Image Enhancement	HP Si_3N_4	Inclusions • High Density • Low Density	 ~25 250	• Sensitivity • Material Thickness Greater than 0.250 in.
	RB Si_3N_4	Inclusions • High Density • Low Density	 125 250	• Low Density Inclusions
Ultrasonics (45 MHz)	HP Si_3N_4	Inclusions • High Density • Low Density	 ~50 ~50	• Component Geometry
	RB Si_3N_4	Inclusions • High Density • Low Density	 125 125	• Component Geometry • Attenuation (RB Si_3N_4)
Dye Penetrants	HP Si_3N_4 Siliconized SiC Sintered SiC	Surface	~250 250-500	• Sensitivity • Surface Conditions

* ¼ in. thick billets, ground, parallel sides midplane detect locations.

HP = Hot-Pressed
RB = Reaction Bonded

Table 6 lists a few of the more promising techniques under development. (3) High frequency ultrasonics have received the most attention and they currently appear to have the greatest potential.

Of all current NDE techniques, ultrasonics appears in principal to be able to satisfy most requirements for defect detection in dense, fine grained, structural ceramics. None of the high frequency ultrasonic techniques are currently advanced enough to be used as engineering tools. Experiments have shown that important defects can be located, and limited experiments have shown that void and solid inclusions can be differentiated. Further work is required to unambiguously determine defect type and shape for correlation with a fracture model. None of the high frequency techniques are yet automated. Automation requires, in part, a surface-coupling method for reducing the ultrasonic technique to engineering practice.

Despite inherent limitations, x-radiography will be an important NDE tool. Today, this tool is more automated than others, and it can be used on materials (porous and large-grained) where the ultrasonics cannot.

Other NDE techniques, i.e., penetrants, which may not be as sensitive to small cracks and flaws, will find important application in screening components prior to critical evaluation.

An acoustic-surface-wave-technique for detecting surface cracks in hot-pressed Si_3N_4 has been devised. (12) The technique has been demonstrated to detect cracks at least as small as 60 μm in depth on polished specimens, and 120 μm in depth on roughly-ground specimens. The amplitude of the scattered background is attributed to surface microcracks produced during grinding.

Although current suppliers and users of structural ceramics routinely use x-radiography (and, in some cases, conventional ultrasonics) to detect defects, the present criteria for rejecting either a billet or a component is ambiguous at best. In general, components, i.e., reaction-sintered vanes, containing large detectable fabrication voids and cracks are rejected, whereas, judgements concerning smaller defects are less definitive. It is obvious that, the development of accept/reject criteria must be concurrent with the development of NDE tools. That is, as NDE tools are developed to locate and identify cracks and inclusions, concurrent effort must be spent to identify the effect of each inclusion type on strength in order to develop a sound accept/reject criteria.

As shown in Figure 6, studies have already been initiated to evaluate the effect of defect size on the strength of hot-pressed

Table 6. Ceramic NDE Status -- Emerging Development.

NDE Methods	Material System	Defect Type	Defect Size (Micron)	Potential Advantages	Current Limitations
Ultrasonics (250 MHz)	HP Si_3N_4	Inclusions • High Density • Low Density	25	• Surface Defects • Improved Sensitivity • Defect Characterization	• Contact Techniques
Scanning Laser Acoustic Microscope[26]	HP Si_3N_4	Surface	100	• Airfoil Shape Evaluation • Minimum Subjective Interpretation	Undefined
	Injection Molded RB Si_3N_4	Subsurface Laminations	100	• Thin Section Evaluation • Airfoil Shape Evaluation	• Material Porosity
Laser[27] Photoacoustic Spectroscopy	RB Si_3N_4	Surface Near Surface	50-100 100	• Improved Resolution -- Near Surface Evaluation • Geometry Insensitive	Undefined

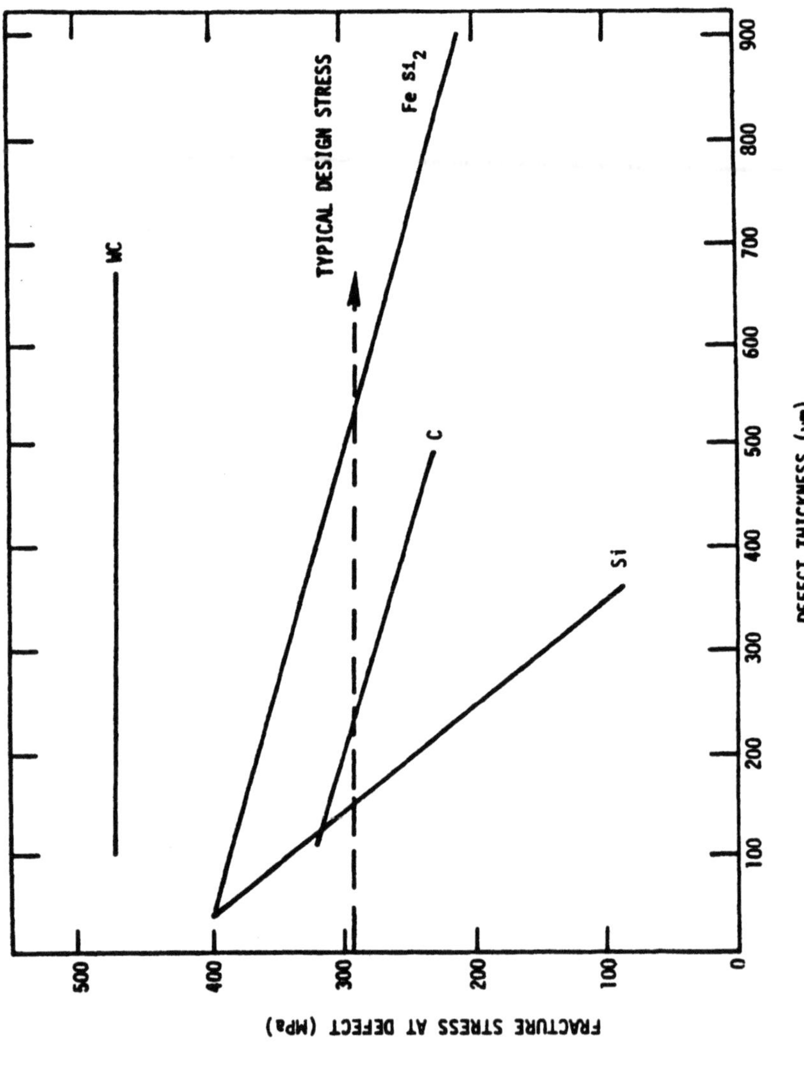

Figure 6. Flexural Strength of Hot-Pressed Si_3N_4 Seeded with Different Types of Inclusions vs. the Size of the Inclusion.

silicon nitride containing silicon, carbon, tungsten carbide and iron silicide inclusions. (13) Additional experimental and theoretical approaches are necessary to determine the relation between specimen strength, defect size, and defect shape for each type of defect. Nondestructive evaluation techniques should be used to measure the size of defects that have been introduced intentionally into specimens intended for strength measurements. The defect size estimated from nondestructive evaluation techniques can then be compared with that determined from fractographic analyses of the specimens after they have been broken, so that the accuracy of the nondestructive technique can be evaluated. Finally, the severity of the defect should be determined by comparing the strength calculated from the defect size (measured by fractography) with that determined in the strength test. Once the severity of the defect and the accuracy of the nondestructive techniques have been evaluated, they can be combined with information on defect location and stress distribution to accept or reject a component that contains a defect.

In practice, both the fracture models and the NDE techniques will contain certain errors which will result in some components being either falsely accepted or falsely rejected. Since both false-accept and false-reject will have adverse economic, political and social results, proper accept/reject criteria must be developed to minimize these problems. A solution to the proper accept/reject criteria involves probability theory, (14) which indicates that three probability functions must be defined: one covering the fracture model for a defect of a given dimension/shape, one covering the NDE estimate of the defect dimension/shape, and one covering the defect size range expected. Knowledge of these three functions can define the inspection level required for a given defect/stress and the probabilities that the component has been either falsely accepted or rejected.

This area has received little attention except for the formalization of the theory and preliminary fracture models for pertinent defect types. Effort is required to experimentally defind the three probability functions for each defect type and test the theoretical model for practical application.

In proof testing, ceramic components are subjected to stresses that are greater than those expected in service in order to break the weak components and, thus, truncate the low end of the strength distribution. In this manner, weak components are eliminated before they can be placed in service. To ensure the validity of proof testing, certain precautions must be followed. (15)

For the proof test to be most effective, it should be conducted in a relatively inert environment and rapid unloading rates should be used. Also, the proof test must duplicate the actual state of stress in the component as expected in service. After proof testing, the component must be protected from any subsequent damage. If these precautions are followed, proof testing can be a practical method for assuring the mechanical reliability of ceramics. Proof testing has been applied to spacecraft windows, (16) electrical porcelain insulators (17) vitrified grinding wheels, (18) and optical glass fibers. (19)

Evans, Wiederhorn, and Fuller (20, 21) have proficed a mathematical foundation for the selection of proof-test stresses and the establishment of the proof-test conditions. Their analysis is based on the assumption that failure of ceramics occurs from the growth of preexisting flaws. By characterizing this crack growth and coupling crack growth kinetic parameters with proof testing, they derived the strength after proof testing assuming flaw growth during the proof-stress, load-unload cycle. The resulting theory indicated that crack growth must be minimized to have effective proof testing. This can be achieved by rapid unloading from the proof stress and good environmental control during the proof test. When this is done, crack growth on unloading from the proof stress is minimized (or eliminated), and the minimum fast-fracture strength after proof testing is the proof stress σ_p. Thus, the corresponding minimum lifetime after proof testing (t_{min}) at an applied stress (σ_a) is: (22)

$$t_{min} = B\, \sigma_p^{\,n-2} \sigma_a^{\,-n}$$

where B and n are measured parameters related to the crack growth kinetics and fracture model.

The proof testing results have important implications. First, good proof-test conditions (relatively inert environment and rapid unloading) must be utilized to help assure effective proof testing. Secondly, it is important to demonstrate experimentally that proof testing is effective for a given set of conditions by comparing the after-proof strength distribution with the initial distribution, (Figure 7). Only in this way, can one be assured that proof testing under a given set of conditions is eliminating the weak components, thereby, truncating the low end of the strength distribution. Thirdly, to minimize any uncertainty in proof testing, some means, such as acoustic emission, should be used to insure that no crack growth occurs in unloading. Finally, the results show that the theoretical analysis of the effects of proof testing must be further developed.

NDE and proof testing are of value only if the flaws detected during the test are, in fact, the flaws that will cause failure

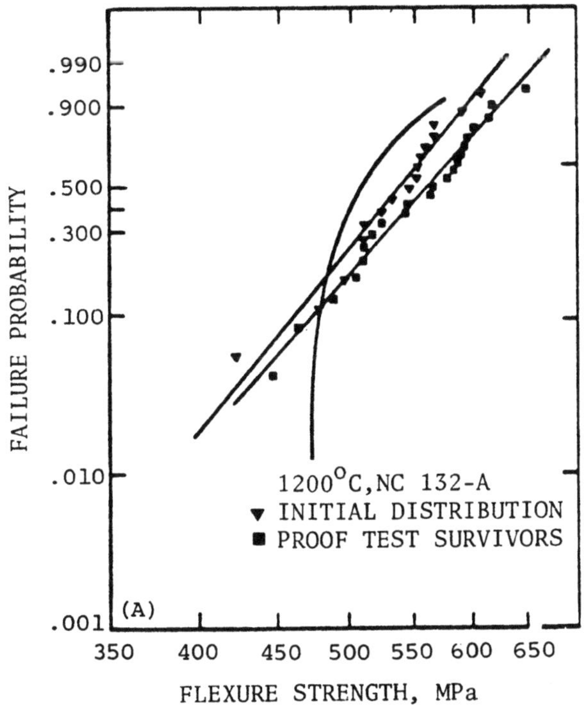

Figure 7. The Effect of a Room Temperature Proof Test on the High Temperature Strength of Hot-Pressed Silicon Nitride. The solid curve gives the expected theoretical distribution for the high temperature strengths after proof testing. (22)

Table 7. Highlights of Heat-Engine Ceramic Technology Demonstration Program.

PROGRAM		ACHIEVEMENT	COMMENT
DARPA/Ford Gas Turbine Test Engine	1974	50-h demonstration of an engine with ceramic stationary components and metal rotors 1930°F (TIT)	First integration of multiple ceramic components into an operating engine
	1977	36.5-h single-state engine demonstration 2200° - 2500°F+ (TIT) 45-50,000 rpm with ceramic rotor	First integration and successful demonstration of an uncooled all-ceramic flow path at temperatures beyond superalloy capability
DoE/Ford Regenerator Program	1977-1978	200-h rig demonstration of all-stationary hardware to 2500°F -- rotor to approximately 2200°F	200-h capability demonstration
	1978	10,000-h regenerator life demonstrated in engines, 1900°F (TIT)	Regenerator problem solved
Army/Solar 10 KW Turboalternator	1978	50-h demonstration of engine with hybrid RBSN/HPSN/RSSIC nozzle 1750°F (TIT)	First ceramic-configured engine providing electric power at required performance levels
DoE-NASA/ Detroit Diesel Allison GT-404 Test Engine	1978	1200+ - h demonstration of KT SiC stator vanes in engine 1900°F (TIT)	No obvious deterioration in the engine environment: first 1000-h+ engine test of flow path component. Engine operating as a prime mover in a truck.
Army/Cummins Adiabatic Diesel	1978	250-h full engine performance demonstration of HPSN Piston cap. max. temp. approx. 1750°F	First successful demonstration of a dynamic component in a heat engine to meet performance goals.
NAVAIR/Garrett	1978	Integration of more than 100 ceramic components into an engine. Engine run at design conditions 2200+°F	Multistage ceramic turbine largest engine run to date. Most components successfully integrated into all engines.

TIT = Turbine Inlet Temperature
RBSN = Reaction Bonded Silicon Nitride
RSSIC = Reaction Sintered Silicon Carbide
HPSN = Hot Pressed Silicon Nitride

during service. Modification of the flaw population by oxidation can invalidate NDE and proof testing procedures either by eliminating the machining flaws that cause failure at room temperature, or by changing their severity. Recent research (23-25) involving proof testing has shown that high temperature (1200°C) modifies the flaw population sufficiently that truncation of the normal flaw population (due to machining) is no longer observed after exposure times as short as 1/2 hr. (Figure 7). Therefore, although the small flaws that result from machining operations can be truncated by proof testing, truncation will service little practical value since these flaws are probably not the ones that cause failure after high temperature exposure. Thus, a need clearly exists to develop a failure prediction scheme accounting for flaw generation while a component is in long-term service. Before a failure prediction methodology can be developed fully, the effect of stress on the oxidation/flaw generation phenomenon must be understood.

CONCLUSIONS

Performance goals for military engines and societal needs for fuel economy and environmental quality at affordable costs are responsible for the current emphasis on ceramics for heat engines. Some of the major ceramic technology demonstrations recently achieved are listed in Table 7. Such successes have been instrumental in changing the outlook among heat engine designers; i.e., from being considered a rather "peculiar notion" in 1971, ceramics are now real engineering options for a wide variety of heat engines. Such a major turnaround in outlook by the design community in an eight-year period constitutes a major success of the existing ceramic demonstration programs with regard to ceramic properties, new ceramic development, and manufacturing techniques. These programs have also defined problems to be resolved in the next phase of development and demonstration; namely, the need for a component and engine reliability sufficient to justify a commitment to manufacturing.

The problems which must be addressed to provide requisite levels of reliability, require further developments in design methodology, materials processing, component fabrication, materials development, quality assurance, and testing procedures including engine tests. Perhaps the most crucial problem is providing mechanisms to maintain iterative test and development programs which integrate all of the above technology areas and assure effective feedback loops amongst them. It is important to note that, the reliability issues and requirements vary from component to component dependent upon duty cycle, power level, etc.

REFERENCES

1. W. H. Dukes, "Handbook of Brittle Material Design Technology," AD 719712, published by the Advisory Group for Aerospace Research and Development, Paris, France, 1971.

2. A. Paluszny and W. Wu, "Probabilistic Aspects of Designing with Ceramics," Journal of Engineering for Power, Vol. 99, No. 4, Oct. 1977, pp. 617-630.

3. C.G. Trantina and H.G. DeLorenzi, "Design Methodology for Ceramic Structures," ASME Paper No. 77-GT-40, 1977.

4. K. Palaniswami and W.G. Knauss, in Mechanics Today, Vol. 4, ed. by S. Nemat-Nasser, Pergamon Press (1978).

5. J. J. Petrovic and M. G. Mendiratta, J. Amer. Ceram. Soc., 59, 163 (1976).

6. J.J. Petrovic and M.G. Mendiratta, J. Amer. Ceram. Soc., 60, 463 (1977)

7. G.K. Bansal, J. Amer. Ceram. Soc., 59, 87 (1976).

8. G.K. Bansal, W.H. Duckworth and D.E. Niesz, J. Amer. Ceram. Soc., 59, 472 (1976)

9. C.A. Johnson and S. Prochaska, "Investigation of Ceramics for High-Temperature Components," General Electric Company, Quarterly Report, No. 3, Navy Contract N72269-76C-0243, January, 1977.

10. G.K. Bansal, W.H. Duckworth, and D.E. Niesz, Bul. Amer. Ceram. Soc., 55, 289 (1976).

11. J.J. Schuldies, AiResearch, Private communications.

12. B.T. Khuri-Yakub, A.G. Evans and G.S. Kino, Sub. to J. Amer. Ceram. Soc.

13. A.G. Evans and H. R. Baumgartner, to be published.

14. A.G. Evans, G.S. Kino, P.T. Khuri-Yakub and B.R. Tittman, Materials Evaluation, p. 85, April 1977.

15. S.M. Wiedernorn, "Reliability, Life Preduction and Proof Testing of Ceramics," pp. 635-65 in Ceramics for High Performance Applications, ed. by J.J. Burke, A.E. Gorum and R.N. Katz, Brook Hill Pub. Co., Chestnut Hill, MA (1974).

16. S.M. Wiederhorn, et.al., "Application of Fracture Mechanics to Space-Shuttle Window," J. Amer. Ceram. Soc. 57, 319-23, (1974).

17. A.G. Evans, et.al., "Proof Testing of Porcelain Insulators and Application of Acoustic Emission," Am. Ceram. Soc. Bull. 54, 576-81 (1975).

18. J.E. Ritter, Jr., and S.A. Wulf, "Evaluation of Proof Testing to Assure Against Delayed Failure," Am. Ceram. Soc. Bull., 58, 186-89 (1978)

19. B.K. Tariyal, et.al., "Proof Testing of Long Length Optical Fibers for a Communication Cable," Am. Ceram. Soc. Bull, 56, 204-05 (1977).

20. A.G. Evans and S.M. Wiederhorn, "Proof Testing of Ceramic Materials--An Analytical Basis for Failure Prediction," Int. J. Fract., 10, 379-92 (1974).

21. A.G. Evans and E.R. Fuller, "Proof Testing - The Effects of Slow Crack Growth," Mat. Sci. and Engr. 19, 69-77 (1975)

22. J.E. Ritter, Jr., "Engineering Design and Fatigue Failure of Brittle Materails," pp. 667-86 in Fracture Mechanics of Ceramics, Vol. 4, ed. by R.C. Bradt, D.P.H. Hasselman, F.F. Lange, Plenum Press (1978).

23. S.M. Wiederhorn and N.J. Tighe, "Effect of Flaw Generation on Proof Testing," (NBSIR 77-1408) Proceedings Ceramic Gas Turbine Engine Demonstration Program Review, Castine, Maine (1977), from "Fracture of Brittle Materials at High Temperature," AFML-TR-78-83 (July 1978).

24. S.M. Wiederhorn and N.J. Tighe, "Proof Testing of Hot-Pressed Silicon Nitride," J. Mat. Sci., 13, 1781-93 (1978)

25. S.M. Wiederhorn and N.J. Tighe, "Application of Proof Testing to Silicon Nitride," pp. 247-258 in Workshop on Ceramics for Advanced Heat Engines, ERDA Division of Conservation Research and Technology, Orlando, Florida (January 24-26, 1977). CONF770110.

STRUCTURAL DESIGN WITH BRITTLE MATERIALS

A.G. Evans

Materials Science and Mineral Engineering
University of California
Berkeley, California 94720

ABSTRACT

Structural design with brittle materials requires that the stress level in the component correspond to a material survival probability that exceeds the minimum survival probability permitted in that application. This can be achieved by developing failure models that fully account for the probability of fracture from defects within the material (including considerations of fracture statistics, fracture mechanics and stress analysis) coupled with non-destructive techniques that determine the size of the large extreme of critical defects. Approaches for obtaining the requisite information are described in this paper. The results provide implications for the microstructural design of failure-resistant brittle materials by reducing the size of *deleterious* defects and enhancing the fracture toughness.

INTRODUCTION

The design of structural components from brittle solids is, in concept, quite straightforward. It simply requires that the stress level in the component should not exceed the strength of the material, at the permissible level of survival probability. The implementation of this concept is, however, very involved. It requires the combination of information derived from the disciplines of fracture statistics, fracture mechanics and flaw detection (or non-destructive evaluation). The description of the general scientific framework for structural design, utilizing these disciplines, and of the future prospects for this class of materials, are the primary intents of the present paper.

Ultimately, design might take the form of a computer simulation of crack growth in real microstructures, coupled with microstructural characterization techniques (such as acoustic scattering). Presently, however, useful progress is being achieved using a partially decoupled approach. The evolution of failure from defects and the crack extension mechanisms are studied separately, and merged where possible. This approach has influenced the structure of the paper, which includes separate considerations of fracture initiating flaws, crack propagation and defect characteristics.

The character of the design problem is illustrated in Fig. 1a, which plots the probability of fracture of a ceramic (measured, say, in flexure) as a function of stress level. It might be construed that for design purposes, it is simply necessary to superimpose the permissible level at failure onto this figure, to obtain a maximum allowable stress in the component: and then to design the component accordingly. The limitations of this approach are exposed when it is appreciated that the fracture probability curve can be substantially perturbed by a wide variety of phenomena. These include: the incidence of slow crack growth (Fig. 1b), the occurrence of undetected flaw populations in the inevitable region of extrapolation (Fig. 1c), and the effects of stress state (Fig. 1d). Because of the problems associated with the direct use of statistical design procedures, alternate approaches have been sought which attempt to effectively truncate the strength distribution at a level above the design stress (Fig. 1e). One such approach, involving the characterization of fracture-initiating defects and of the evolution of failure, is emphasized in the present paper.

Fracture in brittle solids usually occurs either by the direct extension of a single pre-existent flaw (from the large extreme of the flaw population) or by the coalescence of small flaws. The level of stress needed to activate these flaws relates to the size of the flaw in a manner that depends upon the interaction of the flaw with the surrounding microstructure. The character of these interactions and the resultant strength, flaw size, and probability relations are described in the first part of the paper for each of the prevalent flaw types: inclusions, voids, surface cracks, and microcracks. Immediate implications for microstructural design derive from these descriptions of strength.

The ultimate survival of a brittle structural component at an acceptable survival probability requires the use of a flaw characterization technique in conjunction with a failure model. Such techniques involve the detection and analysis of waves scattered or absorbed by defects. The most versatile and sophisticated mode of flaw characterization involves the use of acoustic waves: either bulk waves or surface waves. The utility of acoustic waves

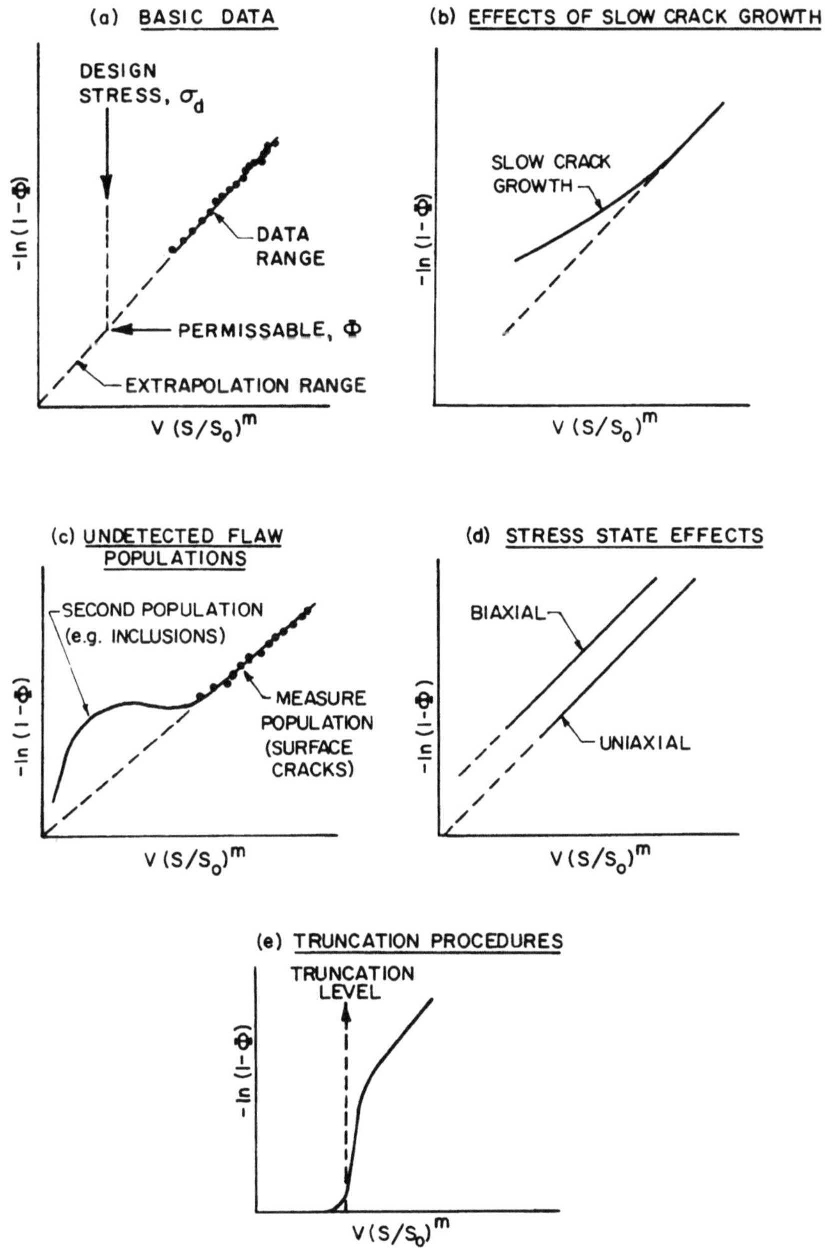

Fig. 1. A schematic illustration of some of the issues that limit the use of a direct statistical approach for structural design. The diagrams relate the fracture probability Φ to the strength level S and the sample volume V.

for providing the requisite survival information, including the combination of the measurement and fracture results to derive optimum accept/reject decision schemes, is described in the third part of the paper.

Finally, some prospects for further advancement of our comprehension of the failure process are discussed.

FRACTURE INITIATING DEFECTS

The flaws that ultimately initiate a fracture in brittle solids can be conveniently classified as intrinsic or extrinsic. The intrinsic flaws are introduced during the fabrication and are predominantly inclusions or voids. The extrinsic flaws are stress induced cracks, such as the surface cracks introduced during machining and the microcracks that result from large residual stresses (e.g., due to thermal contraction anisotropy). Each class of defect will be discussed separately.

The only available analyses of fracture from defects that provide a consistent description of effects of defect size, type and shape invoke the existence of pre-existent microflaws, activated by the concentrated stress fields around and within the defects.[1-3] However, the character of these small flaws is not well defined. It is supposed that the flaws are the small voids (or precipitates) that typically occur at grain boundaries (e.g., at triple points). These flaws are prone to activation at relatively small levels of applied stress because of the large residual stresses that can exist at grain boundaries due to thermal contraction anisotropy[4] (Fig. 2); such flaws located in high energy boundaries would be particularly susceptible to microcrack formation. Direct evidence of this mode of microcracking has not been obtained, however, and the concept must be treated as phenomenological at this juncture.

The statistical character of the microcracking process can be conveniently posed by commencing with the premise that the microflaws exhibit an extreme value size distribution that leads to the probablistic relation;[5,6]

$$\Phi(a) = 1 - \exp[-(A/A_o)(a_o/a)^k] \qquad (1)$$

where $\Phi(a)$ is the probability of finding a microflaw larger than $\underline{a}$ on a grain boundary of area A, a_o is a scale parameter, k is a shape parameter, and A_o is a normalizing constant. Noting that a flaw will extend under the condition that the stress intensity factor K reaches the critical value for grain boundary fracture, $K^c_{g.b}$, then gives the approximate result;[7,8]

$$\frac{\pi\left(K^c_{g.b}\right)^2}{4a} \approx \sigma^2 + \frac{4\tau^2}{(2-\nu)^2} \qquad (2)$$

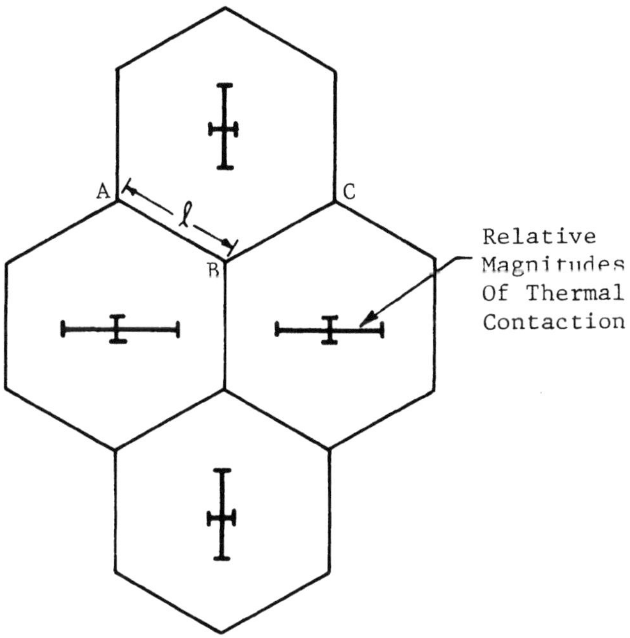

(a) Four Grain Array

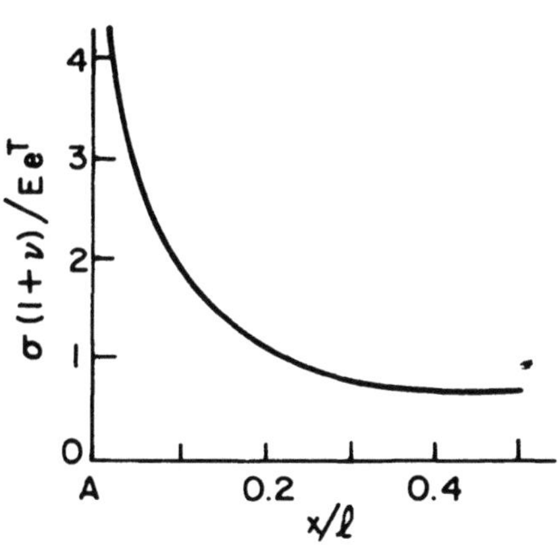

(b) Stress Along Facet AB

Fig. 2. Stresses caused by thermal expansion anisotropy.

where σ is the total stress (applied plus residual) normal to the boundary and τ is the total in-plane shear stress needed to induce crack extension. Substituting $\underline{a}$ from Eq. (2) into Eq. (1) gives the probability of microcracking as a function of applied stress $(\sigma_\infty, \tau_\infty)$ as:

$$\Phi(\sigma_\infty, \tau_\infty) = 1 - \exp\left[-\left(\frac{1}{A_o}\right)\left(\frac{4a_o}{\pi}\right)^k \int_A \left(\frac{(2-\nu)^2(\sigma_\infty+\sigma_R)^2 + 4(\tau_\infty+\tau_R)^2}{K^c_{g.b}(2-\nu)^2}\right)^k dA\right] \tag{3}$$

It should be noted that, since the residual stresses and the toughness are variables, the microcrack probability associated with a specific boundary is not uniquely related to the applied stress; rather, a distribution of probabilities generally exists. This effect allows the origination of a crack tip microcrack zone (i.e., microcracks do not necessarily initiate first at the most highly stressed boundary contiguous with the crack tip). For many other problems, the probability of crack formation averaged over many grains is more pertinent. For this situation, the average residual stress must be zero, and Eq. (3) reduces to the simple form:

$$\Phi(\sigma_A) = 1 - \exp\left[-\frac{1}{V_o}\int_V \left(\frac{\sigma_A}{S_o}\right)^m dV\right] \tag{4}$$

where σ_A is the applied stress, m is the shape parameter (=2k), S_o is the scale parameter (which includes $K^c_{g.b}$, a_o and ν as well as a coefficient that reflects an averaging of the normal and shear stresses over the grain boundaries), and V is the volume of material. A relation similar to Eq. (4) is also generally assumed to describe microcrack extension, except that S_o will have a different significance.

Intrinsic Defects

Voids. The probability of fracture from a void can be ascertained by combining the void stress field with the appropriate statistical relation for extension of microcracks existing in the vicinity of the void. If the microcracks are very much smaller than the void radius, a direct statistical analysis using Eq. (4) suffices.[1,2] For example, when the microcracks predominate at the void surface (Fig. 3), the surface stress field;[9]

$$\frac{\sigma_\theta}{\sigma_A} = [3/2(7-5\nu)]\,[(4-5\nu) + 5\cos 2\theta] \tag{5a}$$

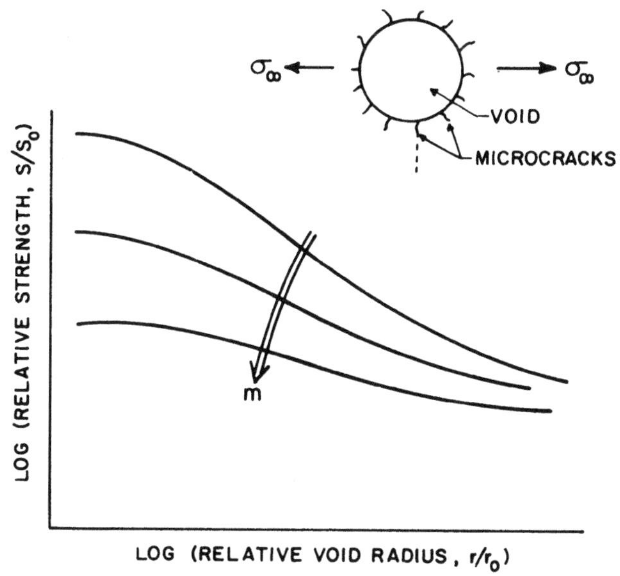

Fig. 3. A schematic illustrating the flaws distributed at the void surface, and the fracture strength, void size relation derived at constant probability.

$$\frac{\sigma_\alpha}{\sigma_A} = [3/2(7-5\nu)]\,[5\nu\cos 2\theta - 1] \tag{5b}$$

can be combined with Eq. (4) and integrated over the tensile portion of the void surface to yield:[2]

$$\xi \equiv -\ln[1-\Phi] \approx 8r^2\left(\frac{\sigma_A}{S_o}\right)^m \exp[0.52m - 1.4] \tag{6}$$

where r is the void radius. A stronger dependence on r emerges for volume distributed microflaws, viz., $\xi \propto r^3$, as deduced by Vardar, et al.[1]

When the microcracks are not small, vis-a-vis the void radius, a stress gradient correction derived from fracture mechanics solutions must be applied.[2] This correction arises because the stress intensity factor for a flaw located in a rapidly varying stress field depends sensitively on the exact flaw location and on its size relative to the gradient. The effect is especially manifest for surface-located microcracks, which are subject to the following approximate peak stress intensity factor:[2]

$$\hat{K} = \frac{2\sigma_A\sqrt{a}}{\sqrt{\pi}\,[1 + 0.3(0.2 + \frac{a}{r})]} \tag{7}$$

This relation for $\hat{K}$ can be used to obtain an effective stress, σ_{eff}, that replaces the applied stress in Eq. (16), given by:

$$\sigma_{eff} = \sigma_A\left\{0.3 + 0.7\left[1 + \left(\frac{\alpha}{2}\right)^2\right]^{-1}\right\} \tag{8}$$

where $\alpha = 1/r\,(K_c/\sigma_A)^2$. Typical void-size strength relations at constant probability, predicted by this analysis, are plotted in Fig. 3.

The basic pertinence of the statistical approach for describing strength in the presence of voids, has been substantiated for voids in silicon nitride[3] and in PZT.[1] A detailed statistical analysis was conducted for the experiments performed on silicon nitride. This analysis revealed a maximum likelihood estimate of the void radius dependence of 2.1 and demonstrated that the coefficients m and S_o were independent of the void radius, with maximum likelihood estimates of $\hat{m} = 4.6$ and $\hat{S}_o = 106$ MPa.

Inclusions. Several modes of failure have been associated with the presence of inclusions. The first distinguishing feature is the tendency for cracking due to thermal contraction mismatch[10,11] (Fig. 4). If the thermal expansion coefficient of the inclusion is appreciably lower than that for the matrix, radial matrix cracks

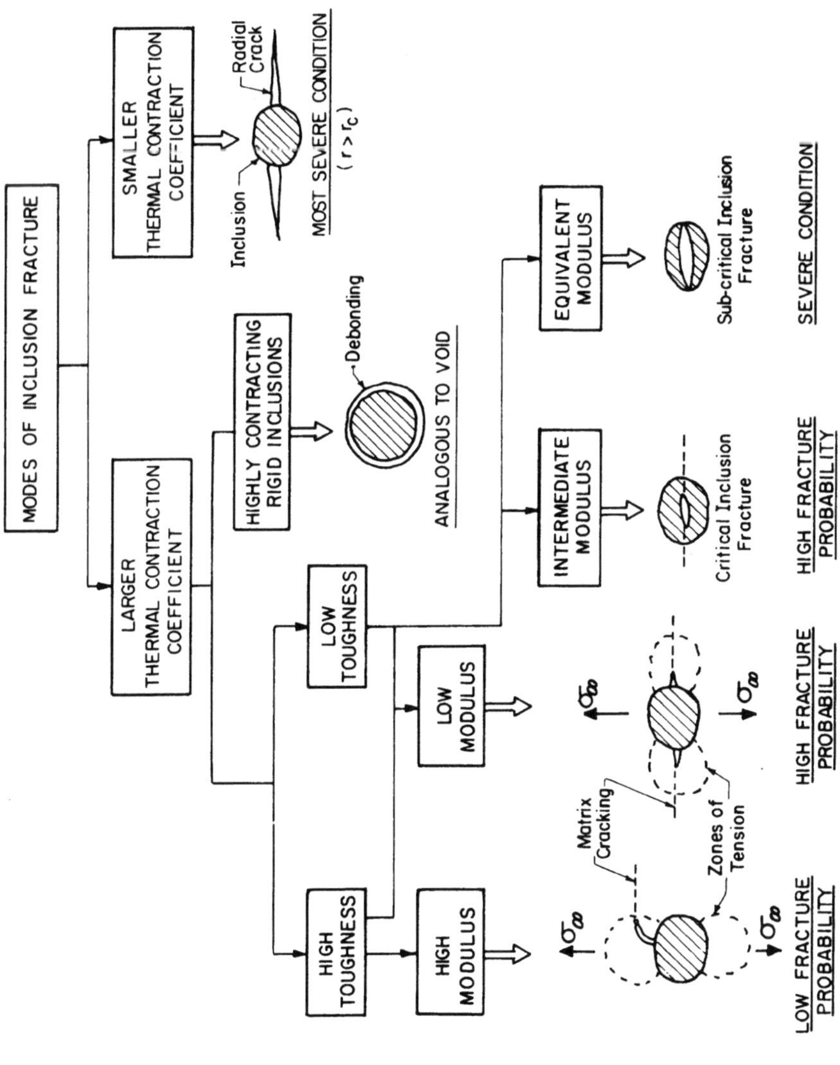

Fig. 4. A schematic indicating the various cracking responses that can occur in the presence of inclusions.

can initiate when the defect exceeds a critical size. This situation can produce severe degradation. This is, however, an unusual condition for structural brittle materials, which must have an intrinsically low thermal expansion coefficient in order to resist thermal shock. Alternately, if the expansion coefficient of the inclusion exceeds that of the matrix, several possibilities can result. Highly contracting, high modulus inclusions will tend to detach from the matrix, tending to produce a defect comparable in character to a void. Inclusions that are more compliant or exhibit smaller relative contractions, remain attached to the matrix. Thereupon, several modes of failure are possible, as exemplified by the results for several types of inclusion in the silicon nitride[3] (Fig. 5). The expected failure mode depends upon the elastic modulus and fracture toughness of the inclusion, vis-a-vis the matrix. When the inclusion has a larger toughness than the matrix (an unusual occurrence) fracture initiates within the matrix, usually from microflaws located within (or adjacent to) the interface. The process then resembles the void fracture problem. However, one additional distinction must be introduced. When the bulk modulus of the inclusion exceeds that of the matrix, the tensile stresses (in a direction suitable for continued extension of the crack due to the applied stress) are confined to a relatively small zone near the poles of the inclusion (Fig. 4). The fracture probability can then be anticipated to be relatively small, as exemplified by the high survival probability for WC inclusions in silicon nitride. Alternatively, when the modulus of the inclusion is smaller than that for the matrix, the maximum tensile stresses occur near the equatorial plane. The fracture condition is then comparable to that for a void, modified by a stress coefficient λ that depends on the modulus ratio

$$\lambda = 1 + \frac{2[(\kappa_i/\kappa_m) - 1](1-2\nu)}{3(1-\nu)} \left[\frac{4\mu_m + 3\kappa_m}{4\mu_m + 3\kappa_i}\right] \tag{9}$$

where κ is the bulk modulus and μ is the shear modulus. This case is expected to be an important one in ceramics, because the inclusions are often porous[3] (following high temperature mass transport driven by thermal contraction anisotropy) and hence, of low effective modulus.

Most inclusions typically encountered in brittle matrices are of low toughness, because they are usually the friable product of chemical reaction with the matrix (Fig. 6). If such an inclusion also has a relatively high modulus (approaching that of the matrix), the inclusion can fracture sub-critically to create a crack of dimensions comparable to the cross-section of the inclusion. The ultimate fracture strength is then dictated by the usual fracture mechanics relation for an internal crack[13]

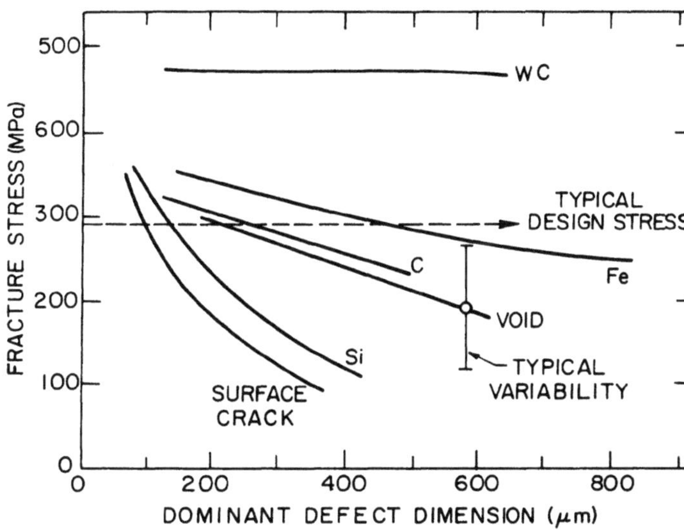

Fig. 5. Strength, size relations for various fracture-initiating defects in silicon nitride.

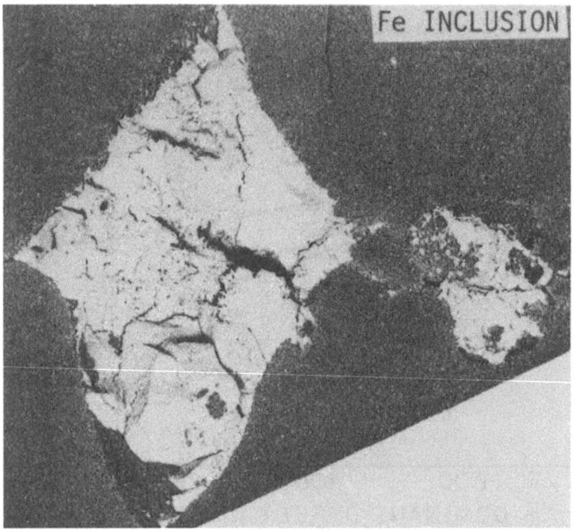

Fig. 6. A scanning electron micrograph of an iron silicide inclusion in silicon nitride.

$$\sigma = z\left(\frac{a}{c}\right) K_c^M a^\beta F\left(\frac{\kappa_m}{\kappa_i}\right) \qquad (10)$$

where a and c are the dimensions of the crack, K_c^M is the effective toughness of the matrix phase neighboring the defect, and β is an exponent (≈ 0.5) that depends on the modulus ratio. This type of defect is the most deleterious of the high expansion defects (Fig. 4). Defects in this category are exemplified by Si inclusions in Si_3N_4 (Fig. 5). When the modulus of the defect becomes very small because of extensive porosity (Fig. 6), the stresses do not attain a sufficient level to induce defect fracture (despite their friability); the situation is then identical to that of low modulus, high toughness inclusions. However, an intermediate condition is also possible; wherein fracture can initiate within the defect and then propagate directly into the matrix to cause complete failure. In this situation, fracture is dictated by the probability of activating microflaws within the inclusion, and the fracture probability becomes

$$\Phi = 1 - \exp\left[-V_i\left(\frac{\lambda\sigma_\infty + \sigma_\alpha}{p_o}\right)^\gamma\right] \qquad (11)$$

where p_o is the scale parameter, γ is the shape parameter, and V_i is the inclusion volume. Fracture results obtained for iron silicide inclusions in silicon nitride satisfy a joint fracture relation involving a combination of the critical defect fracture model (Eq. 11) and the matrix fracture model pertinent to low modulus inclusions.[3] This is a plausible situation considering the potential for a transition, with decrease in size, from inclusion-initiated fracture (volume dependent) to matrix fracture (area dependent).

Extrinsic Defects

Extrinsic defects are usually cracks produced by large transient or localized stress states. The most common sources of extrinsic defects are surface cracks produced by machining,[14] impact[15] or thermal shock. The machining induced cracks are the most prevalent (and comparable in character to the cracks introduced by projectile impact[15]). The evolution of the cracks and their resultant influence on strength is analogous to the cracking that occurs during indentation[16] (Fig. 7). The ultimate dimensions of the cracks are dictated by the residual indentation field, as controlled by the hardness, H, toughness, K_c, and modulus, E, of the material. A specific relation recently derived for the strength controlling radial cracks is:[16]

$$a^{3/2} = 2.10^{-2}(\cot^{2/3}\psi)(E/H)^{2/3} K_c^{-1} P_N \qquad (12)$$

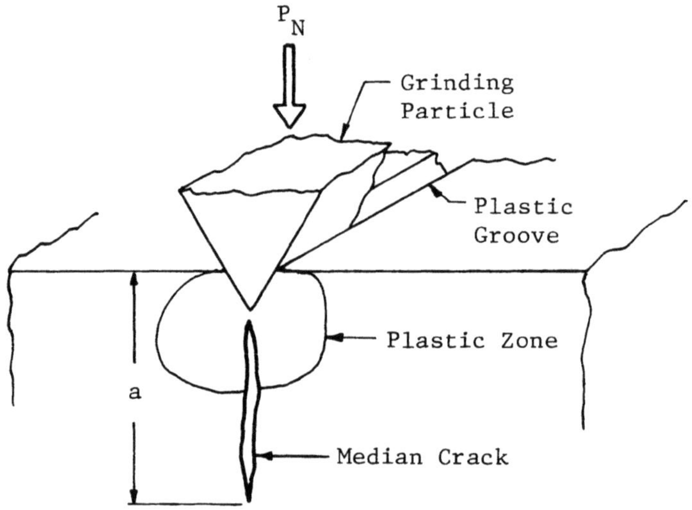

Fig. 7. The median cracking that accompanies the grinding of ceramic surfaces.

where ψ is the included angle of the grinding particle and P_N is the normal force applied to the particle. The term E/H arises because fracture is a residual stress dominated process.[17] The radial cracks are usually semi-circular because of the symmetry of the residual field.

The extension of the surface cracks introduced by grinding is explicable using standard fracture mechanics relations for mode I,[13]

$$K = F(\theta)\frac{2}{\sqrt{\pi}}\sigma\sqrt{a} \tag{13}$$

where $F(\theta)$ is the function, plotted in Fig. 8, that describes the variation in K around the crack periphery. Extension to the mixed mode fracture of inclined cracks appears to be adequately described over an appreciable range by the simple coplanar strain energy release rate criterion.[18] However, the effective stress that produces crack extension can include a significant residual component; particularly in coarse grinding situations where the plastic zone is not removed by subsequent fine grinding. Consequently, the *applied* stress at fracture can exhibit both systematic and random variations from that anticipated by direct application of Eq. (13) (with the peak stress intensity factor $\hat{K}$ equated to K_c). Available test results suggest a probability of fracture given by the normal distribution[19]

$$\Phi(\sigma_\infty|\sigma_\rho) = \frac{1}{v\sqrt{2\pi}}\int_{-\infty}^{\sigma_\infty}\exp\left[\frac{\sigma_\infty - \alpha + \beta\sigma_\rho}{\sqrt{2}\,v}\right]^2 d\sigma_\infty \tag{14}$$

where σ_ρ is the predicted strength obtained from Eq. (13),

$$\sigma_\rho = \frac{K_c}{\hat{F}(\theta)}\left(\frac{\sqrt{\pi}}{2\sqrt{a}}\right) \tag{15}$$

$\hat{F}(\theta)$ corresponds to the value at $\hat{K}$, v is the variance in σ_ρ, β is a systematic error coefficient, and α is a parameter related to the mean strength.

Microstructural Design

Several direct implications for microstructural design emerge from the fracture models. A reduction in the size of the large extreme of voids and inclusions is the most obvious. Inclusions derive from several different sources, usually in connection with the powder preparation and compaction stages of fabrication. The

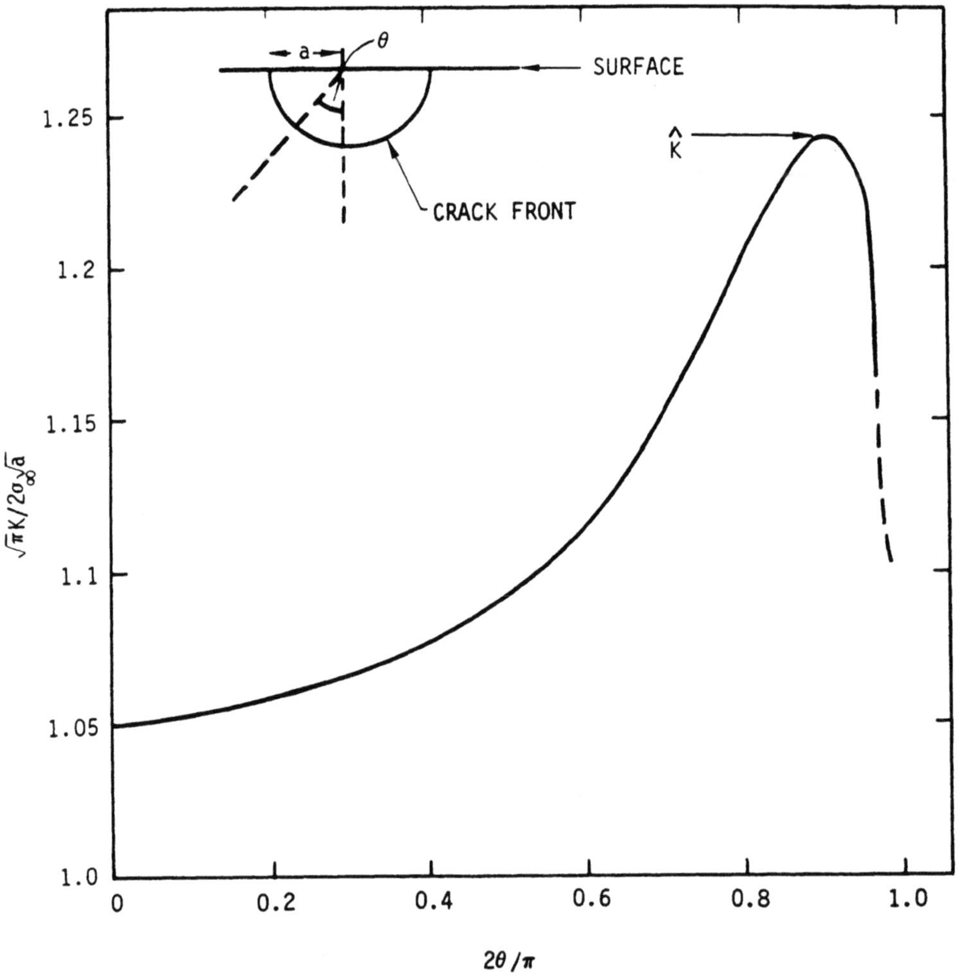

Fig. 8. The variation of stress intensity factor with peripheral location for semi-circular surface crack.

implementation of more stringent powder handling controls (both chemical and size distribution) provides direct benefits. Voids are more difficult to minimize. Large voids form due to coarsening phenomena (driven by surface diffusion or evaporation/condensation) that can occur concurrent with densification. Their existence can be minimized by avoiding domination of the mass transport by surface diffusion control or evaporation condensation control: as ascertained by reference to initial stage sintering maps for the material and the sintering environment.[20]

The formation of extrinsic cracks is determined by those microstructural parameters that influence the toughness and hardness, as indicated by Eq. (12).

NON-DESTRUCTIVE DEFECT CHARACTERIZATION

Accept/Reject Criteria

Accept/reject decisions based on a non-destructive measurement of scattering from a defect must recognize the probabilistic character of the problem.[21] At least three probabilities enter the analysis: the failure probability, given the defect dimensions (discussed above) $\phi(\sigma_\infty^c|a)d\sigma$; the joint probability of identifying the defect type and of estimating its size, $\phi(a_{es}|a)da_{es}$; the a priori distribution of defect sizes, $\phi(a)da$. These probabilities are combined and integrated to various inspection levels, a_{es}^*, to obtain two interrelated probabilities: the false-accept probability Φ_A and the false-reject probability Φ_R (Fig. 9a)

$$\Phi_A = \int_0^{\sigma_A}\int_0^{a_{es}^*}\int_0^{\infty} [\phi(\sigma_\infty|a)d\sigma_\infty]\,[\phi(a_{es}\ a)da_{es}]\,[\phi(a)da] \quad (16)$$

$$\Phi_R = \int_{\sigma_A}^{\infty}\int_{a_{es}^*}^{\infty}\int_0^{\infty} [\phi(\sigma_\infty^c|a)d\sigma_\infty]\,[\phi(a_{es}|a)da_{es}]\,[\phi(a)da] \quad (17)$$

where σ_A is the level of the applied tension in the volume element containing the defect. The inspection level a_{es}^* refers to the estimated defect dimension(s) selected for the rejection or acceptance of the component, e.g., all components with an estimated maximum dimension less than a_{es}^* are accepted and all components with an estimated dimension greater than a_{es}^* are rejected. The false-accept probability Φ_A is thus the probability

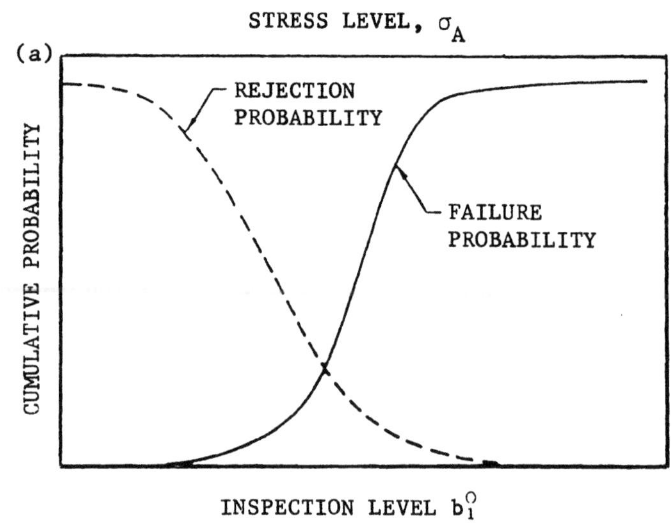

Fig. 9a. False-accept, false-reject curves for two hypothetical measurement techniques, A and B.

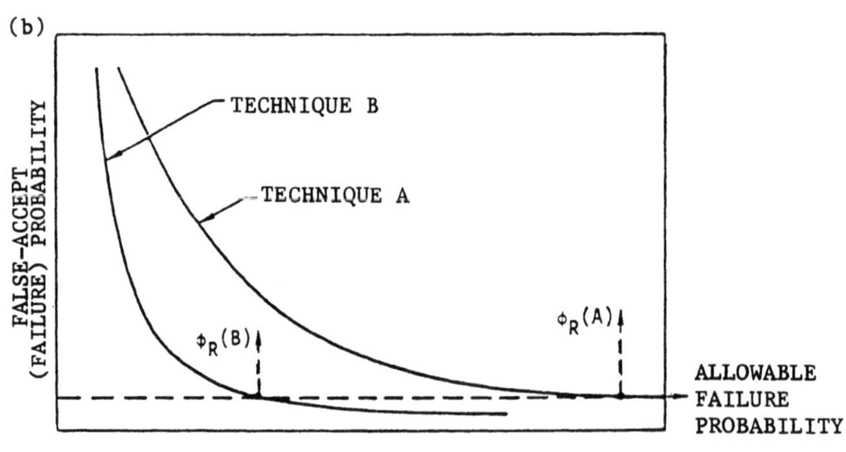

Fig. 9b. The variations of failure probability and rejection probability with inspection level.

that components accepted in accord with the specified inspection level will contain defects more severe than indicated by the estimate, and will actually fail in service (i.e., related to the failure probability). This probability decreases, of course, as the inspection level decreases (Fig. 9b). The false-reject probability Φ_R is the (related) probability that rejected components would, in fact, have performed satisfactorily in service, because the defect severity has been overestimated by the selected inspection level. This probability increases as a_{es}^* decreases (Fig. 9b). However, it is crucial to recognize that these probabilities are interrelated, i.e., they merely represent different ranges of integration of the same combination of probability functions (Eq. 16). This interdependence is exemplified in Fig. 9a, which is a typical plot relating the false-accept and false-reject probabilities: once one of these probabilities has been selected, the other probability, as well as the association inspection level, are *necessarily* defined. It is now apparent from Fig. 9a that the inspection technique, or combination of techniques, that would be preferred is that which yields a curve as close as possible to the probability axes. For example, technique B is preferred over technique A, because the rejection of satisfactory components required to satisfy the failure probability requirements is much lower. Such curves thus represent a quantitative method for characterizing the failure prediction capabilities of various inspection techniques for a given material and service condition.

Scrutiny of the available inspection methods pertinent to ceramics indicates that acoustic methods are preferred, because acoustic waves are appreciably scattered by all of the critical defect types encountered in structural ceramics. The most promising measurement algorithms and their future potential are thus briefly reviewed.

Acoustic Measurement Algorithms

Surface Waves. The most directly successful acoustic method is the use of surface acoustic waves to predict failure from surface cracks; in particular, the use of long wavelength, $\lambda \gg a$, surface waves.[22,23] In the long wavelength limit, the scattering of an acoustic wave (stress wave) by a crack is closely analogous to the interaction of the crack with an applied stress field. In particular, both the scattering coefficient S_1 and the strain energy release rate $\mathcal{G}$ are related to the crack surface integral:[22]

$$\int_{A_s} \sigma_{ij} \Delta u_j' \, n_i \, dA_s$$

where σ_{ij} is the stress across the crack plane in the absence of the crack, $\Delta u_j'$ is the displacement of the crack surfaces, and A_s

is the crack surface area. Hence, it is straightforward to demonstrate that the scattering coefficient is directly related to the crack extension stress σ_c by:[23]

$$\frac{K_c}{\sigma_c} = 2\left[\frac{6(1-\nu)\lambda^2 S_i w}{\pi^5 f_z \eta}\right]^{1/6} \tag{18}$$

where w is the beam width, η is the transducer efficiency, and $f_z \sim 0.4$. This result is strictly correct when both the acoustic wave and the applied stress are normal to the crack plane, as exemplified by recent results[19] summarized in Table 1. However, since the coplanar $\mathcal{G}$ criterion also appears to afford a reasonably satisfactory description of surface crack extension in ceramics,[18] at least over the angular range of interest, the approach appears to be of general applicability. Surface waves also have the advantage that they propagate over curved surfaces, so that complex shapes can be readily probed.[23]

Bulk Waves. The characterization of bulk defects is more complex. Information over a wide range of frequencies appears to be needed to obtain a highly probable defect type classification and hence, a size estimation. Appropriate techniques are available including: the scanning laser acoustic microscope,[24] 200 MHz ZnO transducers,[25] and conventional (2-50 MHz) transducers. Rapid scanning methods for defect location have also been developed. The most critical issue, therefore, concerns the appropriate choice of algorithms to obtain the most reliable defect characterization. A typical set of algorithms and their interaction are illustrated in Table 2, using low and high frequency information as well as acoustic microscopy. This set has not yet been fully evaluated so many redundancies may exist. Four algorithms are employed in this scheme: (i) long wavelength scattering,[26] (ii) intermediate wavelength Born approximation,[27] (iii) high frequency spectroscopy[28] and (iv) cross-sectional information from acoustic microscopy. The impulse response functions (Fig. 10a) are used first to determine whether the defect is a void or an inclusion; the void has an impulse response function (Fig. 10a) characteristic of the transducer, while inclusions have more complex functions (Fig. 10b). Thereafter voids can be analyzed straightforwardly using a variety of algorithms. For example, a long wavelength algorithm similar to that described for surface cracks may be employed. In the long wavelength limit the scattered amplitude is related to the void volume V by[21]

$$A = \frac{V\omega^2}{(4\pi c^2)^2}\left[1 + \frac{1+\nu}{7-2\nu} + \frac{10(1-2\nu)}{7-5\nu}\right]^2 \tag{19}$$

where ω is the frequency and c is the elastic wave speed in the

TABLE I Comparison of Measured Surface Crack Sizes with These Predicted using Long Wave Length Surface Acoustic Waves

Sample	σ_F Actual MPa	Acoustic a μm	σ_F Acoustic MPa	%
5 kg: 1	338.45	56	350	3.3
2	365	51	367	.54
10 kg: 1	298.5	67	320	6.72
2	275.4	66	322.7	14.6
20 kg: 1	159.22	274	158.4	.52
2	179.13	262	159.7	12.17
3	189	255	164.2	15.1

Table 2. Measurement Scheme.

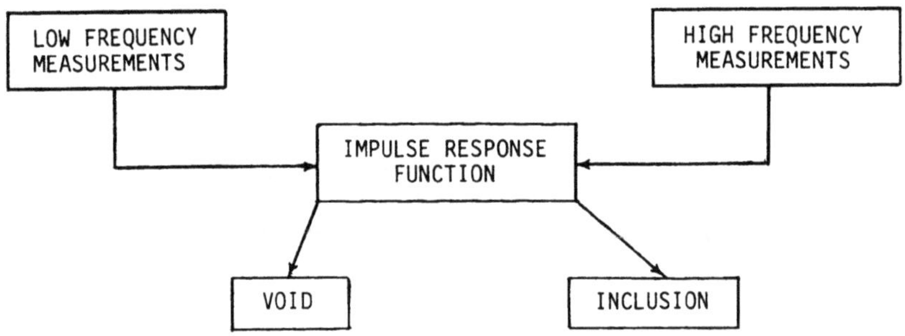

VOID (CRACK)

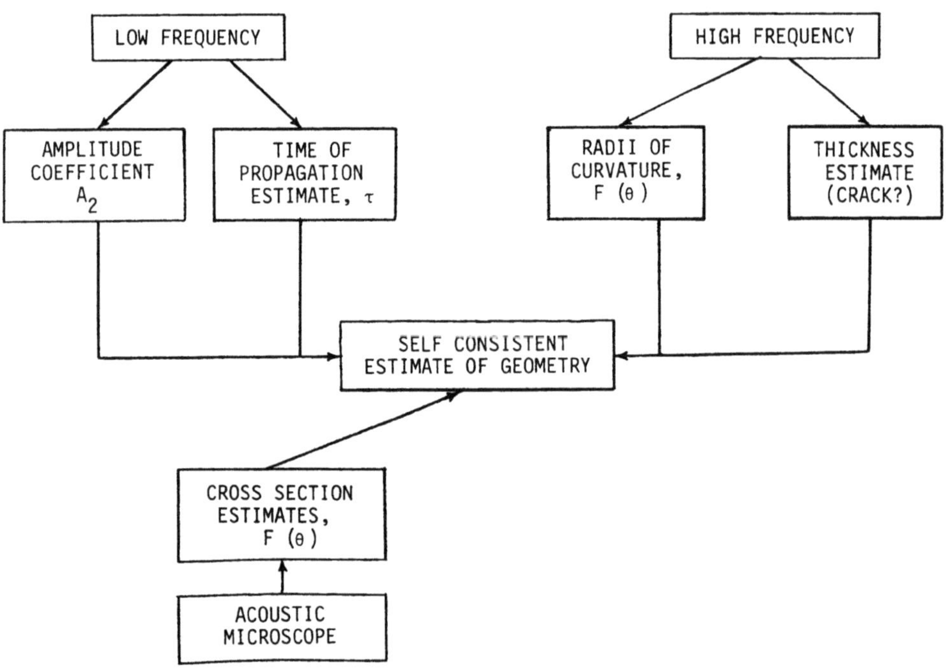

Table 2 (continued). Measurement Scheme Inclusion.

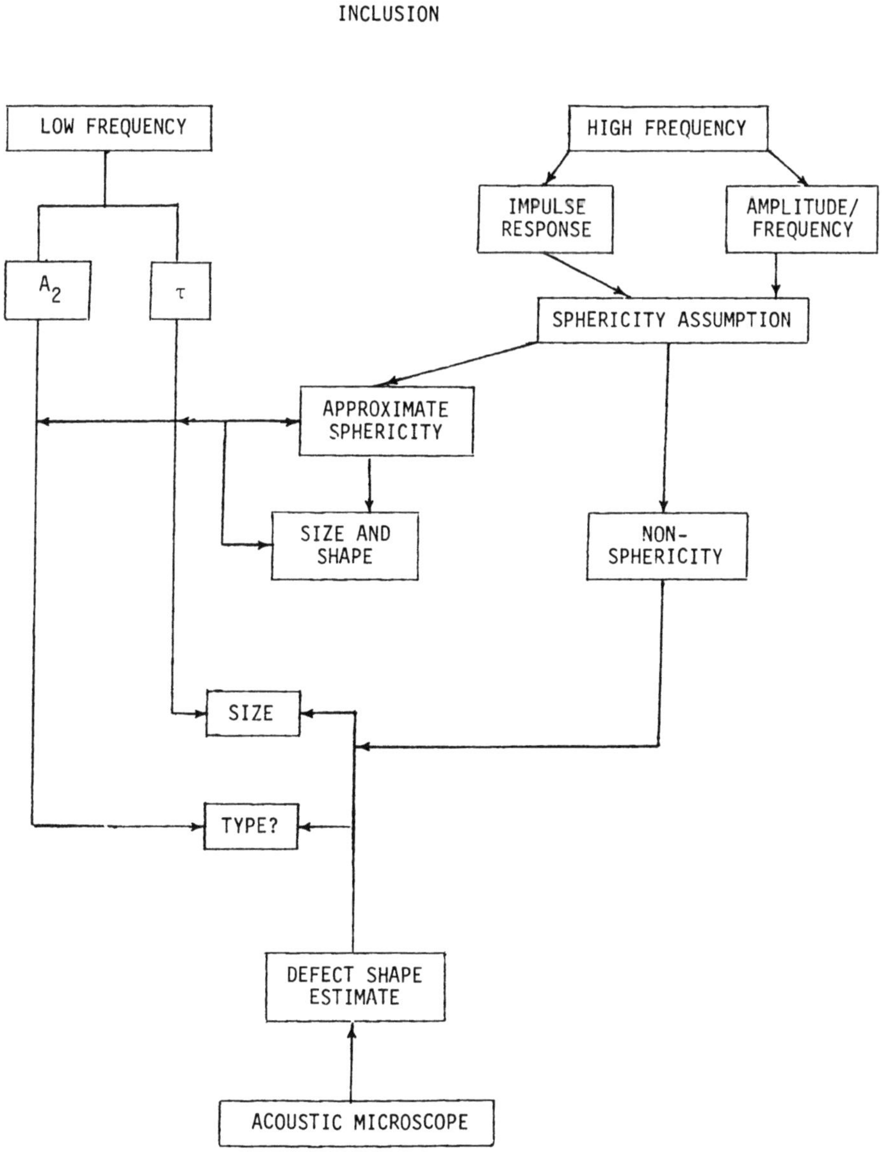

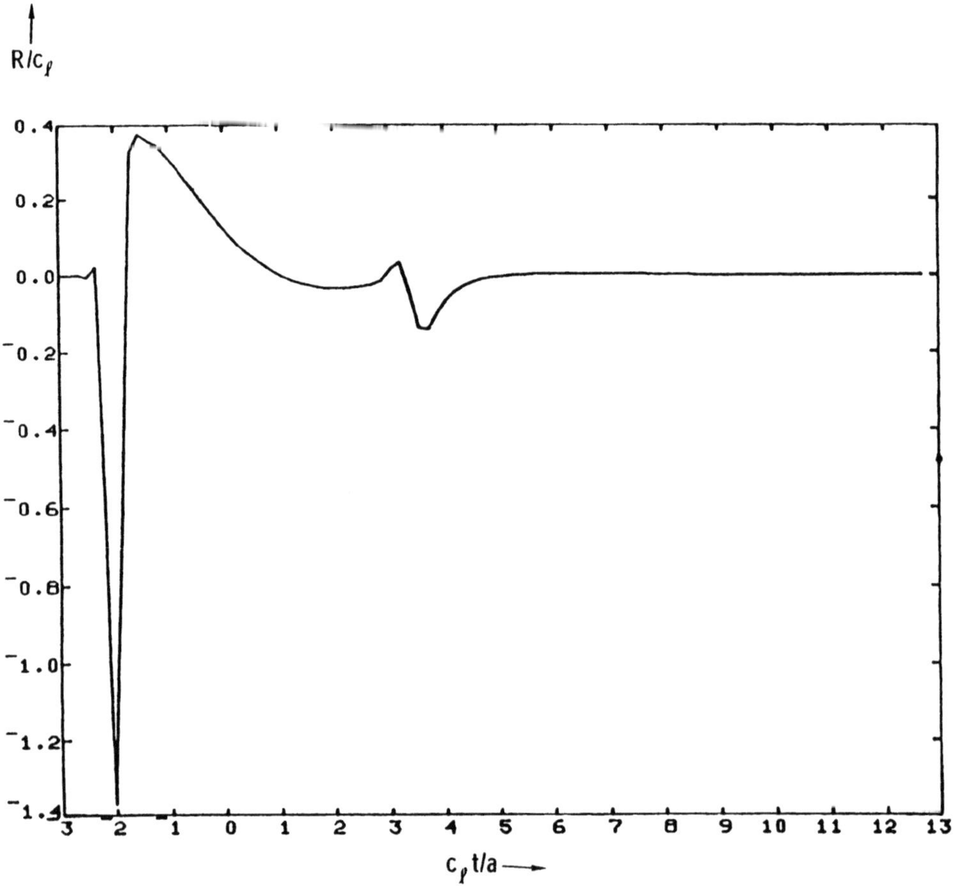

Fig. 10a. Impulse response functions for defects in ceramics, a void in Si_3N_4.

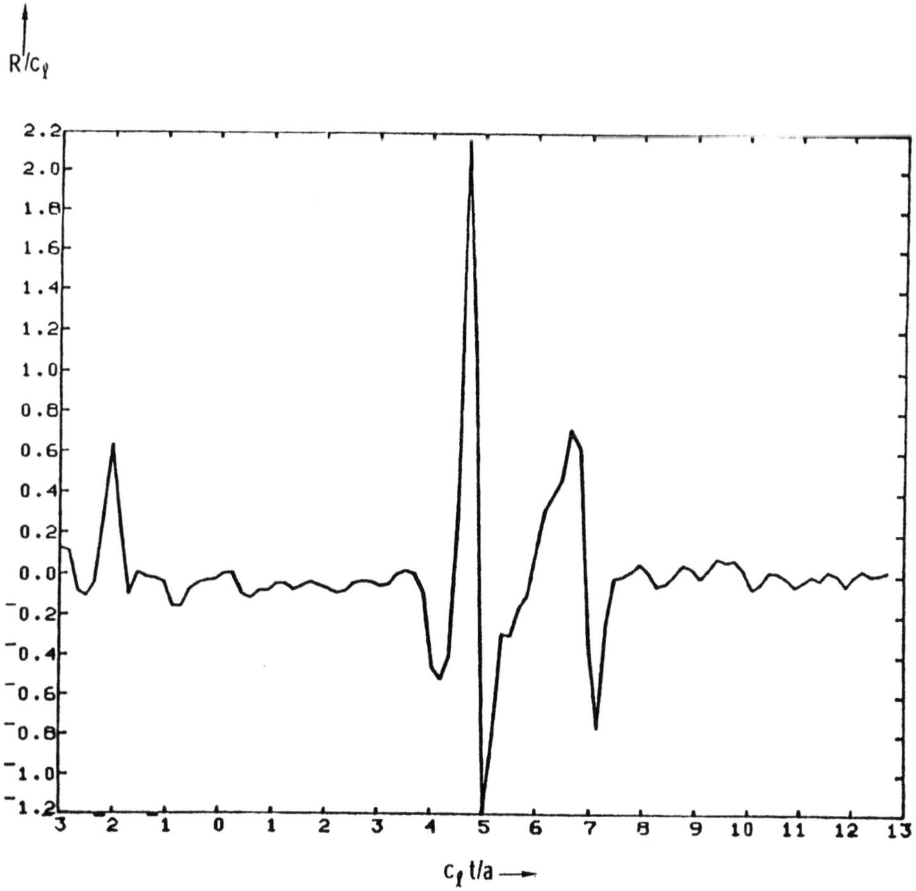

Fig. 10b. Impulse response functions for defects in ceramics, a WC inclusion in Si_3N_4.

host. Inclusions are more difficult to analyze; the combined use of several algorithms is almost certainly required. For nearly spherical inclusions, the interpretation is relatively straightforward. For example, a combination of the long wavelength algorithm (which contains coupled volume and type information) and the Born approximation (which provides an independent estimate of the distance from the geometric center to the back face of the inclusion) can yield the requisite size and type information. A typical result, obtained for a 100 µm radius Si inclusion in Si_3N_4, is illustrated in Fig. 11, wherein the joint probability of the defect type and size is plotted as a function of the estimated size. Alternatively, high frequency measurements displayed in the frequency domain would provide close estimates of the defect size and type.

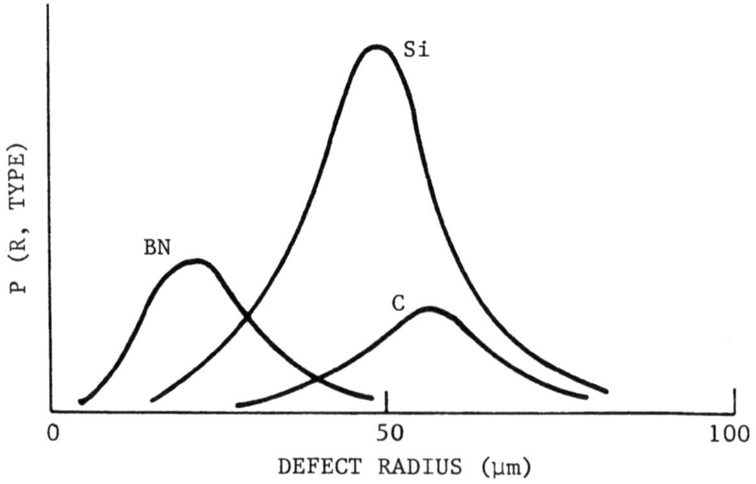

Fig. 11. The joint probability of defect type and size deduced from a coupled long wavelength, Born approximation alogrithm.

FUTURE PROSPECTS

It is hoped that this paper conveys the impression that a positive start has been made in establishing the scientific framework for microstructural design with brittle materials. Certain rewarding research directions have emerged and several exciting near-term and more remote prospects seem viable.

Further studies aimed at characterizing models of fracture from defects are very pertinent. The incisive combination of inputs from mechanics, materials and statistics demonstrated on the limited set of problems addressed thus far should provide some direction and scope for continued activity. Important defects not yet considered include: void clusters, sub-surface inclusions, and surface crack arrays. Progress toward the comprehension of fracture from these defects could utilize existing (or marginally extended) stress analyses coupled with advanced statistical methods and fracture mechanics solutions.

More immediate advances can be anticipated in ultrasonic flaw characterization. A comprehensive set of inversion algorithms already exist, and initial results imply that good estimates of defect size and type are possible, using combinations of these algorithms. Future prospects for devising effective accept/reject schemes pertinent to ceramics are thus very exciting.

Acknowledgments. The author wishes to thank the Division of Materials Sciences, Office of Basic Energy Sciences, U.S. Dept. of Energy under Contract W-7405-ENG-48 for funding the work on toughening mechanisms, the Advanced Research Projects Agency for funding the work on fracture models and accept/reject criteria under Contract F 33615-74-C-5180, and the Office of Naval Research for supporting the work on surface crack dilation.

REFERENCES

1. O.Vardar, I.Finnie, D.R.Biswas and R.M.Fulrath, Intl. Jnl. Frac. 13 (1977) 218.
2. A.G.Evans, D.R.Biswas and R.M.Fulrath, Jnl. Amer. Ceram. Soc. 62 (1979) 101.
3. A.G.Evans, G.Meyer, K.Fertig, B.I.Davis and H.R.Baumgartner, Jnl. of Non-Destructive Evaluation, in press.
4. A.G.Evans, Acta Met. 26 (1978) 1845.
5. A.M.Freudenthal, "Fracture," (H.Liebowitz, ed.), Vol. 2 (Academic Press, New York, 1969), p.592.
6. A.G.Evans, B.R.Tittmann, L.Ahlberg, G.S.Kino and B.T.Khuri-Yakub, Jnl. Appl. Phys. 49 (1978) 2669.

7. S.B.Batdorf and H.L.Heinish, Jnl. Amer. Ceram. Soc. 61 (1978) 355.
8. A.G.Evans, Jnl. Amer. Ceram. Soc. 61 (1978) 302.
9. S.Timoshenko and J.M.Goodier, "Theory of Elasticity," (McGraw Hill, New York, 1951).
10. A.G.Evans, Jnl. Mater. Sci. 9 (1974) 1145.
11. R.W.Davidge and T.J.Green, Jnl. Mater. Sci. 3 (1968) 629.
12. H.R.Baumgartner and D.Richerson, "Fracture Mechanics of Ceramics," (R.C.Bradt, D.P.H.Hasselman and F.F.Lange, eds.), Vol. 1 (1974), p.367.
13. G.C.Sih, "Handbook of Stress Intensity Factors," (Lehigh University Press, 1973).
14. M.V.Swain, "Fracture Mecahnics of Ceramics," ibid, Vol. 3 (1978), p.257.
15. A.G.Evans, M.E.Gulden and M.Rosenblatt, Proc. Roy. Soc. A361 (1978) 343.
16. B.R.Lawn, A.G.Evans and D.B.Marshall, Jnl. Amer. Ceram. Soc., in press.
17. D.B.Marshall and B.R.Lawn, Jnl. Mater. Sci., in press.
18. J.J.Petrovic and M.G.Mendiratta, Jnl. Amer. Ceram. Soc. 59 (1976) 163.
19. B.T.Khuri-Yakub, J.Tien, G.S.Kino and A.G.Evans, Jnl. Amer. Ceram. Soc., to be published.
20. M.F.Ashby, Acta Met. 22 (1974) 275.
21. J.M.Richardson and A.G.Evans, Jnl. of Non-Destructive Evaluation, to be published.
22. B.Budiansky and J.R.Rice, Jnl. Appl. Mech. 45 (1978) 2.
23. B.T.Khuri-Yakub, G.S.Kino and A.G.Evans, Jnl. Amer. Ceram. Soc., in press.
24. L.W.Kessler, work performed at Sonoscan.
25. B.T.Khuri-Yakub and G.S.Kino, Appl. Phys. Lett. 30 (1977) 2.
26. W.Kohn and J.R.Rice, Jnl. Appl. Phys., to be published.
27. J.H.Rosi and J.A.Krumhansl, Materials Science Center Report 2846 (Cornell University).
28. R.B.Thompson and A.G.Evans, Sonics & Ultrasonics 23 (1976) 292.

APPLICATION OF FRACTURE MECHANICS IN ASSURING AGAINST FATIGUE FAILURE OF CERAMIC COMPONENTS

J. E. Ritter, Jr.

University of Massachusetts
Amherst, MA. 01003

S. M. Wiederhorn, N. J. Tighe and E. R. Fuller, Jr.

Fracture and Deformation Division
National Bureau of Standards
Washington, D. C. 20234

INTRODUCTION

The use of ceramics in high performance applications offers a challenge to scientists and engineers; the challenge of designing structural components with brittle materials. Design problems arise for two reasons when brittle materials are used as structural components: 1. The strength of brittle materials is not a well defined quantity, but can vary widely depending on the material; 2. The strength of brittle materials is time dependent so that these materials often exhibit a time delay to failure. This time dependence and scatter of strength so typical of most ceramic materials occurs because of the presence of defects such as cracks or crack-like flaws in these materials. When subjected to an applied tensile stress, these defects act as stress concentrators and fracture occurs when the applied stress intensity factor reaches a critical value. Scatter in the strength of ceramic materials is a consequence of the scatter in the size of the most critical defect in the ceramic. The time dependence of strength results from subcritical crack growth, which gradually lengthens the crack until it reaches critical dimensions, at which time failure occurs. The time delay to failure is the time required for the crack to go from a subcritical to a critical size.

Methods of dealing with design problems involving fatigue of ceramic materials have been developed over the past 10 years through the application of the techniques and concepts of fracture mechanics (1-3). Since fracture mechanics techniques can be used to characterize both the conditions for subcritical crack growth and the conditions for crack instability, they can be used for purposes of design to estimate the allowable applied stress and the expected lifetime for a given component. This is accomplished by estimating the initial crack size in a ceramic component and the time required for the crack to grow from its initial size to a final critical size. Several mutually independent techniques have been developed recently to provide these types of estimates. Since these techniques promise to revolutionize the way in which structural components made of ceramic materials are designed, a complete understanding of these techniques, their application, and their limitations will be necessary to use them correctly.

This paper will review the application of fracture mechanics theory to the prevention of delayed failure of ceramics. Three successful applications of this theory in assuring the mechanical reliability of ceramics are discussed in order to demonstrate the viability of the theory for purposes of engineering design. Finally, a description is presented of practical limitations of the theory with regard to heat engine application. Methods of overcoming these limitations through modification of test procedures, and application of statistical theory are then presented.

THEORY

Since fracture mechanics concepts can be used to characterize both the conditions for subcritical crack growth and the conditions for crack instability, they can be used to predict the failure of ceramic materials under given service conditions. This is accomplished by estimating the initial crack size in a ceramic component and the time for the crack to grow from its initial size to a final critical size. As the crack grows to its critical size, the crack velocity (V) is assumed to be dependent on the applied stress intensity factor (K_I) by (1):

$$V = A\ K_I^N \tag{1}$$

where A and N are crack growth constants that depend on the environment and material composition. From Equation (1) and the relation $\sigma = K_{IC}/Y\sqrt{a}$, it can be shown that the time to failure (t_f) under constant applied tensile stress (σ_a) is (3):

$$t_f = B\ S^{N-2}\ \sigma_a^{-N} \tag{2}$$

where $B = 2/(AY^2(n-2)K_{IC}^{N-2})$, K_{IC} = critical stress intensity factor, and S = fracture strength in an inert environment where no subcritical crack growth occurs prior to fracture. From its definition, B is a crack propagation constant that depends on the environment and material composition.

In Equation (2), t_f represents the time required for a flaw to grow from an initial, subcritical size, to dimensions critical for catastrophic propagation; B and N are the constants that characterize this subcritical crack growth. The initial flaw size is characterized in Equation (2) by the fracture strength in an inert environment*. From Equation (2) it is seen that the time to failure decreases with increasing stress, i.e. fatigue under a static stress.

Also from Eq. (1), a relationship can be derived between the fracture strength σ_f and stressing rate $\dot{\sigma}$. In this case flaws grow from subcritical to critical size under constantly increasing stress. The results of this analysis is (1-3):

$$\sigma_f^{N+1} = B\ (N+1)\ S^{N-2}\ \dot{\sigma} \tag{3}$$

where B and N are the same fatigue constants as in Eq. (2). From Equation (3) it is seen that fracture strength decreases with decreasing stressing rate since the flaws have more time to grow. This behavior is known as dynamic fatigue, i.e. fatigue under constant stressing rate conditions.

The probability of failure (F) for a given lifetime and applied stress can be obtained from Equation (2) by expressing the inert strength in terms of its failure probability distribution. Generally, the inert strength distribution of ceramics can be approximated by the Weibull relationship (3):

$$\ln \ln \frac{1}{1-F} = m \ln \frac{S}{S_o} \tag{4}$$

where m and S_o are empirical constants evaluated by a fit of the strength data. Likewise, the strength distribution at a fixed stressing rate can be obtained by substituting Equation (4) into Equation (3).

Because ceramics exhibit a wide spread in strength values (m is typically 4-8), the allowable stress in service is quite

---

*The initial crack size, a, can be calculated, if desired, from the well known fracture mechanics relationship: $a = K_{IC}^2/Y^2S^2$, where Y is a constant that is determined by the geometry of the specimen and the crack. In subsequent discussion, a is assumed to be small relative to component dimensions so that Y is a single valued constant.

low. If low failure probabilities are required, proof testing offers one means of increasing the design stress for ceramics (1-3). The value of proof testing is that it characterizes the largest effective flaw possible in a tested component, since any larger flaws would have caused failure during the proof test.

Assuming that flaw growth during the proof test cycle is given by Equation (1) (i.e. one region of crack growth) and that the initial inert strength distribution can be characterized by Equation (4), the inert strength after proof testing (S_a) is given by (4):

$$(S_a/S_o)^{N_p-2} = (-\ln(1-F_a)-\ln(1-F_p))^{(N_p-2)/m} - (-\ln(1-F_p))^{(N_p-2)/m} + (S_{min}/S_o)^{N_p-2} \quad (5)$$

where N_p is the crack propagation constant appropriate for the proof test environment, F_a is the failure probability after proof testing, F_p is the failure probability of the proof test and S_{min} is the minimum inert strength of a sample that just passes the proof test cycle, i.e. the truncation strength. It is significant to note that the inert strength after proof testing is truncated at S_{min} and will be greater than the initial inert strength at all levels of failure probability if $m < N_p-2$ (4).

For a given material and proof test environment, S_{min} is determined only by the unloading rate ($\dot{\sigma}_u$) from the proof stress (4):

$$S_{min} = [\dot{\sigma}_u B_p(N_p-2)]^{1/3} [3/(N_p+1)]^{1/(N_p-3)} \quad (6)$$

where B_p is the crack propagation constant appropriate for the proof test environment. If $\dot{\sigma}_u > 0$, Equation (6) shows that proof testing always truncates the strength distribution. The higher $\dot{\sigma}_u$, the greater is the strength level, S_{min}, at which truncation occurs. However, since S_{min} cannot be greater than the maximum stress, σ_p, during the proof test, σ_p is an upper bound for S_{min}. Substitution of σ_p for S_{min} in Equation (6) gives the minimum unloading rate for which σ_p is approximately equal to the truncation strength. Equation (6) also shows that good proof test conditions (high N_p) result in a high S_{min}.

The failure probability of the proof test (F_p) can be determined directly from the number of specimens that break during the proof test or can be predicted from (4):

$$-\ln(1-F_p) = (S_o{}^m)^{-1} (D_p/B_p)^{m/(N_p-2)} \quad (7)$$

where $D_p = \int_0^t \sigma_p^{N_p}(t)\,dt$ represents the amount of strength degradation during the proof stress cycle. For a typical proof test, the component is loaded at a constant rate, $\dot{\sigma}_1$, held at the maximum proof stress, σ_p, for a time t_p, and then unloaded at a constant rate, $\dot{\sigma}_u$. Integrating the D_p^p equation for this typical proof stress cycle gives:

$$D_p = \sigma_p^{N_p} t_p + [\sigma_p^{N_p+1}/(N_p+1)]\,(1/\dot{\sigma}_u + 1/\dot{\sigma}_1) \qquad (8)$$

By coupling Equation (5) with Equations (2) and (3), the t_f and σ_f distribution after proof testing can be predicted for any service environment (4). Of particular interest is the minimum lifetime (t_{min}) in service of a component after proof testing. This is obtained by substituting S_{min} for S in Equation (2) so that:

$$t_{min} = BS_{min}^{N-2}\,\sigma_a^{-N} \qquad (9)$$

where now B and N are the appropriate crack parameters for the service environment.

Equations (5) - (9) summarize failure predictions for ceramic materials after proof testing. These failure predictions are dependent on the crack propagation parameters B and N. These parameters must be determined in a test environment that simulates the appropriate proof-test and service conditions and can be obtained from one of three type of experiments: crack velocity experiments, stress rupture experiments, and stressing rate experiments (3). From crack velocity experiments, Equation (1) is used to determine A and N; K_{IC} is determined in a separate experiment. Stress rupture data (t_f vs. σ_a) and the inert strength data are used to determine B and N from Equation (2), whereas stressing rate data (S vs. $\dot{\sigma}$) and median inert strength data are used to calculate B and N from Equation (3). The stress rupture and stressing rate techniques are commonly referred to as static fatigue and dynamic fatigue, respectively.

APPLICATION

The fracture mechanics principles described in the previous section have been experimentally verified for soda-lime glass (3) and alumina (5) in a moist environment. Specifically, these studies showed that the three experimental techniques for measuring the crack propagation parameters B and N give equivalent results, that the time-to-failure and strength failure probability distributions can be predicted from Equations (2) and (3) coupled with Equation (4), and that strength distributions after proof testing can be predicted on the basis of Equation (5). The purpose of this section is to describe three applications of fracture mechanics principles in assuring against the fatigue failure of ceramic components in service.

Failure Statistics

For alumina substrates fabricated by cofired metallurgy, stress corrosion cracking can be a serious problem. The metals most commonly cofired with alumina are either molybdenum or tungsten. Because these metals possess lower thermal expansion coefficients than alumina, their cosintering can cause the ceramic to retain residual tensile stresses of considerable magnitude. Subsequent cleaning and plating baths, necessary to render the metallurgy solderable, expose the substrate to conditions that promote stress corrosion cracking in the alumina. To statistically assess the susceptibility of polycrystalline alumina to stress corrosion cracking under processing conditions, the stressing rate technique was used to characterize the crack

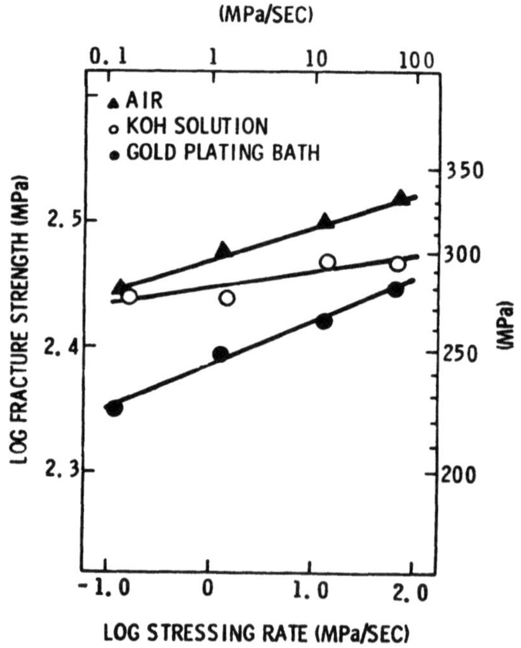

Figure 1. Dynamic fatigue data of polycrystalline alumina in a KOH cleaning solution at 100°C, in a gold-plating bath at 78°C, and in moist air at 30°C and 80% RH (from Ref. 6).

propagation parameters for alumina in a KOH cleaning solution at 100°C and a gold plating solution at 78°C (6).

Figure 1 gives the experimental results of strength as a function of stressing rate of alumina in the KOH and gold solutions (6). For comparison, the fatigue strength results determined for the same alumina in a moist air environment are included (5). From this data and a knowledge of the inert strength of the samples, the crack propagation parameters B and N were determined from Equation (3) for the various environments.

To validate the applicability of fracture mechanics principles in predicting failure probability in the processing solutions, the actual fatigue strength distributions were compared to predictions based on Equation (3) using the appropriate values for B and N and the inert strength distribution of the samples. Figure 2 gives these comparisons and it is evident that agreement between the theoretical predictions and the actual fatigue strength distributions

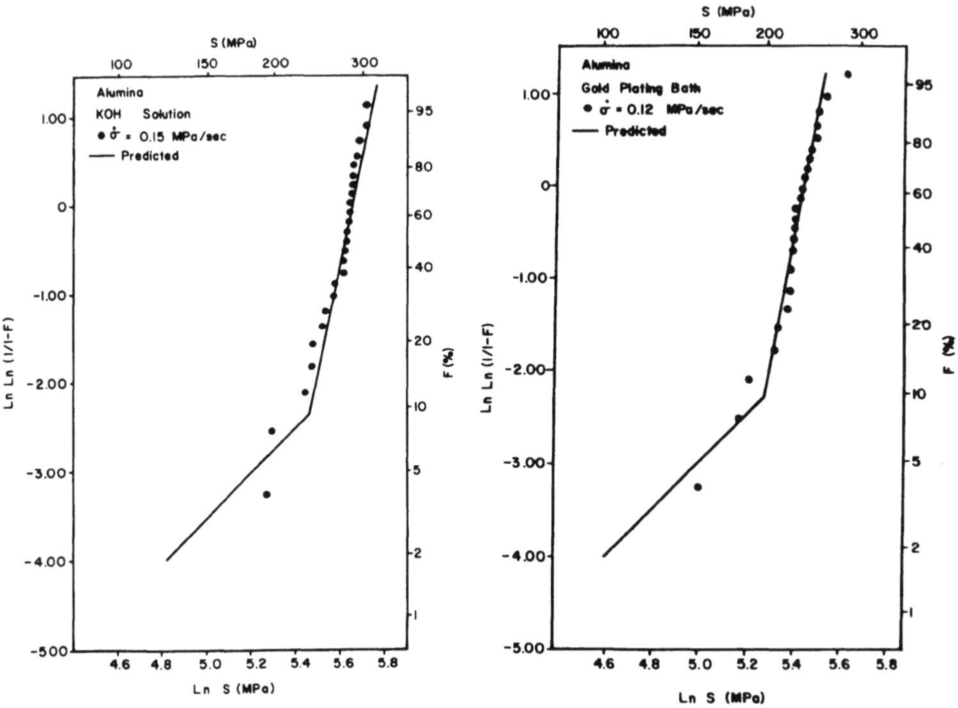

Figure 2. Comparison to theoretical predictions of the strength distributions of polycrystalline alumina in a) KOH cleaning solution at 100°C and b) gold plating bath at 78°C (from ref. 6).

is good. It should also be noted that these strength distributions are bimodal since the low strength regime of the distribution was caused by gross flaws induced in the samples during grinding.

The severity of the processing conditions on the alumina substrates can be most easily assessed by means of a lifetime prediction diagram (Figure 3) based on Equation (2) and the inert strength distribution of the substrates. Although the strengths of the substrates could not be measured directly, they were estimated from the strength distribution of the samples and the Weibull size relation:

$$S_1 = S_2 \ (A_2/A_1)^{1/m} \tag{10}$$

where S_1 and S_2 are the strengths of the substrates and samples, respectively, and A_1 and A_2 are the surface area of the substrates and samples, respectively. Figure 4 shows the estimated inert strength distribution of the substrates. It should be noted that it was assumed that the inert strength distribution of the

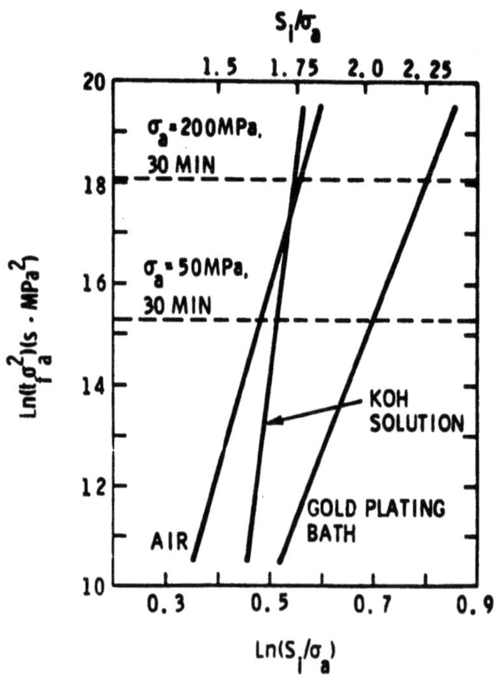

Figure 3. Lifetime prediction diagram for polycrystalline alumina in wet processing environments and moist air (from ref. 6).

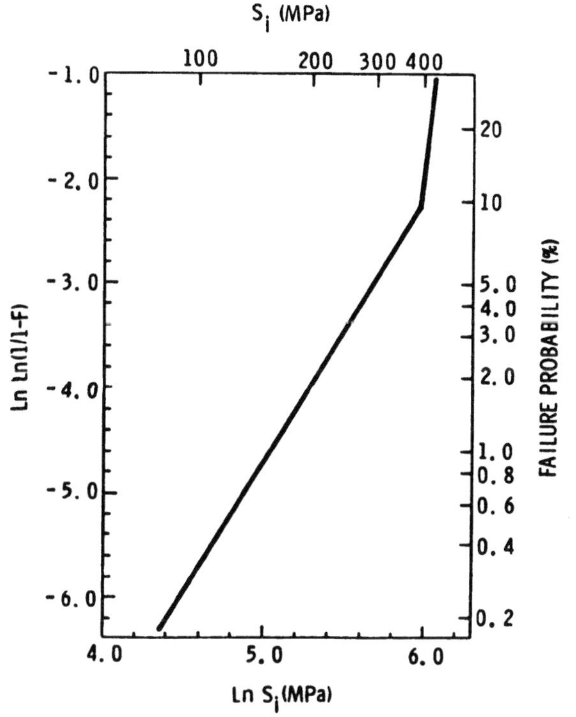

Figure 4. Estimated Weibull inert strength distribution for alumina substrates (from ref. 6).

substrates was not altered through any interaction between the metal and alumina during sintering.

The residual tensile stress in a substrate having cofired metallurgy is dependent on the thermal expansion mismatch between the metallurgy and the ceramic, the temperature difference between the set point of the system and room temperature, and the geometry of the metallurgy in the system. In molybdenum-alumina substrates residual stresses of 200 MPa can be generated. If substrates having such stresses are subjected to the KOH cleaning solution or the gold-plating solution for 30 minutes, respective S/σ_a ratios of 1.75 and 2.27 are predicted (Figure 3). Since σ_a is known, S can be calculated and a failure probability determined from Figure 4. With a 200 MPa residual stress, an 8% failure rate would be expected for the substrates in the 10% KOH solution, and 40% of the substrates in the gold-plating bath would crack after 30 minutes. Also, substrates having a residual stress of this magnitude would have a failure rate of 8% in hot, humid air. Clearly, failure rates of this magnitude are unacceptable.

The failure probability of the substrates can be lowered by: (1) using a stronger alumina to make the substrates, (2) reducing the processing times, or (3) reducing the residual stress in the substrates. The obvious choice is to reduce the residual stress in the ceramic. This can be accomplished by reducing the thermal expansion mismatch between the metal and the ceramic, by altering the geometry of the system, or by a combination of both. Since the thermal expansion differential can be reduced only by altering the metal configuration and/or the ceramic, this choice alone is not particularly attractive. Similarly, if the stress is lowered only through geometric redesign, the potential geometric density of the metallurgy in the system becomes limited. Thus, a combination of both approaches is generally considered most effective in reducing stress.

Obviously, the ideal solution would be to completely eliminate any residual tensile stress in the substrates; this, however, is not always possible without drastically altering the material system. The problem then is to reduce the failure probability of the substrates to a magnitude compatible with good product reliability.

Regardless of the technique used, if the stress in the ceramic is lowered, for instance, to 50 MPa, the lifetime prediction diagram and the inert strength distribution can again be used to assess the probability of stress corrosion cracking of the substrates. If substrates having a residual stress of 50 MPa are placed in the KOH cleaning solution or in the gold plating bath for 30 minutes, it can be determined from Figure 3 and 4 that the failure probabilities are 0.2 and 0.34% respectively. While the possibility of stress corrosion cracking of substrates during processing is not completely eliminated by reducing the residual stress to 50 MPa, the resultant failure rates are now low enough so as to offer a minimal impact on product reliability.

The above examples of lifetime predictions serve to illustrate the importance of fracture mechanics theory in designing electronic substrates fabricated by cofired metallurgy, namely, that substrates of various designs can be compared for the probability of stress corrosion cracking due to the residual tensile stress in the substrates. In this way, fracture mechanics theory is useful in making more rational design decisions. These predictions can also be checked by fabricating substrates of several designs with different residual stresses, and then subjecting the substrates, for instance, to the gold-plating bath for a given period of time. By comparing the incidence of stress corrosion cracking with that predicted by fracture mechanics theory, the theory can be experimentally validated. This was done in the present research program, and the experimental results generally agreed with those theoretically predicted (6).

Proof Testing

Overspeed proof testing is used in the grinding wheel industry to assure individual wheel safety by providing protection against excessive variability in wheel strength and fatigue failure. Empirical standards have been established for overspeeding on the basis of measurements of relative strengths and experience from wheel usage. To place overspeed proof testing on a more sound, fundamental basis, the fracture mechanics principles described in the previous section were applied to verifying the use of overspeed proof testing in assuring against the fatigue failure of vitrified grinding wheels (7). Two vitrified wheel specifications were used in this study: 17A90-L5-VX2 and 2A601-K4-V9676. For purposes of identification these two specifications will be labelled VGW1 and VGW2, respectively.

To measure the crack growth parameters appropriate for VGW1 and VGW2 in a commercial coolant, stress rupture and stressing rate experiments were used (7). Figure 5 compares minimum lifetime predictions for these two vitrified grinding wheels based on the two test techniques (in this case labelled static fatigue and dynamic fatigue). This figure is based on Equation (9) where S_{min} is taken to be equal to σ_p. Since the maximum applied stress in service for these two vitrified grinding wheels was calculated to be 1280 psi and the desired minimum lifetime in service was 6 weeks (7), it is evident from Figure 5 for a given vitrified grinding wheel composition the σ_p/σ_a ratio based on the two test techniques are not significantly different. From the σ_p/σ_a ratio the overspeed proof test could then be calculated.

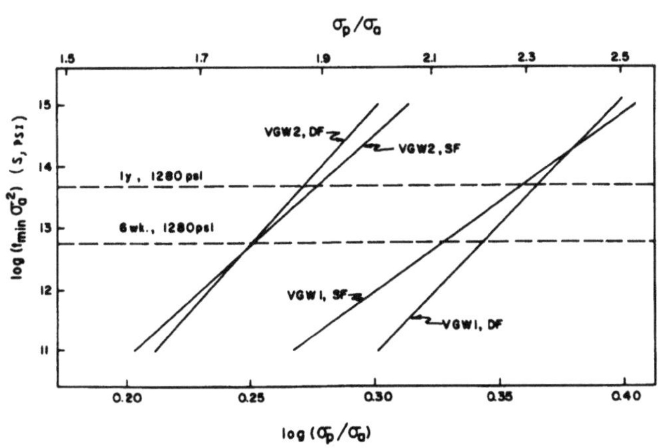

Figure 5. Minimum lifetime predictions comparing static fatigue (SF) and dynamic fatigue (DF) techniques for VGW1 and VGW2 in aqueous coolant at 35°C (from reference 7).

Table 1

Prediction of the Spin-Proof-Ratio (SPR) Necessary to Insure Against Delayed Failure in Service of Vitrified Grinding Wheel for a 6 Week Lifetime at 1280 psi. (after Reference 7)

Material	Test Technique	σ_p (psi)	SPR
VGW1	Static Fatigue	2715	1.57
VGW1	Dynamic Fatigue	2825	1.60
VGW2	Static Fatigue	2285	1.45
VGW2	Dynamic Fatigue	2280	1.45

Table I summarized these results where the spin-proof-ratio (SPR) is defined to be the overspeed proof divided by operating speed. These predicted SPR for the vitrified grinding wheels evaluated in this study can be compared to those actually used: 1.80 for VGW1 and 1.50 for VGW2. It is evident from these results that the SPR actually used for these wheel specifications is greater than that predicted on the basis of fracture mechanics theory. Therefore, it is believed that the overspeed proof test currently used for these vitrified grinding wheel compositions is adequate for preventing fatigue failure in service. In support of this conclusion it is significant to note that there has never been a known fatigue failure in service for these wheel specifications.

To further demonstrate the effectiveness of proof testing, the distribution of initial wheel bursting strengths for the wheel specification VGW2 was compared to that after overspeed proof testing (7). The tested wheels had dimensions 20 x 1 x 5 in. For the proof test the wheels were accelerated in air up to the proof speed of 4550 RPM and then quickly decelerated. During proof testing 3 of the 20 wheels burst. Figure 6 compares the initial wheel bursting strengths to the after-proof strengths. The wheel bursting strength was taken to be the tensile stress present at the wheel arbor when reached its burst speed. It is quite evident that proof testing eliminated the weak samples, resulting in the after-proof strength distribution being stronger than the initial distribution. It is also important to note the good agreement between theory and experiment.

Characterization of High Temperature Fatigue Behavior

The increasing demand for materials that will perform under temperature conditions too severe for metals has led to many possible applications for ceramics. Thus, it is important to

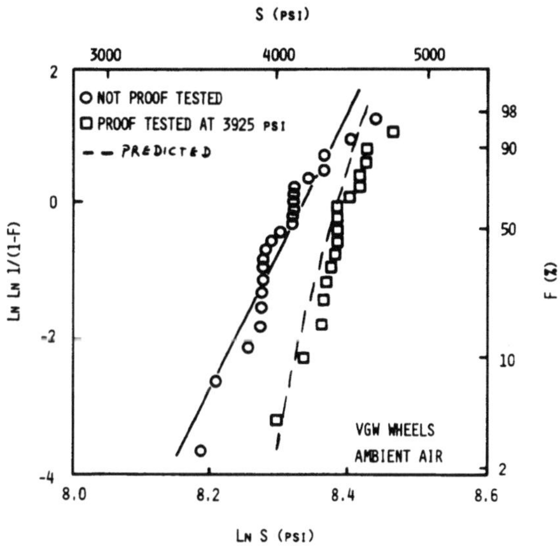

Figure 6. Comparison of wheel bursting strengths before and after proof testing for wheel specification VGW2 (from ref. 7).

characterize the strength and fatigue behavior of ceramics at elevated temperatures to establish allowable stress levels for component design. To study the high temperature strength and fatigue behavior of high purity alumina (Wesgo AL 995), the stressing rate test technique was used because of its simplicity (8).

The effect of temperature on the strength and fatigue behavior of high purity alumina is given in Figure 7 (8). The temperature dependence of both strength and fatigue behavior is small up to 500°C. From 800 to 1100°C, both the strength and fatigue resistance decreases markedly. A convenient way to show the effects of fatigue is to compare minimum lifetime predictions, based on Equation (9) taking S_{min} to be equal to σ_p, as a function of temperature. Figure 8 gives these predictions and clearly illustrates that there is little difference in the fatigue behavior of high purity alumina from 23 to 500°C, however, at higher temperatures fatigue effects become very important. For example, to assure a minimum lifetime of one year under a service stress of 100 MPa the required proof-stress ratio is 2.27 for a temperature of 23°C, but for 1000°C it is 6.90 and for 1100°C it is 14.91. These elevated temperature proof stress ratios would result in such large proof stresses that no sample would be able to pass the proof test. To get more reasonable proof stresses, the expected minimum life or allowable stress in service must be decreased for these elevated temperatures. For example, if the allowable stress is reduced to 30 MPa, then the proof stress to insure a minimum lifetime of 1 year at 1000°C is 179 MPa and 361 MPa at 1100°C.

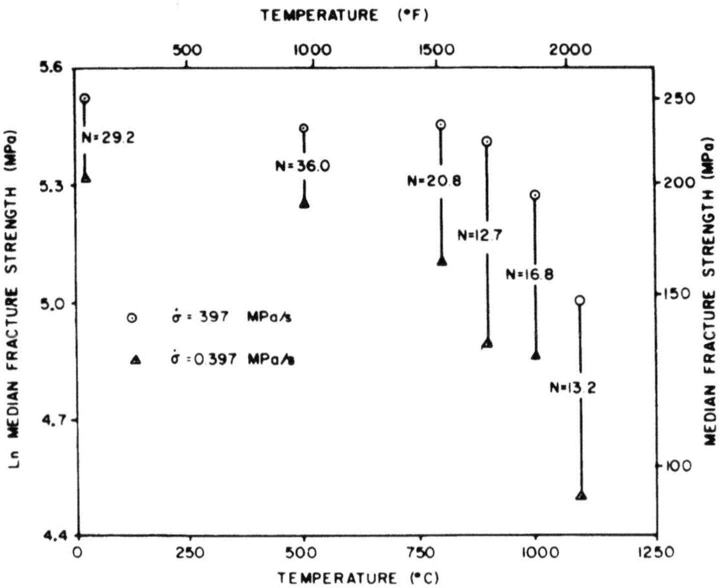

Figure 7. Median fracture strength of alumina as a function of temperature and stressing rate (from ref. 8).

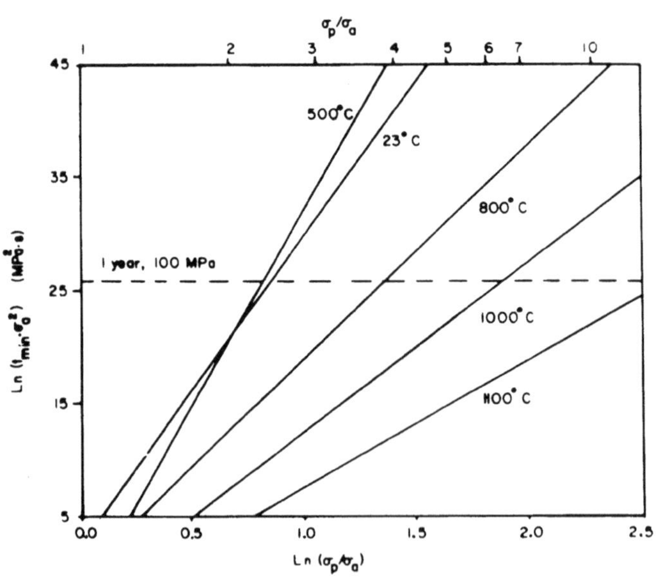

Figure 8. Minimum lifetime prediction diagram for alumina as a function of temperature (from ref. 8).

Scanning electron micrographs showed that the fracture in this high purity alumina changed from transgranular at 23°C to intergranular at temperatures above 800°C (8). In addition, a glassy phase was found at the fracture origins of the samples tested at above 800°C, giving evidence that the glassy phase is playing a major role in determining the strength and fatigue behavior of high purity alumina at elevated temperatures. However, proof test results (8) showed that the preexisting flaw population is not altered at elevated temperatures. Figure 9 compares the strength distribution of alumina at 23 and 1100°C before and after proof testing at room temperature. It is evident, that this proof test eliminated the weak samples so that the after proof strength distribution at low failure probabilities are stronger than the initial distributions and that agreement between theory and experiment is good. On the other hand, Figure 10 shows that proof testing alumina at 1000°C is not effective in improving the strength distribution at 1000°C, however, the good agreement between experiment and theory gives evidence that fracture mechanics theory applies. In this case weak samples are eliminated by proof testing at 1000°C, but flaw growth during the proof test weakens the survivors to such a degree that proof testing is not effective in improving the strength distribution.

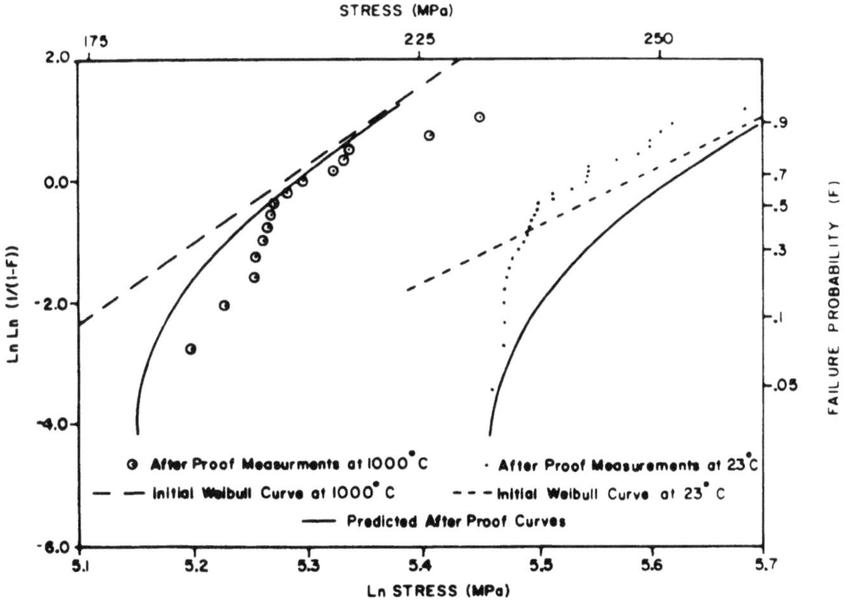

Figure 9. Effect of room temperature proof testing on the strength distribution of alumina at 23°C and 1000°C (from ref. 8).

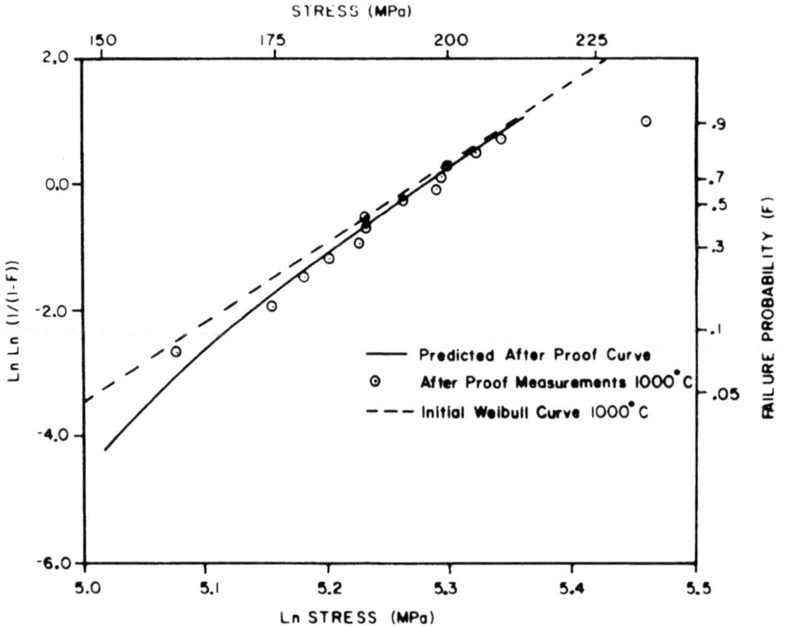

Figure 10. Effect of proof testing at 1000°C on the strength distribution of alumina at 1000°C (from ref. 8).

In summary, it is believed that at moderate temperatures the fatigue failure of alumina is caused by moisture assisted subcritical crack growth; whereas, at elevated temperatures it is caused by subcritical crack growth enhanced by the presence of a glassy phase. It is also thought that the same pre-existing flaws control the strength both at room temperature and at elevated temperatures.

LIMITATIONS OF THE LIFETIME PREDICTION METHOD

The lifetime prediction theory presented in this paper is based on the assumption that subcritical crack growth from pre-existing flaws is the only mechanism of strength degradation. Using this assumption, component lifetime can be predicted once the initial flaw population, the crack growth rate and the critical crack size have been determined. The theory is deterministic because the element of chance is eliminated from the lifetime prediction once these parameters have been evaluated. The accuracy of the lifetime prediction will depend on both the accuracy with which the pertinent parameters can be evaluated (9-11), and the adequacy of the crack growth equation (12). This requirement does not detract from the basic deterministic nature of the method. The prediction scheme can, however, be invalidated when the basic assumptions of the theory are not satisfied. Thus, fracture processes, other than subcritical crack growth, can invalidate failure predictions based on this theory. Unfortunately, other fracture processes do occur for materials currently under consideration for use in heat engines, so that alternate mechanisms of failure must be considered for any scheme of predictive failure for heat engine components. As will be shown, proper modeling of these factors permits extension of the theory to handle fracture situations that are not currently included in the method of lifetime prediction.

As with most other ceramic materials, silicon nitride and silicon carbide fail at low temperatures from preexisting flaws. As discussed by Rice et al. (13,14) and by Richerson and Yonushonis (15), these flaws are usually surface cracks introduced into silicon nitride and silicon carbide components by machining when the components are manufactured. For hot-pressed silicon nitride at temperatures less than ∿1000°C, fracture originates from surface flaws in a completely brittle manner (13,14). At higher temperatures, however, plasticity of components and effects resulting from the reactive nature of the test environment intervene to alter the mode of fracture (13). The same types of effects occur for hot-pressed silicon carbide at higher temperatures (∿1400°C). In oxidizing environments, the effects of temperature on mechanical behavior can be attributed to: (1) enhanced plasticity at the crack tip which results in subcritical crack growth by creep fracture (16,17), and (2) surface oxidation which results in the healing of surface machining cracks, and for some materials, the generation of new flaws that act as new origins for fracture (13-15, 18-20). Both processes have to be considered to accurately predict the lifetime of high temperature structural components. A variety of other processes (thermal shock, particle impact, chemical corrosion, component contacts, etc.) can also intervene to effect the fracture behavior of heat engine ceramics. Although a complete theory of failure would require all of these processes to be considered, such extensive considerations are clearly beyond the scope of this paper. The discussion in the remainder of this paper will be more limited in scope, dealing with pit formation, crack healing and crack growth. Materials to be considered are magnesia-doped, hot-pressed, silicon nitride and reaction bonded silicon nitride for which the processes that occur at elevated temperatures are best understood.

High temperature oxidation has been shown to be particularly effective in altering the initial flaw population in hot-pressed, silicon nitride (NC132). Investigations of the strength and oxidation behavior of this material have shown that surface induced machining flaws are reduced in severity by high temperature oxidation of the ceramic surface (13-15, 18-20). The oxidation process is complex involving mass transport of impurities from the matrix, and transport of oxygen and nitrogen through the glass oxide layer that forms on the surface of the ceramic (17,21). As the oxide scale grows, the initial silicon nitride surface recedes, reducing the size of the surface cracks and thus, their severity (Figure 11). If oxidation continues long enough, the surface cracks can be completely removed from the ceramic surface. In addition to crack removal by dissolution, plastic deformation at elevated temperatures also permits blunting of machining flaws and relief of surface stresses, thus, enhancing the strength of the component (22,23). The effectiveness of oxidation in reducing flaw severity can be illustrated by the introduction of fresh cracks into the surface of silicon nitride by indentation (20,24). As illustrated in Table 2 on hot-pressed silicon nitride (billet B), exposure at 1200°C in air increased the strength from 408 MPa at room temperature to 488 MPa after 100 hours exposure, largely as a result of decreasing the severity of the flaw associated with the indentation. Because of the beneficial effect of oxidation, high temperature exposure in air is being used as a means of improving the finish of hot-pressed silicon nitride ceramic blades (15,23). Similar strengthening results are observed for reaction bonded silicon nitride (NC350) (23,24).

EFFECT OF OXIDATION ON

FLAWS IN Si_3N_4

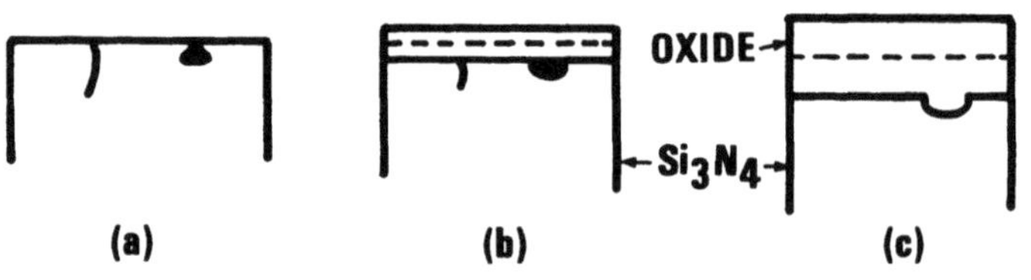

Figure 11. Effect of exposure at 1200°C on flaws in silicon nitride: (a) surface crack and subsurface inclusion before exposure; (b) after exposure crack size is reduced by surface dissolution and subsurface flaw reacts with environment. At the same time, crack blunting or crack healing occurs; (c) oxide growth has removed the subsurface crack and enlarged the subsurface flaw to produce a pit in the silicon nitride (from ref. 34).

Table 2

Strength of Indented Specimens: 2kg Load (34)

Material	Test-Temp. °C	Exposure Conditions	Strength MPa	Fracture at indentation
NC132	25	as indented	397 ± 12	Yes
Billet A	25	16 hr 1200°C	432 ± 27	Yes
	25	100 hr 1200°C	461 ± 42	No
	1200	1/2 hr 1200°C	438 ± 15	Yes
	1200	16 hr 1200°C	396 ± 42	Yes
	1200	100 hr 1200°C	402 ± 12	No
NC132	25	as indented	408 ± 17	Yes
Billet B	25	100 hr 1200°C	524 ± 47	Yes
	1200	1/2 hr 1200°C	441 ± 10	Yes
	1200	100 hr 1200°C	488 ± 11	Yes
NC350	25	as indented	115 ± 24	Yes
	25	33 hr 1200°C	181 ± 22	Yes
	25	100 hr 1200°C	225 ± 38	Yes
	1200	100 hr 1200°C	188 ± 64	Yes
NCX-34	25	as indented	491 ± 16	Yes
Y_2O_3	25	100 hr 1200°C	616 ± 51	No
	1200	100 hr 1200°C	744 ± 25	Yes

If flaw removal were the only process occurring during oxidation, then the lifetime prediction methods presented earlier in this paper could be used as a conservative basis for component design. In this case, the starting flaws would decrease in size as oxidation occurred, and the component would become stronger, not weaker. Research on hot-pressed silicon nitride (NC132) has shown that flaw generation during long term (>100 hr.) exposure results in a decrease in the strength of this material (15, 20, 22). The strength decrease results from pit formation at the ceramic surface due to rapid localized oxidation. Although the exact cause of localized oxidation is not fully understood, it probably results from the presence of impurities in the ceramic which increase the oxidation rate as the oxide interface approaches the location of the impurity (Figure 11; 20,26). The formation of glass-like oxide mounds over the pits with holes in the center of the mounds suggests that rapid gas generation is associated with the nucleation and growth of these pits (20).

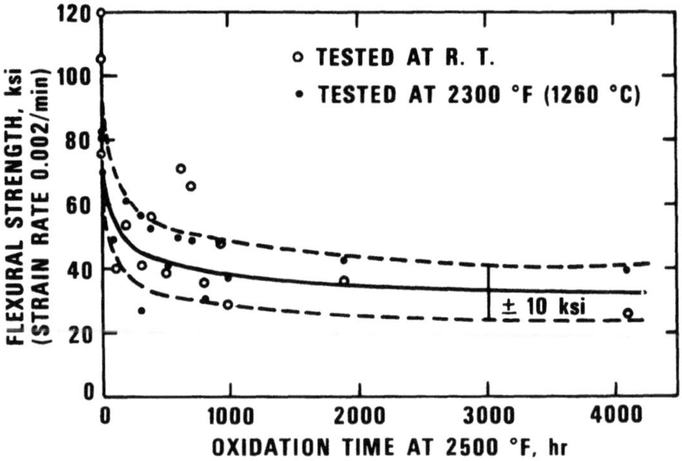

Figure 12. Effect of oxidation at 1371°C (2500°F) on the strength of silicon nitride (NC132). Note that at time zero the scatter in strength is considerably greater than that occurring after substantial periods of exposure (from ref. 25).

The effect of pit formation on the long term strength is illustrated in Figure 12 which shows that after 1000 hours exposure magnesia-doped, hot-pressed silicon nitride has lost approximately 50 percent of its strength (25). This strength decrease seems to occur most rapidly during the initial stages of exposure, suggesting rapid formation of pits initially, and then stabilization of the pit structure after periods of about 1000 hours. Whether or not the pit structure has reached a state of static equilibrium, or is in a state of dynamic equilibrium has not yet been determined.

In freshly machined specimens the combined effect of crack healing, and flaw generation should result in a strength enhancement followed by a strength degradation as pits form in the material (20,24). This type of behavior has been confirmed by a number of investigators. Richerson and Yonushonis (15) have shown, for example, that the strength of transverse ground silicon nitride is increased by exposure to temperatures of 982°C or 1066°C for 50 hours (Table 3). At higher temperatures or for longer periods of time, a strength decrease is observed as pits are formed in the surface. Wiederhorn and Tighe (20,24) showed that this type of behavior depended on the particular billet used for study, since some billets increased in strength

Table 3

Room Temperature Strength of Hot-Pressed, Silicon Nitride (NC132) Transverse Ground Specimens Tested in 4-Point Bending (15)

Exposure °C (°F) time	Strength and Standard Deviation MPa (KSI)	Samples Tested	Predominant Fracture Region
Control	450 ± 36 (63.0 ± 5.0)	12	Grinding Groves
982 (1800) 50 hours	657 ± 61 (92.0 ± 8.6)	50	Surface (Flaws not obvious)
1066 (1950) 50 hours	636 ± 41 (89 ± 5.7)	11	Surface (Flaws not obvious)
1129 (2065) 140 hours	593 ± 36 (83 ± 5.0)	6	Surface Pits
1129 (2065) 240 hours	443 ± 58 (62.0 ± 8.1)	6	Surface Pits
1204 (2200) 24 hours	529 ± 19 (74.0 ± 2.6)	12	Surface Pits
1371 (2500) 24 hours	457 ± 29 (64 ± 4.1)	12	Surface Pits

with exposure at 1200°C (for times of 100 hr.) while others decreased for the same exposure conditions. Richerson and Yonushonis (15) have made similar observations on hot-pressed, silicon nitride. More recently, the combined effect of flaw healing and pit formation was demonstrated by Jones and Rowcliff (22), for silicon nitride exposed at temperatures of 1370°C for periods up to 64 hours. For both indented (knoop; 1 Kg) and transverse ground specimens, the strength first increased then decreased as the exposure time increased (Figure 13).

Reaction bonded silicon nitride behaves in a manner quite different from that of the hot-pressed material. Flaw healing results in a significant strength increase in this material at elevated temperatures. When the material is cooled to room temperature, however, much of this increase is lost as a result of surface cracking caused by a phase transformation of cristobalite in the oxide coat that forms on the surface at high temperatures (27). The absence of strength degradation (at 1200°C)

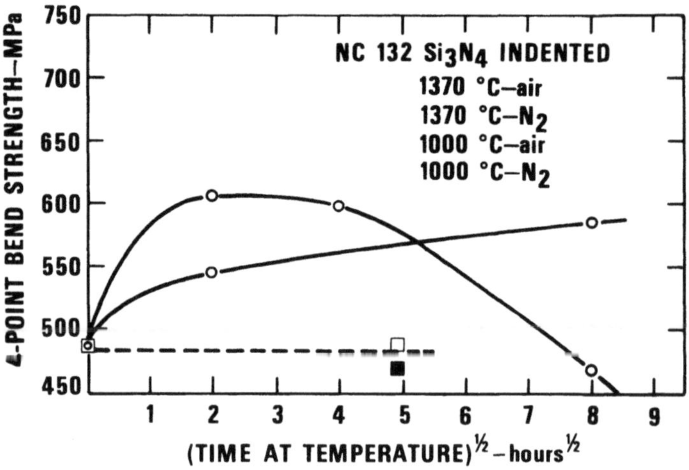

Figure 13. Effect of high temperature exposure on the strength of silicon nitride (NC132) measured at room temperature (from ref. 22).

after sustained (1000 hr.) exposures (28), suggests that pit formation and subcritical crack growth are not significant factors limiting the lifetime of this material.*

LIFETIME ESTIMATE: EFFECT OF FLAW HEALING AND PIT FORMATION

From the above discussion, we conclude that flaw healing, pit formation and subcritical crack growth must be considered simultaneously in order to establish an accurate method of lifetime prediction. The development of a failure prediction method requires each of these processes to be evaluated independently. Once they have been characterized, mathematical techniques can be used to determine the relative contribution of each of these processes to the total failure time.

Subcritical crack growth can be characterized either by strength or fracture mechanics techniques. However, when flaw healing or pit formation is occurring, the use of strength techniques for evaluating crack growth parameters may not be a good procedure, since both of these processes have an independent effect on the material's strength. The feasibility of quantifying the crack growth rate by fracture mechanics techniques has been demonstrated by Evans and Wiederhorn (30), who have shown that

*Recently, Richerson et al. (29) have observed pit formation in heavily oxidized reaction bonded silicon nitride.

the crack velocity can be represented as a power function of the applied stress intensity factor (Equation 1). Details on the application of fracture mechanics techniques to crack growth data can be found in ref. 31.

A second method of collecting crack velocity data was illustrated recently by Rowcliff (22), who used 3 point bend specimens to study the static fatigue of silicon nitride and silicon carbide. The specimens were all indented to form well-characterized cracks in the tensile surface of the specimens. The technique is based on the assumption that equation 1 is the correct form of the crack velocity equation. By rearranging Equation 2, the following equation is obtained for the failure time, t, of a component that is subjected to a constant applied load:

$$t = \frac{2a_i}{(n-2)v_i} \tag{11}$$

Thus, by measuring the initial crack size, a_i, and the time to failure, t, the initial crack velocity, v_i, can be determined. The stress intensity factor that corresponds to v_i is obtained directly from a_i ($K_{Ii} = \sigma Y\sqrt{a_i}$)*.

Data on flaw healing or on pit formation can be obtained from strength measurements. Data on magnesia-doped, hot-pressed silicon nitride and reaction bonded silicon nitride suggest that crack healing occurs relatively quickly and is completed before pits begin to form in the component surface. Therefore, strength changes during the early stages of component exposure can be used to characterize the crack healing process; strength changes occurring after extended exposure can be used to characterize the pit formation process. For normally machined surfaces, quantification of both of these processes requires the collection of extensive strength data so that the strength distribution can be determined accurately as a function of exposure time.

Data on hot-pressed silicon nitride suggests that both the mean strength and its standard deviation change with exposure time when flaw healing occurs (20,24,25). In terms of a Weibull distribution, Equation 4, both the Weibull strength, S_o, and the shape parameter, m, are a function of time. Data on reaction bonded silicon nitride suggests that only S_o is a function of

---

*The validity of the data obtained by this technique is limited by the assumption that the crack size at failure is small relative to the specimen dimensions. As has already been noted, this same limitation applies to earlier parts of this paper.

time (28). The effect of flaw healing on specimens that contain large initial cracks can be determined from studies on specimens that have been indented so that relatively large cracks are introduced into the specimen surface. Studies of this type have been conducted recently by Tighe and Wiederhorn (20,24) and by Cubicciotti et al. (22). The investigation by Singhal et al. (25) of strength degradation after long term exposure suggests that for both crack healing and pit formation S_o and m seem to approach fixed values after a certain amount of time, so that both distributions eventually become stable with regard to further exposure.

Once crack growth, flaw healing, and pit formation have been quantified, the combined failure probability, and the total time to failure can be estimated. To obtain these estimates it is necessary to develop a procedure to determine which mechanism controls failure at a given probability level. Flaw healing and pit formation as a combined failure mechanism will be discussed first. For simplicity it is assumed that the Weibull slope, m_1, for crack healing is less than the Weibull slope, m_2, for pit formation, and that m_1 and m_2 do not depend on time. The Weibull strength for crack healing, S_1, and that for pit formation, S_2, are assumed to follow exponential decay curves that can be fit to the following formulae:

(flaw healing) $$S_1 = S_{o1} + \Delta S_{o1}(1-\exp(-\alpha t)) \quad (12)$$

(pit formation) $$S_2 = S_{o2} - \Delta S_{o2}(1-\exp(-\beta t)) \quad (13)$$

where S_{o1}, S_{o2}, ΔS_{o1}, ΔS_{o2}, α and β are determined from experimental strength data. The initial Weibull strengths for flaw healing and pit growth are S_{o1} and S_{o2} respectively, while ΔS_{o1} and ΔS_{o2} are the changes in strength for these two processes after extended exposure times. α and β determine the rate at which these changes in strength occur. If Equations 12 and 13 are substituted into Equation 4, time dependent strength distributions are obtained that describe the effect of exposure time on strength.

The type of behavior expected from Equations 12 and 13 is shown schematically in Figure 14. We see that during the initial stages of exposure, failure is determined entirely by the initial flaw population, while in the later stages of exposure, pit formation dominates the strength distribution. At any point in time, the combined failure probability, F, is given by the following relation (32):

$$(1-F) = (1-F_1)\,(1-F_2) \quad (14)$$

where F_1 and F_2 are the time dependent failure probabilities for flaw healing and pit formation respectively. Thus, a unique

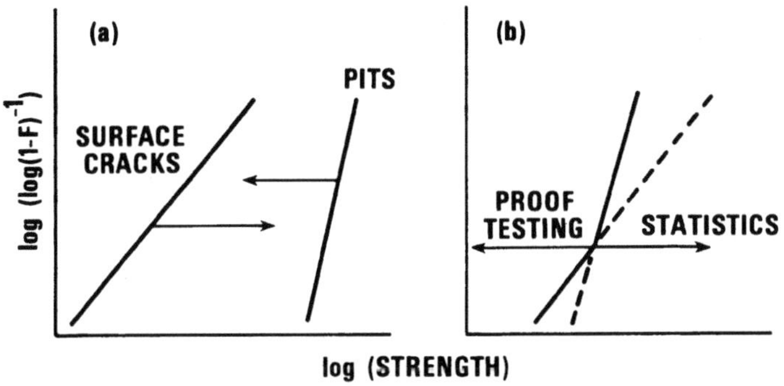

Figure 14. Schematic representation of the effect of high temperature exposure on the strength distribution: (a) initial strength distribution. Curves are assumed to move in the direction of the arrows as the exposure time is increased; (b) distribution after the flaw structure has reached a state of "equilibrium". Proof testing or nondestructive evaluation can be used to improve reliability at stresses lower than the point of intersection of the two lines, statistics can be used to the right of the point of intersection.

equation is obtained that relates the failure strength, S, to the failure probability, F.

With regard to lifetime prediction, the above discussion is sufficient to draw several important conclusions with regard to proof testing and nondestructive testing. At any time during exposure, the strength of components to the left of the point of intersection of the two curves shown in Figure 14 will be controlled by the initial flaw population. Those specimens that lie to the left of the intersection will contain the largest flaws and consequently will be the ones to be eliminated by proof testing or nondestructive inspection. Since the initial flaw population controls the strength in this regime, proof testing and nondestructive evaluation techniques can be used to truncate the population, and the theory presented earlier in this paper can be used to provide a conservative estimate of the component lifetime. The intersection of the two curves shown in Figure 14 provides an upper limit to the proof test load that can be usefully employed to truncate the initial flaw population. Above this load, component strength depends not on the flaws initially present in the component, but on those that are generated by exposure to elevated temperatures (20,24).

After long periods of exposure, proof testing or nondestructive evaluation can be applied again over the entire range of strengths provided the flaw distribution (pits and cracks) has reached a static state. If, however, the distribution to the right of the intersection in Figure 14 represents a dynamic state of equilibrium, then new pits will be continuously generated as the old ones are removed. As a consequence, nondestructive evaluation and proof testing will not be a viable lifetime prediction technique. For the dynamic equilibrium situation, statistical techniques such as those described earlier in this paper would have to be used for purposes of lifetime prediction. Pits generated by the oxidation process would serve as the initial flaw distribution for the lifetime prediction.

The combined effects of subcritical crack growth and pit formation on strength can be determined by considering the rate at which each of these processes degrade the strength. As long as strength degradation due to pit formation exceeds that resulting from crack growth, subcritical crack growth will not play a major role in the strength degradation process; when the converse is true, crack growth will dominate the strength degradation process. The switch from one controlling mechanism to the other will occur when the two strength degradation rates are equal.

The rate of strength degradation due to pit formation can be determined by substituting S_2 of equation 13 for S_o of Equation 4, and differentiating with respect to time:

$$dS/dt = -\beta \, \Delta S_{o_2} [\exp(-\beta t)] \, [\ln(1-F)^{-1}]^{1/m} \tag{15}$$

The rate of strength degradation due to crack growth has been derived by Fuller et al. (4) and is given by the following expression:

$$dS/dt = - (AY^2 K_{IC}{}^{n-2}/2) \, (\sigma/S)^n S^3 \tag{16}$$

where A and n are defined by equation 1; Y is a geometric constant; σ is the applied stress and S is the instantaneous strength. Equating Equations 15 and 16, and substituting Equations 4 and 13 for S, an equation can be obtained which gives the time, t_p, at which the mechanism of strength degradation changes from pit formation to crack growth. The total time to failure is then equal to t_p, plus the time, t_c, for the crack to grow to a critical size. t_c can be determined from Equation 2, (using Equations 4 and 13 to evaluate the strength at which the degradation mechanism changes from pit formation to crack growth).

Although the mathematical scheme for obtaining the total time to failure is straightforward, the equations are not included here because of their complexity. Never-the-less, the procedure does provide a rational method of obtaining the total time to failure when pit formation plays a critical role in determining component lifetime. It is worth noting that the final equation can be expressed in terms of strength, probability and failure time. This type of representation was first suggested by Davidge et al. (33) as a method of design with ceramic materials.

SUMMARY

In this paper, a review is presented of current techniques used to evaluate the lifetime capability of structural ceramics. Three successful applications of these techniques are presented and discussed in order to demonstrate the viability of the techniques for purposes of engineering design. Finally, limitations of the techniques are discussed with regard to heat-engine applications. An extension of theory is recommended to treat crack healing and pit formation resulting from high temperature exposure.

The theory presented in this paper is based on the theory of linear-elastic, fracture mechanics. Failure of structural ceramics is assumed to be the result of subcritical crack growth from pre-existing flaws in the structural components. Lifetime predications require evaluation of the initial flaw size and of the rate at which cracks grow when subjected to applied loads. The three techniques that can be used to evaluate the initial flaw size are nondestructive evaluation, proof-testing and statistics, while the three techniques that can be used to evaluate the rate of crack growth are fracture mechanics techniques, static fatigue techniques (stress rupture) and dynamic fatigue techniques (constant loading rate to failure). The critical equations required for purposes of failure prediction are presented in the paper. Successful applications of the lifetime prediction method are illustrated for electronic-substrate ceramics, for vitreous grinding wheels and for high-purity aluminum oxide at elevated temperatures.

Potential limitations in applying the technique to heat engine ceramics, such as silicon nitride and silicon carbide, are discussed. These limitations arise because of flaw generation and flaw healing that is observed to occur in these materials at elevated temperatures. Flaw generation can be caused by a number of processes: oxidation or other environmental attack which results in surface pit formation, particle impact or component contact which results in localized stresses that nucleate cracks, and thermal shock which results in transient stresses that can result in crack formation. Flaw healing can be attributed to surface dissolution caused by oxidation and silicate glass formation. These processes change the initial flaw population so that failure no longer occurs from machining flaws originally present in the ceramic. In this situation the development of a model for failure prediction requires the consideration of flaw generation processes.

Extension of the theory presented in earlier sections of the paper is illustrated for the combined processes of flaw healing, pit formation and subcritical crack growth. The method independently evaluates these three processes and then determines the relative contribution of each of these processes to the total failure time. Conditions are established for proof testing of as-received components. Because of the occurrence of pit formation, a maximum value for the proof test load is established, above which proof testing does not contribute to component reliability. After long term exposure, proof testing or nondestructive evaluation can be applied again provided the flaw (pit) distribution has reached a state of static equilibrium. If, however, the flaw distribution has reached a state of dynamic equilibrium, then statistical techniques must be applied for lifetime prediction. Assuming that the strength distribution of pits can be represented by a two parameter Weibull distribution, the application of statistics is discussed for the combined failure modes of pit formation and crack growth. For these conditions, equations are developed for determining the total failure time.

ACKNOWLEDGEMENTS:

One of the authors (JER) is pleased to acknowledge the support of the Office of Naval Research, Metallurgy and Ceramics Program, the other authors are pleased to acknowledge the support of the Department of Energy, Division of Fossil Fuel Utilization.

REFERENCES

1. A.G. Evans and S.M. Wiederhorn, "Proof Testing of Ceramic Materials-An Analytical Basis for Failure Prediction," Int. J. Fract., 10, 379-92 (1974).
2. S.M. Wiederhorn, "Reliability, Life Prediction, and Proof Testing of Ceramics," pp. 635-65 in Ceramics for High Performance Applications, ed. by J. J. Burke, A.E. Gorum, and R.N, Katz, Brook Hill Pub. Co., Chestnut Hill, MA. (1974).

3. J.E. Ritter, Jr., "Engineering Design and Fatigue Failure of Brittle Materials," pp. 667-686 in Fracture Mechanics of Ceramics, Vol. 4, ed. by R.C. Bradt, D.P.H. Hasselman, and F. Lange, Plenum Press, NY (1978).
4. S.M. Wiederhorn, E.R. Fuller, Jr., J.E. Ritter, Jr., and P.B. Oates, "Proof Testing of Ceramics: I. Experiment" NBSIR 79-1934; "II Theory," NBSIR 79-1944, (December 1979).
5. J.E. Ritter, Jr. and J.N. Humenik, "Static and Dynamic Fatigue of Polycrystalline Alumina," J. Mat. Sci., 14, 626-632 (1979).
6. J.N. Humenik and J. E. Ritter, Jr., "Susceptibility of Alumina Substrates to stress Corrosion Cracking During Wet Processing," to be published, Bull. Am. Ceram. Soc.
7. J.E. Ritter, Jr. and S.A. Wulf, "Evaluation of Proof Testing to Assure Against Delayed Failure," Bull. Am. Ceram. Soc. 57, 186-190 (1978).
8. K. Jakus, T. Service, and J.E. Ritter, Jr., "High Temperature Fatigue Behaviour of Polycrystalline Alumina," to be published, J. Am. Ceram. Soc.
9. S.M. Wiederhorn, E.R. Fuller, Jr., J. Mandel, and A.G. Evans, "An Error Analysis of Failure Prediction Techniques Derived from Fracture Mechanics," J. Am. Ceram. Soc. 59, 403-11 (1976).
10. D. F. Jacobs and J.E. Ritter, Jr., "Uncertainty in Minimum Lifetime Predictions," J. Am. Ceram. Soc. 59, 481-486 (1976).
11. P.N. Thorby, "Experimental Errors in Estimating Times to Failure," J. Am. Ceram. Soc. 59, 514-16, (1976).
12. S.M. Wiederhorn, "Dependence of Lifetime Predictions on the Form of the Crack Propagation Equation," pp. 893-901 in Fracture 1977, Vol. 3, D.M.R. Taplin, Ed., University of Waterloo Press, Waterloo, Ontario, Canada (1977).
13. R.W. Rice, S.W. Freiman, and J.J. Mecholsky, Jr., "Fracture Sources in Si_3N_4 and SiC," pp. 665 in Proceedings of the 1977 DARPA/NAVSEA Ceramic Gas Turbine Demonstration Engine, J.W. Fairbanks and R.W. Rice, Eds., Metal and Ceramics Information Center Report MCIC-78-36, Battelle Columbus Laboratories, Columbus, Ohio (1978).
14. S.W. Freiman, C. Cm. Wu, K.R. McKinney, and W.J. McDonough, pp 655-663 in ref. 13.
15. D.W. Richerson and T.M. Yonushonis, "Environmental Effects on the Strength of Silicon-Nitride Materials," pp. 247-71 in ref. 13.
16. F.F. Lange, "Evidence for Cavitation in Crack Growth," J. Am. Ceram. Soc. 62 222-3 (1979).
17. N.J. Tighe, "Structure of Slow Crack Interfaces in Silicon Nitride," J. Mater. Sci. 13, 1455-63 (1978).

18. S.W. Freiman, A. Williams, J. J. Mecholsky and R.W. Rice, "Fracture of Si_3N_4 and SiC," pp. 824-34 in Ceramic Microstructures '76, R.M. Fulrath and J.A. Pask, Eds., Westview Press, Boulder, Colo. (1977).
19. S.W. Freiman, J.J. Mecholsky, W.J. McDonough, and R.W. Rice, "Effects of Oxidation on the Room Temperature Strength of Hot-Pressed Si_3N_4-MgO and Si_3N_4-ZrO_2," pp. 1069-76 in Ceramics for High Performance Applications," J.J. Burke, E.M. Lenoe, and R.N. Katz, Brook Hill Publishing Co., Chestnut Hill, Mass. (1978).
20. S.M. Wiederhorn and N.J. Tighe, "Proof-Testing of Hot-Pressed Silicon Nitride," J. Mat. Sci. 13, 1781-93 (1978).
21. D. Cubicciotti and K.H. Lau, "Kinetics of Oxidation of Hot Pressed Silicon Nitride Containing Magnesia," J. Am. Ceram. Soc., 61 512-7 (1978).
22. D.D. Cubicciotti, R.L. Jones, K.H. Lau, and D.J. Rowcliffe, "High Temperature Oxidation and Mechanical Properties of Silicon Nitride," Interim Scientific Report 5522-2, Nov. 15, 1978, Prepared for Air Force Office of Scientific Research/NE, SRI International.
23. T.M. Yonushonis and D.W. Richerson, "Strength of Reaction-Bonded Silicon Nitride," pp.219-233 in ref. 13.
24. S.M. Wiederhorn and N.J. Tighe, "Effect of Flaw Generation on Proof-Testing," pp 689-700 in ref. 13.
25. A.F. McLean, E.A. Risher, R.J. Bratton and D.G. Miller, pp 133-134 in Brittle Materials Design, High Temperature Gas Turbine, technical report AMMRC CTR 75-28 to the Army Materials and Mechanics Research Center, Watertown, Mass., October 1975.
26. J.A. Rubin, "Probable Causes of Pitting in Hot-Pressed Si_3N_4 Ceramics," pp. 739-743 in ref. 13.
27. R.W. Davidge, A.G. Evans, A.G. Gilling and P.R. Wilgman, "Oxidation of Reaction Sintered Silicon Nitride and Effects on Strength," in Special Ceramics, 5, 329-44 (1972).
28. N.J. Tighe, to be published.
29. D.W. Richerson, private communication.
30. A.G. Evans and S.M. Wiederhorn, "Crack Propagation and Failure Prediction in Silicon Nitride at Elevated Temperatures," J. Mat. Sci., 9, 270-278 (1974).
31. A.G. Evans, "Fracture Mechanics Determinations," pp.17-48 in Fracture Mechanics of Ceramics, Vol. 1, R.C. Bradt, D.P.H. Hasselman and F.F. Lange, eds., Plenum Press, New York (1974).
32. N.R. Mann, R.E. Shafer and N.D. Singpurwalla, Methods for Statistical Analysis of Reliability Data, John Wiley and Sons, New York (1974).

33. R.W. Davidge, J.R. McLaren and G. Tappin, "Strength-Probability-Time (SPT) Relationships in Ceramics," J. Mat. Sci. 8, 1699-1705 (1973).

34. N.J. Tighe and S.M. Wiederhorn, "Fracture of Brittle Materials at High Temperatures," Air Force Materials Laboratory Technical Report: AFML-TR-78-83, July 1978.

MATERIAL PARAMETERS FOR LIFE PREDICTION IN CERAMICS

R. K. Govila

Ceramic Materials Department
Scientific Research Staff
Ford Motor Company
Dearborn, Michigan 48121

ABSTRACT

A detailed experimental study of critical material parameters needed for life prediction methodology in hot-pressed silicon nitride, NC132, has been made. The primary experimental techniques were double torsion and indentation induced flaw methods to determine the relationship between crack velocity, V, and stress intensity, K, during subcritical crack growth.

The subcritical crack growth exponent 'n' was determined using flexural stress and strain rate methods and stress rupture methods, and showed a wide scatter in magnitude. When all the relevant life prediction parameters such as inherent flaw size, strength, critical stress intensity factor, and K-V relationship for slow crack growth are known, an estimate of time to failure for a given applied stress, temperature and environment can be made using the numerical relationships outlined by Evans and Wiederhorn earlier. Care should be taken in selecting the appropriate parameters since these parameters are a function of evaluation technique otherwise the predicted time to failure will show a large variation. Future work in the current program will be designed to verify this life prediction methodology by comparing data obtained from simple uniaxial tensile stress rupture testing done in the temperature regimes of fast fracture ($<$ 1200°C) and slow crack growth ($>$ 1200°C).

The life prediction methodology as outlined in this study should be equally applicable to other ceramic materials which show a time dependent fracture behavior.

1. INTRODUCTION

The use of ceramic materials for structural, high temperature, engineering components is being investigated for a number of applications, in particular, heat engines such as the gas turbine and diesel. There are at least two unusual characteristics with respect to the strength of brittle ceramics. The first is that the material is inductile (macroscopically) and its strength is controlled by the largest inherent microcrack or flaw in a given stress field; also, because there is typically a distribution of such flaws, there is a resulting scatter in material strength. The second is that inherent flaws within the material can exhibit the phenomenon of slow (or subcritical) crack growth (SCG) under load at high temperature implying that the strength is time dependent. As a result, one of the most critical factors in the structural application of ceramics is the ability to predict the reliability of a ceramic component as well as its useful life. The useful life of a component under an applied stress is usually referred to as "Lifetime Reliability or Life Prediction" and basically involves determining the times to failure at a given temperature.

In this paper an effort has been made to determine the material parameters usually associated with life prediction for hot-pressed silicon nitride, Norton NC132 material, which is probably the most mature high-strength turbine ceramic currently commercially available. In particular, the various material parameters are: (i) the inherent flaw size, a_o; (ii) fracture stress and its variation with temperature; (iii) temperature for the onset of SCG; (iv) stress intensity factor, K, and its variation with temperature; (v) crack velocity, V, the corresponding stress intensity factor, K, and the associated parameters A and n for SCG. Furthermore, it should be noted that the parameters K, V, and n can be determined by several different methods as illustrated in this study, and the quantitative values obtained often differ considerably resulting in a wide scatter in life prediction time. Therefore, care should be taken in choosing the appropriate data base in order to determine the ceramic component life prediction time. Finally, the future work in continuation of this study involves the actual verification of the analytically obtained life prediction times with experimental results obtained from simple uniaxial tensile stress-rupture tests at varying applied stress levels at 1200 and 1300°C. A brief illustration of the results obtained from tensile stress-rupture tests at 1204°C will also be included in this study.

2. MATERIAL, METHODS AND ANALYSIS

2.1 Material

The material used in this study was a nominally fully-dense, hot-pressed silicon nitride commercially known as Norton NC-132 and was obtained in the form of a billet 6 in. (~ 150 mm) long by 6 in. (~ 150 mm) wide by 1 in. (~ 25 mm) thick. It is believed that at high temperatures, the strength limiting impurities segregate to the grain boundaries in the form of a less refractory phase. Interpretation [1-6] to date considers that the grain boundary phase is a viscous glass at high temperatures, probably predominately unstable, comprised chemically of a combination of processing aids, principally MgO and CaO, as well as SiO_2 residual to the Si_3N_4 powder used in fabrication.

2.2 Methods and Analysis

Various methods for the determination of the critical stress intensity factor, K_{IC}, have been reviewed by Evans [7]. In this study a brief description of the three methods, namely, the double torsion (DT), the indentation induced flaw (IIF) and the uniaxial tensile stress-rupture testing will be given.

2.3 Double Torsion Technique

The double torsion (DT) specimens used were approximately 3 in. (~ 75 mm) long, 1 in. (~ 25 mm) wide and 0.050 in. (~ 1.25 mm) thick. A schematic of the DT specimen and loading configuration is shown in Fig. 1. A prerequisite of the DT test is the presence of a precrack in the test specimen with a crack length of approximately one-half the width of the specimen [8]. For crack velocity and corresponding stress intensity measurements, it is the precrack which is to be repropagated at test temperature. Normally, a very fine notch or slit is made in the center of a DT specimen in order to facilitate the initiation of a precrack when loaded in the DT fixture. The notch or slit is usually accompanied by a narrow groove along the central length of the specimen to guide crack propagation. The groove depth was varied from 0.005 in. (~ 0.127 mm) to 0.010 in. (~ 0.254 mm). The DT testing fixture used, Fig. 1, was similar in design to that used by Evans and Wiederhorn [9] and made from hot-pressed SiC.

2.3.1 Stress Intensity, K_I, Evaluation

The stress intensity factor, K_I, for a DT specimen [10] of an elastic material under plane stress conditions is given by:

$$K_I = P W_m \left[\frac{3(1+\nu)}{W d^3 d_n} \right]^{1/2} \quad \cdots\cdots\cdots\cdots (1)$$

where P is the applied load, ν is Poisson's ratio, W_m, W, d and d_n are specimen dimensions shown in Fig. 1. Note that K_I depends only on the applied load P, specimen dimensions and Poisson's ratio, and

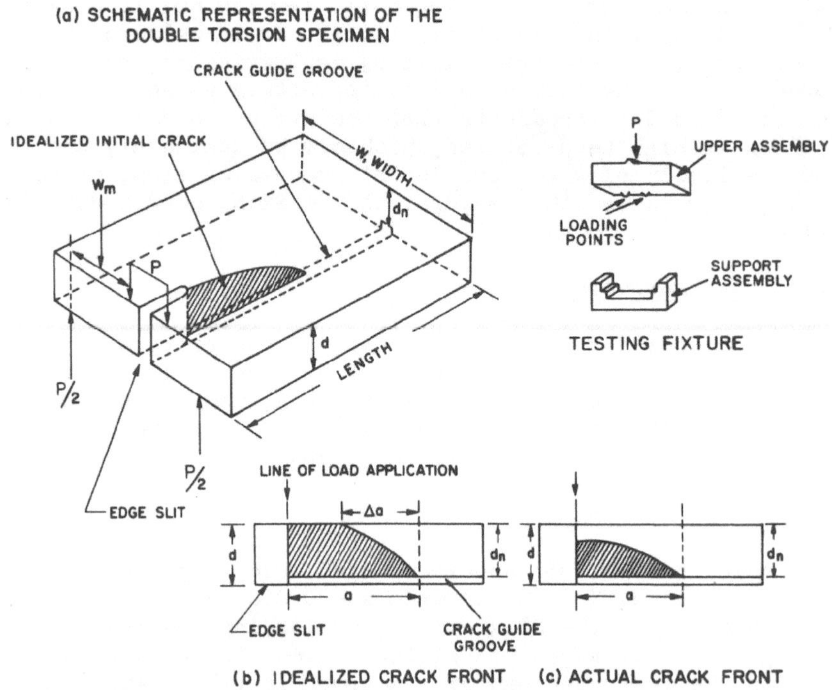

Fig. 1. Schematic display of the double torsion specimen and testing fixture.

is independent of crack length. It is this feature that makes the DT test very popular for making SCG Studies.

2.3.2 Crack Velocity, V, Evaluation

The stress intensity - crack velocity data were obtained by the load-relaxation method as applicable to the DT technique. In this, a sharp crack is initiated, the DT specimen is reloaded in the testing fixture in the Instron testing machine at a fast speed, the machine crosshead is stopped, and the load-relaxation recorded on the chart recorder. Following the work of Evans [11] and others [9, 10, 12, 13], the crack velocity, V = da/dt at a particular load P is computed from the rate of load relaxation (at constant displacement dy/dt = 0, y is the displacement of loading points or the elastic specimen deflection) is given by:

$$V = -\frac{da}{dt} = -\phi \frac{P_{i,f}}{P^2}\left(a_{i,f} + \frac{C}{B}\right)\left(\frac{dP}{dt}\right)_y \quad \cdots\cdots (2)$$

where $P_{i,f}$ is the initial (or final) load, P is the instantaneous load, $a_{i,f}$ is the initial (or final) crack length, and B and C are constants relating to the slope and intercept of a compliance analysis calibration curve of the DT specimen and ϕ is a geometrical factor relating to crack front profile [11], Fig. 1. Assuming that the initial crack is large and contributions due to various constants (as indicated in Eq. 2) are small, the crack velocity expression can be further simplified and given as follows:

$$V = -\frac{P_{i,f}\, a_{i,f}}{P^2}\left(\frac{dP}{dt}\right)_y \quad \cdots\cdots (3)$$

In general, the following method was used for measuring K_I and V from the DT specimen:

(a) After placing the DT specimen in the test fixture, it is loaded in the Instron machine at a slow crosshead speed of 0.001 in. (~ 0.0254 mm) per min. at room temperature (or higher temperature) in air. The Instron machine is kept running until a sudden load drop occurs indicating the nucleation or pop-in of a precrack and immediately the crack is arrested by stopping the machine crosshead speed. The specimen is unloaded at a fast crosshead speed in order to examine the precrack on the surface. The corresponding loads for crack pop-in and arrest are designated P_O and P_A, respectively. When the initial crack is nucleated at a higher temperature such as 1300 or 1400°C, the load-deflection curve does not show a sudden load drop but simply rounds off at the maximum load and a rapid decrease in load occurs indicating the nucleation of a crack. The precracking method is shown schematically in Fig. 2.

(b) For K_{IC} determination, the precracked specimen is reloaded in the Instron machine at a fast crosshead speed of 0.10 in. (~ 2.54 mm) or 0.5 in. (~ 12.7 mm) per min. at the desired test temperature and the fracture load, P_C, for catastrophic crack propagation is measured. This load is usually slightly smaller than the load P_O. Schematic representation is shown in Fig. 2d.

2.4 Indentation Induced Flaw Technique

Test specimens used in this technique were of dimensions 1.25 in.

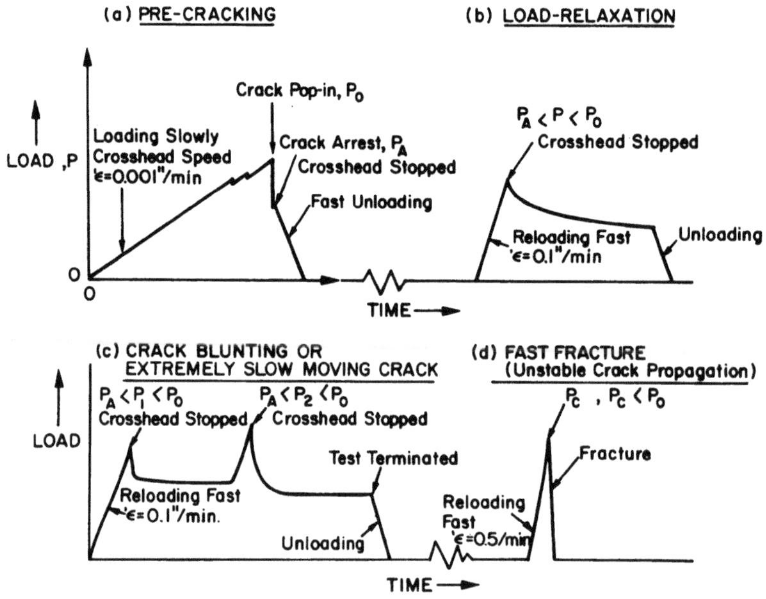

Fig. 2. Schematic representation of steps involved in testing a DT specimen.

(~ 32 mm) long x 1/4 in. (~ 6 mm) wide x 1/8 in. (~ 3 mm) thick. They were machined from the Si_3N_4 billet such that the tensile face was perpendicular to the hot pressing direction [14] (strong direction). All faces were ground in a lengthwise direction using 220-grit diamond wheels and the edges chamfered to prevent notch effects. One surface (the tension face in bending) was carefully ground and wet polished to 6 micron diamond paste finish. The specimens were then precracked using the Diamond Pyramid and Knoop microhardness indenter.+ A self aligning ceramic fixture [14] made from hot-pressed SiC was used for four-point bend tests and specimens were tested in an Instron machine* at a constant crosshead speed of 0.005 in. (~ 0.127 mm) per min. The outer and inner knife edges of the testing fixture were spaced 3/4 in. (~ 19 mm) and 3/8 in. (~ 9.5 mm) apart, respectively. The high-temperature bend tests were conducted in air using a rapid temperature furnace** attached to the Instron machine head. In high-temperature tests, specimens were held at the test temperature for about 15 mins. in order to achieve equilibrium

\+ Wilson Instrument Division of ACCO, Bridgeport, Conn.

* Instron Corp., Canton, Mass.

** CM, Inc., High Temperature Furnaces, Bloomfield, N. J.

before testing is started. No preload was applied on test specimens either at room temperature or at high-temperature tests.

Indenting this material with a Diamond pyramid or Knoop indenter resulted in the introduction of surface cracks of reproducible geometry and size (depth of crack) controlled by choice of indenter load. The resultant crack geometry, Fig. 3a, and typical examples of a crack produced on the polished surfaces of test specimens using Diamond and Knoop indenters are shown in Figs. 3b and 3c, respectively. Great care was taken to make sure that the crack AB, Fig. 3b, or the long diagonal of the Knoop indentation (AB), Fig. 3c, is aligned perpendicular to the direction of the tensile axis. All indentations were made at room temperature. Fracture surfaces were examined by optical microscopy and scanning electron microscopy (SEM).

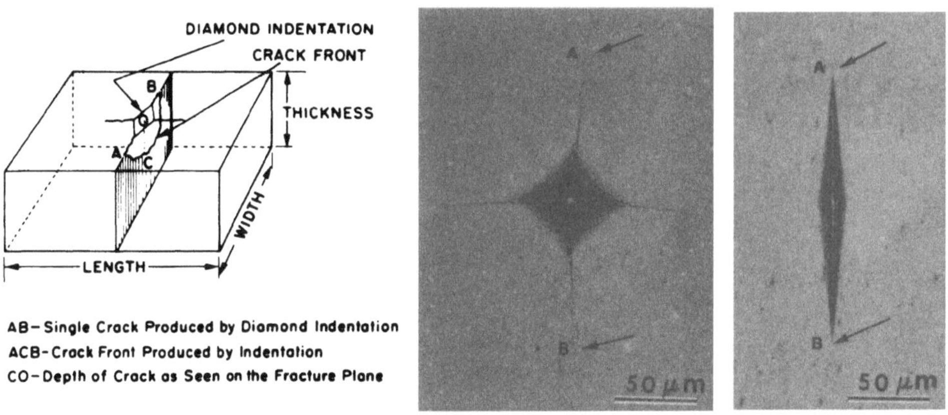

a. b. 4000 gm c. 2600 gm

Fig. 3. (a) Schematic representation of crack geometry as produced by microhardness indentation. (b) Typical example of a crack produced on the polished surface of a NC132 Si_3N_4 test specimen using a Diamond pyramid indenter with 4000 gm load. (c) Typical example of a crack produced on the polished surface of a NC132 Si_3N_4 test specimen using a Knoop indenter with 2600 gm load.

2.4.1 Fracture Mechanics of Surface Flaws

The stress intensity, K_I, for a surface flaw in bending [15] is given by:

$$K_I = \sigma M \left[\frac{\pi c}{Q} \right]^{1/2} \quad \text{..........(4)}$$

where σ is the maximum outer fiber tensile stress, c is the flaw depth, and M and Q are numerical factors related to flaw and specimen geometry. M is a function of the ratio of flaw depth to specimen thickness, the ratio of flaw depth to width, and location along the flaw front [16]. For semi-circular flaws [16], M is 1.03. Also,

$$Q = \Phi^2 - 0.212\,(\sigma/\sigma_{ys})^2 \quad \ldots\ldots\ldots\ldots(5)$$
$$\approx \Phi^2$$

where σ_{ys} is the tensile yield stress and 0.212 $(\sigma/\sigma_{ys})^2$ is a plastic-zone correction factor and for brittle failure in ceramics, this factor is negligible. Finally, Φ is the elliptic integral [15] given by:

$$\Phi = \int_0^{\pi/2} \left[\sin^2\theta + (c/b)^2 \cos^2\theta\right]^{1/2} d\theta \quad \ldots\ldots\ldots\ldots(6)$$

where 2b is the total flaw length at the free surface (length of the flaw AB as seen on the polished surface, Fig. 3). This integral is tabulated in some mathematical tables and in Ref. 15:

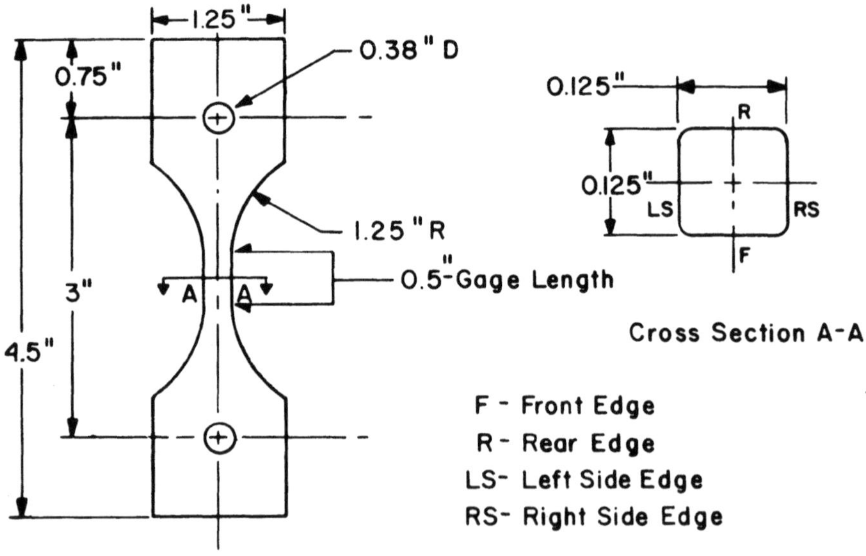

Fig. 4. Tensile stress-rupture test specimen dimensions.

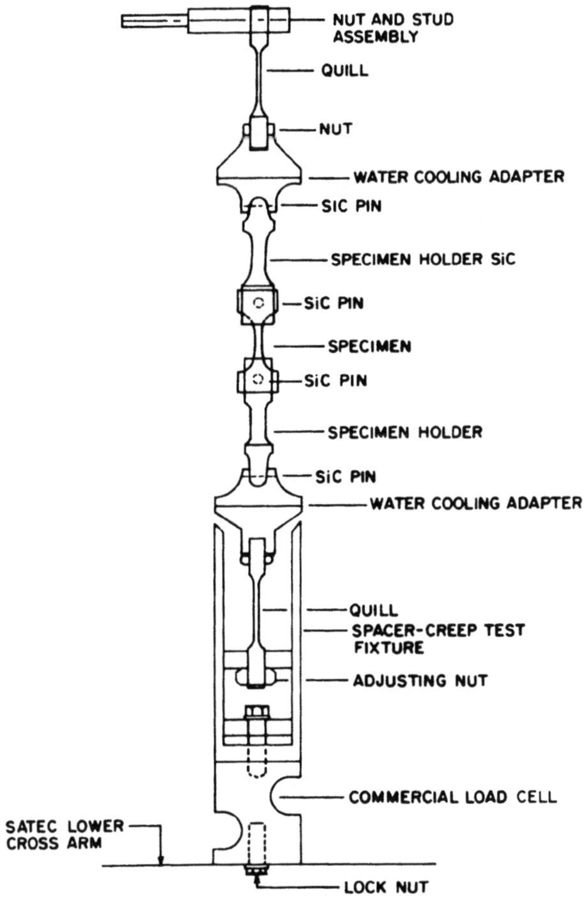

Fig. 5. Schematic representation of the load train assembly used for tensile stress rupture testing at high temperatures.

2.5 Tensile Stress-Rupture Testing

The tensile stress specimen consists of a simple rectangular geometry with a narrow cross-section in gage length and large radii at the shoulders necessitated by the sensitivity of ceramic materials to stress concentrations. The specimen geometry and dimensions are shown schematically in Fig. 4. A Standard creep testing Satec machine with a modified load train assembly was used for stress-rupture testing. The specimen is retained in two slotted SiC holders by large SiC pins. The SiC holders are retained in water cooled metal adaptors which in turn are attached to the standard Satec machine head which includes crossed (90°) knife edges. The assembly procedure includes hanging the load train parts from the Satec head as influenced by gravity. At this point, the lower Satec crossarm is lowered to snub the train in this position. The load train assembly is shown schematically in Fig. 5. Each test specimen contained a total of eight strain gages (two on each face in the gage-length area) for axial alignment and efforts were made to keep the bending stresses below 3% at full load. Initial alignment of the specimen is done at 20°C with full load. Typical test specimen with strain gages attached in the gage section is shown in Fig. 6. A high temperature furnace capable of reaching 1400°C temperatures is mounted on the side area of Satec machine and encloses the full area between the water cooled metal adaptors, Fig. 5. Complete details regarding the load train instrumentation and continuous monitoring of applied stress, test temperature and time are given elsewhere [14].

3. RESULTS AND DISCUSSION

3.1 Inherent Flaw Size Determination

A series of specimens was prepared and precracked using Diamond pyramid indenter with 500 gm, 1000 gm, 2000 gm, and 4000 gm indentation loads. The variation of fracture stress σ_F, with crack depth (actually inverse square root of crack depth) for NC132 Si_3N_4 specimens tested at room temperature is shown in Fig. 7. Although there is considerable scatter in the data, a reasonably good linear relationship is evident which correlates with the Griffith [17] criterion for brittle fracture. Data for a few Knoop indented precracked specimens are also shown, Fig. 7, and behaved in a similar fashion as Diamond pyramid indented precracked specimens. However, from a practical point of view, perhaps the most useful application of the data shown in Fig. 7 is as a means of estimating the inherent flaw+ size in hot pressed

\+ Inherent flaw size refers to specifically the size of flaws which occur in the virgin (or as processed) material and does not refer to surface flaws caused by machining.

NC132 Si_3N_4. Extrapolation of the data in Fig. 7 to the fracture stress of uncracked samples (shown as the approximate scatter band of materials strength) indicates that flaws of the order of 10 to 20 microns in depth are present in the as received (uncracked) Si_3N_4. To date, this is perhaps the best estimate of the inherent flaw size in NC132 Si_3N_4. Numerous investigations [5, 9, 18-25] relating to mechanical properties and fracture mechanics aspects of a similar type of hot pressed Si_3N_4 are on record but none made an attempt to measure or estimate the inherent flaw size. Lange[26] and Mecholsky, Freiman and Rice [27] have also estimated the flaw size to be around 90 and 53 microns in HS-130 Si_3N_4, from their surface energy and mirror constant measurements, respectively.

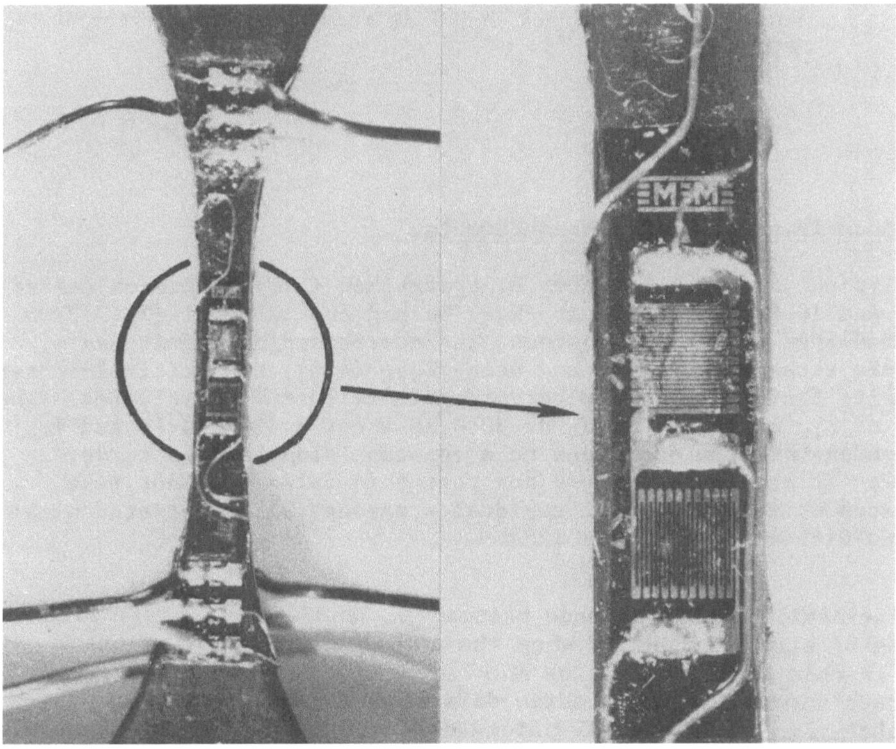

Fig. 6. Typical tensile stress-rupture test specimen with strain gages attached to the gage length.

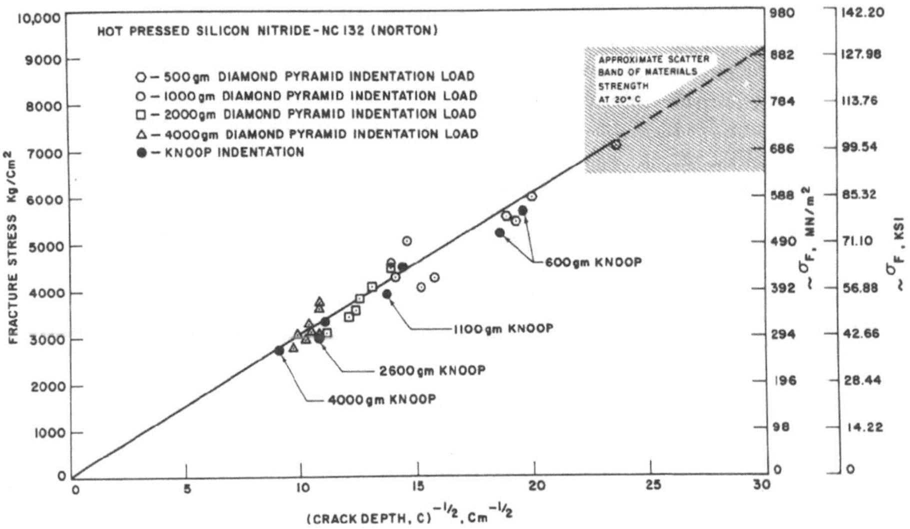

Fig. 7. Variation of fracture stress as a function of inverse square root of crack depth at room temperature for NC132 Si_3N_4.

3.2 Room Temperature K_{IC} Measurements

Typical fracture surfaces of precracked (IIF) specimens tested at 20 and 1000°C are shown in Figs. 8a and 8b, respectively. From the knowledge of flaw dimensions, the corresponding magnitudes of fracture stresses, Fig. 7, and using Eqs. (4-6), the critical stress intensity factor, K_{IC}+, for catastrophic failure was evaluated. The value of K_{IC} for NC132 Si_3N_4 at 20°C is about 3.49 $MN/m^{3/2}$ and is independent of flaw depth due to a constant slope of the curve, Fig. 7. It should be pointed out that this value does not take into account the effects of residual stresses [21] associated with IIF and discussed elsewhere [28].

+A distinction should be made between K_{IC} and K_I. K_{IC} refers to that value of stress intensity when the normal stress at the crack tip reaches a critical value and leads to a catastrophic failure without undergoing any plastic deformation. Under these circumstances, K_{IC} is a material parameter. At high temperatures, when plastic deformation or subcritical crack growth (SCG) occurs prior to final catastrophic failure, stress intensity measured will be referred as K_I.

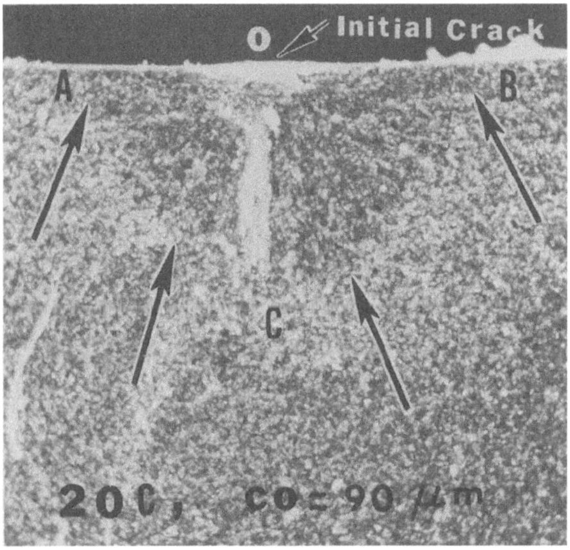

(a)

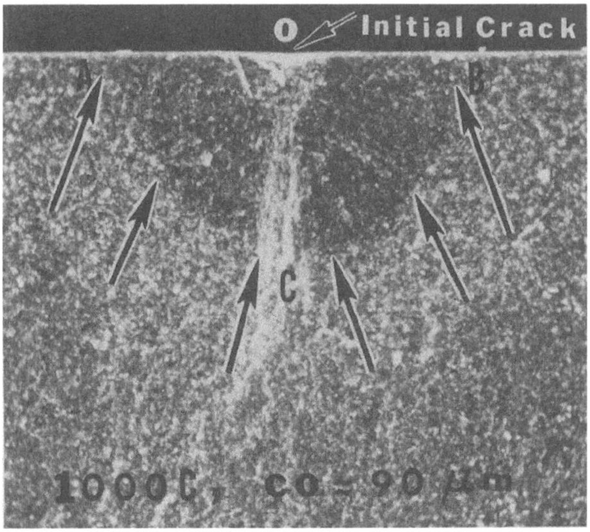

(b)

Fig. 8. Typical SEM micrographs of fracture surfaces of precracked specimen of NC132 Si_3N_4 tested at 20 and 1000°C. Specimen were precracked with 4000 gm load indentation (Diamond pyramid). Arrows indicate the boundary of the precracked region ACB.

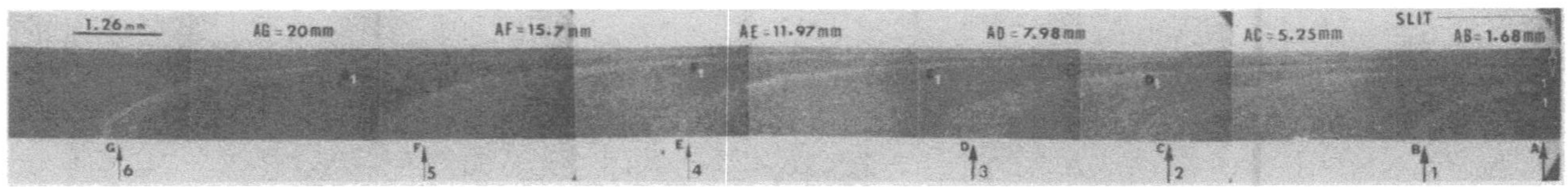

a

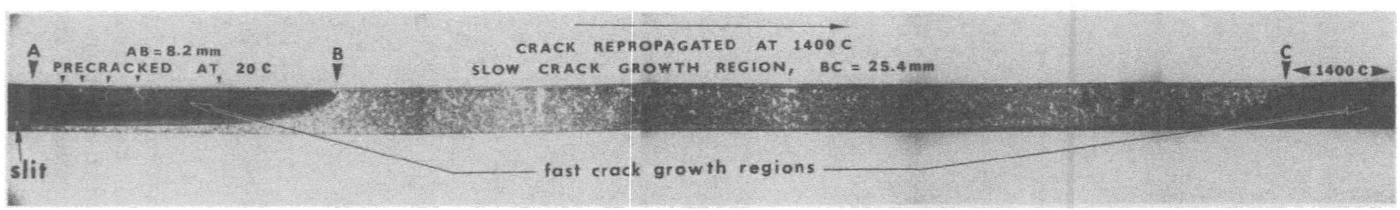

b

Figure 9. (a) Typical fracture surface of a DT specimen of NC132 Si_3N_4 at 20°C. Crack is moving from right to left.

(b) Typical fracture surface of a double torsion specimen of NC132 Si_3N_4 in which the initial crack, AB, was produced at 20°C and subsequently repropagated at 1400°C. Crack in moving from left to right.

The DT technique was also used to evaluate K_{IC} and typical fracture surfaces* of DT specimen tested at 20 and 1400°C are shown in Figs. 9a and 9b respectively. Values of K_{IC} for NC132 Si_3N_4 using Eq. (1) are given in Table 1, and vary from 3.9 to 4.4 $MN/m^{3/2}$. The average value of K_{IC} is about 4.1 $MN/m^{3/2}$. Evans and Wiederhorn [9] using this technique in a similar type of material, HS-130 Si_3N_4, found K_{IC} at 20°C to vary from 4.2 to 5.3 $MN/m^{3/2}$ with an average value of 4.7 $MN/m^{3/2}$, while Lange [26] reported a value of 5.1 $MN/m^{3/2}$ using the double cantilever beam method. Considering the fact that the two hot pressed silicon nitrides NC-132 and HS-130 differ slightly in chemical composition and strength, the difference in K_{IC} values determined using the common DT technique is within reasonable limits.

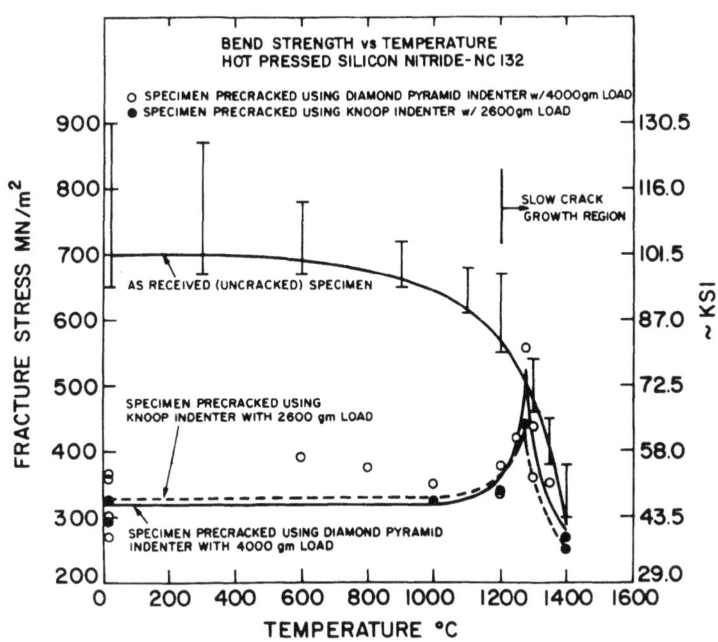

Fig. 10. Temperature dependence of the fracture stress for uncracked and precracked specimens of NC132 Si_3N_4.

* Note that the fracture surface at 20°C shows six crack fronts, the first five were due to small load drops which occurred during initial loading, Fig. 2a, and the crack front G_1G represents the large load drop corresponding to crack pop-in at which time the crack is arrested.

Table 1. Fracture Toughness Measurements in Hot Pressed Silicon Nitride NC132 at 20°C and at constant Crosshead Speed.

Specimen No.	Grooved Specimen: Yes, Groove Depth	Grooved Specimen: No	Precracking: Initiation Temp. °C	Precracking: Initial Length mm	Precracking: Repropagation Temp. °C	Fast Fracture Load P_c pounds	Stress Intensity Factor, $MN/m^{3/2}$: K_I	Stress Intensity Factor, $MN/m^{3/2}$: K_{IC}	Remarks
3 A		X	20	20	20	16		3.9	Crack repropagated centrally all through the length
4 A		X	"	9	"	18		4.4	"
5 A		X	"	12	"	17		4.15	"
10 A	X 0.127 mm		"	15	"	16		4.06	"
8 B		X	1350	16	"	28	6.63		"
11 C	X 0.127 mm		20	8	1400	31	7.87		Load-Relaxation curve obtained
21 D	X 0.315 mm		1350	-	1350	23	6.4		"

3.3 Effect of Temperature and K_I Measurements

The temperature dependence of the fracture stress, σ_F, for uncracked (virgin) material and for precracked with a Diamond Pyramid indenter (containing cracks of about 95 ± 5μm deep) and Knoop indenter (containing cracks of about 80 ± 5μm deep), and tested at temperatures between 20 and 1400°C in air is shown in Fig. 10. It is clear that σ_F for precracked specimens remains essentially constant and independent of temperature from 20 to 1100°C. The constancy of σ_F suggests that no change in fracture mechanism occurs and indicates little or no blunting of the crack tip by plastic deformation in this temperature region. Note that the mode of fracture in the precracked region (inside ACB) and in the repropagated region (outside ACB), Fig. 8, is similar, i.e., a mixed mode of fracture consisting of transgranular and intergranular crack growth. The phenomenology of fracture in a similar type of material (HS130 Si_3N_4) has been discussed in greater detail by Govila, Kinsman, and Beardmore [29]. The sudden and subsequent increase in σ_F at temperatures between 1100 and 1250°C, Fig. 10, is interpreted as indicative of blunting of the micro crack due to the viscous flow. It is believed that the softening of the glassy phases [6, 29] at the grain boundary is a more likely mechanism of "crack blunting." Typical fracture surfaces of precracked specimens (containing cracks of about 95 ± 5μm deep) tested at 1250, 1275, and 1350°C are shown in Fig. 11. The first signs of slow-crack growth (SCG) of the initial micro crack were observed on the fracture face in tests made at 1275°C and higher temperatures, Fig. 11b, as indicated by the change in reflectivity. The extent of SCG prior to catastrophic failure increased with increasing temperature (cf. Fig. 11b-c). The SCG region is distinct in its appearance characterized by bright whitish regions. The nature and mode of fracture during SCG is completely intergranular as discussed elsewhere [29].

The fracture surface appearance at 1250°C, Fig. 11a, is similar to that observed at lower temperatures, Fig. 8, and does not show the presence of SCG. Above 1250°C, SCG starts and σ_F decreases abruptly, Fig. 10, with a proportional increase in SCG region. The two curves (data for virgin and precracked specimens), Fig. 10, seem to merge and show a decreasing strength above 1275°C. The temperature at which the two curves merge characterizes the importance of SCG and points out that failure is governed by the extent of SCG and not by initial crack size. It is fair to conclude from the study of precracked specimens that in NC-132 Si_3N_4 the temperature for the onset of SCG is around 1250°C.+

+ This conclusion is primarily based on the fact that the speed of testing used in data given in Fig. 10 was 0.005 in./min. However, it is likely that the presence of SCG can occur at lower temperatures such as 1200°C if the crosshead speed of testing is one or two orders of magnitudes slower or the material contains large amounts of impurities such as Ca, Fe, etc.

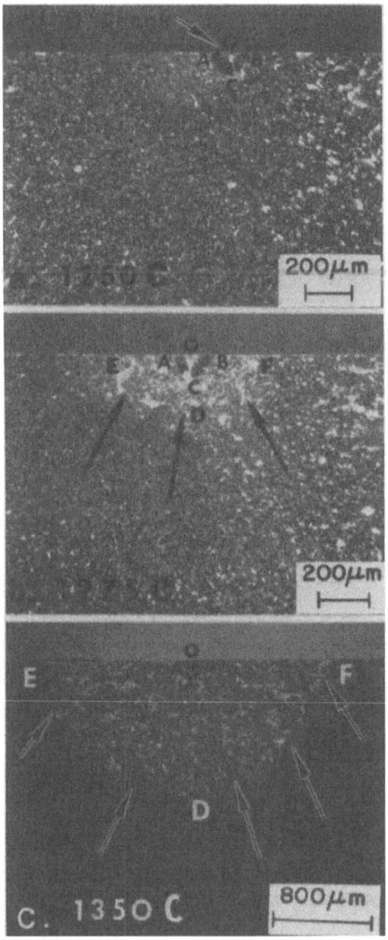

Fig. 11. Typical fracture surfaces of precracked (crack depth $\approx$ 95 $\pm$ 5 μm) NC132 Si_3N_4 specimens tested at a machine head speed of 0.005"/min. at various temperatures. Pictures taken with plane polarized light. Note the absence of subcritical crack growth (SCG) at 1250°C, subsequent appearance of SCG surrounding the initial crack ACB at 1275°C and higher temperatures.

High temperature K_I values were determined using both the DT and IIF methods and the variation of K_I as a function of temperature are shown in Fig. 12. The K_I values for the IIF method, Fig. 12, were calculated using the total depth of flaw, DO, Fig. 11 (initial flaw size, CO + slow crack growth, CD) at the onset of fast fracture and are only approximate in magnitude because plasticity correction factors and shape change in crack front in Eqs. (4-6) have not been taken into account. The variation of K_I as a function of temperature for both methods, Fig. 12, shows a similar form of behavior except values of K_I by DT method at high temperatures are slightly smaller in magnitude relative to IIF method. The reason for this discrepancy is simply due to the fact that faster machine head speeds, 0.2 - 0.5 in./min., were used in evaluating K_I by DT method while orders of magnitude slower machine head speeds (0.005 in./min.) were used with IIF method. It is clear from Fig. 12, that K_I remains essentially constant from 20 to 1100°C. In this temperature region, K_I is independent of plastic yielding effects and thus represent the critical stress intensity factor K_{IC}.

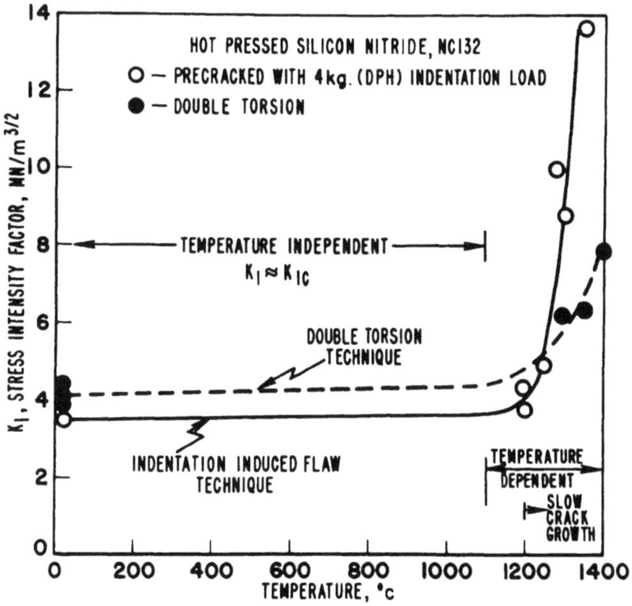

Fig. 12. Variation of stress intensity, K_I, as a function of temperature for NC132 Si_3N_4 using Indentation Induced Flaw and Double Torsion techniques.

3.4 Crack Velocity (V) and Stress Intensity (K) Relationship

A functional relationship between crack velocity (V) and the corresponding stress intensity K_I for subcritical crack growth (SCG) has been approximately described by the following relationship:

$$V = A K_I^n \cdots\cdots\cdots\cdots (7)$$

where A and n are constants for a given temperature and environment. Using the DT method and Eqs. (1) and (3), and the procedure outlined in Section 2.3, the V-K relationship and its variation with temperature is shown in Fig. 13. The curves represent the region of slowly

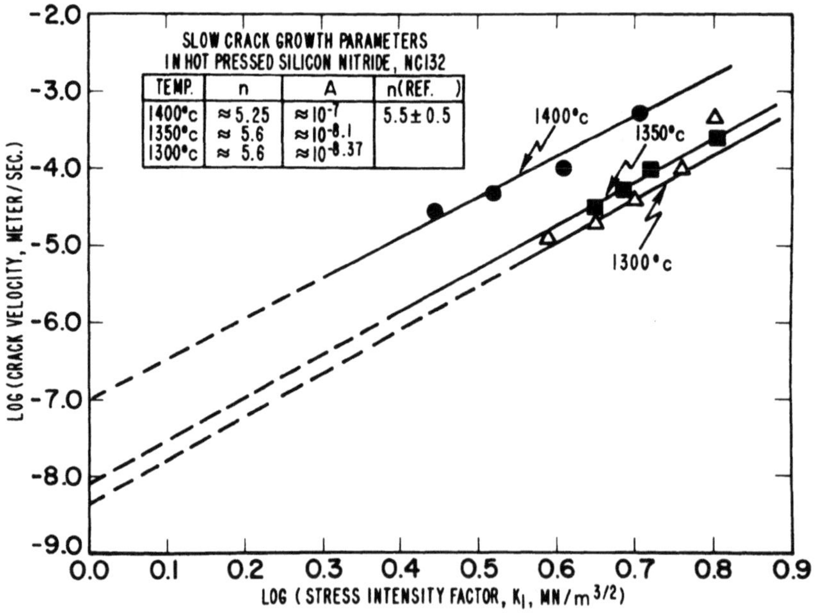

Fig. 13. Subcritical crack growth rate data for NC132 Si_3N_4 as a function of temperature using the double torsion technique. The value of n = 5.5 ± 0.5 at 1400°C as shown (inset) above, is for HS130 Si_3N_4 and is from Ref. 22.

varying velocity and similar behavior has been observed in HS-130 Si_3N_4 [9] at the same temperatures. It should be noted that the curves at 1300, 1350, and 1400°C are essentially parallel and characterized by a constant slope. The individual values of n and A for SCG at these high temperatures are listed in Fig. 13. The constant slope of the curves indicate that once large amounts of slow-crack growth occur, n does not vary significantly for a given environment and is independent of temperature for the SCG temperature regime (1300 - 1400°C or higher). The approximate value of n at 1400°C is about 5.25. This value of "n" is in good agreement with work reported by others [9, 22] in HS-130 Si_3N_4 at the same temperature using similar technique. Typical fracture surface of a DT specimen tested at 1400°C is shown in Fig. 9b.

Recently, Mendiratta and Petrovic [25] used the IIF method using multiple cracks and propagated them in vacuum under a constant stress for different times at temperatures from 1100 to 1300°C, and measured average velocities which were about two orders of magnitude higher than the DT method [9, 22] reported in the same material (HS-130 Si_3N_4). Some stress rupture studies in HS-130 Si_3N_4 have also been reported by us [29] and it is clear from the micrographs that the path taken by SCG region is pseudo elliptical in shape and therefore, velocities measured from surface extensions of crack could be significantly different than those from depth of crack. Furthermore, even if the crack growth or crack front is semi-circular in nature, it is believed that using the IIF method would result in average crack velocities for SCG for two reasons. First, the load-deflection curves for precracked specimens showed linear elastic behavior to the point of fracture (especially for the machine head speeds of 0.005 in./min. and higher) even though large amounts of SCG or extension of the initial crack occurred. From such tests, only the total time to failure can be measured from the Instron recorder chart and would result in much lower than actual velocities. Secondly, the load-deflection curve does not indicate the exact instant at which the crack moves and, therefore, the time to travel a given length of crack cannot be measured accurately.

3.5 Determination of Subcritical Crack Growth Exponent, n

The subcritical crack growth exponent, n, can be determined in a number of ways as outlined below:

a. The Double Torsion

b. Flexural Stress Rate

c. Flexural Strain Rate

d. Flexural Stress Rupture

e. Uniaxial Tensile Stress Rupture

The most direct method is to measure crack velocity as a function of stress intensity as done in the DT method discussed earlier. Results obtained from the other methods (b-e) are presented below:

b. Flexural Stress Rate

Flexural stress rate testing [4, 30] is one simple method for determining the exponent 'n' and utilizes the materials fracture strength as a function of stressing rate as given below for a given temperature and environment:

$$\log \sigma_F = \log D + \frac{1}{n+1} \log \dot{\sigma}_F \quad \cdots\cdots\cdots (8)$$

where σ_F is the fracture stress at a stressing rate $\dot{\sigma}_F$ and D is a constant. A log-log plot of σ_F vs stressing rate $\dot{\sigma}_F$ would yield a linear relationship (straight line) and from the slope of the plot, the parameter 'n' can be determined.

The stress rate measurements were done using the four-point bend test specimens because of simplicity and relatively inexpensive nature of the test compared to other tests such as uniaxial tension and complete details regarding the statistical data and technique are given elsewhere [14]. The temperature dependence of the fracture stress for uncracked (virgin) samples as a function of stressing rate on a log-log scale is shown in Fig. 14. Up to about 1040°C, the fracture stress, σ_F, is almost independent of the stressing rate $\dot{\sigma}_F$, indicating no evidence of SCG. At higher temperatures, the log-log plot showed a clear deviation from the horizontal line indicating presence of SCG and at 1204 and 1371°C the values of n

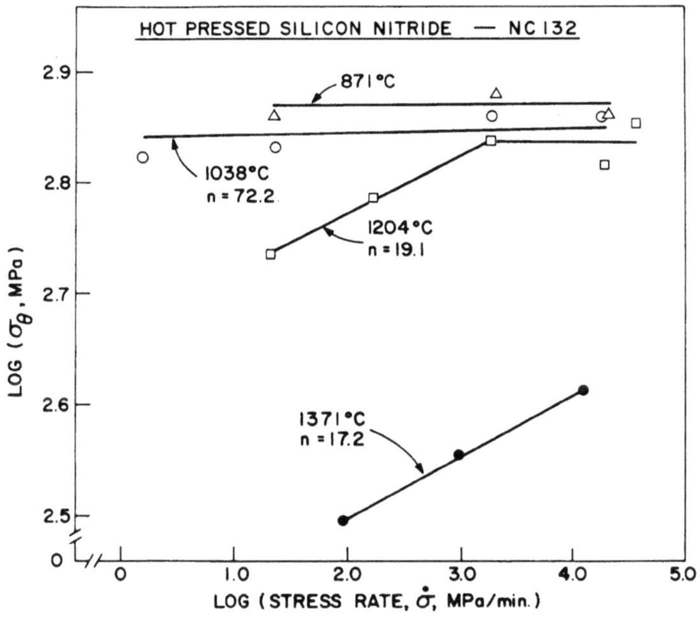

Fig. 14. Log-log plots of flexural strength vs stressing rate at various temperatures. Data presented in this figure was obtained by R. M. Williams, Ford Motor Company.

were 19.1 and 17.2, respectively. Typical fracture surfaces at 1204°C corresponding to the three stress rates (inclined portion of log-log plot) are shown in Fig. 15, and only at the lowest stress rate the presence of SCG was observed, Fig. 15a. Metallographic examination of fracture surfaces for specimens tested at 1371°C (Fig. 14) showed the presence of SCG. It should be noted that the curves at 1204°C and 1371°C are essentially parallel and the value of 'n' is independent of temperature in the range 1200 - 1400°C for a given environment. Similar behavior has been observed by Lange [4] in HS-130 Si_3N_4 (specimen configuration, weak direction [14]), and the values of 'n' determined were 12.7, 10.6, and 11.2 at 1200, 1300 and 1400°C, respectively.

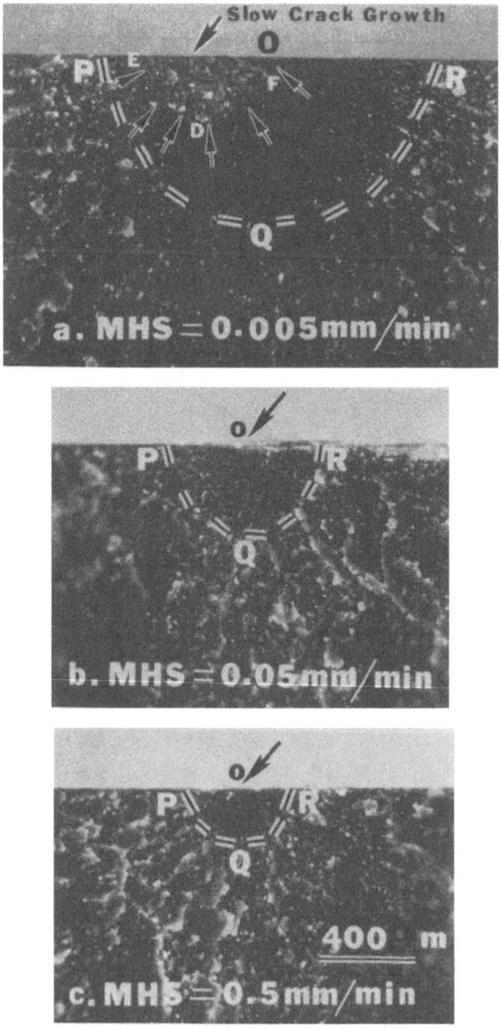

Fig. 15. Typical fracture surface appearance as seen in uncracked NC132 Si_3N_4 flexural specimens tested in air at 1204°C as a function of stressing rate (or machine head speed). PQR is approximately the semi-circular mirror region. Note the presence of slow crack growth, region EDF, was noticed only at extremely slow stress rates.

c. Flexural Strain Rate

Davidge et al. [31] suggested the use of flexural strain rate testing (similar to stress rate testing) in which the ratio of materials fracture strengths (σ_F) at two strain rates ($\dot{\epsilon}$) for equal probability of failure is given by:

$$\frac{\sigma_{F1}}{\sigma_{F2}} = \left(\frac{\dot{\epsilon}_1}{\dot{\epsilon}_2} \right)^{\frac{1}{n+1}} \quad \cdots\cdots\cdots\cdots (9)$$

where σ_{F1} and σ_{F2} are the fracture strengths at strain rates $\dot{\epsilon}_1$ and $\dot{\epsilon}_2$, respectively, and 'n' is the crack velocity exponent for SCG. The use of precracked specimens (IIF method) is particularly suitable in revealing the strain rate sensitivity of SCG qualitatively and also insuring equal probability of failure. This was done using specimens containing a constant crack size (about 95 $\pm$ 5 μm deep) and tested at varying temperatures and strain rates (or machine head speeds). Values of 'n' determined using Eq. (9) were 11 $\pm$ 1 at 1350 and 1400°C and ranged as high as 18 - 20 at 1300°C. Typical fracture surfaces in tests made at 1350°C are shown in Fig. 16. Increasing the strain rate (or machine head speed) makes the material more brittle and initially raises the fracture stress. In agreement with this view, the extent (or depth) of SCG decreased with increasing the strain rate (see Fig. 16) until it disappeared at a machine head speed in the range of 0.2 - 0.5 in./min. Note that in all cases, Fig. 16, fracture was catastrophic.

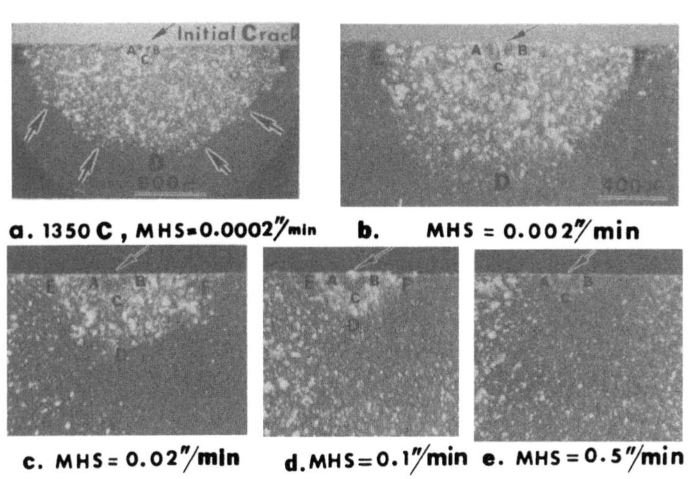

Fig. 16. Successive stages in the slow crack growth of precracked specimen tested at 1350°C as a function of machine head speed. All specimens were precracked with 4Kg load indentation. Pictures taken with plane polarized light.

d. Flexural Stress Rupture

Another simple, but laborious, method has been suggested by Davidge et al. [31] to estimate the value of 'n' and utilizes time to failure measurements at constant applied stresses by flexural stress rupture testing. Under delayed fracture conditions, the ratio of failure times t_1 and t_2 under constant applied stresses σ_1 and σ_2 at a given temperature for a given environment is given by:

$$\frac{\sigma_1}{\sigma_2} = \left(\frac{t_2}{t_1}\right)^{1/n} \qquad \text{..........(10)}$$

A plot of log σ vs log t would result in a straight line with a slope of 1/n.

The stress rupture measurements at 1204°C and at two different stress levels were made using the four-point bend specimens (uncracked) from three different billets of NC-132 Si_3N_4 and the data are shown in Table 2. At the higher stress level of 415 MN/m^2 ($\sim$ 60,187 psi), majority of the specimens failed within three hours. As the applied stress decreased to 298 MN/m^2 ($\sim$ 43,200 psi), the time to failure increased considerably. Typical fracture surfaces for specimens tested at a constant stress of 415 MN/m^2 are shown in Fig. 17 and show the presence of SCG. Furthermore, the extent of SCG is about the same for varying time to failure.

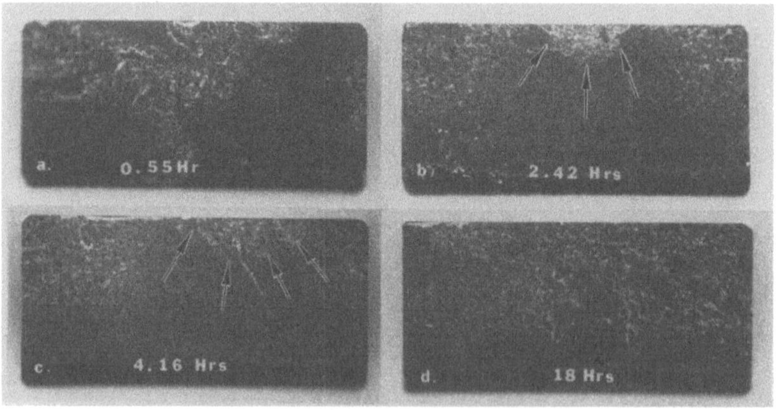

Fig. 17. Typical fracture surfaces of flexural stress-rupture specimens tested at 1204°C in air at a constant stress of $\sim$ 415 MN/m^2 showing the presence of slow crack growth.

TABLE 2 - MOR (FOUR-POINT BENDING) STRESS RUPTURE DATA AT 1204°C FOR HOT PRESSED SILICON NITRIDE (NORTON-NC 132).

SPECIMEN	APPLIED STRESS MN/m^2	APPLIED STRESS PSI	TIME TO FAILURE, HRS.
NC 132-A_1	415	60,187	17.97
-A_2	415	60,187	1.15
-A_3	415	60,187	4.16
-A_4	415	60,187	1.75
-A_5	415	60,187	0.55
NC 132-B_1	415	60,187	2.42
-B_2	415	60,187	0.75
-B_3	415	60,187	1.33
-B_4	415	60,187	2.16
-B_5	415	60,187	0.58
NC 132-C_1	298	43,200	12
-C_2	298	43,200	46
-C_3	298	43,200	195
-C_4	298	43,200	46
-C_5	298	43,200	148
-C_6	298	43,200	59

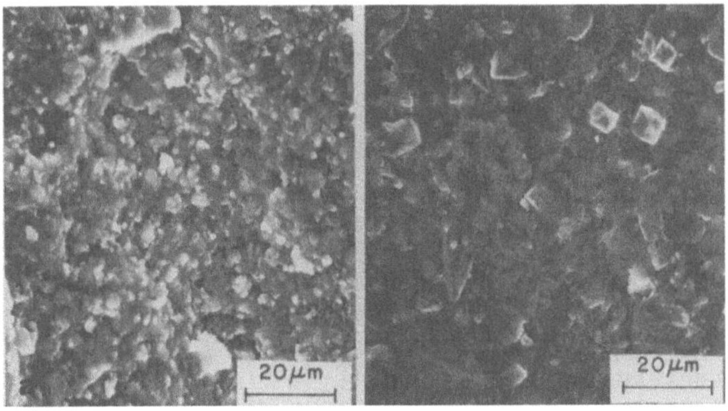

(a) (b)

Fig. 18. Nature and mode of crack propagation as revealed by SEM micrographs during slow crack growth (a) and fast fracture (b).

The nature and mode of crack propagation during SCG and fast fracture are revealed in Fig. 18 (same specimen as shown in Fig. 17c) and consists of intergranular crack propagation (Fig. 18a) and a mixed mode of fracture consisting of transgranular and intergranular crack growth (Fig. 18b), respectively. The results for the flexural stress-rupture testing at 1204°C are shown in Fig. 19 and the value of 'n'$\simeq$ 9.5 was obtained. Quinn and Katz [32] using the similar technique reported n = 9.2 at 1200°C in NC132 Si_3N_4.

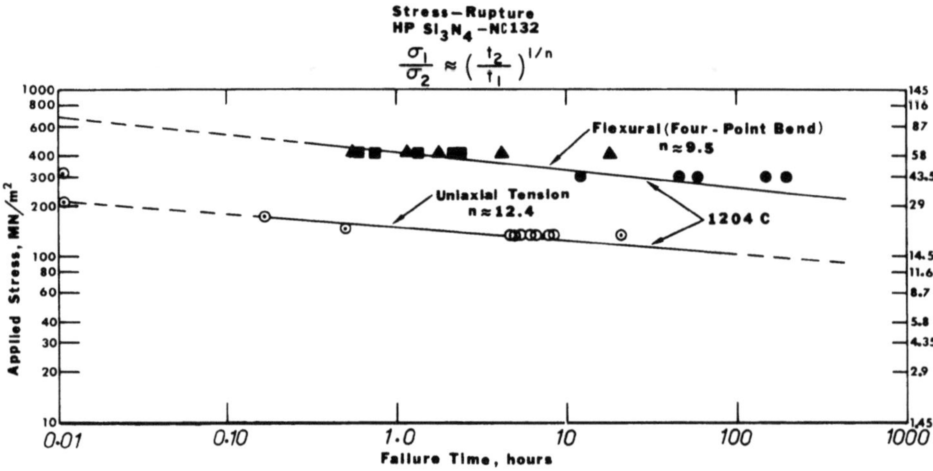

Fig. 19. Flexural and Uniaxial tensile stress rupture data for NC 132 Si_3N_4 at 1204°C.

e. <u>Uniaxial Tensile Stress Rupture</u>

Uniaxial tensile stress rupture testing is one method by which the actual verification of the analytically obtained life prediction times using the various parameters as described above can be made and this was the main objective for carrying out the tests. Considerable time and efforts were spent in carrying out the tensile stress rupture testing of NC-132 Si_3N_4 specimens. A total of sixteen tensile stress rupture specimens were tested at 1204°C at various stress levels and the data are given in Table 3. The tensile stress rupture data can also be used in a similar fashion as the flexural stress rupture data (method 3.5d) and the results are shown in Fig. 19. An approximate value of n $\simeq$ 12.4 at 1204°C is obtained from the slope of the curve and is comparable to the value obtained by flexural stress rupture method. Typical fracture surfaces for specimens subjected to an applied stress of 19,200 psi at 1204°C are shown in Fig. 20. In both cases, failure originated inside the cross-section of the specimen and the crack initiation sites are distinctly visible. Values for the crack growth parameter 'n' as obtained by the various methods are summarized in Table 4. Currently, work is in progress to carry out similar studies at 1300°C.

TABLE 3. - TENSILE STRESS - RUPTURE DATA AT 1204°C FOR HP Si_3N_4 - NC 132

SPECIMEN NO.	APPLIED STRESS PSI	MN/m^2	TIME TO FAILURE
1	19,200	132.5	4.7 Hrs.
2	19,200	132.5	8.4 Hrs.
3	19,200	132.5	6.1 Hrs.
4	19,200	132.5	21.0 Hrs.
5	19,200	132.5	5.3 Hrs.
6	19,200	132.5	6.1 Hrs.
7	19,200	132.5	8.5 Hrs.
8	19,200	132.5	5.1 Hrs.
9	19,200	132.5	7.7 Hrs.
10	19,200	132.5	6.5 Hrs.
11	19,200	132.5	4.9 Hrs.
12	19,200	132.5	7.4 Hrs.
13	21,200	146.3	30 Min.
14	25,000	172.5	10 Min.
15	30,400	210	Fast Failure
16	45,760	316	Fast Failure

3.6 Failure Prediction and Experimental Verification

The relationship between flaw size, a, applied stress, σ_A, and stress intensity factor, K_I, is simply given as:

$$K_I = \sigma_A \, Y \sqrt{a} \quad \cdots\cdots\cdots\cdots (11)$$

where Y is a constant depending on crack geometry. The crack velocity-stress intensity relationship can also be expressed as follows:

$$V = \frac{da}{dt} = A K_I^n \quad \cdots\cdots\cdots\cdots (12)$$

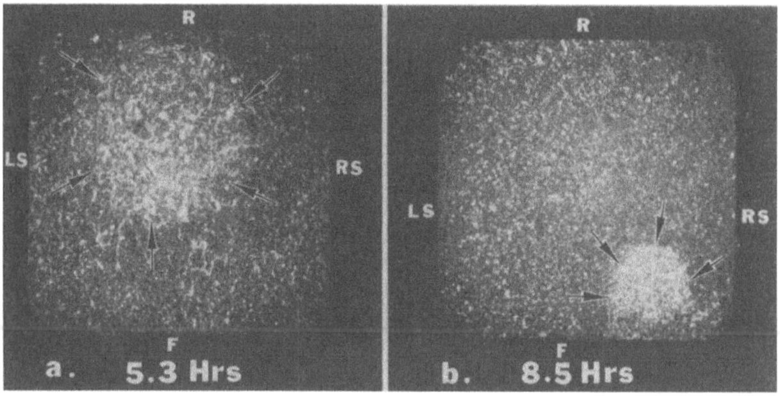

Fig. 20. Typical fracture surfaces showing internal crack initiation sites as seen in tensile stress rupture tests at 1204°C. Specimens shown in (a) and (b) are number 5 and 7, respectively as shown in Table 3.

TABLE 4. - EXPERIMENTALLY DETERMINED CRACK - GROWTH PARAMETERS FOR NC-132 Si_3N_4

METHOD	TEMP. °C	n	A
Double Torsion	1400	5.25	10^{-7}
	1350	5.6	$10^{-8.1}$
	1300	5.6	$10^{-8.37}$
Flexural Stress Rate	1371	17.2	-
	1204	19.1	-
Flexural Strain Rate	1400	10-12	-
	1350	10-12	-
	1300	18-20	-
Flexural Stress Rupture	1204	9.5	-
Tensile Stress Rupture	1300	-	-
	1204	12.4	-

Following the work of Evans and Wiederhorn [19] and combining the above two equations, the time to failure under a given applied stress is given by:

$$t_F \approx \frac{2}{\sigma_A^2 Y^2} \int_{K_{Ii}}^{K_{IC}} \left(\frac{K_I}{V}\right) dK_I$$

$$\approx \frac{2}{(n-2) A \sigma_A^2 Y^2} \left[K_{Ii}^{2-n} - K_{IC}^{2-n} \right]$$

$$\approx \frac{2}{(n-2) A \sigma_A^2 Y^2 K_{Ii}^{n-2}} \quad \cdots\cdots\cdots\cdots (13)$$

where K_{Ii} is the initial stress intensity factor at the most serious flaw in the test specimen. From Eq. 13, it is apparent that the theoretical time to failure or life prediction can be estimated if all the life prediction parameters such as a_o, σ_o, n, and A are known. It is clear from Table 4, that a theoretical estimate of life prediction time cannot be made at 1200°C as the parameter A is not available. However, the uniaxial tensile stress rupture data, Fig. 19, can be used approximately to estimate time to failure at a given stress. For example, at an applied stress of 100 MN/m^2 (~14, 500 psi) at 1204°C, the time to failure is about 100 hrs., while at 150 MN/m^2 (~21,750 psi), the time to failure is only about 48 mins.

The crack propagation data as shown in Fig. 13 and Table 4 can be used in determining a theoretical life prediction curve at a given applied stress and temperature (1300 - 1400°C) for a given environment (air) using Eq. 13. Presently efforts are being made to make analytical life predictions at 1300°C (and other temperatures as well) and compare with the results obtained from tensile stress rupture tests.

SUMMARY

This investigation has presented a cohesive experimental approach for determining the life prediction material parameters in hot-pressed silicon nitride NC132. The various parameters such as the inherent flaw size a_o, the fracture strength, σ_F, and its variation with temperature, the critical stress intensity factor, K_{IC}, for catastrophic failure and effect of temperature on stress intensity factor, K_I, stress intensity - crack velocity relationship for subcritical crack growth and associated parameters A and 'n' were determined using the double torsion and indentation induced flaw methods. Alternative methods such as flexural stress rate, strain rate and stress rupture were also used to determine the magnitude of the subcritical crack growth exponent 'n' in the temperature range of 1200 - 1400°C. Fractography is the principal interpretive agent to reveal the presence of subcritical crack growth and was used extensively. These parameters can be used to estimate the time to failure at a given applied stress, temperature and environment. Actual verification of the life prediction will be done by comparing the experimental data obtained from uniaxial tensile stress rupture testing of Si_3N_4 specimens in the temperature regime of fast and slow crack growth.

ACKNOWLEDGMENTS

The author is thankful to R. C. Elder for doing the tensile stress rupture testing and R. M. Williams for providing the data used in Fig. 14. Thanks are also due to Peter Beardmore and A. F. McLean for encouragement throughout the work. This work was supported in part by Department of Energy under Contract DAAG-46-77-C-0028.

REFERENCES

1. A. G. Evans and J. V. Sharp, J. Mater. Sci., 6, 1292 (1971).
2. R. Kossowsky, J. Am. Ceram. Soc., 56, 531 (1973).
3. D. W. Richerson, Am. Ceram. Soc. Bull., 52, 560 (1973).
4. F. F. Lange, J. Am. Ceram. Soc., 57, 84 (1974).
5. S. D. Hartline, R. C. Bradt, D. W. Richerson and M. L. Torti, J. Am. Ceram. Soc., 57, 190 (1974).
6. D. R. Clarke and G. Thomas, J. Am. Ceram. Soc., 60, 491 (1977).
7. A. G. Evans, Fracture Mechanics of Ceramics, Vol. 1, Plenum Press, New York, pp. 17-48 (1974).
8. G. G. Trantina, J. Am. Ceram. Soc., 60, 338 (1977).
9. A. G. Evans and S. M. Wiederhorn, J. Mater. Sci., 9, 270 (1974).
10. D. P. Williams and A. G. Evans, J. Test. Eval., 1, 264 (1973).
11. A. G. Evans, J. Mater. Sci., 7, 1137 (1972).
12. P. H. Hodkinson and J. S. Nadeau, J. Mater. Sci., 10, 846, 1975).
13. P. W. R. Beaumont and R. J. Young, J. Mater. Sci., 10, 1334 (1975).
14. R. K. Govila, "Methodology for Ceramic Life Prediction and Related Proof Testing," Interim Tech. Rept. AMMRC TR 78-29, July 1978.

15. R. H. Keays, "Review of Stress Intensity Factors for Surface and Internal Cracks," Structures and Materials Report 343, Dept. of Supply, Australian Defense Scientific Service, Aeronautical Research Laboratories, April, 1973.

16. R. C. Shah and A. S. Kobayashi; Stress Analysis and Growth of Cracks. Am. Soc. Test. Mater., Spec. Tech. Publ. No. 513, pp. 3-21 (1972).
17. A. A. Griffith, Phil. Trans. Roy. Soc. London, Ser. A, 221, 163 (1920).
18. J. A. Coppola, R. C. Bradt, D. W. Richerson and R. A. Alliegro, J. Am. Ceram. Soc. Bull., 51, 847 (1972).
19. A. G. Evans and S. M. Wiederhorn, Int. J. Fract. Mech., 10, 379 (1974).
20. J. J. Petrovic and L. A. Jacobson, Ceramics for High Performance Applications, Chestnut Hill, Mass., pp. 397-414 (1974).
21. J. J. Petrovic, L. A. Jacobson, P. K. Talty and A. K. Vasudevan, J. Am. Ceram. Soc., 58, 113 (1975).
22. A. G. Evans, L. R. Russell and D. W. Richerson, Met. Trans., 6A, 707 (1975).
23. N. J. Tighe, J. Mater. Sci., 13, (1978).
24. S. M. Wiederhorn and N. J. Tighe, J. Mater. Sci., 13, 1781, (1978).
25. M. G. Mendiratta and J. J. Petrovic, J. Am. Ceram. Soc., 61, 226 (1978).
26. F. F. Lange, J. Am. Ceram. Soc., 56, 518 (1973).
27. J. J. Mecholsky, S. W. Freiman and R. W. Rice, J. Mater. Sci., 11, 1310 (1976).
28. R. K. Govila and K. R. Kinsman, Am. Ceram. Soc. Bull., 57, 316 (1978).
29. R. K. Govila, K. R. Kinsman and P. Beardmore, J. Mater. Sci., 14, 1095 (1979).
30. R. J. Charles, J. Appl. Phys., 29, 1657 (1958).
31. R. W. Davidge, J. R. McLaren and G. Tappin, J. Mater. Sci., 8, 1699 (1973).
32. G. D. Quinn and R. N. Katz, Am. Ceram. Soc. Bull., 57, 1057 (1978).

SESSION Va

TEST, EVALUATION, AND DEVELOPMENT OF COMPONENTS

Chairman: Mr. V. N. Saffire
General Electric Company

Vice Chairman: Dr. R. Kamo
Cummins Engine Company

COMBUSTION RIG DURABILITY TESTING OF TURBINE CERAMICS

W. D. Carruthers, D. W. Richerson, K. W. Benn

AiResearch Manufacturing Company of Arizona
A Division of The Garrett Corporation
Phoenix, Arizona

INTRODUCTION

A necessary prerequisite to accurate prediction of component life and achievement of system reliability is an understanding of the oxidation/corrosion mechanisms and kinetics for ceramic gas turbine components. In many ways, the situation is analogous to metals technology. The mechanism and rate of oxidation/corrosion can vary markedly, depending upon the composition of the environment, the temperature, and the composition and structure of the material. However, the mode of failure is different for ceramics than for metals. Metals are life-limited in oxidation and corrosion by material removal and the resulting loss in performance. The material-removal rate is much slower for ceramic materials, but the life appears to be limited by formation of surface flaws that reduce the effective material strength.

The oxidation/corrosion studies on ceramic materials performed at AiResearch have been guided by the following philosophy:

(a) Testing should be conducted so that the specimens are exposed to high-velocity combustor discharge gases of the same fuel that will be used for the engine application, plus addition of relevant vapor and condensed gas stream impurities.

(b) Specimens should be exposed to the complete temperature cycle expected in the engine.

(c) A specimen configuration should be selected that permits strength measurement after exposure, in addition to weight and dimensional change.

(d) Surfaces, microstructure, and fracture surfaces should be examined by optical microscopy, scanning electron microscopy and, where necessary, by such techniques as X-ray diffraction, electron microprobe, or others that can detect composition or phase variations and help define the oxidation/corrosion mechanism.

This paper discusses the facilities used to perform this work, and summarizes the results of the evaluations.

INITIAL OXIDATION/CORROSION TESTING

Initial oxidation/corrosion testing at AiResearch was conducted under in-house funding in 1974 and under the ARPA/Navy Ceramic Engine Demonstration Program in 1975 and 1976. An existing burner rig facility designed for metals was used for testing. This rig, shown schematically in Fig. 1, has the following features:

(a) Automatic temperature measurement and control with an Ircon radiation-pyrometer system that can control temperature to within $\pm 15^{\circ}C$ ($\pm 25^{\circ}F$)

(b) Automatic burner cycling between two temperature set-points, including controlled cycling to room temperature by airblast

(c) Controlled addition of sea salt water or other desired contaminants to the burner flame

(d) A control system that provides continuous, unattended cyclic testing with automatic shutoff if undesirable conditions develop during a test

The existing metals test fixture was modified to hold flexure test-suitable ceramic specimens with a cross section of 0.32 x 0.64 cm (0.125 x 0.25 in.). Initially, set-screws were used to hold the ceramic specimens in position. A layer of asbestos cloth was placed between the ceramic and the set-screw to minimize point loading. This was not reliable, and specimens tended to work out of the holder after only a few hours of cyclic testing. Replacing the asbestos with a metal shim and applying a greater clamping load did not resolve this problem. The problem was finally resolved by a notch being machined in the ceramic specimen, in which the set-screw was gently snugged. A high-temperature ceramic cement was also applied to the exposed set-screw threads for protection and to keep the set-screws from loosening.

Initial testing was conducted within the range of 900° to $1100^{\circ}C$ (1650° to $2010^{\circ}F$). The rig and fixture worked well under these conditions. However, much higher temperatures were required

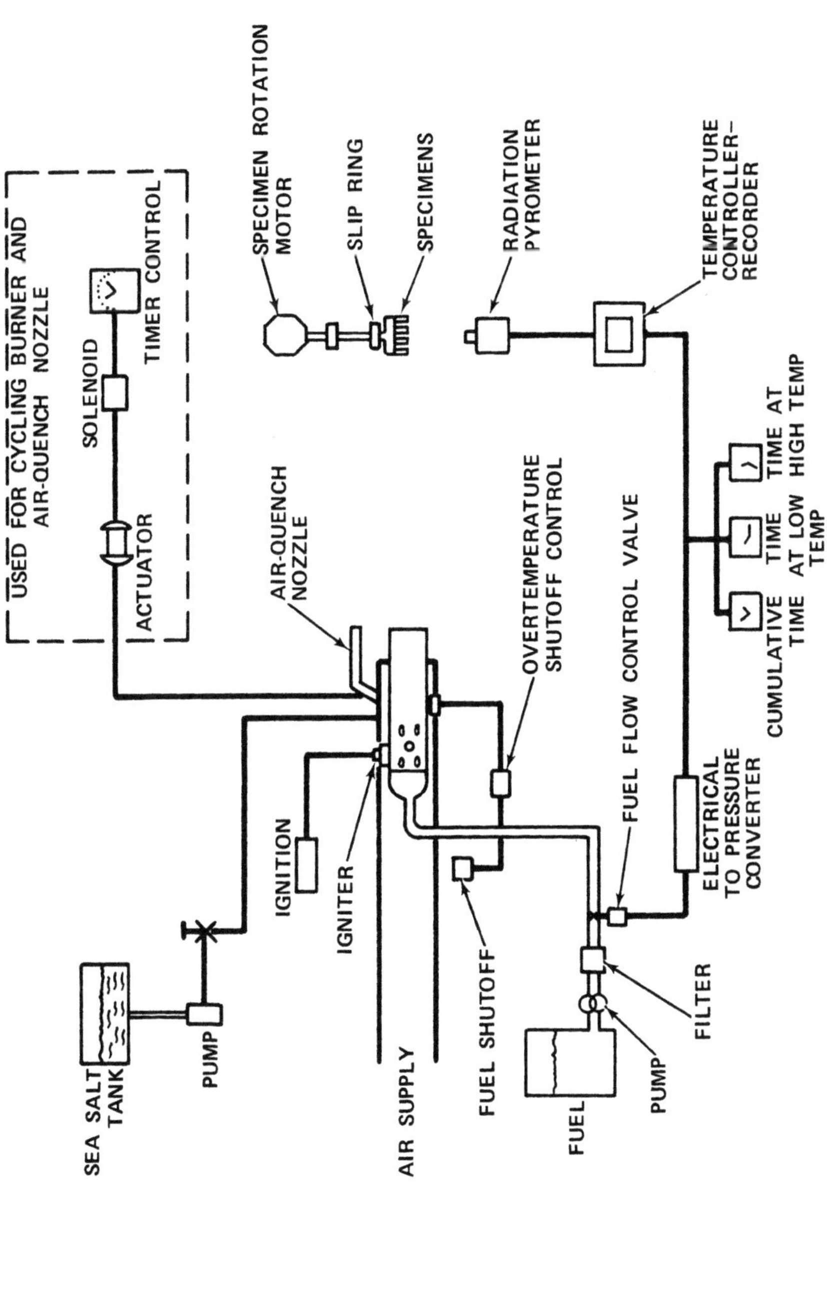

Figure 1. Schematic of Test Rig Used For Initial Ceramics Oxidation/Corrosion Studies.

under the ARPA/Navy program, and both rig and fixture modifications were needed. Even with modifications, testing was still limited to 1230°C (2250°F). To achieve the desired temperature of 1400°C (2550°F), the rig would have to be rebuilt and a cooled fixture designed; neither of which were in the scope of the program task.

Table 1 summarizes test results obtained under the ARPA/Navy/ program.[1*] In each case, cyclic exposure under combustor discharge conditions with no impurities added (other than those already present in the jet aircraft fuel) resulted in increased strength. Fractography indicated that the increase resulted from decrease in severity of surface flaws. Addition of sea salt resulted in the formation of a glassy buildup on the surface, accompanied by strength decrease. Fractography demonstrated that Si_3N_4 had been dissolved locally at the surface in the region of exposure. Figure 2 compares the appearance of specimens tested with and without 5 ppm of sea salt.

Testing conducted under the ARPA/Navy program indicated that Si_3N_4 ceramic materials have good potential for gas turbine conditions, but longer time, higher temperature exposure would be required to determine durability and reliability of these types of materials.

NASA 3500-HOUR DURABILITY OF COMMERCIAL CERAMICS PROGRAM

Recent studies at AiResearch have been performed under a DOE-funded NASA contract (DEN3-27) and have been directed toward evaluation of commercially available ceramic materials for application to automotive gas turbine engines. Conditions selected for this evaluation, including cyclic thermal exposures to 1370°C (2500°F) and 3500 hr, were selected to simulate combustor discharge environment encountered by hot-flow-path components in this type engine.

Four ceramic materials to be tested were selected initially by NASA as representing a cross section of existing and developing Si_3N_4 and SiC material technology. The materials were Norton NCX-34 hot-pressed silicon nitride, Carborundum pressureless sintered, Alpha ( ) SiC, Norton NC-435 siliconized SiC, and AiResearch Air Ceram 101 reaction-bonded silicon nitride (ACC RBN-101). Delivery problems were encountered with the NC-435, which was replaced in the test matrix with Refel siliconized SiC obtained from British Nuclear Fuels, Ltd. (BNFL), through Pure Carbon Company. These materials were to be evaluated as test bars in the as-machined condition.

The cyclic thermal exposure program was constructed so materials would be selectively exposed to successively longer cyclic thermal environments. The NASA-specified testing sequence is summarized in the following paragraphs.

---
*Elevated numbers are to References at end of paper.

Table 1. Oxidation/Corrosion Results from ARPA/NAVY Program.

Material	Surface Preparation	Exposure	Strength, ksi*	
			Before Exposure	After Exposure
NC-132 Hot-Pressed Silicon Nitride (HPSN)	320-grit transverse diamond ground	50 cycles: 1065°C (1950°F)/1 hr, airblast quench/5 min	63	99
	320-grit longitudinal diamond ground	100 cycles: 1120°C (2050°F)/5 min, airblast quench/3 min	97	108
		25 cycles: 900°C (1650°F)/1.5 hr, 1120°C (2050°F)/0.5 hr, airblast quench/5 min		71
		25 cycles with 5 ppm sea salt: 900°C (1650°F)/ 1.5 hr, 1120°C (2050°F)/0.5 hr, airblast quench/ 5 min		100
NXC-34 HSPN		50 cycles: 1065°C (1950°F)/1 hr, airblast quench/5 min	139	150
NC-350 RBSN	As-nitrided			36
	320-grit longitudinal diamond ground			43
	320-grit transverse diamond ground			42
	As-nitrided	70 cycles: 1230°C (2250°F)/5 min, airblast quench/3 min		34
	320-grit transverse diamond ground	70 cycles: 1230°C (2250°F)/5 min, airblast quench/3 min	23	35
	320-grit transverse diamond ground	135 cycles with 5 rpm sea salt: 1065°C (1950°F)/ 3 min, 900°C (1650°F)/5 min, airblast quench/3 min	23	26
	As-nitrided	25 cycles: 900°C (1650°F)/1.5 hr, 1120°C (2050°F)/0.5 hr, airblast quench/5 min	31	30
	As-nitrided	25 cycles with 5 ppm sea salt: 900°C (1650°F)/ 1.5 hr, 1120°C (2050°F)/0.5 hr, airblast quench/ 5 min	31	17

*Measured in four-point bending with an outer span of 4.11 cm (1.5 in.) and an inner span of 1.91 cm (0.75 in.)

A.

25 CYCLES: 900°C (1650°F)/1.5 HR, 1120°C (2048°F)/0.5 HR, AIRBLAST QUENCH/5 MIN. NO SEA SALT ADDITION.

B.

25 CYCLES: 900°C (1650°F)/1.5 HR, 1120°C (2048°F)/0.5 HR, AIRBLAST QUENCH/5 MIN. 5 PPM SEA SALT ADDITION.

Figure 2. Si_3N_4 Specimen Exposure without (A) and with (B) Sea Salt Addition to the Combustion Gases.

Baseline and pretest material evaluations were performed on all test bars including dimensional, density, visual, surface penetrant, surface finish, X-ray radiography, specific damping, and room temperature and 1200°C (2200°F) four-point flexure testing. After inspection and baseline testing, 12 specimens of each material were exposed at 1370°C (2500°F) and 12 at 1200°C (2200°F) for 350 hr/1750 cycles, in a gas burner corrosion/erosion test rig. Testing was conducted so that five cycles were completed each hour with test bars held at test temperature for at least 11 min, then cooled to below 200°C (400°F).

Following exposure, all specimens were again inspected and measured for specific damping capacity changes and then strength-tested at room temperature in four-point flexure. Materials showing greater than 50-percent degradation in strength (compared to the four-point flexure strength of unexposed specimens) were eliminated from further testing.

Materials showing less than 50-percent degradation in strength after 350 hr/1750 testing-cycles were exposed to cyclic testing at 1370° and 1200°C (2500° and 2200°F) for 1050 hr/5250 cycles. Materials showing less than 50-percent degradation in the room temperature strength after 1050 hr/5250 exposure-cycles were then exposed in groups of 12 specimens at 1370° and 1200°C (2500° and 2200°F) for 2100 hr/10,500 cycles.

One material showing less than 50-percent degradation in room temperature strength after 2100 hr/10,500 exposure-cycles will be selected for exposure in groups of 24 specimens at 1370° and 1200°C (2500° and 2200°F) for 3500 hr/17,500 cycles as the program continues. After exposure, 12 specimens will be strength-tested at room temperature and 12 at 1200°C (2200°F), and results compared with comparable tests of unexposed control specimens.

The material selected for 3500-hr thermal cycling will also be used for isothermal exposure in the test rig at 1370° and 1200°C (2500° and 2200°F) for 350 and 1050 hr, respectively. Twelve specimens from each exposure variation will then be strength-tested at room temperature and 1200°C (2200°).

Test Facility Design

Results of prior burner rig tests conducted at AiResearch showed that new specimen fixture and rig facility designs would be required to accomplish the NASA long-term, high-temperature test objectives. Specifically, the high-temperature test cycle would require an air-cooled specimen holder with a positive mechanism for gripping the test bars through the complete temperature cycle and an automatic fail-safe test facility and control system.

Specimen Holder

A successful specimen-holder concept was devised with a spring-loaded holder design that wedged the test bars between compliant packing material and a flat metal surface. This specimen holder design (Figs. 3 and 4) incorporates a circular HS-188 housing, in which a concentric HS-188 body is positioned. The perimeter of this body is slotted to provide space for 24 ceramic test bars of 0.32 x 0.64 cm (1/8 x 1/4 in.) cross section. A circular, slotted specimen spacer, located between the housing and body, provides additional support for the test bars and also provides a cavity for a compressible woven ceramic packing material, such as HitCo Refrasil. When compressed between the body and specimen spacer, this packing material provides a uniform radial pressure on each ceramic test bar. Thus, the test bars are held firmly between the packing material and the outer housing. The compressive radial load on the packing material is maintained between the housing and body by conical spring washers and an adjustment nut mounted on the concentric housing and body shafts outside the high-temperature area. These spring washers apply a force that compresses the packing material between the conical spacer, the tapered inner body, and the inner surface of the specimens. To date, this holder concept has proven very successful and, after an adjustment during the early part of each test, retains the specimens in the holder without damage.

Air cooling and instrumentation passages are provided through the hollow body and housing shaft. Coolant air travels through the body shaft, through coolant passages in the specimen spacer, between and around the test bars, and exits through ports in the housing. Coolant air is directed away from the high-temperature environment by the housing skirt. This airflow direction is noted on Fig. 3.

Test Facilities

To accomplish the required volume of test bar exposures within the 2 years scheduled for the NASA program, four oil burner rigs and control systems were fabricated. With the use of the 24-specimen capacity holders and four test rigs, concurrent evaluations at both 1370° and $1200^{o}C$ (2200° and $2500^{o}F$) on as many as four materials (12 test bars each) is possible.

Each of the four burner rigs consist of a Trane Thermal Co., Model 2085, high-velocity oil burner, a furnace-type enclosure, a specimen drive and actuation system, and associated cycle and temperature controls (Fig. 5).

With this system, the elevated temperature atmosphere is produced by the oil burner and furnace-type enclosure. Thermal cycling is accomplished by translating the specimen holder in and out of the furnace enclosure through a specimen port. After specimen with-

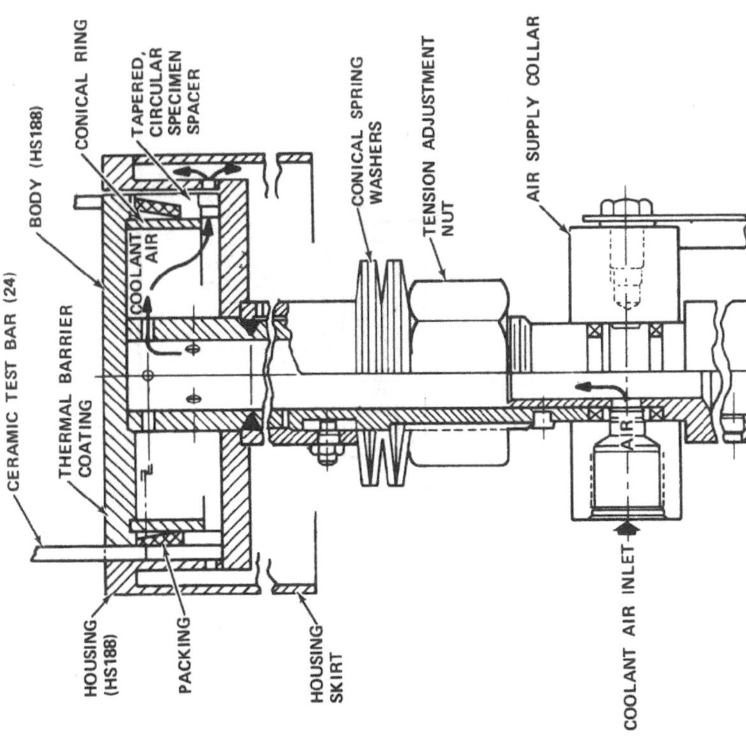

Figure 3. Air-Cooled Specimen Holder for Ceramic Test Bars.

Figure 4. Assembled Specimen Holder.

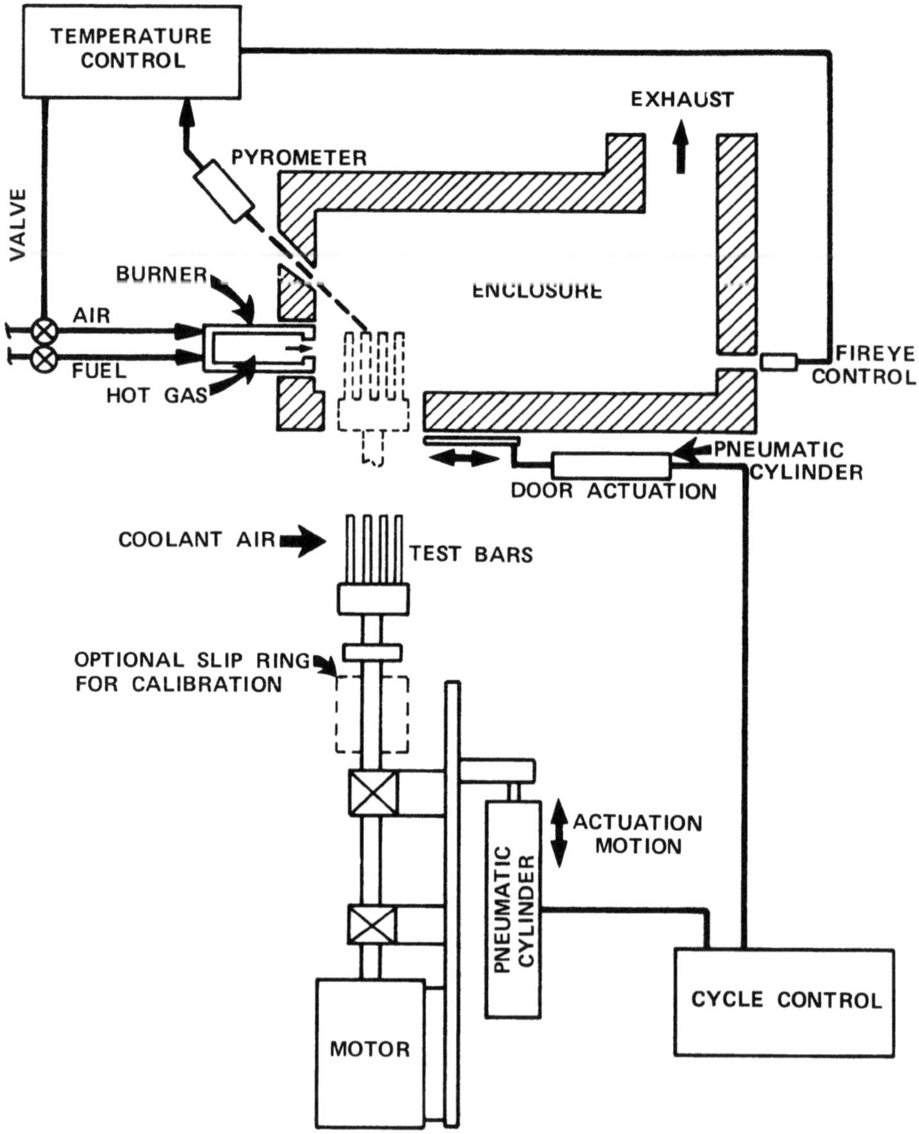

Figure 5. Schematic of Ceramic Test Bar Durability Test Facility.

drawal, coolant air is directed onto the test bars to reduce the temperature to below 200°C (400°F). Constant specimen holder rotation ensures that all specimens are exposed to similar conditions.

Furnace Enclosure. The furnace enclosure is fabricated from 6-in.-thick firebrick encased by stainless steel. Ports are provided in the enclosure assembly for burner installation, specimen holder insertion, pyrometer sighting, burner control, and gas exhaust. After specimen withdrawal, a sliding door on the specimen port closes to prevent furnace heat loss and to allow rapid specimen cooling.

Control System. A control system has been incorporated into the burner rig design that allows continuous, unattended cyclic testing with automatic specimen withdrawal if out-of-limits temperatures develop during a test.

Specimen temperature control is maintained continuously by a Mikron Instrument Co., Model 66, infrared pyrometer, which is focused on the section of the specimen in the direct burner discharge flow. The output from this pyrometer is fed through converters into a Honeywell proportional control unit that varies the air and fuel into the burner. Temperature limits are set at ±25°C (±50°F) and, if these limits are exceeded, an alarm sounds. The specimen holder is withdrawn from the furnace, cooled to below 200°C (400°F), and held in this condition until manual reset is accomplished.

Additional fail-safe features include:

(a) A Fireye Combustion Controls Ultraviolet scanner on the burner flame to initiate shutdown if flameout occurs.

(b) A rotation monitor, so the specimen holder will not be inserted into the furnace if rotation should stop.

(c) A door-position monitor, so the specimen holder cannot be activated into the furnace unless the door is open.

(d) Other normal external fuel shutoffs and safety features included in the facility design.

Specimen Drive System. The specimen holder actuation drive system provides a direct drive to the specimen holder and allows an instrumented specimen holder and slip ring assembly to be mounted on the drive shaft for temperature calibration. The gearbox and motor drive are mounted on a base plate that is supported by two slide shafts and ball bearing slides. For specimen actuation into or out of the enclosure, the base plate is translated by a pneumatic actuator.

EXPERIMENTAL PROCEDURES

All ceramic test bars were machined to 10.2 x 0.64 x 0.32 cm (4.0 x 0.250 x 0.125 in.) with chamfered edges by means of No. 320 diamond-grit grinding wheels. Prior to testing, all bars were pre-exposure inspected by optical, dye-penetrant (except RBSN), and microfocus X-ray techniques. Test bars with detectable indications were omitted from the test matrix and measured for density, weight, and dimensions, as well as specific damping capacity and surface finish. Those not meeting surface finish requirements of 8 μin. rms for NCX-34 and Refel, 15 μin. for Carborundum sintered SiC, and 24 μin. for ACC RBN-101 were also eliminated from further testing. Baseline flexure strengths for all materials were obtained at both room temperature and 1200°C (2200°F) in four-point bending. As in all testing, a minimum of 12 test bars of each material were used.

All inspection bars to be exposed were cleaned and assembled in the specimen holders. Prior to each exposure sequence, rig temperature controls were calibrated, using a special instrumented specimen holder assembled with four instrumented test bars (one of each test material). These were instrumented with Pt-Pt 10 percent Rh thermocouples, and a Lebow slip ring for data retrieval. This assembly fits over the actuator shaft just as the actual bar holders and can be used to measure both peak and minimum temperature values during exposure cycles. With this assembly, accurate pyrometer and control system calibration is possible before each test sequence and can easily be repeated at appropriate intervals during exposure testing, as desired.

A separate furnace temperature monitoring system, using a digital recorder, is also calibrated at this time. This recorder is in operation during the entire exposure period to give a complete historical record of the furnace temperature.

After calibration, loaded specimen holders were mounted on burner rigs adjusted to operate at either 1200° or 1370°C (2200° or 2500°F), at 5 cycles/hr. Upon completion of the desired exposure duration, specimen holders were removed and disassembled. Post-exposure inspection again included visual and dye-penetrant examination, as well as optical and SEM photography. Test bars were also measured for weight, dimensional, and specific damping capacity changes. Four-point flexure strengths at room temperature were subsequently obtained, and fracture surfaces and origins examined by SEM.

Based on the post-exposure room temperature flexure strength results, the materials for additional exposure evaluation were selected. In general, those materials maintaining 50 percent or more of the room temperature baseline flexure strength were selected for further evaluation.

In summary, these rigs have accumulated over 12,500 hr of operation in the last 8 months without any major problems or indication of specimen exposure outside the set temperature limits. Maintenance time, spare parts replacement, and operator time have been well within the expected estimates. Although several commercial power failures have resulted in some lost test time, the fail-safe features have worked satisfactorily in all cases. Specimen and test-time loss has also been well within expectations, and continuous utilization 7 days a week, 24 hr a day, has been demonstrated.

EXPERIMENTAL RESULTS

Baseline flexure strenth results for the four materials tested are summarized on Table 2. Both Weibull and normal statistics have been used to evaluate these results. At 1200°C (2200°F), all materials except Norton NCX-34 exhibited strength improvements over room temperature results. Examination of bar surfaces at both temperatures revealed that bars tested at 1200°C (2200°F) had blunted surface features, when compared with the as-machined surfaces. This as-machined surface blunting during exposure to 1200°C (2200°F) is credited for the flexure strength improvements.

Examinations of the fracture origins of the Carborundum sintered -SiC test bars substantiated the surface flaw-blunting concept for improved strength. Exposed bars more often fractured from internal sites, whereas all unexposed bars fractured at surface origins. Additional examination of baseline ACC RBN-101 test bar fracture origins, however, indicated that for both room temperature and 1200°C (2200°F) testing, over two-thirds of the fractures originated from subsurface sites. These results suggest that in this porous material, the flaw blunting may also affect near-surface, and internal flaws for improved strength.

Fracture origins for the Refel siliconized SiC were not clearly identifiable for either room temperature or 1200°C (2200°F) tests. However, the predominantly tensile face-initiated fractures and improved strength and surface features, when tested at 1200°C (2200°F), support the belief that the flaw blunting was also occurring. The decrease in strength of NCX-34 at 1200°C (2200°F) is similar to that reported by Larsen and Walther[2].

Results of NCX-34 Exposure

Testing of the Norton NCX-34 material at both 1200° and 1370°C (2200° and 2500°F) was terminated prior to achievement of the first 350-hr exposure due to test bar degradation, as observed within 50 hr of exposure at low-temperature regions of the test bars. This degradation was typified by local cracking, expansion, and light brown discoloration in the attachment region. Similar material degradation was observed by Lange[3] and others for Si_3N_4 with

Table 2. Summary of Baseline Four-Point Flexure Strength Data.

Material	Room Temperature					1200°C (2200°F)			
	Average Density, g/cm^3	Average* Strength, ksi	Standard* Deviation, ksi	Weibull Modulus	Characteristic Strength,** ksi	Average* Strength, ksi	Standard* Deviation, ksi	Weibull Modulus	Characteristic Strength,** ksi
Norton NCX-34	3.35	121.0	7.3	17.9	124.4	78.8	9.4	9.0	83.0
Carborundum Sintered SiC	3.11	45.8	10.9	4.8	50.0	50.8	7.3	7.0	54.8
Pure Carbon Refel	3.08	55.4	10.2	4.9	60.4	69.3	6.9	10.4	73.2
ACC RBN-101	2.82	37.4	2.8	12.8	38.6	46.4	3.5	14.5	48.0

NOTES:

1. Outer Span 1.5 in.
2. Inner Span 0.75 in.
3. Cross head speed 0.02 in./min

*Based on normal distribution statistics

**Defined as the strength level at 63.2 percent cumulative percent failure, Weibull statistics

13 percent Y_2O_3, in which oxidation and subsequent volume changes at 1000°C (1830°F) were responsible for cracking and disintegration of unstable phases (notably the J and H) of the Si-Y-O-N system.

X-ray diffraction analysis was conducted for reacted regions of one test bar and a reference, unreacted region of the same test bar. This study revealed several items of interest:

(a) No notable change in the H-phase was observed.

(b) No change in either the position or intensity of the -Si_3N_4 X-ray diffraction peak was observed between the reacted and unreacted areas of the NCX-34.

(c) The unreacted area showed peaks for WSi_2, whereas the reacted material showed none.

(d) No new phase appeared after exposure.

(e) No changes in background were noted, suggesting that no amorphous phases were formed during exposure.

These preliminary results suggest that a reaction other than that identified by Lange may be occurring.

To determine the temperature sensitivity of the degradation, specimens of the NCX-34 were oxidized for 24 hr at 925°, 980°, and 1040°C (1700°, 1800°, and 1900°F); no degradation occurred (Fig. 6). Specimens were then placed in a gradient furnace for 29 hr. Degradation resulted in the temperature range from 580° to 820°C (1080° to 1504°F). A second gradient furnace exposure for 190 hr resulted in degradation between 545° and 890°C (1015° and 1630°F), as shown in Fig. 7. The dark areas are non-degraded material, indicating inhomogeneity within the material and suggesting composition or possibly impurities as a controlling factor.

These oxidation results indicate that a spontaneous material degradation was responsible for the phenomenon observed during durability testing. Based on these results, NCX-34 was suspended from further durability testing.

Results For Carborundum Sintered α-SiC

Cyclic exposure of Carborundum sintered α-SiC to 1200° and 1370°C (2200° and 2500°F) resulted in improved flexure properties over baseline, as indicated in Figs. 8 and 9. At 1200°C (2200°F) after 350 hr of exposure, the average flexure strength rose by 13.5 percent and increased to 16.4 percent after 1050 hr. Both exposure durations resulted in decreased data scatter, as also indicated by the Weibull moduli and data upper and lower values (Fig. 8).

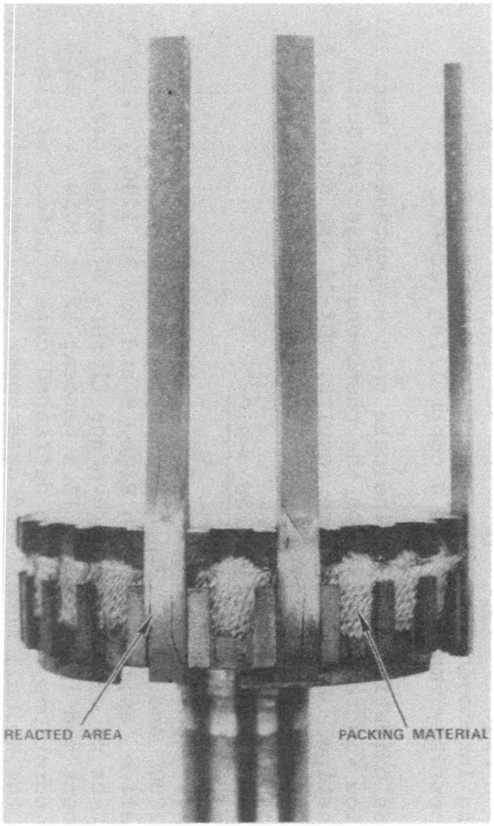

Figure 6. NCX-34 Test Bars After Exposure to 1370^{0}C for 47.1 Hr.

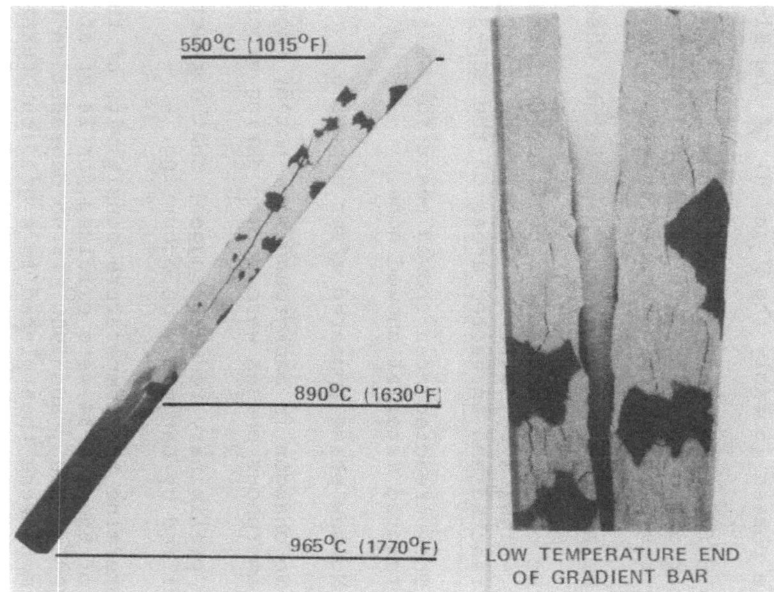

Figure 7. NCX-34 Test Bar After 190-Hr Gradient Furnace Exposure.

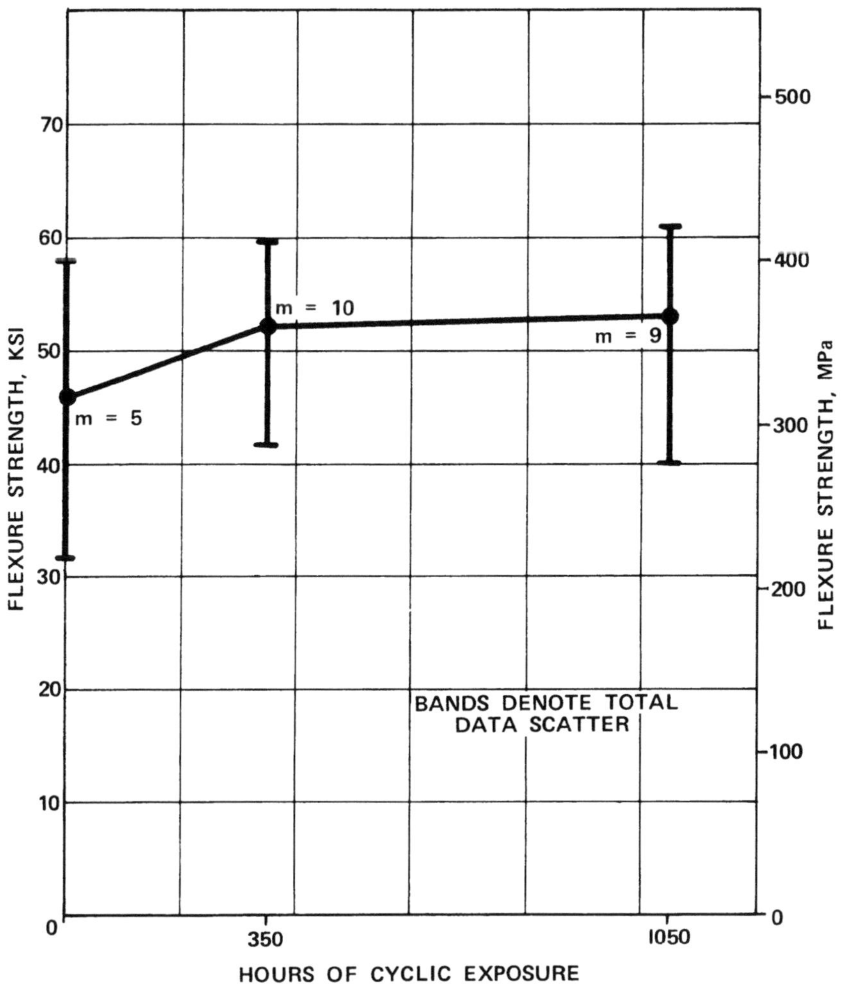

FIG. 8. CARBORUNDUM SINTERED SiC – ROOM TEMPERATURE FLEXURE STRENGTH AFTER CYCLIC EXPOSURE TO 1200°C (2200°F).

Figure 8. Carborundum Sintered SiC - Room Temperature Flexure Strength After Cyclic Exposure to 1200°C (2200°F).

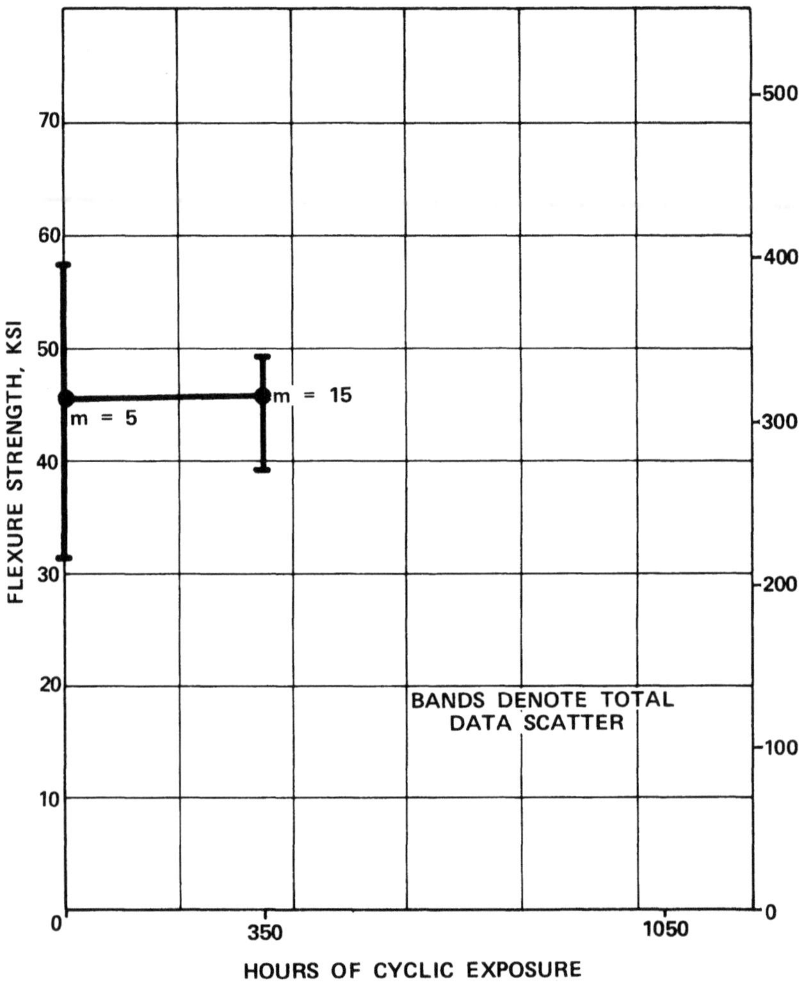

Figure 9. Carborundum Sintered SiC - Room Temperature Flexure Strength After Cyclic Exposure to 1370°C (2500°F).

Data also indicate that the average strength improvements are generally due to a reduced number of low-strength fractures rather than an increase in the maximum flexure strength. These results suggest that the property improvement is a result of surface flaw healing. Examination of exposed surfaces and fracture surfaces on test bars exposed for 350 hr supports this concept, in that most of these test bar fractures originated at subsurface porosity sites. These same results, however, were not observed on bars exposed to 1200°C (2200°F) for 1050 hr. The primary fracture origins for these test bars were surface oxide reaction sites. The fact that a strength improvement was still obtained after 1050 hr of exposure indicates that the flaw healing mechanism is still active, and/or the oxide reaction sites are less severe surface flaws than the as-machined surface defects.

Measurements of dimensional and weight changes with exposure to 1200°C (2200°F) indicate that no dimensional change and a slight weight loss (0.05 percent) occurred after exposure for 350 hr. Additional exposure to 1050 hr resulted in an increase in dimensions and weight of 1.7 and 0.03 percent, respectively. These slight gains are attributed to possible oxide and combustion product build-up during exposure.

After 350 hr and 1750 cycles of exposure to 1370°C (2500°F), the α-SiC flexure strength rose by 0.2 percent, and a significant decrease in scatter was obtained (Fig. 9.) Nearly all test bar fractures originated at surface pore locations. The improvement in data scatter might be attributed to a more uniform surface flaw distribution, produced from flaw blunting with flaws slightly less severe than those induced during the machining operation. Although no dimensional change was detected, a weight loss of 0.35 percent was measured for these bars after 350 hr.

Results for ACC Reaction Bonded Si_3N_4

Cyclic exposure to 1200° and 1370°C (2200° and 2500°F) produced a decrease in flexure strength of 9.0 and 21.4 percent, respectively, for the ACC reaction-bonded material, as illustrated in Figs. 10 and 11. Based on results obtained after 1050 hr of exposure at 1200°C (2200°F), this strength degradation appeared to occur in the first 350 hr of exposure, after which the strength leveled off. No significant changes in the data scatter were noted, as indicated by the Weibull m-values noted on these figures. Weight and dimensional change trends for the reaction-bonded material are similar to those discussed for the sintered SiC; i.e., no dimensional changes were noted after 350 hr, whereas after 1050 hr at 1200°C (2200°F), a 1.5 percent dimensional increase was measured. Likewise, after 350 hr at 1200° and 1370°C (2200° and 2500°F), weight losses of 0.12 and 0.17 percent were recorded, whereas after 1050 hr at 1200°C

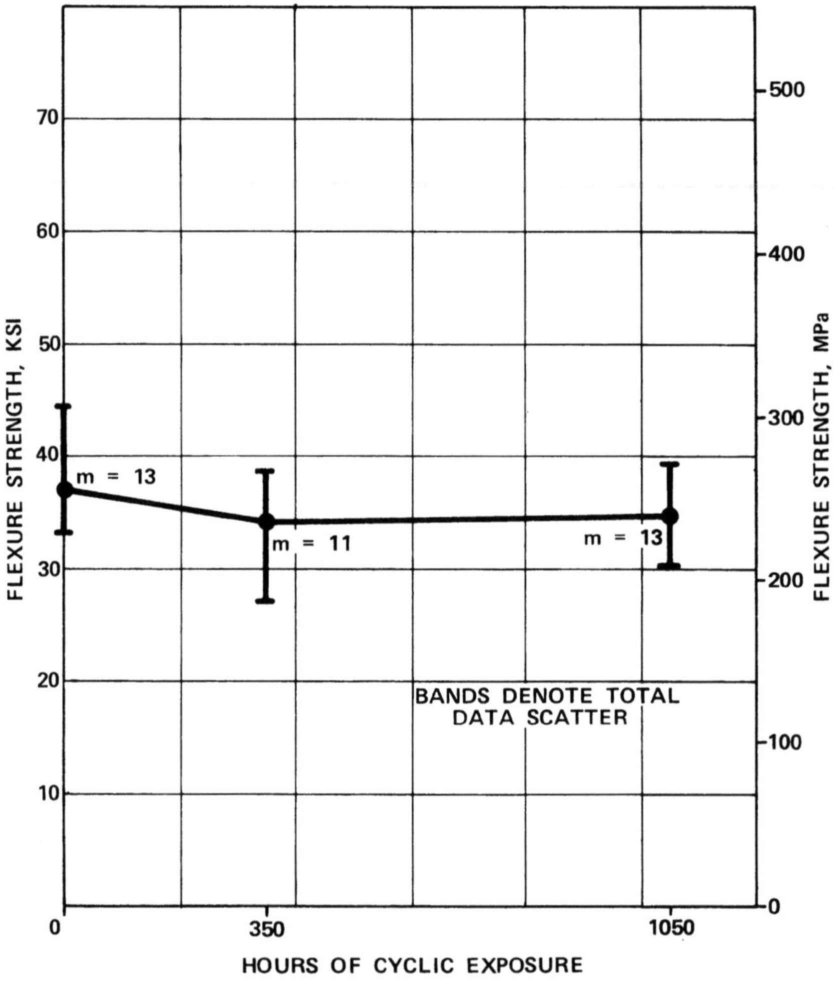

Figure 10. ACC RBN-101 Room Temperature Flexure Strength After Cyclic Exposure to 1200°C (2200°F).

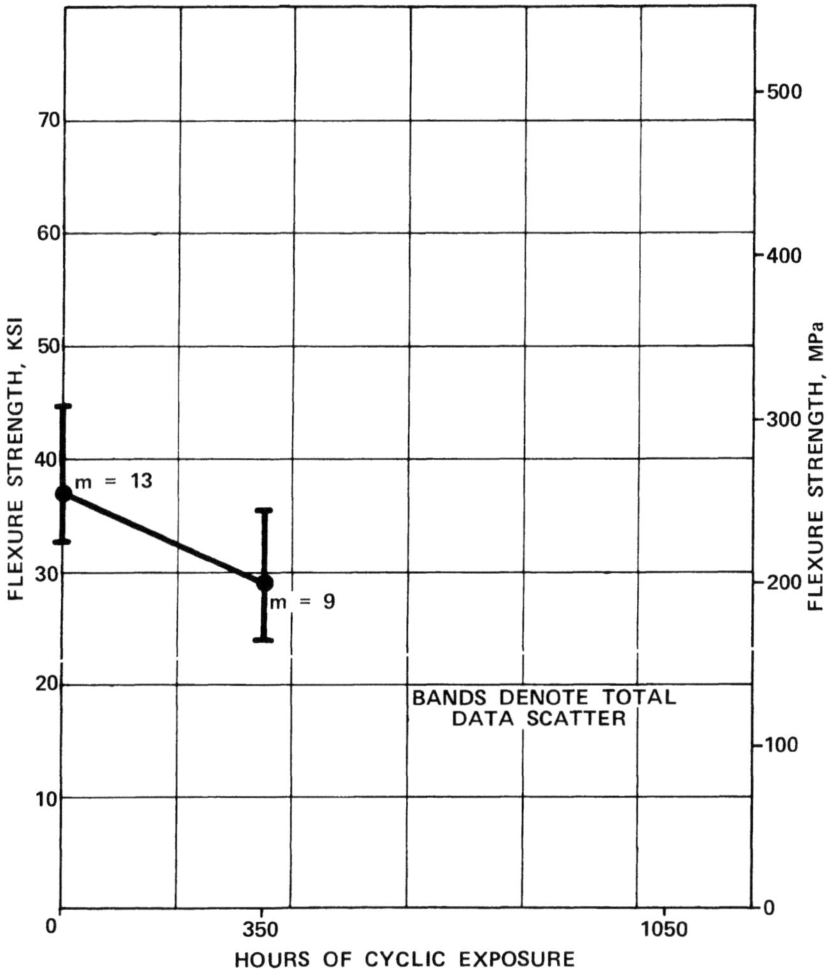

Figure 11. ACC RBN-101 Room Temperature Flexure Strength After Cyclic Exposure to 1370°C (2500°F).

(2200°F), a weight gain of 0.15 percent was obtained. Oxidation and combustion product buildup were attributed to these gains.

Examination of the ACC RBN-101 fractures indicated that the primary fracture origins for all exposures were surface oxidation reactions. In several instances, the oxide reaction penetrated the surface to existing pores or inclusions, which resulted in strength decrease.

Results for Pure Carbon Refel

Cyclic exposure testing of Refel siliconized silicon carbide was initially performed at both 1200° and 1370°C (2200° and 2500°F); however, after 168 hr/840 cycles of exposure at 1370°C (2500°F), test bar cracking in exposed areas and silicon exuding below exposed areas (Fig. 12) became evident. Based on this obvious degradation, exposure testing at that temperature was discontinued. Testing at 1200°C (2200°F) was continued.

Results of cyclic exposure to 1200°C (2200°F) on flexure strength are summarized in Fig. 13. The figure shows a continued decrease in average flexure strength for time and cycles, with little changes in data scatter. Although the average flexure strength decreased by 20.4 percent with 1050 hr of exposure, the lowest flexure strength values for exposed test bars were higher than the baseline low values.

Fracture origins for Refel test bars are often difficult to identify, due to lack of typical "hackle" lines and fracture mirrors. However, when identifiable, both subsurface and surface fracture origins were observed in test bars exposed for 350 hr, whereas bars exposed for 1050 hr, typically fractured from surface origins.

Dimensional inspection of Refel test bars showed a 0.3-percent growth after 350 hr and 2.8 percent growth after 1050 hr. This dimensional growth was greater than that seen in the ACC RBN-101 and the sintered SiC. Weight measurements showed an initial loss of 2.2 percent at 350 hr and a gain of 0.08-percent after 1050 hr.

Discussion of Specific Damping Capacity Measurements

Specific damping capacity measurements were obtained for each test bar before and after thermal cyclic exposures as a means of monitoring possible thermal shock-induced strength reductions; however, no clear correlation with strength changes has been noted. Although specific damping capacity as great as +16.7 and -30.5 percent has been measured after exposure, these changes are within the scatter obtained on reference test bar measurements. These results, as well as those obtained during other investigations, raise doubt

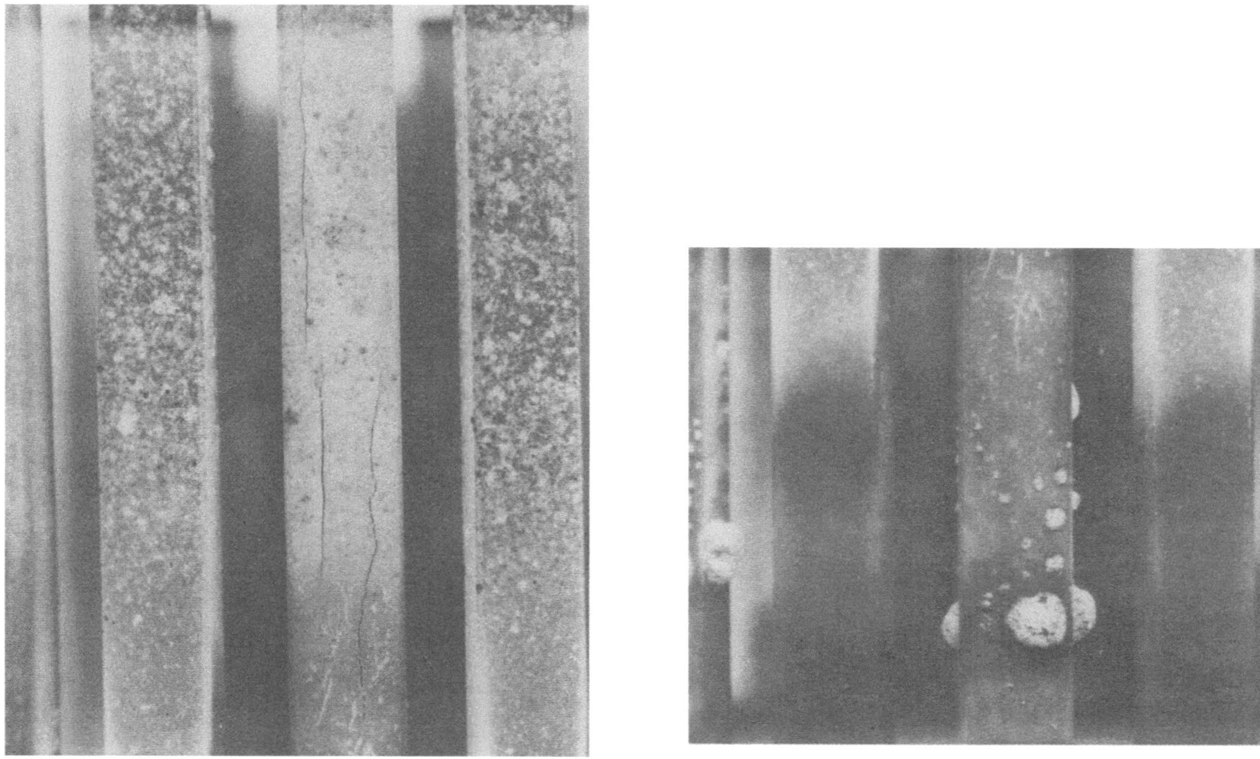

(A) CRACKING (B) Si EXUDING

Figure 12. Close-Up Photo of Refel Test Bars After 168-Hr Exposure to 1370°C (2500°F), Showing Cracking and Si Exuding.

as to the usefulness of the specific damping capacity measurements on test bars exposed to the current test conditions.

SUMMARY/CONCLUSION

A test facility for cyclic exposure of ceramic test bars to 1370°C (2500°F) has been designed and built and has accumulated over 12,500 hr of operation without major problems or indications of specimen exposure outside the set temperature limits.

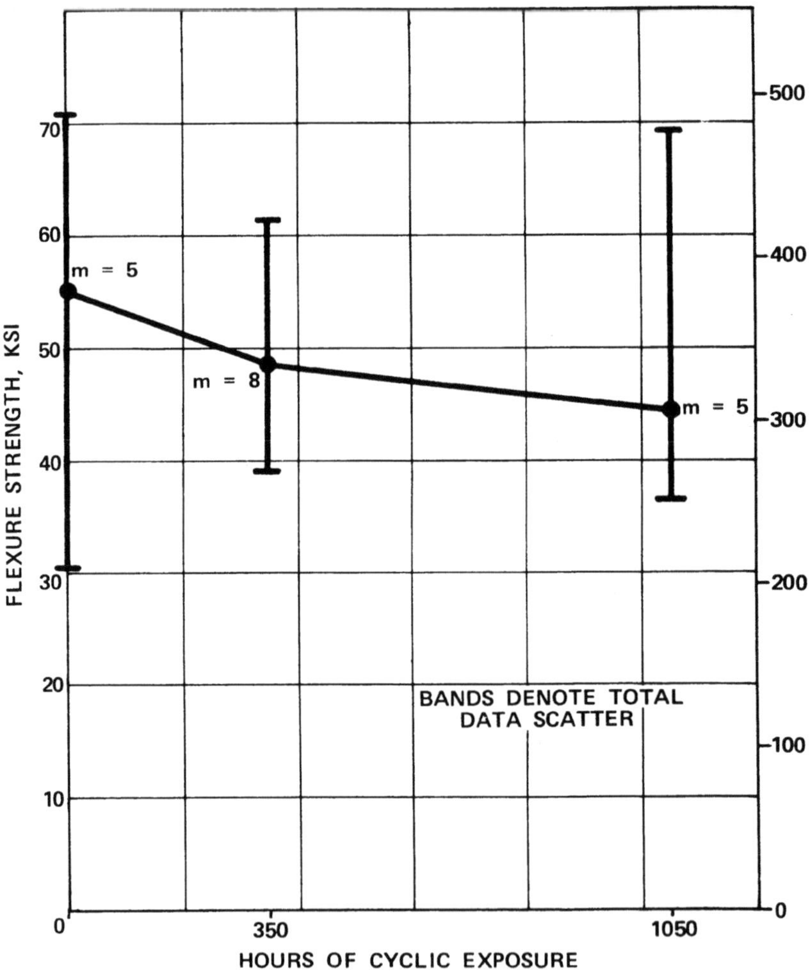

Figure 13. Pure Carbon Refer Room Temperature Flexure Strength After Cyclic Exposure to 1200°C (2200°F).

Ceramic material test results have been encouraging. The Carborundum sintered α-SiC specimens have shown a 16.4-percent increase in strength after 1050 hr/5250 cycles of 1200° (2200°F) exposure and 0.2-percent increase after 350 hr/1750 cycles of 1370°C (2500°F) exposure.

The ACC RBN-101 reaction bonded Si_3N_4 specimens showed a 9.0-percent decrease in strength during the first 350 hr/1750 cycles of 1200° (2200°F) exposure, but no further decrease after 1050 hr/5250 cycles. The RBN-101 has shown a 21.4-percent decrease in strength during the first 350 hr/1750 cycles at 1370°C (2500°F), but results are not yet available for the 1050-hr exposure.

The Pure Carbon Company Refel siliconized SiC decreased in strength by 20.4 percent after 1050 hr/5250 cycles of 1200°C (2200°F) exposure. The 1370°C (2500°F) exposure of Refel SiC was terminated after silicon exuded and material cracked within the first 168 hr/ 840 cycles.

The Norton NCX-34 hot-pressed Si_3N_4 material testing was terminated after less than 50-hr exposure, due to cracking and degradation of the material in a low-temperature region. Separate gradient furnace tests determined the degradation that occurred in the temperature range 545° to 890°C (1015°F to 1630°F), but the mechanism of degradation has not been isolated.

Combustion rig testing will continue for the Carborundum sintered α-SiC, the ACC RBN-101 reaction bonded Si_3N_4, and the Pure Carbon Refel SiC. The rig appears to be a useful and effective means of evaluating the durability of ceramic materials in the high-temperature cycle environment and can be an important tool for evaluating new materials, as available, for heat-engine applications.

REFERENCES

1. D. W. Richerson and T. M. Yonushonis, "Environmental Effects on the Strength of Silicon Nitride Materials," in: Proceedings of the 1977 DARPA/NAVSEA Ceramic Gas Turbine Demonstration Engine Program Review, MCIC-78-36, 247-271 (1978).
2. D. C. Larsen and G. C. Walther, "Property Screening and Evaluation of Ceramic Turbine Engine Materials," AFML Contract F33615-75-C-5196, Interim Technical Report No. 5 (1978).
3. F. F. Lange, S. C. Singhal, and R. C. Kusnicki, "Phase Relations and Stability Studies in the Si_3N_4-Y_2O_3 Pseudoternary Systems," J. Amer. Ceram Soc. 60 [5-6], 249-252 (1977).

DUO-DENSITY CERAMIC TURBINE ROTOR DEVELOPMENT

Robert R. Baker and Lewis R. Swank

Engineering and Research Staff
Ford Motor Company
Dearborn, Michigan 48121

INTRODUCTION

Ford Motor Company has been developing ceramics for gas turbine applications for over 10 years. The replacement of high cost nickel-chrome superalloys with low cost, higher temperature capability ceramics offers significant advantages in efficiency, exhaust emissions, power per unit weight, cost and materials utilization. Successful application of ceramics to gas turbines would therefore not only have military significance but would also greatly influence our national concerns of air pollution, critical materials availability[(1)] and utilization, and the energy crisis.

In July, 1971, the Defense Advanced Research Projects Agency (DARPA) joined Ford to develop and encourage the use of brittle materials for engineering applications. The major program goal of the "Brittle Materials Design - High Temperature Gas Turbine" program was to establish the usefulness of brittle materials in demanding high temperature structural applications by designing and developing the entire ceramic hot flow path of a high temperature (2500°F Turbine Inlet Temperature) vehicular gas turbine engine (Figure 1).

In fiscal year 1977, the Department of Energy (DOE) joined DARPA in partial support of the program (Figure 2). DARPA continued to support fabrication and testing of stationary and rotating components while DOE started to support process development to improve the quality of silicon nitride turbine rotors in addition to some work on ceramic material characterization and development of nondestructive evaluation techniques. The Army

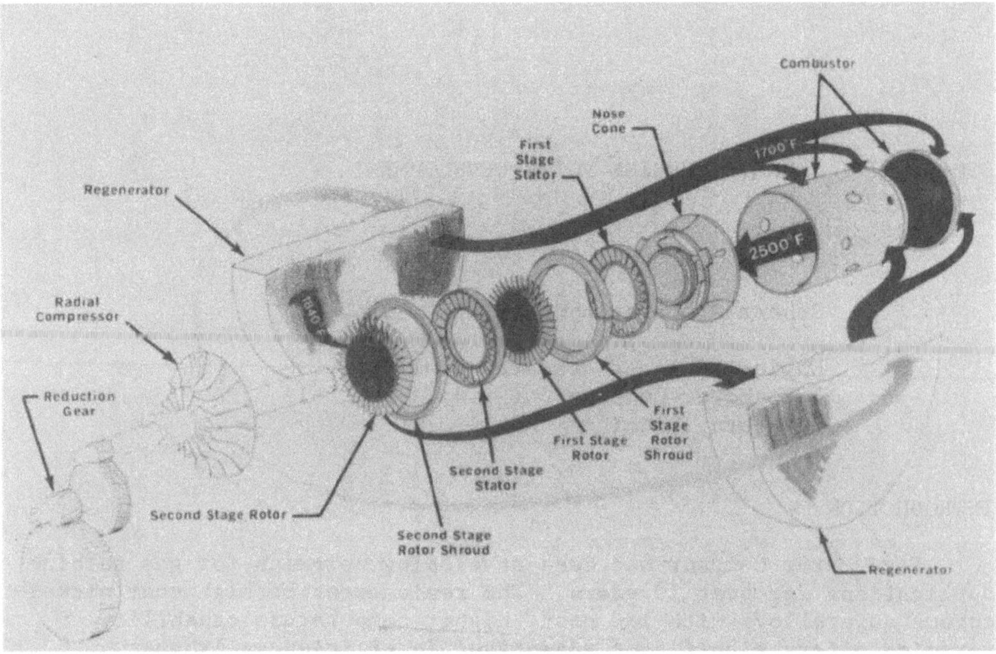

Figure 1 Schematic View of the Vehicular Gas Turbine Engine Flowpath

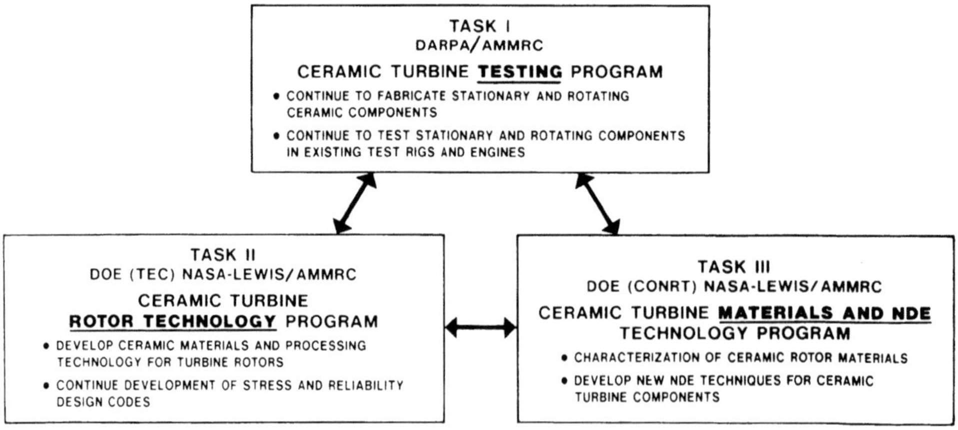

Figure 2 DARPA/DOE Supported Tasks in the "Brittle Materials Design, High Temperature Gas Turbine" Program

Materials and Mechanics Research Center (AMMRC) continued to function as the technical monitor of the overall program.

The stationary ceramic hot flowpath components were designed, developed, and have successfully met the program durability goal of 25 hours of 2500°F plus 175 hours at 1930°F Turbine Inlet Temperature (TIT). The test and development of these components was presented at the "Ceramics for High Performance Applications - II" conference in Newport, Rhode Island, in March, 1977,(2) and updated in the 12th interim report on the program(3).

The fabrication, test and development of all ceramic turbine rotors was presented at the Newport conference(4,5). A brief review of the processes utilized to fabricate turbine rotors, hot spin rig development and test results, and process improvements since Newport are presented in this paper.

ROTOR FABRICATION

The duo-density silicon nitride turbine rotor concept (Figure 3) utilizes high strength hot pressed silicon nitride for the hub where stresses are highest but temperatures are moderate and; therefore, creep resulting from the use of a densification additive is minimized. The complex shaped airfoils of reaction bonded silicon nitride, which has superior high temperature strength, are formed by injection molding or slip casting. The development of the fabrication process used to make duo-density silicon nitride rotors was covered at Newport from its inception(4). The simplified two-piece hot press bonding configuration was utilized to fabricate rotors from December, 1976, to August, 1977 (later referred to as 1977 vintage rotors) and subsequent testing of these rotors is presented in this paper.

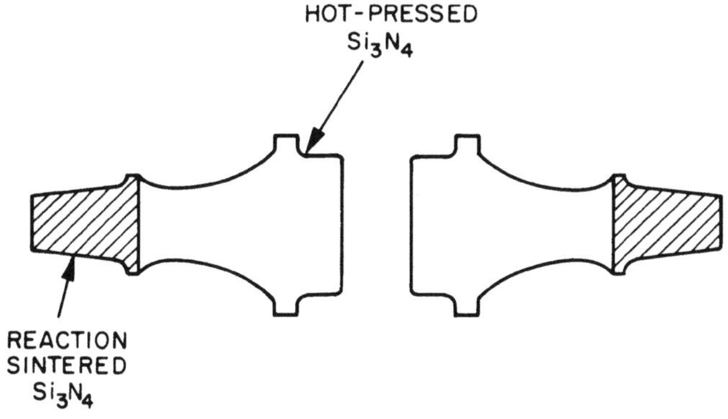

Figure 3 Duo-Density Ceramic Turbine Rotor

The reaction bonded silicon nitride blade ring was injection molded (Figure 4) in one shot forming the 36 complex shaped airfoils and the blade ring rim as an integral part. The green blade ring was burned out to remove the organic binders and subsequently nitrided to 2.7g/cc density reaction bonded silicon nitride. The development of the injection molding process, including the solid state injection molding system, was covered in detail at Newport[6].

The blade ring was encapsulated in a blade fill of slip cast silicon nitride to support the rim during hot pressing and protect the blades from the carbonaceous hot pressing environment[7]. The blade fill process uses boron nitride as a lubricant for subsequent blade fill removal and as a barrier material which prevented blade ring to blade fill bonding during nitriding. Controlling the thickness of the boron nitride layers (Figure 5) was determined to be critical in order to prevent cracking of the blades during hot press bonding[8].

The simplified two-piece hot press bonding configuration (Figure 6) was used to densify the silicon nitride powder to form the hot pressed hub while simultaneously press bonding the hub to the blade ring. A pressure of 1000 psi was applied to the powder for 2-1/2 hours at a temperature of 1715°C. The graphite wedge system shown in Figure 6 was used to support the blade filled blade ring during press bonding. After cooling the blade fill was removed and the rotor machined to the contour previously shown in Figure 3.

Figure 4 Injection Molded Blade Ring

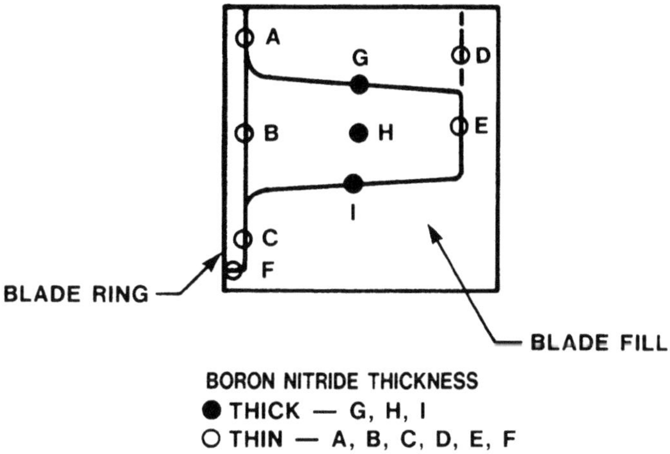

Figure 5 Blade Filled Blade Ring Showing Boron Nitride Layers

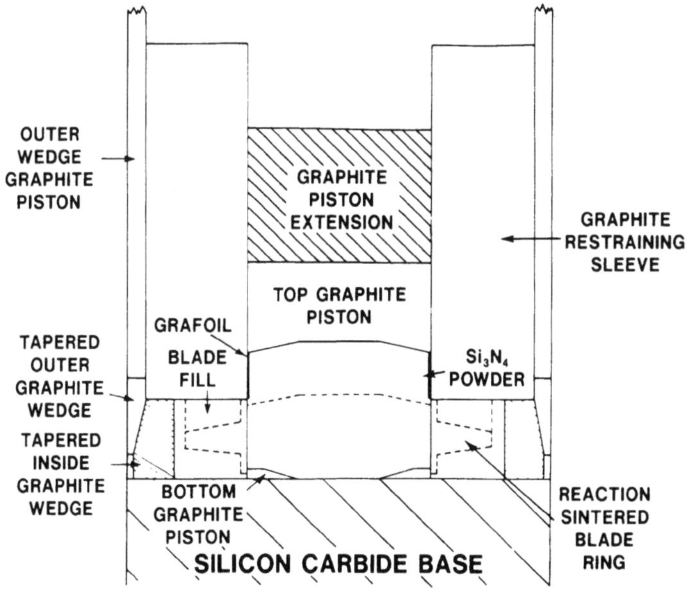

Figure 6 Simplified Two Piece Hot Press Bonding Configuration

COLD SPIN TESTING

All rotors which were to be hot tested in the hot spin rigs or engines were cold spin tested to eliminate flawed blades of the type shown in Figure 7. Two levels of cold spin qualification testing were set; 55,000 rpm for rotors to be hot tested to 50,000 rpm and 70,000 rpm for rotors to be hot tested to 64,240 rpm. Nine out of 16 rotors achieved 55,000 rpm or higher without blade failures; however, 14 out of 17 failed one or more blades while in the process of being qualified to 70,000 rpm. Figure 8 shows the Weibull distribution of 1977 vintage rotors. Also shown in this figure is the Weibull distribution representing the state-of-the-art in 1975[9]. The data, while still not at desired levels, shows significant improvements in rotor processing as the 1977 vintage rotors demonstrated a cold spin reliability almost two orders of magnitude greater than the state-of-the-art in 1975.

HOT SPIN RIG DEVELOPMENT

The purpose of the Hot Spin Test Rig (HSTR) is to evaluate ceramic turbine rotors in an environment similar to that of a turbine engine. The results are fed back to the ceramic material and fabrication engineer to guide him in his development effort. In order to expedite the evaluation of ceramic turbine rotors, the HSTR was designed to be a simple rig with a minimum of ceramic and metal parts in order to minimize the time required to rebuild the rig after failure.

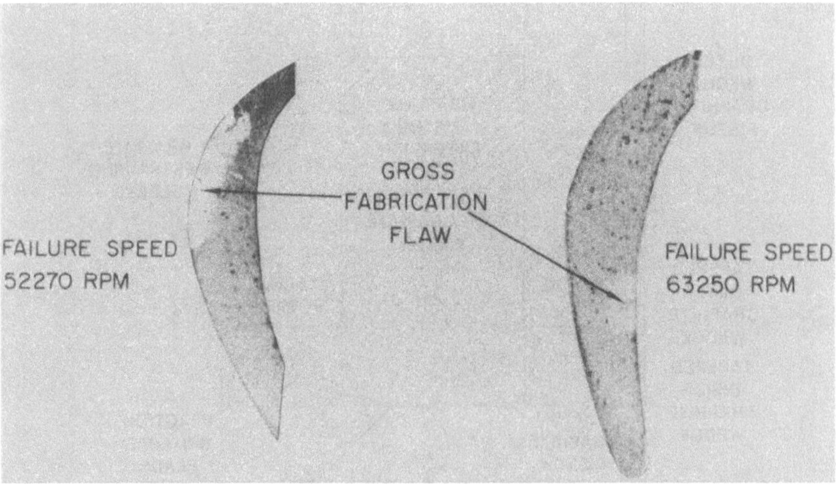

Figure 7 Typical Fracture Surfaces With Gross Fabrication Flaws

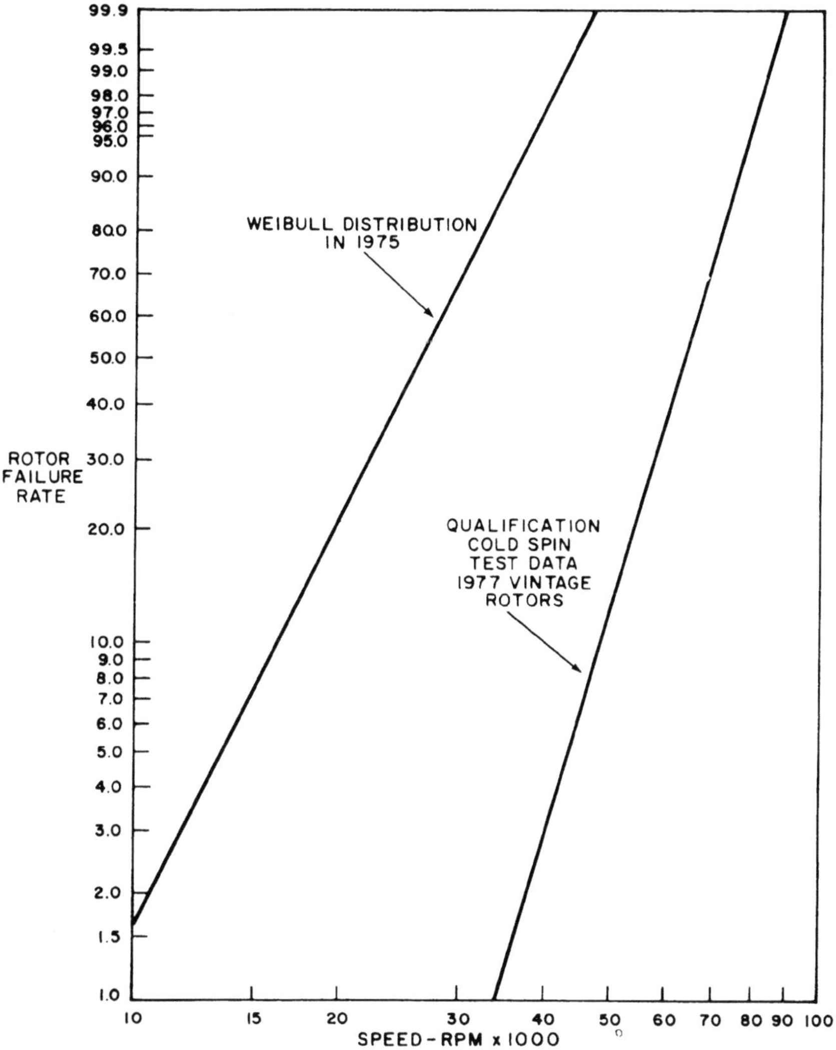

Figure 8 Weibull Distributions of Rotor Failures

The Hot Spin Test Rig configuration previously reported[10] has undergone considerable modification; therefore, this section of the report will review the earlier configuration of the rig, which is shown in Figure 9. The combustor had a set of twelve flame tubes arranged in a circle on the combustor cover at a diameter matching the mid-span of the rotor blades. The flame tubes were evenly spaced on the circle and tube inlets were encased in a plenum so that the air flow to the flame tubes could be controlled. The air-fuel ratio was controlled so that the mixture was maintained at near

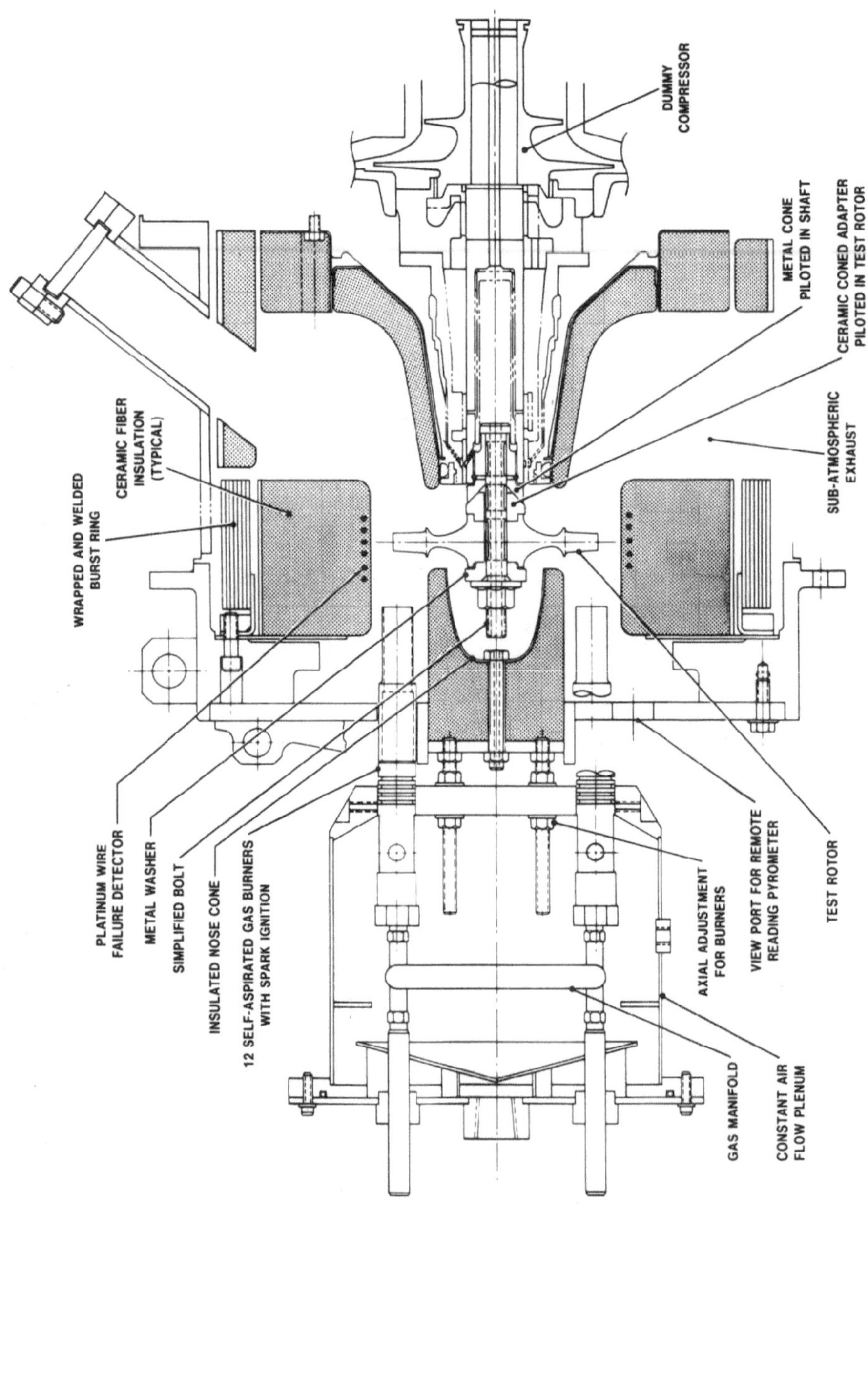

Figure 9 Hot Spin Test Rig Early Configuration

stoichiometric. Mounted in the center of the flame tubes on the combustor cover was an insulated nose cone. This part shielded the hub of the rotor and the rotor attachment hardware from direct radiation from the flames. The insulation used on the nose cone was a premolded shape of alumina ceramic fiber. A thermocouple was mounted in the nose cone attachment bolt to monitor the tie bolt cooling air temperature. The bead of the thermocouple was adjusted so that it was in close proximity to the end of the tie bolt.

The rotor was attached with a ceramic coned adaptor system. This system relied on mating ceramic and metal cones to maintain the rotor concentric and square with the shaft through the required temperature excursions. At the upstream end of the rotor, a metal washer was used to maintain the tie bolt concentric with the rotor. Relative motion between the rotor and washer due to temperature excursions was made possible by gold plating the metal washer, with the gold acting as a solid lubricant. The ceramic coned adaptor system utilized a short tie bolt of low ductility H-11 tool steel.

The early HSTR configuration utilized an alumina ceramic fiber shroud. The shroud had a thin inner ring with platinum wire wrapped around the ring to act as a failure detector. The inner ring was then cemented to the inside of the shroud body. When a rotor failed, the fragments sliced through the wire activating a shutdown cycle.

The temperature of the rotor is one of the important test parameters. The measuring device used to determine the rotor temperature was a radiation pyrometer, which is a practical device for measuring elevated temperatures on a rotating body from a remote location. The pyrometer was mounted on a bracket to hold it in a fixed position relative to the rotor and a viewing port was located in the combustor cover.

Several modifications have been made to the Hot Spin Test Rig and this section describes an intermediate configuration with which a series of 50,000 rpm durability tests were conducted. The intermediate configuration is shown in Figure 10.

The number of flame tubes was reduced from twelve to five. This was done because it was found that twelve flame tubes delivered too much heat at the flow rates required for stable combustion per tube. Reduction of the number of tubes increased the flow rate per tube and put the operation in a stable combustion region. Further improvement in the stability of the combustor operation was achieved by supplying the combustor with excess air, and these two modifications eliminated flame out problems.

The rotor attachment system used in the HSTR has been completely changed. Tests on the simple ceramic coned adaptor attachment system

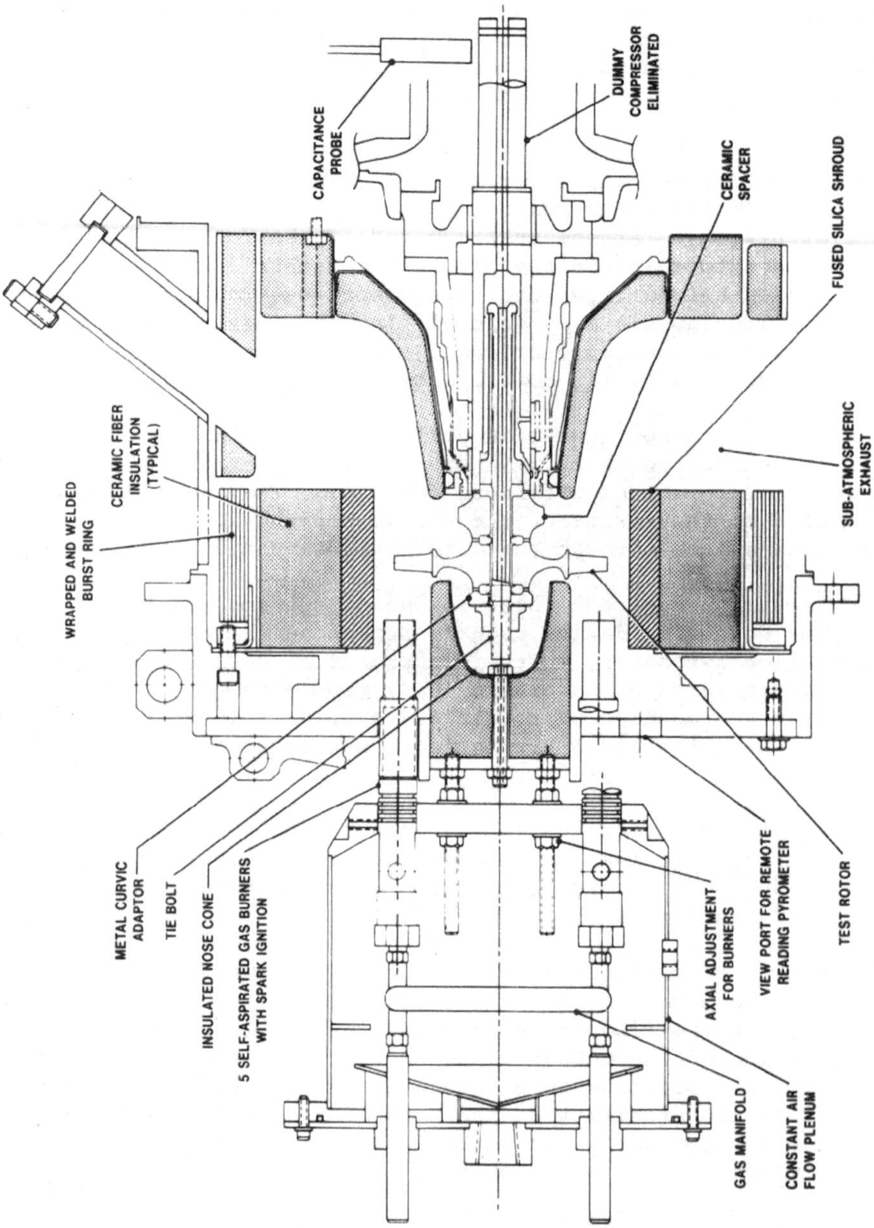

Figure 10 Hot Spin Test Rig Intermediate Configuration

showed that it did not adequately pilot the turbine wheel resulting in increased unbalance with increasing temperature, as a result it was abandoned in favor of a Curvic™ coupling system, which was designed for and proven successful in engine testing. Curvic couplings have the ability to maintain the concentricity of mating parts while accommodating relative radial motion due to differences in thermal expansion. The present attachment system uses a ceramic spacer to replace the second stage turbine to allow testing of a single rotor stage. A metal curvic adaptor pilots the tie bolt relative to the rotor and shaft. A long tie bolt is used to accommodate the differences in thermal expansion between the tie bolt and the ceramic hubs. The tie bolt material was changed from low ductility H-11 tool steel to Inconel 718 to minimize the possibility of bolt failure. All metal curvic surfaces were electroplated with .0001 to .0002 inches of gold to act as a solid lubricant between the ceramic and metal.

Initially, the HSTR shroud was made of a ceramic fiber insulation material. This material worked well for short time tests, but as the rig was developed for longer running times, it became apparent that the ceramic fiber required protection from high velocity gases since it was being eroded. A fused silica shroud was substituted for the ceramic fiber shroud as shown in Figure 10. The fused silica tube nests in the ceramic fiber insulation and is held in place by mechanically packing the insulation around it and bonding with a high temperature cement. This arrangement has proven to be very durable, but the construction does not permit the use of a wire wound failure detector. This has been compensated for by carefully monitoring a capacitance probe sensing shaft vibrations.

The early version of the HSTR used a dummy wheel simulating the compressor, and a compressor nut on the shaft. These two parts were used to duplicate compressor inertia effects on the shaft dynamics. A critical shaft frequency was noted near 35,000 rpm. The dummy compressor and compressor nut were eliminated from the shaft assembly and this eliminated critical frequencies in the operating range. In addition, this modification had two important benefits: one, the unbalanced load was reduced on the shaft after-rotor failure, which eliminated damage to the shaft bearing journals as had been previously reported[10]; two, the shaft and rotor are installed in the test rig without disassembly so that testing is conducted with the same balance as achieved in the balancing operation. This removes the balance uncertainty inherent in shaft systems requiring disassembly of the rotor and shaft for installation in the test rig.

ROTOR TEST RESULTS

The intermediate configuration of the HSTR was used for a series of 25 hour durability tests. The rotor platform temperature for these tests was 1800°F and the speed was 50,000 rpm as planned. Once the test conditions were attained, they were held continuously for 25 hours. The test schedule, Figure 11, shows the rotor platform temperature and speed versus time during the start up and shut down sequences of the test. The results of the 25 hour durability tests are presented in Table 1. The rotor and blade ring serial numbers are given so that the test results can be related to the fabrication process and materials development. The rotors had from 27 to 35-1/2 blades during the test. Those blades that were removed generally contained gross flaws, though some blades were removed for balancing. The rotors went through the start up, testing, and shut down procedures without incident, but during disassembly three rotors gave audible indications of cracking as the tie bolt was unloaded. In two cases, cracks were found, using Zyglo techniques, in the curvic teeth of the rotor on the side mating with the metal curvic adaptor. This indicated that the gold film lubricant was not functioning as required. Radiation pyrometer and temperature sensitive paint measurements indicated that the metal curvic adaptor temperature was 1650°F during the tests. At this temperature, the gold may be diffusing into the silicon nitride rotor and the metal curvic adaptor, reducing its lubricating effectiveness. It was also suspected that a gold-silicon nitride reaction might occur at this temperature forming gold silicides that would reduce the gold's lubricating effec-

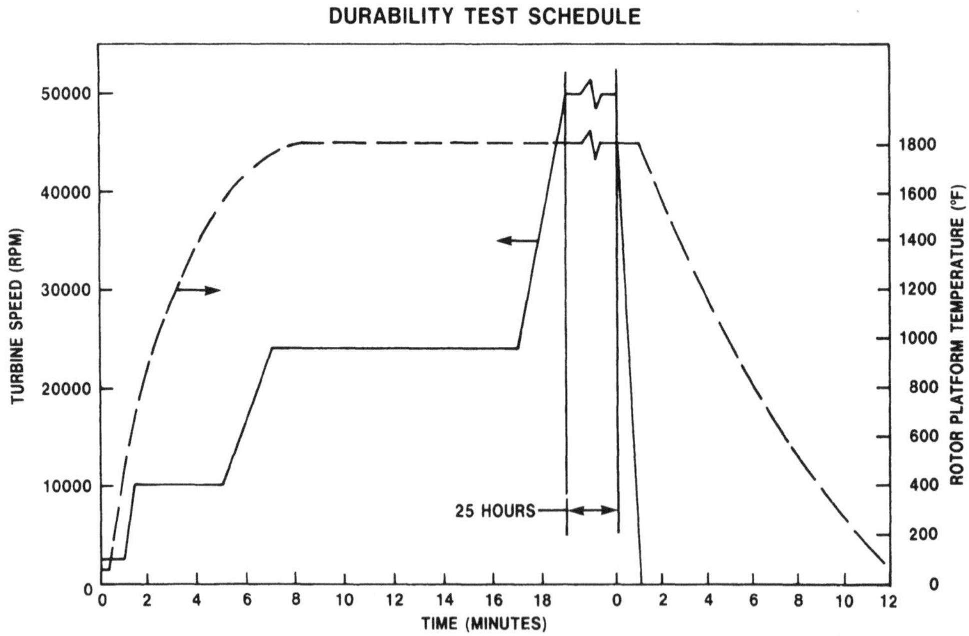

Figure 11 25 Hour Durability Test Schedule

tiveness. X-ray diffraction studies were performed on the curvic teeth of rotor 1324, but the results did not indicate the presence of any gold-silicon compounds.

Figure 12 shows the HSTR in its present configuration. It is the same as the intermediate configuration except an air deflector was added to the nose cone to direct the rotor bolt cooling air against the metal curvic adaptor. Temperature measurements indicated that the temperature was reduced to 1000°F with this configuration when the rotor platform temperature was 1800°F with a rotor speed of 50,000 rpm. This HSTR configuration has been used for testing all rotors since the 25 hour durability tests of Table 1.

The first test conducted with the present configuration was a 175 hour durability test on Rotor 1306, which had been previously tested for 25 hours, to complete a major program objective of 200 duty cycle hours on a rotor. The speeds, time at speed, and percent time of the duty cycle are given in Table 2. The 86.5 percent speed point and below were run as a nonstop 175 hour test. The start-up procedure was the same as shown in Figure 11 except that terminal speed was 43250 rpm. Running and shutdown were without incident, which was very encouraging as this was the first attempt to complete a 200 hour durability test on a rotor.

The next program objective was to test a rotor for 25 hours at 64,240 rpm and 1800°F rotor platform temperature. A series of tests, summarized in Table 3, have been conducted in an attempt to meet this objective. These tests have been conducted with the present rig configuration as shown in Figure 12, and with a test schedule as shown

Table 1

25 Hour Ceramic Rotor Tests

Intermediate Rig Configuration

Rotor Serial Number	Blade Ring Serial Number	Number Of Blades During Test	Rim Temperature °F	Speed RPM	Audible Indications Of Cracking On Disassembly	Zyglo Indications After Disassembly
1324	2032	27	1800	50000	Yes	Yes
1287	2045	27	1800	50000	Yes	Yes
1294	1794	31	1800	50000	Yes	No
1306	2033	35 1/2	1800	50000	No	No

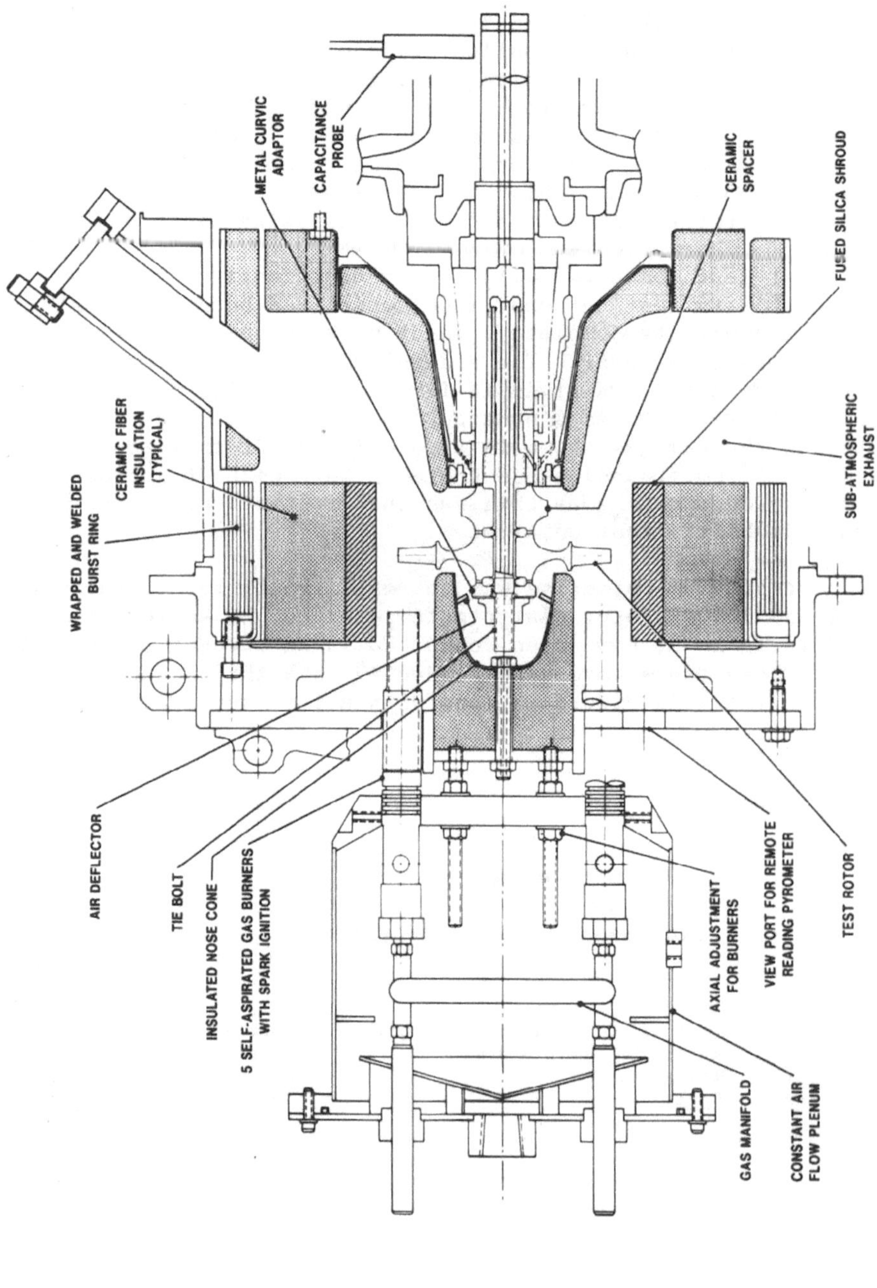

Figure 12 Hot Spin Test Rig Present Configuration

in Figure 11, except with a maximum speed of 64,240 rpm. Only one of the rotors has reached 64,240 rpm, rotor 1312, and it was destroyed when a flame tube vibrated loose and worked forward striking the rotor. Four of the other rotors were completely destroyed during the acceleration from 50,000 rpm to 64,240 rpm. The three rotors which failed at low speeds, 10,000 to 23,000 rpm, were different in nature as the hubs remained intact after the test while the blades and blade ring rim spalled off as shown in Figure 13.

ANALYSIS OF TEST RESULTS

A time dependent reliability analysis was conducted on the rotor configuration and operating conditions of the four rotor tests,[11] which successfully completed 25 hours of operation at 50,000 rpm and 1800°F rim temperature (Table 1). The estimated time dependent reliability was 0.95525 at 25 hours. Although no obvious time dependent failures have been encountered in rotor testing, it is possible to estimate the demonstrated reliability using Bayes formula applied to the success run theorem[12].

$$Rc = (1 - C)^{\frac{1}{N+1}}$$

where Rc = minimum demonstrated reliability
C = confidence level of prediction
N = number of successful trials with no failures

The minimum demonstrated reliability by four successes and no time dependent failures is 0.87055 at 50% confidence. It is interesting to note that in order to demonstrate a minimum reliability of 0.95525, 14 successive tests would be required without encountering any failures[12]. Thus correlation between time dependent reliability predictions and actual test results could be confirmed if 10 additional rotors were tested to 25 hours each without any failures.

A correlation analysis of cold spin rotor hub failures with the calculated failure distribution was conducted in 1976[7]. Fourteen hot pressed silicon nitride rotor hubs were made under as identical conditions as possible. Five of the hubs were sectioned to provide the material strength data for the analytical prediction while the remaining nine hubs were cold spun to failure. The results are shown in Figure 14. The calculated failure distribution (characteristic speed 103,800 rpm, Weibull modulus m = 16.8) was within the 90 percent confidence band of the experimental data (characteristic speed 108,500 rpm, Weibull modulus m = 14.8). This confirmed the use of fast fracture Weibull theory in predicting cold spin failures.

Table 2

200 Hour Test

Rotor 1306

Present Rig Configuration

Rim Temperature °F	Percent Speed	Turbine Speed RPM	Time at Speed Percent	Time at Speed Hours	Total Time Hours
1800	100	50000	12	25	25
1800	86.5	43250	5	10	35
1800	77.5	38750	5	10	45
1800	69	34500	7	14	59
1800	59	29500	25	50	109
1800	55	27500	46	91	200

Table 3

64240 rpm Speed Tests

Present Rig Configuration

Rotor Serial Number	Blade Ring Serial Number	Number Of Blades During Hot Test	Rim Temperature °F	Speed RPM	Time Hours	Comments
1312	2142	28	1800	64240	1 3/4	Flame tube fell into rotor causing failure
1329	2274	24	1800	52800	-	Complete rotor failure
1357	2317	19	1800	57100	-	Complete rotor failure
1364	2336	24	1800	55400	-	Complete rotor failure
1368	2350	36	1800	10000	-	Blade/rim spalling
1382	1323	29 1/2	1800	23000	-	Blade/rim spalling
1392	2371	23	1800	10000	-	Blade/rim spalling
1395	2353	25	1800	60600	-	Complete rotor failure

Figure 13 Blade-Rim Spalling of Rotor 1392

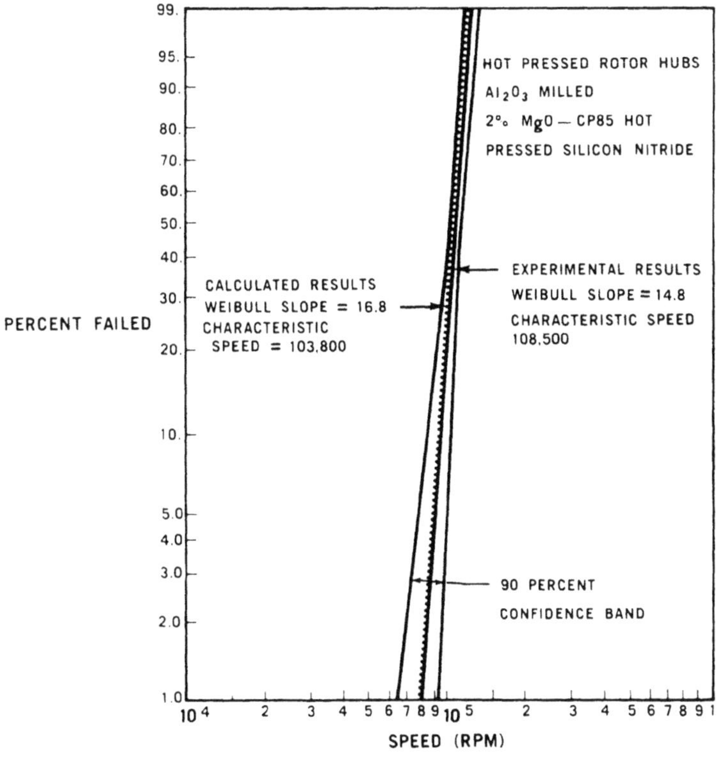

Figure 14 Predicted and Actual Weibull Distributions of Cold Burst Rotor Hubs

The prediction of fast fracture failures at elevated temperatures is more complex due to the combined thermal and centrifugal stresses in the rotor. Speed can be controlled precisely but thermal gradients and hence thermal stresses are much more difficult to predict and control. The measurement of high temperature material properties is also more difficult; however, fast fracture reliability analyses were conducted for the rotors tested in the hot spin rig. The estimated reliability was 0.985 at 50,000 rpm[11] and 0.846 at 64,240 rpm, hence the predicted failure rates were 1.5% at 50,000 rpm and 15.4% at 64,240 rpm. The Weibull distribution of the test results previously presented in Tables 1 and 3, with the low speed blade/rim spalling failures eliminated, is shown in Figure 15. The predicted failure rates fall within the 99 percent confidence band of the experimental data indicating a reasonable correlation.

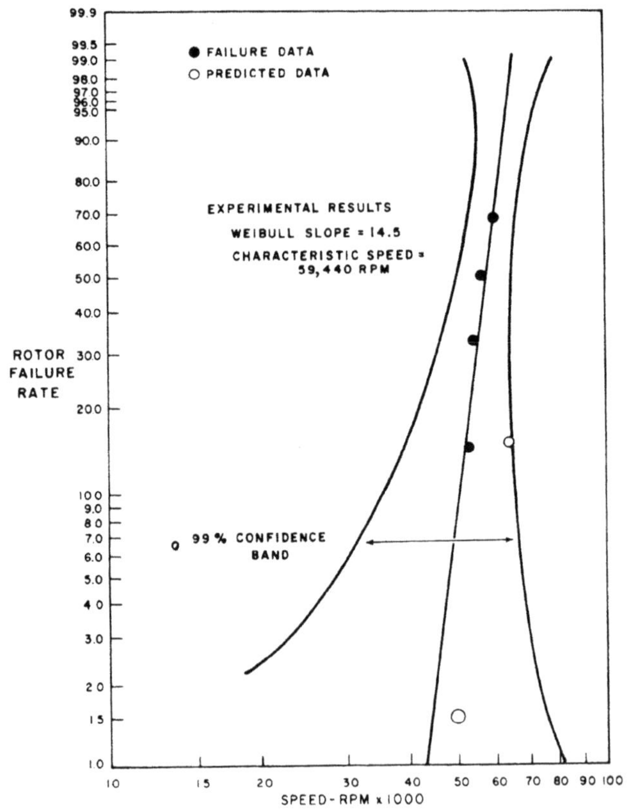

Figure 15 Predicted and Actual Weibull Distributions of Hot Burst Rotors

PROCESSING IMPROVEMENTS

Some rotors produced during the development of the simplified two-piece hot press bonding concept were cold spin tested to failure. The test results of one rotor were particularly interesting as 10 unflawed blades failed in the 87,100 to 96,900 rpm speed range. Typical fracture surfaces are shown in Figure 16. The Weibull distribution representing this data is shown in Figure 17 along with the qualification test results of 1977 vintage rotors previously shown in Figure 8. The potential for unflawed, improved process rotors is shown dramatically by comparing the failure rates at 70,000 rpm for 1977 rotors (80% failure rate) to that for unflawed, improved process rotors (8% failure rate), a tenfold reduction in failure probability. As a result, a program was initiated to improve the rotor fabrication process by improving the injection molding and hot press bonding processes.

Two types of blade flaws were revealed during cold spin testing to qualify rotors for further hot testing. Subsurface voids, shown in Figure 18, and planar flaws, both subsurface and surface (shown in Figure 19), were responsible for the majority of low speed cold spin failures. In the past, blade rings were injection molded, visually inspected, burned out, nitrided and visually inspected again along with conventional x-ray. Recently, conventional and microfocus x-ray techniques were used to detect flaws in the green, as-molded, state prior to burnout and nitriding. This step was taken to eliminate the time required for burnout and nitriding, thereby providing timely feedback to the injection molding engineers who were conducting parametric experiments to eliminate molding flaws.

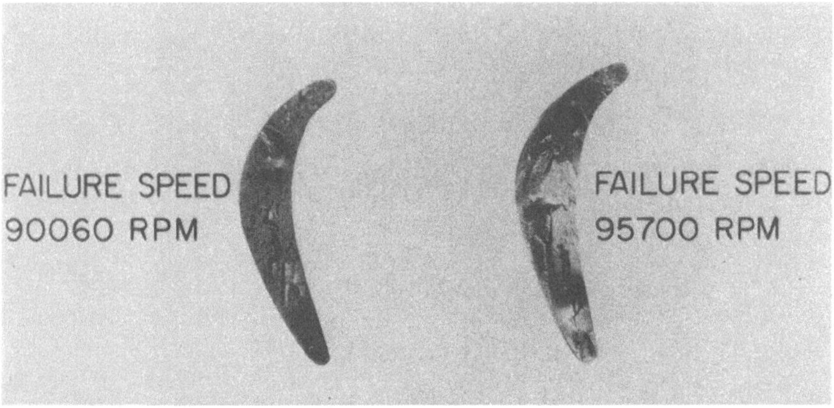

Figure 16 Typical Fracture Surfaces Without Gross Fabrication Flaws

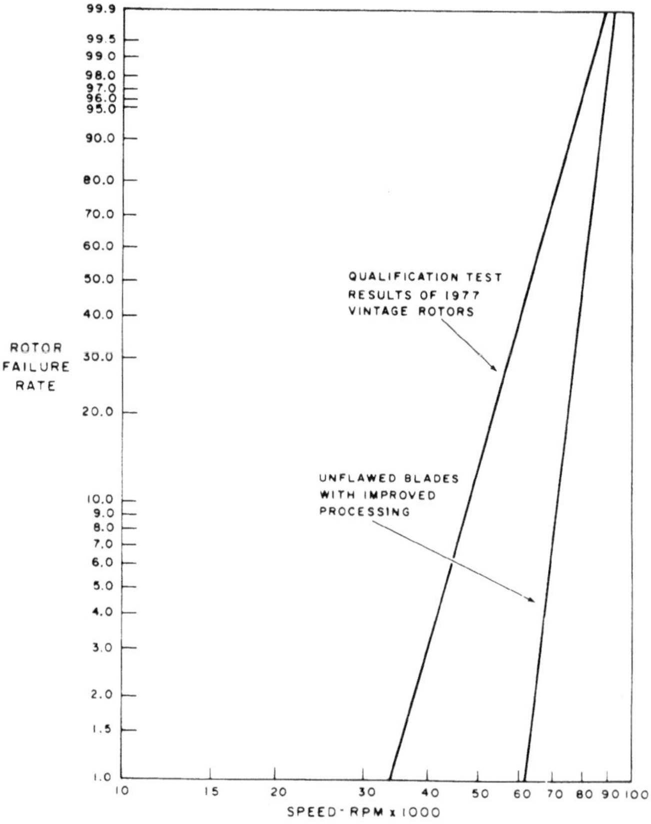

Figure 17 Weibull Distributions of Rotor Failures

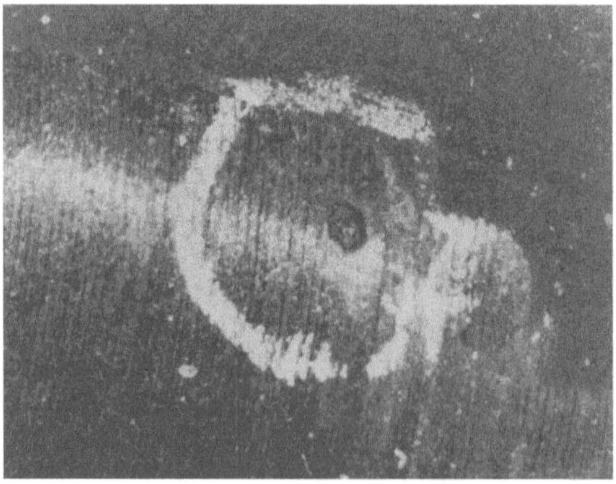

Figure 18 Subsurface Void Uncovered by Grinding

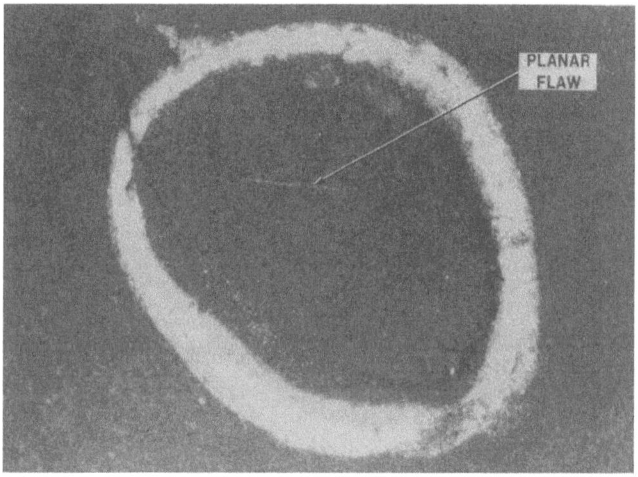

Figure 19 Planar Type Defect Detected by Visual Inspection

A Hunkar flow control unit was coupled to the existing automated injection molding system described by Johnson at Newport(6). This new unit controls and monitors the injection velocity and die cavity pressure during the injection and hold portions of the molding cycle (Figure 20). Ten parametric molding studies were conducted with systematically varied injection profiles and hold pressures. Microfocus x-ray results indicated that high injection flow rates combined with low hold pressures in the die cavity reduced the number of subsurface void-type flaws in the outer half of the blades and completely eliminated voids in the inner, more highly stressed portion of the blades.

Conventional and microfocus X-ray techniques were used to try and detect planar flaws of the type previously shown in Figure 19 in both the green as-molded and nitrided conditions. Up to 18x magnification was used with the microfocus x-ray equipment in addition to multiple orientation of the parts. Planar type flaws were only detectable after nitriding, indicating that this type of flaw may only occur after burn out and/or nitriding.

Flaws in the rim region of the blade ring were detected using the panoramic microfocus X-ray setup shown in Figure 21. The panoramic tube head fits inside the blade ring while the film strip is located outside the blade tips in one of a series of concentric grooves in the plexiglass disk. The result is a radial panoramic X-ray of the rotor blades and rim at up to 10x magnification.

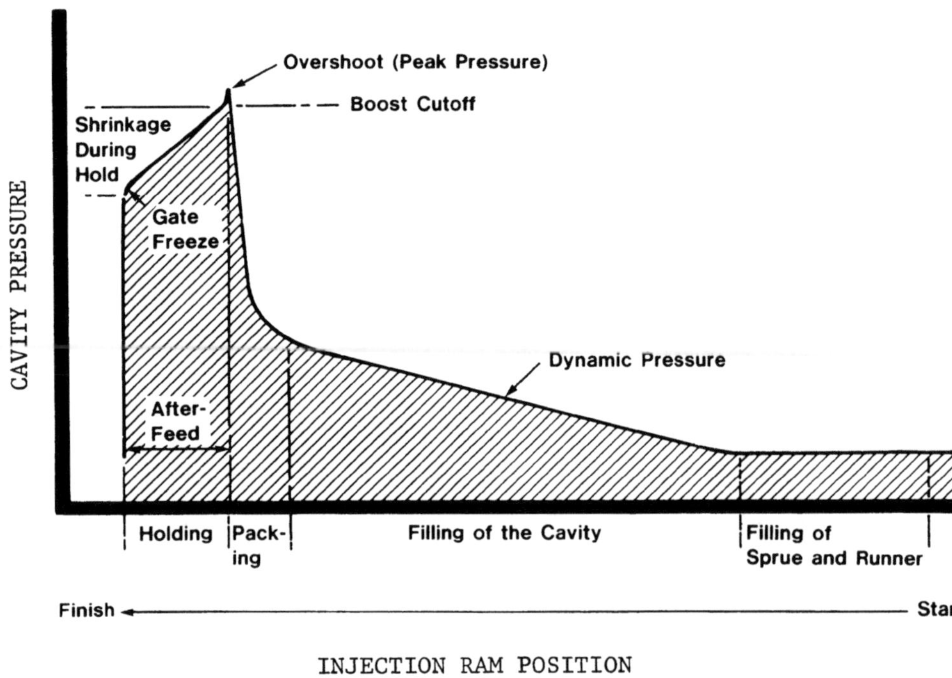

Figure 20 Schematic of Cavity Pressure Versus Ram Position

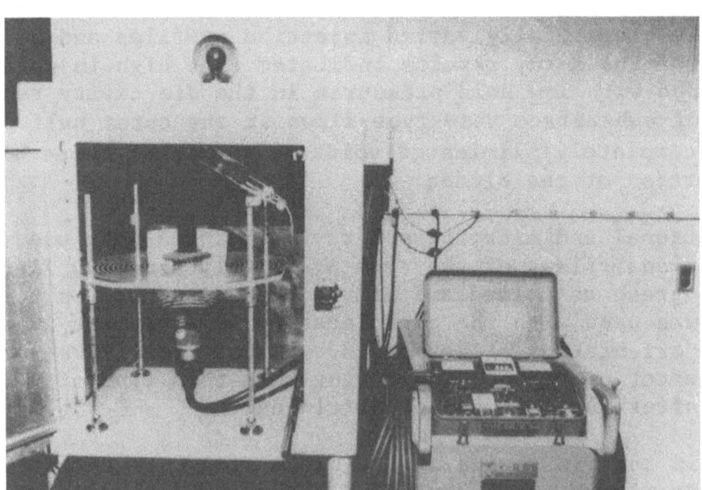

Figure 21 Panoramic Microfocus Equipment and Fixture for Radial X-ray

Several of the 1977 vintage rotors were sectioned and examined with some interesting results with respect to the reaction bonded silicon nitride blade rings. The blade rings varied in color, from light gray to shiny black, and microstructure, from 70% to 0% alpha silicon nitride. A subsequent investigation revealed that the reaction bonded silicon nitride was dissociating, during hot press bonding, resulting in an increase in porosity, hence the gray color. A few rotors were subjected to blade bend testing to determine if blade strength was affected by hot press bonding. The results, Table 4, show the characteristic loads after press bonding were 14 to 38% lower than those for the "as-nitrided" or before hot press bonding state. As a result a degradation study was conducted, which defined the nature of the changes and identified the hot pressing parameters which could be controlled to eliminate this degradation of microstructure and strength. The processing steps in the degradation study are shown in Figure 22. Every other blade of an as-nitrided blade ring was loaded to failure (A). The blade ring was then blade filled (B) and hot press bonded. After hot press bonding (C), the remaining blades were loaded to failure. This technique provided blades from before and after hot pressing from the same blade ring, hence the same nitriding cycle, which were examined with respect to color, microstructure, hardness, and phase composition.

The ranges of hot pressing temperature and time at temperature which were investigated in this degradation study encompassed the nominal conditions of 1715°C and 150 minutes used when fabricating 1977 vintage rotors. The color changes which occurred during hot pressing are shown in Figure 23. As expected, the degree of color change, from black to light gray, increased with hot pressing time and temperature. Comparison of the microstructure before and after hot pressing, for degraded parts, showed that the "after" microstructure contained much more fine interconnected porosity, hence the gray color. This increased porosity degraded the microstructure, which resulted in a strength degradation. Figure 24 shows the relationship derived between change in strength and hot pressing time and

Table 4

Summary of Blade Bend Test Results

NITRIDING	CHARACTERISTIC LOAD (POUNDS)		%
NUMBER	AS-NITRIDED	AFTER HOT PRESSING	CHANGE
48	89.9	77.0	-14
67	85.7	53.0	-38
78	79.1	61.6	-22

temperature[11]. It should be noted that the expected strength degradation at the nominal hot pressing conditions used to fabricate rotors in 1977 (1715°C - 150 minutes) is 10-20%. Hence, future improved process rotors can be fabricated under hot pressing conditions, which would result in no strength degradation effectively increasing their strength 10-20% over 1977 vintage rotors.

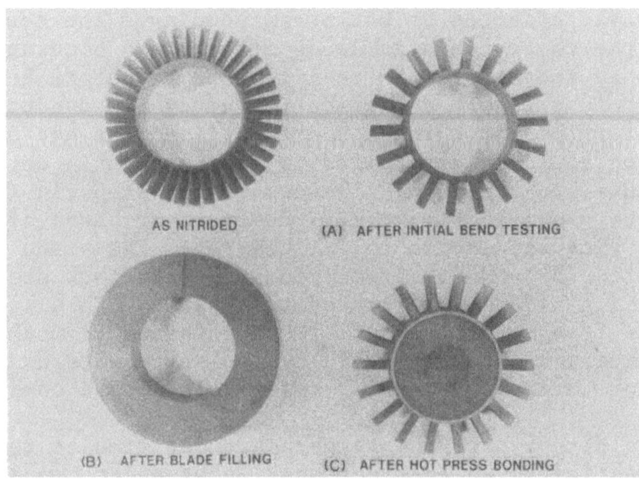

Figure 22 Processing Steps for Degradation Study

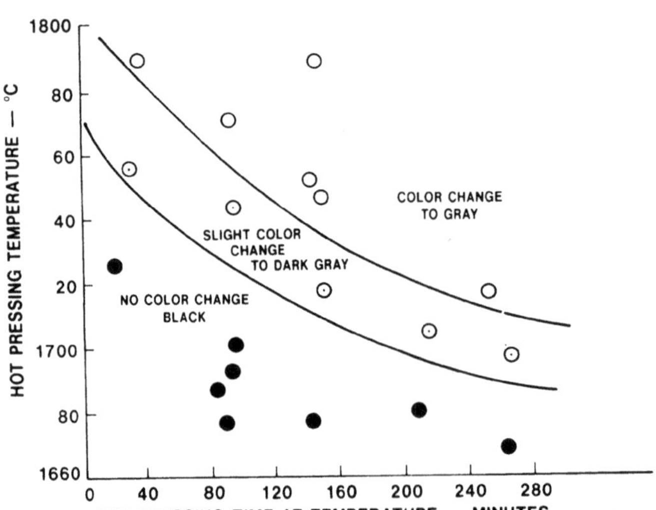

Figure 23 Color Change Versus Hot Pressing Parameters

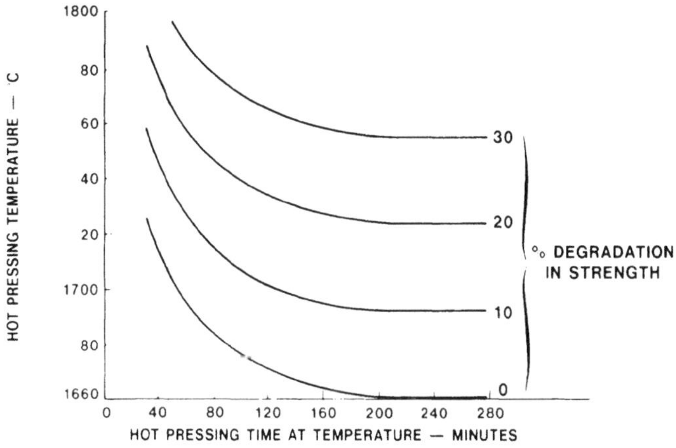

Figure 24 Percent Degradation In Strength Versus Hot Pressing Parameters

SUMMARY

Duo-density silicon nitride turbine rotors fabricated in 1977 have demonstrated a significant improvement in cold spin reliability over state-of-the art rotors fabricated in 1975. Hot test results (of 1977 vintage rotors) in hot spin rigs are very encouraging in that the first attempt to complete a 200 hour durability test was successful. Ceramic turbine rotors have now been run at turbine inlet temperatures of 2000-2500°F at speeds of 27,500-64,240 rpm for periods up to 200 hours.

Reasonable correlation was demonstrated between fast fracture reliability predictions based on Weibull theory and actual test results both on cold spin disks and hot turbine rotors. Correlation between time dependent reliability predictions and actual test results has not been confirmed nor disproved as more test results are required.

Process improvements in injection molding were identified, which eliminated subsurface voids in the inner, more highly stressed portions of the blades. Microfocus X-ray techniques were developed to produce magnified X-rays in addition to radial view panoramic X-rays of the rotor blade ring rim. Microstructure and strength degradation of reaction bonded silicon nitride blade rings, which previously occurred during hot press bonding, was determined to be controllable. Improved process rotors can now be fabricated with no degradation utilizing the hot pressing parameters of time and temperature recently defined; this will result in a significant increase in rotor reliability. Further improvements to better detect and eliminate planar type flaws in blades would be of added benefit.

ACKNOWLEDGEMENTS

The authors acknowledge the Defense Advanced Research Projects Agency for partial support of the fabrication, test and evaluation, the Department of Energy for partial support of process improvements and the Army Materials and Mechanics Research Center, the technical monitor of the overall program.

REFERENCES

1. R. G. Salter, C. Dziter, E. D. Harris, W. E. Mooz, K. A. Wolf, "Strategic Defense Materials: A Case Study of High Temperature Engines", ARPA Order No. 189-1, February, 1977.
2. R. R. Baker, J. H. Buechel, P. H. Havstad, D. L. Hartsock, "Test and Development of Ceramic Combustors, Stators, Nose Cones, and Rotor Tip Shrouds", Ceramics for High Performance Applications - II, J. J. Burke, E. N. Lenoe, and R. N. Katz, editors, Brook Hill Publishing Company, p. 291-315.
3. A. F. McLean, R. R. Baker, "Brittle Materials Design, High Temperature Gas Turbine", AMMRC TR 78-14, 12th Interim Report, March, 1978.
4. R. R. Baker, A. Ezis, M. U. Goodyear, D. L. Hartsock, "Developments in Press Bonding of Duo-Density Rotors", Ceramics for High Performance Applications - II, J. J. Burke, E. N. Lenoe, and R. N. Katz, editors, Brook Hill Publishing Company, p. 207-230.
5. R. R. Baker, J. C. Caverly, P. H. Havstad, "Ceramic Turbine Rotors - Engine Test and Development", ibid, p. 273-290.
6. C. F. Johnson, T. G. Mohr, "Injection Molding 2.7g/cc Silicon Nitride Turbine Rotor Blade Rings Utilizing Automatic Control", ibid, p. 193-206.
7. A. F. McLean, R. R. Baker, "Brittle Materials Design - High Temperature Gas Turbine", AMMRC CTR 76-31, 10th Interim Report, October, 1976.
8. R. R. Baker, "Crack Protection Method", U. S. patent 4,127,684, November 28, 1978.
9. A. F. McLean, E. A. Fisher, R. J. Bratton, D. G. Miller, "Brittle Materials Design - High Temperature Gas Turbine", AMMRC CTR 75-28 8th Interim Report, October, 1975.
10. G. C. DeBell, J. G. LaFond, W. E. Meyer, J. R. Secord, "Development of a Ceramic Turbine Rotor Hot Spin Test Rig", Ceramics for High Performance Applications, - II, J. J. Burke, E. N. Lenoe, and R. N. Katz, editors, Brook Hill Publishing Company, p. 243-258.
11. A. F. McLean, R. R. Baker, "Brittle Materials Design - High Temperature Gas Turbine", AMMRC TR 79-11 13th Interim Report, February, 1979.
12. L. G. Johnson, General Motors Research Reliability Manual, GMR 302, Chapter 2.

THE EVALUATION OF CERAMIC TURBINE STATORS

Eugene A. Fisher and Walter Trela

Engineering and Research Staff
Research
Ford Motor Company
Dearborn, Michigan 48121

INTRODUCTION

This program is supported by the Department of Energy (DOE) as part of the Component Development Technology Program. Administered by the NASA Lewis Research Center (Contract DEN3-00019), this program's objectives are to evaluate the durability of ceramic stators over a highly dynamic automotive gas turbine duty cycle and to assess the capability of current ceramic design methodology to predict the durability of the stators over the selected duty cycle for times in excess of 200 hours. The program also provides an assessment of the ability of the ceramic industry to fabricate complex geometry monolithic stators by near-net-shape forming processes.

Figure 1 illustrates the Ford Model 820 stator which was selected as the test component for this evaluation. This stator is representative of the size and design of components required for automotive gas turbines. This stator was designed, fabricated, and evaluated during the Ford/ARPA/ERDA Program "Brittle Materials Design-High Temperature Gas Turbine" and significant aspects of its development and testing are noted in interim reports [1-6] issued under that program. The stator is 5 inches in overall diameter and consists of 25 airfoil vanes attached to both an integral outer shroud and a fully segmented inner shroud. This latter feature provides stress relief from uneven heating and cooling of adjacent vanes during transient operation.

The program consists of four major technical tasks, as follows:

Task I — The development of the stator reliability prediction model and the computation of expected reliability over the duty cycle based upon ceramic material property data.

Task II — The fabrication and/or procurement of the ceramic stators, test specimens for material property data, and the ceramic support hardware needed for test rig operation.

Task III — The determination of material property data required for the reliability calculation in Task I.

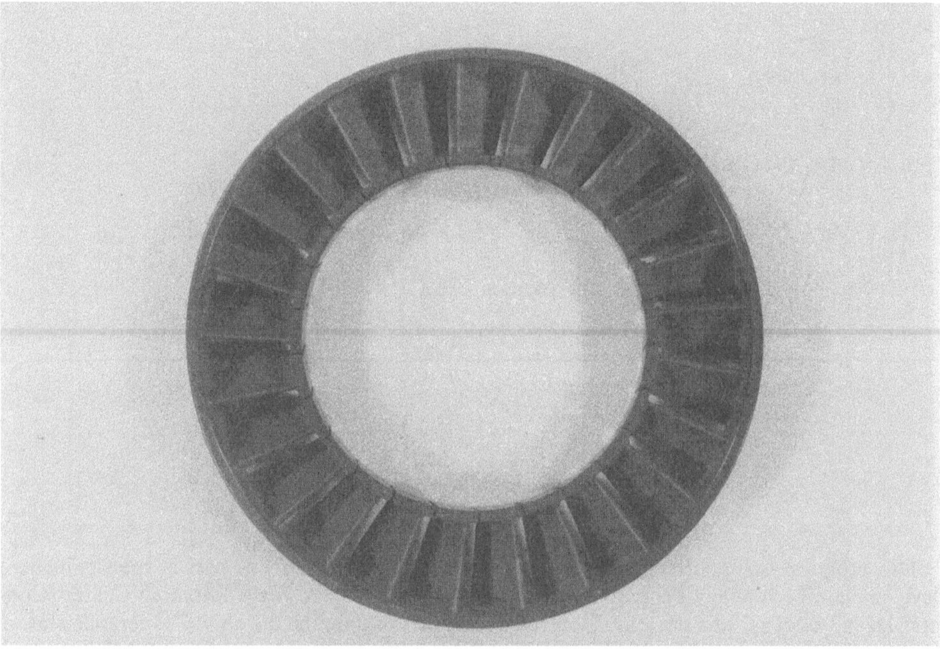

FIGURE 1 — FORD MODEL 820 MONOLITHIC CERAMIC GAS TURBINE STATOR

Task IV — The evaluation of the stators, using previously developed qualification tests (2,3,6) followed by the duty cycle testing. The durability goal is 500 hours of duty cycle operation at temperatures up to 1204°C (2200°F).

PROGRAM PLAN

Task I — Reliability Prediction

The reliability prediction program, illustrated schematically in Figure 2, will be prepared based on 3-D finite element thermal and stress analyses and statistical distribution of material strength properties for the reliability calculations.

A 3-D finite element model of the Ford Model 820 one-piece ceramic stator will be prepared. Thermal analyses will be conducted to define the temperature gradients for the various combinations of airflow and transient temperature conditions of the testing program. The aerodynamic loads imposed on the stator vanes will be determined for inclusion in the stress calculations.

The transient stress distributions within the model resulting from the combined thermal and aerodynamic loads will be generated for a cold light-off qualification test and the thermal transients imposed during the durability cycle at the maximum air flow condition. The resulting stress data along with ceramic material properties will then be used as input to the reliability program to determine the probability of survival of the particular stator for the testing conditions imposed.

3-D FINITE ELEMENT MODEL
- ONE BLADE SEGMENT OF STATOR
  - SEGMENT OF OUTER SHROUD
  - COMPLETE VANE
  - INNER SHROUD

THERMAL AND STRESS ANALYSIS
- THERMAL GRADIENTS
- AERODYNAMIC LOADS
- TEMPERATURE DEPENDENT PROPERTIES
- BOUNDARY CONSTRAINTS

RELIABILITY PREDICTION
- WEIBULL MATERIAL PROPERTIES

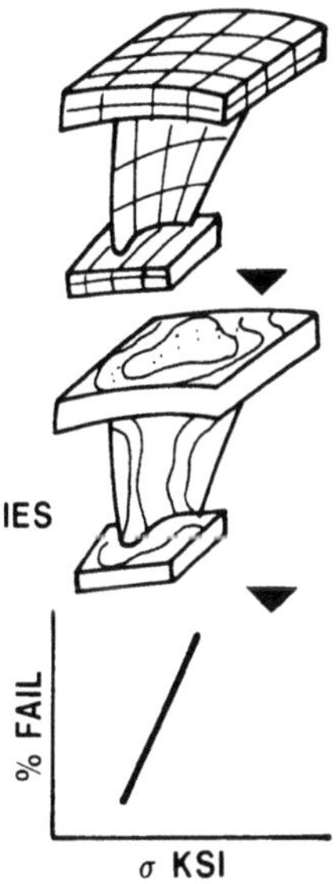

FIGURE 2 — SCHEMATIC ILLUSTRATION OF THE RELIABILITY PREDICTION PROGRAM, TASK I

Task II — Ceramic Stator Fabrication

As specified by NASA, silicon nitride and silicon carbide stators were to be fabricated to the Ford Model 820 Gas Turbine Engine design by at least three ceramic component suppliers. Fabrication of both Si_3N_4 and SiC stators by each supplier was not required, but fabrication processes were to be selected having potential for near-net-shape, low-cost production. A minimum of 12 stators were to be provided by each supplier. In addition, test samples required for material property determination for use in the reliability predictions were to be fabricated from the same starting materials utilizing as closely as possible the identical processing conditions as the stators.

Task III — Material Property Characterization

This task comprises the determination of material properties of each of the stator materials for use in the reliability prediction under Task I. The inputs required are:

Shear Modulus	Thermal Conductivity	Four-Point MOR
Young's Modulus	Thermal Diffusivity	Stress Rate Dependence
Specific Heat	Thermal Expansion	

All of these properties except the stress rate dependence are required over a temperature range from ambient to 1204°C (2200°F), the maximum temperature of the stator testing portion of the program.

The stress rate dependence test on each material will be initially conducted at 1204°C. If, at this temperature, the material behavior indicates the absence of slow-crack growth, it is assumed that no slow-crack growth exists at lower temperatures and, therefore, no time-dependent considerations are needed for the reliability program. If, on the other hand, slow-crack growth is indicated at 1204°C, then the stress rate dependency test must be repeated at several temperatures in order to provide the necessary data for time-dependent reliability prediction.

In order to obtain the MOR data, a minimum of 30 samples will be broken at room temperature, 1000°, 1100°, and 1204°C. Data will be reduced using the Maximum-Likelihood Estimator (MLE) (6) technique to fit the Weibull model, with the resulting distribution described in terms of characteristic strength and Weibull slope. Thermal expansion measurements to 1204°C will be done using a Theta Industries differential dilatometer.

Thermal diffusivity will be determined by the laser pulse technique from ambient to 1204°C. Specific heat will be determined over the same temperature range using a drop calorimeter technique. Thermal conductivity will be calculated using thermal diffusivity, specific heat, and density data. Both Young's and shear modulii will be determined by resonant frequency techniques over a range of temperatures from ambient to 1204°C.

Task IV — Ceramic Stator Evaluation

The duty cycle durability testing is to be carried out in an existing test rig at Ford. The rig, the Hot Flowpath Qualification Rig, is shown in section in Figure 3.

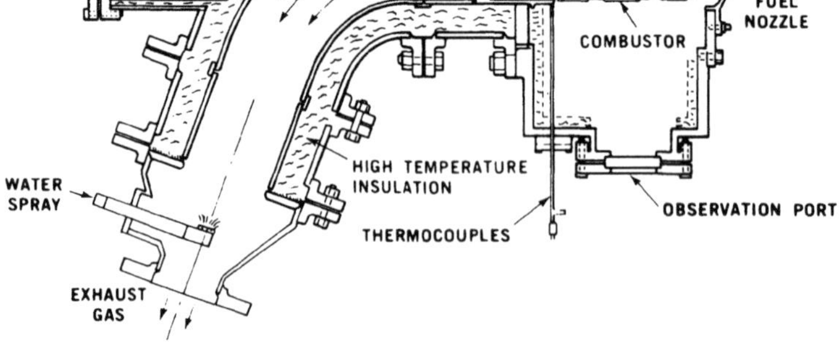

FIGURE 3 — HOT FLOWPATH QUALIFICATION TEST RIG

Before initiation of durability testing, stators will be subjected to NDE and qualification tests. The NDE procedures consist of microscope-aided visual defect inspection, dimensional checks, and radiographic inspection. Qualification tests are stator vane bend testing, stator outer shroud pressure testing, and light-off qualification testing in a Ford 820 turbine engine operated without turbine rotors. Unleaded gasoline with a research octane number of 90-92 is the fuel used for both rigs. The test cycle which was proposed for durability testing was based in general upon the airflow, equivalent engine speed, amount of time at any one test condition, and the number of accelerations and decelerations of the EPA Composite Driving Cycle with a maximum turbine inlet temperature of 1204°C (2200°F). This cycle will be described later in greater detail.

SUMMARY OF RESULTS TO DATE

Task I — Reliability Prediction

The 3-D model and the thermal and stress analyses used in this program were based on an earlier analysis performed by Lawrence Livermore Laboratories.(7) This previous study, however, had several shortcomings relative to the scope of the current program. It was based on an earlier version of the stator, did not include temperature-dependent material properties or aerodynamic loads, employed simplified boundary constraints, and did not include reliability prediction.

A new 3-D model was prepared and is shown in an exploded view, for clarity, in Figure 4. The model includes a 1/25th section of the outer shroud, a complete vane and the attached inner shroud segment.

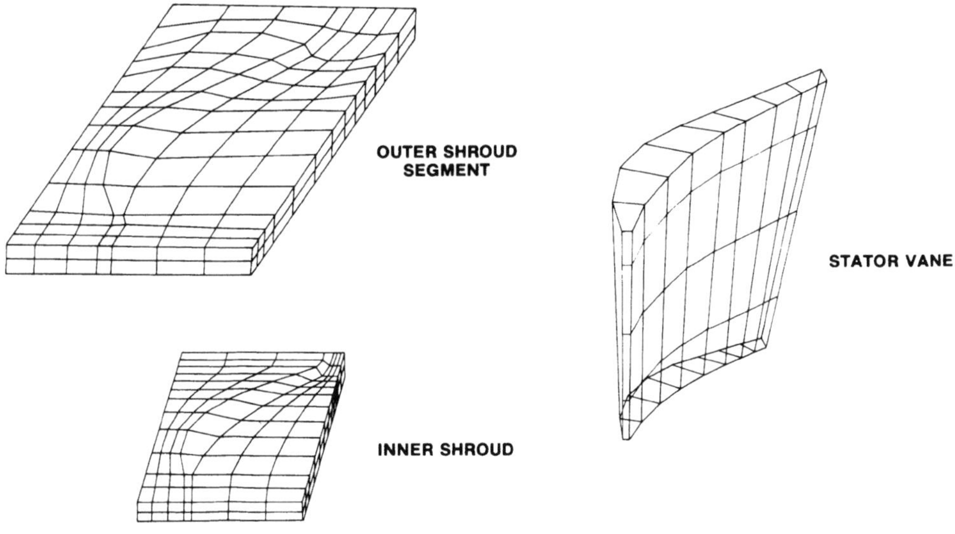

FIGURE 4 — EXPLODED VIEW OF STATOR MODEL PREPARED FOR THREE DIMENSIONAL THERMAL AND STRESS ANALYSIS

Modifications of the thermal and stress analysis programs were completed to allow the use of temperature-dependent material properties. Aerodynamic loads on the vane have been calculated for the idle speed air flow and the maximum air flow condition of the durability cycle. These loads have been added to the stress analysis program. Continuity of the outer shroud ends has been achieved in both the thermal and stress analysis programs.

As the refinements were being added to the analyses, stresses were computed for the thermal up-shock portion of the cold light-off transient (a step change in inlet temperature from 15°C (60°F) to 1054°C (1930°F) followed by a ramp to "idle speed" air flow rate). Material properties used thus far in these calculations are those previously determined for the Ford injection molded reaction bonded 2.7 g/cc density Si_3N_4 material. All stress calculations resulted in maximum principal tensile stresses occurring in the outer shroud. Calculated maximum tensile stresses have been approximately 3400 psi. However, it was noted that compressive stresses in the vane reached 11,000 psi and may possibly be reversed during the downshock. In addition, still higher stresses may be produced during the one-minute thermal cycle of the durability test.

Temperature gradient data, required to continue the stress analysis, has been generated for the 30-second hold at 1054°C (1930°F) during the light-off test and for the upshock and 40 second hold at 1204°C (2200°F) which occurs during the maximum airflow condition of the durability cycle. Computer input has been prepared but not yet run for the downshock portion from both of these conditions.

Task II — Ceramic Stator Fabrication

As noted earlier, at least three ceramic component suppliers were to be involved in the fabrication of stators for evaluation. A survey of suppliers was initiated, resulting in the following list which was submitted to NASA for approval.

Company	Process & Material
AiResearch Casting Co.	Slip Cast RBSN (Airceram. RBN-101)
The Carborundum Co.	Injection Molded Sintered α SiC
Ford Motor Company	Injection Molded RBSN
The Norton Company	Slip Cast RSSC (Noralide NC-433)

Following NASA approval, subcontracts were issued by Ford to ACC, Carborundum, and Norton.

Each of the suppliers is to fabricate a minimum of 12 stators plus 300 MOR bars for material strength testing, and 25 physical property test samples. The current status of each stator fabrication program, including preparation of the required test samples, will be discussed in the remainder of this section.

AiResearch Casting Company (ACC)

ACC is fabricating stators of Airceram RBN-101 RBSN by the slip casting process, with the stators being the product of the mold with final machining done by Ford. The various test samples would be cast in separate molds but would be made from the same slip as the stators and would be processed with the stators. The test bars would be delivered in the as-cast condition while the various physical property specimens would be supplied fully machined. Molds were prepared using wax patterns which were molded using Ford's injection molding tooling. The patterns were molded by a vendor using wax supplied by ACC.

All of the ACC stator castings, MOR bars, and test samples contracted for have been delivered. One of the stator castings is shown in Figure 5.

The Carborundum Company

Carborundum is fabricating SiC stators by injection molding and is assuming responsibility for tooling design and construction. Stators will be supplied in the as-sintered condition to Ford, who will be responsible for final machining. This tooling has been built, incorporating a test bar cavity upstream of the stator cavity, and is currently being used for development of the fabrication process.

The Ford Motor Company

Fabrication of Ford molded RBSN stators and test samples has been completed. After finish machining, the stators and all of the test samples were then given the Ford flash-oxidation

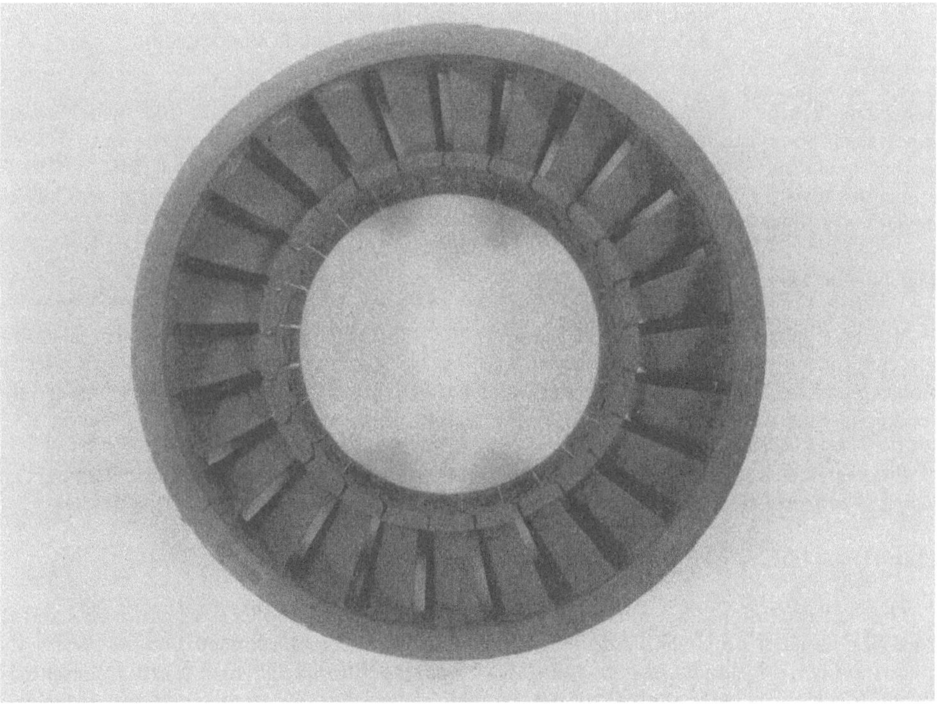

FIGURE 5 — REACTION BONDED SILICON NITRIDE STATOR CASTING FABRICATED BY AIRESEARCH CASTING COMPANY

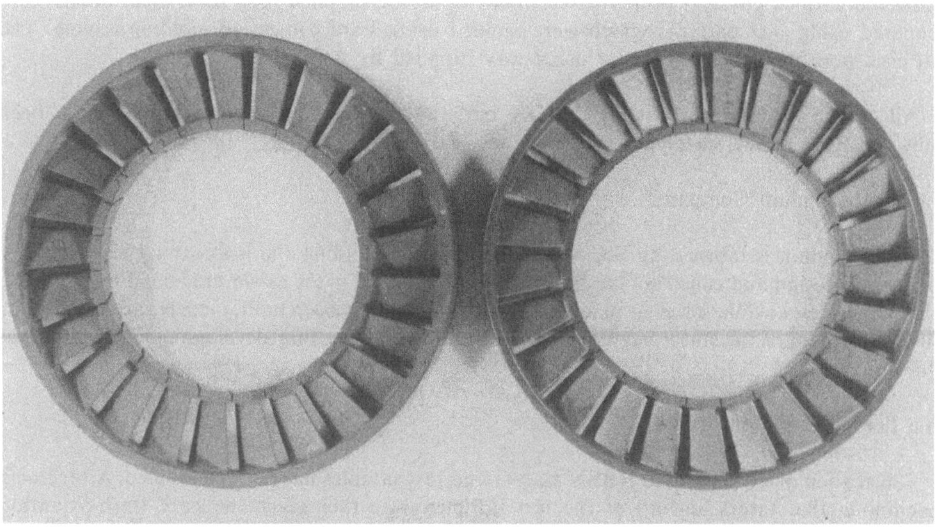

FIGURE 6 — REACTION BONDED SILICON NITRIDE STATORS FABRICATED BY FORD MOTOR COMPANY; (LEFT), AS-MOLDED AND NITRIDED, AND (RIGHT), AFTER FINISH MACHINING AND FLASH OXIDIZING

treatment, a heat treatment process to produce a thin siliceous surface film which retards penetration of oxygen into the pores of the RBSN. This treatment has been shown to decrease further oxidation of molded RBSN stators during steady-state testing at 1054°C (1930°F). Figure 6 shows a finish-machined flash oxidized stator next to an as-nitrided stator. Only the shrouds are finish-machined.

The Norton Company

Norton is fabricating stators from Noralide NC-433, one of a family of RSSC materials. The slip cast forming technique was selected, and wax patterns were molded using Ford's injection molding tooling. The wax materials were supplied by Norton and the patterns were molded by a vendor. Norton also proposed to supply the stators finish machined to Ford's drawing.

Thus far, seven stators have been received from Norton, along with all of the MOR bars and physical property test samples. Figure 7 is a photograph of one of the Norton stators.

Task III — Material Property Characterization

Thermal expansion determinations have been completed on the Carborundum SiC and the Ford Si_3N_4 materials.[8] Strength measurements have also been completed on the Ford Si_3N_4 material. Figure 8 is a plot of four-point MOR vs. temperature. This data is also represented in Table 1, which includes the calculated Weibull slope value for each set of test samples as determined by the MLE technique.[6]

The results of the stress rate test are shown in Figure 9. As indicated, the strength was insensitive to stress rate at 1200°C, leading to the conclusion that this material does not exhibit

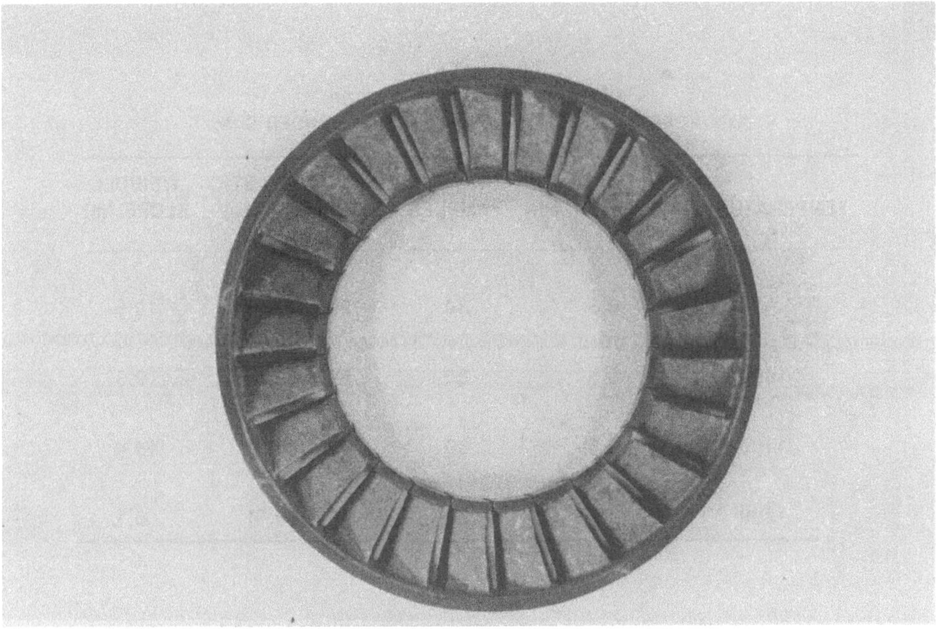

FIGURE 7 — REACTION SINTERED SILICON CARBIDE STATOR FABRICATED BY THE NORTON COMPANY

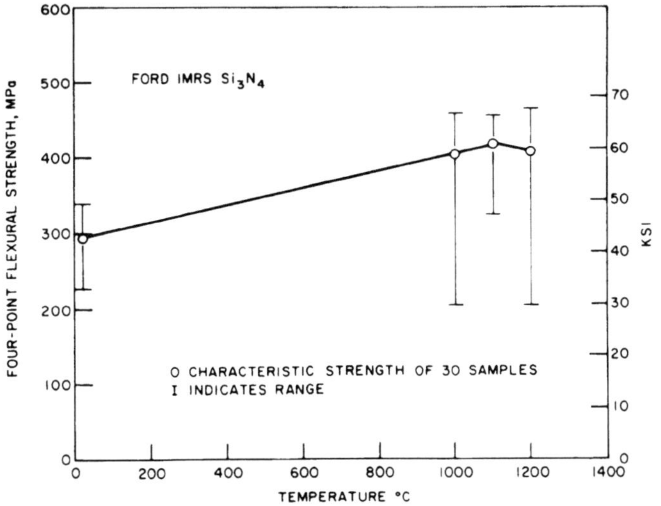

FIGURE 8 — FOUR-POINT FLEXURAL STRENGTH VS. TEMPERATURE FOR FORD INJECTION MOLDED REACTION BONDED SILICON NITRIDE

TABLE 1
MOR STRENGTH DATA FOR
FORD INJECTION MOLDED REACTION BONDED Si_3N_4

TEMPERATURE °C	CROSS HEAD SPEED MM/MIN	NO. OF SAMPLES	CHARACTERISTIC STRENGTH (σ_θ) MPa (KSI)	WEIBULL SLOPE (m)
20	0.5	30	295 (42.8)	10.6
1000	0.5	30	406 (58.9)	9.3
1100	0.5	30	420 (60.8)	15.4
1200	0.5	30	409 (59.3)	8.1

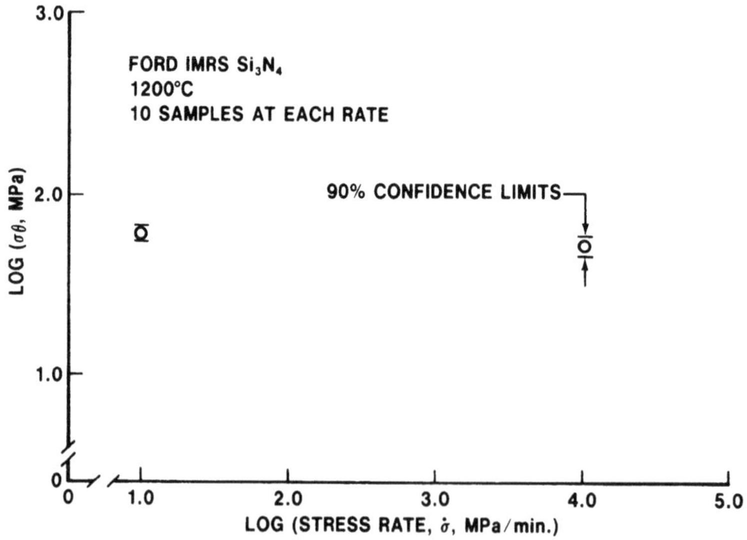

FIGURE 9 — RESULTS OF STRESS RATE DEPENDENCE TESTS ON FORD INJECTION MOLDED REACTION BONDED SILICON NITRIDE slow-crack growth tendencies at this temperature. Therefore, for this material, time dependency need not be considered in the calculation of stator reliability up to 1200°C in Task I.

Task IV — Ceramic Stator Evaluation

A. Qualification Testing

After all processing and inspection/NDE of stators had been completed, they were next subjected to a series of qualification tests. These qualification tests consist of a stator vane bend test, a stator outer shroud pressure test, and a 10 light test in a Ford Engine Simulator Rig. These tests are also described in a previous report.[6]

1. Stator Vane Bend Test

As part of the Ford/ARPA/ERDA program, a hydraulically-loaded device was utilized for simultaneously applying a variable load to each of the 25 vanes. It was found that a load of 19 pounds, when applied normal to the flat surface of the segmented vane inner shroud, would cause fracture if defects were present at the most highly stressed region. This load corresponds to a maximum tensile stress in the vane leading edge, at the junction of the outer shroud, of 28,080 psi in the case of loading from the leading edge side. Parts which withstood this test successfully, when loaded from both leading and trailing edge sides, did not experience failures in the vane roots during subsequent steady state durability testing. This particular test is now considered routine for Ford injection molded RBSN stators of 2.7 g/cc density. The stators selected for this program successfully survived the vane bend test after nitriding. Six of the original 19 stators were lost during flash oxidation due to the build-up of unusually heavy glassy coatings. This was traced to the presence of sodium nitrite in the particular coolant used during finish grinding. The remaining 13 stators were subjected to careful washing with distilled water prior to flash oxidation, which was apparently successful in that the remaining stators exhibited a normal appearance after flash oxidation. After machining and flash-oxidation, the stators were again subjected to the vane bend test in order to eliminate any parts which might have been damaged by the additional processing. Since the hydraulic fixture was designed for evaluating an as-nitrided part having larger shroud dimensions than a finish-machined part, it was decided to conduct this second vane bend test manually. The additional design and fabrication work required to modify the hydraulic fixture would have been too time-consuming.

The vane bend test was conducted on the remaining 13 stators, loading each vane on the inner shroud at a steadily increasing load up to the 19 pound per vane limit using a calibrated loading device. Each vane was loaded in both the fore and aft direction, as was done earlier using the hydraulic fixture. One vane in one of the 13 stators failed just as the 19 pound load limit was reached. Examination of the fracture surfaces revealed a small pinhole at the leading edge of the vane, which was the origin of fracture.

2. Stator Outer Shroud Pressure Test

The remaining 12 stators were then subjected to the stator outer shroud pressure test. In the outer shroud pressure test, as shown in Figure 10, the stator is contained but not clamped between two heavy steel plates and is sealed by means of a rubber lip seal. Fluid is then admitted to the cavity and the pressure increased until the outer shroud is subjected to a tensile stress of

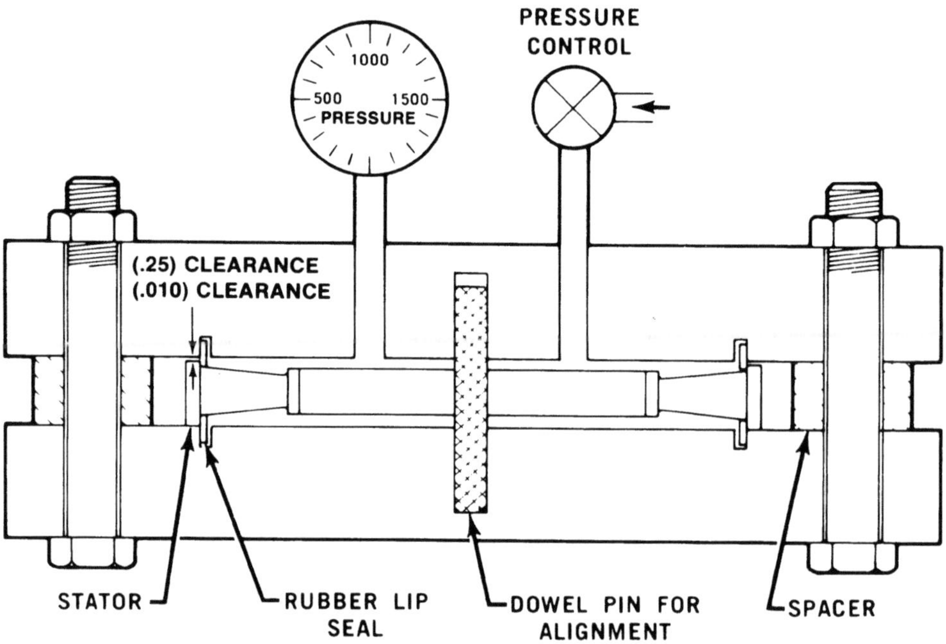

FIGURE 10 — SCHEMATIC ILLUSTRATION OF STATOR OUTER SHROUD PRESSURE TEST

6000 psi. This level had been selected during the Ford/ARPA/ERDA program so that reasonable assurance of survival could be provided for the subsequent light-off qualification testing.

Some initial difficulties with the rubber lip seal were probably to blame for failure of two stators. Upon examination of fracture surfaces, there was no evidence of flaws. After improvements were made, the remaining parts were successfully evaluated to the 6000 psi stress level.

3. Ten Light Test

The final test in the qualification sequence, the ten light test, was also developed during the Ford/ARPA/ERDA program. A modified Ford 820 engine, called The Engine Simulator Rig (ESR), is used for this test. Operating without turbine rotors, one stator at a time is placed in the first stage position and subjected to 10 lights. When engine light off occurs, equivalent engine idle airflow is attained while the temperature is maintained at 1054°C (1930°F). After the required hold time, fuel is shut off and the ESR is force-cooled by compressor discharge to 66°C (150°F) stator inlet temperature before the remaining light-offs are initiated. A typical light off transient is shown schematically in Figure 11.

All the remaining 10 stators survived the 10 light sequence with no damage.

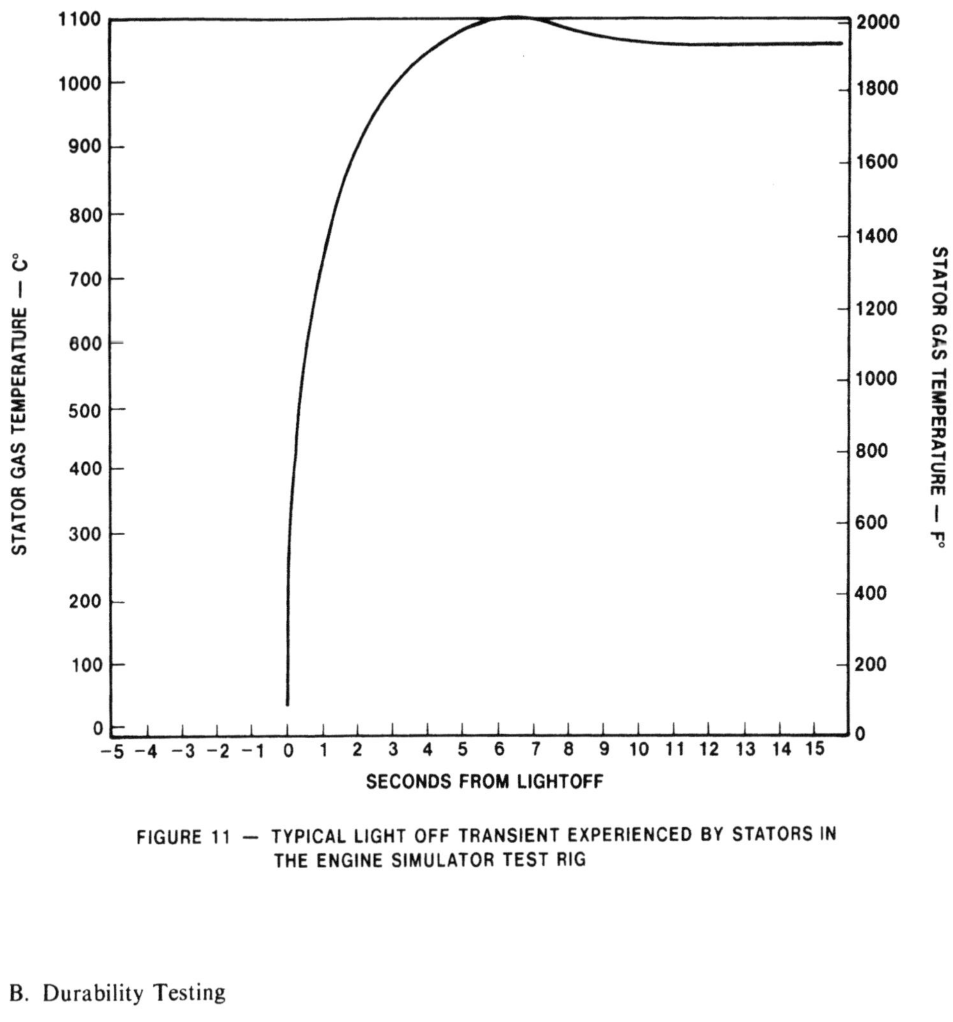

FIGURE 11 — TYPICAL LIGHT OFF TRANSIENT EXPERIENCED BY STATORS IN THE ENGINE SIMULATOR TEST RIG

B. Durability Testing

1. The Duty Cycle

The durability test was established to evaluate stator life in a highly cyclic manner under simulated gas turbine engine conditions. The test section is composed of an injection-molded RBSN nose cone, two stators, and associated slip-cast RBSN shroud and centering rings (Figure 3).

The durability duty cycle is illustrated in Figure 12, and consists entirely of 1 minute cycles. Three different airflow rates are used in the cycles, with a different peak inlet temperature corresponding to each airflow. As shown, the first three hours of the durability cycle are the same. Each hour consists of 40 one-minute cycles from 704°C (1300°F) to 1121°C (2050°F) at an airflow of 0.32 Kg/sec (0.70 lbs/sec) and 20 one-minute cycles from 704°C to 1177°C (2150°F) at an airflow of 0.41 Kg/sec (0.93 lbs/sec).

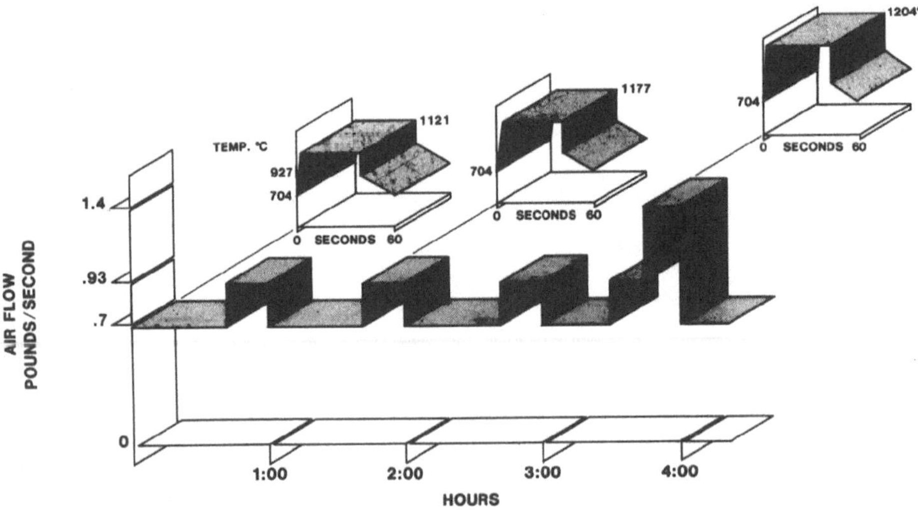

FIGURE 12 — STATOR DURABILITY DUTY CYCLE

The fourth hour of the durability cycle is different from the first three and consists of 29 and 14 one minute cycles respectively at the same conditions noted earlier plus 17 one-minute cycles at the maximum conditions varying from 704°C to 1204°C (2200°F) at an airflow of 0.64 Kg/sec (1.40 lbs/sec). This basic four hour segment is then repeated over and over again to accumulate the durability objective. Note that each individual one-minute cycle features a very rapid temperature increase from 704°C to the maximum temperature desired, a 40 second hold at that temperature, a sharp temperature drop to 927°C (1700°F) and a slow temperature decrease to the 704°C starting point.

In order to provide testing representative of actual engine usage, additional engine light-offs are needed. These will be provided at the rate of one light-off per accumulated hour of durability testing. This testing will be done in the ESR used for the 10 light qualification test. The light-offs will be done at various temperatures ranging from ambient to 593°C (1000°F) in order to simulate a range of both cold and hot starts.

As noted earlier, the objective of this test was to achieve 500 hours of duty cycle durability attaining a maximum temperature of 1204°C (2200°F) and to survive 500 light-offs in the ESR. During the durability testing, the stators will be removed from the rig at various intervals for inspection and measurement of weight and dimension changes.

In order to insure that the various cycles making up the durability test will be as reliable and reproducible as possible, a computerized control system was developed. The programmable analog system provides fully automatic control of the cycle, and includes provisions for shutting down the rig should off-design or safety problems arise. In addition, other instrumentation provides strip chart records of airflow and temperature conditions.

Some difficulties were experienced with the control system, and considerable time and effort were expended in trouble-shooting. Eventually all problems except one were cleared up. After reaching the end of each 40 second hold at maximum temperature, the fuel flow must be sharply reduced in order to achieve the desired rapid cool-down. The system had difficulty in distinguishing between this condition and a flame-out, resulting in a number of shut-downs at this point. The operator would then need to re-start the rig (at 538°C) and return the unit to fully automatic control in order to resume the duty cycle. Occasionally a real flame-out would occur due to the reduced fuel flow. As a result of all this, a number of unscheduled stops and starts were made during duty cycle testing. These were all recorded and constitute some form of hot starts which may relate to hot starts in engine operation. Additional programming experiments are being conducted in an attempt to eliminate this problem while still retaining the desired cool-down ramp.

2. Durability Test Results

The 10 Ford injection molded RBSN stators, discussed earlier in the section on Qualification Testing, were the first components available for testing. Two of the ten parts were selected, installed, and testing was initiated. The testing involved four hours of duty cycle operation, through one complete sequence as shown in Figure 12, before stopping for inspection and measurement.

However, before the initial 4 hour test could be completed, the stator in the second stage position experienced a failure of the outer shroud. Another set of stators were then inserted and testing resumed. The first 4 hour test sequence was completed successfully, but another second stage stator outer shroud failure was experienced just before completion of a second 4 hour run. Still another set of stators was installed, with approximately the same results.

These three second stage failures had thus occurred early in the program and in a somewhat similar fashion, even though the first stage stator is believed to be exposed to more severe test conditions. An investigation pointed to the inner shroud extension ring, shown in Figure 13. This ring, made from slip-cast RBSN, is not a normal 820 engine component but was added in the absence of a rotor for the purpose of ducting the hot gases to the second stage stator, preventing bypass of the hot gases to the inside of the stator. Since it was difficult to properly mount this ring, it was wired to the vane-inner shroud junctions of the second stage stator, using platinum wire tied through holes drilled in the ring. Some of this wire had come off during testing and there was evidence that the ring had shifted. It is likely that this ring applied unusual loads to the stator, therefore it was decided to resume testing without using the ring. This appears to have eliminated the problem, although some hot gas flow probably by-passes the second stage stator.

Testing was then resumed using the new second stage stator and a first stage stator which had accumulated 10.6 hours, 636 cycles, during the previous testing. Further testing continued without incident for 64.1 hours, 3846 cycles, when some chipping of the second stator outer shroud occurred at the point where the stator contacted the second stage shroud ring. The first stage stator and the remainder of the ceramic components were not damaged.

A new second stage stator was then installed and testing was resumed. Attention was primarily focused on the first stage stator, since it was now the high time part. The testing proceeded smoothly at this point, and the testing schedule was expanded to operate around the clock in 50 hour increments before stopping for inspection and measurement. In this manner, the first stage

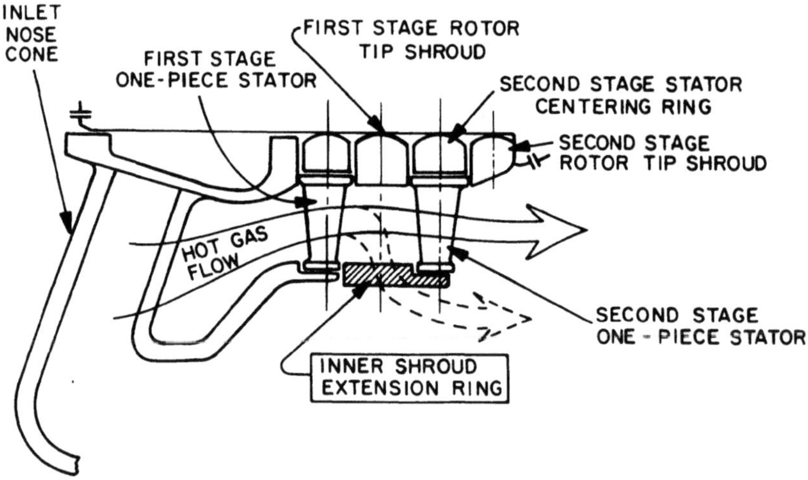

FIGURE 13 — SCHEMATIC ILLUSTRATION OF CERAMIC TEST SECTION. NOTE LOCATION OF INNER SHROUD EXTENSION RING

stator had now accumulated 202.2 hours. 12,134 cycles of duty cycle testing and a total hot running time of 224.5 hours. The latest second stage stator had accumulated 110.8 hours, 6647 cycles of duty cycle testing and a total hot running time of 124.4 hours.

At this point, another around the clock run was initiated, this time with the goal of achieving 100 hour increments of testing. At 69 hours into this run, a failure was observed. A visual check 30 minutes earlier showed no sign of distress. Upon disassembly, it was found that the nose cone and both stators had failed. At the point of shut down, the first stage stator had accumulated 266.6 hours, 15,998 cycles of duty cycle testing and a total hot running time of 293.5 hours. The second stage stator had accumulated 175.2 hours, 10,511 cycles of duty cycle testing and a total hot running time of 193.6 hours.

The nose cone bell section had broke away behind the struts and two large fractured sections were found in position against the first stage stator inner shroud. This is a non-typical nose cone failure and is shown in Figure 14. Seven vanes of the first stage stator were damaged, as shown in Figure 15. Segments of two of the vanes were found resting between the two stators upon disassembly. The second stage stator, shown in Figure 16, had seven vanes missing completely, two additional vanes broken away at about mid span, and one crack through the outer shroud. Virtually all the remaining second stage vanes had impact damage on the leading edge.

An extensive visual fractographic analysis of the failure surfaces was conducted. No evidence of internal or surface flaws was observed on any of the fracture surfaces. The nose cone showed no evidence of foreign object damage. The bell fracture surfaces indicate that the bell failure was initiated at the strut/bell junction area. Six of the seven vane failures on the first stage

FIGURE 14 — REAR VIEW OF NOSE CONE FAILURE. THE TWO LARGER FRACTURED SECTIONS CAME FROM THE INNER BELL SECTION

stator show evidence of impact damage from the trailing edge. The seventh vane was completely missing. All failure surfaces of the second stage stator show evidence of impact crack initiation and are considered as secondary failures.

Although the failure analysis is continuing, the preliminary conclusion is that the nose cone bell failure precipitated one vane failure in the first state stator and the subsequent damage was entirely secondary in nature.

Weight change data in the two stators is shown in Figure 17 as percent weight gain vs total hot running time up to the last inspection point before failure. The weight-gain curve is parabolic in nature, typical of the oxidation behavior of RBSN. The actual weight gain is higher than that experienced on flash-oxidized Ford injection molded RBSN stators when engine tested under steady state conditions at 1054°C (1930°F) as shown in Figure 18. Also shown in the weight gain behavior of untreated stators made from the same material, which showed very rapid weight gain and failure of the outer shroud upon reaching approximately 2% weight gain. The weight gain data shown in Figure 17 may be somewhat conservative, since the stator was covered with a dark red deposit on all surfaces which were exposed to the hot gases during

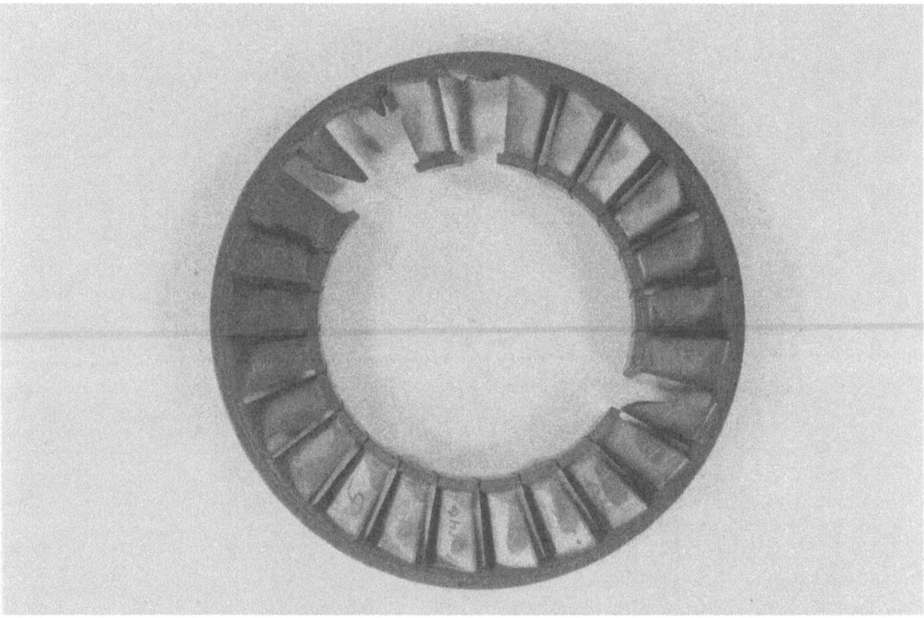

FIGURE 15 — DAMAGED FIRST STAGE STATOR, VIEWED FROM TRAILING EDGE SIDE

testing. A preliminary X-ray fluorescence analysis of one of these vanes indicates the presence of iron, so the deposit is primarily iron oxide. Since 1% weight gain of these stators is only 0.9 grams, some of the measured weight gain may be due to the deposit.

Eight of the remaining undamaged vanes of the first stage stator were individually loaded to failure in the vane bend test fixture described earlier, resulting in a characteristic breaking load of 34.8 pounds with a Weibull slope of 6.1. This test was repeated on another stator from the same batch which had accumulated only 10 hours of rig testing during control development work, which exhibited a characteristic breaking load of 35.8 pounds with a Weibull slope of 10.0. A statistical analysis, comparing the results of these two tests, indicated that at a 90% confidence level there is no statistically significant difference between either the characteristic strength values or the slopes. Thus, it appears that the Ford RBSN material is not degraded by the durability test cycle for the testing times achieved to date.

Future Plans

As a result of the amount of secondary damage which resulted from the presence of the second stage stator, which may prevent definitive establishment of the cause of failure, continued testing will be conducted with only one stator which will occupy the first stage position. Another factor in this decision is the suspected difference in exposure conditions between the two stators, making comparison of test results uncertain. Visual observation of the first stage stator during testing will also be greatly improved by removal of the second stage stator.

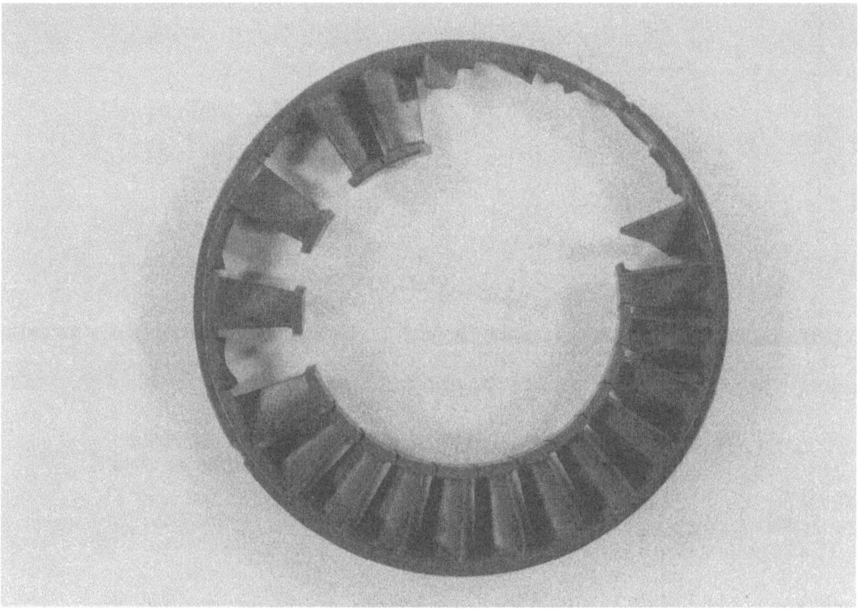

FIGURE 16 — DAMAGED SECOND STAGE STATOR. VIEWED FROM LEADING EDGE SIDE

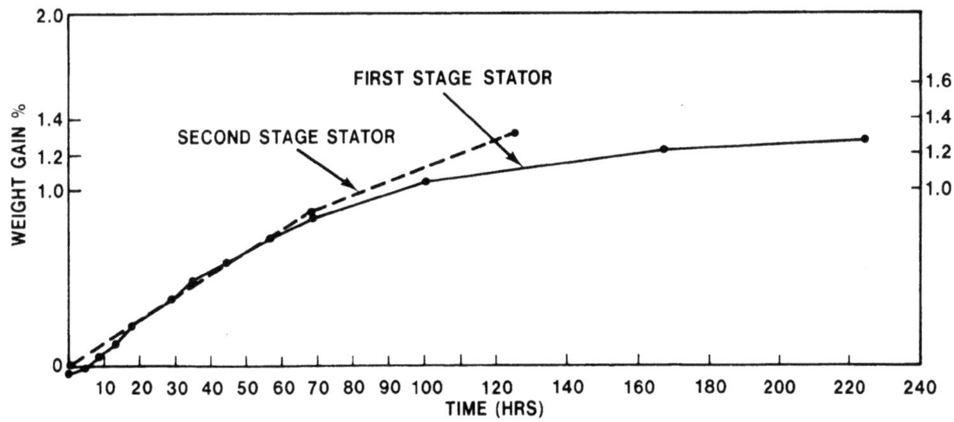

FIGURE 17 — WEIGHT CHANGE VS TEST TIME FOR BOTH STATORS

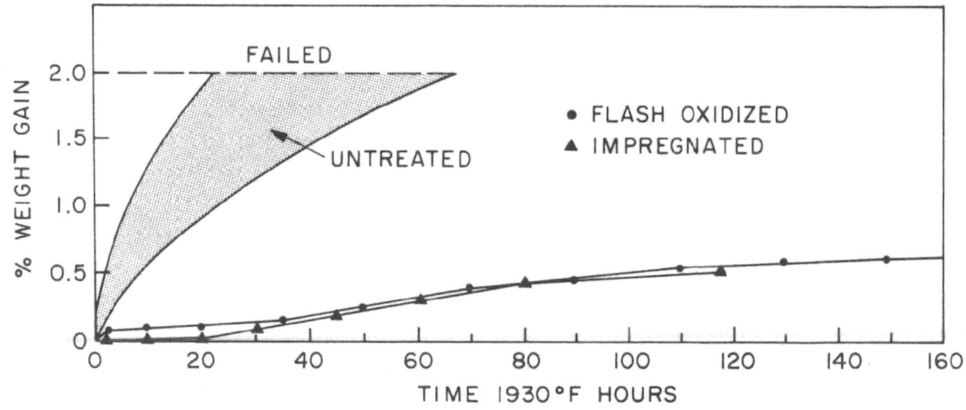

FIGURE 18 — PREVIOUS HISTORY OF STATOR WEIGHT CHANGE WHEN ENGINE-TESTED AT STEADY STATE CONDITIONS AT 1054°C.

A color television camera will be positioned at the rig's viewing port, and connected to a monitor at the operator's console. This will permit continual visual observation of the ceramic parts during the test. A taping system will also be utilized, so that permanent visual records of any future failures could be obtained.

Since finished stators made by Norton from NC-433 RSSC are now available, these parts will be durability tested, alternating Norton and Ford RBSN stators in the first stage position. As stators from the remaining vendors become available for testing, they will likewise be subjected to the durability test cycle. Longer range plans contemplate the addition of other turbine ceramic materials to this program.

REFERENCES

1. McLean, A. F., Fisher, E. A., Bratton, R. J., "Brittle Materials Design, High Temperature Gas Turbine." AMMRC-CTR-74-59, Interim Report, September, 1974.

2. McLean, A. F., Fisher, E. A., Bratton, R. J., Miller, D. G., "Brittle Materials Design, High Temperature Gas Turbine." AMMRC-CTR-75-8, Interim Report, April, 1975.

3. McLean, A. F., Fisher, E. A., Bratton, R. J., Miller, D. G., "Brittle Materials Design, High Temperature Gas Turbine." AMMRC-CTR-75-28, Interim Report, September, 1975.

4. McLean, A. F., Baker, R. R., Bratton, R. J., Miller, D. G., "Brittle Materials Design, High Temperature Gas Turbine." AMMRC-CTR-76-12, Interim Report, April, 1976.

5. McLean, A. F., Baker R. R., "Brittle Materials Design, High Temperature Gas Turbine." AMMRC-CTR-76-31, Interim Report, October, 1976.

6. McLean, A. F., Fisher, E. A., "Brittle Materials Design, High Temperature Gas Turbine." AMMRC-CTR-77-20, Interim Report, August, 1977.

7. Norris, D. M. and Grantham, P. V., "Thermal Stress Analysis of a Ceramic Gas Turbine Stator Blade," Report UCRL-51923, Lawrence Livermore Laboratory, October 3, 1975.

8. W. Trela, "Evaluation of Ceramics for Stator Applications — Gas Turbine Engines", Sixth Quarterly Report, Contract DEN 3-00019, May-July, 1979.

DEVELOPMENT AND CHARACTERIZATION OF CERAMIC TURBINE COMPONENTS

P.W. Heitman and P.K. Khandelwal

Detroit Diesel Allison
Division of General Motors Corporation
Indianapolis, Indiana

INTRODUCTION

Polycrystalline silicon carbide and silicon nitride materials are currently being utilized in the development of ceramic hot flow-path components for vehicular and industrial gas turbine engines. The ultimate success of these efforts, as reflected in marketable products, is tied directly to the realization of consistent, reproducible properties in actual components. The current state-of-the-art of structural silicon base ceramics, while developing rapidly, is still judged to be in a state of infancy. At present, engineering ceramic materials, with sufficiently well-developed materials and process specifications, that meet component reliability goals are not available. Considering the susceptibility of silicon base ceramics to failure from flaws of the order of 100 microns or less in size, it is clear that new, sophisticated process control procedures, heretofore, not required in commercial applications, and new, highly sensitive, nondestructive evaluation (NDE) techniques must be developed.

Detroit Diesel Allison (DDA) is currently involved in a major government-sponsored program intended to develop the technology needed to introduce ceramic components into the vehicular gas turbine engine. Faced with the current situation regarding the evolution of structural silicon base ceramic, DDA formulated a component development plan designed to yield, in a timely fashion, complex components with acceptable properties that could be consistently produced for commercial use. The development plan including an example of its application to a specific component, the turbine tip shroud, is reviewed. Also discussed is the current effort toward developing improved inspection techniques for ceramic materials.

COMPONENT MATERIALS DEVELOPMENT PLAN

The development of reliable ceramic components requires an iterative systems approach, which addresses the interplay between component design, evaluation/test, and manufacturing methodology. This necessitates a high degree of cooperation between the component user and manufacturer. The general scheme employed at DDA under the Ceramic Applications in Turbine Engines (CATE) program is illustrated in Fig. 1. First, materials are selected upon consideration of (1) component requirements, (2) potential for cost-affordable processing, and (3) manufacturing source capability. Materials that are acceptable in terms of these considerations are then evaluated on a preliminary basis to confirm their suitability. Such materials are processed, when possible, with appropriate fabrication techniques compatible with end usage. Structure and strength are generally evaluated and, on occasion, other critical tests deemed necessary in view of specific component operating requirements are performed.

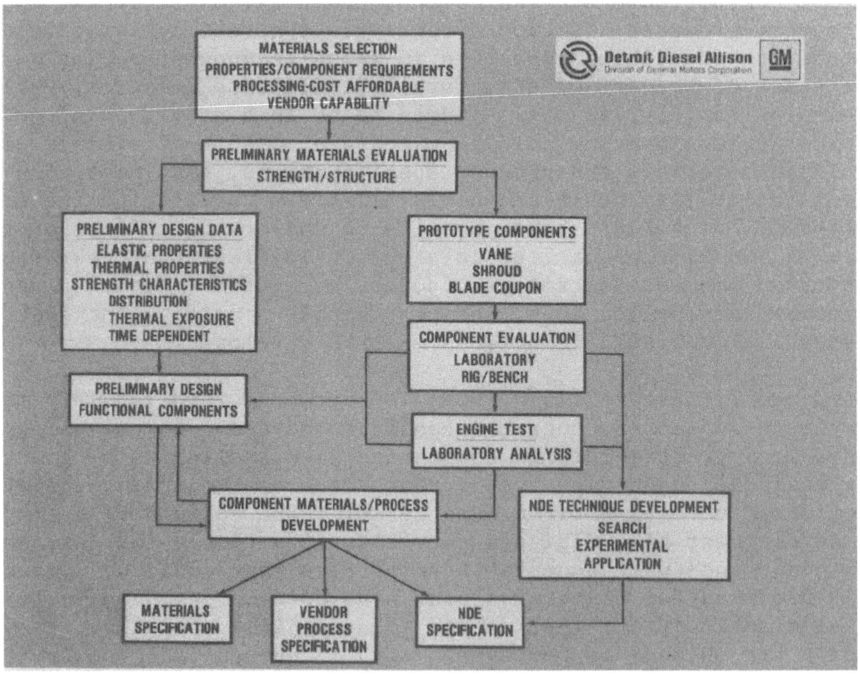

Fig. 1. Component materials development plan.

At this juncture, viable candidates are further evaluated to establish a data base for use in the preliminary design and analysis of functional components. Concurrently, fabrication of prototype components is initiated. These components are usually simplified versions of their function counterparts, which retain required critical features. Emphasis is placed on developing fabrication experience and establishing a base-line process routing. Further, such components are designed such that they may be readily and inexpensively evaluated in existing engines. The detailed evaluation and testing of these components in conjunction with the fabrication studies and preliminary design work are then used to structure a detailed component manufacturing development program intended to supply components with acceptable, reproducible properties. The results of tests and laboratory evaluations of first generation function components together with in-depth data developed on a mature material can be used to establish the final design configuration.

The initial materials and component characterization work also provides a basis for developing new inspection techniques. The evaluation of the size and nature of strength-controlling defects sets the requirements for the development program.

TURBINE TIP SHROUD DEVELOPMENT

Component Fabrication

Prototype shrouds of both silicon carbide and silicon nitride have been fabricated, evaluated, and successfully engine tested. The silicon carbide shrouds were produced by the Carborundum Company using a reaction-bonded grade of material. The silicon nitride shrouds were fabricated from 3502 material by GTE Sylvania. In addition to these materials, work is now proceeding with Corning Glass Co. (LAS) and Coors Porcelain Co. (R.B. Silicon Carbide). However, to illustrate the component materials development approach outlined earlier, the work in sintered silicon nitride shrouds will be taken as an example.

Four prototype silicon nitride shrouds were fabricated by GTE Sylvania at their Towanda facility. All shrouds were fabricated from GTE 3502 silicon nitride powder. The fabrication sequence, depicted schematically in Fig. 2, consisted of first isopressing simple cylindrical shells. Each shell was then sliced radially, thus producing two rings of approximately square cross section. One ring was then green machined to near-net shape. Next, both rings were simultaneously pressureless sintered to densities in excess of 95% theoretical. Finally, the near-net shaped ring was finish ground to final dimensions, while the remaining companion ring was sectioned into qualification test bars (10 bars per ring).

The shrouds were received in lots of two. The first lot, composed of S/N 35-25 and -35, was followed three months later by the second lot, S/N 35-45 and -55.

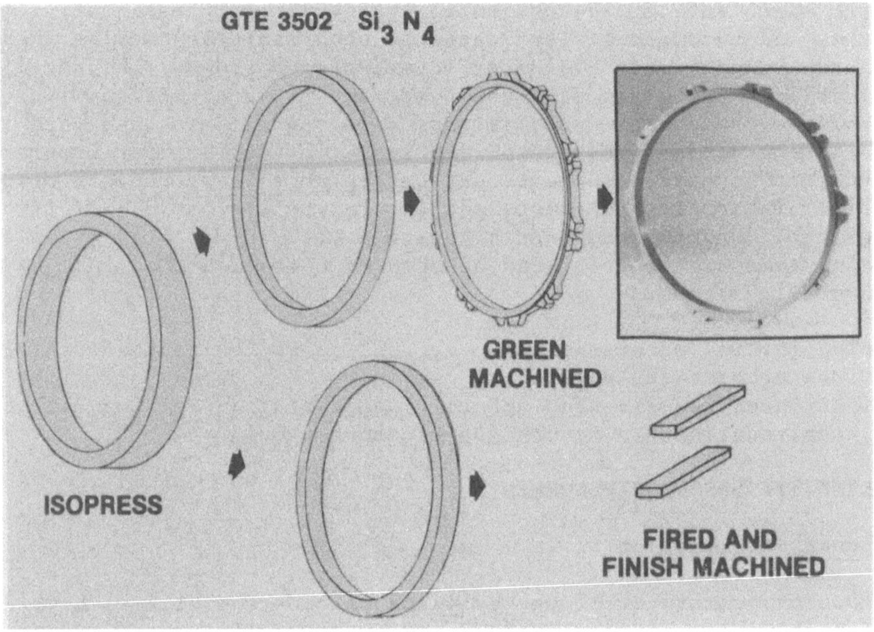

Fig. 2. Fabrication sequence for GTE 3502 Si_3N_4 shrouds.

SHROUD CHARACTERIZATION

Upon receipt, each shroud was subjected to detailed characterization. Both the shrouds and qualification test bars were examined nondestructively using standard aerospace X-ray radiographic, ultrasonic (1-10 MHz) and dye penetrant procedures. In addition, density and surface finish were also catalogued. Qualification test bars were used to establish material strength and microstructural characteristics. These results were correlated with burst strengths and microstructures of actual shrouds. Posttest fracture surface analyses were used to determine the size and nature of strength-controlling defects. The results of these investigations, were then compared with experience on early laboratory material to assess quality and estimate performance potential.

Nondestructive Evaluation

Bulk densities of both the components and the companion qualification test bars were determined by water immersion. The results are shown in Table 1.

Table 1. Density of GTE 3502 Si_3N_4 Shrouds

Lot	S/N	Shroud Density, g/cm^3	Test Bar Density, g/cm^3
1	35-25	3.15	3.17
	35-35	3.20	3.21
2	35-45	3.16	3.16
	35-55	3.23	3.26

The theoretical density of the 3502 composition, estimated from the rule of mixture, is approximately 3.45 g/cm^3. Thus, these shrouds ranged in density from 91 to 94% theoretical. However, since both alumina and yttria dissolve in silicon nitride the measured densities should be significantly closer to theoretical than the estimates in Table 1 indicate.

The variability in density from shroud to shroud is significant. It is independent of the time of manufacture and is thought to be related to variability in the starting powder. Further, as indicated below this variation is reflected in a significant variation in component strength.

Radiographic inspection revealed the presence of occasional large high-density inclusions in both the shrouds and the qualification test bars. A greater number of such inclusions were found to be present in the two higher density components (S/N 35-35 and -55). As discussed below these inclusions were found to contain both molybdenum and iron.

Standard ultrasonic inspection techniques using 1-10 MHz revealed no material defects. Dye penetrant inspection, using both standard (ZL-22) and filtered particle techniques, were somewhat ineffectual in detecting small defects, because of a high background level. This suggested a relatively high level of porosity consistent with bulk density measurements. However, the results clearly indicate a lack of any gross surface cracking or crazing.

SURFACE STRUCTURE

Examination of the surface of qualification test bars revealed the presence of several types of small pores which gave the material a salt and pepper appearance (Fig. 3). Of particular interest are those dark pores surrounded by a black halo. Observation at higher magnification shows the presence of bright metallic particles around each pore (Fig. 4). Figure 5 shows both the inside and edge region of this same pore as seen in the scanning electron microscope (SEM). Energy dispersive X-ray analysis shows that the small metallic particles contain molybdenum and silicon with small amounts of yttrium, iron, and sulfur.

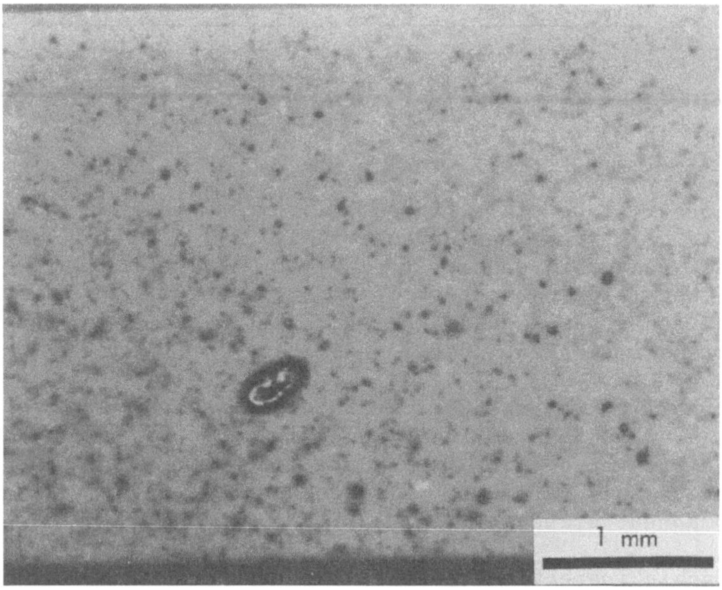

Fig. 3. As-ground surface structure of GTE 3502 Si_3N_4.

MICROSTRUCTURE

The general microstructure of GTE 3502 silicon nitride is shown in Fig. 6. The structure consisted of relatively large plate-like grains each surrounded by a matte of much finer grains.

COMPONENT MATERIAL STRENGTH CHARACTERISTICS

Material strength characteristics were established from the companion MOR test specimens accompanying each shroud. These specimens have the following dimensions: 3.18 mm x 6.36 mm x 50.8 mm. The measurements themselves were made using the fixture shown in Fig. 7. The loading geometry is quarter point with an inner span of 19.05 mm. The roller support structure is free to rotate in two directions thus keeping parasitic stresses to a minimum. A constant cross head speed of 0.5 mm/min was used for all tests.

Fig. 4. Pore structure of GTE 3502 Si_3N_4.

The strength and bulk density of each of the four shrouds is summarized in Table 2. There is a significant variation strength which is consistent with the variation in bulk density. The lower density shrouds (35-25 and -45) have an average strength of 430 MPa, while the higher density shrouds (35-35 and -55) display strengths of 507 and 518 MPa, respectively.

Table 2
Strength and Density of GTE 3502 Shrouds

S/N	Shroud Density, g/cm^3	Test Bar Density, g/cm^3	MOR, MPa
35-25	3.15	3.17	428
35-35	3.20	3.21	507
35-45	3.16	3.16	433
35-55	3.23	3.26	518

Fig. 5. Electron micrographs of pore structure of GTE 3502 Si_3N_4 shown in Fig. 4 (magn: 1800X).

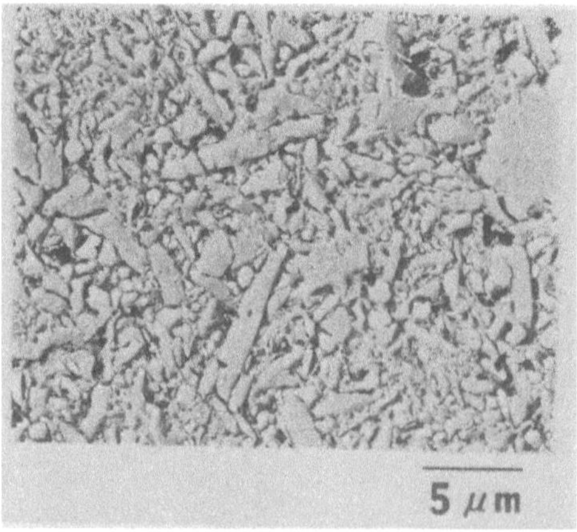

Fig. 6. General microstructure of GTE 3502 Si_3N_4.

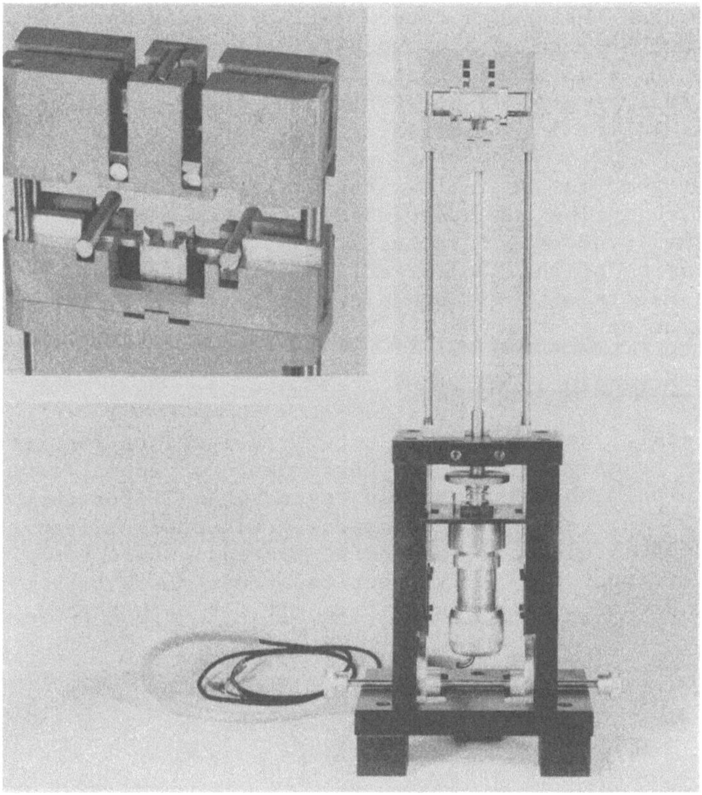

Fig. 7. Test fixture used to measure modulus of rupture.

An evaluation of fracture surface topography of the test specimens indicated that fracture was primarily intergranular at ambient temperature. Cleavage of large plate grains, however, was common. Fracture was caused by large pores, either internal or surface connected, in approximately 90% of the tests conducted. In the remaining specimens, fracture resulted from machining flaws. Generally, machining flaws were associated with the higher strength specimens, whereas pores initiated failure over the entire range of strengths observed.

Three types of pore structures were found to initiate failure of the test specimens. Most often failure occurred either from a smooth surface spherical pore or an irregularly shaped pore with a granular interior surface. The latter tended to be larger and were associated with low strength. Both types of pores, however, have a rather broad size range. The third type of pore, which caused failure in about 10 to 20% of the specimens, was the black pore structure described previously. It should be noted that pores larger than that which caused failure were generally present within the gage section of a given specimen. This clearly indicates, as would be expected, that, in addition to gross defect size, other factors such as pore shape and the surrounding structure are influential in establishing flaw severity.

The flaws that result in low fracture strengths generally have molybdenum rich inclusions associated with them. Molybdenum is not intentionally added but rather finds its way into the material during processing of the SN 3502 silicon nitride powder. During sintering it reacts with silicon nitride forming a silicide phase an accompanying pore structure.

Shroud Burst Strength

The successful development of ceramic components requires that a correlation be made between the performance of actual components, both in the laboratory and in engines, and laboratory test specimens. To this end, two nonabradable GTE 3502 silicon nitride shrouds (S/N 35-45 and -55) were burst tested. While both shrouds were processed under nominally identical conditions, bulk density and inclusion count differed significantly for each shroud.

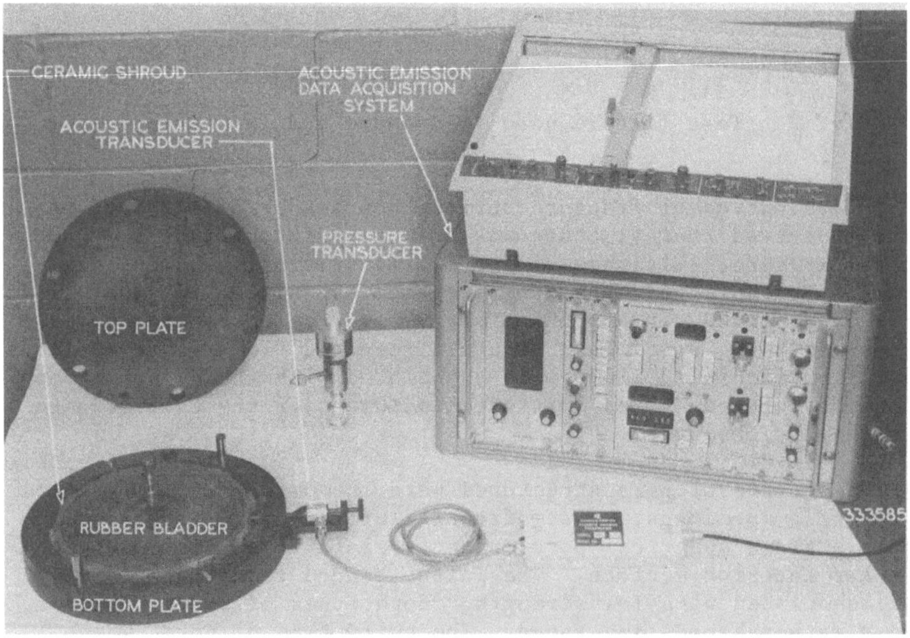

Fig. 8. Test fixture used to pressure test ceramic shrouds.

The test arrangement used for pressure testing is shown in Fig. 8. The fixture applies an internal pressure to the inside diameter of the shroud by means of a rubber bladder pressurized with hydraulic fluid. There was millimeter clearance between the shroud and retaining plates. MoS_2 lubricant was used to ensure freedom of movement between the bladder and retaining plates during pressurization. A wax ring was placed around the shroud to catch fragments of the burst. Acoustic emission instrumentation was attached to the shroud through a coupling device in hopes of assessing any crack growth or other damage prior to failure. Stress state at failure was established with the aid of a 3-D finite element analysis.

Component burst strengths, along with associated density and qualification test bar strength (MOR), are summarized in Table 3. A direct correlation between density, MOR, and burst strength is evident.

Table 3. Summary of Burst Test Results of GTE 3502 Shrouds

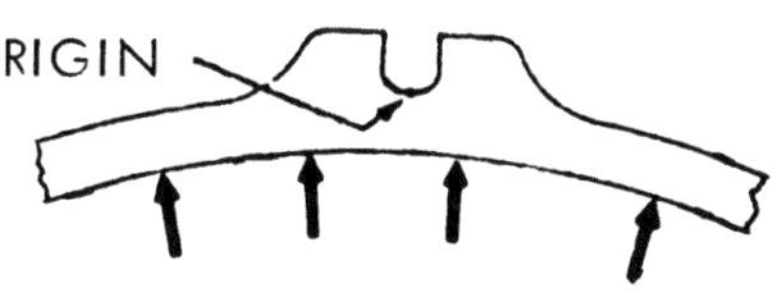

S/N	MOR,* MPa	Density, (g/cm^3)	Burst Strength, MPa
35-45	433	3.16	267
35-55	518	3.23	323

*Test bars

Fig. 9 shows one of the two shrouds after test. The arrow denotes the location of the initial fracture. Examination of the fracture surfaces showed that failure originated from a machining flaw located in the high stress region of the lug (Fig. 10). The remaining shroud (35-45) also failed at a similar location from a machining defect.

Close examination of the slot surfaces indicated the presence of rather rough grinding marks not typical of the general surface finish. This was particularly true in the case of S/N 35-45, the lower strength shroud. The qualification bars failed primarily from pores with only about 10% of the failures originating from machining defects. Further high strengths usually were associated

Fig. 9. GTE 3502 Si_3N_4 shroud after burst testing. Arrow denotes the location of the initial fracture.

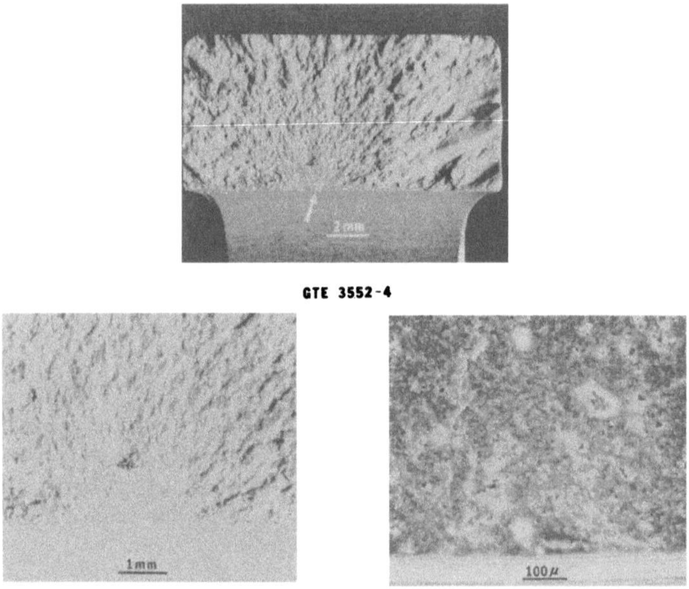

Fig. 10. Fracture surface of GTE 3502 Si_3N_4 shroud after burst testing. Arrow denotes the failure origin.

with surface flaws introduced by grinding. The surface area under high stress is a factor of 2 to 3 greater than that of a simple test bar. Assuming similar stress states, a 10% decrease in strength from the test bar value of 450 MPa would be expected for shrouds. The significantly lower value of strength observed must be related to excessive machining damage. This points out the difficulty in trying to assess component performance from laboratory test data. Similar experience has been noted with silicon carbide shrouds.

FLAW DETECTION STUDIES

Two classes of flaws have been found to be of importance in establishing the performance capability of silicon base structural ceramics. What shall be called class I flaws are those large in degree but small in extent (e.g., pores, inclusions, foreign matter and microcracks). Such flaws directly control material strength characteristics, particularly fast fracture strength.

Conversely, class II flaws are those small in degree and large in extent. Examples of this type of flaw include long range variations in microstructure (e.g., phase content, porosity, composition and grain size). This class of flaws also effects strength through control of thermal and elastic properties as well as fracture tightness and oxidation behavior. The nature and extent of both classes of flaws are highly dependent on material processing.

New or improved techniques for detecting and classifying both classes of defects in ceramic material are needed. In addressing this task under the CATE program, an assessment of available and emerging techniques was made. McGillen and co-workers, under subcontract to DDA, conducted a state-of-the-art survey of NDE techniques that might be applicable for class I flaws in silicon base ceramics. They concluded that reflective ultrasound for bulk defects and photoacoustic spectroscopy for surface defects show the maximum promise for near-term exploitation. Further, with regard to class II flaws, our own assessment indicated that ultrasonic velocity/attenuation and differential spectroscopy were most attractive for near-term work.

Based on this assessment an experimental program was initiated to examine and evalute the most promising techniques that might impact inspection of CATE hardware. These techniques are listed in Table 4. Examples of some of the ongoing work are given in subsequent subsections.

Table 4
Selected NDE Flaw Detection Techniques.

Class	Flaw Type	Technique
I	Voids, inclusions, cracks	Photoacoustic spectroscopy Reflective high-frequency ultrasound Acoustic microscopy
II	Microstructural variations	Ultrasonic velocity Ultrasonic attenuation Differential spectroscopy

Scanning Photoacoustic Spectroscopy (SPAS)

Laser scanned photoacoustic spectroscopy shows excellent potential for detecting surface and near-surface flaws in opaque ceramics. In this technique, the material to be examined is placed in a closed cell and scanned with modulated laser light. The periodic optical excitation of the sample results in a periodic heating of the specimen and subsequent heat flow into the gas layer immediately adjacent to the specimen surface. This creates pressure variation in the gas, at the modulation frequency, which can be detected by a microphone as an acoustic signal. The detected photoacoustic signal depends on the optical and thermal properties of the material and, in addition, upon the details of local microstructure and topography.

Fig. 11 shows the experimental setup used in work conducted at Gilford Instrument. A 2-mW He-Ne laser (λ = 0.6328 micron) was rigidly mounted to an optical bench. The laser light was modulated by a mechanical chopper at 40 Hz and focused on the surface of the specimen with a 76-mm focal length lens. The specimens were placed in a photoacoustic cell, which was situated in a 2-D translational stage. The stage was motorized in one direction to scan with one of the three available speeds (3.5, 7.5, and 15 microns) and incremented manually in the other direction using a micrometer. The signal was detected by a sensitive microphone and processed in the amplifier by a phase tracking lock and displayed on a strip chart recorder.

Two specimens--one each of sintered silicon nitride (SN 3502) and alpha silicon carbide--have been evaluated. The specimens dimensions were 12.7 mm x 6.35 mm x 1.5 mm. A 2.5-kg Knoop indentation was induced in each specimen. A black line was also placed on the surface of each specimen as a reference line. Figure 12 shows the results of the flaw detection study. The indentation

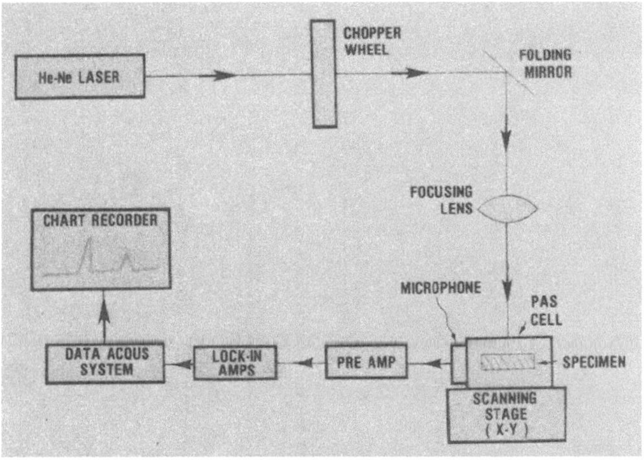

Fig. 11. Schematic diagram for flaw detection using photoacoustic spectroscopy.

was detected in both the silicon nitride and silicon carbide specimens. The amplitude of the photoacoustic signal, however, was small in alpha-silicon carbide compared to silicon nitride both for the black line and the indentation. This is attributed primarily to the differences in the thermal conductivity of these two materials since the density and specific heat (C_p) are similar. Thermal conductivity of sintered silicon nitride is 0.097 W/cm/^{o}C and that of alpha-silicon carbide is 1.531 W/cm/^{o}C. Therefore, silicon nitride is heated locally to a much higher temperature than silicon carbide. Consequently, the photoacoustic signal is much higher for nitride.

Reflective Ultrasound

High frequency ultrasonic flaw detection studies in silicon based ceramics are currently being conducted at DDA. Fig. 13 shows the present experimental setup for flaw detection (A-scan) in flat ceramic specimens. A broad band 75.0-MHz pulser/receiver, a stepless gate (for echo isolation), and a 50.0-MHz nominal frequency focused transducer were used. The transducer had a focal length of 4.13 cm in water. The waterpath was maintained at 6.4 mm.

Flaw detection studies on sintered silicon carbide seeded (voids) specimens have been initiated. The specimens consist of five circular disks 25.4 mm in diameter and 2.44 mm thick. These disks were seeded with voids of five size ranges (38 to 1400 microns) by Carborundum.

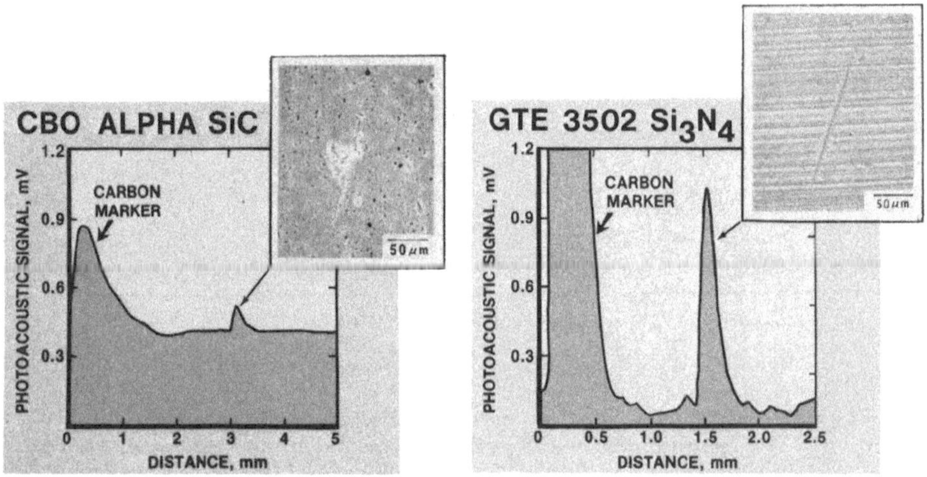

Fig. 12. Photoacoustic signal from surface on CBO alpha-SiC and GTE 3502 Si_3N_4

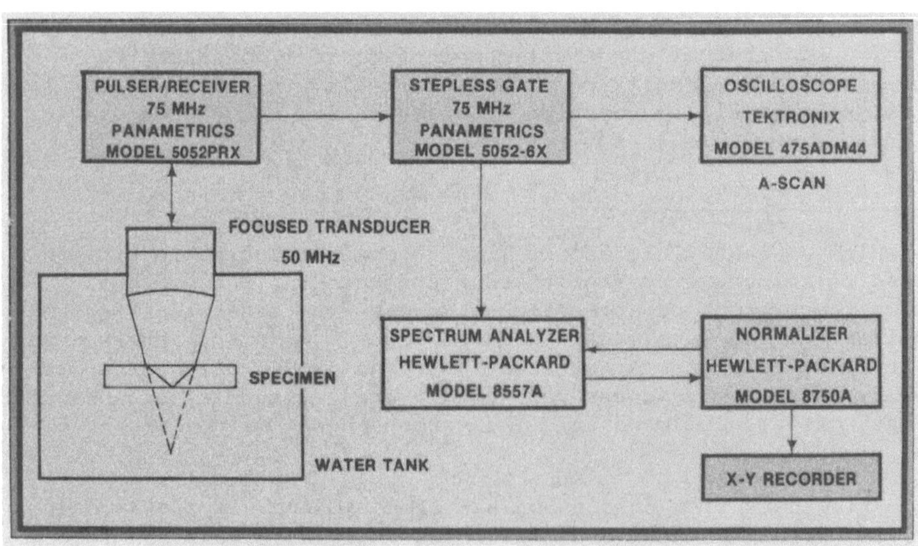

Fig. 13. Schematic diagram of high-frequency ultrasonic test arrangement.

The seeded disks were evaluated for amplitude detection (A-scan) in the longitudinal mode. Figure 14 shows the frequency spectrum of the wave pulse as it enters the coupling medium (water), the front surface reflection and the first back surface echo of the specimen. The output of the transducer shows high frequency contents from 30 to 65 MHz with significant power level. Attenuation of the high frequency portion of this spectrum does occur during transmission through water. However, the center frequency of the pulse as it enters the specimen is still about 38 MHz with a mechanical Q of 2.27. The center frequency of the back surface echo was also about 38 MHz with substantially less amplitude than the front surface echo because of very large impedance mismatch between water and alpha-silicon carbide. It was also observed that the ultrasonic attenuation of this material is very small at these test frequencies.

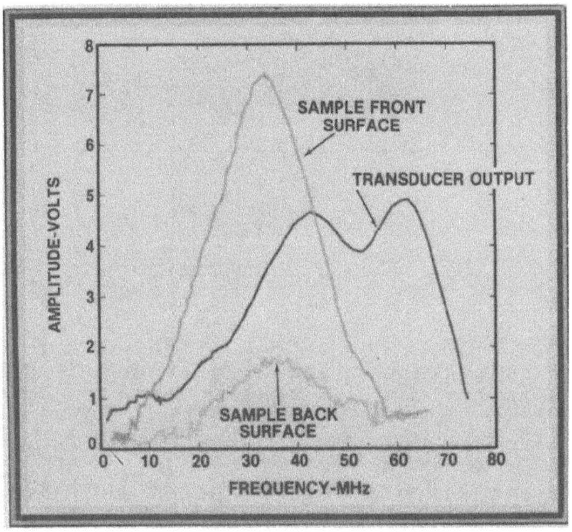

Fig. 14. Frequency spectrum of three reflected echoes. (1) Delay line-water interface, (2) front surface of specimen, and (3) first back surface of specimen.

The results of flaw detection studies on the disks are shown in Fig. 15. The upper portion of this figure shows the reflections of both the front and back surfaces of a defect free section and the position of the stepless gate. The lower portion shows the gated output from each disk along with a positive of the corresponding radiograph taken. The front surface dead zone varied between 70 and 100 ns due to variability in the specimen surface conditions while that at the back surface was about 30 ns. Maximum noise resulting from the material microstructure and electronics was observed to be about 7 mV. The signal-to-noise ratio was greater than 3 to 1 in all cases.

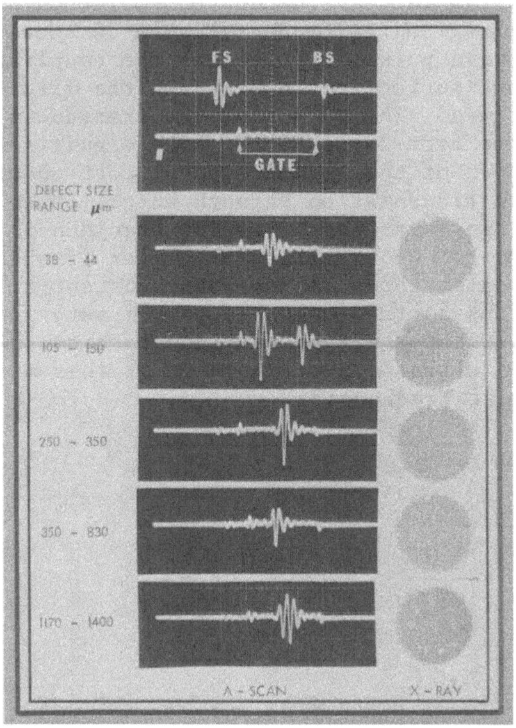

Fig. 15. X-Ray radiographic and high-frequency ultrasonic flaw detection of defects in alpha-SiC.

Numerous apparent defects were detected as indicated by intermediate reflections and loss of back surface reflections in all specimens, including the specimens where defect size ranges from 38 to 44 microns. Standard radiographic procedures did not detect flaws below 250 microns. The detection of these defects should not be construed, however, to suggest that small individual voids can be detected with certainty. The observed flaw signals may very well represent reflections from a group of small defects.

Acoustic Microscopy (SLAM)

Flaw detection studies using a 100-MHz acoustic microscope (Sonoscan Model 100) are currently underway at the Indianapolis Center for Advanced Research (ICFAR). The technique being employed for optically opaque silicon base ceramics is illustrated schematically in Fig. 16. The specimen is placed on the acoustic stage of the microscope and covered with a mirrored coverslip. An ultrasonic transducer (100 MHz) mounted in the stage insonifies

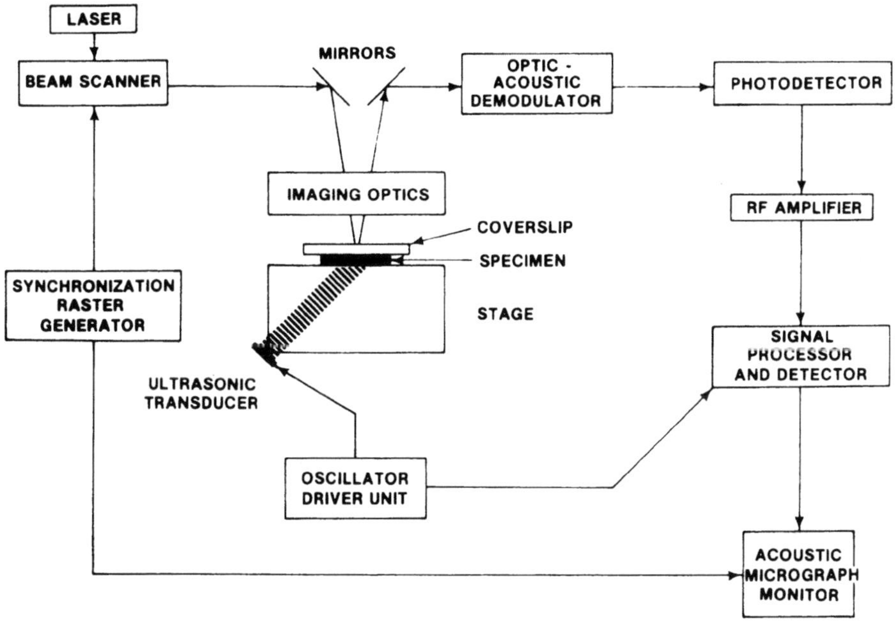

Fig. 16. Schematic diagram for flaw detection in ceramics using acoustic microscopy (SLAM).

the specimen causing a dynamic ripple on the surface of the coverslip. A laser beam continuously scans the rippled surface of the coverslip and forms an acoustic image. Localized variations (inhomogeneity and flaws) in the ultrasonic attenuation and velocity of the material modulates the amplitude and phase of the surface perturbations and cause the image to change. The image can be displayed on the TV monitor both as an acoustic amplitude micrograph and an acoustic interferogram.

Preliminary work has included the evaluation of two of the seeded alpha-silicon carbide disks (38 to 44 and 105 to 150 micron voids) discussed previously. Each specimen was interrogated at two locations: first, where there was no defect and transmission was good. Figure 17A shows a typical acoustic interferogram of such an area in one disk (38-44 microns). The interference lines are straight. Second, the specimen was moved to a new location and examined. This area showed voids across the entire viewing area of the stage. Figures 17B and C show a typical acoustic interferogram of such a defective area. A line of discontinuities extending almost the entire width of the picture is seen. The defect at one location was about two line spacings wide (190

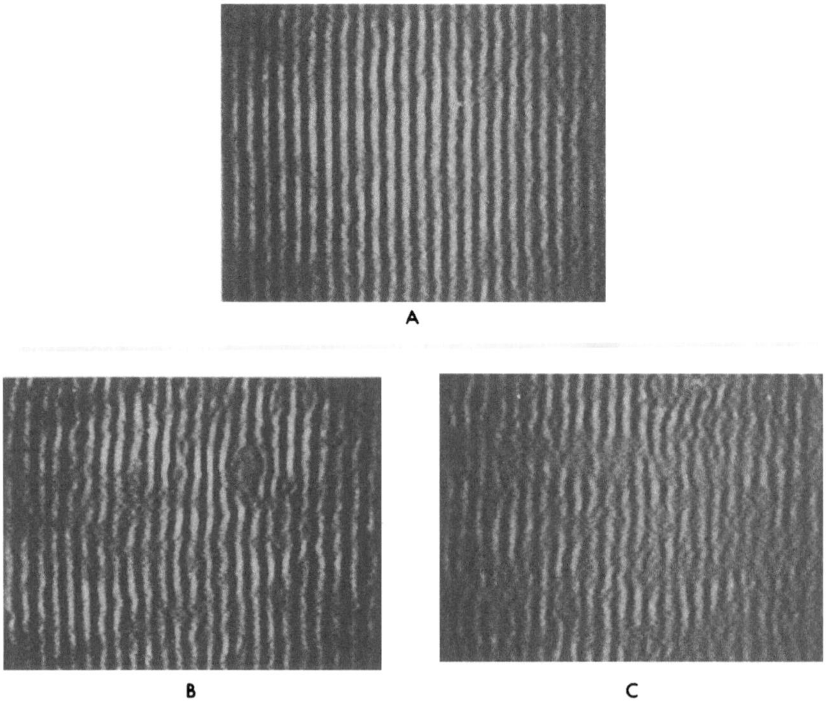

Fig. 17. Acoustic interferogram of alpha-SiC seeded disks (nominal defect size, 38-44 microns). (A) Base material, (B and C) areas with seeded defects.

microns). The examination of the disk at various other locations revealed that voids were as much as several millimeters long. The foregoing experiments have established that seeds in the disks are generally coalesced and not individual. Our ultrasonic A-scan inspection described earlier, therefore, must have detected groups of small defects.

SESSION Vb

TEST, EVALUATION, AND DEVELOPMENT OF SPECIMENS

Chairman: Prof. W. Love
University of Washington

Vice Chairmen: Prof. A.S. Kobayashi
University of Washington

Dr. R. Chait
Army Materials and Mechanics Research Center

HIGH TEMPERATURE STATIC FATIGUE TESTING OF TURBINE CERAMICS

George D. Quinn

Army Materials and Mechanics Research Center

Watertown, MA 02172

ABSTRACT

An experimental program was conducted to investigate the degradation in mechanical properties of silicon based ceramics exposed to high temperature oxidizing environments. This paper will present the results of testing on three grades of reaction bonded silicon nitride (RBSN): Norton NC 350, Kawecki Berylco Industries RBSN, and Ford 2.7 RBSN. Testing included flexural stress rupture at 1200°C and stepped temperature stress rupture (STSR) experiments. The latter test was devised for this study and is a variation of the stress rupture test such that a range of temperatures are employed. The purpose of the STSR test is to explore the potential stress-temperature regimes of static fatigue failure. In addition, a combined cycle durability sequence was applied to the three RBSN's. This procedure is a crude service simulation involving static heat soaks and rapid thermal cycling on bend bars. Extensive fractography was conducted to identify causes of failure.

I. INTRODUCTION

Silicon based ceramics have considerable potential for usage in structural applications at high temperature and promise to play an important role in energy conversion devices. A severe impediment to the use of ceramics is the deficiency of mechanical property data after long term exposure to oxidizing environments, particularly with temperature cycling. For example, data on materials such as hot pressed silicon nitride indicate that strength could degrade by as much as fifty percent after a few hundred hours exposure at 1371°C.[1] Due to the lack of property data after environmental exposure, the present generation of ceramic components for gas turbines have been designed using virgin material properties.

A program of research was undertaken to address these durability related issues. Static fatigue testing was conducted to identify mechanical property degradation under the influence of stress in an oxidizing environment. High temperature testing in furnaces was employed. Since the passive environment in a test furnace may not adequately represent actual service conditions, an alternate procedure was utilized as well. Unstressed samples were exposed to a crude simulated service environment and subsequent property degradation was measured. Extensive fractography was used to identify the cause of material failure or degradation.

Thirteen materials were chosen for evaluation:

* Norton NC 132 hot pressed silicon nitride
* Norton NCX 34 hot pressed silicon nitride
* Norton NC 136 hot pressed silicon nitride
* Norton NC 350 reaction bonded silicon nitride
* Kawecki Berylco Industries reaction bonded silicon nitride
* Ford 2.7 reaction bonded silicon nitride
* Norton NC 203 hot pressed silicon carbide
* Norton NC 435 reaction bonded silicon carbide
* Norton NC 433 reaction bonded silicon carbide
* Carborundum 1977 sintered alpha silicon carbide
* Carborundum 1978 sintered alpha silicon carbide
* General Electric Silcomp CC siliconized silicon carbide
* General Electric Silcomp CRC siliconized silicon carbide

This paper will discuss the results obtained for the three grades of reaction bonded silicon nitride. While the mechanical properties of these three materials will be compared, it is important to note that they are at different stages of development, and that they were furnished in different forms.

II. MATERIALS

Bend bar samples were carefully ground by a surface grinder such that surface striations ran parallel to the long axis. All four long faces were machined with either a 220 or 280 grit diamond wheel with a final finish removal rate of about 1.5×10^{-4} cm (0.00005 inch) per pass. In this manner about 0.008 cm (.003 inch) of material were removed from each face in the final finishing phase. A wheel speed of 150 surface meters per minute was used (500 feet/minute). All four long edges were chamfered approximately 45° to a depth of about 0.015 cm, and specifications were for striation markings to be parallel to the long axis.

The NC 350 tested was fabricated by Norton in mid 1975 and was in the form of a plate 19.1 cm in diameter by 0.46 cm thick. Sixty samples of size 0.203 x 0.280 x 5.08 cm were prepared and had an average density of 2.53 g/cm^3 with a standard deviation of 0.019

g/cm^3. Quantitative analysis by x-ray diffraction[2] (on a powdered sample) indicated 74 weight percent alpha silicon nitride with the balance beta phase. Impurity content was quantitatively determined by emission spectroscopy and is shown in Table 1. The distribution of porosity is illustrated by a typical polished section shown in Figure 1. The pores are well distributed with no excessively large pores or zones of porosity. No residual silicon was evident.

The Kawecki Berylco Industries RBSN was procured in January 1977 in the form of a large plate 30.5 cm square by 1.27 cm thick. This size is much larger than the billets of the other two RBSN's evaluated. The billet was formed by isopressing a large billet of silicon powder, slicing the 1.27 cm slab from the billet, and by nitriding the slab. One hundred bend bars of size 0.203 x 0.280 x 5.08 cm were cut from the billet with at least two samples coming out of the thickness of the billet. The average bend bar density was 2.59 g/cm^3 with a standard deviation of 0.40. A density gradient through the plate thickness was evident as shown in Figure 2 and probably accounted for the high standard deviation in bend bar density. The billet was least dense in the interior. Polished sections indicated the porosity is irregular in shape with a wide distribution of size. Occasional silicon grains associated with pores were also observed (Fig. 3a,b). X-ray diffraction identified 72 percent alpha silicon nitride and the remainder beta silicon nitride. Emission spectroscopy showed Al and Fe to be the primary impurities or additives (Table 1).

Ford injection molded RBSN, nominal density 2.7 g/cm^3, was obtained in late 1977. The material was furnished in many small slabs of size 0.3 x 0.6 x 7.6 cm which had been injection molded and then nitrided. From thirty-two of these slabs, sixty-four samples of size 0.216 x 0.280 x 5.08 cm were prepared. Average density was 2.77 g/cm^3 with a standard deviation of 0.016 g/cm^3. Porosity was uniformly distributed except for large pockets that were often observed (Fig. 4). Very little free silicon was observed, although the shape of some pores suggested some meltout may have occurred.[3] X-ray diffraction indicated 80 weight percent alpha and 20 percent beta silicon nitride. A few samples had large cracks and were not used. A relatively high Fe content (Table 1) detected by emission spectroscopy is the result of a three percent iron oxide addition.[4]

III. EXPERIMENTAL PROCEDURE

Four test procedures were employed in the program:

* Room temperature (RT) flexural strength testing
* 1200°C flexural stress rupture testing
* 1000-1400°C stepped temperature stress rupture (STSR) testing
* Combined cycle thermal exposure testing

TABLE 1

Metallic Element Content of RBSN's
(Weight Percents)

	Al	B	Ca	Cr	Cu	Fe	Mg	Mn	Mo	Ni	Ti	V	Zr
Ford 2.7 RBSN	.07	∿.005	.06	<.04	.01	1.0	.003	.02	--	<.04	<.06	<.03	--
KBI RBSN	.12	∿.005	.07	<.04	.01	0.23	.004	.03	∿.05	<.04	<.06	<.03	∿.01
Norton NC 350	.08	∿.005	.07	<.04	.01	.14	.003	.01	--	<.04	<.06	<.03	--

Notes: -- Not Detected or Scanned for.
∿ Semiquantitative

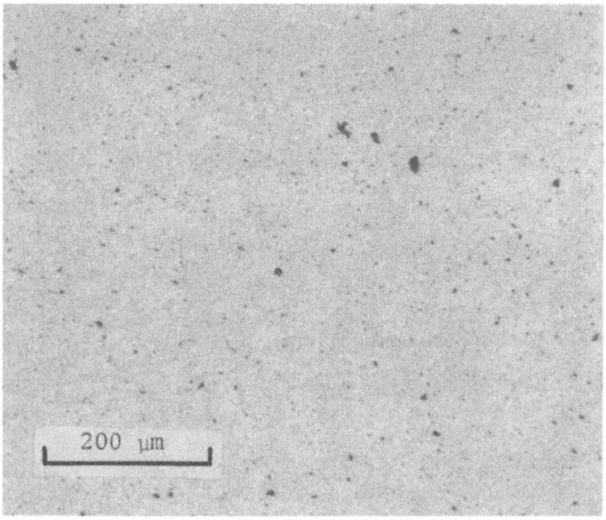

Fig. 1. A polished section of Norton NC 350 showing a uniform distribution of porosity

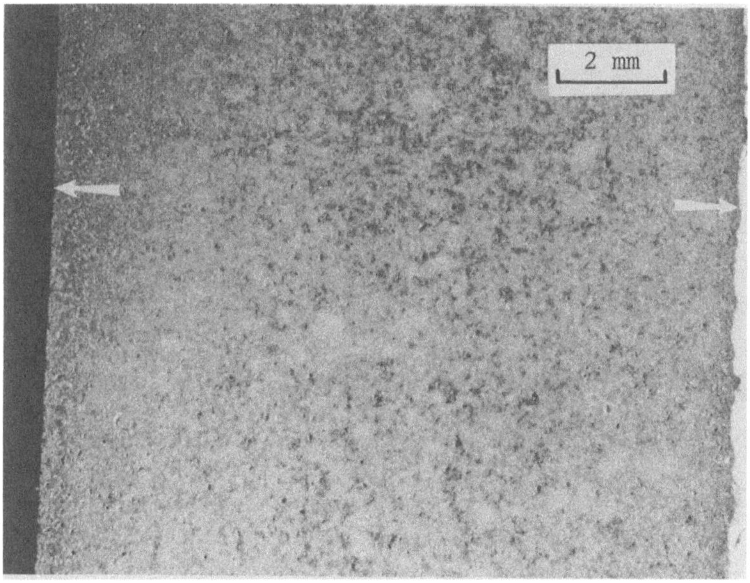

Fig. 2. A macrophotograph of a section through the thickness of the KBI RBSN billet. The billet was 1.27 cm thick (arrow to arrow) and a gradient in porosity is evident

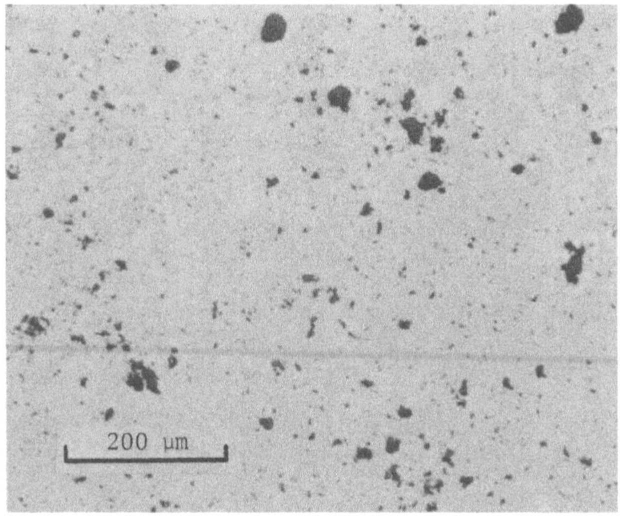

Fig. 3a. A polished section of KBI RBSN showing a wide distribution of pore size and shape

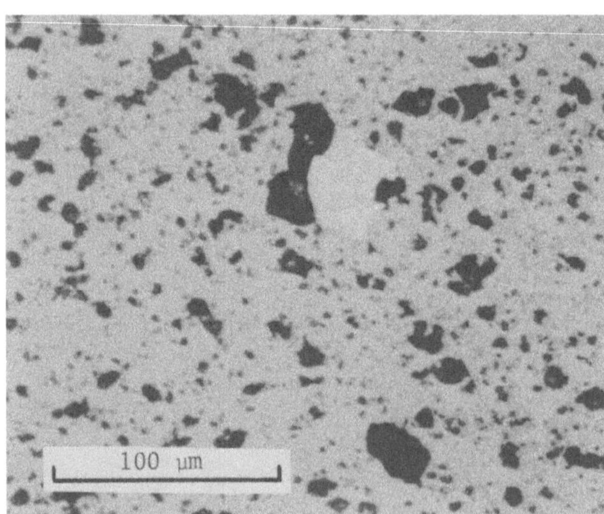

Fig. 3b. A higher magnification section showing residual silicon (white)

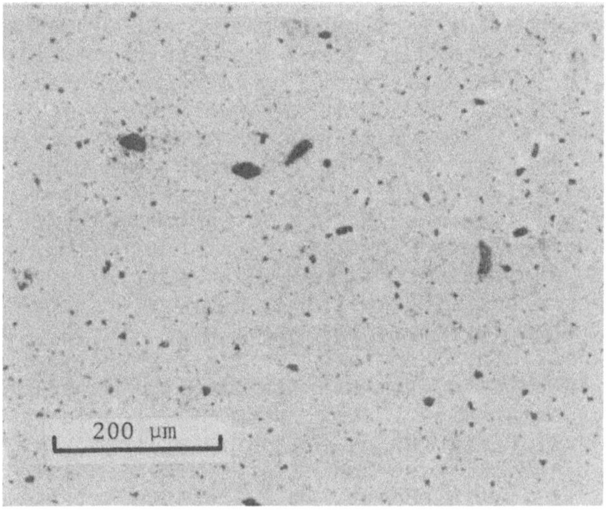

Fig. 4. A polished section of Ford 2.7 RBSN. Most of the porosity was finely distributed but some large pores were evident as shown in this figure

Fig. 5. The mechanical test furnace used for the stress rupture and STSR testing

A reference level of strength was determined by breaking a control lot of samples at room temperature in four point bending. Fixture spans were 3.04 and 1.52 cm. A universal testing machine* was used with a crosshead speed of 5 x 10^{-3} cm/minute. Care was taken to insert the combined cycle exposed samples into the fixtures (for retained strength measurement) such that the most severely damaged face was in tension inside the inner load pins. The stress rupture survivors (also for retained strength measurement) were mounted so that the zone most highly stressed during the high temperature test was again loaded in a tensile manner within the inner load pins of the room temperature fixture. The control samples, being 5.08 cm long, occasionally permitted two breaks per sample. In such case, care was taken to insure the previously stressed portion of the sample was not reinserted into the inner gauge length. Stress at failure was calculated from the elastic beam equation for the maximum outer fiber stress. No adjustments were made for subsurface flaws or for curvature in the case of samples that survived high temperature testing (and were broken at RT for retained strength). Weibull parameters for the data will be reported elsewhere.[5]

High temperature flexural stress rupture trials were conducted at 1200°C as the second major experimental procedure. This temperature is representative of working temperatures to be expected in ceramic gas turbine passages and other energy conversion devices. A bank of furnaces was constructed for this purpose and permitted experiments to run for several hundred hours. Tests were usually terminated at 300 hours if samples were still intact, although in some instances, tests of much longer duration were conducted. Sufficient samples were tested to discern any trend toward diminishing load carrying capacity with time. In the cases where static fatigue failures were confined to narrow stress ranges (samples fail on loading or survive intact), only enough samples were used to establish an approximate critical stress level.

A small and inexpensive mechanical test furnace was designed and constructed to perform the flexural stress rupture tests (Fig. 5). The furnace is made with refractory firebrick and utilizes silicon carbide heating elements to attain temperatures up to 1500°C in air. The fixtures are machined from hot pressed silicon carbide and have spans of 3.81 x 1.91 cm. Loading is effected by a dead weight lever arm mechanism on the top of the furnace which transmits a load to the fixtures. Complete details of the design, construction and operation of the furnaces as well as the fixtures are in Reference 6.

Static fatigue failures over a broader range of temperatures were investigated via a stepped temperature stress rupture (STSR) procedure specifically devised for this program.[7] The STSR

*Instron Corp., Canton, MA

procedure is very similar to the common stress rupture test except that a range of temperatures are employed upon each sample. This test was conducted over the range of 1000 to 1400°C and is intended to detect the temperature and stress ranges where time dependent failure can occur. The temperature cycle chosen for our requirements is illustrated in Figure 6. A flexural sample is loaded into the standard furnace with four point bend fixture and the furnace is heated to 1000°C in air with no load applied to the sample. At 1000°C a deadweight load is applied. Should the sample survive twenty-four hours at that temperature, the furnace is then heated (approximately 1/2 hour) to 1100°C and again allowed to soak for twenty-four hours. This cycle is repeated for 1200, 1300 and 1400°C, but in the last case, the soak is maintained for 60 hours. This cycle was arbitrarily chosen, but the range encompasses temperatures that critically stressed ceramic components such as nozzles and blades will experience in vehicular gas turbines with inlet temperatures up to 1400°C. Throughout the test, the same sample is subjected to a constant deadweight load. In the event a sample breaks, the furnace is immediately cooled and the time of failure is denoted by an arrow on the STSR plot. The arrow is labelled with the stress that was applied to the sample. A series of trials are executed with differing loads corresponding to the stress levels calculated from the elastic beam formula. The STSR test allows one to quickly focus on temperatures and stresses where a material may exhibit time dependent failure.

The fourth procedure applied was a combined cycle thermal exposure pattern wherein a group of samples were alternately heat treated in an oxidizing environment (furnace in air) and then thermally cycled in a flame heating compressed air quench apparatus. The combined thermal exposure and thermal cycling sequence consists of five repetitions of 24 hour soaks in air at 1000, 1200 and 1371°C with a subsequent 100 thermal shocks from room temperature (RT) up to 1371°C and then down to RT. In this manner, specimens accumulate 360 hours of exposure and 500 thermal shock cycles, as shown in Figure 7. While not intended to match any specific application, the purpose of this procedure is to impart a more abusive treatment to the samples than that generated by a static oxidation treatment. The thermal soaks were carried out in air in a furnace (similar to the stress rupture furnace) equipped with a specially constructed Si_3N_4 muffle to protect samples from the possibility of contamination with the furnace refractories. Thermal shock cycling was performed in the AMMRC thermal fatigue rig shown in Figure 8. This machine is modeled after the one initially described by Johnson and Hartsock.[8] Samples were placed on an indexing table which rotates samples into a flame heating station. Once steady state conditions were achieved (maximum sample temperature is 1371°C) the table was rotated, thereby indexing the sample to a quench station where compressed air cools the sample. Cycle time was set to 20 seconds. Samples were allowed to accumulate 100 such cycles each and then removed from the

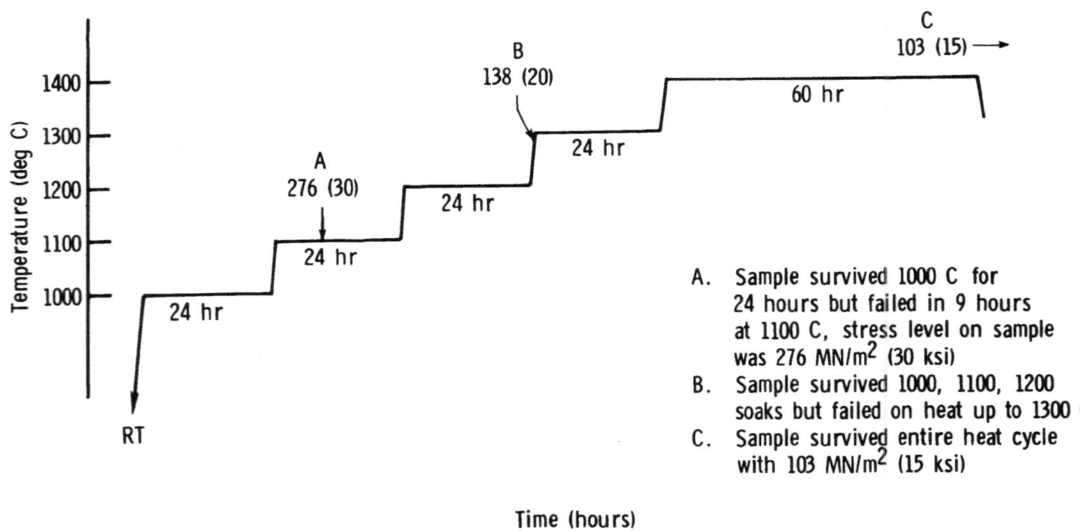

Figure 6. The stepped temperature stress rupture sequence with three examples.

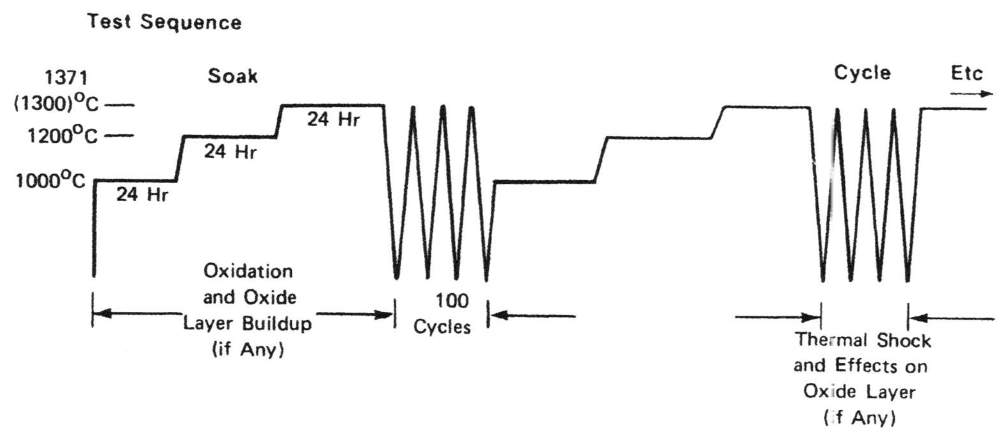

Figure 7. The combined cycle exposure sequence. The soak cycling procedure is repeated five times to give a total of 360 hours soak time and 500 shock cycles to each of twelve samples.

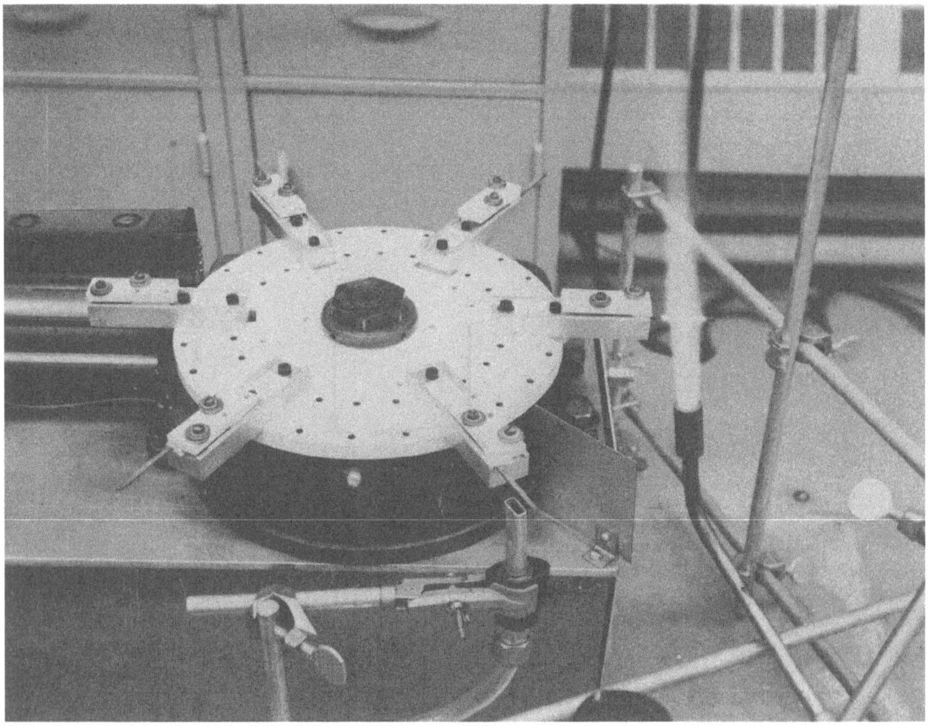

Fig. 8. The thermal fatigue rig. An indexing table cycles samples from a flame heating station to a compressed air quench station

rig for subsequent furnace exposure or final strength testing. Care was taken to ensure the same portion (and side) of each sample was exposed to the flame heating-quench cooling zone on each sequence. Complete details of the operation of the thermal fatigue rig, the heat transfer conditions and the thermal stresses are discussed in References 5 and 9.

IV. RESULTS

A. Reference Strength

Sixteen NC 350 reference fractures yielded an average strength of 294 MPa with a 41 MPa standard deviation. Distinct mirrors indicated the origins of failure. Strength limiting flaws were large volume distributed pores (Fig. 9a,b) and also machining defects at the surface (Fig. 9c).

Twenty-five KBI reference fractures resulted in an average strength of 206 MPa with a standard deviation of only 15 MPa. Fractography revealed two strength limiting flaws: large pores and also large pores with inclusions that are probably unreacted silicon. Both flaw types were volume distributed and are illustrated in Figure 10.

Sixteen Ford 2.7 reference fractures gave an average strength of 288 MPa with a standard deviation of 38 MPa. Strength controlling flaws were usually pores or defects associated with pores as shown in Figure 11. These flaws were volume distributed.

These average strengths and standard deviations are illustrated in Figure 12.

B. Stress Rupture 1200°C

None of the nine NC 350 stress rupture trials failed in a time dependent manner (Figure 13). The cut-off stress level is 345 MPa, a value considerably higher than the reference RT strength (294 MPa). The six survivors (samples which did not break and were removed from the furnace intact) were broken at RT to measure retained strength. Table 2 shows the results and the strength limiting flaws. The samples with the highest retained strengths tended to fail from defects that were at the sample surface and as such may have been healed by the formation of the coherent alpha cristobalite oxide layer. The two samples that failed from subsurface defects had strengths similar to the reference group.

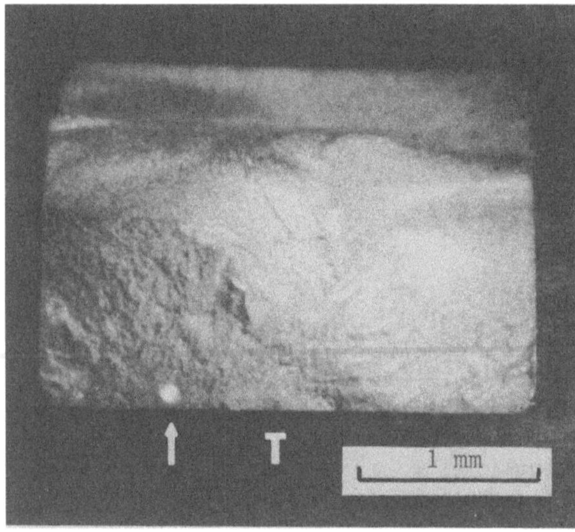

Fig. 9a. A light photmacrograph of the fracture surface of a NC 350 reference strength sample. T marks the tensile edge. The strength controlling defect is the light spot indicated by the arrow. Fracture stress was 316 MPa

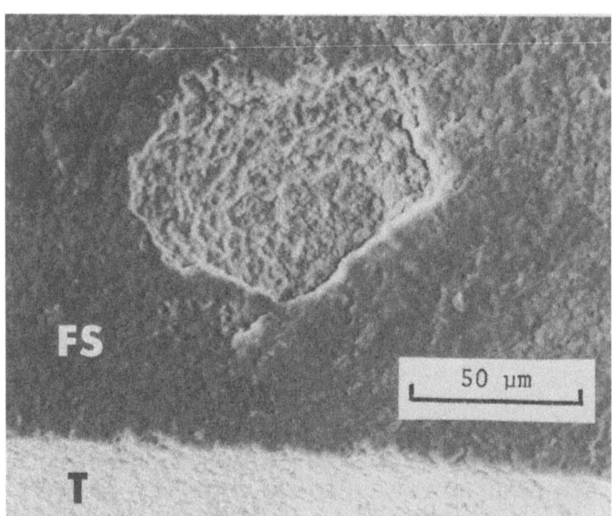

Fig. 9b. A SEM closeup of (a) shows the defect is a subsurface pore

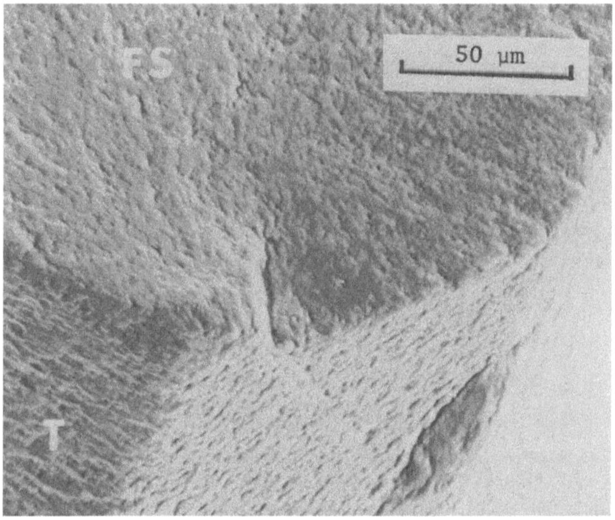

Fig. 9c. A SEM photo of the fracture surface of another NC 350 referecne strength sample. T marks the tensile flat face of the sample, and FS indicates the fracture surface. The sample has been tilted back to show the strength limiting flaw which is machining damage related to a chamfer machining striation. The fracture stress was 264 MPa

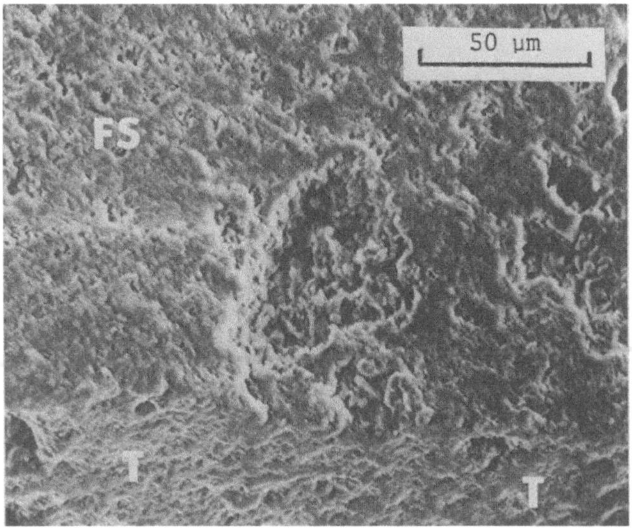

Fig. 10a. A SEM photo of the strength limiting flaw in a KBI RBSN reference strength sample. The flaw is a large pore near the surface of the sample. Fracture stress was 214 MPa

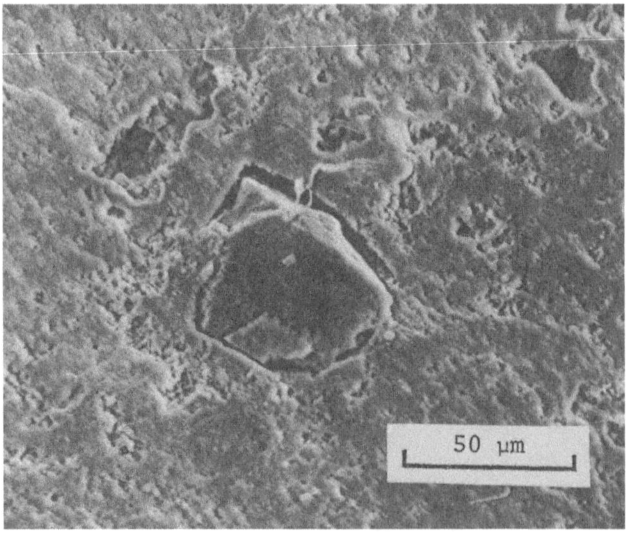

Fig. 10b. A SEM photo of another strength limiting flaw in a sample that had a strength of 225 MPa. The flaw is a subsurface pore with inclusion that may be excess silicon

Fig. 11a. Fracture surfaces of a Ford 2.7 RBSN reference strength sample showing a subsurface strength controlling defect. The fracture halves are mounted back to back along the tensile edge. Fracture stress was 307 MPa

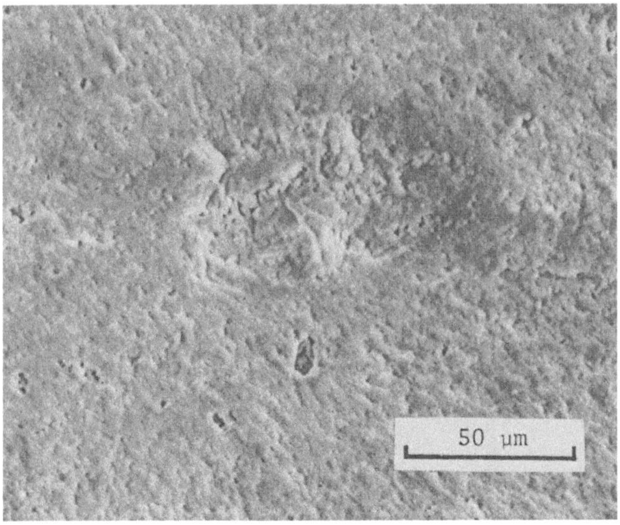

Fig. 11b. A SEM closeup of the defect in (a). The flaw is a porous zone

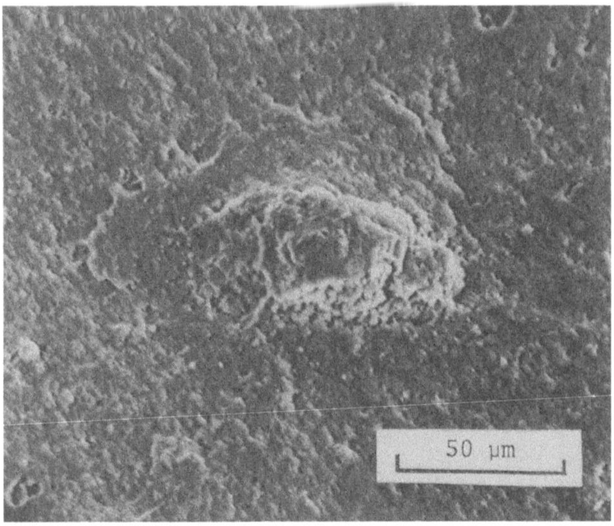

Fig. 11c. A SEM photo of the strength limiting flaw in a different sample which failed at 273 MPa. The flaw was subsurface and is a porous zone

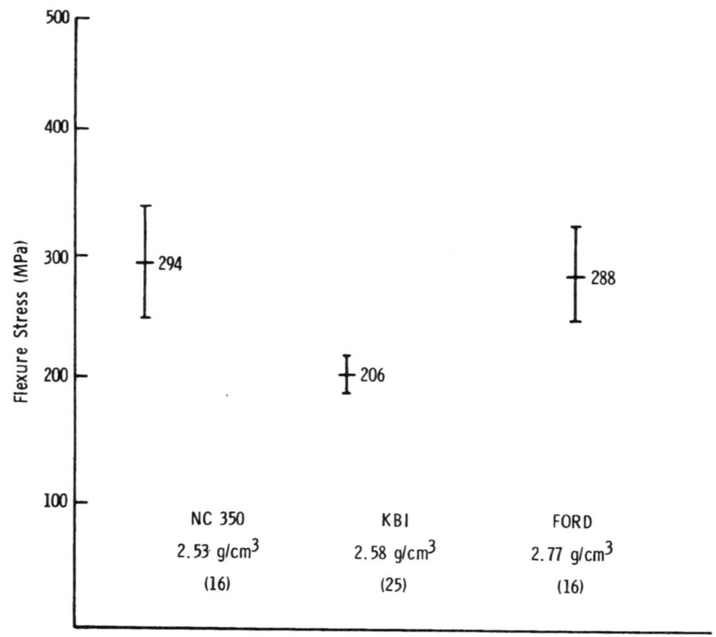

Fig. 12. Comparative chart showing the RT reference strengths of the three RBSN's. The error bars are one standard deviation. The number of fractures is in parenthesis

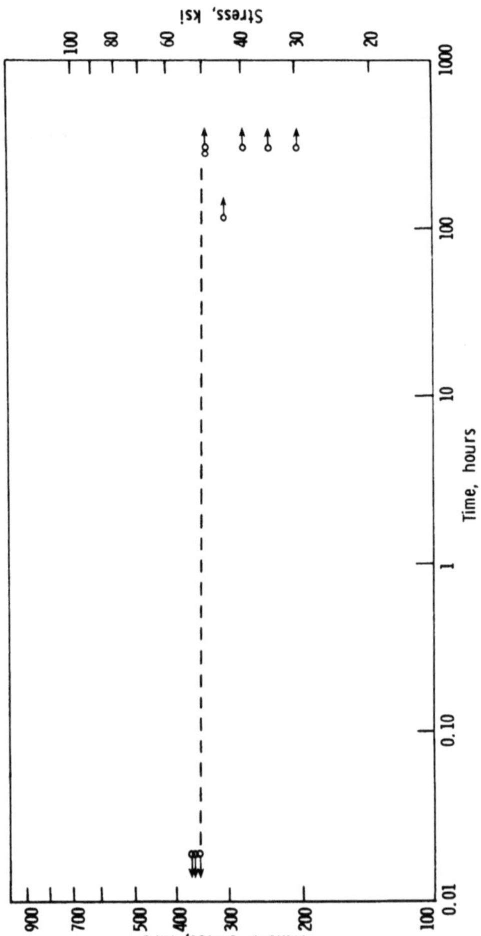

Fig. 13. Stress rupture at 1200°C for Norton NC 350 RBSN

TABLE 2

NC-350 Static Fatigue Survivors

Stress Rupture Loading	Survival Time	Room Temperature Retained Strength	Room Temperature Failure Source
345 MPa	300 hours	470 MPa	*Surface Connected Porosity
345	267	427	*Machining Damage
310	118	454	*Machining Damage
276	300	373	Subsurface Porosity
241	300	257	*Surface Porosity
207	300	256	Subsurface Porosity

*Flaw healing potential

The results of fourteen KBI RBSN stress rupture trials are shown in Figure 14. The cut-off stress is approximately 190 MPa, a value slightly less than the reference RT strength (206 MPa). The failure origins for the two samples which failed in a time dependent manner could not be determined. The retained strength of the 500 hour survivors were 219, 208 and 191 MPa for the samples loaded to 207, 193 and 150 MPa respectively. Failures initiated from an unidentified site, a pore and an unidentified site respectively. The 200 MPa sample which survived 300 hours had a retained strength of 198 MPa and failed from a large pore. These retained strength values are not very different from the reference RT strength and indicate little or no degradation.

Of twelve Ford 2.7 RBSN tests, three failed on loading, five survived intact and four failed in a time dependent manner (Figure 15). The cut-off stress cannot be positively specified as there is some overlap, but it is between 250 and 300 MPa. No major strength degradation compared to the RT strength is evident. Three time dependent failures occurred from surface connected defects: two from large pores and one from a corner pit-pore. The fourth sample, 0.028 hours at 228 MPa, shattered into many pieces on failure. All but the samples which failed on loading had oxide layers that sealed over most porosity. The oxide was identified by x-ray diffraction as alpha cristobalite. The retained strengths and sources of failure for the Ford 2.7 RBSN survivors are shown in Table 3. The 241 MPa sample did not fail during the stress rupture test despite a large (longitudinal) crack which ran the length of the sample. On subsequent retained strength testing, the initiation site was at the intersection of the crack with tensile surface. The sample was similar to a few other samples that were rejected for testing and apparently the crack was overlooked. The consistency of the retained strengths of the survivors is remarkable. The average is 340 MPa with a standard deviation of 3 MPa. This average is well above the reference RT strength (288 MPa).

The 1200°C cut-off stresses and the retained strengths of the survivors are shown in Figure 16 for the three RBSN's. Whereas the 1200°C static fatigue cut-off stress for NC 350 is higher than the reference RT strength, the other two RBSN's showed either a slight decrease or no change. The average retained RT strength of the survivors was considerably greater than the reference RT strength in the cases of NC 350 and Ford RBSN's. Such strength enhancement was not observed for the KBI material. In the case of the NC 350, although the average retained strength is higher, the scatter is greater as well. The samples with surface connected defects (flaw healing potential) had improved strength, but those samples with internal defects probably had no strength improvement, thus leading to a dichotomy of strength results which in turn led to the large standard deviation. On the other hand, the five Ford RBSN survivors had very consistent retained strength.

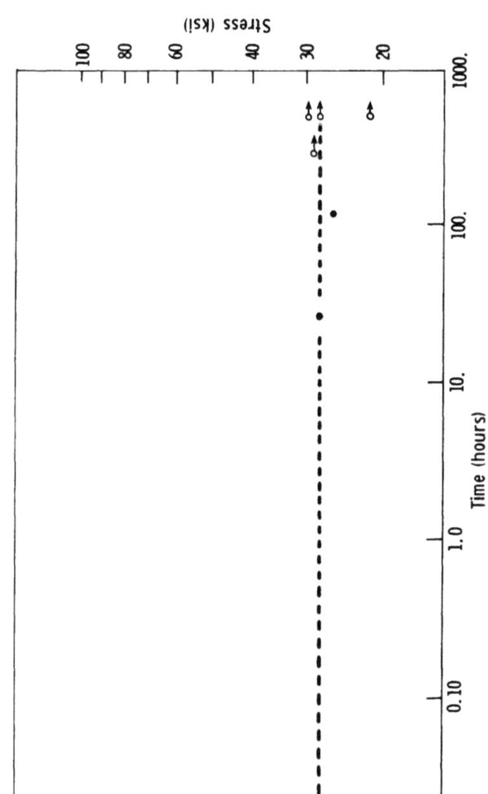

Fig. 14. Stress rupture at 1200°C for KBI RBSN

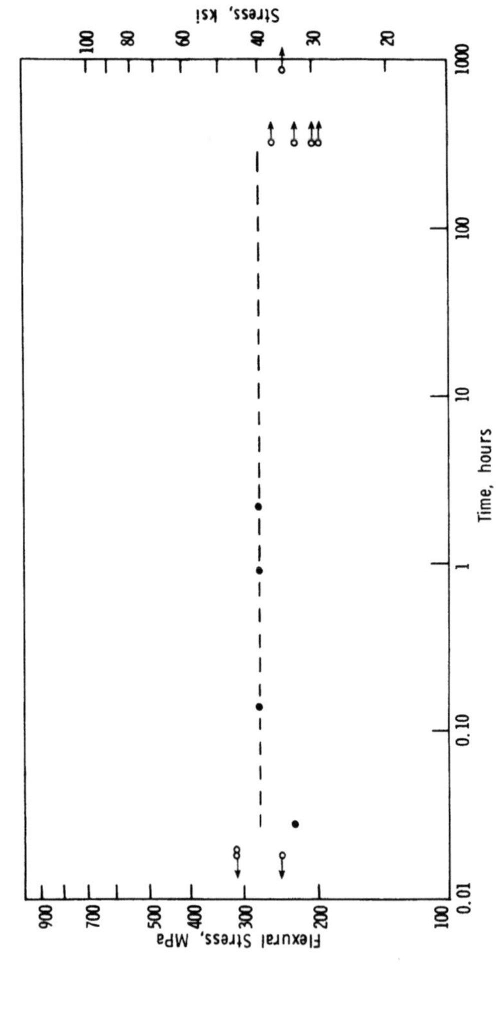

Fig. 15. Stress rupture at $1200^{o}C$ for Ford 2.7 RBSN

TABLE 3

Ford 2.7 RBSN 1200°C Stress Rupture Survivor Data

Stress Rupture Loading	Survival Time	Room Temperature Retained Strength	Room Temperature Failure Source
200 MPa	308 hours	342 MPa	Surface Pore/pit
207	308	336	Surface Pore/pit
228	307	337	?
241	847	343	Crack
255	300	340	Subsurface Inclusion (silicon?)

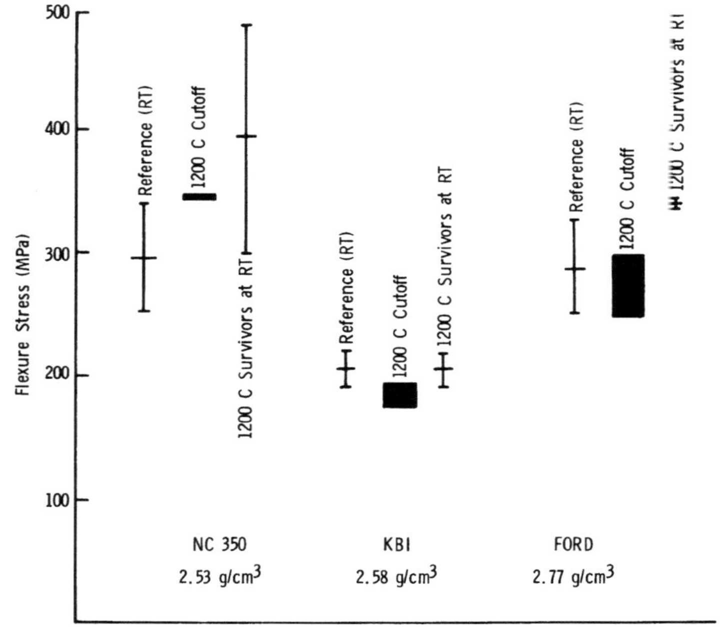

Fig. 16. Comparative chart showing the room temperature reference strengths of the three RBSN's. Also shown is the 1200°C cut-off stress in each case relative to the reference strength. The cutoff stress is represented as bars since the exact value (or range) was not precisely determined. In addition, the room temperature strengths of the stress rupture survivors are shown in each case.

V. STEPPED TEMPERATURE STRESS RUPTURE

STSR results for NC 350 indicate the material has a high resistance to static fatigue failure (Figure 17). Only two samples out of fourteen failed in a time dependent manner. Both failures were initiated from surface connected porosity. All eight survivors had retained strengths higher than the reference RT strength as shown in Table 4. Seven had strength controlling defects at the surface. As with the 1200°C stress rupture survivors, the higher retained strengths are probably due to the healing of the surface connected flaws.

The twelve STSR tests conducted on KBI RBSN did show a susceptibility to static fatigue failure (Figure 18). Eight samples loaded to 172 MPa tended to fail either in the 1300-1400°C range or at the low temperature end of the pattern. That no failures occurred at 1200°C is in accord with the stress rupture trials conducted at that temperature. The actual stress level for survival of the STSR sequence was not precisely determined, but lies between 138 and 172 MPa. This is substantially less than the 1200°C stress rupture cut off level (190 MPa), and also less than the reference strength (206 MPa). No common defect was observed for all the failures although in three cases a small glassy formation associated with a pore was detected at the initiation area. This feature was visible only when viewed by low power optical microscopy (with any type of lighting). SEM examination failed to discern this feature, however, even when light photomicrographs were used to guide SEM examination. In one other sample, a large version of this feature (a white band) extended through a significant portion of the fracture surface. Two other failure samples, 14 hours at 1300°C and 44 hours at 1400°C, had unidentified failure initiation sites. The retained strength of one survivor was 196 MPa (very similar to the reference strength) and it failed from a surface connected pore. The other survivor was broken in handling during removal from the furnace. The oxide layer was identified as alpha cristobalite by x-ray diffraction and covers nearly all of the surface of the survivors with a smooth transparent layer. All but the largest pores are sealed over. The two 1000°C failures were initiated at surface connected pores. The third sample which failed at low temperature (on heatup to 1100°C) failed from an unidentified source.

STSR test results for Ford 2.7 RBSN are shown in Figure 19 and indicate the material is susceptible to time dependent failure at 1000-1100°C. This is common with the KBI RBSN and to a lesser extent the NC 350 RBSN, but unlike the latter two, no Ford sample failed in the 1300-1400°C range. Some reduction in load carrying capacity compared to the reference strength is evident. Eight samples survived, with seven intact (the eight, loaded to 241 MPa failed during furnace cooldown). Retained strengths (in order of lowest to highest applied stress) were 254, 276, 240, 279, 284, 287 and 248 MPa. These average

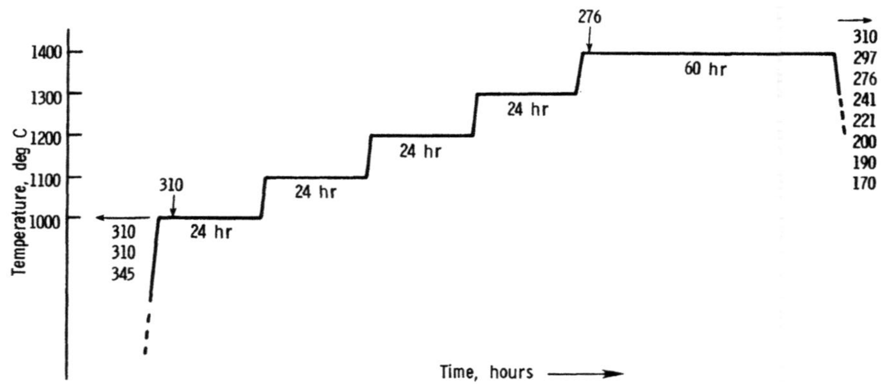

Fig. 17. STSR results for Norton NC 350 RBSN

TABLE 4

NC 350 STSR Survivor Data

Stress Rupture Loading	Room Temperature Retained Strength	Room Temperature Failure Source
310 MPa	343 MPa	*Chamfer machine mark
310	Broke on Cooldown	*Surface pore?
297	349	*Surface pore filled with oxide
276	387	*Surface pore?
241	Broke on Cooldown	*Corner defect
221	319	*Surface pore filled with oxide
200	343	*Surface pore filled with oxide
190	312	Shiny spot in mirror Silicon?
170	305	*Chamfer machine mark
Average =	337 MPa	

*Flaw healing potential

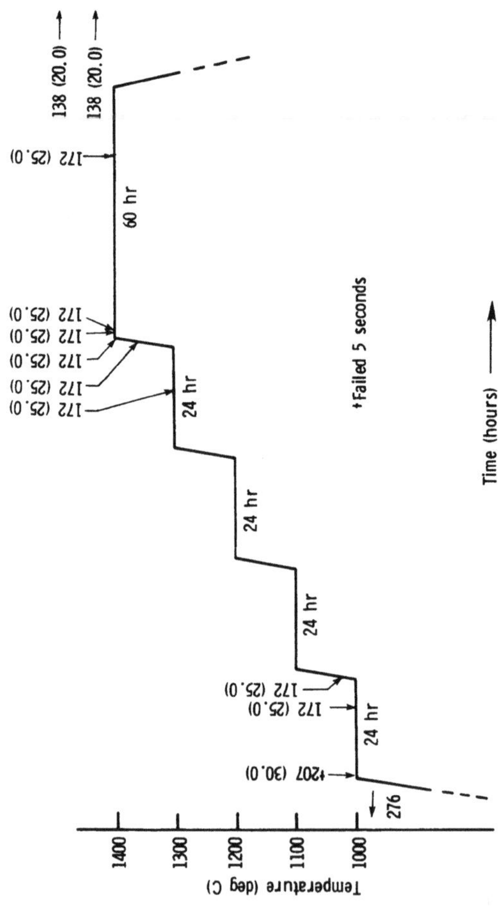

Fig. 18. STSR results for KBI RBSN

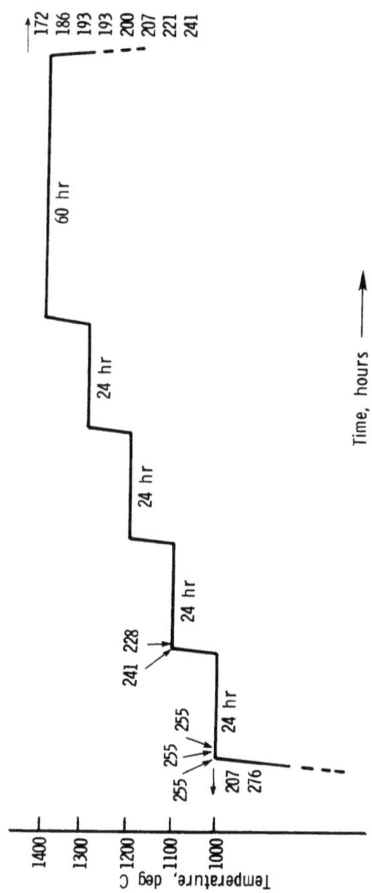

Fig. 19. STSR results for Ford 2.7 RBSN

to 267 MPa (with a standard deviation of 19 MPa) which is slightly less than the 288 MPa control strength. At least six of these samples failed from surface connected pores that were partially or completely oxide filled. The oxide layer was smooth and glossy in appearance and was identified by x-ray diffraction to be alpha cristobalite. Time dependent failures were initiated from surface or near surface defects. Four failed from porous zones that were difficult to observe with SEM. The specific mechanism that involved these porous regions and results in failure could not be identified. The fifth samples, which failed after 16 minutes at 1100°C, failed from a subsurface porous zone which had a dark black zone (light microscopy) near it. The dark zone is less evident when viewed with SEM, but close examination shows it probably is a melted silicon zone (from processing). A coherent oxide layer formed in patches on the sample.

VI. COMBINED CYCLE PROCEDURE

Twelve NC 350 combined cycle samples survived intact with very little change in appearance compared to their original condition. Some slight pitting and erosion was evident at the edges where the flame and quench air washed over the sample. The retained flexural strength of the eleven samples* averaged 243.5 MPa with a standard deviation of 37.2 MPa. This represents a 17% strength loss compared to the reference strength. A shift in failure sites relative to the reference group was apparent. Whereas the strength limiting pores were volume distributed in the reference samples, eight of the eleven combined cycle samples failed from surface connected pores. The pores were often sealed over by the oxide layer, but the layer itself was irregular and was unlike the smooth stress rupture oxide layers. One other sample failed from a pore beneath the surface that may or may not have been surface connected. The remaining two samples failed from unidentified sources, but were probably initiated at the corners.

Eleven KBI RBSN samples were subjected to the combined cycle procedure.* The average retained strength was 166 MPa with a standard deviation of 11.5 MPa. The average is twenty percent lower than the reference strength. Fractography determined that surface connected porosity caused most failures (at least seven). The mechanism of strength degradation is the enlargement of the surface connected porosity already in the material. This mechanism is consistent with the observation that the distribution of strengths of the combined cycle and control samples are nearly identical (same Weibull moduli and standard deviation).[5] The oxide layer that formed was quite irregular and damaged in the region which was subjected to the flame attack - air quench. At the ends of the samples (away from the flame - air streams) the oxide layer was glossy, smooth and coherent.

*One was broken in handling.

The average retained strength of twelve Ford 2.7 RBSN combined cycle samples was 246 MPa and the standard deviation was 25.1 MPa. The average is 15 percent less than the reference strength (288 MPa). A Weibull plot of the strength distribution shows a striking similarity to the control distribution.[5] Fractography identified all of the strength controlling defects. All but one were surface or surface connected flaws. Five failures occurred from surface pits (or enlargened pores) located at the corner of the sample. The greatest erosion attack occurred at the corners and led to a significant rounding of the corners. Six other samples failed from porosity that appeared as a white patch when viewed with the binocular microscope. These pores were sometimes sealed over by oxide and the oxide at least partially filled the pore. Away from the flame impinged zone, the oxide layer was smooth, coherent and glossy in appearance. At the impinged zone, significant attack occurred resulting in an irregular and damaged oxide layer. The corners were well rounded and more pores (pits) were evident. At the corners the oxide layer was gone exposing matrix material to the attack.

VII. DISCUSSION

The results of this program highlight the relationship between microstructure and mechanical properties in RBSN. Although the Ford 2.7 RBSN has a higher density than the Norton NC 350 RBSN, it has a nearly identical RT strength and standard deviation. This points out the desireability of maintaining a uniform porosity distribution rather than simply a higher density. The strength controlling defects in the Ford 2.7 RBSN are large pores that can assume a variety of sizes and shapes. Samples were cut from a number of separate small pieces, unlike the NC 350 and KBI materials tested in this program (which were cut from a single billet each). The Kawecki Berylco Industries RBSN material tested in this study was sampled from a relatively large plate. Despite the thickness, a density of 2.58 g/cm^3 was achieved, yet a gradient in structure did exist. Polished sections of the microstructure indicated a wide distribution of pore size and shape. Large pores, sometimes associated with silicon, were strength limiting at room temperature and elevated temperature. Despite these factors, the RT strength scatter was quite small as evidenced by the low standard deviation and a high Weibull modulus.[5] Evidently, although the overall pore distribution is large, the high end (strength controlling) of the distribution is narrow.

Norton NC 350 reaction bonded silicon nitride has a high resistance to static fatigue failure at temperatures of 1000-1400°C. The material is able to sustain higher stress levels at elevated temperature than at room temperature. Subsequent strength at room temperature is also significantly improved for many of the survivor samples. These observations are consistent with reports that the strength of NC 350 increases with temperature[10-14] and with high temperature oxidation exposure (room temperature strength).[15-18] The fractographic evidence indicates the samples which had significantly improved strengths during or after exposure to high temperature had strength controlling defects at the surface. These defects were either machining related or surface connected porosity. Samples which showed little or no improvement in strength compared to the control group had strength controlling defects, usually porosity, in the interior. Thus an oxidizing heat treatment cannot be relied upon to improve the strength of all samples. Examination of the oxide layer that forms as a result of the stress rupture procedure shows it is a coherent, smooth and thin layer of alpha cristobalite. In some cases the oxide layer filled the surface porosity. Thus the evidence indicates that the oxide formation is a strong aid in crack blunting or flaw healing. Sufficient time for oxidation was presumably available (during heat up to temperature) for the stress rupture trials at 1000°C, that an improvement in strength is attained. A further improvement is experienced for stress rupture testing at 1200°C probably due to the longer time to reach temperature (more oxidation time). Although the bulk of the data generated indicates passive oxidation of surface defects leads to an improvement in strength, the converse may also be true. In the two cases where NC 350 static fatigue failure occurred, the failure site was surface connected porosity. Just what was occurring to these porous regious that led to sample failure is not clear, but the possibility that oxidation was a contributing factor cannot be ruled out.

The KBI RBSN exhibited a greater sensitivity to static fatigue failure. A tendency of the KBI RBSN to fail in the 1000-1100°C as well as the high temperature range of 1300-1400°C was noted. Failures for both regimes were from surface or surface connected porosity although the mechanisms of failure may be different. Little or no strength degradation in the KBI RBSN occurs at temperatures lower than 1300°C under docile thermal conditions. This is unlike the Norton NC 350 grade which tended to improve in strength with exposure. These observations on the KBI material are similar to those reported in Reference 11 and 13 which reported high temperature flexural MOR results. The strength controlling defects for the KBI material are pores that were much larger than those in the NC 350 grade. Thus the KBI pores are harder to seal or fill, and therefore are not an ameanable to the healing effect of the oxidizing environment.

Ford 2.7 RBSN was also susceptible to static fatigue failure at the lower temperatures. Stress rupture trials showed static fatigue failure can occur at 1200°C although the STSR results do not so indicate. Failures at 1200°C most often occurred for samples loaded to 276 MPa. Based upon Figure 19, similarly loaded samples in the STSR trials would likely fail immediately or in the 1000-1100°C range. Thus the lower temperature failures may in the STSR trials mask the 1200°C failure tendency. The material may have less load carrying capacity at 1000-1100°C than at 1200°C. No failures were sustained at higher temperatures despite eight STSR trials. It may also be true that the same STSR masking effect applies to the higher temperatures as well. Four point flexural stress rupture trials conducted on the same material are reported in Reference 3. No time dependent failures were attained in tests at 1038, 1149 and 1204°C, but the stresses employed (138-207 MPa) were substantially lower than those which caused failure in this study. In this study retained strengths after 1200°C stress rupture trials were very consistent and well above reference strengths. The surface oxide formation tended to seal over and partially fill the surface connected pores which were strength controlling. Alternately, the retained strengths of the STSR survivors were less consistent and showed no improvement relative to the reference strength. This occurred despite the oxide surface layer which again sealed and filled surface pores.

The origin of failure was surface connected porosity for both the low (1000-1100°C) and high temperature (1300-1400°C) RBSN static fatigue samples. The mechanisms of failure may be different for these two temperature regimes however. Regarding the lower temperature failures, it is reported that the oxidation mass gain on some grades of RBSN can be far greater at 1000-1100°C than at higher temperatures.[19,20] This effect is due to the failure to develop a protective oxide layer thus permitting oxidation into the interior via continuous porosity.[19-23] Oxidation attack may act to enlarge the surface connected pores leading to static fatigue failure. Whereas negligible creep deformation was noted for the RBSN samples during the 1000 and 1100°C exposure, noticeable permanent curvature was observed on samples that were exposed to the higher temperatures. Creep deformation was such that the maximum tensile strain was of the order of 0.1 to 0.2 percent.[5] The high temperature failures may be related to this macroscopically observed creep. Creep along oxide grain boundary layers may act to enlarge the surface connected pores. Grain boundary sliding has been observed in RBSN in the 1300-1450°C range[21,22,24] and at lower temperatures from earlier 2.3 g/cm^3 grade material.[25] On the other hand, recent work suggests the creep and oxidation phenomena are interrelated.[21,22] It is possible that the low temperature failures involve the formation of oxide grain boundary phases throughout the interior. The creep rates in the bulk material may not be significant enough to cause overall sample deformation, but in a microscopic region near a surface connected pore, locally higher creep rates (possibly due to a higher impurity concentration) could cause pore enlargement. Additional stress rupture testing on these materials, but at 1000°C, is discussed in Reference 5.

The combined cycle results for all three RBSN's show the hazard in applying conclusions regarding flaw healing from static oxidizing environments to a thermal environment more likely to be encountered in service. The combined cycle process leads to an oxide layer that is not smooth, but porous, irregular and possibly cracked by the cycling. At sample edges much of the oxide is eroded away. When retained strength tests were performed at RT, nearly all of the samples tested (all three RBSN's) failed from surface connected porosity or machining damage that had grown in size due to repeated oxidation attack and the failure to retain a protective oxide layer. Unlike the samples from static oxidation experiments, the combined cycle samples thus experienced a measurable (15-20%) decrease in strength. This result reinforces the contention that proper materials evaluation requires exposure to non-docile, dynamic oxidizing environments. A similar strength decrease (20%) in RBSN due to thermal exposure with cycling has been reported elsewhere.[26]

VIII. CONCLUSIONS

The strength of RBSN was shown to be strongly dependent upon structure rather than merely density. Strengths limiting flaws in flexural samples were surface machining damage and volume distributed porosity. Static fatigue failures were obtained on all three RBSN's, but NC 350 was the most resistant to failure. The materials seemed more susceptible to static fatigue failure at 1000-1100°C than at higher temperatures. Although the mechanism of failure could not be determined, the origins of failure were usually surface connected pores. Samples which survived the stress rupture procedure intact often had higher retained strengths as the result of flaw healing although the three RBSN's showed different propensities for this effect. The cutoff stress at 1200°C for NC 350 was substantially higher than the RT reference strength whereas the KBI and Ford RBSN's had cutoff stresses similar to their RT reference strengths. The more aggressive combined thermal exposure - thermal shock sequence caused degradation of surface oxide layers and led to attack upon surface connected pores. This led to a strength degradation of 15-20%.

A final report of this work is in preparations and will have detailed fractographic analysis as well as a Weibull parameter analysis.[5]

IX. ACKNOWLEDGEMENTS

Special thanks are in order to Messrs. Valante Ociepka, Robert Cicerone and Michael Slavin who performed the bulk of the stress rupture, STSR and combined cycle testing.

Dr. R.N. Katz, Mr. G. Gazza, Dr. D. Messier and Dr. E. Lenoe of AMMRC were helpful in discussions regarding the data interpretation.

This program was supported by the Department of Energy through DOE/AMMRC Interagency Agreement EC-A-1017.

REFERENCES

1. D. Miller, C. Anderson, S. Singhal, F. Lange, E. Diaz, R. Kossowsky and R. Bratton, "Brittle Materials Design, High Temperature Gas Turbine Materials Technology," AMMRC CTR 76-32, Volume IV, December 1976.
2. C. Gazzara and D. Messier, "Determination of Phase Content of Si_3N_4 by X-ray Diffraction Analysis," Am. Cer. Soc. Bull., 56, #9, 1977, p. 777.
3. A. McLean and R. Baker, "Brittle Materials Design, High Temperature Gas Turbine," Interim Report #10, AMMRC CTR 76-31, October 1976.
4. A. McLean, E. Fisher, R. Bratton and D. Miller, "Brittle Materials Design, High Temperature Gas Turbine," Interim Report #7, AMMRC CTR 75-8, April 1975.
5. G. Quinn, "Characterization of Turbine Ceramics after Long-Term Environmental Exposure," Final Report for AMMRC/DOE Interagency Agreement EC-76-A-1017, AMMRC TR 80-15, April 1980.
6. G. D. Quinn, "Guide to the Construction of a Simple 1500°C Test Furnace," AMMRC TN 77-4, August 1977.
7. G. D. Quinn and R. N. Katz, "Stepped Temperature Stress Rupture Testing of Silicon Based Ceramics," Am. Cer. Soc. Bull., 51, #11, 1978, p. 1057.
8. C. F. Johnson and D. L. Hartsock, "Thermal Response of Ceramic Turbine Stators," Ceramics for High Performance Applications, Proceedings of the Second Army Materials Technology Conference, November 1973, Eds. J. Burke, A. Gorum and R. N. Katz, Brook Hill Publishing Co., Chestnut Hill, MA, 1974, p. 549.
9. G. D. Quinn, R. N. Katz and E. M. Lenoe, "Thermal Cycling Effects, Stress Rupture and Tensile Creep in Hot Pressed Silicon Nitride," Proceedings of the DARPA/NAVSEA Ceramic Gas Turbine Demonstration Engine Program Review, Castine, ME, MCIC 78-36, August 1977, p. 715.
10. M. Washburn and H. Baumgartner, "High Temperature Properties of Reaction Bonded Silicon Nitride," Ceramics for High Performance Applications , 1974, p. 479.
11. D. Larsen, S. Bortz, R. Ruh and N. Tallan, "Evaluation of Four Commercial Si_3N_4 and SiC Materials for Turbine Applications," Ceramics for High Performance Applications II, Proceedings of the Fifth Army Materials Technology Conference, March 1977, Eds. J. Burke, E. Lenoe and R. Katz, Brook Hill Publishing Co., Chestnut Hill, MA, 1978, p. 651-68.
12. T. Yonushonis and D. Richerson, "Strength of Reaction Bonded Silicon Nitride," Proceedings of the 1977 DARPA/NAVSEA Ceramic Gas Turbine Demonstration Engine Program Review, August 1977, Castine, ME, MCIC 78-36, p. 219.
13. D. Larsen and G. Walther, "Property Screening and Evaluation of Ceramic Vane Materials," IITRI/AFML Contract F33615-75-C-5196, Interim Report #4, October 1977.

14. A. McLean, E. Fisher, and R. Bratton, "Brittle Materials Design, High Temperature Gas Turbine," Interim Report #6, AMMRC CTR 74-59, September 1974.
15. D. Richerson, J. Schuldies, T. Yonushonis and K. Johansen, "ARPA/NAVY Ceramic Engine Materials and Process Development Summary," Ceramics for High Performance Applications II, 1978, p. 625.
16. D. Richerson and T. Yonushonis, "Environmental Effects on the Strength of Silicon Nitride Materials," Proceedings of the 1977 DARPA/NAVSEA Ceramic Gas Turbine Demonstration Program Review, August 1977, Castine, ME, MCIC 78-36, p. 247.
17. D. Larsen and G. Walther, "Property Screening and Evaluation of Ceramic Vane Materials," ITTRI/AFML Contract F33615-75-C-5196, Interim Report #5, January 1978.
18. M. Mendiratta and H. Graham, "Effect of Oxidation Upon the Room Temperature Strength of Reaction Bonded Si_3N_4," Presented at the Application of Advanced Materials Conference, American Ceramic Society, Merritt Island, FL, 22 January 1979.
19. A. McLean, R. Baker, R. Bratton and D. Miller, "Brittle Materials Design, High Temperature Gas Turbine," Interim Report #9, AMMRC CTR 76-12, April 1976.
20. R. Davidge, "Mechanical Properties of Reaction Bonded Silicon Nitride," Nitrogen Ceramics, Ed. F. Riley, Proceedings of NATO Advanced Study Institute on Nitrogen Ceramics, Cantebury, UK, August 1976, Noordhoff International Publishing, 1977, p. 541.
21. G. Grathwohl and F. Thummler, "Creep of Reaction Bonded Silicon Nitride," J. Mat. Sci., 13, 1978, p. 1177.
22. G. Grathwohl and F. Thummler, "Creep of Reaction Bonded Silicon Nitride," Ceramics for High Performance Applications II, p. 573.
23. J. Warburton, J. Antill and R. Hawes, "Oxidation of Thin Sheet Reaction Sintered Silicon Nitride," J. Am. Cer. Soc., 61, #1, 1978, p. 67.
24. S. U. Din and P. S. Nicholson, "Creep Deformation of Reaction Sintered Silicon Nitride," J. Am. Cer. So., 58, #11-12, 1975, p. 500.
25. J. Mangels, "Development of a Creep Resistant Reaction Sintered Si_3N_4," Ceramics for High Performance Applications I, p. 195.
26. K. Benn and W. Carruthers, "3500 Hour Durability Testing of Commercial Ceramic Materials," AiResearch/NASA Contract DEN 3-27, Interim Report #5, June 1979.

A COMPARATIVE STUDY OF THE MECHANICAL STRENGTH OF REACTION-BONDED SILICON NITRIDE

H. Fessler*, D. C. Fricker* and D. J. Godfrey+

* University of Nottingham, England
\+ Admiralty Marine Technology Establishment
Holton Heath

INTRODUCTION

The main objective of this investigation was to determine the room-temperature fracture strength of RBSN[1] for a wide range of multiaxial stress ratios; it was compared with that for uniaxial tension, using Weibull statistics to describe the strength variability of the material, and the 'independent-action' failure criterion[2] in preference to other criteria[3].

The failure probability of a brittle component depends on the mean strength and the variability of strength of the material, the shape and size of the component, as well as the type and magnitude of the loads. All these effects are incorporated in the four-function Weibull equation[4],

$$P_f = 1 - \exp\left\{-\left[\frac{1}{m}!\right]^m \left[\frac{\sigma_{nom}}{\sigma_{fv}}\right]^m \frac{V}{v}\,\Sigma(V)\right\} \qquad (1)$$

where
$$\Sigma(V) = \int_V \left[\left(\frac{\sigma_1}{\sigma_{nom}H(\sigma_1)}\right)^m + \left(\frac{\sigma_2}{\sigma_{nom}H(\sigma_2)}\right)^m + \left(\frac{\sigma_3}{\sigma_{nom}H(\sigma_3)}\right)^m\right]\frac{dV}{V} \qquad (2)$$

The first function $[(1/m)!]^m$ is associated with strength variability only; it varies from 0.65 for m = 5 to 0.58 for m = 20. The second function $(\sigma_{nom}/\sigma_{fv})^m$ combines the effects of material strength and applied load, and is the only function affected by the magnitude of the load. The mean unit strength allows the size-dependence of the strength of brittle components to be conveniently accounted for by the inclusion of the third function V/v. The effects of component shape and type of loading are represented in the fourth function, the stress-volume integral; the non-dimensional nature of $\Sigma(V)$, with respect to both stress and volume, is an important feature. Using the alternative assumption that all failures originate at the surface of the component leads to analogous equations in terms of areas instead of volumes.

Because previous strength tests[5] suggest the possibility of an inherent difference in strength between beams and discs, most of the tests use circular disc shapes. As the stress ratio in most practical specimens varies, a high value of Weibull modulus $(m > 10)$ of the material is necessary to assign a particular critical multiaxial stress ratio to a test.

NOTATION

- a — unit surface area
- A — component surface area
- b — half-width of Hertzian line contact
- c — half-width of square bars
- F — total radial force per unit thickness on discs and open rings
- H — step function whose value depends on the sign of each principal stress
- i — stress index = principal stress/nominal stress
- I — value of an integral
- L — supported span of bar bending tests and length between grips in torsion test
- m — Weibull modulus
- $\frac{1}{m}!$ — gamma function of $(\frac{1}{m} + 1)$
- M_t — torque applied to square bars
- N — number of specimen tests
- p — uniform radial pressure applied to Brazilian discs
- p_{max} — Hertzian line contact peak pressure
- P_f — failure probability
- r — radial coordinate
- R — support radius of disc bending and compression
- R_i — bore radius of open rings
- R_o — outer radius of discs and open rings
- t — thickness of discs and open rings
- T — transverse friction force per unit length in disc bending
- v — unit volume
- V — component volume
- W — total force
- α — modulus of the ratio of uniaxial compressive to uniaxial tensile strength
- β — half-angle of the Brazilian disc loading arc
- Δ — standard error
- θ — angular coordinate
- μ — Coulomb coefficient of sliding friction
- ν — Poisson's ratio
- σ — normal stress at any point (+ve value denotes tensile stress)
- $\sigma_{1\,2\,3}$ — principal stresses at any point

σ_{fa} mean failure stress of unit area of the material under uniform, uniaxial tension

σ_{fv} mean failure stress of unit volume of the material under uniform, uniaxial tension

σ_{nom} nominal stress

$\bar{\sigma}_{nom}$ mean nominal failure stress

$\Sigma(A)$ stress-area integral

$\Sigma(V)$ stress-volume integral

$\Sigma(V)_t$ contribution to $\Sigma(V)$ from tensile principal stresses alone

$\Sigma(V)_c$ contribution to $\Sigma(V)$ from compressive principal stresses alone

τ shear stress at any point

SHAPES AND LOADING

Biaxial stressing was used except to evaluate α, the ratio of compressive to tensile strength; as inadvertent, tensile stresses can cause premature failure in biaxial compressive tests[6], triaxial compression was carried out. For uniaxial tension, square bars 4.6 mm x 4.6 mm were loaded in 3-point bending over a 19 mm span. Discs, slotted as shown in Fig. 2, called open rings, also give convenient, uniaxial tensile loading when compressed on their perimeter. As fracture initiates near the outer edge where tensile bending stresses exist, the surface finish of the hole and slit are unimportant. Closed rings have the greatest tensile stresses at the inner edge and it is very difficult to calculate the exact stresses theoretically. For equibiaxial tension, ring-loaded, ring-supported plate bending of the discs (see Fig. 3) was chosen as the highly stressed volume is greater than for ball-loading. Making the load radius half the support radius gives a large region under equibiaxial stress but avoids significant "thick plate" shearing stresses. To avoid variations due to wear of knife-edged contacts and to reduce friction, rounded toroidal contacts (2 mm radius) were used for face loading.

Although Hertzian ball contact produces pure shear at the surface outside the contact area, the tensile stresses are so localised that the pure shear strength could not be obtained from such tests because of grain size and surface roughness effects. Twisting of discs by two loading and two support points symmetrically spaced on the same pitch circle produces pure shear at the centre of the disc, but fractures did not originate there and the pure shear strength could not be obtained. Instead of these tests, square bars 4.6 mm x 4.6 mm x 51 mm long were loaded in torsion.

The well-known Brazilian disc test gives a nominal stress ratio $\frac{1}{3}$:0:-1 at its centre. A mean ratio 1:0:-3.5 for the highly stressed region was calculated and confirmed by the positions of fracture origins. Discs and open rings were 25 mm diameter 4 mm thick. For triaxial compression, discs were compressed between equal toroidal rings, a Hertzian line contact problem. As no tensile

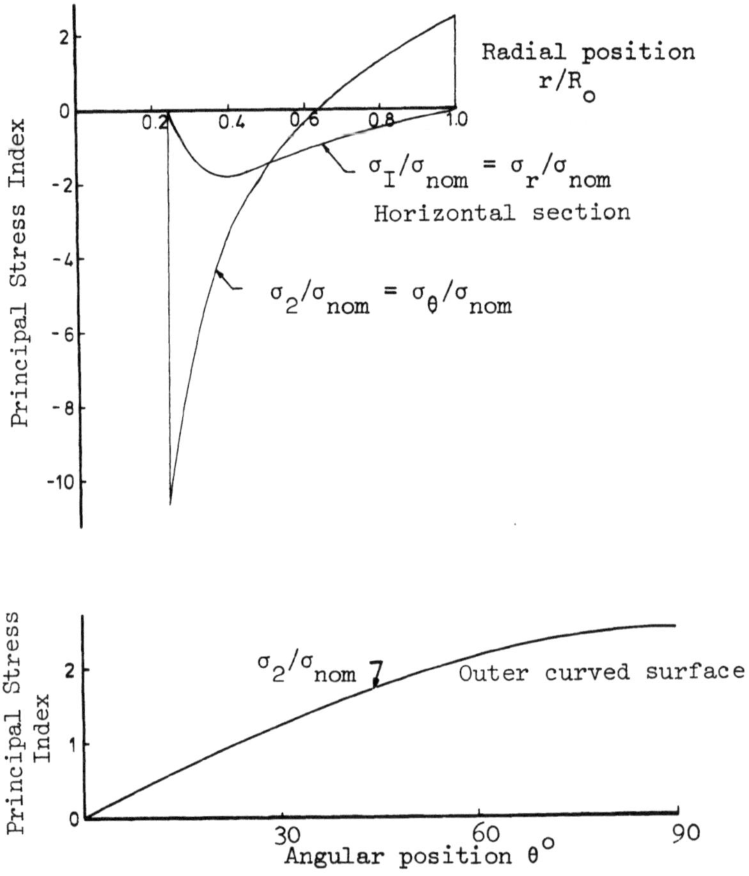

Figure 1. Principal stress indices in a thick curved bar $R_i/R_o = 0.24$

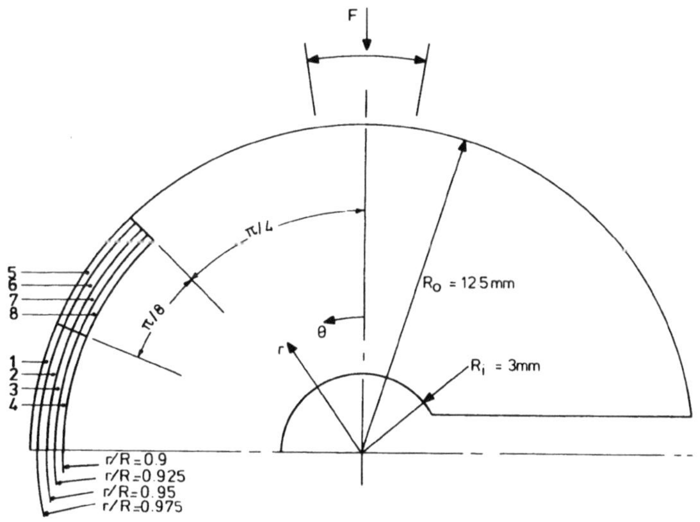

Region	tensile min i	compressive max |i|	% contribution to $\Sigma(V)_t$			
			m = 5	m = 10	m = 15	m = 20
1	2.203	-0.059	19.4	39.2	54.8	66.6
2	2.069	-0.120	14.0	20.9	21.7	19.5
3	1.933	-0.183	9.8	10.6	7.9	5.2
4	1.796	-0.248	6.7	5.1	2.7	1.2
5	1.687	-0.046	9.0	9.3	7.2	5.1
6	1.588	-0.096	6.5	5.0	2.9	1.5
7	1.489	-0.150	4.6	2.5	1.1	0.4
8	1.392	-0.208	3.2	1.2	0.4	0.1
1 to 8	-	-	73.2	93.8	98.5	99.6

Figure 2. Contributions of sub-regions to $\Sigma(V)_t$ for the open ring for four Weibull moduli, m.

stresses exist, this test was chosen to study compressive fracture. Careful alignment of the loading and supporting rings was required to avoid bending and shearing stresses.

THEORETICAL CONSIDERATIONS

The high stresses in each shape-loading combination are analysed theoretically and integrated to give the stress integrals used in the four-function Weibull equation. For single integrals a computer subroutine accepts the stress function, limits of integration, and a required relative accuracy. It successively applies a series of Gaussian quadrature formulae with up to 255 points until two results differ by less than the required accuracy (usually 0.01% or 0.001%). This subroutine is suited to 'difficult' problems with rapidly varying integrands. Accidental premature convergence of the integral is unlikely. Numerical integration of double integrals of the form

$$I = \int_{y_1}^{y_2}\int_{x_1}^{x_2} f(x,y)dx\,dy = \int_{y_1}^{y_2}\left[\int_{x_1}^{x_2} f(x,y)dx\right]_y dy \qquad (3)$$

was performed using two (necessarily different) computer subroutines, the output of one subroutine being the input for the other. The 'difficult' inner quadrature was performed as above, and the current value for y in the integrand was defined by the second subroutine which performed the outer quadrature. It uses a series of Chebyshev's quadrature formulae with up to 128 points and is best suited to 'well-behaved' problems. The output from this subroutine is the value of the double numerical integral. The inner limits of integration x_2 and x_1 can be explicitly stated or defined by functions of y.

Square Bar Bending Tests

The stress analysis, stress-volume and stress-area integrals for frictionless 3-point bending of square bars[7] are applicable because free roller supports were used. With a nominal stress equal to the maximum theoretical tensile bending stress, the stress integrals are

$$\textstyle\sum(V) = \frac{1}{2(m+1)^2} \qquad \sum(A) = \frac{m+2}{4(m+1)^2}$$

The effects of rolling friction and all contact stresses are likely to be negligible.

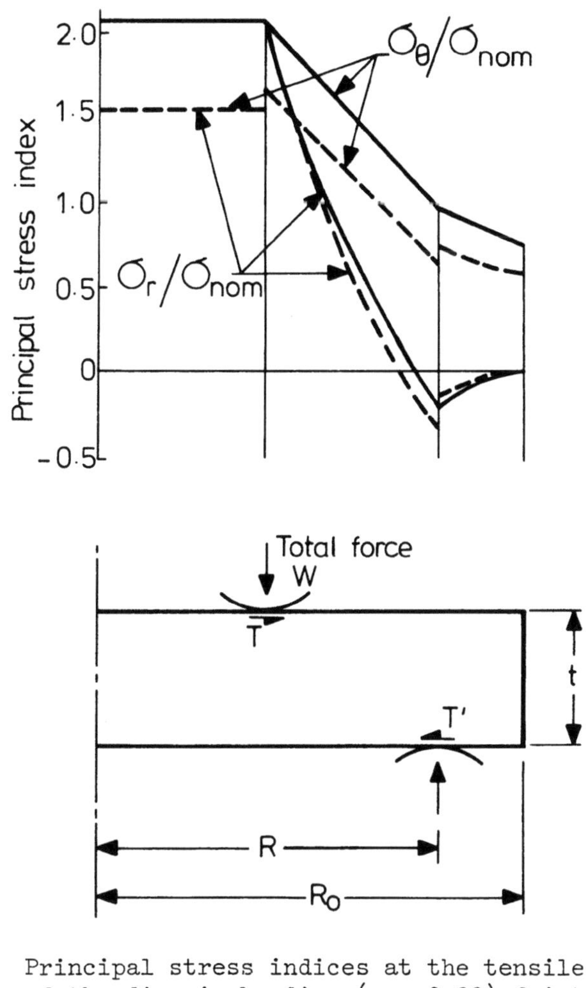

Figure 3. Principal stress indices at the tensile face of the disc in bending ($\nu = 0.23$) frictionless shown by full and Coulomb friction $\mu = \mu' = 0.3$ by dotted lines.

Open Ring Bending Tests

The stress analysis of a radially thick, axially 'thin', curved bar[8] is applicable to the highly stressed region, by Saint-Venant's principle. The stresses are calculated as stress indices, i.e. multiples of σ_{nom} the average compressive stress across the plane of symmetry, independent of size. To investigate the possibility of compressive failure, the stress integrals were divided into tensile and compressive contributions, i.e.

$$\nleq(V) = \nleq(V)_t + \frac{\nleq(V)_c}{\alpha^m} \quad \text{and similarly for } \nleq(A) \qquad (4)$$

making the ratio of compressive to tensile strength α explicit in the stress integral evaluations which were carried out separately.

Table 1 Greatest tensile and compressive principal stress indices in thick curved bars of different shapes

Shape	tensile			compressive			ratio
	Value	Position		Value	Position		
R_i/R_o	i_{max}	r/R_o	θ	i_{min}	r/R_o	θ	$\frac{\lvert i_{min}\rvert}{i_{max}}$
0.5	6.4	1.0	$\pi/2$	-12.9	0.5	$\pi/2$	2.00
0.25	2.6	1.0	$\pi/2$	-10.5	0.25	$\pi/2$	4.00
0.125	1.9	0.3	0	-12.2	0.125	$\pi/2$	6.28
0.0625	1.8	0.25	0	-16.7	0.0625	$\pi/2$	9.07

It may be seen from Table 1 that, depending on the shape and the relative strength of the material in different biaxial stress states, failure may be tensile at the outside, compressive at the inside or shear beneath the load. $R_i/R_o = 0.24$ was chosen to avoid shear failure, after evaluation of $\nleq(V)_c/\nleq(V)_t$ had shown compressive failure to be unlikely. For this shape

Weibull modulus m	5	10	15	20
Stress integral contribution $\nleq(V)_t$	.319E1	.156E3	.112E5	.932E6
Stress integral contribution $\nleq(V)_c$	.323E3	.131E8	.887E12	.728E17

The typical distributions in Fig. 1 show that the tensile stress index reduces more rapidly in the radial direction than in the circumferential direction. Fig. 2 shows that for $R_i/R_o=0.24$ this is effectively a uniaxial tensile strength test when m and α

are large, as the probability of failure initiating in a region where the stress state is significantly biaxial is negligible.

Disc Bending Tests

The theoretical stress analysis[9] was extended to account for 'thick' plate bending and contact friction. A parabolic shear stress distribution whose resultant force at any radius was equal to the unit shearing force was added; these shear stresses have little effect on the stress integrals. Friction modifies the bending moments predicted from simple theory; each friction force alone was replaced by an equivalent radial bending moment about the mid-plane of the disc acting at the loading or support radii. The boundary conditions of continuity of displacement and slope were observed and the two separate stress distributions were superposed on the 'thick' plate stresses. Replacing a friction force by a bending moment causes stress discontinuities; it was hoped that these would not significantly affect the stress integrals. Constant but possibly different coefficients of Coulomb friction μ and μ' at the ring-load and the ring-support respectively were assumed. The nominal stress $\sigma_{nom} = 3W/4\pi t^2$ takes disc thickness variations into account. The disc and the principal stress indices at the tensile face are shown in Fig. 3. The maximum tensile principal stresses occur near the ring-load. The part of the disc outside the support, the upper half of the specimen and the compressive stresses near the support were neglected in calculating stress-volume and stress-area integrals. It can be seen from Table 2 that, as friction increases, $\xi(A)$ and $\xi(V)$ decrease and failure becomes more likely in the outer region where the biaxial stress ratio varies. The effect of 'thick' plate shear stresses can be seen by comparison with the 'thin' plate frictionless pure bending ratio[10]

$$\frac{\xi(A)}{\xi(V)} = \frac{R_o(m+1)}{R_o + t} = 8.33 \text{ for } m = 10$$

Table 2 Stress integrals for disc bending tests

Weibull modulus m = 10, Poisson's ratio ν = 0.23. Central region is inside loading ring, Outer region is outside loading ring

Coefficient of friction $\mu = \mu'$	Stress-area integral $\xi(A)$	Stress-volume integral $\xi(V)$	$\frac{\xi(A)}{\xi(V)}$	$\frac{\text{Central contrib.}}{\text{Outer contrib.}}$
0	225	27.1	8.28	3
0.1	99.5	12.15	8.22	2½
0.2	43.6	5.39	8.10	1¾
0.3	19.8	2.51	7.87	1
0.4	10.2	1.35	7.55	⅓
0.5	6.47	0.891	7.26	1/10
0.6	4.90	0.693	7.08	1/40

$\xi(A)$ is independent of 'thick' plate shear stresses whilst $\xi(V)$ is not. As friction increases, the bending stresses reduce relative to these shear stresses. Table 3 shows that the effect of these shear stresses on frictionless stress integrals reduces for higher m values. Finite element stress integrals[5] for $\mu = \mu' = 0.3$ show good agreement. For example when $m = 10$ $\xi(V) \simeq 3$ and $\xi(A) \simeq 20$. The accuracy of the stress integrals depends on the accuracy with which friction is known.

Table 3 Stress integrals for frictionless disc bending tests. $\nu = 0.23$.

Weibull modulus m	Stress-area integral $\xi(A)$	Stress-volume integral $\xi(V)$	$\frac{\xi(A)}{\xi(V)}$	Central contrib. / Outer contrib.	Pure bending $\frac{\xi(A)}{\xi(V)}$
5	.696 E1	.157 E1	4.42	2	4.55
10	.225 E3	.271 E2	8.28	3	8.33
15	.807 E4	.667 E3	12.09	6	12.12
20	.299 E6	.188 E5	15.88	8	15.91

Square Bar Torsion Tests

The exact elastic, pure shear solution[11] with a nominal stress $\sigma_{nom} = \tau_{max} = (\sigma_1)_{max}$ is presented in Fig. 4. It is convenient that the torsional stresses are low near the corners, where clamping by the jaws may cause tensile stresses.

Table 4 Stress integrals for a square bar in torsion for $\alpha = 13.5$ or 15

Weibull modulus m	5	10	15	20
Stress-volume integral $\xi(V)$	0.108	0.0420	0.0238	0.0158
Stress-area integral $\xi(A)$	0.446	0.334	0.278	0.244

Stress integrals for the volume and surface area between the jaws are given in Table 4. Fig. 5 shows the variation of the stress function at the surface, with a linear approximation. As the stress-area integral $\xi(A)$ equals the mean value of this function, for small failure probabilities, the likelihood of any failure initiating within a region from the mid-face to any position is directly the proportion of the corresponding area under the graph to the total area under the graph.

Brazilian Disc (edge compression) Tests

Using an exact theoretical stress analysis[12] and a nominal stress $\sigma_{nom} = 2\beta p/\pi = F/\pi R_o$, makes the stress indices almost independent of 2β, the angle over which the force is assumed uniformly distributed (see Fig. 7). The recommended[13] half-angle β

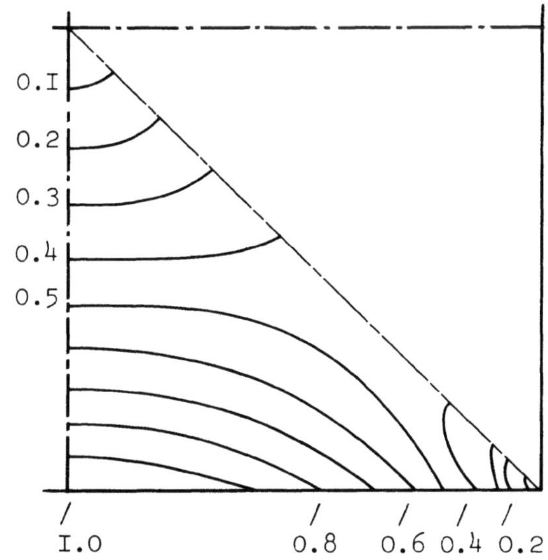

Figure 4. Contours of equal tensile and compressive principal stress index σ_1/σ_{nom} and σ_2/σ_{nom} within 1/8 sector of a square bar in torsion.

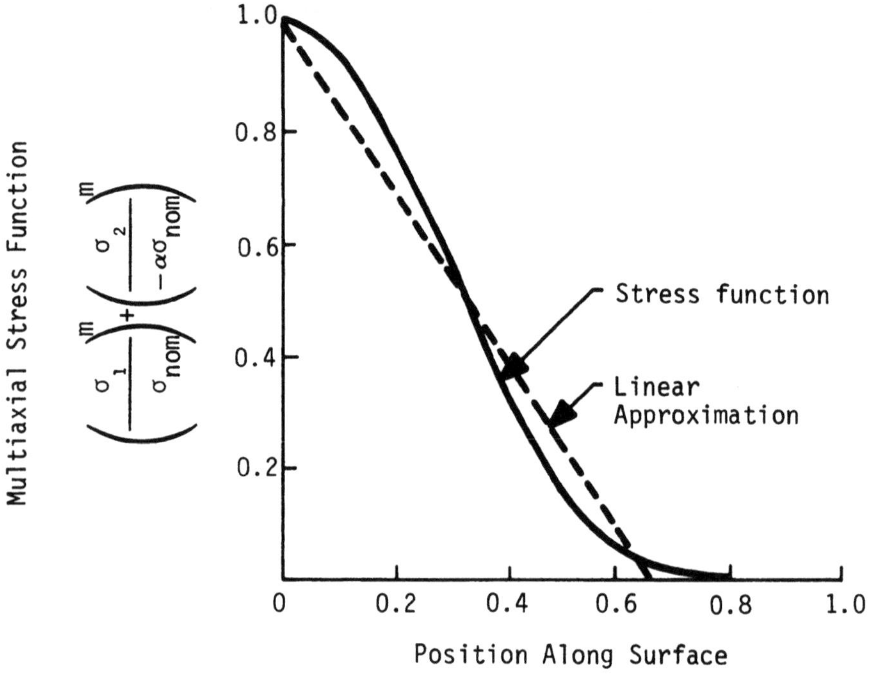

Figure 5. Stress function for a square bar in torsion ($m = 10$, $\alpha = 13.5$).

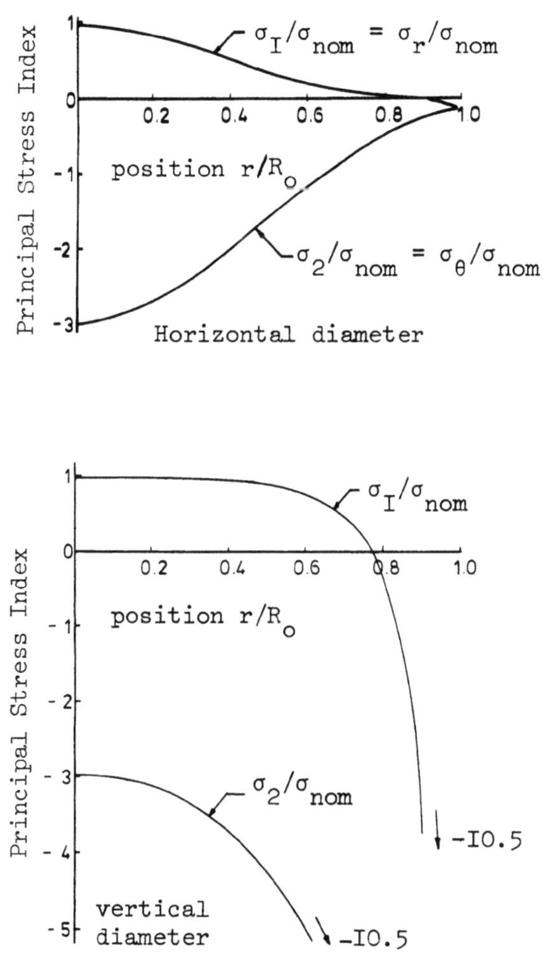

Figure 6. Principal stress indices along the horizontal and vertical diameters of a Brazilian disc (β = 0.15).

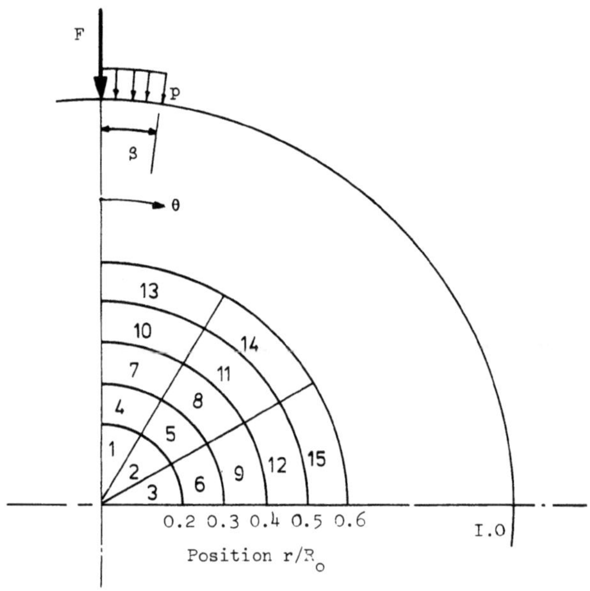

Region	Principal stress ratio		% contribution to Σ (V) for β = 0.15			
	β = 0.28	β = 0.15	m = 5	m = 10	m = 15	m = 20
1	3.2 to 3.5	3.1 to 3.2	10.3	16.3	21.2	25.4
2	3.2 to 3.4	3.1 to 3.2	8.9	12.3	14.2	15.4
3	3.2 to 3.4	3.1 to 3.2	7.8	9.6	10.1	10.2
4	3.4 to 3.8	3.2 to 3.5	11.4	16.0	18.6	20.1
5	3.4 to 3.8	3.2 to 3.5	6.9	6.2	4.8	3.5
6	3.4 to 3.6	3.2 to 3.5	4.4	2.5	1.3	0.6
7	3.8 to 4.5	3.5 to 4.0	13.1	15.6	15.7	15.0
8	3.6 to 4.3	3.5 to 3.9	5.0	2.6	1.2	0.6
9	3.6 to 4.0	3.5 to 3.8	2.1	0.4	0.0	0.0
10	4.3 to 6.1	3.9 to 4.7	12.3	11.4	9.3	7.3
11	4.0 to 5.1	3.8 to 4.5	2.5	0.6	0.1	0.0
12	4.0 to 4.6	3.8 to 4.4	0.6	0.0	0.0	0.0
13	5.1 to 11.4	4.5 to 6.3	8.8	5.4	3.1	1.7
14	4.6 to 6.5	4.4 to 5.5	0.8	0.0	0.0	0.0
15	4.5 to 5.5	4.4 to 5.3	0.1	0.0	0.0	0.0
1 to 15	-	-	95.0	98.9	99.6	99.8

Figure 7. Biaxial stress ratios and contributions to Σ(V) of sub-regions for a Brazilian disc for four Weibull moduli, m.

for brittle rock material is 0.15 radians. The principal stress indices for the measured mean effective value of 0.28 are similar to those shown in Fig. 6 for $\beta = 0.15$. As the biaxial stress ratio varies near the centre of the disc, the critical central portion of the disc was sub-divided as shown in Fig. 7, which also gives computed values of the biaxial stress ratios.

Tensile and compressive stress-volume integral contributions show that tensile failure will occur if failure at the loading arcs (where no tensile stresses exist) is avoided by the insertion of soft interlayers between disc and platens.

Weibull modulus m	5	10	15	20
Stress integral contribution $\Sigma(V)_t$	.106 E0	.546 E-1	.344 E-1	.236 E-1
Stress integral contribution $\Sigma(V)_c$	.344 E4	.270 E9	.288 E14	.340 E19

The % contributions to $\Sigma(V)$ shown in Fig. 7 were computed for $\beta = 0.15$ and will be similar to those for the measured value of $\beta = 0.28$. The ratio of principal stresses at the origin of fracture is likely to be between 3.2 and 3.8 as regions 1 to 6 contain between 50% (for m = 5) and 75% of $\Sigma(V)$ (for m = 20).

Disc Face Compression Tests

Hertzian theory[14] is applicable because

1) the stressed volume is large enough for both bodies to be effectively homogeneous and isotropic
2) the thickness of the discs is a sufficient number of contact widths for the bodies to be "semi-infinite" (see Fig. 8 and Table 5)
3) the curvature of the contact is small compared to its width, i.e. generalised plane strain exists
4) the roughness of the contact surfaces will have no effect
5) interfacial friction is not important if sliding is prevented by ensuring that the disc surfaces are perpendicular to the direction of the compressive force.

No tensile stresses are produced anywhere in the bodies if the line contacts are exactly opposite. The triaxial compression differs at different points. By using a nominal stress equal to the maximum Hertzian pressure, the stress index equations form a unique solution for any normally loaded, plane-strain, line contact. The stress indices under the centre of contact are shown in Fig. 8.

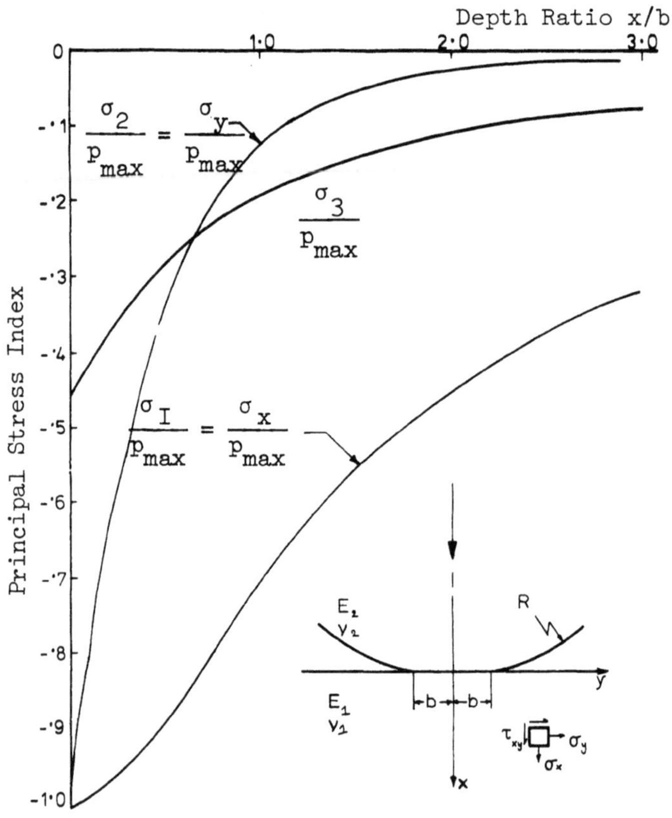

Fig. 8 Principal stress indices along axis of symmetry y/b = 0 for a Hertzian line contact (ν_I = 0.23). $p_{max} = W/\pi^2 Rb$, where

$$b^2 = 2WR_I \mid (I-\nu_I^2)/E_I + (I-\nu_2^2)/E_2 \mid /\pi R$$

As the fracture behaviour of brittle materials under compressive loading is fundamentally different from their behaviour under tensile loading, the critical flaw basis for the statistical approach to tensile failure is inapplicable, but it was also used for comparison. Compressive stress intergral contributions are derived and evaluated[15] using characteristic volume $V = b^2$ times contact length and characteristic area A = b contact length: -

Weibull modulus m	5	10	15	20
Stress integral contribution $\Sigma(V)_c$	0.889	0.387	0.249	0.183
Stress integral contribution $\Sigma(A)_c$	1.984	1.478	1.234	1.081

Table 5 Contributions to Hertzian line contact stress-volume integral from layers at different depths ($\nu = 0.23$)

depth of region x/b	% contributions to $\Sigma(V)$			
	m = 5	m = 10	m = 15	m = 20
0.0 to 0.5	61.2	83.1	91.7	95.7
0.5 to 2.0	35.9	16.9	8.3	4.3
2.0 to 5.0	2.9	0.1	0.1	0.1

Summary

The expressions needed to calculate failure probabilities from other test results for the shapes of specimens used in this paper are collected in Table 6.

Table 6 Expressions for volumes and surfaces and nominal stresses

Shape	loading	definition	volume for calculation V	surface for calculation A	nominal stress σ_{nom}
square bar	bending	in text	$4c^2L$	8cL	$3WL/16c^3$
open ring	bending	Fig. 2	$\pi t(R_o^2 - R_i^2)$	$2\pi[(R_o^2 - R_i^2) + t(R_o + R_i)]$	$F/(R_o - R_i)$
disc	bending	Fig. 3	πtR_o^2	$2\pi R_o(R_o + t)$	$3W/4\pi t^2$
square bar	torsion	notation	$4c^2L$	8cL	$M_t/1.665c^3$
Brazilian disc	edge compression	Fig. 7	πtR_o^2	$2\pi R_o(R_o + t)$	$F/\pi R_o$
disc	face compression	Fig. 8	$2\pi Rb^2$	$2\pi Rb$	p_{max} (see Fig. 8)

EXPERIMENTAL WORK

'As-fired' RBSN bars, open rings and discs were manufactured at the Admiralty Marine Technology Establishment, Holton Heath, and fired in four batches. Batches I to III were manufactured from 25 μm silicon powder whereas batch IV was from 12 μm powder. All the specimens achieved an acceptable percentage weight gain during nitriding. 20 discs for bending were lapped on both faces using 40 μm carborundum powder and polished on the tensile face with ½ μm diamond paste. The widths of bars, open ring and disc diameters and thicknesses were measured so that actual dimensions could be used to calculate each nominal failure stress. As-fired, square bar torsion, disc bending and face compression specimens were buffed to minimise friction and wear. Re-measuring the buffed specimen dimensions showed no significant size reduction. 20 discs were considered adequate for each test. Discs were allocated to the different tests so that visible blemishes should be outside critically stressed regions and Brazilian test discs, which could be oversize on diameter, had minimum thickness variations between discs. The loading fixtures and details of the methods used to achieve accurate alignment will be described in a subsequent paper.

Three point bending of span 19 mm was used so that three fracture loads could be obtained from each bar. These tests were carried out for each batch of material for comparison of batch properties and quality control. After the tests there was uniform transverse metal deposition at the support rollers but not at the loading rollers, evidence of uniform loading and some friction.

For open ring and Brazilian disc tests, 0.15 mm thick lead foil was located between the greased steel anvils and the specimens to avoid fracture in the contact region. The lead foil extruded equally on both sides into the gaps between the anvils and the clamping blocks; it remained unbroken during tests. These specimens could only be very lightly clamped to avoid failure at the locating blocks. In the testing machine all possible combinations of position of specimens, anvils and locating blocks were used to avoid systematic alignment errors. Plastic flow of the steel anvils occurred in every Brazilian test. Three test results were rejected because the specimen assembly could not easily be removed from the loading fixture after the test because of jamming; these invalid results indicated little friction under these extreme conditions. In the Brazilian disc tests the half angle of the loading arc was estimated from the profiles of the anvils and the widths of the lead foil adhering to some surviving disc contact areas. The latter varied from 6.13 mm after test 5 to 7.57 after test 23. A mean effective value for β of 0.28 radians was estimated; stress integrals are not greatly affected by the exact value of β.

For bending tests of as-fired discs, the least blemished face was loaded in tension. To ensure that they were in contact with unworn loading and support rings, the polished discs were tested first. The loading and support rings for disc bending and face compression were coated with a thin layer of molybdenum disulphide grease and new pieces of PTFE tape. The PTFE tape and the disc surfaces showed that contact was uniform along the lengths of both the line contacts. Very little damage or wear of the loading fixture occurred.

In square bar torsion tests some angular movement occurred due to the deformation of the lead foil, but it was never broken. Fracture never initiated at the clamps. Very little wear of the fixture occurred.

In disc face compression tests the PTFE tape and the disc surfaces showed that contact was uniform along the lengths of both the line contacts. Slight wear of the bore of the fixture occurred because, at failure, the discs expanded in diameter and had to be pressed out after each test. The mean contact widths of 3 discs were measured to be 0.28, 0.22 and 0.28 mm.

FRACTURES

Visual inspection of specimen fracture surfaces showed that the origin of fracture (often a visible and sometimes 'bright' inclusion) lay at the centre of a circle, whose diameter was of the order of 1 mm, with a smooth surface. Beyond this circle, the fracture surface was rough and there were relief lines radiating outwards. Further out still, the surface again became and continued smooth. These features have been observed before with ceramics, rocks and glass, e.g. Baratta et al[16] and Gramberg[17] have given detailed descriptions of similar fracture surfaces. Failure origins could easily be located in square bars in bending and torsion, open rings, and Brazilian discs.

For square bars in bending and open rings, results of microscopic analyses are presented in Table 7. The two histograms in Fig. 9 show only one failure origin close to the surface of the bars. The gradients of failure origin distributions away from the surface for bars and open rigns are similar, despite normal stress gradient for bars three times greater than for open rings; this is explained by the much lower measured Weibull modulus for the bars (presented later). The significantly different frequencies of void failure origins away from and near unnitrided silicon (Table 7) also indicate that the size of the silicon powder affects the behaviour of RBSN. Scanning electron microscopy of void defects revealed that around some voids there was unnitrided silicon, suggesting that unnitrided silicon may produce voids; Fig. 10 shows photographs of both types.

In square bar bending tests failure originated near the mid-span position, 4 of 21 fractures starting in the middle third of the width; depths are shown in Fig. 9. The fracture paths were in planes normal to the axes of the bars.

Table 7 Results of microscopic analysis of fractures

shape	powder particle max. size μm	mean flaw size μm	loading	probable cause of failure		
				voids without u.s. %	voids near u.s. %	machining defects %
square bar	12	60	bending	29	71	0
open ring	25	40	bending	58	30	12

Note: u.s. = unnitrided silicon

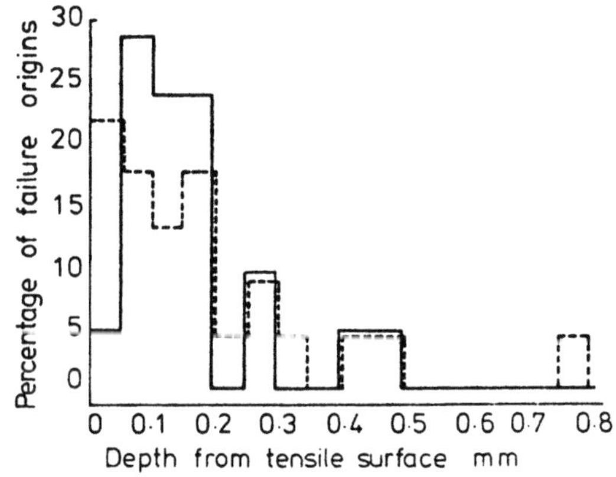

Figure 9. Distribution of failure origin positions
——— refers to 21 square bar bending tests
----- refers to distance from curved surface of 26 open ring tests

The angular distribution of failure origins of <u>open rings</u> agreed well with the predictions in Fig. 2 for the appropriate Weibull modulus. 61% of failure origins occurred within 0.5 mm of flat surfaces (3 of the 26 were due to machining defects). Radial positions are shown in Fig. 9; none initiated at the bore. In the typical fracture pattern of an open ring shown in Fig. 11a, X is the failure origin and XB the primary fracture path. Secondary cracks EF and GH were often absent or occurred after the slit had closed.

<u>Disc bending</u> fracture origins could not be established with certainty; since crack bifurcation is likely to proceed at an acute angle, the failure origin positions were traced to the primary failure crack marked as AB in Fig. 11b. The radial positions of 15 of these cracks were measured; only the shortest (0.1 mm long) occurred completely outside the central region. The mean radial position of the mid-point of these initial cracks was 4.0 mm from the centre, (77% of the mean loading radius). Their mean length was 1.2 mm. The orientation of the primary crack in Fig. 11b is in the hoop direction but there was no preferred direction. These results for both as-fired and polished discs indicate that the fracture origins were in a region of sensibly constant equibiaxial tension, in agreement with theoretical predictions for the measured low friction which occurred in these tests.

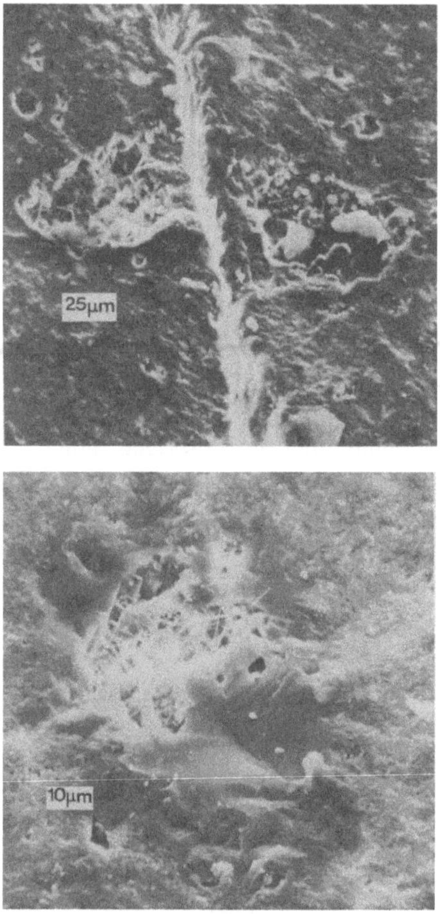

Fig. 10 Failure origins in Batch IV bend-tested bars (upper, simple void defect, lower, void with unnitrided silicon)

Square bar torsion fracture origins were never at the clamping jaws and there was no bias to the axial position of the origins. Failure origins were at voids or unnitrided silicon near the middle of a face, as shown in Fig. 11c, and mostly between 0.1 mm and 0.5 mm from the surface; there was a slight bias to failure initiating near two particular adjacent faces. The orientation of the primary fracture paths corresponded to the direction of the predicted tensile principal stresses, i.e. at 45° to the bar axis and normal to the face.

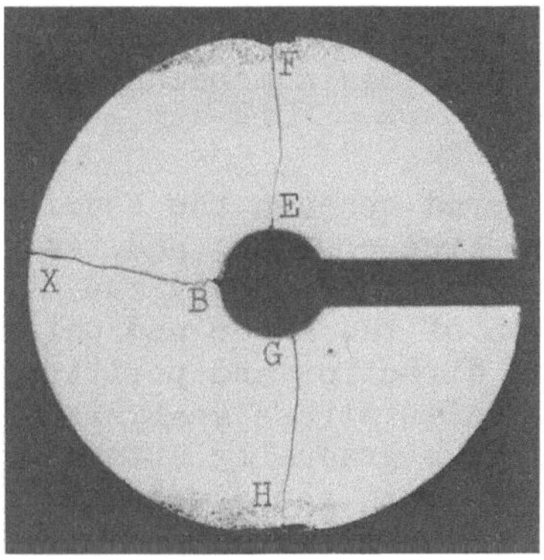

(a) Open ring

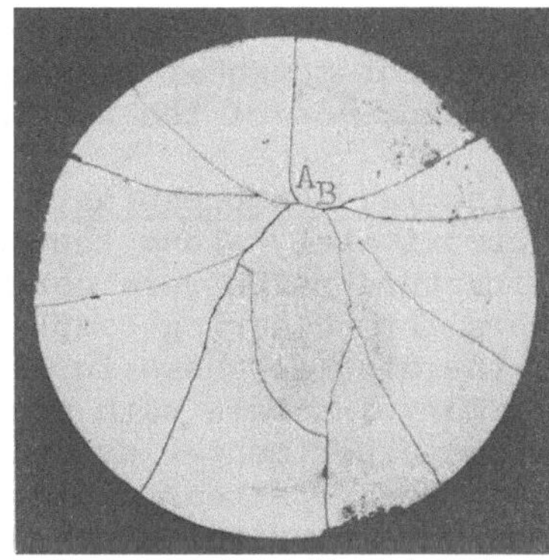

(b) Ring-loaded ring-supported disc bending (tensile face)

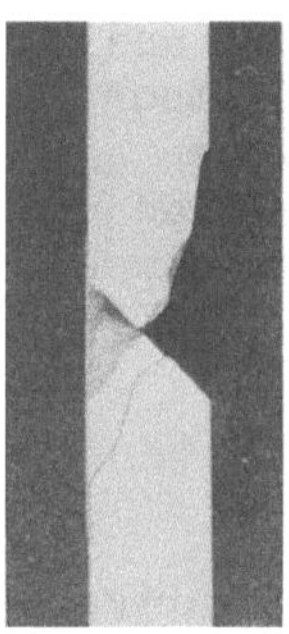

(c) Square-bar torsion

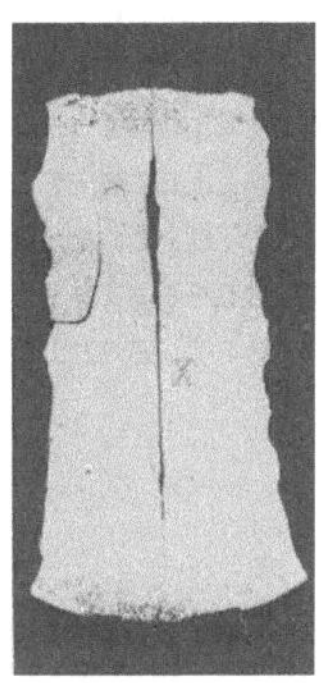

(d) Brazilian disc

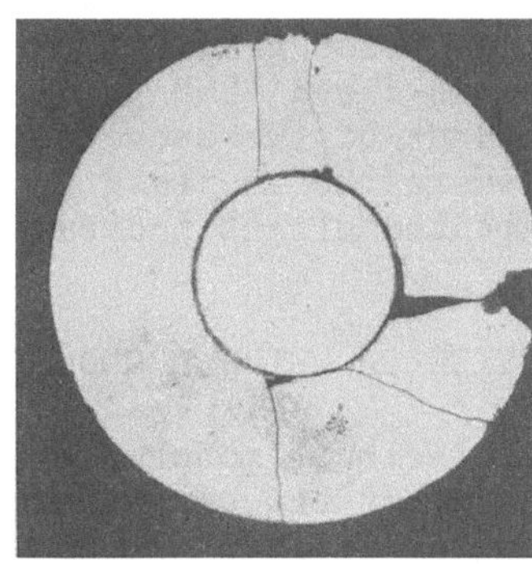

(e) disc face compression

Fig. 11 Typical fracture patterns of as-fired specimens

The Brazilian disc fracture origin was usually at some interior position, always in the centre of the discs, never at the loading arcs. The secondary cracks usually met the edge of the disc outside the contact arc; the fracture pattern shown in Fig. 11d has been predicted using finite element analysis[18] and observed previously[19]. Point X is the failure origin position. Examination of the positions of fracture origins showed that there was no significant bias to the axial positions, indicating that accurate loading alignment had been achieved and that the central region of the discs had been subjected to axially-uniform stressing. Inadvertent axial non-uniformity of diametral loading for 'thick' specimens, as suggested by previous results[19], was thus avoided by this choice of specimen shape. The angular distribution of failure origins, either voids or unnitrided silicon, could not be measured accurately but the radial distribution agreed with the predictions in Fig. 7 for the appropriate Weibull modulus. The orientation of the primary fracture paths corresponded to the tensile principal stresses and not to shear stresses despite their maximum being over twice the magnitude of the tensile stresses.

In disc face compression, fracture occurred beneath the loading rings, approximately along the centre of the line contact, and the primary fracture paths were normal to the surface; secondary radial cracks were formed during complete separation of the inner and outer parts of the disc as shown in Fig. 11e. The direction and position of the primary fracture path agrees with Hertzian stress analysis and previous experimental evidence[20] that cracks gradually extending in compressive stress fields tend to curve into the direction of the maximum principal stress.

The consistent result that, for tensile loading, sub-surface failure origins are usual for as-fired RBSN specimens means that failure predictions based on an area-size effect are invalid and that the stress gradient normal to the critical surface is important for as-fired RBSN predictions. The failure stresses were independent of the sequence of testing for all except Brazilian tests where the increase in β, due to successive yielding, caused the expected slight increase during the test series.

ESTIMATION OF WEIBULL MODULI

The Weibull modulus m is a 'material' property, which varies for different batches of ceramics, and different machining operations after firing. It was estimated using least-squares, straight-line fits to the individual, test-ranked, nominal failure stress distributions $\sigma_{nom}/\bar{\sigma}_{nom}$ [21], i.e. the slopes of $\ln \ln \left[\frac{1}{1 - P_{fi}}\right]$ against $\ln(\sigma_{nom}/\bar{\sigma}_{nom})$ graphs, as shown in Fig. 12. These values are given in Table 8 with the standard errors $\Delta m = m/\sqrt{2N}$, and values of 1.2/coefficient of variation, which is another measure[22] of m. Fig. 13 gives a linear graph of failure probability P_f against the as-fired distribution of ranked failure strengths for batch II with the standardised, two-parameter, Weibull curve.

Although the least-squares straight line method of estimating m values is biased towards low $(\sigma_{nom}/\bar{\sigma}_{nom})$ values (especially when there are few test results), the results [23] show that the Weibull two-parameter distribution is an adequate description of the experimental failure results because there is no trend of the lower ranked strengths to curve downwards away from the straight lines. It also describes the failure results of RBSN with a polished surface under biaxial tensile loading and surprisingly, of RBSN failing under purely compressive triaxial loading. The least-squares straight line ordinate values when $\sigma_{nom} = \bar{\sigma}_{nom}$ are very close to the theoretical values $m \ln(1/m\,!)$. This indicates the slight skewness of the

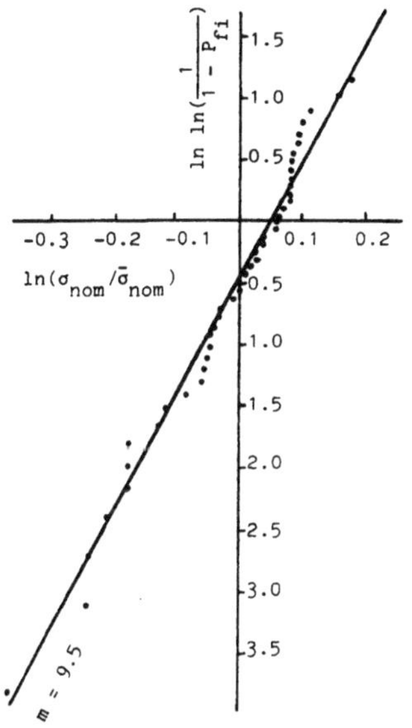

Figure 12. Weibull modulus estimation from pooled as-fired batch IV distribution of ranked strengths.

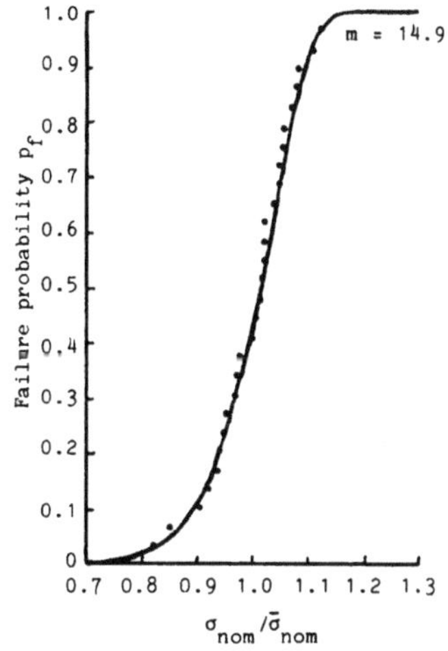

Figure 13. Two-parameter Weibull distribution fit to the pooled as-fired batch II data.

Table 8. Weibull moduli for the materials used.

batch	number of tests N	surface condition	$m \pm \Delta m$	$\frac{1.2}{cv}$	use
I	72	as fired	18.9 ± 1.6	19.9	Open ring, Brazilian disc, disc & square bar bending
II	28	" "	14.9 ± 2.0	16.4	Disc & square bar bending
II	20	polished	8.5 ± 1.4	10.5	Disc bending
II	15	as fired	19.6 ± 3.5	23.9	Disc face compression
III	15	" "	23.6 ± 4.3	27.4	Square bar bending
IV	45	" "	9.5 ± 1.0	10.4	Square bar torsion & bending

$$cv = \sqrt{\frac{\sum\left(\frac{\sigma_{nom}}{\bar{\sigma}_{nom}} - 1\right)^2}{N - 1}}$$, called coefficient of variation

experimental data in that $P_f < 0.5$ when $\sigma_{nom} = \bar{\sigma}_{nom}$ and so the Weibull distribution is likely to be a better overall fit to the data than the normal distribution.

For the Brazilian discs only, m is 13.5, which is lower than the other estimates from batch I, because loading variability is superimposed on the inherent material strength variability.

Table 8 shows that the different firing batches of the same ceramic produce different m values. The considerably lower m value for the batch IV RBSN from 12 μm powder than those ($m \geq 15$) for the other batches from 25 μm powder has been observed frequently; there is no apparent reason for this behaviour. Polishing the 'as-fired' surfaces reduces m below 15. The Weibull modulus is a measure of component variability, not a true material property. Other previous results[24] have shown some material variability increases with certain surface production processes but no general trend was indicated. The Weibull modulus in compression is higher than the pooled batch II value, because compressive failure requires a very large number of inherent flaws to join[20]. Most of the approximate m values from the coefficients of variation of the nominal failure stress distributions are less than one standard error Δm higher than the least-squares straight line estimates.

UNIT TENSILE STRENGTHS

It has been shown[22] that

$$\sigma_{fv} = \bar{\sigma}_{nom}\left[v/V \leq (V)\right]^{-1/m} \quad \text{and} \quad \sigma_{fa} = \bar{\sigma}_{nom}\left[a/A \leq (A)\right]^{-1/m} \qquad (5)$$

These expressions were evaluated for $v = 1\ mm^3$ and $a = 1\ mm^2$, using the mean actual sizes and are shown in Table 9. The sampling errors are too big to distinguish between the area- and volume-flaw assumptions.

For batch I, the bar modulus of rupture is 235 MN/m^2 whereas the open ring value is 212 MN/m^2, showing a size-weakening effect and the short-comings of using a 'modulus of rupture'. The lower value is expected from Weibull statistics because there is more critically stressed material in open rings than in bars.

Table 9 also shows significant batch-to-batch variations in unit tensile strengths. Batch IV RBSN from 12 μm silicon powder has typically higher unit strengths and lower m value than RBSN manufactured from 25 μm powder. For batches of RBSN produced in the same way, the consistency in material properties is high. The independence of Weibull moduli and unit strength values can be seen from comparison with Table 8.

Table 9 Unit volume and unit area tensile strengths from as-fired, uniaxial bending

batch	type of test	number of tests N	mean nominal failure stress $\bar{\sigma}_{nom}$ MN/m^2	Standard error $\Delta\bar{\sigma}_{nom}$ MN/m^2	Unit strengths volume σ_{fv} MN/m^2	area σ_{fa} MN/m^2
I	bar	6	235	±6	227	255
I	open ring	26	91.1	±1.1	236	254
II	bar	7	216	±7	213	243
III	bar	15	236	±3	225	249
IV	bar	21	254	±7	279	320

$\Delta\bar{\sigma}_{nom} = 1.2\ \bar{\sigma}_{nom}/m\sqrt{N}$

Mean values of batch I unit strengths σ_{fv} = 234 MN/m^2 and σ_{fa} = 255 MN/m^2 were obtained by weighting the bar and open ring strengths inversely proportional to their percentage standard errors.

MULTIAXIAL FAILURE STRENGTHS

Unit strengths were obtained using Equations 5 with the appropriate values from Tables 6 and 8. Equations 5 use the independent-action failure criterion, the validity of which is tested here. The ratio of compressive to tensile strength, α is needed to evaluate $H(\sigma)$ and hence $\leq(V)$ and $\leq(A)$ from Equation 2; it was obtained from the disc face compression results and the tensile results from the same batch of material.

Sines and Adams[6, 20, 25, 26] have shown that the biaxial compressive strength of alumina is almost independent of the magnitude of the smaller compressive stress. As a maximum shear stress criterion for RBSN in compression is not applicable[27], the maximum principal stress failure assumption, which is also consistent with the above results is used; the disc face compression results then give the uniaxial compressive strengths shown in Table 10. For disc bending, uniaxial tensile strength was derived from Equations 5, with the further assumption that the same equibiaxial/uniaxial strength ratio as measured from batch I applied. The values of σ_{fv} and σ_{fa} 240 and 267 MN/m^2 in Table 10 are not exactly equal to the ratios of unit and multiaxial strengths (e.g. 231 x 234/222 MN/m^2 for the volume-flaw criterion) because there exist regions of non-equibiaxial tension outside the ring-load of disc bending and the m values of batches I and II differ.

Table 10 Compressive strength ratios α from batch II results

uniaxial tensile strengths from bending			uniaxial compressive strengths from discs		
type	volume flaw σ_{fv} MN/m^2	area flaw σ_{fa} MN/m^2	volume flaw σ_{fvc} MN/m^2	area flaw σ_{fac} MN/m^2	modulus of rupture $\bar{\sigma}_{nom}$ MN/m^2
			2980	3640	3260
bar	213	243	14.0	14.9	15.3 13.4
disc	240	267	12.7	14.1	13.6 12.2
mean			13.8		13.2

The mean values of α were obtained by weighting the individual values inversely proportional to the percentage standard errors on corresponding unit tensile strengths. An overall mean value $\alpha = 13.5$ was used to evaluate $\not\leq(V)$ and $\not\leq(A)$ for the Brazilian disc and disc face compression tests to obtain the multiaxial strengths shown in Table 11. (The compressive contribution to the stress integrals for torsion is negligible; strengths for Brazilian discs are negligible dependent on the exact value of α).

The independent-action criterion (see Equation 2) predicts for the volume and area-flaw criteria that (equibiaxial strength)/(uniaxial strength) = $(\frac{1}{2})^{1/m}$. For the large m values obtained here the predicted biaxial unit strengths for torsion and Brazilian discs are equal to the unit tensile strength, as shown in Table 12.

The measured ratios of biaxial to uniaxial strengths in Table 12 were calculated as the product of the predicted ratio and the

Table 11 Unit volume and unit area strengths from biaxial tests

type of test	stress ratio	batch	number of tests N	mean nominal failure stress $\bar{\sigma}_{nom}$ MN/m^2	multiaxial strengths for 1 mm^3 σ_{mfv} MN/m^2	for 1 mm^2 σ_{mfa} MN/m^2
disc bending	1:1:0	I	20	99.7	222	251
" "	"	II	21	97.8	231	266
" " (1)	"	II	20	90.4	275	330
torsion	1:0:-1	IV	24	192	281	343
Brazilian disc	1:0:-3.5	I	20	198	220	212

(1) polished on tensile surface

Table 12 Ratios of biaxial to uniaxial strengths

stress ratio	batch	predicted by independent-action	measured	
			volume-flaw	area-flaw
1:1	I	0.964	0.914	0.949
1:1	II	0.955	1.037	1.043
1:-1	IV	1.000	1.043	1.074
1:-3.5	I	1.000	0.940	0.834

ratio of multiaxial to uniaxial strengths: if the independent-action criterion were exact, this latter ratio would always be unity and the measured strength ratio would equal that predicted. The ratios of multiaxial to uniaxial strengths for batch II are subject to significantly greater errors than the others because there were only 7 uniaxial tests. This explains the discrepancy in Table 10 between unit strengths from bar and disc bending results and the apparent equibiaxial strengthening shown in Table 12 for batch II.

Table 12 shows that the independent-action failure criterion applies for as-fired RBSN for the important tensile-tensile ratios and is safe for most practical applications. An exponent form of failure criterion[5] was previously proposed to explain equibiaxial weakening significantly greater than that predicted by the independent-action criterion but that apparent greater weakening could be due to an overestimate of friction in those disc bending tests.

As no polished uniaxial tests were carried out the polished biaxial results can only be compared with unpolished ones. Rows 3 and 2 in Table 11 show a reduction in mean failure stress, probably due to the polishing scratches, but increases in unit strengths due to the lower m value (see Table 8 and Equations 5) associated with the polishing scratches.

The slight equibiaxial weakening for as-fired RBSN agrees with theoretical predictions[2] and roughly with previous results for other materials as reported[2]. The slight pure shear strengthening agrees roughly with previous results for other materials[28]. The 1:-3.5 tension-compression weakening measured here for as-fired RBSN lies between previously measured weakenings for some and strengthenings for other ceramics[29, 13].

CONCLUSIONS

To measure uniaxial tensile strengths, bar bending tests are easy to perform whilst open ring tests are not subject to important contact friction effects. Radially thinner open rings would be easier to test and would be useful to obtain the strength of ceramic tube-like components. To measure equibiaxial tensile strength the disc bending test is easy to perform but the results are subject to important contact friction effects. Pure shear strength is easily obtained from square bar torsion tests.

The Brazilian disc test measures specimen strengths in a stress state between 1:-3 and typically 1:-3.5, depending on the loading contact conditions. It needs low material variability, high compressive strength and careful experimental design and procedure. The Hertzian line contact test measures the -0.4:-1:-1 triaxial compressive strength and is easy to perform. Inadvertent tensile stresses are unlikely but the test relies on contact stresses which can be subject to significant effects which cannot be accounted for theoretically. The value $\alpha = 13.5$ is probably conservative for as-fired RBSN at room-temperature. Previous results indicate that α is likely to be reduced at high temperatures because the compressive strength is likely to fall.

The two-parameter Weibull probability distribution adequately describes all the test failure strength distributions. Population material property estimates from small samples are often poor and it is recommended that at least 20 nominally identical samples be used for typical ceramic materials.

Weibull moduli m are material batch properties but subsequent machining and the form of testing can change this property. In particular polishing as-fired surfaces has been shown to reduce m and material tested in purely compressive stress states shows less variability than when tested with at least one tensile principal stress. This indicates the inapplicability of the weakest-link hypothesis to compression failure.

Experimental loading variability superimposed upon material variability gives inaccurately low material Weibull moduli estimates. The importance of this apparently obvious conclusion must be remembered whenever ceramic specimen strength tests are designed and performed, and the results analysed.

The independent-action multiaxial failure criterion is probably adequate for calculations of the failure probabilities of most ceramic components.

Fractography, i.e. the description of failure origins and crack paths, provides valuable information about the failure of ceramic specimens and components. For as-fired RBSN specimens, sub-surface failure origins were clearly seen to be typical; for polished RBSN specimens, the increased strength variability measured indicated that the polishing process had introduced surface flaws from which failures initiated. Since sub-surface failure origins are typical for as-fired RBSN, the volume-flaw criterion should be used. For polished RBSN, and most other post-firing machined ceramics, the area-flaw criterion is appropriate. Leaving non-critical RBSN component surfaces in the as-fired condition is likely to give lower failure probability than machining them.

ACKNOWLEDGEMENTS

This work was carried out with the support of the Procurement Executive, Ministry of Defence. The authors also thank the technicians at Nottingham University for their skilful assistance.

REFERENCES

1. D.J. Godfrey and M.W. Lindley, The strength of reaction-bonded silicon nitride ceramics, Proc. Brit. Ceram. Soc. 22:229 (1973).
2. S.B. Batdorf, Fundamentals of the statistical theory of fracture, in: "Fracture mechanics of ceramics Vol. 3", R.C. Bradt et al., eds., Plenum Press, New York (1978).
3. A.D. Sivill, "The development and application of a statistical basis for the design of brittle components", Ph.D. Thesis, Nottingham University (1974), pp. 212-219.
4. P. Stanley, H. Fessler and A.D. Sivill, An engineer's approach to the prediction of failure probability of brittle components, Proc. Brit. Ceram. Soc., 22:453 (1973).
5. P. Stanley and A. D. Sivill, On the biaxial fracture strength of reaction-bonded silicon nitride, Proc. Brit. Ceram. Soc., 26:97 (1978).
6. G. Sines and M. Adams, Compression testing of ceramics, in: ibid. Ref. 2.
7. P. Stanley, A.D. Sivill and H. Fessler, The unit strength concept in the interpretation of beam test results for brittle materials, Proc. Instn. Mech. Engrs. 190 49/76:585 (1976).
8. S.P. Timoshenko and J.N. Goodier, "Theory of elasticity", McGraw-Hill Kogakusha Ltd., Tokyo (1970), p. 83.
9. Ibid. Ref. 3, pp. 340-341.
10. Ibid. Ref. 3, p. 342.
11. Ibid. Ref. 8, pp. 309-313.
12. G. Hondros, Poisson's ratio and elastic modulus by Brazilian test, Australian J. of Appl. Sci., 10:265 (1959).
13. Ö. Vardar and I. Finnie, An analysis of the Brazilian disk fracture test using the Weibull probabilistic treatment of brittle strength, Int. J. Fracture, 11:495 (1975).
14. E.I. Radzimovsky, Stress distribution and strength condition of two rolling cylinders pressed together, Univ. of Illinois Eng. Exp. Station, Bulletin No. 408 (1953).

15. D. C. Fricker, "Failure predictions for ceramic gas turbine components under mechanical loading", Ph.D. Thesis, Nottingham University (1979), pp. 262-266.
16. F.I. Baratta, G.W. Driscoll and R.N. Katz, The use of fracture mechanics and fractography to define surface finish requirements for Si_3N_4, in: "Ceramics for high performance applications", J.J. Burke et al., eds., Brook Hill Publishing Co., Chestnut Hill, Mass. (1974).
17. J. Gramberg, Axial cleavage fracture, Eng. Geol. 1:31 (1965).
18. H.L. Price and K.H. Murray, Finite element analysis of the diametral compression test of polymer moldings, J. Eng. Mat. Tech., 95H:186 (1973).
19. R.H. Marion and J.K. Johnstone, A parametric study of the diametral compression test for ceramics, Amer. Ceram. Soc. Bulletin, 56:998 (1977).
20. G. Sines and M. Adams, The use of brittle materials as compressive structural elements, ASME Paper No. 75-DE-23 (1975).
21. Ibid. Ref. 3, pp. 105-134.
22. P. Stanley, A.D. Sivill and H. Fessler, The application and confirmation of a predictive technique for the fracture of brittle components, in: "Proc. 5th Int. Conf. on Experimental Stress Analysis", CISM, Udine, Italy (1974).
23. Ibid. Ref. 15, pp. 183-198.
24. R.W. Rice, Machining of ceramics, in: ibid. Ref. 16.
25. M. Adams and G. Sines, Determination of biaxial compressive strength of a sintered alumina ceramic, J. Amer. Ceram. Soc., 59:300 (1976).
26. M. Adams and G. Sines, Methods for determining the strength of brittle materials in compressive stress states, JTEVA, 4:383 (1976).
27. Ibid. Ref. 15, pp. 215-219.
28. G. Tappin, R.W. Davidge and J.R. McLaren, The strength of ceramics under biaxial stresses, in: ibid. Ref. 2.
29. H.W. Babel and G. Sines, An improved method of uniaxial and biaxial testing of brittle materials, JMLSA, 3:134 (1968).

STRENGTH AND FRACTURE TOUGHNESS OF SILICON CARBIDE

Jochen Kriegesmann, Alfred Lipp, Klaus Reinmuth and
Karl A. Schwetz

Elektroschmelzwerk Kempten GmbH
Herzog-Wilhelm-Straße 16
D-8000 München 33

ABSTRACT

Flexural strength of four different grades of dense silicon carbide was measured at different temperatures. It could be demonstrated by scanning electron microscopy that strength remains constand up to high temperatures, if the fracture mechanism turns out to be transgranular. In the case of intergranular fracture the decrease of strength with temperature is governed by subcritical crack growth.

Static fracture toughness vs. temperature characteristics have also been determined. The static fracture toughness data at high temperatures are assumed to be dependent upon subcritical crack growth in the case of intergranular fracture mechanism.

INTRODUCTION

In recent years there has been considerable interest in ceramics for high temperature structural applications. Most ceramic materials, however, exhibit significant strength degradation at temperatures above 900°C. For mechanical applications at high temperatures materials are needed that show no or little decrease of strength at increasing temperatures.

This paper demonstrates that it is possible for a structural ceramic material like silicon carbide to deep strength constant up to high temperatures, if a particular fracture criterion is fullfilled.

FRACTURE MECHANISM

The steady decline of strength at rising temperatures in structural ceramics like Si_3N_4 and SiC is governed by subcritical crack growth.[1,2,3].

Evans [4] has proposed a model - schematically shown in fig 1 - for slow crack growth due to grain boundary sliding near preexisting microcracks. Regarding this model the rigid grains will be sliding along each other starting adjacent to the crack tip, if a load is applied. This sliding causes a time-dependent displacement that may become critical at the triple points of the grains. Thus secondary cracks can be formed. Since these secondary cracks also give rise to grain boundary sliding, a network of secondary cracks (process zone) can be built up around the crack tip. In this way the microstructure of the material is gradually weakened. Crack growth becomes critical when the grains are beginning to separate.

If this model is valid, a material characterized by an intergranular fracture mechanism should exhibit a more or less distinct subcritical crack growth depending essentially upon the concentration and the viscosity of the supposed glassy phase in the grain boundary. In this case the strength of the material will decrease with increasing temperature. On the other hand the influence of subcritical crack growth on strength should be small if the fracture mechanism of the material turns out to be transgranular. Transgranular and intergranular fracture mechanism are schematically shown in fig 2.

It has been demonstrated elsewhere [5,6] that both mechanisms of fracture are possible in hot pressed silicon carbide depending on the kind and the quantity of the sintering aid.

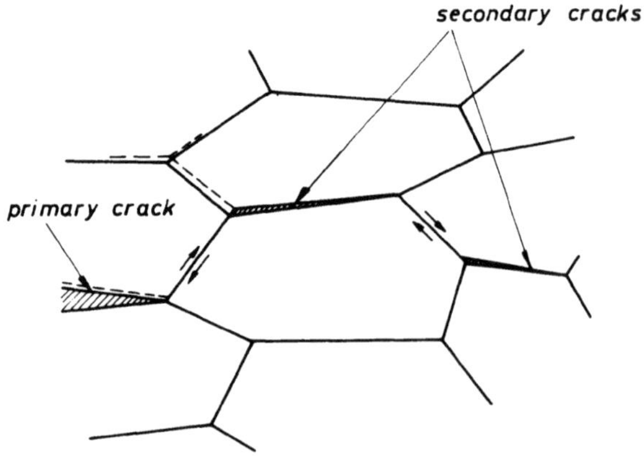

Fig. 1 Schematic of subcritical crack growth due to grain boundary sliding (Evans [4])

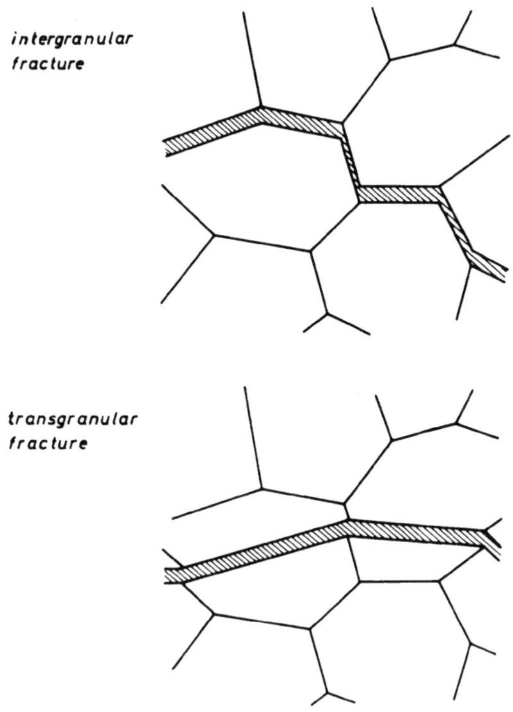

Fig. 2 Schematic of intergranular and transgranular fracture mechanisms

EXPERIMENTAL PROCEDURE

Three grades (HP-I, HP-T, HP-TI) of hot pressed silicon carbide and one grade of pressureless sintered silicon carbide (S-T) have been chosen for this investigation. In all cases except one alpha-SiC powder has been used as starting powder for sintering. Beta-SiC powder has been taken for manufacturing of one hot pressed grade (HP-TI). The average density of the hot pressed grades was at least 3.19 g/cm³ and the density of the sintered grade turned out to be 3.11 g/cm³ that is 97 % of the theoretical density of silicon carbide.

For the determination of the four-point flexural strength (outer span 30 mm, inner span 15 mm) the silicon carbide bodies were cut into rectangular test bars of the size 3.15 x 4.1 x 35 mm³ by diamond tools. The apparatus used in this study has been described elsewhere[5]. The chosen stressing rate has been 3 MN/m²s.

The fracture toughness was determined using the single-edge-notched-beam technique[7]. The size of the specimens have been 3.5 x 7 x 35 mm³. Notches of radius $\rho<50$ µm and average depth of 1.8 mm were introduced into the specimens by a high precision diamond cut. Those notches were supposed to be equivalent to a zero volume crack [8]. The fracture toughness data were determined by four-point loading (outer span 30 mm, inner span 7,5 mm). For these measurements the same apparatus has been used as for the flexural strength determinations.

MICROSTRUCTURES

Specimens for scanning electron microscopy were polished down to 1 μm surface finish and etched in a boiling Murakami solution. The time for etching was about 3 min. Fig. 3 to fig. 6 show the microstructures of the studied materials. The hot pressed grades HP-I and HP-T exhibit uniform equiaxed grain structures whereas the microstructure of SiC HP-TI is composed of large tabular grains. The sintered grade S-T is characterized by uniformly fine, elongated grains.

Fig. 3 Scanning electron micrograph of the microstructure of SiC HP-I

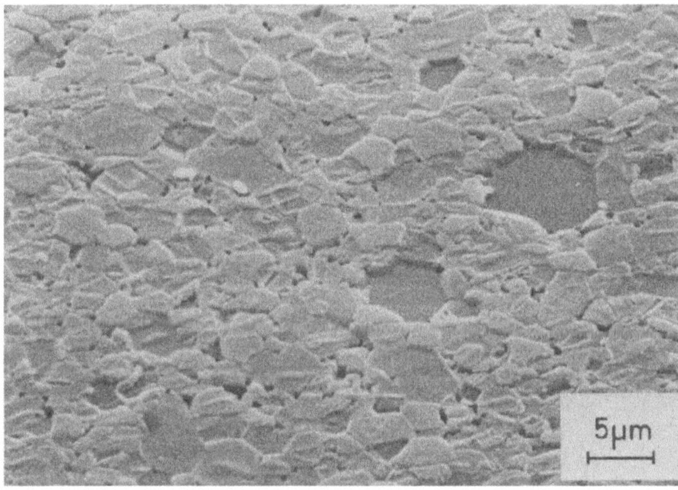

Fig. 4 Scanning electron micrograph of the microstructure of SiC HP-T

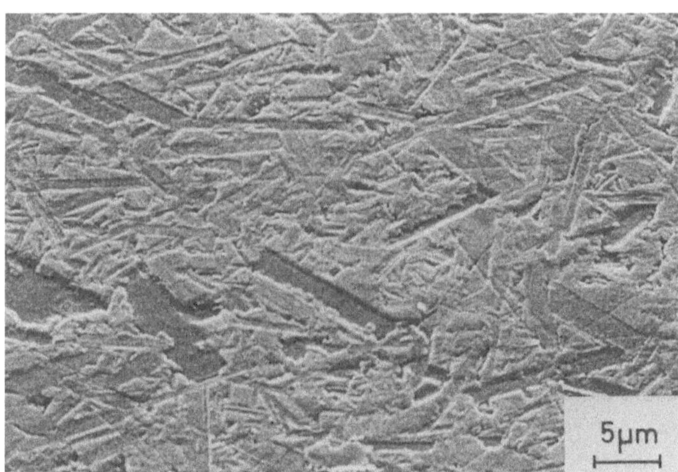

Fig. 5 Scanning electron micrograph of the microstructure of SiC HP-TI

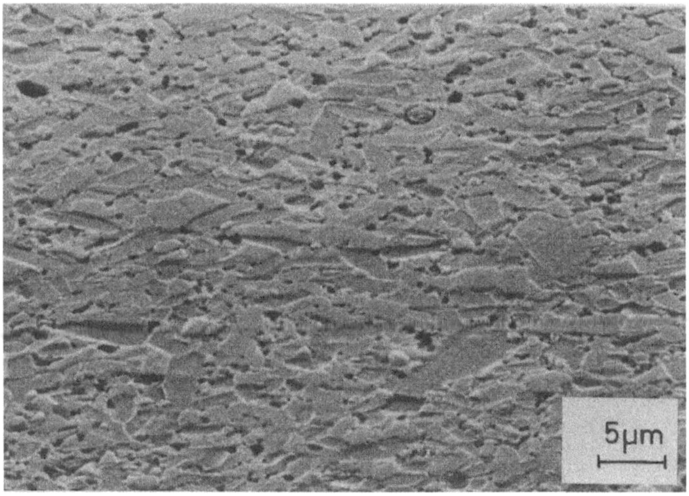

Fig. 6 Scanning electron micrograph of the microstructure of SiC S-T

STRENGTH AND FRACTURE TOUGHNESS RESULTS

The results of flexural strength measurements and fracture toughness determinations are summerized for the four materials in table 1 and table 2 respectively.

Table 1. Flexural Strength of SiC, MN/m^2

	grade			
Temperature	HP-I	HP-T	HP-TI	S-T
Room temp.	634 (10) ± 44	657 (11) ± 61	736 (6) ± 54	430 (12) ± 49
800 °C	638 (5) ± 61	652 (7) ± 67	732 (4) ± 83	436 (8) ± 38
1000 °C	589 (8) ± 52	648 (10) ± 56	743 (4) ± 38	445 (10) ± 51
1200 °C	496 (7) ± 31	663 (10) ± 72	683 (6) ± 46	453 (10) ± 41
1370 °C	319 (8) ± 21	660 (8) ± 49	580 (6) ± 61	450 (10) ± 45

Number of specimens tested given in perentheses

Table 2. Static Fracture Toughness of SiC, MN/m^{2} '^{3}

Temperature	HP-I	HP-T	S-T
		grade	
Room temp.	5.47 (5) ± 0.32	4.84 (5) ± 0.39	4.80 (5) ± 0.27
800 °C	5.40 (4) ± 0.19	4.95 (4) ± 0.43	-
1000 °C	4.78 (5) ± 0.26	5.00 (5) ± 0.37	-
1200 °C	4.20 (5) ± 0.22	4.90 (5) ± 0.20	-
1370 °C	3.58 (5) ± 0.22	5.06 (5) ± 0.36	-

Number of specimen tested given in perentheses

Though the microstructure of the grades HP-I and HP-T is similar, the strength behaviour of these two silicon carbides is quite different. Strength and static fracture toughness of HP-I distinctly depend upon temperature, whereas the data of HP-T are kept almost constant up to high temperatures. Strength and static fracture toughness vs. temperature results are plotted in fig. 7 and 8 respectively.

Though the microstructure of grade HP-TI shows exaggerated grains, its room temperature strength is the highest of all studied materials. The strength degrades only slightly with temperature. The strength data of this material is presented in fig. 9. Fracture toughness values of this grade have not been measured.

The room temperature strength of the presented sintered silicon carbide (fig. 10) is a bit lower compared with the hot pressed grades because of the minor density. But the strength of this material even shows little increase in temperature. The fracture toughness has only been measured at room temperature, its value (table 2) is comparable to that of SiC HP-T.

FRACTURE SURFACES

It was pointed out above that strength vs. temperature relationship of structural ceramics like silicon carbide should mainly be influenced by the fracture criterion. Regarding the strength results grade HP-I should be featured by an intergranular and grade HP-T by a transgranular kind of fracture. Fig. 11 and 12 show that fracture mechanisms of these two grades are indeed just like they were expected. These figures contain only the fracture surfaces at room

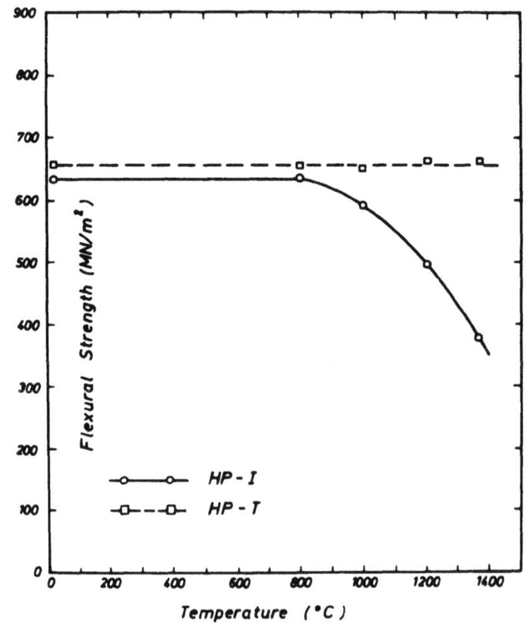

Fig. 7 Flexural strength as a function of temperature for SiC HP-I and HP-T

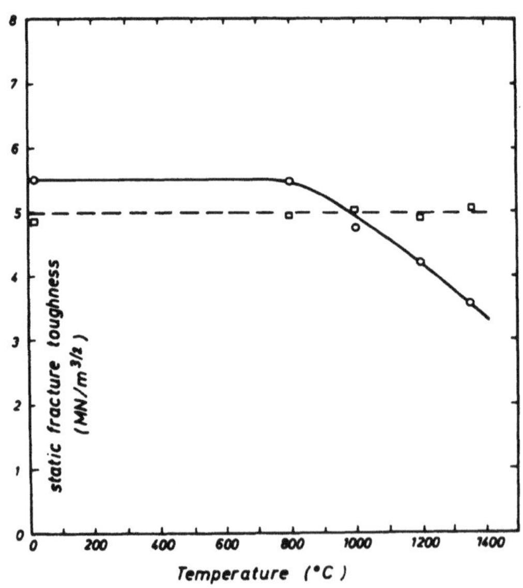

Fig. 8 Static fracture toughness as a function of temperature for SiC HP-I and HP-T

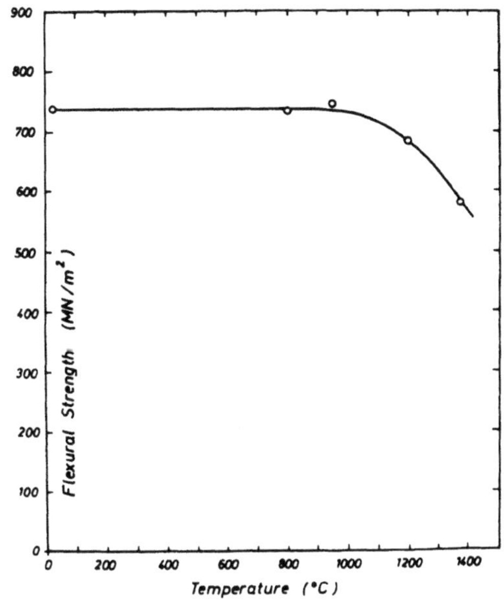

Fig. 9 Flexural strength as a function of temperature for SiC HP-TI

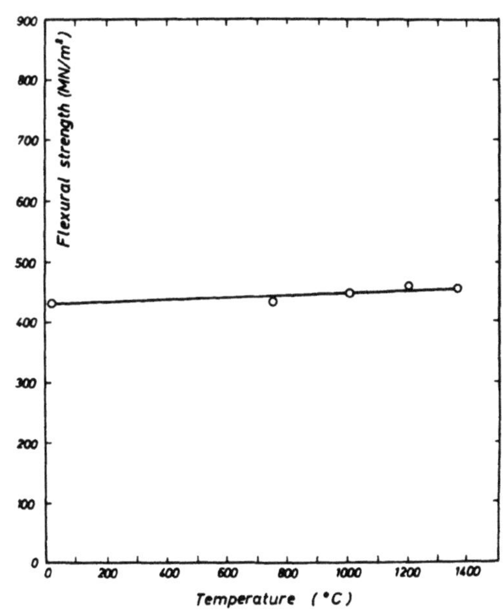

Fig. 10 Flexural strength as a function of temperature for SiC S-T

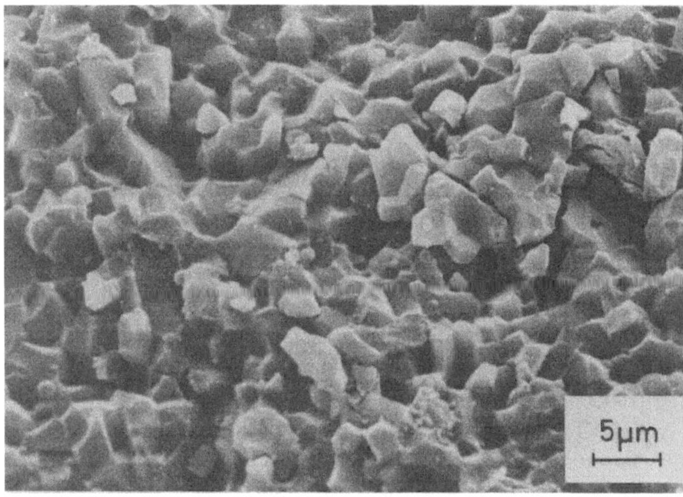

Fig. 11 Scanning Electron Micrograph of the fracture surface of SiC HP-I at room temperature

Fig. 12 Scanning Electron Micrograph of the fracture surface of SiC HP-T at room temperature

temperatures since there is no principle change in fracture surface up to high temperatures.

This is different with grade HP-TI which is characterized by a mainly transgranular kind of fracture near room temperature, whereas the fracture behaviour changes into intergranular at high temperatures. This is demonstrated in fig. 13.

It is not expected that the desirable transgranular kind of fracture that is responsible for a good high temperature strength is restricted only to one special technology of sintering. Fig. 14 shows that the pressureless sintered grade S-T as well is characterized by a transgranular fracture mechanism which may explain the strength results.

DISCUSSION

The basic idea of this investigation is the statement that subcritical crack growth gives rise to a time-dependent fracture stress. This means that the measured fracture stress for a given structural ceramic is supposed to be higher if the stressing rate is inclined. Thus the strength of the material showing an intergranular kind of fracture is not a constant parameter for a given temperature but is strongly depending on the stressing rate. If the stressing rate is high enough the strength of the material exhibiting an intergranular kind of fracture at quasistatic stressing rates will not decrease with temperatures. Unfortunately it was not possible during this investigation to demonstrate this. But since all strength results reported in this paper could be explained by the fracture mechanisms, the static fracture toughness data of SiC HP-I cannot be explained in the same manner. Since fracture toughness of a material should definitely be independent on time effects, the fracture toughness data should be kept constant up to high temperatures. Therefore the measured static fracture toughness of grade HP-I cannot be regarded as independent on time effects. Thus only the fracture toughness data of SiC HP-T can be regarded as valid. Mendiratta et al.[9] have measured static and dynamic strength and fracture toughness in hot pressed silicon nitride. They found out that the static strength and fracture toughness are distinctly dependent on temperature while the dynamic parameters are not. Thus high temperature fracture toughness data of ceramics reported in literature should be examined critically.

Going back to normal quasistatic stressing rates it has not been pointed out enough which is the reason for either an inter- or a transgranular fracture mechanism. It has already been reported[5,6] that a boron doped silicon carbide shows a transgranular kind of fracture while for an aluminium doped silicon carbide an intergranular fracture mechanism is favoured.

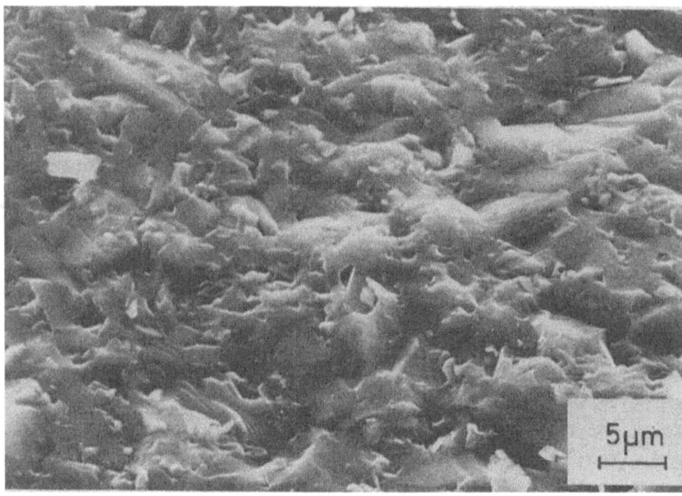

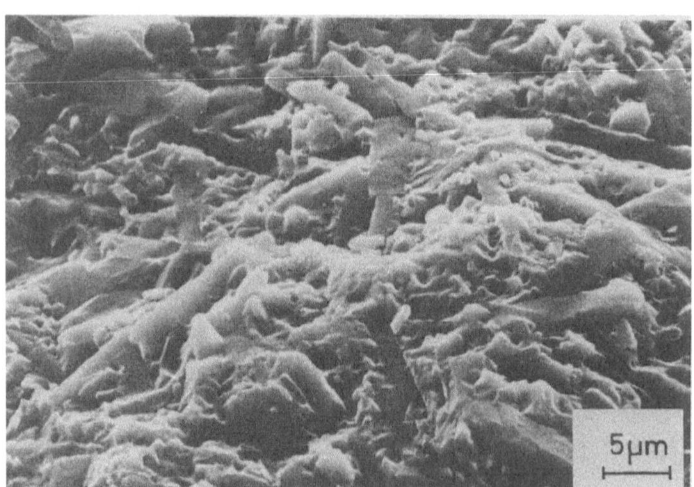

Fig. 13 Scanning Electron Micrograph of fracture surface of SiC HP-TI

above: at room temperature (unetched)
below: at 1370 °C (etched with conc. hydrofluoric acid)

Fig. 14 Scanning Electron Micrograph of fracture surface of SiC S-T

Since for boron doped silicon carbide a solid boron carbide grain boundary phase can be supposed, the grain boundary phase of the aluminium doped silicon carbide is thought to be an aluminium silicate glass. Thus a glassy phase at the grain boundaries seems to favour an intergranular kind of fracture. On the other hand a transgranular fracture should be favoured if the grain boundaries are free of glassy phase.

Evans and Lange [3] have measured high temperature slow crack growth of a boron doped sintered silicon carbide and a hot pressed silicon carbide with an aluminium containing additive. Since for the aluminium doped SiC an extensive crack growth was apparent at 1400 °C the corresponding experiments for the boron doped grade did not show any significant slow crack growth. Though fracture surfaces have not been shown in the cited paper, it can be supposed that generally the predicted connection between fracture mechanism and subcritical crack growth at high temperatures is correct.

CONCLUSION

A silicon carbide ceramic characterized by a transgranular fracture mechanism under quasistatic conditions exhibits little subcritical crack growth even at high temperatures. Such a material does not show any distinct drop of strength at elevated temperatures and is thus extremely resistent against delayed fracture.

REFERENCES

1. F. F. Lange, and J.L. Iskoe, High-Temperature Strength Behavior of Hot-Pressed Si_3N_4 and SiC: Effect of Impurities, in "Ceramics for High Performance Applications", J.J. Burke, A.E. Gorum, and R.N. Katz. eds., Brook Hill Publishing Company, Chestnut Hill (1974)
2. F.F. Lange, High-Temperature Strength Behavior of Hot-Pressed Si_3N_4: Evidence for Subcritical Crack Growth, J. Amer. Ceram., Soc. 57:84 (1974)
3. A.G. Evans, and F.F. Lange, Crack Propagation and Fracture in Silicon Carbide, J. Mat. Sci, 10: 1659 (1975)
4. A.G. Evans, High-Temperature Slow Crack Growth in Ceramic Materials, in the same source as ref. 1
5. J. Kriegesmann, Biegefestigkeitsuntersuchungen an heißgepreßtem Siliciumcarbid, Ber. Dt. Keram. Ges., 55:391 (1978)
6. J. Kriegesmann, K. Reinmuth, and A. Lipp, Die Herstellung und Eigenschaften von heißgepreßtem und heißisostatisch gepreßtem SiC, in "Keramische Komponenten für Fahrzeug-Gasturbinen" W. Bunk, and M. Böhmer eds.,Springer Verlag, Berlin, Heidelberg, New York (1978)

7. L.A. Simpson, Use of the Notched-Beam Test for Evaluation of Fracture Energies, J. Amer. Ceram. Soc. 57:151 (1974)
8. R. Pabst, G. Popp, and L. Li, private communication
9. M.G. Mendiratta, J. Wimmer, and I. Bransky, Dynamic K_{IC} and Dynamic Flexural Strength in HS-130 Si_3N_4, J.Mat. Sci. 12:212 (1977)

DEVELOPMENT OF PLANE STRAIN FRACTURE TOUGHNESS TEST FOR CERAMICS USING CHEVRON NOTCHED SPECIMENS

R. T. Bubsey, J. L. Shannon, Jr.,
NASA Lewis Research Center, Cleveland, Ohio, 44135

and D. Munz
DFVLR, Cologne, FR Germany

SUMMARY

The chevron-notch specimen is especially useful for measuring the plane strain fracture toughness, K_{Ic}, of ceramics. During test, a crack initiates at the tip of the chevron notch and extends stably as load is increased. For materials with flat crack growth resistance curves, maximum load for a given specimen configuration will always occur at the same relative crack length independent of the material. K_{Ic} is calculated from maximum load and the corresponding (minimum) value of the dimensionless stress intensity factor coefficient (Y_m^*), with no need for crack length measurement.

Y_m^* values have been determined for the short bar and four-point-bend specimens and have been used in fracture toughness measurements on hot pressed silicon nitride and sintered aluminum oxide. K_{Ic} for Si_3N_4 was essentially independent of specimen size and chevron notch configuration, with values ranging only from 4.6 to 4.9 $MNm^{-3/2}$. In contrast, significant specimen size and notch geometry effects were observed for Al_2O_3, with K_{Ic} values ranging from 3.1 to 4.7 $MNm^{-3/2}$. These effects are believed due to a rising crack growth resistance curve for the Al_2O_3 tested.

INTRODUCTION

At the present time there is no standard test for the plane-strain fracture toughness (K_{Ic}) of brittle nonmetallic materials. A number of specimen types are used which have either blunt machined notches or are pre-cracked by wedge loading or thermal shock (1-7). Those with machined notches can overestimate the fracture toughness if the notch width exceeds a critical size (1, 8-11). Precracked specimens are difficult to prepare in a reproducible manner, and the initial crack front often cannot be seen on the fracture surface after testing, making it difficult, if not impossible, to measure the initial crack length. Chevron notch specimens, such as the short rod specimen proposed by Barker (12), overcome these difficulties. A stable crack initiates at the tip of the triangular notch, and for materials with a flat crack growth resistance curve the maximum load (P_{max}) occurs at a relative crack length that is dependent only on the specimen geometry (and not the material).

Three types of chevron notch specimens have been given attention: bend, short rod and short bar. Three-point-bend specimens have been used in work-of-fracture studies (13,14). These studies were not totally successful because of difficulty in obtaining stable crack growth much beyond maximum load. The effective surface energy obtained did not always agree with the surface energy determined with other fracture mechanics specimens (9, 15-18).

Barker (12) has used a short-rod specimen of a specific geometry. The relation between P_{max} and K_{Ic} was estimated for this specimen geometry by a comparison procedure involving materials having well documented K_{Ic} values obtained in accordance with ASTM Standard Method E 399-78 for Plane-Strain Fracture Toughness of Metallic Materials. Such a calibration is subject to material variability and test procedure, and also assumes an independence of specimen geometry.

The present authors (19) have used the chevron-notched specimen in four-point-bend tests, applying the results of Bluhm (20, 21) for the P_{max}-K_{Ic} relation. They have also developed the P_{max}-K_{Ic} relation for the double-cantilever loaded short bar specimen analytically and from experimental compliance measurements (22), and are currently making experimental compliance measurements on the short rod specimen.

The four-point-bend and short-bar specimens (Figure 1) have been used to determine the fracture toughness of aluminum oxide and silicon nitride.

STRESS INTENSITY FACTOR FOR THE CHEVRON NOTCH SPECIMEN

The chevron notch (Figure 1) is characterized by the notch length at the specimen surface, a_1; the notch length to the tip of the chevron, a_o; and the specimen width, W, and thickness, B.

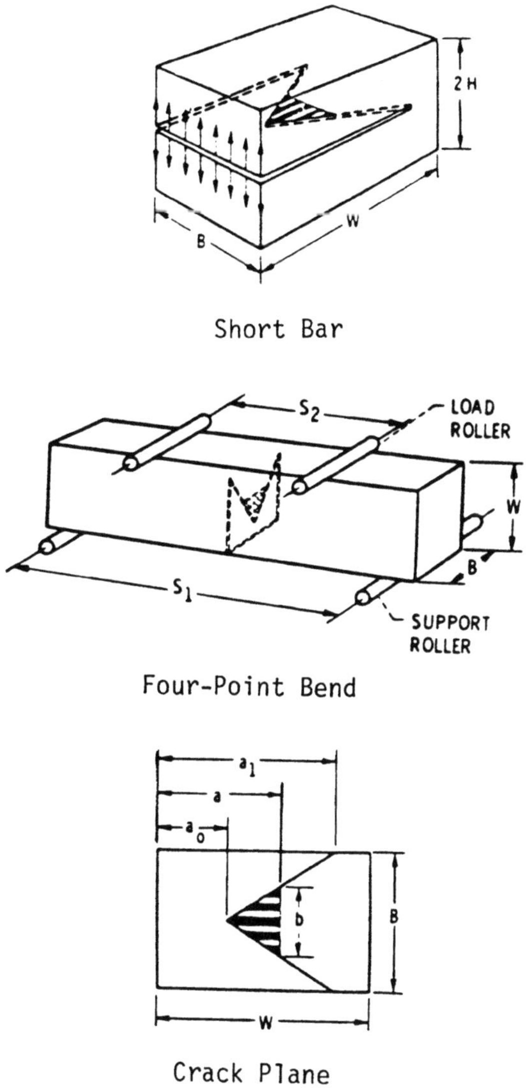

Figure 1. Chevron-notch specimens.

The shaded (or striped) area in Figure 1 represents a crack of length a. The crack front length b is

$$b = B \frac{a - a_o}{a_1 - a_o} = B \frac{\alpha - \alpha_o}{\alpha_1 - \alpha_o} \tag{1}$$

where $\alpha = a/w$, $\alpha_o = a_o/w$, and $\alpha_1 = a_1/w$

The relation between load P and fracture toughness K_{Ic} is obtained by considering the available and the necessary energies for crack propagation. The energy available to extend the crack a small increment Δa is

$$\Delta U = \frac{P^2}{2W} \frac{dC}{d\alpha} \Delta a \tag{2}$$

where C is the compliance of the specimen. The energy necessary to extend the crack Δa is

$$\Delta\overline{W} = G_{Ic}\, b\, \Delta a = \frac{K_{Ic}^{2}}{E'}\, b\, \Delta a \tag{3}$$

with elastic modulus $E' = E$ for plane stress and $E' = E/(1-\nu^2)$ for plane strain.

During crack extension, $\Delta U = \Delta\overline{W}$, leading to

$$K_{Ic} = P \left[\frac{(dC/d\alpha)\,E'}{2Wb}\right]^{\frac{1}{2}} = \frac{P}{B\sqrt{W}} \left[\frac{1}{2} \frac{dC'}{d\alpha} \frac{\alpha_1 - \alpha_o}{\alpha - \alpha_o}\right]^{\frac{1}{2}} \tag{4}$$

where $C' = E'BC$ is the dimensionless compliance. The term

$$Y^* = \left[\frac{1}{2} \frac{dC'}{d\alpha} \frac{\alpha_1 - \alpha_o}{\alpha - \alpha_o}\right]^{\frac{1}{2}} \tag{5}$$

is the dimensionless stress intensity factor coefficient.

There are no available analytical solutions for the chevron notch specimen stress intensity factor coefficient. However, as a first approximation it can be assumed that $dC'/d\alpha$ for the chevron notch with a crack is identical to that for a straight-through crack (referred to as the "straight-through crack assumption", STCA). Using the relation for a straight-through crack

$$\frac{dC}{d\alpha} = 2Y^2 \tag{6}$$

where

$$Y = \frac{KB\sqrt{W}}{P},$$

Eqs. (4) and (5) can be rewritten

$$K_{Ic} = \frac{P}{B\sqrt{W}} \, Y \left[\frac{\alpha_1 - \alpha_o}{\alpha - \alpha_o}\right]^{\frac{1}{2}} \tag{7}$$

and

$$Y^* = Y \left[\frac{\alpha_1 - \alpha_o}{\alpha - \alpha_o}\right]^{\frac{1}{2}} \tag{8}$$

Short Bar

The dimensionless stress intensity factor coefficient for the short bar specimen with a straight-through crack is (22):

$$Y = \frac{\alpha}{(1-\alpha)^{3/2}} \ln\left(\exp\left[\frac{2.702}{\alpha} + 1.628\right] + \exp\left\{\left[12\,\frac{W^3(1-\alpha)^3}{H^3}\right]^{\frac{1}{2}}\left[1 + \frac{0.679}{\alpha(W/H)}\right]\right\}\right) \tag{9}$$

To determine the validity of the straight-through crack assumption, a series of short bar experimental compliance measurements were made. The compliance specimens were machined from 7075-T651 aluminum plate to the dimensions shown in Figure 2 for W/H ratios of 3 and 4. A 17.8 mm slot was machined into the front face to accommodate loading knife edges. The displacement measurements were made with a modified ASTM E-399 clip-in gage. Hardened steel cones were fastened to the inner surfaces of the gage arms

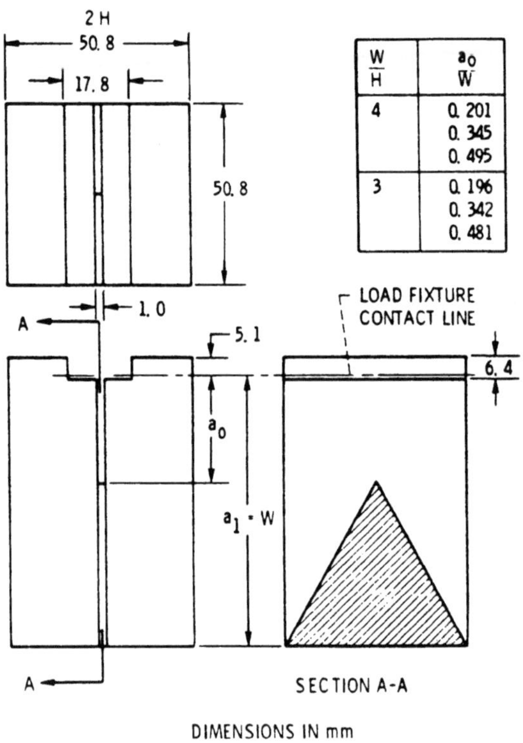

Figure 2. Short bar chevron-notch compliance specimen.

and set into small indentations in the top and bottom of the specimen (22). A 0.6 mm slot was extended incrementally to simulate an advancing crack in the chevron notched compliance specimens (24).

Compliance results (substituted into Eq. (5)) are compared in Figure 3 with analytical results from Eqs. (8) and (9) for the W/H = 4 specimen proportions. In the region of the curve minima the experimental and analytical results agree. As expected, the minima decrease and broaden with decreasing initial crack length, α_0. Companion results on the influence of α_1 on Y* are shown in Figure 4. Decreasing α_1 lowers the curves and, at $\alpha = \alpha_1$ the chevron-notch specimen curves join the curve for a straight-through crack specimen. Y_m^* for the entire range of short bar specimen geometries examined are given in Table 1. The greatest difference reported is less than 4 percent. The authors have a preference for a chevron notch geometry of $\alpha_0 = 0.2$ and $\alpha_1 = 1$. For this geometry the agreement in Y_m^* is within 1 percent.

Four-Point-Bend

For a specimen with a straight-through crack subjected to pure bending, the dimensionless stress intensity coefficient from (23) is:

$$Y = \frac{S_1 - S_2}{W} \quad \frac{3\Gamma m \sqrt{\alpha}}{2(1-\alpha)^{3/2}} \tag{10}$$

with

$$\Gamma m = 1.9887 - 1.326\alpha - \frac{(3.49 - 0.68\alpha + 1.35\alpha^2)\ \alpha(1-\alpha)}{(1+\alpha)^2} \tag{11}$$

An alternate analysis for the dimensionless stress intensity factor coefficient for the four-point-bend specimen has been developed by Bluhm (21) and used by the present authors (22). The specimen is divided into slices of uniform thickness. Compliance of a slice is taken to be that of a straight-through crack specimen of the same thickness. The compliance of the full chevron-notch specimen is obtained by summing the compliance of the slices. Interlaminar shear stresses are accounted for by effectively increasing the slice thickness (reducing the compliance) based on a curve-fitting of experimental compliance measurements.

For materials that have a flat crack growth resistance curve, the chevron-notch specimen stress intensity factor during crack extension is constant and equal to K_{Ic}. It is computed from:

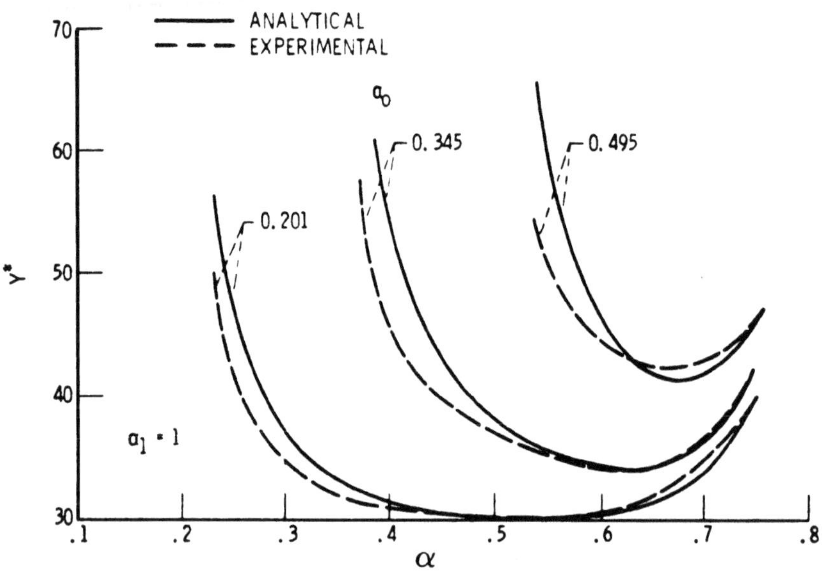

Figure 3. Comparison of Y* determined experimentally and analytically for short bar chevron-notch specimens of W/H = 4 and $\alpha_1 = 1$.

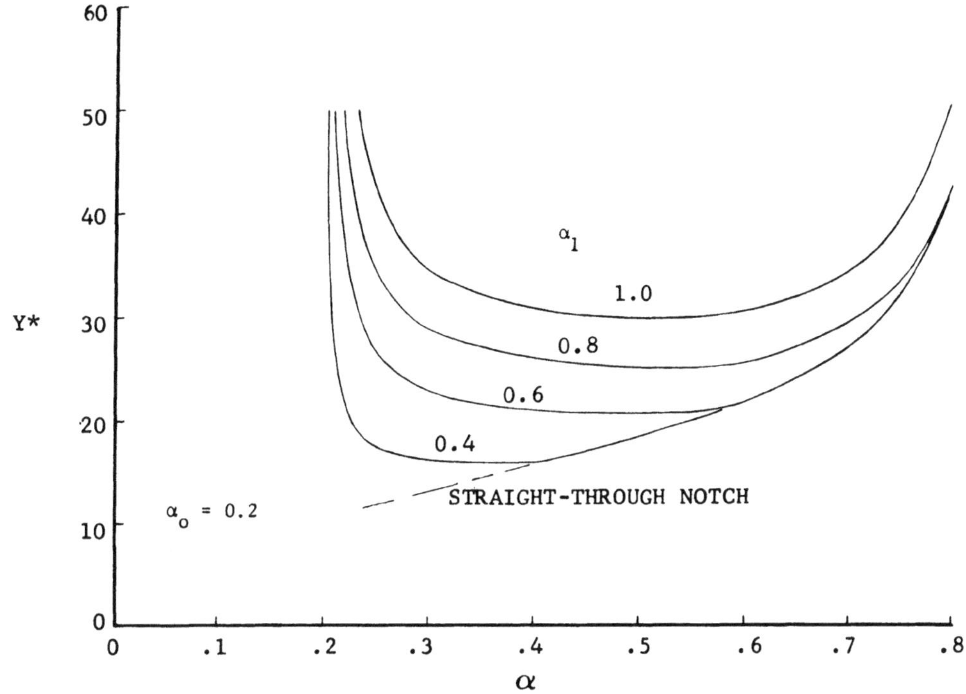

Figure 4. Experimentally determined Y* for short bar specimens of W/H = 4 containing either a chevron notch of $\alpha_o = 0.2$ or a straight-through notch.

Table 1. Dimensionless stress intensity factor coefficient minimum value, Y^*_m, for short bar chevron-notch specimens.

W/H	α_o	α_1	Y^*_m		DIFFERENCE
			ANALYTICAL	EXPERIMENTAL	%
3	0.196	1.000	21.50	21.40	0.4
	0.342	1.000	25.70	25.95	1.0
	0.481	1.000	33.56	32.54	3.0
4	0.201	1.000	30.21	29.95	0.9
	0.198	0.800	26.11	25.15	3.7
	0.198	0.600	21.34	20.71	3.0
	0.199	0.403	15.87	15.90	0.2
	0.345	1.000	34.14	34.14	0.0
	0.495	1.000	41.48	42.93	3.5

$$K_{Ic} = \frac{P}{B\sqrt{W}} Y^* \tag{12}$$

Substituting P_{max} for P and Y^*_m for Y^*, K_{Ic} is obtained from maximum load with no need for crack length measurement. Y^*_m for the short bar specimen, determined with the STCA, may be expressed (22):

$$Y^*_m = \left\{4.08 + 3.95\left(\frac{W}{H}\right) + 0.50\left(\frac{W}{H}\right)^2 + \left[-23.15 + 1.15\left(\frac{W}{H}\right) + 1.30\left(\frac{W}{H}\right)^2\right]\alpha_o + \left[172.5 - 43.5\left(\frac{W}{H}\right) + 3.0\left(\frac{W}{H}\right)^2\right]\alpha_o^2\right\}\left\{\frac{\alpha_1 - \alpha_o}{1 - \alpha_o}\right\}^{\frac{1}{2}} \tag{13}$$

for $3 \leq W/H \leq 4$ and $0.2 \leq \alpha_o \leq 0.4$. For four-point-bend specimens, Bluhm's slice model yields (25):

$$Y^*_m = \left[3.08 + 5.00\,\alpha_o + 8.33\,\alpha_o^2\right]\frac{S_1 - S_2}{W} \cdot \left[1 + 0.007\left(\frac{S_1 S_2}{W^2}\right)^{\frac{1}{2}}\right]\left[\frac{\alpha_1 - \alpha_o}{1 - \alpha_o}\right] \tag{14}$$

With the STCA, the four-point-bend relation is (25):

$$Y^*_m = \left[2.92 + 4.52\,\alpha_o + 16.14\,\alpha_o^2\right]\frac{S_1 - S_2}{W}\left[\frac{\alpha_1 - \alpha_o}{1 - \alpha_o}\right]^{\frac{1}{2}} \tag{15}$$

for $1 \leq W/B \leq 1.25$, $0.12 \leq \alpha_o \leq 0.24$, and $0.90 \leq \alpha_1 \leq 1.00$.

FRACTURE TOUGHNESS DETERMINATIONS

Materials and Procedures

Two materials were investigated: Norton Company NC-132 hot pressed silicon nitride (Si_3N_4) and 3M Company Alsimag-614 sintered aluminum oxide (Al_2O_3). The silicon nitride was received in the form of a 150 mm x 150 mm x 10 mm hot pressed plate. Short bar and four-point-bend chevron notch specimens were machined with their crack planes either parallel or perpendicular to the plate surface. The aluminum oxide was received in the form of 25.4 mm x 25.4 mm x 54.6 mm, 12.7 mm x 12.7 mm x 29.7 mm, and 4.1 mm x 5.1 mm x 50.8 mm sintered blanks. Short bar chevron notch specimens were machined from the 12.7 mm and 25.4 mm square sintered blanks. Four-point-bend chevron-notch specimens were machined from the 25.4 mm square and the 4.1 mm x 5.1 mm sintered blanks. Chevron notches were produced by wire sawing with diamond impregnated wires, or by slotting with diamond coated wheels.

The test fixtures and alignment procedures used in testing the short bar and four-point-bend specimens are described in (22) and (19), respectively. Fracture toughness was computed from the maximum load with Y_m^* values obtained using the STCA for the short bar specimens and Bluhm's slice model for the four-point-bend specimens.

RESULTS

Silicon nitride short bar and four-point-bend specimen results are presented in Table 2 and Figures 5 and 6. K_{Ic} appears to be independent of the chevron notch parameters α_0 and α_1, specimen size, and crack plane orientation. The bend specimen results are only slightly different (+3 percent) from the short bar specimen results. All values fall in the range 4.64 to 4.85 $MNm^{-3/2}$.

The results for aluminum oxide also show K_{Ic} independent of α_0 (Figures 7 and 8). However, a definite size effect is indicated by the short bar data in Figure 7. Doubling the specimen size (increasing thickness B from 12.7 mm to 25.4 mm) produced a 15% to 18% increase in K_{Ic}. Increasing the width-to-height ratio, W/H, from 3 to 4 raises the computed K_{Ic} by 6% to 10%. These size effects are more clearly represented in Figure 9 where K_{Ic} is plotted as a function of the specimen width, W. The average K_{Ic} increased from 3.74 $MNm^{-3/2}$ to 4.71 $MNm^{-3/2}$ over the range investigated. While α_0 had minimal effect on the computed K_{Ic}, α_1 had a pronounced effect (Figure 10). Decreasing α_1 from 1.0 to 0.4 reduced the computed K_{Ic} by over 25%. Total variation in K_{Ic} observed for Al_2O_3 in this study was 3.07 $MNm^{-3/2}$ to 4.71 $MNm^{-3/2}$.

DISCUSSION

Chevron notch specimens are well suited for the determination of K_{Ic} for ceramic materials having a flat crack growth resistance curve. This is demonstrated by the observed independence of the computed K_{Ic} on specimen type, size and notch geometry for the hot pressed silicon nitride.

The aluminum oxide K_{Ic} results summarized in Table 2 varied considerably with specimen type, size, and notch geometry. To ensure that material variation was not a contributing factor in

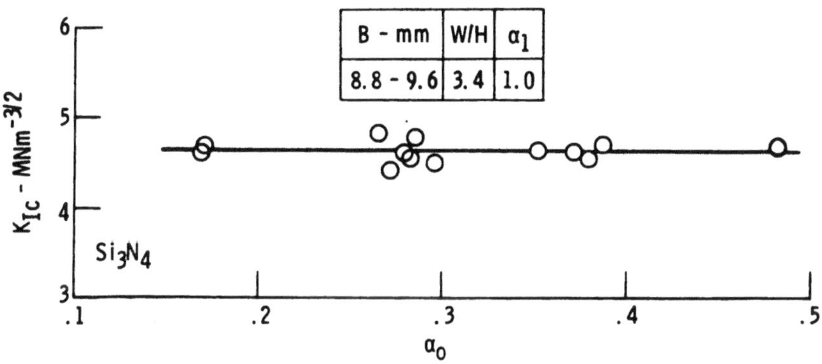

Figure 5. Effect of α_o on K_{Ic} of hot pressed silicon nitride (NC-132) determined with short bar chevron-notch specimens.

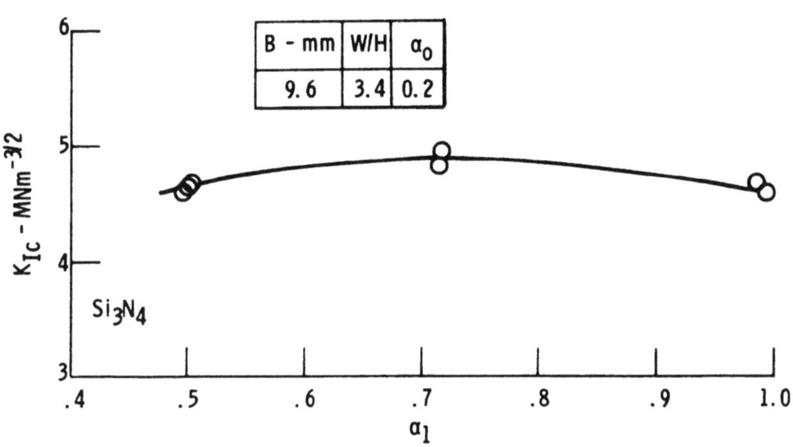

Figure 6. Effect of α_1 on K_{Ic} of hot pressed silicon nitride (NC-132) determined with short bar chevron-notch specimens.

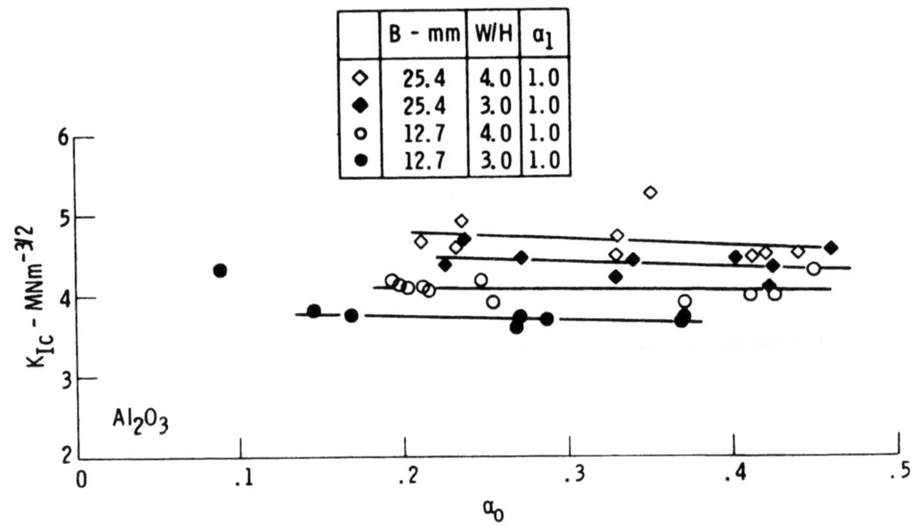

Figure 7. Effect of α_o on K_{Ic} of sintered aluminum oxide (Alsimag-614) determined with short bar chevron-notch specimens.

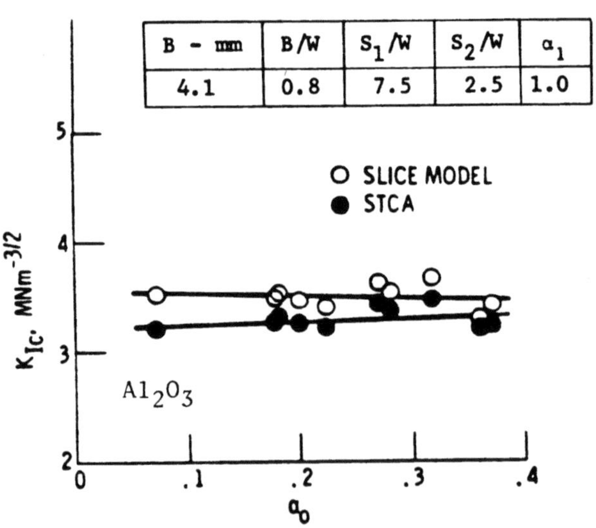

Figure 8. Effect of α_o on K_{Ic} of sintered aluminum oxide (Alsimag-614) determined with four-point-bend chevron-notch specimens of α_1 = 1 and calculated using both the slice model of Bluhm and the straight-through crack assumption.

Table 2. Summary of plane strain fracture toughness results obtained with chevron-notch specimens.

MATERIAL	SPECIMEN TYPE	SPECIMEN ORIENTATION (1)	NUMBER OF TESTS	B mm	W mm	W/H	S_1/W	S_2/W	α_0 (2)	α_1 (2)	K_{Ic} $MNm^{-3/2}$
Si_3N_4	Short Bar	Parallel	13	8.8-9.6	15.2	3.4			0.2-0.5	1.0	4.64
	Short Bar	Parallel	7	9.6	15.2	3.4			0.2	0.5-1.0	4.72
	Short Bar	Perpendicular	9	9.0	13.3	3.0			0.2-0.4	1.0	4.71
	4-Pt Bend	Perpendicular	2	9.0	12.8		3.1	0.8	0.2	1.0	4.83
	4-Pt Bend	Perpendicular	2	3.5	5.0		4.0	2.0	0.2	1.0	4.85
Al_2O_3	Short Bar		9	25.4	50.8	4.0			0.2-0.4	1.0	4.71
	Short Bar		9	25.4	38.1	3.0			0.2-0.5	1.0	4.43
	Short Bar		11	12.7	25.4	4.0			0.2-0.5	1.0	4.10
	Short Bar		8	12.7	19.1	3.0			0.2-0.4	1.0	3.74
	Short Bar		12	12.7	25.4	4.0			0.2	0.4-1.0	3.07-4.13
	4-Pt Bend		10	4.1	5.1		7.5	2.5	0.1-0.4	1.0	3.49
	4-Pt Bend		7	4.0	5.0		4.0	2.0	0.2	1.0	3.56
	4-Pt Bend		5	8.0	10.0		4.0	1.0	0.2	1.0	3.60

(1) Relation between crack plane and plate surface.
(2) Values rounded off to nearest tenth.

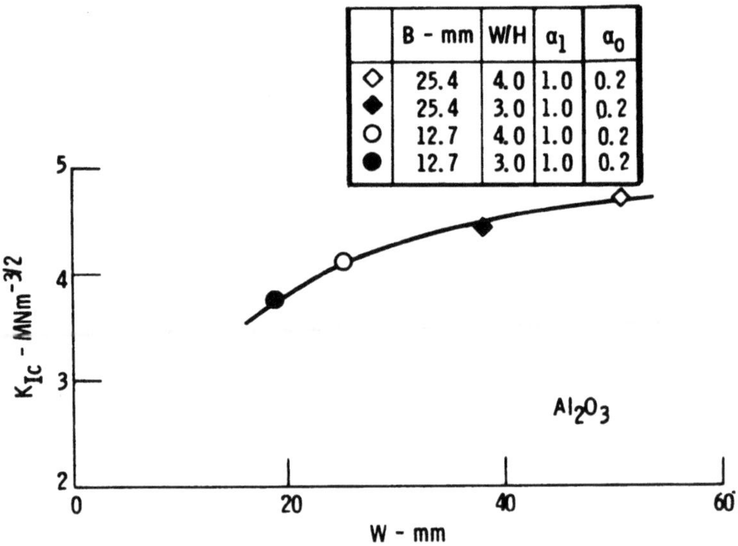

Figure 9. Effect of specimen width on K_{Ic} of sintered aluminum oxide (Alsimag-614) determined with short bar chevron-notch specimens.

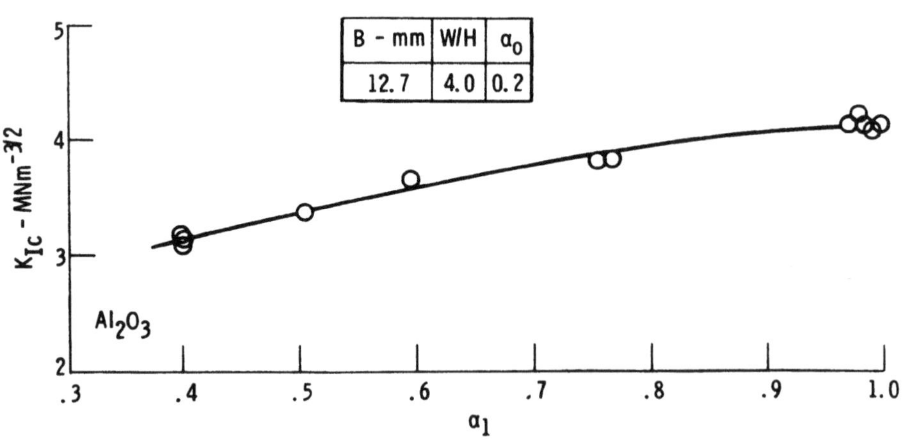

Figure 10. Effect of α_1 on K_{Ic} of sintered aluminum oxide (Alsimag-614) determined with short bar chevron-notch specimens.

the observed K_{Ic} variation, small 12.7 mm thick short bar specimens were machined from the center of the 25.4 mm square sintered blanks. The K_{Ic} for these specimens ranged from 4.02 to 4.08 $MNm^{-3/2}$, agreeing well with the 4.10 $MNm^{-3/2}$ value obtained from the same size specimens machined from the smaller 12.7 mm sintered blanks. Therefore, material variation was not a factor in the variation of K_{Ic} for the aluminum oxide.

The stress state varies along the crack front from one of plane strain at the center to plane stress at the specimen surface. This transition might be expected to influence the computed K_{Ic}. However, this stress state transition would be expected to produce a trend opposite to that observed (26). Therefore, this effect cannot explain the observed K_{Ic} variation.

While most ceramic materials are considered to have flat crack growth resistance curves, some may not (25-27). Since crack extension at maximum load is proportional to specimen size, a rising crack growth resistance curve would cause the computed K_{Ic} to increase with increasing specimen width. This is in agreement with the results shown in Figure 9. Therefore, it may be concluded that the variation in computed K_{Ic} observed for aluminum oxide is due to a rising crack growth resistance curve for this material.

REFERENCES

1. W. G. Clark and W. A. Logsdon, Fracture Mechanics of Ceramics, Vol 2, R. C. Bradt, D. P. H. Hasselman, and F. F. Lange, eds., Plenum, New York, 1974, pp. 843-861.

2. R. W. Davidge and A. G. Evans, "The Strength of Ceramics", Mater. Sci. Eng. 6 (1970), pp. 281-298.

3. F. F. Lange, "Relation Between Strength, Fracture Energy, and Microstructure of Hot-Pressed Si_3N_4", J. Am. Ceram. Soc., 56 (1973), pp. 518-522.

4. L. A. Simpson, "Use of the Notched-Beam Test for Evaluation of Fracture Energies of Ceramics", J. Am. Ceram. Soc. 57:4 (1974), pp. 151-154.

5. G. Berry, "The Fracture-Toughness Testing of Cemented Carbides", Metal. Sci., 10 (1976), pp. 361-366.

6. L. A. Simpson, T. R. Hsu, and G. Merrett, "Application of the Single-Edge Notched Beam to Fracture Toughness Testing of Ceramics", J. Test. Eval. 2:6, (1974), pp. 503-509.

7. J. L. Henshall, D. J. Rowcliffe and J. W. Edington, "Fracture Parameters in Refel Silicon Carbide", J. Mater. Sci., 9:9 (1974), pp. 1559-1561.

8. R. L. Bertolotti, "Fracture Toughness of Polycrystalline Al_2O_3", J. Amer. Ceram. Soc., 56:2 (1973), p. 107.

9. J. L. Cermant, A. Deschanvres, and A. Iost, Fracture Mechanics of Ceramics, Vol. 1, R. C. Bradt, D. P. H. Hasselman, and F. F. Lange, eds., Plenum, New York (1974), pp. 347-366.

10. R. F. Pabst, "Fracture Mechanics of Ceramics", Vol. 2, R. C. Bradt, D. P. H. Hasselman, and F. F. Lange, eds., Plenum, New York (1974), pp. 555-565.

11. N. Claussen, R. Pabst and C. P. Lahmann, "Influence of Microstructure of Aluminum Oxide and Zirconium Oxide on K_{Ic}", Proc. British Ceramic Soc., 25 (1975), pp. 139-149.

12. L. M. Barker, "A Simplified Method for Measuring Plane Strain Fracture Toughness", Eng. Fracture Mech., 9:2 (1977) pp. 361-369.

13. J. Nakayama, "Direct Measurement of Fracture Energies of Brittle Heterogeneous Materials", J. Am. Ceram. Soc., 48:11 (1965), pp. 583-587.

14. H. G. Tattersall and G. Tappin, "The Work of Fracture and Its Measurement in Metals, Ceramics and other Materials", J. Mater. Sci., 1:3 (1966), pp. 296-301.

15. R. W. Davidge, and G. Tappin, "The Effective Surface Energy of Brittle Materials", J. Mater. Sci., 3:2 (1968), pp. 165-173.

16. L. A. Simpson, "Effect of Microstructure on Measurements of Fracture Energy of Al_2O_3", J. Am. Ceram. Soc., 56:1 (1973), pp. 7-11.

17. D. J. Green, P. S. Nicholson, and J. D. Embury, "Fracture Toughness of a Partially Stabilized ZrO_2 in the System CaO, ZrO_2", J. Am. Ceram. Soc., 56:12 (1973), pp. 619-623.

18. H. Hubner, "The Determination of the Specific Work of Fracture of Two Cemented Carbide Alloys in Controlled Fracture Experiments," Z. Metallkd, 67:8 (1976), pp. 507-513.

19. D. Munz, R. T. Bubsey and J. L. Shannon, Jr., "Fracture Toughness Determination of Al_2O_3 Using Four-Point-Bend Specimens with Straight Through and Chevron Notches", to be published in J. Am. Ceram. Soc., 1980.

20. J. I. Bluhm, "Slice Synthesis of a Three Dimensional 'Work of Fracture' Specimen -- For Brittle Materials Testing", Eng. Fracture Mech., 7 (1975), pp. 593-604.

21. J. I. Bluhm, 4th International Conference on Fracture, D. M. R. Taplin, ed., Univ. Waterloo Press, Waterloo, Ontario (1977), pp. 409-417.

22. D. Munz, R. T. Bubsey and J. E. Srawley, "Compliance and Stress Intensity Coefficients for Short Bar Specimens with Chevron Notches Useful for Fracture Toughness Testing of Ceramics", to be published in Int. J. Fracture, 1980.

23. R. T. Bubsey, D. M. Fisher, M. H. Jones and J. E. Srawley, "Compliance Measurements", Experimental Techniques in Fracture Mechanics, Soc. for Experi. Stress Anal. Monograph 1, A. S. Kobayashi, ed., Iowa State Univ. Press and Soc. for Experimental Stress Analysis, Cambridge (1976), pp. 76-95.

24. J. E. Srawley, and B. Gross, Cracks and Fracture, ASTM STP 601, Amer. Soc. for Testing & Materials, Philadelphia (1976), pp. 559-579.

25. D. Munz, R. T. Bubsey and J. L. Shannon, Jr., "Fracture Toughness Calculations from Maximum Load in a Four-Point-Bend Test with Chevron Notched Specimens", to be published.

26. D. Munz, R. T. Bubsey, and J. L. Shannon, Jr., "Performance on Chevron-Notch Short Bar Specimen in Determining Fracture Toughness of Si_3N_4 and Al_2O_3", to be published in J. Test. Eval., May 1980.

27. H. Hubner, and W. Jillek, "Subcritical Crack Extension and Crack Resistance in Polycrystalline Alumina", J. Mat. Sci. 12 (1977), pp. 117-125.

OXIDATION BEHAVIOUR AND MECHANICAL PROPERTIES OF INJECTION MOLDED Si_3N_4

Ulf Dworak, Hans Olapinski

Feldmuehle Aktiengesellschaft
Fabrikstrasse 23 - 29
7310 Plochingen

INTRODUCTION

The present work deals with the improvement of oxidation behaviour and mechanical properties of injection molded and reaction-bonded silicon-nitride (RBSN). It was found that homogeneous RBSN with small micropores exhibits excellent oxidation resistance and mechanical strength after long-term heat treatment.

CONTENT

1. Oxidation and creep behaviour of RBSN.
2. Preparation and characterisation of investigated material.
3. Porosity-oxidation-creep correlation.
4. RT-strength of investigated RBSN.
5. RT-strength after thermal treatment.
6. RT-strength after long term thermal treatment in air.
7. Summary

1. OXIDATION AND CREEP BEHAVIOUR OF RBSN

It is known, that the oxidation resistance and mechanical high temperature properties of commercial RBSN still need to be improved for high performance applications as for instance gas turbine components.

Thümmler[1,2] and colleagues investigated the high temperature creep of RBSN in vacuum and air (Fig. 1). He demonstrated, that the observed creep in air is due to internal oxidation.

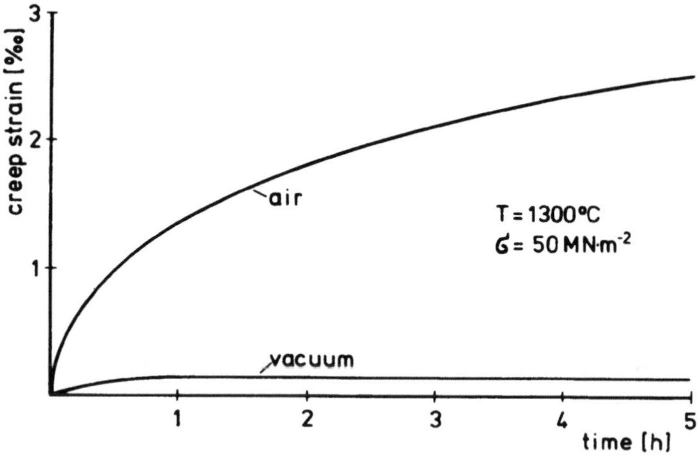

Fig. 1 - Creep of RBSN in vacuum and air

Furthermore he could show, that the differences in the microporosity of various RBSN result in different creep rates and that in conclusion the amount of internal oxidation depends on the pore diameter distribution (Fig. 2).

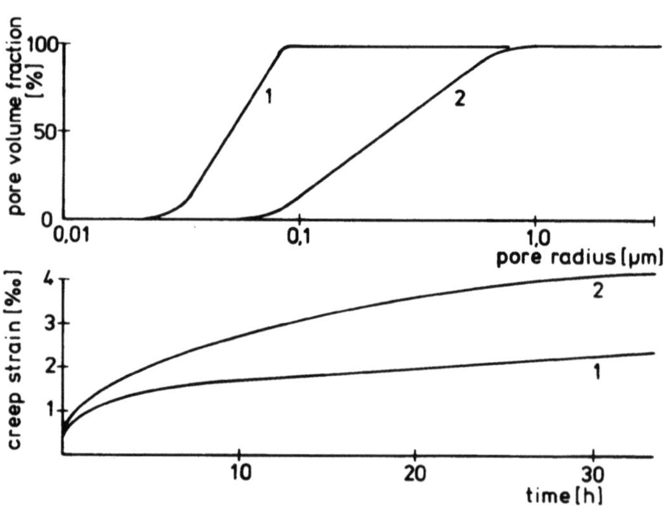

Fig. 2 - Correlation of pore size and creep strain for RBSN

Following these results, it was our objective to reduce the radii of the microporosity drastically and to narrow the range of porosity distribution.

2. PREPARATION AND CHARACTERISATION OF THE INVESTIGATED MATERIAL

In order to realize this goal, we carried out a far-reaching parameter variation of factors influencing the micro-porosity, which included:

- increase of final density by increasing the green density
- variation of the particle size distribution of the starting powder (1,5 to 5μm)
- modifying the nitridation cycle using thermal gravimetric attachments (96% - 95% nitridation).

All test specimens had been injection molded and reaction sintered. The dimensions were 3,5 x 4,5 x 60 mm.

The material is characterized as follows:

- final density	:	2,45 - 2,58 g/ccm
- crystallography	:	70 - 80% α-Si_3N_4
- free silicon	:	5%
- total impurities	:	1%

As a result of the different preparation parameters we obtained very distinct microstructures after nitridation, as shown in Fig. 3 and Fig. 4.

The micrographs taken by means of optical microscopy show a broad variation of pore size and distribution at constant densities. Of course the total porosity is higher and reaches in the first case (Fig. 3) 19%, in the second case 23% (Fig. 4).

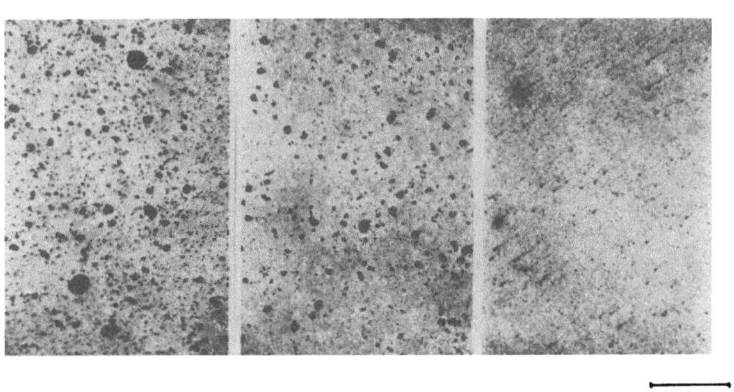

Fig. 3 - Porosity (optical micrograph) of various RBSN's with a density of 2,58 g/ccm

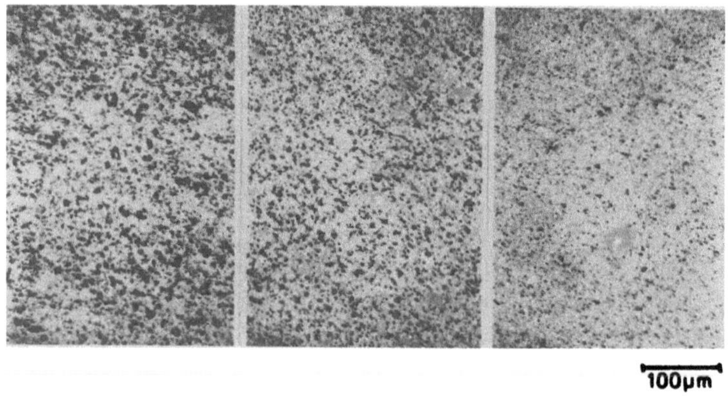

Fig. 4 - Porosity (optical micrograph) of various RBSN's with a density of 2,45 g/ccm

3. POROSITY-OXIDATION-CREEP CORRELATION

The pore fractions with diameters above 1 µm in no case correlate to the oxidation behaviour of the material. (However the correlation to the fracture strength is very good and will be discussed later).

The internal oxidation of the material depends not on the size of the pores but on the diameter of the connecting channels which build-up the so called open porosity. These "channel-diameters" in size much below 1 µm are quite accurately detected by the mercury-porosimeter.

The pore-size distribution of various RBSN-ceramics from different suppliers and our new development materials are shown in Fig. 5. The measurement had been carried out in the "Institut für Werkstoffkunde II" of the University of Karlsruhe by Prof. Thümmler and colleagues. Curves A and B represent a material with a remarkably smaller porosity. The long-term oxidation behaviour of the corresponding samples A and B was studied at 900°C and 1260°C in stationary air, as shown in Fig. 6. As expected, sample A with a very low pore size and a narrow distribution - 95% of all pore radii < 0,017 µm - has excellent oxidation resistance compared to sample B with 95% of all pores radii < 0,03 µm.

As has been observed by other investigators, the oxidation in the case of sample B at 900°C is more severe compared to 1260°C. Because of its unique microstructure sample A is subject to almost no oxidation even at 900°C. Creep experiments at 1300°C have shown, that specimen A and B exhibit no measurable deformation.

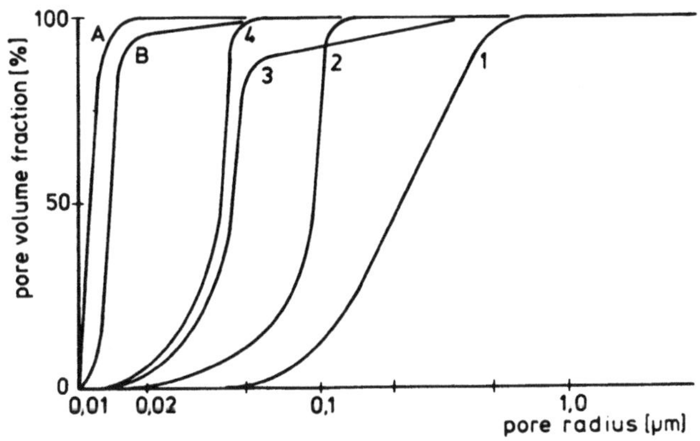

Fig. 5 - Micropore-size distribution of various RBSN's (Hg-porosimeter-measurements)

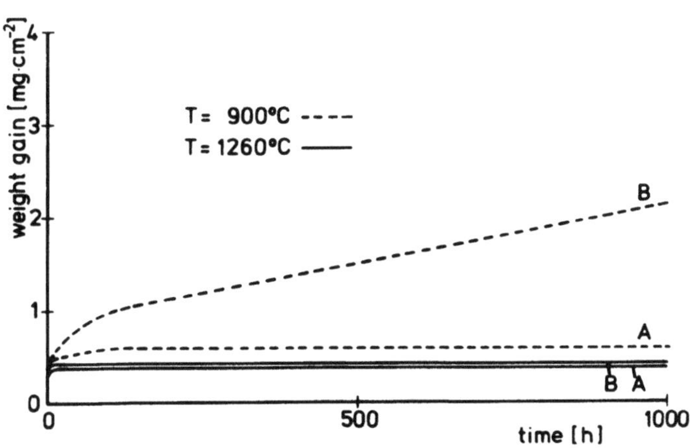

Fig. 6 - Oxidation behaviour of RBSN sample A and B with distinct microporisity (see Fig. 5)

Fig. 7 demonstrates a drastic decrease in creep strain with decreasing pore size, as shown in Fig. 5. In order to differentiate between samples A, B, 3, and 4, a preoxidation at critical conditions with respect to internal oxidation (see Fig. 6) will be carried out in the near future.

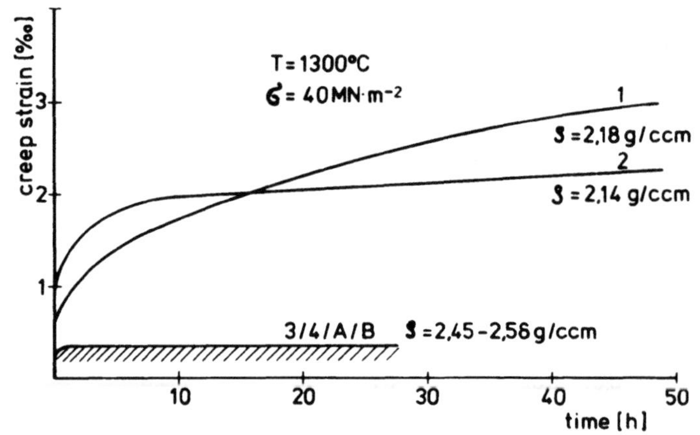

Fig. 7 - Creep behaviour of various RBSN's (see Fig. 5)

4. RT-STRENGTH OF INVESTIGATED RBSN

The bending strength at room temperature of RBSN depends on the total porosity and its stereological arrangement. Larger pores as can be seen by optical microscopy play a dominant role. At a constant density, but different porosity in the pore fraction above 1 μm, an increase in strength with the pore size reduction within this fraction is observed. (Fig. 8 and Fig. 9).

Even at a higher total porosity (g = 2,45) it is possible to establish a higher RT-fracture strength when the pores are smaller than in the case of larger pores at lower total porosity.

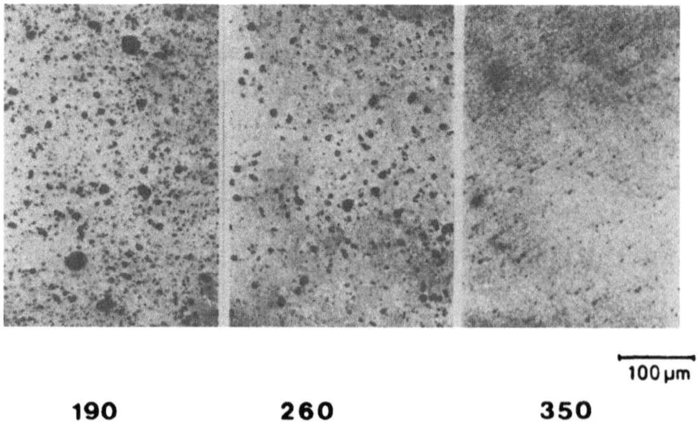

Fig. 8 - Porosity-strength correlation of various RBSN's with a density of 2,58 g/ccm.

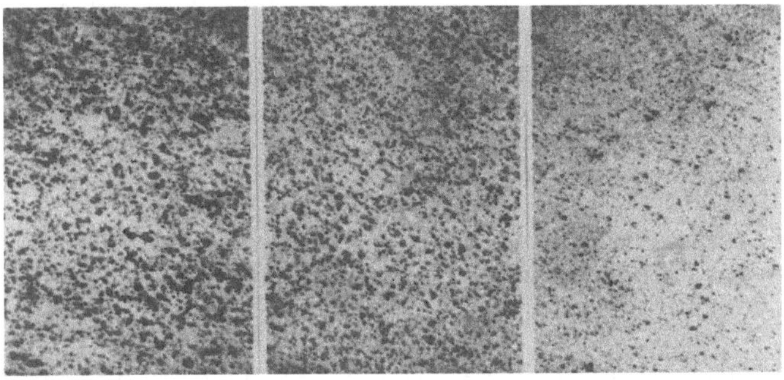

Fig. 9 - Porosity-strength correlation of various RBSN with a density of 2,45 g/ccm

5. RT-STRENGTH AFTER THERMAL TREATMENT

We observed an increase of the RT-bending strength up to 25% after a short term heat treatment between 1300°C and 1400°C for 1 hour in the case of homogeneous materials, that means materials with a bending strength of above σ_β = 2250 MN/m^2 (Fig. 10).

as nitrided material; ground surface
ρ = 2,58 g/ccm
σ_B = 236 MN m^{-2}

thermal treatment 1350°C/1h in	RT-strength [MN m^{-2}] thermal treatment after grinding	before grinding
air	307	243
argon	291	217
vacuum	269	224
N_2 : H_2/90:10	298	254

Fig. 10 - RT-strength of RBSN after thermal treatment

This strengthening is limited to RBSN's with small pore diameters and independent of the materials bulk density. Furthermore strengthening was observed in oxidizing and inert atmospheres (Air and vacuum) which suggests healing effects. The thermal strengthening is lost by subsequent material removal through surface grinding. In conclusion the strength increase is a pure surface effect.

6. RT-STRENGTH AFTER LONG TERM THERMAL TREATMENT IN AIR

It is well known, that all commercial RBSN materials exhibit a drastic strength deterioration after long-term heating in air. This strength decrease occurs after few hours up to 1000 h depending on the material.

In contrast much better mechanical properties after long-term heating were observed, when the initial material again had a bending strength above 220 MN/m^2.

Fig. 11 summarizes the results of RT bending strength as a function of preheating time in air at 900^oC.

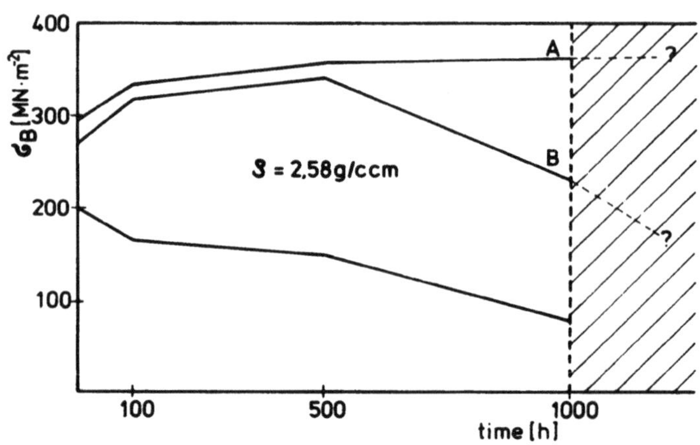

Fig. 11 - RT-strength of various RBSN's after heating in air at 900^oC

Curve A corresponds to RBSN-sample A shown previously in Fig. 5 with a low pore size and narrow distribution where 95% of all pre radii are < 0,017 μm. No strength deterioration can be observed after 1000 hours heating time. Sample B with 95% of all pore radii < 0,03 μm according to the mercury-porosimetry (Fig. 5) exhibits after an initial strength increase a significant drop of strength values, due to a more severe internal oxidation, which generally takes place with RBSN materials.

The amount and depth of oxidation from the surface to the interior depends on time, temperature, and on the pore size diameters of the investigated material. In order to prove experimentally this direct correlation of internal oxidation and material strength after long term heating experiments we plan to carry out creep tests on just these pretreated samples. We expect no measurable creep strain for specimen A after the 1000 h air baking at 900^oC.

SUMMARY

Following the results from Thümmler et al who studied the relations between microporosity and internal oxidation, we developed a new RBSN with excellent oxidation resistance, and high mechanical strength after long term heat treatment under critical conditions.

In the present paper we demonstrated the correlation between material properties, especially its microstructure, and long term mechanical strength which is a fundamental requirement for the application and reliability of RBSN-components in gas turbine engines.

REFERENCES

1) Thümmler, F., Porz, F., Gratwohl, G., and Engel, W., Science of Ceramics 8, 1975.
2) Gratwohl, G., Thümmler, F., Kriechen von Si_3N_4 unter oxidierenden und nichtoxidierenden Bedingungen, Ber.Dt.Keram.Ges. 52 (1975) 268 - 270.
3) Siebels, G., Volkswagenwerk, Entwicklung von keramischen Turbinenbauteilen für Automobilgasturbinen. 2. Fortschrittsbericht Forschungsvorhaben NTS 1006, 1978.

SESSION VI

CRITICAL ISSUES WORKSHOP

Chairman: Prof. J. Mueller
University of Washington

COMMENTS ON CREEP OF REACTION-BONDED SILICON NITRIDE AND HOT-PRESSED SILICON CARBIDE

George Grathwohl

Institut für Werkstoffkunde II

Universität Karlsruhe
D-7500 Karlsruhe, Germany

In recent years studies concerning creep of reaction-bonded silicon nitride (RBSN) (1) and different silicon carbide materials (2) and the production and high-temperature properties of hot-pressed silicon nitride (HPSN) (3) were made in our institute. The early results were partly published at the Newport-conference (4) and elsewhere, e.g. (5,6,7). Some comments on creep were given during this conference relating to material parameters and properties of RBSN and hot-pressed silicon carbide (HPSC) discussed in other conference papers.

The influence of oxidation on the creep behaviour of RBSN as previously reported (4,5) has been further demonstrated in recent experiments. Fig. 1 shows the synopsis of two typical experiments under identical thermal and environmental conditions, one performed in a thermobalance system and the other in a creep machine. The bending creep specimen was loaded to an outer fiber stress of 70 MN/m^2 before heating the system with a low heating rate ($1^oC/min$). The thermogravimetric experiment shows the oxidation taking place mainly during the heating period. After this period the pore channels of RBSN were closed and the surface was sealed by an oxide layer. Any further oxidation proceeds very slowly and is diffusion-controlled through the SiO_2-surface layer. During the period of the high oxidation rate a relatively high creep rate (10^{-4} h^{-1}) was measured especially regarding the low temperature around 1000^oC. The onset of creep was found at a temperature as low as 800^oC. This temperature also represents the onset of the oxidation. However, this oxidation assisted creep at 800^oC is restricted to in-situ-oxidation and may be explained by the atomic mobility due to the transition of atoms into a new structure. In further experiments the dependence of creep on the heating rate,

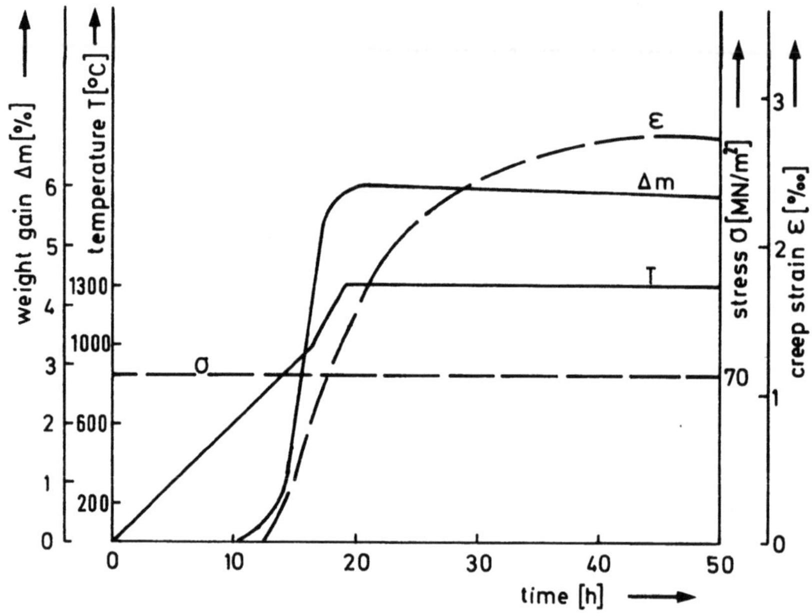

Figure 1: Weight gain by oxidation and creep of RBSN under identical thermal conditions.

any thermal pretreatment, and on material parameters as pore sizes has been shown.

In long-term tests and stress-change and temperature-change experiments (Fig. 2) the influence of stress and temperature on creep of RBSN was evaluated (1). Also strain recovery was observed in these experiments after stress reduction as reported recently for HPSN (8). In the experiment in Fig. 2 it can be seen that ∿6% of the previous creep strain is recovered after a stress reduction from 100 MN/m^2 to 50 MN/m^2 at 1400°C. The recovery rate decreases over a period of 15 h followed by "zero creep" and no "positive" creep rate is regained for the next 50 h. Preliminary results show an influence of temperature, ratio of stresses before and after stress reduction and of the creep strain in the recovery behaviour. As discussed for HPSN (9) this behaviour reflects the viscoelastic deformation mode of these materials, which can also be assumed for RBSN. A further point to explain the transient creep rates after stress changes is to be found in the stress distribution over the cross-section of the bending specimen. With the stress exponent n larger than unity ($\dot{\varepsilon} \sim \sigma^n$) any loading or unloading of the specimen leads to a time-dependent stress redistribution resulting in an non-stationary component of the creep rate.

The creep rupture strain of RBSN correlates in an unusual way with the creep rate. Most commonly a constant rupture strain independent of temperature and stress is expected according to the Monkman-Grant-relation which has also been observed for RBSN (10). This could not be confirmed in our investigation. The measured rupture strains differ widely depending on test conditions and materials quality. It is interesting to see that a material with a lower creep resistance typically allows a higher creep rupture strain and vice versa. The maximum creep strain of a highly creep resistant RBSN is very limited and lies around $1 \cdot 10^{-3}$. Higher rupture strains are achievable with large amounts of intergranular phases. This is expected and explained on the basis of the active creep mechanism which is thought to be a grain boundary separation process (11) accompanied by the relative motion of the elastic grains and by deformation of the original pores. Two adjacent grains with a thick glassy boundary sheet will then show a lower creep strength and sustain a larger strain before this boundary fails. By this way the deformation process implies a progressive destruction of the material which could be expressed in terms of increasing pore sizes. In Fig. 3 the modification of the original pore size distribution in creep specimens (B,C,D) is shown. With higher creep rates and increasing creep strain the frequency of relatively small and large pores increases and a bimodal distribution is found. Up to now the association of these two frequency maxima to the tension (large pores) and compression (small pores) parts of the bending bar has not been proved but

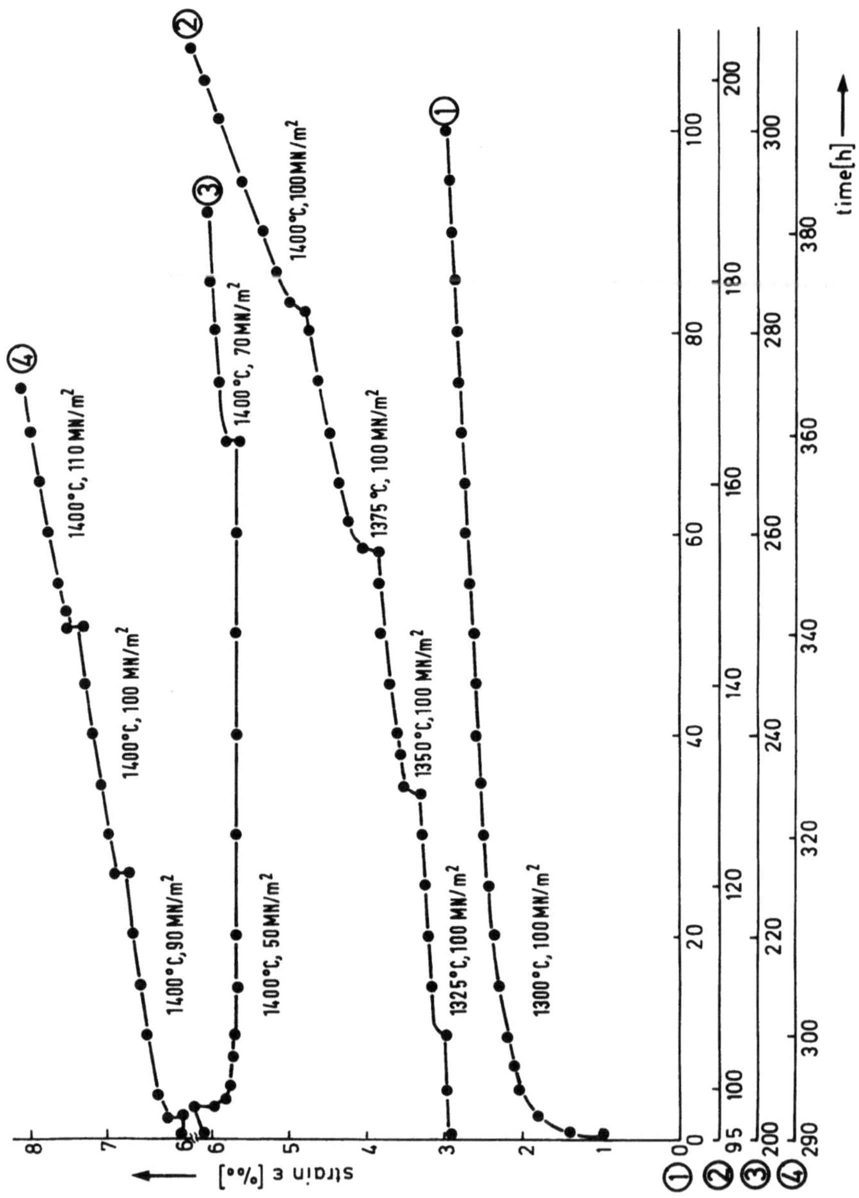

Figure 2: Creep curve of RBSN with temperature and stress change experiments.

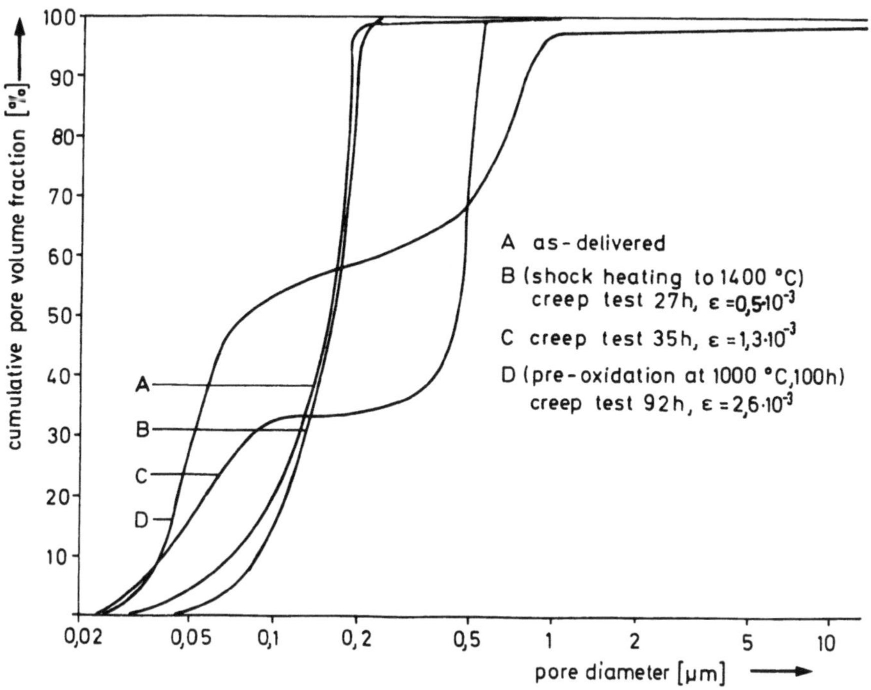

Figure 3: Pore size distribution of RBSN in the as-delivered state (A) and in creep specimens (B,C,D) after creep tests at 1300°C, 70 MN/m^2.

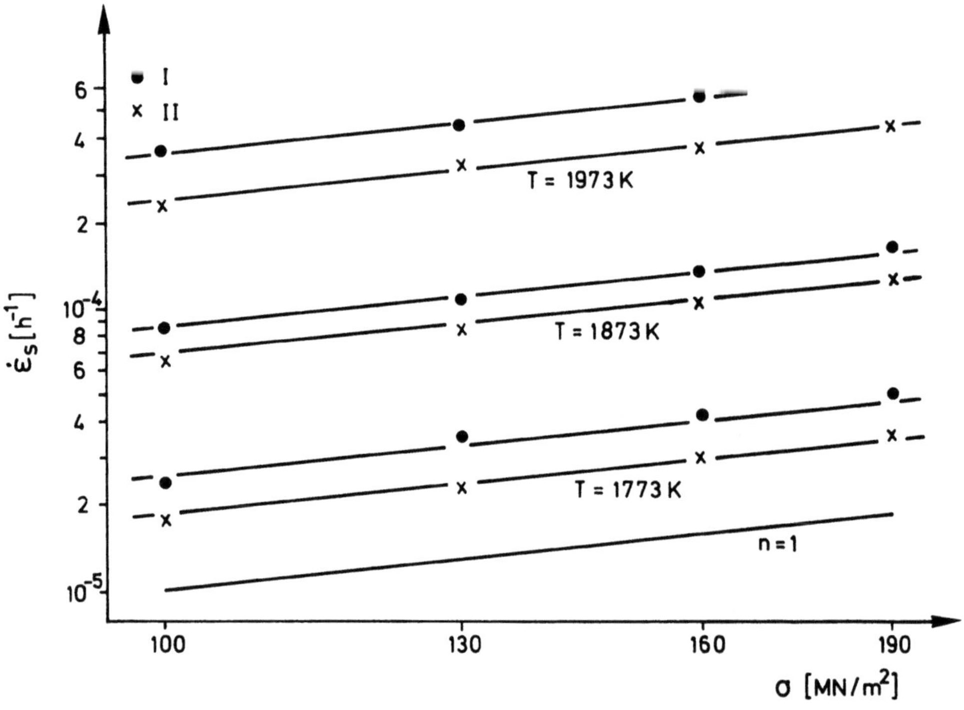

Figure 4: Stress dependence of stationary creep rate of two HPSC materials.

seems to be possible. Eventually specimen failure will occur when the cross-section reaches a critical number of "broken" boundaries or the pore size extends up to a critical value.

With HPSC we investigated the creep behaviour (2,7) of two materials which are characterized in another paper during this conference (12). These two HPSC materials differ from each other by the fracture mode which is intergranular in the material I with a higher impurity content and transgranular up to high temperatures in the material II with a reduced impurity content. The creep behaviour of both materials in vacuum is shown in Fig. 4. Besides a slightly lower creep rate of the material with higher purity no considerable difference in the vacuum creep behaviour was found. According to the measured stress exponent ($n = 1$) a diffusion controlled creep mechanism was determined; a more detailed presentation is given in (2).

In contrast to these results in vacuum considerable differences between both materials were found in the creep experiments in air. Oxidation rates measured by weight change were ten times higher in materials I than in air higher creep rates, an increase of the stress exponent into the region of 2 and a considerable loss of strength compared to the vacuum experiments, while with material II no change at all was observed.

References

(1) Grathwohl,G., "Creep of reaction singered silicon nitride", KfK report 2675, 1978, also: Ph.D. dissertation, Universität Karlsruhe, 1978.

(2) Schnürer,K., "Creep of different SiC-materials in vacuum and in air", Ph.D. dissertation, Universität Karlsruhe, 1979.

(3) Becker,R., "The influence of process parameters and additives on the densification, creep properties and strength of hot-pressed silicon nitride", KfK report 2771, 1979, also: Ph.D. dissertation, Universität Karlsruhe, 1979.

(4) Grathwohl, G. and Thümmler, F., "Creep of reaction bonded silicon nitride", Ceramics for high performance applications - II, ed. by J.J. Burke, E.M. Lenoe and R.N. Katz, Brook Hill Publ.Co., Chestnut Hill, MA, 1979, pp. 573-591.

(5) Grathwohl,G. and Thümmler,F., "Creep of reaction bonded silicon nitride", J. Mater. Sci. 13, 1177-1186 (1978).

(6) Becker,R. and Thümmler,F., "Hot pressed Si_3N_4 with very low amount of binder phase", 4th CIMTEC, St.Vincent, May 28/June 1,1979.

(7) Schnürer, K., Thümmler,F. and Grathwohl,G., "Creep of different SiC-materials", Science of Ceramics 10, Berchtesgaden, Sept. 2 - 4, 1979.

(8) Arons,R.M., Tien,J.K. and Lenoe,E.M., "Observation of visco-elastic strain recovery in hot-pressed silicon nitride", J. Am. Ceram. Soc. 62, 216 f (1979).
(9) Arons,R.M., "Creep and strain recovery in silicon nitride", Ph.D. dissertation, Columbia University 1979.
(10) Ud Din,S. and Nicholson, P.S., "Creep deformation of reaction-sintered silicon nitrides", J. Am. Ceram. Soc. 58, 500-502, (1975).
(11) Lange,F.F., "Non-elastic deformation of polycrystals with a liquid boundary phase", Deformation of ceramic materials, ed. by R.C. Bradt, R.E. Tressler, Plenum Press, New York, 1975, pp. 361-381.
(12) Kriegesmann, J., Lipp, A., Reinmuth, K. and Schwetz, K.A., "Strength and fracture toughness of silicon carbide", this volume.

OXIDATION AND STRENGTH OF SILICON NITRIDE AND SILICON CARBIDE

Johann E. Siebels

Research and Development
Volkswagenwerk AG,
Wolfsburg, West Germany

ABSTRACT

Ceramic components in a gas turbine have to operate in an oxidizing atmosphere at high temperatures. Nonoxides like Si_3N_4 and SiC, easily oxidizing materials, have to withstand this attack for long times without changing their characteristics.

The results of laboratory investigations on oxidation resistance and long time stability of Si_3N_4 and SiC are described. Recent materials show a satisfactory resistance against oxidation. Nevertheless, all materials investigated show changes in strength after annealing. Durability problems of ceramic components may arise from the strength degradation behaviour shown by nearly all materials at certain temperature levels.

INTRODUCTION

The gas turbine puts through several time the amount of air being necessary for combustion. Therefore all components in the hot gas steam are exposed to a heavy oxidation attack. Due to the position of the component the temperatures at which the oxidation occurs partially are differing very much. Furthermore, the oxidation is not generally isothermal. Temperature changes occur not only during the starting and shutdown cycle but also during the driving cycle.

At present, nonoxide ceramic materials like silicon nitride as reaction bonded (RBSN), hot pressed (HPSN) and sintered or pressureless sintered (SSN, PSSN) Si_3N_4 and silicon carbide of various fabrication are being investigated for high temperature application in a small gas turbine. In general these materials can be oxidized easily. the application in oxidizing atmosphere under high temperatures is possible only because of a rapid formation of a SiO_2 protection layer. This layer, if not damaged or removed by any environment, stops or at least slows down further oxidation substantially.

Besides the transformation of Si_3N_4 and SiC into SiO_2 depending on the kind and amount of additives converting into silicates of various composition, an influence on other material characteristics can be expected. These changes, for instance of the strength may be of positive or negative.

Ceramic components in a passenger car gas turbine have to survive more than 3500 hours with several 10,000 lead cycles. This means the materials utilized have to remain stable for a long time in their characteristics being basic for the design of a part.

Early results of long-time tests of silicon nitrides under conditions similar to operation were reported in 1, 2. Those tests have been continued and extended on all ceramic materials being candidates of application in a small gas turbine.

TEST CONDITIONS

The oxidation test conditions were chosen to represent the temperature schedule of an all ceramic turbine rotor. Figure 1 shows the different temperatures cycles applied to the specimens. The 1260° level represents the average maximum temperature in the middle portion of a blade, 900°C the average operation temperature in the rim.

Cycle I is an isothermal test. The time at which specimens are removed from the furnace are marked.

In Cycle II the test is interrupted daily after 20 hours heat soak.

Cycle III combines both temperature levels with different heat soak times derived from a mixed driving cycle. Three 2-hr. cycles with 1260°C are followed by a 16-hr. 900°C cycle. Loading and unloading the furnace causes moderate thermal shocks to the test bars. All heat treatment is done in electric furnaces in air.

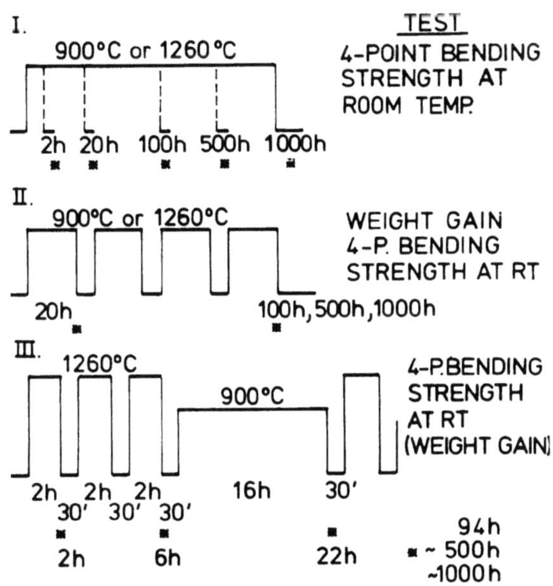

Figure 1. Oxidation cycles for testing ceramics for turbine applications.

Oxidation resistance by means of weight gain per geometric surface area is determined with cycle II.

The evaluation of the long-time-stability of the ceramic materials is done by measurement of 4-point-bending-strength at room temperature after the different exposure times marked in figure 1 in the cycles I, II and III.

TEST RESULTS

RBSN. Most of the investigations up to now were done on the oxidation of RBSN, isostatically pressed and by preference injection molded.

The results of cyclic oxidation (cycle II) of RBSN are shown in figure 2. The weight gain curves cover a wide range of values for both temperatures, 900 °C and 1260 °C. More recent materials however show substancial improvements in oxidation resistance. In general they do not exceed 2 mg/cm^2 weight gain after 1000 hours oxidation time, where hot gas exposure times are counted only. Some injection molded materials of group A show attrition when the test temperature is changed from 1260 °C to 900 °C after long oxidation times.

As a more important fact the effects of oxidizing heat treatment on the strength have shown. Former materials had a drop of strength preferably at the 900 °C level and remained stable at 1260 °C. Improved RBSN with high oxidation resistance and a high initial strength (300 ... 350 N/mm^2, 4-point-bending) show stability or increase of strength now at 900 °C but strength degradation at 1260 °C. Figure 3 shows three selected materials tested according to figure 1. Material A is a RBSN of the early state of development. RBSN B and C are more recent materials. The most promising material for application in a gas turbine is material C, because it shows a strength in the order of 250 N/mm^2 after all annealing cycles.

That the effects on the strength are not only due to the oxidation could be shown by removing the oxidized specimen surface layers down to the bulk material prior to the flexure test (see RBSN B, figure 3). Annealing in vacuum or in a Si_3N_4 powder bed as an oxygen getter shows slightly different results, but those results are not comparable because of changes and sintering effects in the surface area especially at 1260 °C. But this point is of academic interest only as components in a heat engine have to operate in an oxidizing atmosphere anyway.

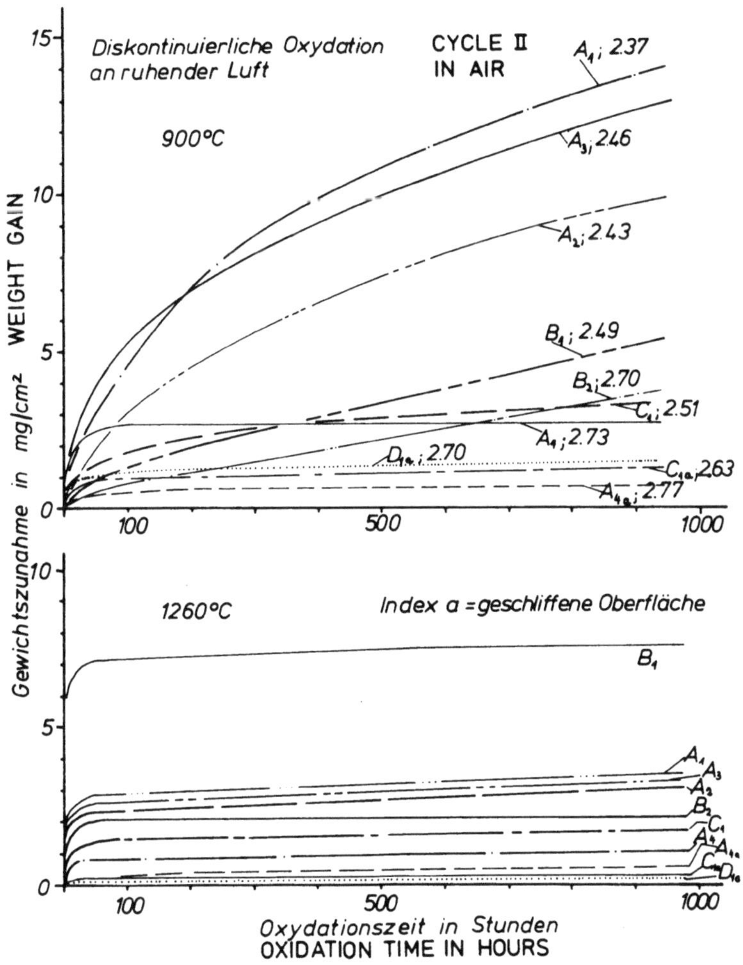

Figure 2. Oxidation of reaction bonded silicon nitride.

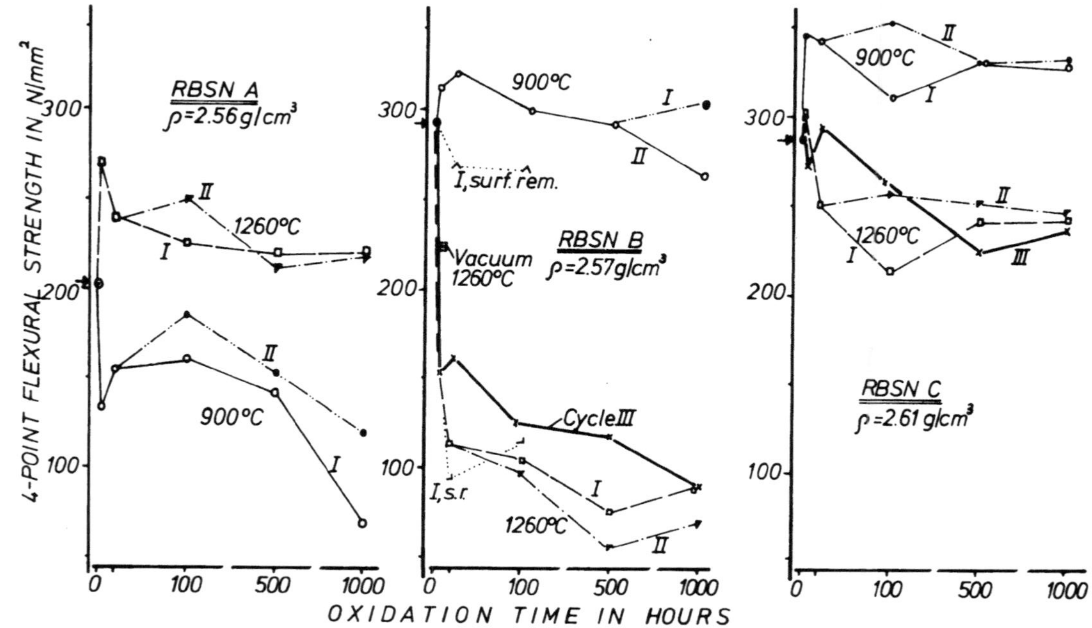

Figure 3. Room temperature strength of injection molded silicon nitride after annealing in air.

An explanation for the strength effects described above has not yet been found. Investigations on density, porosity, microstructure, phase analysis, chemical composition and oxygen content, among others, did not give a clear answer what the reason for this behaviour is.

As an optimum density for getting high strength RBSN $\rho = 2.5 \ldots 2.6$ g/cm³ was found.

With a limited number of test bars it could be demonstrated, that at least some RBSN materials have the same strength behaviour in high temperature strength after longer periods of annealing as they show for room temperature strength.

DENSE SILICON NITRIDE (HPSN, PSSN). Dense silicon nitrides have negligible oxidation rates at 900 °C (see figure 4, top) with weight gains less than 0,1 ... 0.2 mg/cm² after 1000 hours. At 1260 °C two groups of hot pressed materials are found. The different oxidation rates are relatively independent of additives and impurities, but a correlation to the manufacturers is evident.

The shown pressureless sintered Si_3N_4 of a density of 3.1 g/cm³ has due to its high content of additives comparably high oxidation rates at 1260 °C that exceed even those of RBSN.

At 900 °C the strength of HPSN is uninfluenced even after long times of cyclic oxidation (bottom of figure 4). After annealing at 1260 °C (any cycle) all sorts of HPSN show the same strength degradation within a narrow data band. A removal of the oxide layer raises the strength only little. This result agrees with [3].

The tested PSSN shows strength degradation after heat treatment at both temperature levels.

SILICON CARBIDE. Hot pressed (HPSC) and sintered (SiC-DLS) silicon carbide of theoretical density have negligible oxidation rates at 900 °C and 1260 °C (weight gain respectively weigth loss $\leq$ 0,1 mg/cm² after 1000 hours).

It is difficult to get a clear idea how annealing influences the strength of silicon carbide at present. All available kinds of dense, not silicon infiltrated silicon carbide show a wide range of scatter of single flexural strength values up to a magnitude of $\pm$ 100 N/mm^2 and even more. However, some silicon carbides show significant changes of room temperature strength after different oxidizing cycles.

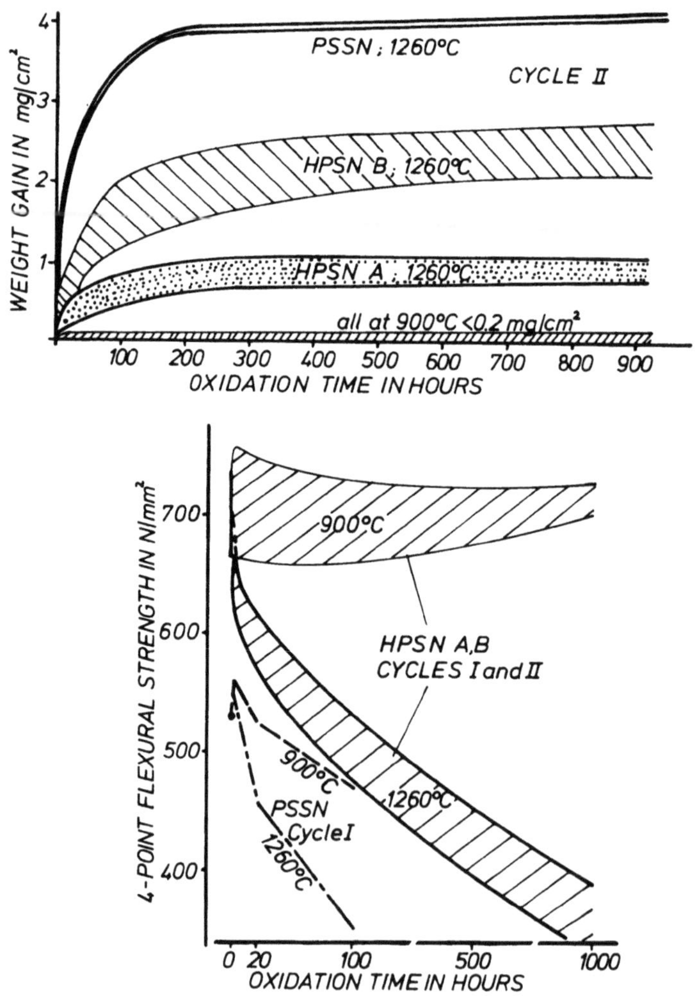

Figure 4. Oxidation and strength of dense silicon nitrides.

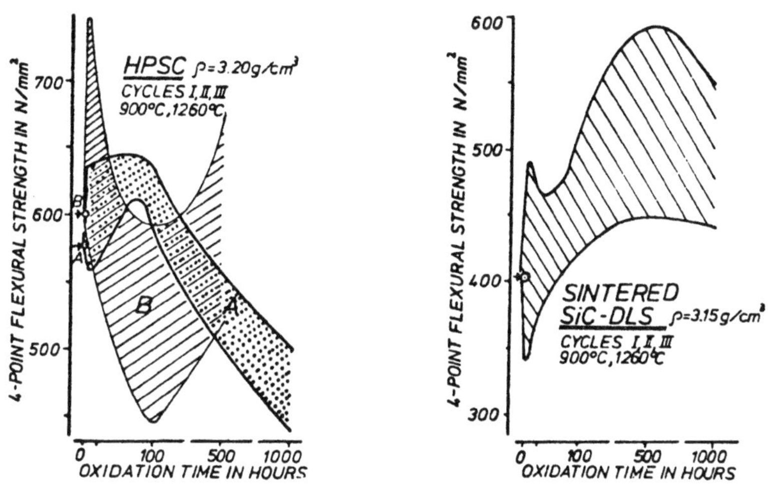

Figure 5. Strength of dense silicon carbide after oxidation.

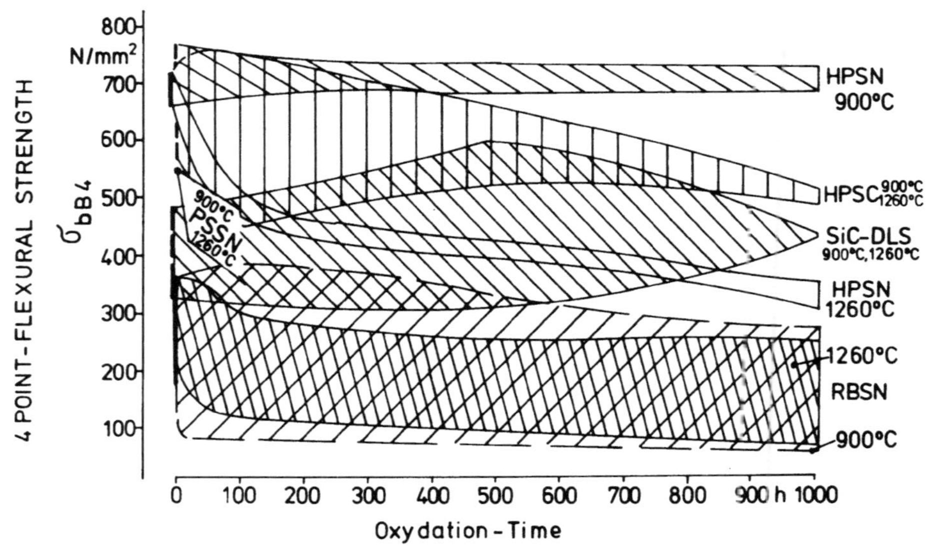

Figure 6. General data bands of room temperature strength after oxidizing heat treatment of nonoxide ceramics (state of may 1979).

Figure 5 shows the data bands of the mean values of two hot pressed silicon carbides and a sinteres SiC. It is evident that there is an influence on the strength similar to silicon nitrides. A valid explanation of this behaviour has not yet been found, as well.

CONCLUDING REMARKS

Figure 6 shows the present situation of the strength behavior of all silicon nitride and silicon carbide materials tested during the last five years.

In addition to the described laboratory investigations turbine components were examined after annealing procedures. This is necessary to prove the observed phenomena of strength changes being valid under states of stresses different to flexural stress in a test bar. In spin tests this could be shown with turbine blades and rotor hubs and in thermal shock tests with a nozzle ring. This means that Si_3N_4 and SiC materials have to be carefully examined according to the desired application.

In many cases of application an attachment of ceramic parts to metals or ceramic parts one to another (example: multidensity rotor) is necessary. Those brazing [4] and bonding techniques in general require elevated temperatures and may damage the ceramic internally, if the joining temperature is critical for strength degradation.

The results reported in this paper show that much effort has to be put into increasing the thermal stability of nonoxide ceramics like Si_3N_4 und SiC prior to raise the initial strength as it was done over the last years.

ACKNOWLEDGEMENTS

The work being reported here has been partially sponsored by the German Ministry for Research and Technology. On behalf of the Volkswagenwerk AG the author would like to acknowledge this support.

REFERENCES

[1] P. Walzer, M. Langer, J. Siebels, "Development of Multi-Density Silicon Nitride Turbine Rotors" Proceedings of the 5th Army Material Technology Conference, 1977

[2] P. Walzer, J. Siebels, "Stand der Entwicklung eines Turbinenlaufrades aus einem Siliziumnitrid-Verbund", Keramische Komponenten für Fahrzeug-Gasturbinen, Springer-Verlag, Berlin; 1978

[3] S. W. Freiman, J. J. Mecholsky, W. J. Mc Donough, R. W. Rice, "Effect of Oxidation on the Room Temperature Strength of Hot-Pressed Si_3N_4-MgO and Si_3N_4-ZrO_2", Proceedings of the 5th Army Materials Technology Conference, 1977

[4] J. Siebels, German patent P 28 22 627.9

THERMAL SHOCK TESTING OF A HOT-PRESSED Si_3N_4 TURBINE ROTOR

K. D. Mörgenthaler and F. Neubrand

Daimler-Benz Aktiengesellschaft, Stuttgart

INTRODUCTION

For a number of years, within the context of the German Ceramics Program, Daimler-Benz has been developing an all ceramic turbine rotor. This is an important activity since it is one of the few efforts to develop an integral all ceramic rotor concept. The monolithic hot pressed silicon nitride rotor is meticulously machined from single billets and subjected to a series of proof tests, which have been described in previous Army Materials Technology conferences on high temperature structural ceramics.

Summary of Current Results

During this presentation, a brief review is provided of current accomplishments regarding evaluation of the performance capabilities of the all ceramic, hot pressed rotor.

In an automotive gas turbine, the components exposed to the hot gas are subjected to very fast changes in gas temperature. One of the characteristics necessary for components in a gas turbine, especially ceramic components, is a resistance to rapid temperature changes. By way of example, Fig. 1 shows a glowing turbine wheel which has been swung out of the hot gas stream and is now exposed to the static ambient air.

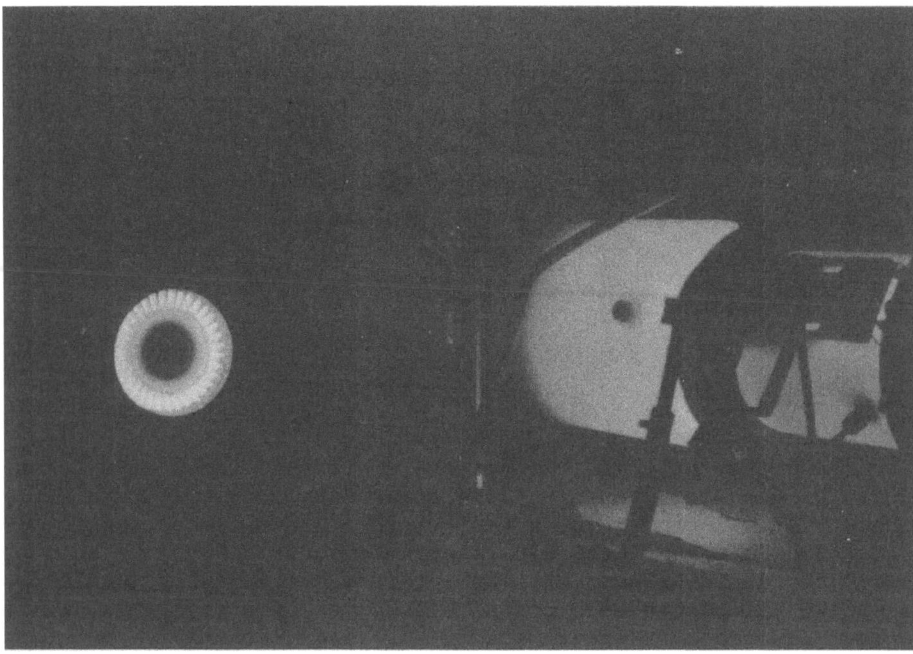

Figure 1: Glowing Turbine Wheel, Swung Out from the Thermal Shock Test Rig

Resistance to rapid temperature changes is examined with the aid of thermal shock tests. Thermal shock tests on ceramic turbine wheels are carried out in a test rig (Fig. 2) which allows cool down and heatup rates in gas temperature up to 600 K/s.

In addition, it is possible to swing the turbine wheel out of the hot gas stream and into the ambient air at room temperature. The tests were carried out with the turbine wheel at a standstill and with a relatively low gas pressure in comparison with actual conditions in the turbine.

Figure 2: Thermal Shock Test Rig

Fig. 3 shows a turbine wheel of hot-pressed silicon nitride as installed in the thermal shock test rig. Wheel and wheel mounting are swung out of the test path.

Fig. 4 shows the different thermal shock cycles to which a turbine wheel of hot-pressed silicon nitride was subjected. During the tests, the gas temperatures altered from a lower temperature of between 400 and 750 °C to a higher temperature between 900 and 1250 °C within a period of 1.5 to 34 seconds.

One turbine wheel was subjected to temperature changes in the gas stream for more than 230 thermal shock cycles. These temperature changes can be seen in Fig. 4.

In addition, more than 400 thermal shock tests were carried out where a wheel was swung out of the hot gas stream at various temperatures and exposed to static ambient air of about 20 °C.

Hot-Pressed Silicon Nitride Turbine Rotor in a Thermal Shock Test Rig

Forschung und Entwicklung
E6SV 0579 10 502

Figure 3: Turbine Wheel of HPSN in the Thermal Shock Test Rig

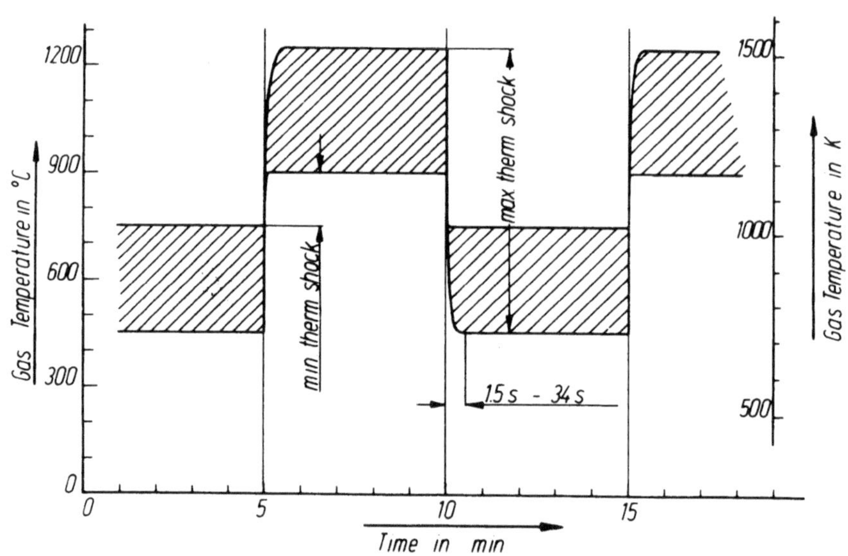

Figure 4: Thermal Shock Cycles

With the aid of the thermal shock test rig, the temperature distribution on the wheel surface was measured by using infrared photography. Fig. 5 shows a temperature distribution pattern established in this way.

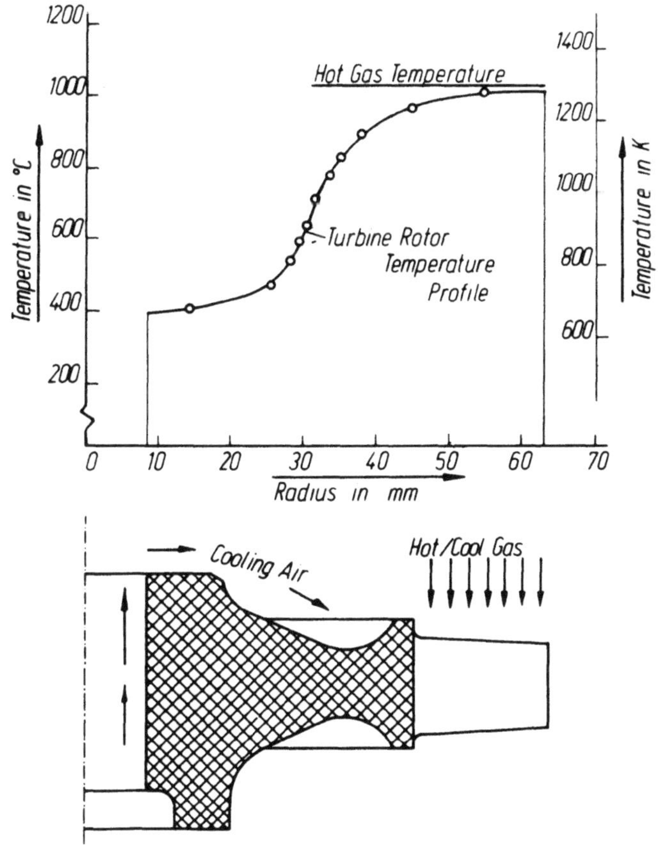

Figure 5: Temperature Distribution on the Ceramic Turbine Wheel

Damage was caused only in one single thermal shock test, where the temperature of the gas streams was reduced from 1250 to 450 °C within 1.5 secs. As Fig. 6 shows, the turbine wheel of hot-pressed silicon nitride broke into a few pieces. In all other cases no damage whatsoever was found to have resulted from the various thermal shock tests.

Fractured Hot-Pressed Silicone Turbine Rotor After Severe Thermal Shock

Forschung und Entwicklung
E6SV 0579 10 505

Figure 6: Damage after Severe Thermal Shock

It is intended to carry out further tests in the thermal shock test stand in order to obtain more information regarding the calculation of the thermal stresses which are fundamental to the understanding of the behavior of the ceramic rotor and of fundamental importance to the proper interpretation of the test results obtained.

INTRODUCTION TO

THE CRITICAL ISSUES WORKSHOP

The Editors

On the last day of the Conference, a Critical Issues Workshop was held. It was the intention of the Conference organizers that this workshop would provide an off-the-record forum at which discussions of those pacing problems impeding the confident use of ceramics in high performance engine applications could be discussed by the attendees. The workshop also provided a forum for brief presentations (both prepared and ad hoc) on specific topics.

In order to set the Workshop in motion, the session Vice Chairmen identified the critical issues raised in (or derived from) the papers presented in their sessions. At the end of the workshop, these presentations were summarized by Prof. J. Mueller, of the University of Washington.

The Workshop proved to be a valuable contribution to the Conference, and produced much animated and lively discussion. Because the workshop was "off-the-record", it is not possible to either completely summarize it or reproduce it in its entirety. However, the following section includes several of the short presentations which were "for-the-record" and the University of Washington Workshop Summary based on inputs from Session Leaders and summarized by Prof. J. Mueller.

WORKSHOP SUMMARY

Professor James I. Mueller

University of Washington

In preparation for the meeting summary, session vice-chairmen were requested to identify topics or issues raised during each session which were deemed critical to the successful utilization of high performance ceramic materials to improve heat engine efficiency. It was anticipated that several common topics would be identified at different sessions but it was agreed in advance that all critical issues would be listed regardless of possible redundancy. The following is a paraphrased recapitulation of those items identified as critical issues by each session vice-chairman:

Session Ia

1. Reliability and durability together with material consistency
2. Lack of confidence in failure prediction results in consideration of temporary restrictions as real barriers
3. Design of the ceramic to metal interface

Session Ib

1. Improved material with consistent properties
2. Extension of data base of materials properties, especially at high temperatures
3. Development of standard test procedures
4. Requirements for NDE or reliability testing

5. Ability to form large and/or complex shapes

6. Design of ceramic to metal or ceramic to ceramic contacts

7. Understanding of surface phenomena, particularly temperature gradients and complex stress states

Session II

1. Development of detailed processing method to improve quality and eliminate defects

2. Detailed attention to manufacturing procedures and in-process inspection to assure consistency

3. A thorough understanding of the chemistry of processing, in particular grain boundary engineering

Session III

1. Improved communication between the mathematical modelers, the materials processors and the design engineers

2. Development of accurate boundary condition statements

3. Development of test procedures which can result in a successful data base of all material properties

4. Improved use of sophisticated optimization techniques

5. Development of correlations between sample test results to operating structures

6. The need of consistency in materials

7. Reliable data and improved modeling of long-life behavior are a necessity for suitable lifetime predictions

Session IV

1. Interdisciplinary communication between the designer and the materials specialist

2. Future development of NDE techniques satisfactory for use in production

3. Development of a rational life prediction capability

Session Va

1. Knowledge of contact stress and surface interaction of ceramic-ceramic and metal-ceramic interfaces

2. Longlife testing to assess the effects of service and environment induced flaws

Session Vb

1. Influences of micro-fracture mechanisms in micro-strength and fracture toughness

2. Need for correct stress intensity factors to generate valid fracture data

3. Improved communication between experts in ceramic materials and those in fracture mechanics

It becomes obvious that the major critical issue confronting each and all of the speakers was life prediction. All addressed the shortcomings or needs in today's technology which, at times, was likened unto Aesop's tale of the blind men describing the elephant, where each, in his own way, addressed part of the whole. A review of the above issues leads to the identification of the most critical issues to be solved prior to the successful utilization of ceramic materials for structural applications involving hostile environments. I have chosen to identify these criteria as the "5-C's":

1. Consistency of material

2. Confidence in data through standardization of testing and the development of an adequate data base

3. Contact studies, including tribology, in order to develop design and fabrication techniques

4. Communication improvement among all the people involved with iterative design

5. Confidence in the material through improved mutual understanding between the mechanics and materials specialists

Although the last criterion was not specifically identified, it is deemed to be crucial to the potential utilization of ceramic materials. We are in the infancy of a new technology, the design with brittle materials. As a result, we are often discouraged when success seems to be in the undeterminable future rather than right around the corner. We must remember that we are working with first, second, or perhaps, third generation materials -- a far cry from the mass produced metals with which most designers, analysts and managers are familiar. Nevertheless, considerable progress has been made in a relatively short-time -- the result not only of considerable government and industrial funding but also of the early enthusiasm of those who spoke at this meeting.

This is not the time to lose that enthusiasm. If we recall Morris Berg's comment [1] that, in his organization, there is a ten to fifteen year span between concept and full production, we can readily see that we are only about one third to one half way toward fruitation. In the past few years, some subtley positive terminology has crept into our lexicon, as we now talk about LIFE PREDICTION rather than failure predictions. Why not in the future refer to PROBABILITY OF SUCCESS rather than probability of failure? Ceramic materials have shown a high probability of success in many engineering applications in the past. There is every reason to expect that the present goals can and will be attained through proper support, continued excellent developments, good judgement and meaningful dialogue and understanding.

REFERENCE

1. M. Berg, "The Spark Plug: A Case History", an unpublished talk presented at the Conference.

INDEX